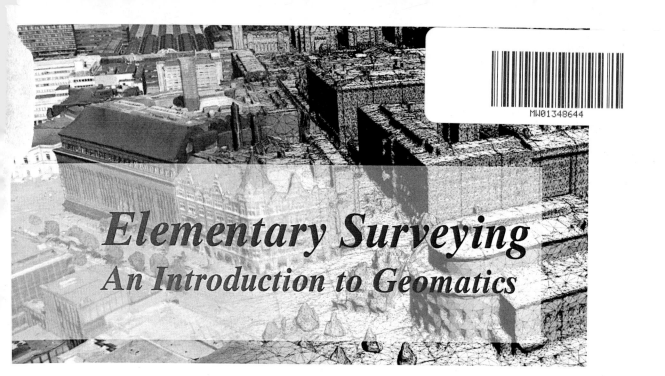

Elementary Surveying
An Introduction to Geomatics

Sixteenth Edition

CHARLES D. GHILANI
Professor Emeritus of Engineering
Pennsylvania State University

Content Management: Dawn Murrin, Erin Ault
Content Development: Coleen Morrison
Content Production: Rajinder Singh, Pallavi Pandit, Kaitlin Smith
Product Management: Holly Stark
Product Marketing: Stacey Sveum, Wayne Stevens
Rights and Permissions: Anjali Singh

Please contact https://support.pearson.com/getsupport/s/contactsupport with any queries on this content

Cover Image: Courtesy of Bentley Systems, Incorporated.

Copyright © 2022, 2018, 2015 by Pearson Education, Inc. or its affiliates, 221 River Street, Hoboken, NJ 07030. All Rights Reserved. Manufactured in the United States of America. This publication is protected by copyright, and permission should be obtained from the publisher prior to any prohibited reproduction, storage in a retrieval system, or transmission in any form or by any means, electronic, mechanical, photocopying, recording, or otherwise. For information regarding permissions, request forms, and the appropriate contacts within the Pearson Education Global Rights and Permissions department, please visit www.pearsoned.com/permissions/

Acknowledgments of third-party content appear on the appropriate page within the text.

PEARSON, ALWAYS LEARNING, and Mastering^{TM} are exclusive trademarks owned by Pearson Education, Inc. or its affiliates in the U.S. and/or other countries.

Unless otherwise indicated herein, any third-party trademarks, logos, or icons that may appear in this work are the property of their respective owners, and any references to third-party trademarks, logos, icons, or other trade dress are for demonstrative or descriptive purposes only. Such references are not intended to imply any sponsorship, endorsement, authorization, or promotion of Pearson's products by the owners of such marks, or any relationship between the owner and Pearson Education, Inc., or its affiliates, authors, licensees, or distributors.

Library of Congress Cataloging-in-Publication Data

Names: Ghilani, Charles D., author.
Title: Elementary surveying : an introduction to geomatics / Charles Daniel Ghilani.
Description: Sixteenth edition. | Hoboken, N.J. : Pearson, 2022. | Includes bibliographical references and index.
Identifiers: LCCN 2020038712 | ISBN 9780136822820 (paperback) | ISBN 9780136822806 (rental edition)
Subjects: LCSH: Surveying. | Geomatics.
Classification: LCC TA545 .W77 2022 | DDC 526.9—dc23 LC record available at https://lccn.loc.gov/2020038712

ScoutAutomatedPrintCode

Rental

ISBN-10: 0-13-682280-0
ISBN-13: 978-0-13-682280-6

Print Offer

ISBN-10: 0-13-682282-7
ISBN-13: 978-0-13-682282-0

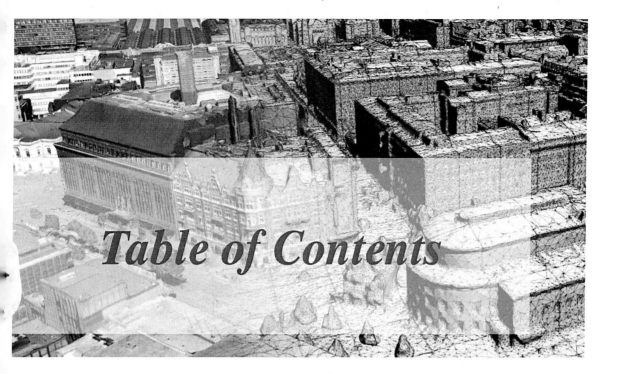

Table of Contents

PREFACE — XIX

Acknowledgments xxi

1 • INTRODUCTION — 1

1.1 Definition of Surveying 1
1.2 Geomatics 3
1.3 History of Surveying 4
1.4 Geodetic and Plane Surveys 9
1.5 Importance of Surveying 10
1.6 Specialized Types of Surveys 11
1.7 Surveying Safety 12
1.8 Land and Geographic Information Systems 14
1.9 Federal Surveying and Mapping Agencies 15
1.10 The Surveying Profession 16
1.11 Professional Surveying Organizations 17
1.12 Surveying on the Internet 18
1.13 Future Challenges in Surveying 18
Problems 20
Bibliography 21

2 • UNITS, SIGNIFICANT FIGURES, AND FIELD NOTES — 22

PART I UNITS AND SIGNIFICANT FIGURES 22

- 2.1 Introduction 22
- 2.2 Units of Measurement 22
- 2.3 International System of Units (SI) 24
- 2.4 Significant Figures 26
- 2.5 Rounding Off Numbers 28

PART II FIELD NOTES 29

- 2.6 Field Notes 29
- 2.7 General Requirements of Handwritten Field Notes 30
- 2.8 Types of Field Books 31
- 2.9 Kinds of Notes 32
- 2.10 Arrangements of Notes 32
- 2.11 Suggestions for Recording Notes 34
- 2.12 Introduction to Survey Controllers 35
- 2.13 Transfer of Files from Survey Controllers 38
- 2.14 Digital Data File Management 40
- 2.15 Advantages and Disadvantages of Survey Controllers 41

Problems 42
Bibliography 43

3 • THEORY OF ERRORS IN OBSERVATIONS — 44

- 3.1 Introduction 44
- 3.2 Direct and Indirect Observations 44
- 3.3 Errors in Measurements 45
- 3.4 Mistakes 45
- 3.5 Sources of Errors in Making Observations 46
- 3.6 Types of Errors 46
- 3.7 Precision and Accuracy 47
- 3.8 Eliminating Mistakes and Systematic Errors 48
- 3.9 Probability 48
- 3.10 Most Probable Value 49
- 3.11 Residuals 50
- 3.12 Occurrence of Random Errors 50
- 3.13 General Laws of Probability 54
- 3.14 Measures of Precision 55
- 3.15 Interpretation of Standard Deviation 57
- 3.16 The 50%, 90%, and 95% Errors 57
- 3.17 Error Propagation 59
- 3.18 Applications 64
- 3.19 Conditional Adjustment of Observations 64
- 3.20 Weights of Observations 65

3.21 Least-Squares Adjustment 66
Problems 67
Bibliography 69

4 • LEVELING—THEORY, METHODS, AND EQUIPMENT   70

PART I LEVELING—THEORY AND METHODS 70

4.1 Introduction 70
4.2 Definitions 70
4.3 North American Vertical Datum 73
4.4 Curvature and Refraction 74
4.5 Methods for Determining Differences in Elevation 76

PART II EQUIPMENT FOR DIFFERENTIAL LEVELING 83

4.6 Categories of Levels 83
4.7 Telescopes 84
4.8 Level Vials 86
4.9 Tilting Levels 88
4.10 Automatic Levels 88
4.11 Digital Levels 90
4.12 Tripods 91
4.13 Hand Levels 91
4.14 Level Rods 92
4.15 Turning Points 94
4.16 Testing and Adjusting Levels 95
Problems 99
Bibliography 101

5 • LEVELING—FIELD PROCEDURES AND COMPUTATIONS   102

5.1 Introduction 102
5.2 Carrying and Setting Up a Level 102
5.3 Duties of a Rod Person 104
5.4 Differential Leveling 105
5.5 Precision 111
5.6 Adjustments of Simple Level Circuits 112
5.7 Reciprocal Leveling 113
5.8 Three-Wire Leveling 114
5.9 Profile Leveling 116
5.10 Grid, Cross-Section, or Borrow-Pit Leveling 120
5.11 Use of the Hand Level 121
5.12 Sources of Error in Leveling 121
5.13 Mistakes 123

5.14 Reducing Errors and Eliminating Mistakes 124
5.15 Using Software 124
Problems 126
Bibliography 128

6 • DISTANCE MEASUREMENT 129

PART I METHODS FOR MEASURING DISTANCES 129

6.1 Introduction 129
6.2 Summary of Methods for Making Linear Measurements 129
6.3 Pacing 130
6.4 Odometer Readings 130
6.5 Optical Rangefinders 130
6.6 Tacheometry 131
6.7 Subtense Bar 131

PART II DISTANCE MEASUREMENTS BY TAPING 131

6.8 Introduction to Taping 131
6.9 Taping Equipment and Accessories 132
6.10 Care of Taping Equipment 133
6.11 Taping on Level Ground 134
6.12 Horizontal Measurements on Sloping Ground 136
6.13 Slope Measurements 137
6.14 Sources of Error in Taping 139

PART III ELECTRONIC DISTANCE MEASUREMENT 143

6.15 Introduction 143
6.16 Propagation of Electromagnetic Energy 144
6.17 Principles of Electronic Distance Measurement 147
6.18 Electro-Optical Instruments 148
6.19 Total Station Instruments 151
6.20 EDM Instruments Without Reflectors 152
6.21 Computing Horizontal Lengths from Slope Distances 152
6.22 Errors in Electronic Distance Measurement 154
6.23 Using Software 161
Problems 162
Bibliography 163

7 • ANGLES, AZIMUTHS, AND BEARINGS 164

7.1 Introduction 164
7.2 Units of Angle Measurement 164
7.3 Kinds of Horizontal Angles 165

7.4 Direction of a Line 166
7.5 Azimuths 167
7.6 Bearings 168
7.7 Comparison of Azimuths and Bearings 169
7.8 Computing Azimuths 171
7.9 Computing Bearings 173
7.10 The Compass and the Earth's Magnetic Field 174
7.11 Magnetic Declination 176
7.12 Variations in Magnetic Declination 176
7.13 Software for Determining Magnetic Declination 178
7.14 Local Attraction 179
7.15 Typical Magnetic Declination Problems 180
7.16 Mistakes 182
Problems 182

8 • TOTAL STATION INSTRUMENTS; ANGLE OBSERVATIONS 185

PART I TOTAL STATION INSTRUMENTS 185

8.1 Introduction 185
8.2 Characteristics of Total Station Instruments 185
8.3 Functions Performed by Total Station Instruments 188
8.4 Parts of a Total Station Instrument 189
8.5 Handling and Setting Up a Total Station Instrument 192
8.6 Servo-Driven and Remotely Operated Total Station Instruments 195

PART II ANGLE OBSERVATIONS 196

8.7 Relationship of Angles and Distances 196
8.8 Observing Horizontal Angles with Total Station Instruments 197
8.9 Observing Multiple Horizontal Angles by the Direction Method 200
8.10 Closing the Horizon 201
8.11 Observing Deflection Angles 202
8.12 Observing Azimuths 204
8.13 Observing Vertical Angles 204
8.14 Sights and Marks 206
8.15 Prolonging a Straight Line 208
8.16 Balancing-in 209
8.17 Random Traverse 209
8.18 Total Stations for Determining Elevation Differences 211
8.19 Adjustment of Total Station Instruments and Their Accessories 212
8.20 Sources of Error in Total Station Work 216
8.21 Propagation of Random Errors in Angle Observations 222
8.22 Mistakes 223
Problems 224
Bibliography 225

9 • TRAVERSING 226

9.1 Introduction 226
9.2 Observation of Traverse Angles or Directions 228
9.3 Observation of Traverse Lengths 229
9.4 Selection of Traverse Stations 230
9.5 Referencing Traverse Stations 231
9.6 Traverse Field Notes 232
9.7 Angle Misclosure 233
9.8 Traversing with Total Station Instruments 234
9.9 Radial Traversing 236
9.10 Sources of Error in Traversing 237
9.11 Mistakes in Traversing 237
Problems 237

10 • TRAVERSE COMPUTATIONS 239

10.1 Introduction 239
10.2 Balancing Angles 240
10.3 Computation of Preliminary Azimuths Or Bearings 242
10.4 Departures and Latitudes 243
10.5 Departure and Latitude Closure Conditions 245
10.6 Traverse Linear Misclosure and Relative Precision 245
10.7 Traverse Adjustment 246
10.8 Rectangular Coordinates 249
10.9 Alternative Methods for Making Traverse Computations 250
10.10 Inversing 254
10.11 Computing Final Adjusted Traverse Lengths and Directions 255
10.12 Coordinate Computations in Boundary Surveys 257
10.13 Use of Open Traverses 259
10.14 State Plane Coordinate Systems 262
10.15 Traverse Computations Using Computers 263
10.16 Locating Blunders in Traverse Observations 263
10.17 Mistakes in Traverse Computations 266
Problems 266
Bibliography 269

11 • COORDINATE GEOMETRY IN SURVEYING CALCULATIONS 270

11.1 Introduction 270
11.2 Coordinate Forms of Equations for Lines and Circles 271
11.3 Perpendicular Distance from a Point to a Line 273
11.4 Intersection of Two Lines, Both Having Known Directions 275

11.5 Intersection of a Line With a Circle 277
11.6 Intersection of Two Circles 280
11.7 Three-Point Resection 282
11.8 Two-Dimensional Conformal Coordinate Transformation 285
11.9 Inaccessible Point Problem 290
11.10 Three-Dimensional Two-Point Resection 292
11.11 Software 295
Problems 296
Bibliography 300

12 • AREA   301

12.1 Introduction 301
12.2 Methods of Measuring Area 301
12.3 Area by Division Into Simple Figures 302
12.4 Area by Offsets from Straight Lines 303
12.5 Area by Coordinates 305
12.6 Area by Double-Meridian Distance Method 309
12.7 Area of Parcels with Circular Boundaries 312
12.8 Partitioning of Lands 313
12.9 Area by Measurements from Maps 317
12.10 Software 319
12.11 Sources of Error in Determining Areas 320
12.12 Mistakes in Determining Areas 320
Problems 320

13 • GLOBAL NAVIGATION SATELLITE SYSTEMS—INTRODUCTION AND PRINCIPLES OF OPERATION   323

13.1 Introduction 323
13.2 Overview of GPS 324
13.3 The GPS Signal 326
13.4 Reference Coordinate Systems 329
13.5 Fundamentals of Satellite Positioning 339
13.6 Errors in Observations 341
13.7 Differential Positioning 349
13.8 Kinematic Methods 352
13.9 Relative Positioning 352
13.10 Precise Point Positioning 355
13.11 Other Satellite Navigation Systems 357
13.12 The Future 359
Problems 361
Bibliography 362

14 • GLOBAL NAVIGATION SATELLITE SYSTEMS—STATIC SURVEYS   363

14.1 Introduction   363
14.2 Field Procedures in Static GNSS Surveys   364
14.3 Planning Satellite Surveys   369
14.4 Performing Static Surveys   380
14.5 Data Processing and Analysis   381
14.6 Things to Consider   391
14.7 A Method for Obtaining Orthometric Height Differences Using GNSS   394
14.8 Sources of Errors in Satellite Surveys   396
14.9 Mistakes in Satelite Surveys   398
Problems   399
Bibliography   401

15 • GLOBAL NAVIGATION SATELLITE SYSTEMS—KINEMATIC SURVEYS   402

15.1 Introduction   402
15.2 Planning of Kinematic Surveys   403
15.3 Initialization Techniques   405
15.4 Equipment Used in Kinematic Surveys   406
15.5 Methods Used in Kinematic Surveys   408
15.6 Performing Post-Processed Kinematic Surveys   411
15.7 Communication in Real-Time Kinematic Surveys   414
15.8 Real-Time Networks   415
15.9 Performing Real-Time Kinematic Surveys   417
15.10 Machine Guidance and Control   419
15.11 Errors in Kinematic Surveys   421
15.12 Mistakes in Kinematic Surveys   422
Problems   422
Bibliography   423

16 • ADJUSTMENTS BY LEAST SQUARES   425

16.1 Introduction   425
16.2 Fundamental Condition of Least Squares   427
16.3 Least-Squares Adjustment by the Observation Equation Method   428
16.4 Matrix Methods in Least-Squares Adjustment   432
16.5 Matrix Equations for Precisions of Adjusted Quantities   434
16.6 Least-Squares Adjustment of Leveling Circuits   436
16.7 Propagation of Errors   440
16.8 Least-Squares Adjustment of GNSS Baseline Vectors   442
16.9 Least-Squares Adjustment of Conventional Horizontal Plane Surveys   447
16.10 The Error Ellipse   456

16.11 Adjustment Procedures 461
16.12 Other Measures of Precision for Horizontal Stations 462
16.13 Software 465
16.14 Conclusions 465
Problems 466
Bibliography 471

17 • MAPPING SURVEYS 472

17.1 Introduction 472
17.2 Basic Methods for Performing Mapping Surveys 473
17.3 Map Scale 474
17.4 Control for Mapping Surveys 475
17.5 Contours 476
17.6 Characteristics of Contours 479
17.7 Method of Locating Contours 480
17.8 Digital Elevation Models and Automated Contouring Systems 482
17.9 Basic Field Methods for Locating Topographic Details 483
17.10 Planning a Laser-Scanning Survey 494
17.11 Three-Dimensional Conformal Coordinate Transformation 496
17.12 Selection of Field Method 498
17.13 Working With Survey Controllers and Field-to-Finish Software 498
17.14 Hydrographic Surveys 501
17.15 Sources of Error in Mapping Surveys 506
17.16 Mistakes in Mapping Surveys 506
Problems 507
Bibliography 508

18 • MAPPING 510

18.1 Introduction 510
18.2 Availability of Maps and Related Information 511
18.3 National Mapping Program 511
18.4 Accuracy Standards for Mapping 512
18.5 Manual and Computer-Aided Drafting Procedures 516
18.6 Map Design 517
18.7 Map Layout 519
18.8 Basic Map Plotting Procedures 520
18.9 Contour Interval 522
18.10 Plotting Contours 523
18.11 Lettering 524
18.12 Cartographic Map Elements 525
18.13 Drafting Materials 527
18.14 Automated Mapping and Computer-Aided Drafting Systems 528
18.15 Migrating Maps Between Software Packages 533
18.16 Impacts of Modern Land and Geographic Information Systems on Mapping 534

18.17 The Importance of Metadata 535
18.18 Sources of Error in Mapping 535
18.19 Mistakes in Mapping 535
Problems 536
Bibliography 538

19 • CONTROL SURVEYS AND GEODETIC REDUCTIONS 539

19.1 Introduction 539
19.2 The Ellipsoid and Geoid 540
19.3 The Conventional Terrestrial Pole 542
19.4 Geodetic Position and Ellipsoidal Radii of Curvature 544
19.5 Geoid Undulation and Deflection of the Vertical 547
19.6 U.S. Reference Frames 549
19.7 Transforming Coordinates Between Reference Frames 556
19.8 Accuracy Standards and Specifications for Control Surveys 561
19.9 The National Spatial Reference System 564
19.10 Hierarchy of the National Horizontal Control Network 564
19.11 Hierarchy of the National Vertical Control Network 565
19.12 Control Point Descriptions 565
19.13 Field Procedures for Conventional Horizontal Control Surveys 569
19.14 Field Procedures for Vertical-Control Surveys 574
19.15 Reduction of Field Observations to their Geodetic Values 579
19.16 Geodetic Position Computations 591
19.17 The Local Geodetic Coordinate System 594
19.18 Three-Dimensional Coordinate Computations 596
19.19 Software 598
Problems 599
Bibliography 601

20 • STATE PLANE COORDINATES AND OTHER MAP PROJECTIONS 603

20.1 Introduction 603
20.2 Projections Used in State Plane Coordinate Systems 604
20.3 Lambert Conformal Conic Projection 608
20.4 Transverse Mercator Projection 609
20.5 State Plane Coordinates in Nad 27 and Nad 83 610
20.6 Computing SPCS 83 Coordinates in the Lambert Conformal Conic System 611
20.7 Computing SPCS 83 Coordinates in the Transverse Mercator System 616
20.8 Reduction of Distances and Angles to State Plane Coordinate Grids 622
20.9 Computing State Plane Coordinates of Traverse Stations 632
20.10 Surveys Extending from One Zone to Another 635
20.11 The Universal Transverse Mercator Projection 636
20.12 Other Map Projections 637

20.13 Ground versus Grid Problem 641
20.14 Proposed Changes to SPCS in 2022 647
20.15 Map Projection Software 652
Problems 654
Bibliography 656

21 • BOUNDARY SURVEYS 657

21.1 Introduction 657
21.2 Categories of Land Surveys 658
21.3 Historical Perspectives 659
21.4 Property Description by Metes and Bounds 660
21.5 Property Description by Block-and-Lot System 663
21.6 Property Description by Coordinates 665
21.7 Retracement Surveys 665
21.8 Subdivision Surveys 668
21.9 Partitioning Land 670
21.10 Registration of Title 671
21.11 Adverse Possession and Easements 672
21.12 Condominium Surveys 672
21.13 Geographic and Land Information Systems 679
21.14 Sources of Error in Boundary Surveys 679
21.15 Mistakes 679
Problems 680
Bibliography 682

22 • SURVEYS OF THE PUBLIC LANDS 683

22.1 Introduction 683
22.2 Instructions for Surveys of the Public Lands 684
22.3 Initial Point 687
22.4 Principal Meridian 688
22.5 Baseline 688
22.6 Standard Parallels (Correction Lines) 690
22.7 Guide Meridians 690
22.8 Township Exteriors, Meridional (Range) Lines, and Latitudinal (Township) Lines 690
22.9 Designation of Townships 692
22.10 Subdivision of a Quadrangle into Townships 692
22.11 Subdivision of a Township into Sections 693
22.12 Subdivision of Sections 695
22.13 Fractional Sections 695
22.14 Notes 696
22.15 Outline of Subdivision Steps 696
22.16 Marking Corners 698
22.17 Witness Corners 698
22.18 Meander Corners 698

22.19 Lost and Obliterated Corners 699
22.20 Accuracy of Public Land Surveys 701
22.21 Descriptions by Township Section, and Smaller Subdivision 702
22.22 BLM Land Information System 703
22.23 Sources of Error 704
22.24 Mistakes 704
Problems 704
Bibliography 706

23 • CONSTRUCTION SURVEYS 707

23.1 Introduction 707
23.2 Specialized Equipment for Construction Surveys 708
23.3 Horizontal and Vertical Control 712
23.4 Staking Out a Pipeline 713
23.5 Staking Pipeline Grades 714
23.6 Computing the Bend Angles in Pipelines 716
23.7 Staking Out a Building 717
23.8 Staking Out Highways 721
23.9 Other Construction Surveys 726
23.10 Construction Surveys Using Total Station Instruments 727
23.11 Construction Surveys Using GNSS Equipment 730
23.12 Machine Guidance and Control 732
23.13 As-Built Surveys with Laser Scanning 734
23.14 Sources of Error in Construction Surveys 735
23.15 Mistakes 735
Problems 735
Bibliography 737

24 • HORIZONTAL CURVES 738

24.1 Introduction 738
24.2 Degree of Circular Curve 739
24.3 Definitions and Derivation of Circular Curve Formulas 741
24.4 Circular Curve Stationing 743
24.5 General Procedure of Circular Curve Layout by Deflection Angles 744
24.6 Computing Deflection Angles and Chords 746
24.7 Notes for Circular Curve Layout by Deflection Angles and Incremental Chords 748
24.8 Detailed Procedures for Circular Curve Layout by Deflection Angles and Incremental Chords 749
24.9 Setups on Curve 750
24.10 Metric Circular Curves by Deflection Angles and Incremental Chords 751
24.11 Circular Curve Layout by Deflection Angles and Total Chords 753
24.12 Computation of Coordinates on a Circular Curve 753
24.13 Circular Curve Layout by Coordinates 755

24.14 Curve Stakeout Using GNSS Receivers and Robotic Total Stations 760
24.15 Circular Curve Layout by Offsets 761
24.16 Special Circular Curve Problems 764
24.17 Compound and Reverse Curves 765
24.18 Sight Distance on Horizontal Curves 766
24.19 Spirals 767
24.20 Computation of "As-Built" Circular Alignments 771
24.21 Sources of Error in Laying Out Circular Curves 774
24.22 Mistakes 775
Problems 775
Bibliography 777

25 • VERTICAL CURVES 778

25.1 Introduction 778
25.2 General Equation of a Vertical Parabolic Curve 779
25.3 Equation of an Equal Tangent Vertical Parabolic Curve 780
25.4 High or Low Point on a Vertical Curve 782
25.5 Vertical Curve Computations Using the Tangent-Offset Equation 782
25.6 Equal Tangent Property of a Parabola 786
25.7 Curve Computations by Proportion 787
25.8 Staking a Vertical Parabolic Curve 787
25.9 Machine Control in Grading Operations 788
25.10 Computations for an Unequal Tangent Vertical Curve 788
25.11 Designing a Curve to Pass Through a Fixed Point 790
25.12 Sight Distance 792
25.13 Sources of Error in Laying Out Vertical Curves 794
25.14 Mistakes 794
Problems 794
Bibliography 796

26 • VOLUMES 797

26.1 Introduction 797
26.2 Methods of Volume Measurement 797
26.3 The Cross-Section Method 798
26.4 Types of Cross Sections 799
26.5 Average-End-Area Formula 800
26.6 Determining End Areas 801
26.7 Computing Slope Intercepts 804
26.8 Prismoidal Formula 806
26.9 Volume Computations 808
26.10 Unit-Area, or Borrow-Pit, Method 809
26.11 Contour-Area Method 811
26.12 Measuring Volumes of Water Discharge 812
26.13 Software 813
26.14 Sources of Error in Determining Volumes 814

26.15 Mistakes 814
Problems 815
Bibliography 817

27 • PHOTOGRAMMETRY   818

27.1 Introduction 818
27.2 Uses of Photogrammetry 819
27.3 Aerial Cameras 820
27.4 Types of Aerial Photographs 822
27.5 Vertical Aerial Photographs 824
27.6 Scale of a Vertical Photograph 825
27.7 Ground Coordinates from a Single Vertical Photograph 828
27.8 Relief Displacement on a Vertical Photograph 830
27.9 Flying Height of a Vertical Photograph 832
27.10 Stereoscopic Parallax 833
27.11 Stereoscopic Viewing 836
27.12 Stereoscopic Measurement of Parallax 837
27.13 Analytical Photogrammetry 839
27.14 Stereoscopic Plotting Instruments 839
27.15 Orthophotos 844
27.16 Ground Control for Photogrammetry 845
27.17 Flight Planning 846
27.18 Airborne Laser-Mapping Systems 848
27.19 Small Unmanned Aerial Systems 849
27.20 Remote Sensing 851
27.21 Software 856
27.22 Sources of Error in Photogrammetry 858
27.23 Mistakes 858
Problems 858
Bibliography 861

28 • INTRODUCTION TO GEOGRAPHIC INFORMATION SYSTEMS   862

28.1 Introduction 862
28.2 Land Information Systems 864
28.3 GIS Data Sources and Classifications 865
28.4 Spatial Data 865
28.5 Nonspatial Data 871
28.6 Data Format Conversions 871
28.7 Creating GIS Databases 874
28.8 Metadata 880
28.9 GIS Analytical Functions 880
28.10 GIS Applications 884
28.11 Data Sources 885
Problems 887
Bibliography 889

APPENDIX A • TAPE CORRECTION PROBLEMS — 890

A.1 Correcting Systematic Errors in Taping 890

APPENDIX B • EXAMPLE NOTEFORMS — 893

APPENDIX C • ASTRONOMIC OBSERVATIONS — 900

C.1 Introduction 900
C.2 Overview of Usual Procedures for Astronomical Azimuth Determination 901
C.3 Ephemerides 903
C.4 Definitions 906
C.5 Time 908
C.6 Timing Observations 910
C.7 Computations for Azimuth from Polaris Observations by the Hour Angle Method 911
C.8 Azimuth from Solar Observations 913
C.9 Importance of Precise Leveling 915

APPENDIX D • USING THE WORKSHEETS FROM THE COMPANION WEBSITE — 916

D.1 Introduction 916
D.2 Using the Files 916
D.3 Worksheets as an Aid in Learning 921

APPENDIX E • INTRODUCTION TO MATRICES — 922

E.1 Introduction 922
E.2 Definition of a Matrix 922
E.3 The Dimensions of a Marix 923
E.4 The Transpose of a Matrix 924
E.5 Matrix Addition 924
E.6 Matrix Multiplication 924
E.7 Matrix Inverse 926

APPENDIX F • U.S. STATE PLANE COORDINATE SYSTEM DEFINING PARAMETERS — 928

F.1 Introduction 928
F.2 Defining Parameters for States Using the Lambert Conformal Conic Map Projection 928
F.3 Defining Parameters for States Using the Transverse Mercator Map Projection 931

APPENDIX G • ANSWERS TO SELECTED PROBLEMS — 933

APPENDIX H • COMMONLY USED CONVERSIONS AND ABBREVIATIONS — 938

INDEX — 944

Preface

This 16th edition of *Elementary Surveying: An Introduction to Geomatics* presents basic concepts and practical material in each of the areas fundamental to modern surveying (geomatics) practice. It is written primarily for students beginning their study of surveying (geomatics) at the college level. Although the book is introductory to the practice of surveying, its depth and breadth also make it ideal for self-study and preparation for licensing examinations. This edition includes more than 400 figures and illustrations to help clarify discussions, and numerous example problems are worked to illustrate computational procedures.

In keeping with the goal of providing an up-to-date presentation of surveying equipment and procedures, total stations are stressed as the instruments for making angle and distance observations. In the chapters on GNSS surveys, discussions have been included to bring awareness to the inaccuracies of directions obtained from a GNSS survey along with the effect of centering errors on horizontal positioning. Chapter 17 introduces the reader to unmanned systems and Chapter 18 has a section on the importance of metadata. Chapter 20 on mapping the ground versus grid problem is discussed including subsections on low-distortion projections and single project factor use.

As with past editions, this book continues to emphasize the theory of errors in surveying work. At the end of each chapter, common errors and mistakes related to the topic covered are listed so that students will be reminded to exercise caution in all of their work. Practical suggestions resulting from the author's many years of experience are interjected throughout the text. Many of the 1,000 end-of-chapter problems have been rewritten so that instructors can create new assignments for their students. An Instructor's Manual is available on the instructor resource center for instructors who adopt the book by contacting their Pearson sales representative.

Updated versions of STATS, WOLFPACK, and MATRIX are available in Mastering Engineering. These programs contain options for statistical computations, traverse computations for polygon, link, and radial traverses; area calculations; astronomical azimuth reduction; two-dimensional coordinate transformations; horizontal and vertical curve computations; and least-squares adjustments. Mathcad® worksheets and Excel® spreadsheets are included on the companion website. These programmed computational sheets demonstrate the solution to many of the example problems discussed herein. For those desiring additional knowledge in map projections, the Mercator, Albers Equal Area, Oblique Stereographic, and Oblique Mercator map projections have been included with these files.

RESOURCES FOR INSTRUCTORS

All instructor resources are available for download at www.pearson.com If you are in need of a login and password for this site, please contact your local Pearson representative.

Instructor's Solutions Manual
Available to adopters of this textbook, it contains the solutions to the homework problems. The Solutions Manual is available in PDF format.

PowerPoint Slides
A complete set of all the figures and tables from the textbook are available in PowerPoint format.

Mastering Engineering
This online tutorial and assessment program allows you to integrate dynamic homework with automated grading of the calculation parts of problems and personalized feedback. MasteringTM Engineering allows you to easily track the performance of your entire class on an assignment-by-assignment basis, or the detailed work of an individual student. For more information visit www.masteringengineering.com.

WHAT'S NEW IN THIS EDITION

- Discussion on use and limitations of small unmanned aerial systems (sUAS).
- Discussion on the importance of metadata.
- Discussion on GNSS Precise Point Positioning.
- Discussion on the new gravimetric- and geometric-based datums for the United States in 2022.
- Discussion on the computational changes to the state plane coordinate systems in 2022.
- Discussion on the planned vertical datum for the North American continent involving Canada, Mexico, and the United States.
- Discussion on the planned new horizontal and vertical datums.

- Discussion on the ground versus grid problem with map projections including low-distortion projections.
- Revised problems sets.

ACKNOWLEDGMENTS

Past editions of this book, and this current one, have benefited from the suggestions, reviews, and other input from numerous students, educators, and practitioners. For their help the author is extremely grateful. In this edition, those professors and graduate students who reviewed material or otherwise assisted include Robert Schultz, Oregon State University; Steven Frank and James Reilly, New Mexico State University; Jeremy Deal, University of Texas-Arlington; Eric Fuller, St. Cloud State University; Loren J. Gibson, Florida Atlantic University; John J. Rose, Phoenix College; Robert Moynihan, University of New Hampshire; Marlee Walton, Iowa State University; Douglas E. Smith, Montana State University; Jean M. Rüeger, The University of New South Wales, Sydney, Australia; Thomas Seybert, Frank Derby, and Brian Naberezny, Pennsylvania State University; Paul Dukas and Bon DeWitt, University of Florida; Preston Hartzell, University of Houston; Ryan White, Univeristy of New Brunswick; Heather Nicholson; Thomas Meyer, University of Connecticut, Craig Rollins, National Geospatial Intelligence Agency; Dru Smith, Daniel Gillins, and Michael Dennis, National Geodetic Survey. The author would like to acknowledge the following professionals for their contributions and suggestions including Charles Harpster, Pennsylvania Department of Transportation; Eduardo Fernandez-Falcon, Topcon Positioning Systems; Joseph Gabor, Thomas Meyer, the University of Connecticut, and Dru Smith, Michael Dennis, and Dan Gillins from the National Geodetic Survey.

In addition, the author wishes to acknowledge the contributions of charts, maps, or other information from the National Geodetic Survey, the U.S. Geological Survey, and the U.S. Bureau of Land Management. Also appreciation is expressed to the many instrument manufacturers who provided images and other descriptive information on their equipment for used herein. To all of those named above, and to any others who may have been inadvertently omitted, the author is extremely thankful.

Combine this...

Hwk: 4.32   Intro to surveying   Annie Ghilani

First Setup

$$\Delta Elev_{AB} = (5.123 + 2\varepsilon) - (5.705 + \varepsilon)$$

Second Setup

$$\Delta Elev_{AB} = (5.123 + \varepsilon) - (5.697 + 2\varepsilon)$$

So $(5.123 + 2\varepsilon) - (5.705 + \varepsilon) = (5.123 + \varepsilon) - (5.697 + 2\varepsilon)$

Solving for ε

$$-0.582 + \varepsilon = -0.574 - \varepsilon$$
$$2\varepsilon = -0.574 + 0.582$$
$$2\varepsilon = 0.008$$
$$\varepsilon = 0.004$$

Check

Setup 1: $[5.123 + 2(0.004)] - (5.705 + 0.004) = -0.578$ ✓

Setup 2: $(5.123 + 0.004) - [5.697 + 2(0.004)] = -0.578$ ✓

With the Power of Mastering Engineering for Elementary Surveying

Mastering™ is the teaching and learning platform that empowers every student. By combining trusted authors' content with digital tools developed to engage students and emulate the office hours experience, Mastering personalizes learning and improves results for each student.

Learn more at www.pearson.com/mastering/engineering

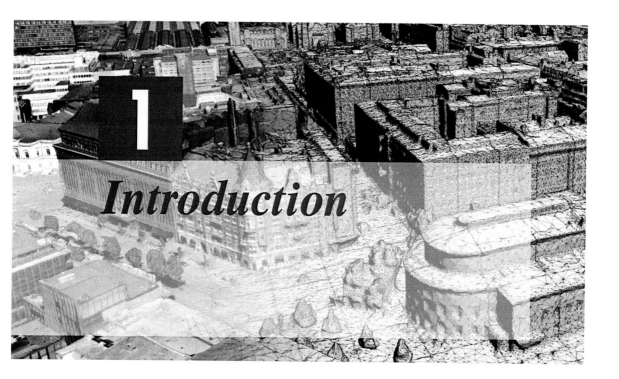

1

Introduction

1.1 DEFINITION OF SURVEYING

Surveying, which is also interchangeably called *geomatics* (see Section 1.2), has traditionally been defined as the science, art, and technology of determining the relative positions of points above, on, or beneath the Earth's surface, or of establishing such points. In a more general sense, however, surveying (geomatics) can be regarded as that discipline that encompasses all methods for measuring and collecting information about the physical earth and our environment, processing that information, and disseminating a variety of resulting products to a wide range of clients. Surveying has been important since the beginning of civilization. Its earliest applications were in measuring and marking boundaries of property ownership. Throughout the years, its importance has steadily increased with the growing demand for a variety of maps and other spatially related types of information, and the expanding need for establishing accurate line and grade to guide construction operations.

Today the importance of measuring and monitoring our environment is becoming increasingly critical as our population expands, land values appreciate, our natural resources dwindle, and human activities continue to stress the quality of our land, water, and air. Using modern ground, aerial, and satellite technologies, and computers for data processing, contemporary surveyors are now able to measure and monitor the Earth and its natural resources on literally a global basis. Never before has so much information been available for assessing current conditions, making sound planning decisions, and formulating policy in a host of land-use, resource development, and environmental preservation applications.

Recognizing the increasing breadth and importance of the practice of surveying, the *International Federation of Surveyors* (see Section 1.11) adopted the following definition:

"A surveyor is a professional person with the academic qualifications and technical expertise to conduct one, or more, of the following activities;

- to determine, measure and represent the land, three-dimensional objects, point-fields, and trajectories;
- to assemble and interpret land and geographically related information;
- to use that information for the planning and efficient administration of the land, the sea and any structures thereon; and
- to conduct research into the above practices and to develop them.

Detailed Functions

The surveyor's professional tasks may involve one or more of the following activities, which may occur either on, above, or below the surface of the land or the sea and may be carried out in association with other professionals.

1. The determination of the size and shape of the earth and the measurements of all data needed to define the size, position, shape and contour of any part of the earth and monitoring any change therein.
2. The positioning of objects in space and time as well as the positioning and monitoring of physical features, structures and engineering works on, above or below the surface of the earth.
3. The development, testing, and calibration of sensors, instruments, and systems for the above-mentioned purposes and for other surveying purposes.
4. The acquisition and use of spatial information from close range, aerial and satellite imagery and the automation of these processes.
5. The determination of the position of the boundaries of public or private land, including national and international boundaries, and the registration of those lands with the appropriate authorities.
6. The design, establishment, and administration of geographic information systems (GIS), and the collection, storage, analysis, management, display and dissemination of data.
7. The analysis, interpretation, and integration of spatial objects and phenomena in GIS, including the visualization and communication of such data in maps, models, and mobile digital devices.
8. The study of the natural and social environment, the measurement of land and marine resources and the use of such data in the planning of development in urban, rural, and regional areas.
9. The planning, development and redevelopment of property, whether urban or rural and whether land or buildings.
10. The assessment of value and the management of property, whether urban or rural and whether land or buildings.
11. The planning, measurement and management of construction works, including the estimation of costs.

In application of the foregoing activities surveyors take into account the relevant legal, economic, environmental, and social aspects affecting each project."

The breadth and diversity of the practice of surveying (geomatics), as well as its importance in modern civilization, are readily apparent from this definition.

1.2 GEOMATICS

As noted in Section 1.1, geomatics is a term that is now commonly being applied to encompass the areas of practice formerly identified as surveying. The principal reason cited for making the name change is that the manner and scope of practice in surveying have changed dramatically in recent years. This has occurred in part because of recent technological developments that have provided surveyors with new tools for measuring and/or collecting information, for computing, and for displaying and disseminating information. It has also been driven by increasing concerns about the environment locally, regionally, and globally, which have greatly exacerbated efforts in monitoring, managing, and regulating the use of our land, water, air, and other natural resources. These circumstances, and others, have brought about a vast increase in demands for new spatially related information.

Historically surveyors made their measurements using ground-based methods using the transit and tape[1] as their primary instruments. Computations, analyses, and the reports, plats, and maps they delivered to their clients were prepared (in hardcopy form) through tedious manual processes. Today's surveyor has an arsenal of tools for measuring and collecting environmental information that includes electronic instruments for automatically measuring distances and angles, satellite surveying systems for quickly obtaining precise positions of widely spaced points, modern aerial digital imaging, and laser-scanning systems for quickly mapping and collecting other forms of data about the Earth, and, as shown in Figure 1.1, small unmanned systems to carry sensors such as digital cameras, laser scanners, and sonar. In addition, computer systems are available that can process the measured data and automatically produce plats, maps, and other products at speeds unheard of a few years ago. Furthermore, these products can be prepared in electronic formats and be transmitted to and from remote locations via telecommunication systems.

Concurrent with the development of these new data collection and processing technologies, *geographic information systems* (GISs) have matured. These

Figure 1.1
Unmanned aerial system. (Courtesy of Leica Geosystems, Inc.)

---

[1]These instruments are described in Appendix A and Chapter 6, respectively.

computer-based systems enable virtually any type of spatially related information about the environment to be integrated, analyzed, displayed, and disseminated.[2] The key to successfully operating GIS is spatially related data of high quality, and the collection and processing of this data placing great new demands upon the surveying community.

As a result of these new developments noted above, and others, many feel that the name surveying no longer adequately reflects the expanded and changing role of their profession. Hence the new term geomatics has emerged. In this text, the terms surveying and geomatics are both used, although the former is used more frequently. Nevertheless, students should understand that the two terms are synonymous as discussed above. Additionally due to the many diverse fields in what was traditionally called surveying, the entire field is now part of what is known as the *geospatial industries*.

1.3 HISTORY OF SURVEYING

The oldest historical records in existence today that bear directly on the subject of surveying state that this science began in Egypt. Herodotus recorded that Sesostris (about 1400 B.C.) divided the land of Egypt into plots for the purpose of taxation. Annual floods of the Nile River swept away portions of these plots, and surveyors were appointed to replace the boundaries. These early surveyors were called *rope-stretchers*, since their measurements were made with ropes having markers at unit distances.

As a consequence of this work, early Greek thinkers developed the science of geometry. Their advance, however, was chiefly along the lines of pure science. Heron stands out prominently for applying science to surveying in about 120 B.C. He was the author of several important treatises of interest to surveyors, including *The Dioptra*, which related the methods of surveying a field, drawing a plan, and making related calculations. It also described one of the first pieces of surveying equipment recorded, the *diopter* [Figure 1.2(a)]. For many years, Heron's work was the most authoritative among Greek and Egyptian surveyors.

Significant development in the art of surveying came from the practical-minded Romans, whose best-known writing on surveying was by Frontinus. Although the original manuscript disappeared, copied portions of his work have been preserved. This noted Roman engineer and surveyor, who lived in the 1st century, was a pioneer in the field, and his essay remained the standard for many years. The engineering ability of the Romans was demonstrated by their extensive construction work throughout the empire. Surveying necessary for this construction resulted in the organization of a surveyors' guild. Ingenious instruments were developed and used. Among these were the *groma* [Figure 1.2(b)], used for sighting; the *libella*, an A-frame with a plumb bob, for leveling; and the *chorobates*, a horizontal straightedge about 20-ft long with supporting legs and a groove on top for water to serve as a level.

---

[2]Geographic information systems are briefly introduced in Section 1.9, and then described in greater detail in Chapter 28.

Figure 1.2
Historical surveying instruments: (a) the diopter and (b) the groma.

One of the oldest Latin manuscripts in existence is the *Codex Acerianus*, written in about the 6th century. It contains an account of surveying as practiced by the Romans and includes several pages from Frontinus's treatise. The manuscript was found in the 10th century by Gerbert and served as the basis for his text on geometry, which was largely devoted to surveying.

During the Middle Ages, the Arabs kept Greek and Roman science alive. Little progress was made in the art of surveying, and the only writings pertaining to it were called "practical geometry."

In the 13th century, Von Piso wrote *Practica Geometria*, which contained instructions on surveying. He also authored *Liber Quadratorum*, dealing chiefly with the *quadrans*, a square brass frame having a 90° angle and other graduated scales. A movable pointer was used for sighting. Other instruments of the period were the *astrolabe*, a metal circle with a pointer hinged at its center and held by a ring at the top, and the *cross staff*, a wooden rod about 4-ft long with an adjustable crossarm at right angles to it. The known lengths of the arms of the cross staff permitted distances to be measured by proportion and angles.

Early civilizations assumed the Earth to be a flat surface, but by noting the Earth's circular shadow on the moon during lunar eclipses and watching ships gradually disappear as they sailed toward the horizon, it was slowly deduced that the planet actually curved in all directions.

Determining the true size and shape of the Earth has intrigued humans for centuries. History records that a Greek named Eratosthenes was among the first to compute its dimensions. His procedure, which occurred about 200 B.C., is illustrated in Figure 1.3. Eratosthenes had concluded that the Egyptian cities of Alexandria and Syene were located approximately on the same meridian, and he had also observed that at noon on the summer solstice, the sun was directly overhead at Syene. (This was apparent because at that time of that day, the image of the sun could be seen reflecting from the bottom of a deep vertical well

Figure 1.3
Geometry of the procedure used by Eratosthenes to determine the Earth's circumference.

there.) He reasoned that at that moment, the sun, Syene, and Alexandria were in a common meridian plane, and if he could measure the arc length between the two cities, and the angle it subtended at the Earth's center, he could compute the Earth's circumference. He determined the angle by measuring the length of the shadow cast at Alexandria from a vertical staff of known length. The arc length was found from multiplying the number of caravan days between Syene and Alexandria by the average daily distance traveled. From these measurements, Eratosthenes calculated the Earth's circumference to be about 25,000 mi. Subsequent precise geodetic measurements using better instruments, but techniques similar geometrically to Eratosthenes', have shown his value, though slightly too large, to be amazingly close to the currently accepted one. (Actually, as explained in Chapter 19, the Earth approximates an oblate spheroid having an equatorial radius about 13.5 mi longer than the polar radius.)

In the 18th and 19th centuries, the art of surveying advanced more rapidly. The need for maps and locations of national boundaries caused England and France to make extensive surveys requiring accurate triangulation; thus, geodetic surveying began. The U.S. Coast Survey (now the National Geodetic Survey of the U.S. Department of Commerce) was established by an act of Congress in 1807. Initially its charge was to perform hydrographic surveys and prepare nautical charts. Later its activities were expanded to include establishment of reference monuments of precisely known positions throughout the country.

Increased land values and the importance of precise boundaries, along with the demand for public improvements in the canal, railroad, and turnpike eras, brought surveying into a prominent position. More recently, the large volume of general construction, numerous land subdivisions that require precise records, and demands posed by the fields of exploration and ecology have entailed an augmented surveying program. Surveying is still the sign of progress in the development, use, and preservation of the Earth's resources.

In addition to meeting a host of growing civilian needs, surveying has always played an important role in our nation's defense activities. World Wars I and II, the Korean and Vietnam conflicts, and the more recent conflicts in the Middle East, Europe, and Afghanistan have created staggering demands for precise measurements and accurate maps. These military operations also provided the stimulus for improving instruments and methods to meet these needs. Surveying also contributed to, and benefited from, the space program where new equipment and systems were needed to provide precise control for missile alignment, and for mapping and charting portions of the moon and nearby planets.

Developments in surveying and mapping equipment have now evolved to the point where the traditional instruments that were used until about the 1960s or 1970s—the transit, theodolite, dumpy level, and steel tape—have now been almost completely replaced by an array of new digital instruments. These include electronic *total station instruments*, which can be used to automatically measure and record horizontal and vertical distances, and horizontal and vertical angles; and *Global Navigation Satellite Systems* (GNSS) such as the *Global Positioning System* (GPS) that can provide precise location information for virtually any type of survey. Laser-scanning instruments combine automatic distance and angle measurements to compute dense grids of coordinated points. Also new aerial digital cameras and remote sensing instruments have been developed, which provide images in digital form, and these images can be processed to obtain spatial information and maps using new *digital photogrammetric restitution instruments* (also called *softcopy plotters*). Figures 1.4, 1.5, 1.6, and 1.7, respectively, show a total station instrument, 3D mobile mapping system, laser-scanning instrument, and modern softcopy plotter. The 3D mobile mapping system in Figure 1.5 is an integrated system consisting of scanners, GNSS-receiver, inertial measurement unit, and a high-quality hemispherical digital camera that can map all items within 100 m of the vehicle as the vehicle travels at highway speeds. The system can capture 1.3 million

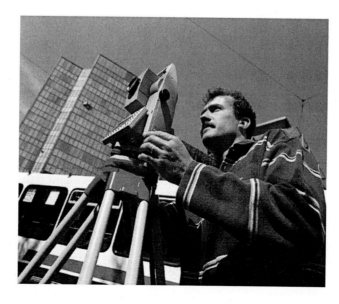

Figure 1.4
LEICA TPS 1100 total station instrument. (Courtesy Leica Geosystems, Inc.)

8 INTRODUCTION

Figure 1.5
A mobile mapping system. (Courtesy Topcon Positioning Systems.)

Figure 1.6
A ground-based laser scanner. (Courtesy of Christopher Gibbons, Leica Geosystems, Inc.)

data points per second providing the end user with high-quality, georeferenced coordinates on all items visible in the images.

One area where laser scanning is finding great acceptance is in *Building Information Management* (BIM). In BIM, scanners are being used to capture the three-dimensional features of both the exterior and interior of existing buildings. The three-dimensional point cloud is then used by engineers, architects, and building contractors in developing plans for reconstruction projects or checking buildings for compliance with codes. Archeologists are also finding uses for laser scanning. By scanning ancient artifacts, not only is the artifact being preserved for future preservation projects, scans have also identified possible clues to how items were constructed and have even identified possible locations of buildings whose locations were previously unknown.

Figure 1.7
Intergraph Image Station Z softcopy plotter. (From *Elements of Photogrammetry: With Applications in GIS*, by Wolf and Dewitt, 2000, Courtesy Intergraph, Inc., and the McGraw-Hill Companies.)

■ 1.4 GEODETIC AND PLANE SURVEYS

Two general classifications of surveys are *geodetic* and *plane*. They differ principally in the assumptions on which the computations are based, although field measurements for geodetic surveys are usually performed to a higher order of accuracy than those for plane surveys.

In geodetic surveying, the curved surface of the Earth is considered by performing the computations on an *ellipsoid* (curved surface approximating the size and shape of the Earth—see Chapter 19). It is now becoming common to do geodetic computations in a three-dimensional, *Earth-centered, Earth-fixed* (ECEF) Cartesian coordinate system. The calculations involve solving equations derived from solid geometry and calculus. Geodetic methods are employed to determine relative positions of widely spaced monuments and to compute lengths and directions of the long lines between them. These monuments serve as the basis for referencing other subordinate surveys of lesser extents. The modern reference systems are four-dimensional including the three-dimensional position of a point on the surface of the Earth with respect to time. These X, Y, H, and t systems take into account a location's movement due to ground motions caused by subsidence and uplift, plate tectonics, and other disturbing sources.

In early geodetic surveys, painstaking efforts were employed to accurately observe angles and distances. The angles were measured using precise ground-based theodolites, and the distances were measured using special tapes made from metal having a low coefficient of thermal expansion. From these basic measurements, the relative positions of the monuments were computed. Later, electronic instruments were used for observing the angles and distances. Although these latter types of instruments are still sometimes used on geodetic surveys, satellite positioning has now almost completely replaced other instruments for these types of surveys. Satellite positioning can provide the needed positions with much greater accuracy, speed, and economy. GNSS receivers enable ground stations to be located precisely by observing distances to satellites operating in known positions along their orbits. GNSS surveys are being used in all forms of surveying including geodetic, hydrographic, construction, and boundary surveying. When combined with a real-time network (RTN), GNSS surveys are capable of providing accuracy within 0.1 ft over a 50-km region with as little as three minutes of data. The principles of operation of the global positioning system are given in Chapter 13, field and office procedures used in static GNSS surveys are discussed in Chapter 14, and the methods used in kinematic GNSS surveys including RTNs are discussed in Chapter 15.

In plane surveying, except for leveling, the reference base for fieldwork and computations is assumed to be a flat horizontal surface. The direction of a plumb line (and thus gravity) is considered parallel throughout the survey region, and all observed angles are presumed to be plane angles. For areas of limited size, the surface of our vast ellipsoid is actually nearly flat. On a line 5-mi line, the ellipsoid arc and chord lengths differ by only about 0.02 ft. A plane surface tangent to the ellipsoid departs only about 0.7 ft at 1 mi from the point of tangency. In a triangle having an area of 75 square miles, the difference between the sum of the three ellipsoidal angles and three plane angles is only about 1 second. Therefore, it is evident that except in surveys covering extensive areas, the Earth's surface can be approximated as a plane, thus simplifying computations and techniques. In general, algebra, plane and analytical geometry, and plane trigonometry are used in plane-surveying calculations. Even for very large areas, map projections, such as those described in Chapter 20, allow plane-surveying computations to be used. This book concentrates primarily on methods of plane surveying, an approach that satisfies the requirements of most projects.

1.5 IMPORTANCE OF SURVEYING

Surveying is one of the world's oldest and most important arts because, as noted previously, from the earliest times it has been necessary to mark boundaries and divide land. Surveying has now become indispensable to our modern way of life. The results of today's surveys are used to (1) map the Earth above and below sea level; (2) prepare navigational charts for use in the air, on land, and at sea; (3) establish property boundaries of private and public lands; (4) develop data banks of land-use and natural resource information that aid in managing our environment; (5) determine facts on the size, shape, gravity, and magnetic fields of the earth; and (6) prepare charts of our moon and planets.

Surveying continues to play an extremely important role in many branches of engineering. For example, surveys are required to plan, construct, and maintain highways, railroads, rapid-transit systems, buildings, bridges, missile ranges, launching sites, tracking stations, tunnels, canals, irrigation ditches, dams, drainage works, urban land subdivisions, water supply and sewage systems, pipelines, and mine shafts. Surveying methods are commonly employed in laying out industrial assembly lines and jigs.[3] These methods are also used for guiding the fabrication

---

[3]These instruments are described in Appendix A and Chapter 6, respectively.

of large equipment, such as airplanes and ships, where separate pieces that have been assembled at different locations must ultimately be connected as a unit. Surveying is important in many related tasks in agronomy, archeology, astronomy, forestry, geography, geology, geophysics, landscape architecture, meteorology, paleontology, and seismology, but particularly in military and civil engineering.

All engineers must know the limits of accuracy possible in construction, plant design and layout, and manufacturing processes, even though someone else may do the actual surveying. In particular, surveyors and civil engineers who are called on to design and plan surveys must have a thorough understanding of the methods and instruments used, including their capabilities and limitations. This knowledge is best obtained by making observations with the kinds of equipment used in practice to get a true concept of the theory of errors and the small but recognizable differences that occur in observed quantities.

In addition to stressing the need for reasonable limits of accuracy, surveying emphasizes the value of significant figures. Surveyors and engineers must know when to work to hundredths of a foot instead of to tenths or thousandths, or perhaps the nearest foot, and what precision in field data is necessary to justify carrying out computations to the desired number of decimal places. With experience, they learn how available equipment and personnel abilities govern procedures and results.

Engineers who design buildings, bridges, equipment, and so on are fortunate if their estimates of loads to be carried are correct within 5%. Then a factor of safety of 2 or more is often applied. However, except for some topographic work, only exceedingly small errors can be tolerated in surveying, and there is no factor of safety. Traditionally, therefore, both manual and computational precision are stressed in surveying.

1.6 SPECIALIZED TYPES OF SURVEYS

Many types of surveys are so specialized that a person proficient in a particular discipline may have little contact with the other areas. Persons seeking careers in surveying and mapping, however, should be knowledgeable in every phase, since all are closely related in modern practice. Some important classifications are described briefly here.

Control surveys establish a network of horizontal and vertical monuments that serve as a reference framework for initiating other surveys. Many control surveys performed today are done using techniques discussed in Chapters 14 and 15 with GNSS instruments.

Topographic surveys determine locations of natural and artificial features and elevations used in map making.

Land, boundary, and *cadastral surveys* establish property lines and property corner markers. The term cadastral is now generally applied to surveys of the public lands systems. There are three major categories: *original surveys* to establish new section corners in unsurveyed areas that still exist in Alaska and several western states; *retracement surveys* to recover previously established boundary lines; and *subdivision surveys* to establish monuments and delineate new parcels of ownership. *Condominium surveys*, which provide a legal record of ownership, are a type of boundary survey.

Hydrographic surveys define shorelines and depths of lakes, streams, oceans, reservoirs, and other bodies of water. *Sea surveying* is associated with port and offshore industries and the marine environment, including measurements and marine investigations made by shipborne personnel.

Alignment surveys are made to plan, design, and construct highways, railroads, pipelines, and other linear projects. They normally begin at one control point and progress to another in the most direct manner permitted by field conditions.

Construction surveys provide line, grade, control elevations, horizontal positions, dimensions, and configurations for construction operations. They also secure essential data for computing construction pay quantities.

As-built surveys document the precise final locations and layouts of engineering works and record any design changes that may have been incorporated into the construction. These are particularly important when underground facilities are constructed, so their locations can be accurately known for maintenance purposes, and so that unexpected damage to them can be avoided during later installation of other underground utilities.

Mine surveys are performed above and below ground to guide tunneling and other operations associated with mining. This classification also includes geophysical surveys for mineral and energy resource exploration.

Solar surveys map property boundaries, solar easements, obstructions according to sun angles, and meet other requirements of zoning boards and title insurance companies.

Optical tooling (also referred to as *industrial surveying* or *optical alignment*) is a method of making extremely accurate measurements for manufacturing processes where small tolerances are required.

Except for control surveys, most other types described are usually performed using plane-surveying procedures, but geodetic methods may be employed on the others if a survey covers an extensive area or requires extreme accuracy.

Ground, aerial, and *satellite surveys* are broad classifications sometimes used. Ground surveys utilize measurements made with ground-based equipment such as automatic levels and total station instruments. Aerial surveys are accomplished using either *photogrammetry* or *remote sensing*. Photogrammetry uses cameras that are carried usually in airplanes to obtain images, whereas remote sensing employs cameras and other types of sensors that can be transported in either aircraft or satellites. Procedures for analyzing and reducing the image data are described in Chapter 27. Aerial methods have been used in all the specialized types of surveys listed, except for optical tooling, and in this area *terrestrial* (ground-based) photographs are often used. Satellite surveys include the determination of ground locations from measurements made to satellites using GNSS receivers, or the use of satellite images for mapping and monitoring large regions of the Earth.

1.7 SURVEYING SAFETY

Surveyors (geomatics engineers) generally are involved in both field and office work. The fieldwork consists in making observations with various types of instruments to either (a) determine the relative locations of points or (b) to set out stakes in accordance with planned locations to guide building and construction operations. The office work involves (1) conducting research and analysis in preparing for surveys, (2) computing and processing the data obtained from field measurements, and (3) preparing maps, plats, charts, reports, and other documents according to client specifications. Sometimes the fieldwork must be performed in hostile or dangerous environments, and thus it is very important to be aware of the need to practice safety precautions.

Among the most dangerous of circumstances within which surveyors must sometimes work are job sites that are either on or near highways or railroads, or that cross such facilities. Job sites in construction zones where heavy machinery is operating are also hazardous, and the dangers are often exacerbated by poor hearing conditions from the excessive noise, and poor visibility caused by obstructions and dust, both of which are created by the construction activity. In these situations, whenever possible, the surveys should be removed from the danger areas through careful planning and/or the use of *offset* lines. If the work must be done in these hazardous areas, then certain safety precautions should be followed. Safety vests of fluorescent yellow color should always be worn in these situations, and flagging materials of the same color can be attached to the surveying equipment to make it more visible. Depending on the circumstances, signs can be placed in advance of work areas to warn drivers of the presence of a survey party ahead, cones and/or barricades can be placed to deflect traffic around surveying activities, and flaggers can be assigned to warn drivers, or to slow or even stop them, if necessary. The *Occupational Safety and Health Administration* (OSHA), of the U.S. Department of Labor,[4] has developed safety standards and guidelines that apply to the various conditions and situations that can be encountered.

Besides the hazards described above, depending on the location of the survey and the time of year, other dangers can also be encountered in conducting field surveys. These include problems related to weather such as frostbite and overexposure to the sun's rays, which can cause skin cancers, sunburns, and heat stroke. To help prevent these problems, plenty of fluids should be drunk, large-brimmed hats and sunscreen can be worn, and on extremely hot days surveying should commence at dawn and terminate at midday or early afternoon. Outside work should not be done on extremely cold days, but if it is necessary, warm clothing should be worn and skin areas should not be exposed. Other hazards that can be encountered during field surveys include wild animals, poisonous snakes, bees, spiders, wood ticks, deer ticks (which can carry Lyme disease), poison ivy, and poison oak. Surveyors should be knowledgeable about the types of hazards that can be expected in any local area, and always be alert and on the lookout for them. To help prevent injury from these sources, protective boots and clothing should be worn and insect sprays used. Certain tools can also be dangerous, such as chainsaws, axes, and machetes that are sometimes necessary for clearing lines of sight. These must always be handled with care. In addition, care must be exercised in

---

[4]The mission of OSHA is to save lives, prevent injuries, and protect the health of America's workers. Its staff establishes protective standards, enforces those standards, and reaches out to employers and employees through technical assistance and consultation programs. For more information about OSHA and its safety standards, consult its website http://www.osha.gov.

handling certain surveying instruments, like long-range poles and level rods, especially when working around overhead wires, to prevent accidental electrocutions.

Many other hazards, in addition to those cited above, can be encountered when surveying in the field. Thus, it is essential that surveyors always exercise caution in their work, and know and follow accepted safety standards. In addition, a first-aid kit should always accompany a survey party in the field, and it should include all of the necessary antiseptics, ointments, bandage materials, and other equipment needed to render first aid for minor accidents. The survey party should also be equipped with cell phones for more serious situations, and telephone numbers to call in emergencies should be written down and readily accessible.

1.8 LAND AND GEOGRAPHIC INFORMATION SYSTEMS

Land Information Systems (LISs) and GISs are areas of activity that have rapidly assumed positions of major prominence in surveying. These computer-based systems enable storing, integrating, manipulating, analyzing, and displaying virtually any type of spatially related information about our environment. LISs and GISs are being used at all levels of government, and by businesses, private industry, and public utilities to assist in the management and decision-making process. Specific applications have occurred in many diverse areas and include natural resource management, facilities siting and management, land records modernization, demographic and market analysis, emergency response and fleet operations, infrastructure management, and regional, national, and global environmental monitoring. Data stored within LISs and GISs may be both natural and cultural, and be derived from new surveys, or from existing sources such as maps, charts, aerial and satellite photos, tabulated data and statistics, and other documents. However, in most situations, the needed information either does not exist, or it is unsatisfactory because of age, scale, or other reasons. Thus, new measurements, maps, photos, or other data must be obtained.

Specific types of information (also called *themes* or *layers* of information) needed for land and GISs may include political boundaries, individual property ownership, population distribution, locations of natural resources, transportation networks, utilities, zoning, hydrography, soil types, land use, vegetation types, wetlands, and many, many more. An essential ingredient of all information entered into LIS and GIS databases is that it is spatially related, that is, located in a common geographic reference framework. Only then are the different layers of information physically relatable so they can be analyzed using computers to support decision making. This geographic positional requirement will place a heavy demand upon surveyors (geomatics engineers) in the future, who will play key roles in designing, implementing, and managing these systems. Surveyors from virtually all of the specialized areas described in Section 1.6 will be involved in developing the needed databases. Their work will include establishing the required basic control framework; conducting boundary surveys and preparing legal descriptions of property ownership; performing topographic and hydrographic surveys by ground, aerial, and satellite methods; compiling and digitizing maps; and assembling a variety of other digital data files.

The last chapter of this book, Chapter 28, is devoted to the topic of land and geographic information systems. This subject seems appropriately covered at the end, after each of the other types of surveys needed to support these systems has been discussed.

1.9 FEDERAL SURVEYING AND MAPPING AGENCIES

Several agencies of the U.S. government perform extensive surveying and mapping. Three of the major ones are:

1. The National Geodetic Survey (NGS), formerly the Coast and Geodetic Survey, was originally organized to map the coast. Its activities have included control surveys to establish a network of reference monuments throughout the United States that serve as points for originating local surveys, preparation of nautical and aeronautical charts, photogrammetric surveys, tide and current studies, collection of magnetic data, gravimetric surveys, and worldwide control survey operations. The NGS plays a major role in coordinating and assisting in activities related to upgrading the national network of reference control monuments, and to the development, storage, and dissemination of data used in modern LISs and GISs.
2. The U.S. Geological Survey (USGS), established in 1879, has as its mission the mapping of our nation and the survey of its resources. It provides a wide variety of maps, from topographic maps showing the geographic relief and natural and cultural features, to thematic maps that display the geology and water resources of the United States, to special maps of the moon and planets. The National Mapping Division of the USGS has the responsibility of producing topographic maps. It currently has nearly 70,000 different topographic maps available, and it distributes approximately 10 million copies annually. In recent years, the USGS has been engaged in a comprehensive program to develop a national digital cartographic database, which consists of map data in computer-readable formats.
3. The Bureau of Land Management (BLM), originally established in 1812 as the General Land Office, is responsible for managing the public lands. These lands, which total approximately 264 million acres and comprise about one eighth of the land in the United States, exist mostly in the western states and Alaska. The BLM is responsible for surveying the land and managing its natural resources, which include minerals, timber, fish and wildlife, historical sites, and other natural heritage areas. Surveys of most public lands in the conterminous United States have been completed, but much work remains in Alaska.

In addition to these three federal agencies, units of the U.S. Army Corps of Engineers have made extensive surveys for emergency and military purposes. Some of these surveys provide data for engineering projects, such as those connected with flood control. Surveys of wide extent have also been conducted for special purposes by nearly 40 other federal agencies, including the Forest Service, National Park Service, International Boundary Commission, Bureau of

Reclamation, Tennessee Valley Authority, Mississippi River Commission, U.S. Lake Survey, and Department of Transportation.

All states have a surveying and mapping section for purposes of generating topographic information upon which highways are planned and designed. Likewise, many counties and cities also have surveying programs, as have various utilities.

1.10 THE SURVEYING PROFESSION

The personal qualifications of surveyors are as important as their technical ability in dealing with the public. They must be patient and tactful with clients and their sometimes-hostile neighbors. Few people are aware of the painstaking research of old records required before fieldwork is started. Diligent, time-consuming effort may be needed to locate corners on nearby tracts for checking purposes as well as to find corners for the property in question.

Land or boundary surveying is classified as a learned profession because the modern practitioner needs a wide background of technical training and experience, and must exercise a considerable amount of independent judgment. Registered (licensed) professional surveyors must have a thorough knowledge of mathematics (particularly geometry, trigonometry, and calculus); competence with computers; a solid understanding of surveying theory, instruments, and methods in the areas of geodesy, photogrammetry, remote sensing, and cartography; some competence in economics (including office management), geography, geology, astronomy, and dendrology; and a familiarity with laws pertaining to land and boundaries. They should be knowledgeable in both field operations and office computations. Above all, they are governed by a professional code of ethics and are expected to charge professional-level fees for their work.

Permission to trespass on private property or to cut obstructing tree branches and shrubbery must be obtained through a proper approach. Such privileges are not conveyed by a surveying license or by employment in a state highway department or other agency (but a court order can be secured if a landowner objects to necessary surveys).

All 50 states, Guam, and Puerto Rico have registration laws for professional surveyors and engineers (as do the provinces of Canada). In general, a surveyor's license is required to make property surveys, but not for construction, topographic, or route work, unless boundary corners are set.

To qualify for registration as either a professional land surveyor (PLS) or a professional engineer (PE), it is necessary to have an appropriate college degree, although some states allow relevant experience in lieu of formal education. In addition, candidates must acquire two or more years of mentored practical experience and must also pass comprehensive written examinations. In most states, common national examinations covering fundamentals and principles and practice of surveying are now used. However, usually two hours of the principles and practice exam are devoted to local legal customs and aspects. As a result, transfer of registration from one state to another has become easier.

Many states also require continuing education units (CEUs) for registration renewal. Typical state laws require that a licensed land surveyor sign all plats, assume responsibility for any liability claims, and take an *active part* in the fieldwork.

■ 1.11 PROFESSIONAL SURVEYING ORGANIZATIONS

There are many professional organizations in the United States and worldwide that serve the interests of surveying and mapping. Generally, the objectives of these organizations are the advancement of knowledge in the field, encouragement of communication among surveyors, and upgrading of standards and ethics in surveying practice. The National Society of Professional Surveyors (NSPS) represents boundary and construction surveyors in the United States. The *American Association for Geodetic Surveying* (AAGS) represents geodetic surveying professionals and the *Geographic and Land Information Society* (GLIS) represent those involved in GISs. AAGS and GLIS publish the biannual journal *Surveying and Land Information Science* (SaLIS).

As noted in the preceding section, all states require persons who perform boundary surveys to be licensed. Most states also have professional surveyor societies or organizations with full membership open only to licensed surveyors. These state societies are generally affiliated with NSPS and concentrate on matters of state and local concern.

The *American Society for Photogrammetry and Remote Sensing* (ASPRS) is an organization also devoted to the advancement of the fields of measurement and mapping, although its major interests are directed toward the use of aerial and satellite imagery for achieving these goals. Its monthly journal *Photogrammetric Engineering and Remote Sensing* regularly features surveying and mapping articles.

The *Geomatics Division* of the *American Society of Civil Engineers* (ASCE) is also dedicated to professional matters related to surveying and publishes quarterly the *Journal of Surveying Engineering*.

The *Surveying and Geomatics Educators Society* (SAGES) holds pedagogical conferences on the instruction of surveying/geomatics in higher educational institutions. These conferences occur biennially at host institutions throughout the North American continent.

Another organization in the United States, the *Urban and Regional Information Systems Association* (URISA), also supports the profession of surveying and mapping. This organization uses information technology to solve problems in planning, public works, the environment, emergency services, and utilities. Its *URISA Journal* is published quarterly.

The *Canadian Institute of Geomatics* (CIG), formerly the *Canadian Institute of Surveying and Mapping* (CISM), is the foremost professional organization in Canada concerned with surveying. This organization disseminates information to its members through its *Geomatica*.

The *International Federation of Surveyors* (FIG), founded in 1878, fosters the exchange of ideas and information among surveyors worldwide. The acronym

FIG stems from its French name, *Fédération Internationale des Géométres*. FIG membership consists of professional surveying organizations from many countries throughout the world. FIG is organized into nine technical commissions, each concerned with a specialized area of surveying. The organization sponsors international conferences, usually at four-year intervals, and its commissions also hold periodic symposia where delegates gather for the presentation of papers on subjects of international interest.

■ 1.12 SURVEYING ON THE INTERNET

The explosion of available information on the Internet has had a significant impact on the field of surveying (geomatics). The Internet enables the instantaneous electronic transfer of documents to any location where the necessary computer equipment is available. It brings resources directly into the office or home, where previously it was necessary to travel to obtain the information or wait for its transfer by mail. Some equipment manufacturers provide cloud access in their instruments to support the transfer of information from and to the field crews. Software, educational materials, technical documents, standards, and much more useful information are available on the Internet. As an example of how surveyors can take advantage of the Internet, data from a *Continuously Operating Reference Station* (CORS) can be downloaded from the NGS website for use in GNSS surveys (see Section 14.3.5).

Many agencies and institutions maintain websites that provide data free of charge on the Internet. Additionally, some educational institutions now place credit and noncredit courses on the Internet so that distance education can be more easily achieved. With a web browser, it is possible to research almost any topic from a convenient location, and names, addresses, and phone numbers of goods or services providers in a specific area can be identified. As an example, if it was desired to find companies offering mapping services in a certain region, a web search engine could be used to locate web pages that mention this service. Such a search may result in over a million pages if a very general term such as "mapping services" is used to search, but using more specific terms can narrow the search.

Unfortunately, the addresses of particular pages and entire sites, given by their *Uniform Resource Locators* (URLs), tend to change with time. However, at the risk of publishing URLs that may no longer be correct, a short list of important websites related to surveying is presented in Table 1.1.

■ 1.13 FUTURE CHALLENGES IN SURVEYING

Surveying is currently in the midst of a revolution in the way data are measured, recorded, processed, stored, retrieved, and shared. This is in large part because of developments in computers and computer-related technologies. Concurrent with technological advancements, society continues to demand more data, with increasingly higher standards of accuracy, than ever before. Consequently, in a few years the demands on surveying engineers (geomatics engineers) will likely be very different from what they are now.

In the future, the National Spatial Reference System, a network of horizontal and vertical control points, must be maintained and supplemented to meet

TABLE 1.1 UNIFORM RESOURCE LOCATOR ADDRESSES FOR SOME SURVEYING RELATED SITES

Uniform Resource Locator	Owner of Site
http://www.ngs.noaa.gov	National Geodetic Survey
http://www.usgs.gov	U.S. Geological Survey
http://www.blm.gov	Bureau of Land Management
http://www.navcen.uscg.mil	U.S. Coast Guard Navigation Center
http://www.usno.navy.mil	U.S. Naval Observatory
http://www.nsps.us.com/	National Society of Professional Surveyors
http://www.aagsmo.org	American Association for Geodetic Surveying
https://www.cig-acsg.ca/	Canadian Institute of Geomatics
http://www.asprs.org	American Society for Photogrammetry and Remote Sensing
http://www.asce.org	American Society of Civil Engineers
http://www.geosages.org/	Surveying and Geomatics Educators Society
http://pipelinesurveyor.org/	Association of Pipeline Surveyors
https://www.asce.org/utility-engineering-and-surveying/utility-engineering-and-surveying-institute/	Utility Engineering and Surveying Institute
https://aagsmo.org/salis-journal/	Surveying and Land Information Science

requirements of increasingly higher-order surveys. New topographic maps with larger scales as well as digital map products are necessary for better planning. Existing maps of our rapidly expanding urban areas need revision and updating to reflect changes, and more and better map products are needed of the older parts of our cities to support urban renewal programs and infrastructure maintenance and modernization. Large quantities of data will be needed to plan and design new rapid-transit systems to connect our major cities, and surveyors will face new challenges in meeting the precise standards required in staking alignments and grades for these systems.

In the future, assessment of environmental impacts of proposed construction projects will call for more and better maps and other data. GISs and LISs that contain a variety of land-related data such as ownership, location, acreage, soil types, land uses, and natural resources must be designed, developed, and maintained. Cadastral surveys of the yet unsurveyed public lands are essential. Monuments set years ago by the original surveyors have to be recovered and remonumented for preservation of property boundaries. Appropriate surveys with very demanding accuracies will be necessary to position drilling rigs as mineral and oil explorations press further offshore. Other future challenges include making precise deformation surveys for monitoring existing structures such as dams, bridges, and skyscrapers to detect imperceptible movements that could be precursors to catastrophes caused by their failure. Timely measurements and maps of the effects of natural disasters such as earthquakes, floods, and hurricanes will be needed so that effective relief and assistance efforts can be planned and implemented. In the space program, the desire for maps of neighboring planets will continue. In addition, we must increase our activities in measuring and monitoring natural and human-caused global changes (glacial growth and retreat, volcanic activity, large-scale deforestation, and so on) that can potentially affect our land, water, atmosphere, energy supply, and even our climate.

These and other opportunities offer professionally rewarding indoor or outdoor (or both) careers for numerous people with suitable training in the various branches of surveying.

PROBLEMS

NOTE: Answers for some of these problems, and some in later chapters, can be obtained by consulting the bibliographies, later chapters, websites, or professional surveyors.

1.1 List 10 uses for surveying in areas other than land surveying.
1.2 Explain the difference between geodetic and plane surveys.
1.3 List some application of surveying in geology, forestry, and mining.
1.4 Why is it important to make accurate surveys of underground utilities?
1.5 Discuss the uses for topographic surveys.
1.6 What are hydrographic surveys, and why are they important?
1.7 Print a view of your location using Google Earth.®
1.8 Briefly explain the procedure used by Eratosthenes in determining the Earth's circumference.
1.9 Describe the steps a land surveyor would need to do when performing a boundary survey.
1.10 What is the name of the state-level professional surveying organization in your state or region?
1.11 List the tasks of a surveyor as defined by the International Federation of Surveyors?
1.12 What organizations in your state furnish maps and reference data to surveyors and engineers?
1.13 List the legal requirements for registration as a land surveyor in your state.
1.14 Briefly describe an Earth-Centered, Earth-Fixed coordinate system.
1.15 List the professional societies representing the geospatial industry in the
 (a) United States. **(b)** Canada. **(c)** International.
1.16 Explain how aerial photographs and satellite images can be valuable in surveying.
1.17 List the various specialized types of surveys.
1.18 What types of office work will a surveyor encounter?
1.19 Visit one of the surveying websites listed in Table 1.1, and write a brief summary of its contents. Briefly explain the value of the available information to surveyors.
1.20 Read one of the articles cited in the bibliography for this chapter, or another of your choosing, that describes an application where satellite surveying methods were used. Write a brief summary of the article.
1.21 Same as Problem 1.20, except the article should be on safety as related to surveying.

BIBLIOGRAPHY

Anderson, J. 2012. "Wow! I'm an LSIT! (Uh, Now What Do I Do?" *The American Surveyor* 11 (No. 5): 23.

Aust, M. 2014. "A Safe Habitat." *Point of Beginning* 38 (No. 2): 36.

Blouin, D. 2014. "An Intro to IoT: Internet of Things and Internet of Where-Things-Are." *XYHT* 1 (No. 3): 33.

Brown, D. and J. Ouellete. 2014. "Aid after Catastrophe." *XYHT Fall Supplement*: 8.

Cantu, O. R. 2014. "The Cloud: Changing the Dynamic of Field-to-Office Communication." *The American Surveyor* 11 (No. 9): 34.

Cheves, M. 2014. "Smithsonian X 3D" *The American Surveyor* 11 (No. 4): 16.

Dahn, R. E. and R. Lumos. 2006. "National Society of Professional Surveyors." *Surveying and Land Information Science* 66 (No. 2): 111.

Duffy, L. 2014. "RFID: Going Below the Surface." *Point of Beginning* 39 (No. 6): 14.

Greenfeld, J. 2006. "The Geographic and Land Information Society and GIS/LIS Activities in the United States." *Surveying and Land Information Science* 66 (No. 2): 119.

Koon, R. 2014. "Safety Sense: How Bad Can It Get after an Accident." *Point of Beginning* 39 (No. 9): 32.

Lathrop, W. and D. Martin. 2006. "The American Association for Geodetic Surveying: Its Continuing Role in Shaping the Profession." *Surveying and Land Information Science* 66 (No. 2): 97.

Luccio, M. 2014. "Captured: From Traditional Photogrammetry to UAS." *XYHT* 1 (No. 2): 42.

_____. 2014. "From Shore to Floor." *XYHT* 1 (No. 4): 16.

_____. 2014. "Extraterrestrial Photogrammetry." *XYHT* 1 (No. 1): 20.

_____. 2014. "Precision Ag: Satellite Signals Steer Farmers Straight." *XYHT* 1 (No. 5): 31.

Morton, C. 2014. "Digging into Archaeology." *The American Surveyor* 11 (No. 6): 38.

Naberezny, B. 2014. "National Surveyors Week 2014." *Professional Surveyor* 34 (No. 5): 8.

Paulk, M. 2016. "One Man Crews." *The American Surveyor* 14 (No. 7): 4.

Plaza, J. 2014. "UAVs in Latin America: All Over the Map." *XYHT* 1 (No. 5): 24.

Silver, M. 2014. "The Dinosaur Surveyors." *The American Surveyor* 11 (No. 8): 18.

Velde, G. 2014. "Out of Africa." *The American Surveyor* 11 (No. 2): 16.

Wilson, B. 2014. "A Dam with a View." *The American Surveyor* 11 (No. 10): 20.

2
Units, Significant Figures, and Field Notes

PART I • UNITS AND SIGNIFICANT FIGURES

■ 2.1 INTRODUCTION

Five types of observations illustrated in Figure 2.1 form the basis of traditional plane surveying: (1) horizontal angles, (2) horizontal distances, (3) vertical (or zenith) angles, (4) vertical distances, and (5) slope distances. In the figure, *OAB* and *ECD* are horizontal planes, and *OACE* and *ABDC* are vertical planes. Then as illustrated, horizontal angles, such as angle *AOB*, and horizontal distances, *OA* and *OB*, are measured in horizontal planes; vertical angles, such as *AOC*, are measured in vertical planes; zenith angles, such as *EOC*, are also measured in vertical planes; vertical lines, such as *AC* and *BD*, are measured vertically (in the direction of gravity); and slope distances, such as *OC*, are determined along inclined planes. By using combinations of these basic observations, it is possible to compute relative positions between any points. Equipment and procedures for making each of these basic kinds of observations are described in later chapters of this book.

■ 2.2 UNITS OF MEASUREMENT

Magnitudes of measurements (or of values derived from observations) must be given in terms of specific units. In surveying, the most commonly employed units are for *length, area, volume,* and *angle.* Two different systems are in use for specifying units of observed quantities, the *English* and *metric* systems. Because of its widespread adoption, the metric system is called the *International System of Units,* and abbreviated *SI.*

The basic unit employed for length measurements in the English system is the foot, whereas the meter is used in the metric system. In the past, two different

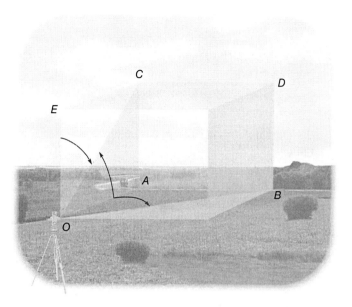

Figure 2.1
Kinds of measurements in surveying.

definitions have been used to relate the foot and meter. Although they differ slightly, their distinction must be made clear in surveying. In 1893, the United States officially adopted a standard in which 39.37 in. was exactly equivalent to 1 m. Under this standard, the foot was approximately equal to 0.3048006 m. In 1959, a new standard was officially adopted in which the inch was equal to exactly 2.54 cm. Under this standard, 1 ft equals exactly 0.3048 m. This current unit, known as the international foot, differs from the previous one by about 1 part in 500,000, or approximately 1 foot per 100 miles. This small difference is thus important for very precise surveys conducted over long distances, and for conversions of high elevations or large coordinate values such as those used in State Plane Coordinate Systems as discussed in Chapter 20. Because of the vast number of surveys performed prior to 1959, it would have been extremely difficult and confusing to change all related documents and maps that already existed. Thus the old standard, now called the *U.S. survey foot (sft)*, is still used. Individual states have the option of officially adopting either standard, which is further discussed in Sections 2.3 and 20.2. The National Geodetic Survey uses the meter in its distance measurements; thus, it is unnecessary to specify the foot unit. However, those making conversions from metric units must know the adopted standard for their state and use the appropriate conversion factor.

Because the English system has long been the officially adopted standard for measurements in the United States, except for geodetic surveys, the linear units of feet and *decimals* of a foot are most commonly used by surveyors. In construction, feet and inches are often used. Because surveyors perform all types of surveys including geodetic, and they also provide measurements for developing construction plans and guiding building operations, they must understand all the various systems of units and be capable of making conversions between them. Caution must always be exercised to ensure that observations are recorded in their proper units, and conversions are correctly made.

A summary of the length units used in past and present surveys in the United States includes the following:

1 foot = 12 inches

1 yard = 3 feet

1 inch = 2.54 centimeters (basis of international foot)

1 meter = 39.37 inches (basis of U.S. survey foot)

1 rod = 1 pole = 1 perch = 16.5 feet

1 vara = approximately 33 inches (old Spanish unit often encountered in the southwestern United States)

1 Gunter's chain (ch) = 66 feet = 100 links (lk) = 4 rods

1 mile = 5280 feet = 80 Gunter's chains

1 nautical mile = 1852 m $\approx$ 6076.10 sft $\approx$ 6076.12 ift (nominal length of a minute for any great circle on the Earth)

1 fathom = 6 feet.

In the English system, areas are given in *square feet* or *square yards*. The most common unit for large areas is the *acre*. Ten square chains (Gunter's) equal 1 acre. Thus, an acre contains 43,560 ft^2, which is the product of 10 and 66^2. The *arpent* (equal to approximately 0.85 acre, but varying somewhat in different states) was used in land grants of the French crown. When employed as a linear term, it refers to the length of a side of 1 square arpent.

Volumes in the English system can be given in *cubic feet* or *cubic yards*. For very large volumes, for example, the quantity of water in a reservoir, the *acre-foot* unit is used. It is equivalent to the area of an acre having a depth of 1 ft, and thus is 43,560 ft^3.

The unit of angle used in surveying is the *degree*, defined as 1/360 of a circle. One degree (1°) equals 60 minutes, and 1 minutes equals 60 seconds. Divisions of seconds are given in tenths, hundredths, and thousandths. Other methods are also used to subdivide a circle, for example, 400 *grad* (with 100 *centesimal min/grad* and 100 *centesimal sec/min*. Another term, *gons*, is now used interchangeably with grads. The military services use *mils* to subdivide a circle into 6400 units.

A *radian* is the angle subtended by an arc of a circle having a length equal to the radius of the circle. Therefore, 2π rad = 360°, 1 rad $\approx$ 57°17'44.8" $\approx$ 57.2958°, and 0.01745 rad $\approx$ 1°.

■ 2.3 INTERNATIONAL SYSTEM OF UNITS (SI)

As noted previously, the meter is the basic unit for length in the metric or SI system. Subdivisions of the meter (m) are the *millimeter* (mm), *centimeter* (cm), and *decimeter* (dm), equal to 0.001, 0.01, and 0.1 m, respectively. A kilometer (km) equals 1000 m, which is approximately five eighths of a mile.

Areas in the metric system are specified using the *square meter* (m^2). Large areas, for example, tracts of land, are given in *hectares* (ha), where 1 ha is equivalent to a square having sides of 100 m. Thus, there are 10,000 m^2, or about 2.471 acres/ha. The *cubic meter* (m^3) is used for volumes in the SI system. Degrees, minutes, and seconds, or the radian, are accepted SI units for angles.

The metric system was originally developed in the 1790s in France. Although other definitions were suggested at that time, the French Academy of Sciences chose to define the meter as 1/10,000,000 of the length of the Earth's meridian through Paris from the equator to the pole. The actual length that was adopted for the meter was based on observations that had been made up to that time to determine the Earth's size and shape. Although later measurements revealed that the initially adopted value was approximately 0.2 mm short of its intended definition related to the meridional quadrant, still the originally adopted length became the standard.

Shortly after the metric system was introduced to the world, Thomas Jefferson who was the then secretary of state, recommended that the United States adopt it, but the proposal lost by one vote in the Congress! When the metric system was finally legalized for use (but not officially adopted) in the United States in 1866, a meter was defined as the interval under certain physical conditions between lines on an international prototype bar made of 90% platinum and 10% iridium, and accepted as equal to exactly 39.37 in. A copy of this bar was held in Washington, D.C. and compared periodically with the international standard held in Paris. In 1960, at the General Conference on Weights and Measures (CGPM), the United States and 35 other nations agreed to redefine the meter as the length of 1,650,763.73 waves of the orange-red light produced by burning the element krypton (Kr-86). That definition permitted industries to make more accurate measurements and to check their own instruments without recourse to the standard meter-bar in Washington. The wavelength of this light is a true constant, whereas there is a risk of instability in the metal meter-bar. The CGPM met again in 1983 and established the current definition of the meter as the length of the path traveled by light in a vacuum during a time interval of 1/299,792,458 seconds. Obviously with this definition, the speed of light in a vacuum becomes exactly 299,792,458 m/sec. The advantage of this latest standard is that the meter is more accurately defined, since it is in terms of time, the most accurate of our basic measurements.

During the 1960s and 1970s, significant efforts were made toward promoting adoption of SI as the legal system for weights and measures in the United States. However, costs and frustrations associated with making the change generated substantial resistance, and the efforts were temporarily stalled. Recognizing the importance to the United States of using the metric system in order to compete in the rapidly developing global economy, in 1988 the Congress enacted the *Omnibus Trade and Competitiveness Act*. It designated the metric system as the *preferred* system of weights and measures for U.S. trade and commerce. The Act, together with a subsequent *Executive Order* issued in 1991, required all federal agencies to develop definite metric conversion plans and to use SI standards in their procurements, grants, and other business-related activities to the extent economically feasible. As an example of one agency's response, the Federal Highway Administration adopted a plan calling for (1) use of metric units in all publications and correspondence after September 30, 1992 and (2) use of metric units on all plans and contracts for federal highways after September 30, 1996. Although the Act and Executive Order did not mandate states, counties, cities, or industries to convert to metric, strong incentives were provided, for example, if SI directives were not complied with, certain federal matching funds could be withheld.

In light of these developments, it appeared that the metric system would soon become the official system for use in the United States. However, again much resistance was encountered, not only from individuals but also from agencies of some state, county, town, and city governments, as well as from certain businesses. As a result, the SI still has not been adopted officially in the United States.

Besides the obvious advantage of being better able to compete in the global economy, another significant advantage that would be realized in adopting the SI standard would be the elimination of the confusion that exists in making conversions between the English System and the SI. The 1999 crash of the Mars Orbiter underscores costs and frustrations associated with this confusion. This $125 million satellite was supposed to monitor the Martian atmosphere, but instead it crashed into the planet because its contractor used English units while NASA's Jet Propulsion Laboratory was giving it data in the metric system. For these reasons and others, such as the decimal simplicity of the metric system, surveyors who are presently burdened with unit conversions and awkward computations involving yard, foot, and inch units should welcome official adoption of the SI. However, since this adoption has not yet occurred, this book uses both English and SI units in discussion and example problems.

Legislatively, Arizona, Michigan, Montana, North Dakota, Oregon, and South Carolina have adopted the international foot definition for conversions. Alabama and Missouri did not legislatively define the foot. All remaining states adopted the survey foot definition adding confusion to which definition of the foot should be used in conversions. This book will use the survey foot definition unless otherwise specified. It should be noted that sometime after January 1, 2023, the United States will be deprecating the survey foot definition, and thus the surveyor will need to adopt this change according to the epoch of the record data.

2.4 SIGNIFICANT FIGURES

In recording observations, an indication of the accuracy attained is the number of digits (significant figures) recorded. By definition, the number of significant figures in any observed value includes the positive (certain) digits plus one (*only one*) digit that is estimated or rounded off, and therefore questionable. For example, a distance measured with a tape whose smallest graduation is 0.01 ft, and recorded as 73.52 ft, is said to have four significant figures; in this case the first three digits are certain, and the last is rounded off and therefore questionable but still significant.

To be consistent with the theory of errors discussed in Chapter 3, it is essential that data be recorded with the correct number of significant figures. If a significant figure is dropped in recording a value, the time spent in acquiring certain precision has been wasted. On the other hand, if data are recorded with more figures than those that are significant, false precision will be implied. The number of significant figures is often confused with the number of decimal places. Decimal places may have to be used to maintain the correct number of significant figures, but in themselves they do not indicate significant figures. Some examples follow:

Two significant figures: 24, 2.4, 0.24, 0.0024, 0.020

Three significant figures: 364, 36.4, 0.000364, 0.0240

Four significant figures: 7621, 76.21, 0.0007621, 24.00.

Zeros at the end of an integer value may cause difficulty because they may or may not be significant. In a value expressed as 2400, for example, it is not known how many figures are significant; there may be two, three, or four, and therefore definite rules must be followed to eliminate the ambiguity. The preferred method of eliminating this uncertainty is to express the value in terms of powers of 10. The significant figures in the measurement are then written in scientific notation as a number between 1 and 10 with the correct number of zeros and power of 10. As an example, 2400 becomes 2.400×10^3 if both zeros are significant, 2.40×10^3 if one is significant, and 2.4×10^3 if there are only two significant figures. Alternatively, a bar may be placed over the last significant figure, as $240\bar{0}, 2\bar{4}00,$ and $2\bar{4}00$ for 4, 3, and 2 significant figures, respectively.

When observed values are used in the mathematical processes of addition, subtraction, multiplication, and division, it is imperative that the number of significant figures given in answers be consistent with the number of significant figures in the data used. The following three steps will achieve this for addition or subtraction: (1) identify the column containing the rightmost significant digit in each number being added or subtracted, (2) perform the addition or subtraction, and (3) round the answer so that its rightmost significant digit occurs in the leftmost column identified in step (1). Two examples illustrate the procedure.

```
       (a)                    (b)
      46.7418                 378.
    +  1.03                  − 2.1
    +375.0                    375.9
     ───────               (answer 376.)
     422.7718
   (answer 422.8)
```

In (a), the digits 8, 3, and 0 are the rightmost significant ones in the numbers 46.7418, 1.03, and 375.0, respectively. Of these, the 0 in 375.0 is leftmost with respect to the decimal. Thus, the answer 422.7718 obtained on adding the numbers is rounded to 422.8, with its rightmost significant digit occurring in the same column as the 0 in 375.0. In (b), the digits 8 and 1 are rightmost, and of these the 8 is leftmost. Thus, the answer 375.9 is rounded to 376.

In multiplication, the number of significant figures in the answer is equal to the least number of significant figures in any of the factors. For example, $362.56 \times 2.13 = 772.2528$ when multiplied but the answer is correctly given as 772. Its three significant figures are governed by the three significant digits in 2.13. Likewise, in division the quotient should be rounded off to contain only as many significant figures as the least number of significant figures in either the divisor or the dividend. These rules for significant figures in computations stem from error propagation theory, which is discussed further in Section 3.17.

In surveying, four specific types of problems relating to significant figures are encountered and must be understood.

1. Field measurements are given to some specific number of significant figures, thus dictating the number of significant figures in answers derived when the measurements are used in computations. In an intermediate calculation, it is

a common practice to carry at least one more digit than required, and then round off the final answer to the correct number of significant figures.

2. There may be an implied number of significant figures. For instance, the length of a football field might be specified as 100 yd. But in laying out the field, such a distance would probably be measured to the nearest hundredth of a foot, not the nearest half-yard.

3. Each factor may not cause an equal variation. For example, if a steel tape 100.00 ft long is to be corrected for a change in temperature of 15°F, one of these numbers has five significant figures while the other has only two. However, a 15° variation in temperature changes the tape length by only 0.01 ft. Therefore, an adjusted tape length to five significant figures is warranted for this type of data. Another example is the computation of a slope distance from horizontal and vertical distances, as in Figure 2.2. The vertical distance V is given to two significant figures, and the horizontal distance H is measured to five significant figures. From these data, the slope distance S can be computed to five significant figures. For small angles of slope, a considerable change in the vertical distance produces a relatively small change in the difference between slope and horizontal distances.

4. Observations are recorded in one system of units but may have to be converted to another. A good rule to follow in making these conversions is to retain in the answer a number of significant figures equal to those in the observed value. As an example, to convert 178 ft 6-3/8 in. to m, the number of significant figures in the measured value would first be determined by expressing it in its smallest units. In this case, 1/8th in is the smallest unit and there are $(178 \times 12 \times 8) + (6 \times 8) + 3 = 17{,}139$ of these units in the value. Thus, the measurement contains five significant figures, and the answer is $17{,}139 \div (8 \times 39.37 \text{ in./m}) = 54.416$ m, properly expressed with five significant figures. (Note that 39.37 used in the conversion is an exact constant and does not limit the number of significant figures.)

2.5 ROUNDING OFF NUMBERS

Rounding off a number is the process of dropping one or more digits so the answer contains only those digits that are significant. In rounding off numbers to any required degree of precision in this text, the following procedures will be observed:

1. When the digit to be dropped is lower than 5, the number is written without the digit. Thus, 78.374 becomes 78.37. Also 78.3749 rounded to four figures becomes 78.37.

2. When the digit to be dropped is exactly 5, the nearest even number is used for the preceding digit. Thus, 78.375 becomes 78.38 and 78.385 is also rounded to 78.38.

Figure 2.2
Slope correction.

3. When the digit to be dropped is greater than 5, the number is written with the preceding digit increased by 1. Thus, 78.386 becomes 78.39.

Procedures 1 and 3 are standard practice. When rounding the value 78.375 in procedure 2, however, some people always take the next higher hundredth, whereas others invariably use the next lower hundredth. However, using the nearest even digit establishes a uniform procedure and produces better-balanced results in a series of computations. It is an improper procedure to perform two-stage rounding where, for example, in rounding 78.3749 to four digits it would be first rounded to five figures, yielding 78.375, and then rounded again to 78.38. The correct answer in rounding 78.3749 to four figures is 78.37.

It is important to recognize that rounding should only occur with the final answer. Intermediate computations should be done without rounding to avoid problems that can be caused by rounding too early. Example (a) of Section 2.4 is repeated below to illustrate this point. The sum of 46.7418, 1.03, and 375.0 is rounded to 422.8 as shown in the "correct" column. If the individual values are rounded prior to the addition as shown in the "incorrect" column, the incorrect result of 422.7 is obtained.

Correct	Incorrect
46.7418	46.7
+ 1.03	+ 1.0
+375.0	+375.0
422.7718	422.7
(answer 422.8)	(answer 422.7)

PART II • FIELD NOTES

2.6 FIELD NOTES

Field notes are the records of work done in the field. They typically contain measurements, sketches, descriptions, and many other items of miscellaneous information. In the past, field notes were prepared exclusively by hand lettering in field books or special note pads as the work progressed and data were gathered. However, survey controllers, also known as data collectors and electronic field books, have been introduced that can interface with many different modern surveying instruments. As the work progresses, they create computer files containing a record of observed data. All surveying controllers provide a mapping feature (see Chapter 17). Some controllers and total stations also provide a camera so that an image of the area where data is being collected can be captured. When these features are absent manually prepared sketches and descriptions often supplement the numerical data they capture. Regardless of the manner or form in which the notes are taken, they are extremely important.

Whether prepared manually, created by a survey controller, or a combination of these forms, surveying field notes are the only permanent records of work done in the field. If the data are incomplete, incorrect, lost, or destroyed, much or all of the time and money invested in making the measurements and records have been wasted. Hence, the job of data recording is frequently the most important and difficult one in a surveying party. Field books and computer files containing information gathered over a period of weeks are worth many thousands of dollars because of the costs of maintaining personnel and equipment in the field.

Recorded field data are used in the office to perform computations, make drawings, or both. The office personnel using the data are usually not the same people who took the notes in the field. Accordingly, it is essential that without verbal explanations notes be intelligible to anyone.

Property surveys are subject to court review under some conditions, so field notes become an important factor in litigation. Also, because they may be used as references in land transactions for generations, it is necessary to index and preserve them properly. The salable "goodwill" of a surveyor's business depends largely on the office library of field books. Cash receipts may be kept in an unlocked desk drawer, but field books are stored in a fireproof safe!

2.7 GENERAL REQUIREMENTS OF HANDWRITTEN FIELD NOTES

The following points are considered in appraising a set of field notes:

Accuracy. This is the most important quality in all surveying operations.

Integrity. A single omitted measurement or detail can nullify use of the notes for computing or plotting. If the project was far from the office, it is time consuming and expensive to return for a missing measurement. Notes should be checked carefully for completeness before leaving the survey site and never "fudged" to improve closures.

Legibility. Notes can be used only if they are legible. A professional-looking set of notes is likely to be professional in quality.

Arrangement. Note forms appropriate to a particular survey contribute to accuracy, integrity, and legibility.

Clarity. Advance planning and proper field procedures are necessary to ensure clarity of sketches and tabulations, and to minimize the possibility of mistakes and omissions. Avoid crowding notes; paper is relatively cheap. Costly mistakes in computing and drafting are the end results of ambiguous notes.

Appendix B contains examples of handwritten field notes for a variety of surveying operations. Their plate number identifies each. Other example note forms are given at selected locations within the chapters that follow. These notes have been prepared keeping the above points in mind.

In addition to the items stressed in the foregoing, certain other guidelines must be followed to produce acceptable handwritten field notes. The notes should be lettered with a sharp pencil of at least 3H hardness so that an indentation is made in the paper. Books so prepared will withstand damp weather in the field (or even a soaking) and still be legible, whereas graphite from a soft pencil, or ink from a pen or ballpoint, leaves an undecipherable smudge under such circumstances.

Erasures of recorded data are not permitted in field books. If a number has been entered incorrectly, a single line is run through it without destroying the number's legibility, and the proper value is noted above it (see Figure 5.5). If a partial or entire page is to be deleted, a single diagonal line in red is drawn through opposite corners, and **VOID** is lettered prominently on the page, giving the reasons.

Field notes are presumed to be "original" unless marked otherwise. Original notes are those taken at the same time the observations are being made. If the original notes are copied, they must be so marked (see Figure 5.11). Copied notes may not be accepted in court because they are open to question concerning possible mistakes, such as interchanging numbers, and omissions. The value of a distance or an angle placed in the field book from memory 10 minutes after the observation is definitely unreliable. Students are tempted to scribble notes on scrap sheets of paper for later transfer in a neater form to the field book. This practice may result in the loss of some or all of the original data and defeats one purpose of a surveying course—to provide experience in taking notes under actual field conditions. In a real job situation, a surveyor is not likely to spend any time at night transcribing scribbled notes. Certainly, an employer will not pay for this evidence of incompetence.

■ 2.8 TYPES OF FIELD BOOKS

Since field books contain valuable data, suffer hard wear, and must be permanent in nature, only the best should be used for practical work. Various kinds of field books as shown in Figure 2.3 are available, but bound and loose-leaf types are most common. The bound book, a standard for many years, has a sewed binding, a hard cover of leatherette, polyethylene, or covered hardboard, and contains 80 leaves. Its use ensures maximum testimony acceptability for property survey records in courtrooms. Bound duplicating books enable copies of the original notes to be made through carbon paper in the field. The alternate duplicate pages are perforated to enable their easy removal for advance shipment to the office.

Loose-leaf books have come into wide use because of many advantages, which include (1) assurance of a flat working surface; (2) simplicity of filing

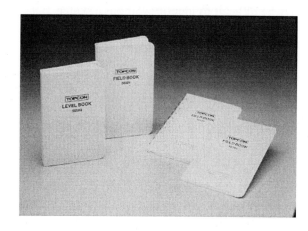

Figure 2.3
Field books.
(Courtesy Topcon America Corp.)

individual project notes; (3) ready transfer of partial sets of notes between field and office; (4) provision for holding pages of printed tables, diagrams, formulas, and sample forms; (5) the possibility of using different rulings in the same book; and (6) a saving in sheets and thus cost since none are wasted by filing partially filled books. A disadvantage is the possibility of losing sheets.

Stapled or spiral-bound books are not suitable for practical work. However, they may be satisfactory for abbreviated surveying courses that have only a few field periods, because of limited service required and low cost. Special column and page rulings provide for particular needs in leveling, angle measurement, topographic surveying, cross-sectioning, and so on.

A camera is a helpful note-keeping "instrument." Moderately priced, reliable, lightweight cameras can be used to document monuments set or found and to provide records of other valuable information or admissible field evidence. Recorded images can become part of the final record of survey. Tape recorders can also be used in certain circumstances, particularly where lengthy written explanations would be needed to document conditions or provide detailed descriptions.

2.9 KINDS OF NOTES

Four types of notes are kept in practice: (1) sketches, (2) tabulations, (3) descriptions, and (4) combinations of these. The most common type is a combination form, but an experienced recorder selects the version best fitted to the job at hand. The note forms in Appendix B illustrate some of these types and apply to field problems described in this text. Other examples are included within the text at appropriate locations. Sketches and digital images often greatly increase the efficiency with which notes can be taken. They are especially valuable to persons in the office who must interpret the notes without the benefit of the notekeeper's presence. The proverb about one picture being worth a thousand words might well have been intended for notekeepers!

For a simple survey, such as measuring the distances between points on a series of lines, a sketch showing the lengths is sufficient. In measuring the length of a line forward and backward, a sketch together with tabulations properly arranged in columns is adequate, as in Plate B.1. The location of a reference point may be difficult to identify without a sketch, but often a few lines of description are enough. Photos may be taken to record the location of permanent stations and the surrounding locale. The combination of a sketch with dimensions and photographic images can be invaluable in later station relocation. Bench marks are usually briefly described, as in Figure 5.5.

In notekeeping this axiom is always pertinent: *When in doubt about the need for any information, include it and make a sketch. It is better to have too much data than not enough.*

2.10 ARRANGEMENTS OF NOTES

Note styles and arrangements depend on departmental standards and individual preference. Highway departments, mapping agencies, and other organizations engaged in surveying furnish their field personnel with sample note forms, similar to those in Appendix B, to aid in preparing uniform and complete records that can be checked quickly.

It is desirable for students to have as guides predesigned sample sets of note forms covering their first fieldwork to set high standards and save time. Sample note forms are placed in appropriate locations throughout this book. The note forms shown in Appendix B are composites of several models. They stress the open style, especially helpful for beginners, in which some lines or spaces are skipped for clarity. Thus, angles observed at a point *A* (see Plate B.4) are placed opposite *A* on the page, but distances observed between *A* and *B* on the ground are recorded on the line between *A* and *B* in the field book.

Left- and right-hand pages are practically always used in pairs and therefore carry the same page number. A complete title should be lettered across the top of the left page and may be extended over the right one. Titles may be abbreviated on succeeding pages for the same survey project. Location and type of work are placed beneath the title. Some surveyors prefer to confine the title on the left page and keep the top of the right one free for date, party, weather, and other items. This design is revised if the entire right page must be reserved for sketches and bench mark descriptions. Arrangements shown in this book and Appendix B demonstrate the flexibility of note forms. The left page is generally ruled in six columns designed for tabulation only. Column headings are placed between the first two horizontal lines at the page top and follow from left to right in the anticipated order of reading and recording. The upper part of the left or right page must contain the following items:

1. *Project name, location, date, time of day* (A.M. *or* P.M.), *and starting and finishing times.* These entries are necessary to document the notes and furnish a timetable as well as to correlate different surveys. Precision, troubles encountered, and other facts may be gleaned from the time required for a survey.
2. *Weather.* Wind velocity, temperature, and adverse weather conditions such as rain, snow, sunshine, and fog have a decided effect on accuracy in surveying operations. Surveyors are unlikely to do their best possible work at temperatures of 15°F or with rain pouring down their necks. Hence, weather details are important in reviewing field notes, in applying corrections to observations due to temperature variations, and for other purposes.
3. *Party.* The names and initials of party members and their duties are required for documentation and future reference. Jobs can be described by symbols, such as $\overline{\wedge}$ for instrument operator, φ for rod person, and *N* for notekeeper. The party chief is frequently the notekeeper.
4. *Instrument type and number.* The type of instrument used (with its make and serial number) and its degree of adjustment affects the accuracy of a survey. Identification of the specific equipment employed may aid in isolating some errors—for example, a particular total station is found to have a 40″ indexing error when was used in trigonometric leveling.

To permit ready location of desired data, each field book must have a table of contents that is kept current daily. In practice, surveyors cross-index their notes on days when field work is impossible.

2.11 SUGGESTIONS FOR RECORDING NOTES

Observing the suggestions given in preceding sections, together with those listed here, will eliminate some common mistakes in recording notes.

1. Letter the notebook owner's name and address on the cover and the first inside page using permanent ink. Number all field books for record purposes.
2. Begin a new day's work on a new page. For property surveys having complicated sketches, this rule may be waived.
3. Employ any orderly, standard, familiar note-form type, but, if necessary, design a special arrangement to fit the project.
4. Include explanatory statements, details, and additional observations if they might clarify the notes for field and office personnel.
5. Record what is read without performing any mental arithmetic.
6. Run notes down the page, except in route surveys, where they usually progress upward to conform with sketches made while looking in the forward direction. (See Plate B.5.)
7. Use sketches instead of tabulations when in doubt. Carry a straightedge for ruling lines and a small protractor to lay off angles.
8. Make drawings to general proportions rather than to exact scale, and recognize that the usual preliminary estimate of space required is too small. Lettering parallel with or perpendicular to the appropriate features, showing clearly to what they apply.
9. Exaggerate details on sketches if clarity is thereby improved, or prepare separate diagrams.
10. Line up descriptions and drawings with corresponding numerical data. For example, a bench mark description should be placed on the right-hand page opposite its elevation, as in Figure 5.5.
11. Avoid crowding. If it is helpful to do so, use several right-hand pages of descriptions and sketches for a single left-hand sheet of tabulation. Similarly, use any number of pages of tabulation for a single drawing. Paper is cheap compared with the value of time that might be wasted by office personnel in misinterpreting compressed field notes, or by requiring a party to return to the field for clarification.
12. Use explanatory notes when they are pertinent, always keeping in mind the purpose of the survey and needs of the office personnel. Put these notes in open spaces to avoid conflict with other parts of the sketch.
13. Employ conventional symbols and signs for compactness.
14. A meridian arrow is vital for all sketches. Have north arrow at the top and on the left side of sketches if possible.
15. Keep tabulated figures inside of and off column rulings, with decimal points and digits in line vertically.
16. Make a mental estimate of all measurements before receiving and recording them in order to eliminate large mistakes.
17. Repeat aloud values given for recording. For example, before writing down a distance of 124.68, call out "one, two, four, point six, eight" for verification by the person who submitted the measurement.

18. Place a zero before the decimal point for numbers smaller than 1; that is, record 0.37 instead of .37.
19. Show the precision of observations by means of significant figures. For example, record 3.80 instead of 3.8 only if the reading was actually determined to hundredths.
20. Do not superimpose one number over another or on lines of sketches, and do not try to change one figure to another, as a 3 to a 5.
21. Make all possible arithmetic checks on the notes and record them before leaving the field.
22. Compare all misclosures and error ratios while in the field. On large projects where daily assignments are made for several parties, completed work is shown by satisfactory closures.
23. Arrange essential computations made in the field so they can be checked later.
24. Title, index, and cross-reference each new job or continuation of a previous one by client's organization, property owner, and description.
25. Sign surname and initials in the lower right-hand corner of the right page on all original notes. This places responsibility just as signing a check does.

■ 2.12 INTRODUCTION TO SURVEY CONTROLLERS

Advances in computer technology in recent years have led to the development of sophisticated automatic data collection systems for taking field notes. These devices are about the size of a pocket calculator and are produced by several different manufacturers. They are available with a variety of features and capabilities. Figure 2.4 illustrates three different survey controllers.

Survey controllers can be interfaced with modern surveying instruments, and when operated in that mode they can automatically receive and store data in computer compatible files as observations are taken. Control of the measurement and storage operations is maintained through the survey controller's keyboard. For

(a) (b) (c)

Figure 2.4
Various survey controllers that are used in practice: (a) Trimble TSC2 data collector (Courtesy Trimble Navigation.), (b) Carlson Explorer data collector (Courtesy of Carlson.) and (c) Topcon FC2500 data collector. (Courtesy of Topcon Positioning Systems.)

clarification of the notes, the operator inputs point identifiers and other descriptive information along with the measurements as they are being recorded automatically. When a job is completed or at day's end, the files can be transferred directly to a computer for further processing. Where cell coverage is available, this transfer can be performed using a data modem that is part of the survey controller. Some manufacturers now provide cloud service to provide field and office personnel a common storage location to exchange data and files.

In using survey controllers, the usual preliminary information such as date, party, weather, time, units, datum, and instrument number is entered manually into the file through the keyboard. For a given type of survey, the survey controller's internal microprocessor is programmed to follow a specific sequence of steps. The operator identifies the type of survey to be performed from a menu, and then follows instructions that appear on the unit's screen. Step-by-step prompts will guide the operator to either (1) input "external" data (which may include station names, descriptions, or other information) or (2) press an icon or key to initiate the automatic recording of observed values. Because data collectors require users to follow specific steps, they are typically referred to as survey controllers. Due to the common usage of both survey controller and data collector for the electronic field book, this book will use both names interchangeably throughout.

Survey controllers store information in either binary or ASCII (American Standard Code for Information Interchange) format. Binary storage is faster and more compact, but usually the data must be translated to ASCII before they can be read or edited. Survey controllers enable an operator to scroll through stored data, displaying them on the screen for review and editing while still at the job site. They also provide a mapped image of the data that is captured. In some controllers, this image can be overlaid with the map features to provide clarity to the user and the office. The organizational structures used by different data collectors in storing information vary considerably from one manufacturer to the next. They all follow specific rules, and once they are understood, the data can be readily interpreted by both field and office personnel. A disadvantage of having varied data structures from different manufacturers is that a new system must be learned with each instrument of different make. The LandXML organization has made an effort toward standardizing the data structures. This structure for surveying data serves a similar function as does the hypertext markup language (HTML) for the Internet. Another example is the *Survey Data Management System* (SDMS), which has been adopted by the *American Association of State Highway and Transportation Officials* (AASHTO) and is recommended for all surveys involving highway work. The example field notes for a radial survey given in Table 17.1 of Section 17.9 are in the SDMS format.

Most manufacturers of modern surveying equipment have developed survey controllers specifically to be interfaced with their own instruments, but some are flexible. The survey controller shown in Figure 2.4(a), for example, can be interfaced with instruments from the same company as well as other company's instruments. In addition to serving as a survey controller, can to perform a variety of timesaving calculations directly in the field. It has a Windows operating system and thus can run a variety of Windows software programs. Additionally, it has Bluetooth technology so that it can communicate with instruments without using cables, Wi-Fi capabilities for connecting to the Internet, and *universal serial bus* (USB) ports for uploading or downloading data from the unit.

Typically survey controllers can also be operated as electronic field books. In the electronic field book mode, the data collector is not interfaced with a surveying instrument. Instead of handwriting the data in a field book, the notekeeper enters observations into the survey controller manually by means of keyboard strokes after readings are taken. This has the advantage of enabling field notes to be recorded directly in a computer format ready for further processing, even though the surveying instruments being used may be older and not compatible for direct interfacing with the survey controller. However, survey controllers provide the utmost in efficiency when they are interfaced with surveying instruments such as total stations that have automatic readout capabilities.

The touch screen of the data collector shown in Figure 2.4(b) is a so-called third-party unit; that is, it is made by an independent company to be interfaced with instruments manufactured by others. It also utilizes a Windows operating system and has both Bluetooth and Wi-Fi capabilities, as well as USB ports. It can be either operated in the electronic field book mode or interfaced with a variety of instruments for automatic data collection.

The survey controller shown in Figure 2.4(c) is a rugged tablet with the Windows operating system. It has a 5.7 in. screen and can perform cloud networking allowing data to be transferred from the field to the office during the survey where Wi-Fi or cell coverage is available. This controller also has Bluetooth technology, a cellular modem, and a USB port. Additionally it has an internal 5.7 megapixel camera to capture photo notes. Like the other units in Figure 2.4, it can be operated as an electronic field book and works with other manufacturers instruments.

Many instrument manufacturers incorporate data collection systems as internal components directly into their equipment. These incorporate many features of external data collectors, including the display panel, within the instrument. The Topcon image station shown in Figure 2.5 is a robotic total station that has the survey controller software built into its operating system. Additionally it has the capability of collecting a set of overlapping images for the entire scene at the job site providing a record of the area surveyed for later use by office personnel. The unit has slots for USB drives and *compact flash* (CF) cards.

Figure 2.5
The Topcon GPT 7500 series total station with internal data collector. (Courtesy Topcon Positioning Systems.)

Many survey controllers use the Windows operating system. A pen and pad arrangement enables the user to point on menus and options to run software. The units shown in Figure 2.4 through 2.7 have this type of interface. A code-based Global Navigation Satellite System (GNSS) antenna can be inserted into a PCMCIA[1] port of several data collectors to add code-based GNSS capabilities (see Chapter 13) to the unit. Most modern survey controllers have the capability of running advanced computer software in the field. They can come with a keyboard or, as shown in Figure 2.7, a smaller unit that comes with a touchpad keyboard. Most have a *secure digital* (SD) port to expand their internal memory and many come with internal digital cameras. As one example of their utility, field crews can check their data before sending it to the office.

As each new series of survey controllers is developed, more sophisticated user interfaces are being designed, and the software that accompanies the systems is being improved. These systems have resulted in increased efficiency and productivity, and have provided field personnel with new features, such as the ability to perform additional field checks. However, the increased complexity of operating surveying instruments with advanced survey controllers also requires field personnel with higher levels of education and training.

■ 2.13 TRANSFER OF FILES FROM SURVEY CONTROLLERS

At regular intervals, usually at lunchtime and at the end of a day's work, or when a survey has been completed, the information stored in files within a data collector is transferred to another device. This is a safety precaution to avoid accidentally losing substantial amounts of data. Ultimately, of course, the files are downloaded to a host computer, which will perform computations or generate maps and plots from the data. Depending on the peripheral equipment available, different procedures

Figure 2.6
Trimble TSC2
with Bluetooth
technology.
(Courtesy Trimble
Navigation
Systems.)

---

[1]A PCMCIA port conforms to the *Personal Computer Memory Card International Association* standards.

Figure 2.7
The Topcon FC250 survey controller. (Courtesy Topcon Positioning Systems.)

for data transfer can be used. In one method that is particularly convenient when surveying in remote locations, data can be returned to the home office via telephony technology using devices called *data modems*. (Modems convert computer data into digital tones for transmission via telephone systems.) Thus, office personnel can immediately begin using the data. In areas with cell phone coverage, this operation can be performed in the field. Another method of data transfer consists in downloading data straight into a computer by direct hookup via an RS-232 cable. This can be performed in the office, or it can be done in the field if a laptop computer is available. In areas with wireless Internet, data can be transferred to the office using wireless connections. Data collectors with Wi-Fi capabilities allow field crews to communicate directly with office personnel, thus allowing data to be transferred, checked, and verified before the crews leave the field.

Some surveying instruments, for example, the Topcon image station shown in Figure 2.5, have computers and cameras built into them. These total stations can capture an image of the work site as evidence. Many survey controllers also come with cameras built into them to provide the same capabilities. Thus field crews can capture images of important features such as evidence of boundary location, monuments occupied, and so on. When a data modem is available these images along with relevant data can be transferred to an office computer. Office personnel can analyze field data, or compute additional points to be staked, in the office, and return the results to the field crews while they are still on the site.

From the preceding discussion, and as illustrated in Figure 2.8, computers are central components of modern computerized surveying systems. In these systems, data flow automatically from the field instrument through the survey controller to the printer, computer, plotter, and other units in the system. The term "field-to-finish systems" is often applied when this form of instrumentation and software is utilized in surveying.

Figure 2.8
The computer—a central component in the modern office. (Images (a)-(d): Courtesy of Topcon Positioning Systems; Image (e): Courtesy of Michal Vitek/ Shutterstock; Image (f): Courtesy of Rawpixel.com/ Shutterstock; Image (g): Courtesy of Reha Mark/ Shutterstock.)

2.14 DIGITAL DATA FILE MANAGEMENT

Once the observing process is completed in the field, the generated data files must be transferred (downloaded) from the survey controller to another secure storage device. Typical information downloaded from a survey controller includes a file of computed coordinates and a raw data file. Survey controllers generally provide the option of exporting these and other types of files. In this case, the coordinate file consists of computed coordinate values generated using the observations and any applied field corrections and their field codes. Field corrections may include a scale factor, offsets, and Earth curvature and refraction corrections applied to distances. Field crews generally can edit and delete information from the computed file. However, the raw data file consists of the original unreduced observations and cannot be altered in the field. The necessity for each type of data file is dependent on the intended use of the survey. In most surveys, it would be prudent to save both the coordinate and raw files. As an example, for projects that require specific closures, or that are subject to legal review, the raw data file is an essential element of the survey. However, in topographic and GNSS surveys large quantities of data are often generated. In these types of projects, the raw data file can sometimes be eliminated to provide more storage space for coordinate files. In GNSS surveys the raw data files are typically stored on the GNSS receiver to save storage space on the survey controller for the coordinates and the field codes of the points captured.

With survey controllers and digital instruments, personnel in modern surveying offices deal with considerably more data than was customary in the past. This increased volume inevitably raises new concerns about data reliability and safe storage. Many methods can be used to provide backup of digital data. Some storage options include removable tapes. Since these tend to be magnetic, there is an inevitable danger that data could be lost due to the presence of external magnetic devices, or from the failure of the surface material due to age. Because of this problem, it is wise to keep two copies of the files for all jobs. Other options to this problem include the use of compact disk (CD) and digital video disk (DVD) writers. These drives will write an optical image of a project's data on a portable disk media. Since CDs and DVDs are small but have large storage capabilities, entire projects, including drawings, can be recorded in a small space that is easily archived for future reference. However, these disks can fail if their surface is scratched. Thus, care must be taken in their handling and storage. Other options include individually maintained network storage devices and cloud storage systems.

2.15 ADVANTAGES AND DISADVANTAGES OF SURVEY CONTROLLERS

The major advantages of automatic data collection systems are that (1) mistakes in reading and manually recording observations in the field are precluded and (2) the time to process, display, and archive the field notes in the office is reduced significantly. Survey controllers can execute some programs in the field, which adds a significant advantage. As an example, the data for a survey can be corrected for systematic errors and misclosures computed, so verification that a survey meets closure requirements is made before the crew leaves a site.

Survey controllers are most useful when large quantities of information must be recorded, for example, in topographic surveys or cross-sectioning. In Section 17.9, their use in topographic surveying is described, and an example set of notes taken for that purpose is presented and discussed.

Although survey controllers have many advantages, they also present some dangers and problems. There is the slight chance, for example, the files could be accidentally erased through carelessness or lost because of malfunction or damage to the unit. Some difficulties are also created by the fact that sketches cannot be entered into the computer. However, this problem can be overcome by supplementing files with sketches made simultaneously with the observations that include field codes. These field codes instruct the drafting software to draw a map of the data complete with lines, curves, and mapping symbols. The process of collecting field data with field codes that can be interpreted later by software is known as a *field-to-finish* survey. This greatly reduces the time needed to complete a project. Field-to-finish mapping surveys are discussed in more detail in Section 17.12. It is important to realize that not all information may be stored in digital form, and thus it is important to keep a traditional field book to enter sketches, comments, and additional notes when necessary. Many modern survey controllers also contain digital cameras that allow field personnel to capture a digital image of the survey. Note that survey controllers should not be used for long-term storage. Rather the data should be downloaded and immediately saved to some permanent storage device once the field collection for a project is complete.

Survey controllers are available from numerous manufacturers. They must be capable of transferring data through varied hardware in modern surveying systems such as that illustrated in Figure 2.8. Since equipment varies considerably, it is important when considering the purchase of a survey controller to be certain it fits the equipment owned or perhaps needed in the future.

PROBLEMS

Asterisks (*) indicate problems that have partial answers given in Appendix G.

2.1 List the five types of measurements that form the basis of traditional plane surveying.

2.2 Give the basic units that are used in surveying for length, area, volume, and angles in
  (a) The English system of units.
  (b) The SI system of units.

2.3 The easting coordinate for a point is 646,284.381 m. What is the coordinate using the
  (a) Survey foot definition?
  (b) International foot definition?
  (c) Why was the survey foot definition maintained in the United States?

2.4 Convert the following distances given in meters to U.S. survey feet:
  *(a) 4129.574 m   (b) 640.137 m   (c) 487,356.009 m

2.5 Convert the following distances given in feet to meters:
  *(a) 537.52 sft   (b) 2,128,453.84 sft   (c) 4809.34 ift

2.6 Compute the lengths in survey feet corresponding to the following distances measured with a Gunter's chain:
  *(a) 10 ch 13 lk   (b) 6 ch 23 lk   (c) 14 ch 33 lk

2.7 Express $487,300$ sft^2 in:
  *(a) acres   (b) hectares   (c) square Gunter's chains

2.8 Convert 9.5438 ha to:
  (a) square survey feet   (b) acres   (c) square international feet

2.9 What are the lengths in feet and decimals for the following distances shown on a building blueprint?
  (a) 42 ft 6-1/2 in.   (b) 30 ft 6-3/4 in.

2.10 What is the area in acres of a rectangular parcel of land measured with a Gunter's chain if the recorded sides are as follows:
  *(a) 9.17 ch and 10.64 ch   (b) 16 ch 78 lk and 52 ch 49 lk

2.11 Compute the area in acres of triangular lots shown on a plat having the following recorded right-angle sides:
  (a) 220.36 ft and 440.02 ft   (b) 60.005 m and 30.006 m

2.12 A distance is expressed as 9756.12 sft. What is the length in
  *(a) international feet?   (b) meters?

2.13 What are the radian and degree–minute–second equivalents for the following angles given in grads:
  *(a) 136.000 grads   (b) 146.845 grads   (c) 202.964 grads

2.14 Give answers to the following problems in the correct number of significant figures:
  *(a) sum of 23.15, 0.984, 124, and 12.5
  (b) sum of 8.106, 34.28, 14.5, and 202.96
  (c) product of 220.08 and 220.1
  (d) quotient of 42,568.83 divided by 220.61

2.15 Express the value or answer in powers of 10 to the correct number of significant figures:
  (a) 135,897.96
  (b) 3284.0
  (c) square of 220.00
  (d) sum of (156.087 + 23.54 + 218.94) divided by 15.3

2.16 Convert the angles of a triangle to radians and show a computational check:
  (a) *39°41′54″, 91°30′16″, 48°47′50″
  (b) 89°45′23″, 46°12′35″, 44°02′02″

2.17 What item must be contained in the upper-right corner of a field book page?

2.18 Explain why one number should not be superimposed over another or the lines of sketches.

2.19 Explain why data should always be entered directly into the field book at the time measurements are made, rather than on scrap paper for neat transfer to the field book later.

2.20 Why should the field notes show the precision of the measurements?

2.21 Explain the reason for item 3 in Section 2.11 when recording field notes.

2.22 Explain the reason for item 9 in Section 2.11 when recording field notes.

2.23 Explain the reason for item 13 in Section 2.11 when recording field notes.

2.24 Why are sketches in field books not usually drawn to scale?

2.25 Justify the requirement to list in a field book the makes and serial numbers of all instruments used on a survey.

2.26 Discuss the advantages of survey controllers that can communicate with several different types of instruments.

2.27 Discuss why data should always be backed up regular intervals.

2.28 Search the Internet and find at least two sites related to
  (a) Manufacturers of survey controllers.
  (b) Manufacturers of total stations.
  (c) Manufacturers of global navigation satellite system (GNSS) receivers.

2.29 Why do many survey controllers contain digital cameras?

2.30 What are the dangers involved in using a survey controller?

2.31 Describe what is meant by the phrase "field-to-finish."

2.32 Discuss the advantages of survey controllers.

BIBLIOGRAPHY

Alder, K. 2002. *The Measure of All Things—The Seven-Year Odyssey and Hidden Error That Transformed the World*. New York, NY: The Free Press.

Billingsley, S. 2014. "Hardware and Software Purchase Decisions." 34 (No. 1): 22.

Brazeal, R. 2014. "Topcon MAGNET V2.0." *Professional Surveyor* 34 (No. 1): 18.

Cavell, J. A. 2014. "Are Your Surveys Accurate?" *The American Surveyor* 11 (No. 9): 22.

Duffy, L. 2014. "A Different Approach to Data Management." *Point of Beginning* 39 (No. 9): 16.

Ghilani, C. D. 2010. *Adjustment Computations: Spatial Data Analysis*. New York, NY: Wiley.

Hockert, C., et al. 2006. *The International System of Units (SI)—Conversion Factors for General Use*. NIST Special Publication 1028. Department of Commerce. Accessed at https://www.nist.gov/sites/default/files/documents/pml/wmd/metric/SP1038.pdf on February 17, 2018.

Meade, M. E. 2007. "The International versus U.S. Survey Foot." *Point of Beginning* 33 (No. 1): 66.

Schrock, G. 2015. "Go Wide." *XYHT* 2 (No. 3): 20.

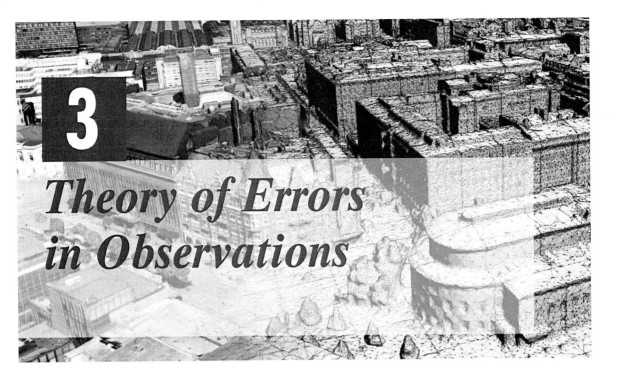

3 Theory of Errors in Observations

■ 3.1 INTRODUCTION

Taking observations (measurements), and subsequent computations and analyses using them, are fundamental tasks of surveyors. Good observations require a combination of human skill and equipment applied with the utmost judgment. However no matter how carefully made, observations are never exact and will always contain errors. Geomatics engineers (surveyors) whose work must be performed to exacting standards, should therefore thoroughly understand the different kinds of errors, their sources and expected magnitudes under varying conditions, and their manner of propagation. Only then can they select instruments and procedures necessary to reduce error sizes to within tolerable limits.

Of equal importance, surveyors must be capable of assessing the magnitudes of errors in their observations so that either their acceptability can be verified or, if necessary, new ones made. Computers and sophisticated software are tools now commonly used by surveyors to plan measurement projects, design measurement systems, investigate, and distribute observational errors after results have been obtained. Section 3.21 and Chapter 16 discuss the method of least squares adjustments that is often used to adjust observations in the modern surveying office.

■ 3.2 DIRECT AND INDIRECT OBSERVATIONS

Observations may be made directly or indirectly. Examples of *direct observations* are applying a tape to a line, fitting a protractor to an angle, or turning an angle with a total station instrument.

An *indirect observation* is secured when it is not possible to apply a measuring instrument directly to the quantity to be observed. The answer is therefore determined by its relationship to some other observed value or values. As an example, we can find the distance across a river by observing the length of a line on one side of the river and the angle at each end of this line to a point on the other side, and then computing the distance by one of the standard trigonometric formulas. Many indirect observations are made in surveying, and since all measurements contain errors, it is inevitable that quantities computed from them will also contain errors. The manner by which errors in measurements combine to produce erroneous computed answers is called error propagation. This topic is discussed further in Section 3.17.

3.3 ERRORS IN MEASUREMENTS

By definition, an error is the difference between an observed value for a quantity and its true value, or

$$E = X - \overline{X} \tag{3.1}$$

where E is the error in an observation, X the observed value, and $\overline{X}$ its true value. It can be unconditionally stated that (1) no observation is exact, (2) every observation contains errors, (3) the true value of an observation is never known, and, therefore, (4) the exact error present is always unknown. These facts are demonstrated by the following. When a distance is observed with a scale divided into tenths of an inch, the distance can be read only to hundredths (by interpolation). However, if a better scale graduated in hundredths of an inch was available and read under magnification, the same distance might be estimated to thousandths of an inch. In addition, with a scale graduated in thousandths of an inch, a reading to ten-thousandths might be possible. Obviously, accuracy of observations depends on the scale's division size, reliability of equipment used, and human limitations in estimating closer than about one-tenth of a scale division. As better equipment is developed, observations more closely approach their true values, but they can never be exact. Note that observations, not counts (of cars, pennies, marbles, or other objects), are under consideration here.

3.4 MISTAKES

These are observer blunders and are usually caused by misunderstanding the problem, carelessness, fatigue, missed communication, or poor judgment. Examples include transposition of numbers, such as recording 73.96 instead of the correct value of 79.36; reading an angle counterclockwise, but indicating it as a clockwise angle in the field notes; sighting the wrong target; or recording a measured distance as 682.38 instead of 862.38. Large mistakes such as these are not considered in the succeeding discussion of errors. They must be detected by careful and systematic checking of all work, and eliminated by repeating some or all of the measurements. It is very difficult to detect small mistakes because they merge with errors. When not exposed, these small mistakes will therefore be incorrectly treated as errors.

■ 3.5 SOURCES OF ERRORS IN MAKING OBSERVATIONS

Errors in observations stem from three sources and are classified accordingly.

Natural errors are caused by variations in wind, temperature, humidity, atmospheric pressure, atmospheric refraction, gravity, and magnetic declination. An example is a steel tape whose length varies with changes in temperature.

Instrumental errors result from any imperfection in the construction or adjustment of instruments and from the movement of individual parts. For example, the graduations on a scale may not be perfectly spaced, or the scale may be warped, elongated, or shrunk. The effect of many instrumental errors can be reduced, or even eliminated, by adopting proper surveying procedures or applying computed corrections.

Personal errors arise principally from limitations of the human senses of sight and touch. As an example, a small error occurs in the observed value of a horizontal angle if the vertical cross hair in a total station instrument is not aligned perfectly on the target, or if the target is the top of a rod that is being held slightly out of plumb.

■ 3.6 TYPES OF ERRORS

Errors in observations are of two types: *systematic* and *random*.

Systematic errors, also known as *biases*, result from factors that comprise the "measuring system" and include the environment, instrument, and observer. So long as system conditions remain constant, the systematic errors will likewise remain constant. If conditions change, the magnitudes of systematic errors also change. Because systematic errors tend to accumulate, they are sometimes called *cumulative errors*.

Conditions producing systematic errors conform to physical laws that can be modeled mathematically. Thus, if the conditions are known to exist and can be observed, a correction can be computed and applied to observed values. An example of a constant systematic error is the use of a 100-ft steel tape that has been calibrated and found to be 0.02 ft too long. It introduces a 0.02-ft error each time it is used but applying a -0.02-ft correction readily eliminates the error. An example of variable systematic error is the change in length of a steel tape resulting from temperature differentials that occur during the period of the tape's use. If the temperature changes are observed, length corrections can be computed by a simple formula, as explained in Section 6.14.

Random errors are those that remain in measured values after mistakes and systematic errors have been eliminated. They are caused by factors beyond the control of the observer, obey the laws of probability, and are sometimes called *accidental errors*. They are present in all surveying observations.

The magnitudes and algebraic signs of random errors are matters of chance. There is no absolute way to compute or eliminate them but they can be estimated using adjustment procedures known as *least squares* (see Section 3.21 and Chapter 16). Random errors are also known as *compensating errors* since they tend to partially cancel themselves in a series of observations. For example, a person interpolating to hundredths of a foot on a tape graduated only to tenths, or reading a level rod marked in hundredths, will presumably estimate too high on some values and too low on others. However, individual personal characteristics may nullify such partial compensation since some people are inclined to interpolate high, others interpolate low, and many favor certain digits—for example, 7 instead of 6 or 8, 3 instead of 2 or 4, and particularly 0 instead of 9 or 1.

■ 3.7 PRECISION AND ACCURACY

A *discrepancy* is the difference between two observed values of the same quantity. A small discrepancy indicates there are probably no mistakes and random errors are small. However, small discrepancies do not preclude the presence of systematic errors.

Precision refers to the degree of refinement or consistency of a group of observations and is evaluated based on discrepancy size. If multiple observations are made of the same quantity and small discrepancies result, this indicates high precision. The degree of precision attainable is dependent on equipment sensitivity and observer skill.

Accuracy denotes the absolute nearness of observed quantities to their true values. The difference between precision and accuracy is perhaps best illustrated with reference to target shooting. In Figure 3.1(a), for example, all five shots exist in a small group, indicating a precise operation; that is, the shooter was able to repeat the procedure with a high degree of consistency. However, the shots are far from the target's center and therefore not accurate. This probably results from misaligned sights. Figure 3.1(b) shows randomly scattered shots that are neither precise nor accurate. In Figure 3.1(c), the closely spaced grouping, in the target's center, represents both precision and accuracy. The shooter who obtained the results in (a) was perhaps able to produce the shots of (c) after aligning the sights. In surveying, this would be equivalent to the calibration of observing instruments or the removal of systematic errors from the observations.

As with the shooting example, a survey can be precise without being accurate. To illustrate, if refined methods are employed and readings taken carefully, say to 0.001 ft, but there are instrumental errors in the measuring device and corrections are not made for them, the survey will not be accurate. As a numerical example, two observations of a distance with a tape assumed to be 100.000 ft long, that is actually 100.050 ft, might give results of 453.270 and 453.272 ft. These values are precise, but they are not accurate, since there is a systematic error

  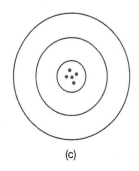

(a) (b) (c)

Figure 3.1
Examples of precision and accuracy.
(a) Results are precise but not accurate.
(b) Results are neither precise nor accurate.
(c) Results are both precise and accurate.

of approximately 4.53 × 0.050 = 0.23 ft in each. The precision obtained would be expressed as (453.272 − 453.270)/453.271 = 1/220,000, which is excellent, but accuracy of the distance is only 0.23/453.271 = 1 part in 2000. Also, a survey may appear to be accurate when rough observations have been taken. For example, the angles of a triangle may be read with a compass to only the nearest 1/4 degree and yet produce a sum of exactly 180°, or a zero misclosure error. On good surveys, precision and accuracy are consistent throughout.

3.8 ELIMINATING MISTAKES AND SYSTEMATIC ERRORS

All field operations and office computations are governed by a constant effort to eliminate mistakes and systematic errors. Of course, it would be preferable if mistakes never occurred, but because humans are fallible, this is not possible. In the field, experienced observers who alertly perform their observations using standardized repetitive procedures can minimize mistakes. Mistakes that do occur can be corrected only if discovered. Comparing several observations of the same quantity is one of the best ways to identify mistakes. Making common sense estimates and analysis is another. Assume that five observations of a line are recorded as follows: 567.91, 576.95, 567.88, 567.90, and 567.93. The second value disagrees with the others, apparently because of a transposition of figures in reading or recording. Either casting out the doubtful value, or preferably repeating the observation can eradicate this mistake. The numbers of the offending observation should never be transposed to match the other four observations since it is possible, even if it is unlikely, that the other four are mistakes.

When a mistake is detected, it is usually best to repeat the observation. However, if enough observations of the quantity are available and in agreement, as in the foregoing example, the widely divergent result may be discarded. Serious consideration must be given to the effect on an average before discarding a value. It is seldom safe to change a recorded number, even though there appears to be a simple transposition in figures. Tampering with physical data is always a bad practice and will certainly cause trouble, even if done infrequently.

Systematic errors can be calculated, and proper corrections applied to the observations. Procedures for making these corrections to all basic surveying observations are described in the chapters that follow. In some instances, it may be possible to adopt a field procedure that automatically eliminates systematic errors. For example, as explained in Section 5.4, a leveling instrument out of adjustment causes incorrect readings, but if all backsights and foresights are made the same length, the errors cancel in differential leveling.

3.9 PROBABILITY

At one time or another, everyone has had an experience with games of chance, such as coin flipping, card games, or dice, which involve probability. In basic mathematics courses, laws of combinations and permutations are introduced. It is shown that events that happen randomly or by chance are governed by mathematical principles referred to as probability.

Probability may be defined as the ratio of the number of times a result should occur to its total number of possibilities. For example, in the toss of a fair die there is a one-sixth probability that a 2 will come up. This simply means that there are six possibilities, and only one of them is a 2. In general, if a result may occur in m ways and fail to occur in n ways, then the probability of its occurrence is $m/(m + n)$. The probability that any result will occur is a fraction between 0 and 1; 0 indicating impossibility and 1 denoting absolute certainty. Since any result must either occur or fail, the sum of the probabilities of occurrence and failure is 1. Thus if 1/6 is the probability of throwing a 2 with one toss of a die then $(1 - 1/6)$ or 5/6 is the probability that a 2 will not come up.

The theory of probability is applicable in many sociological and scientific observations. In Section 3.6, it was pointed out that random errors exist in all surveying work. This can perhaps be better appreciated by considering the measuring process, which generally involves executing several elementary tasks. Besides instrument selection and calibration, these tasks may include setting up, centering, aligning, or pointing the equipment; setting, matching, or comparing index marks; and reading or estimating values from graduated scales, dials, or gauges. Because of equipment and observer imperfections, exact observations cannot be made, so they will always contain random errors. The magnitudes of these errors, and the frequency with which errors of a given size occur, follow the laws of probability.

For convenience, the term error will be used to mean only random error for the remainder of this chapter. It will be assumed that all mistakes and systematic errors have been eliminated before random errors are considered.

■ 3.10 MOST PROBABLE VALUE

It has been stated earlier that in physical observations, the true value of any quantity is never known. However, its *most probable value* can be calculated if redundant observations have been made. *Redundant observations* are measurements in excess of the minimum needed to determine a quantity. For a single unknown, such as the length of a line that has been directly and independently observed a number of times using the same equipment and procedures,[1] the first observation establishes a value for the quantity, and all additional observations are redundant. The most probable value in this case is simply the arithmetic mean, or

$$\overline{M} = \frac{\Sigma M}{n} \qquad (3.2)$$

where $\overline{M}$ is the most probable value of the quantity, ΣM the sum of the individual measurements M, and n the total number of observations. As shown in Section 16.3, Equation (3.2) can be derived using the principle of least squares, which is based on the theory of probability.

---

[1]The significance of using the same equipment and procedures is that observations are of equal reliability or weight. The subject of unequal weights is discussed in Section 3.20.

50 THEORY OF ERRORS IN OBSERVATIONS

As discussed in Chapter 16, in more complicated problems, where the observations are not made with the same instruments and procedures, or if several interrelated quantities are being determined through indirect observations, most probable values are calculated by employing least-squares methods. The treatment here relates to multiple direct observations of the same quantity using the same equipment and procedures.

■ 3.11 RESIDUALS

Having determined the most probable value of a quantity, it is possible to calculate *residuals*. A residual is simply the difference between the most probable value and any observed value of a quantity, which in equation form is

$$v = \overline{M} - M \tag{3.3}$$

where v is the residual in any observation M, and $\overline{M}$ is the most probable value for the quantity. Residuals are theoretically identical to errors, with the exception that residuals can be calculated whereas errors cannot because true values are never known. Thus, residuals rather than errors are the values actually used in the analysis and adjustment of survey data.

■ 3.12 OCCURRENCE OF RANDOM ERRORS

To analyze the manner in which random errors occur, consider the data of Table 3.1, which represents 100 repetitions of an angle observation made with a precise total station instrument (described in Chapter 8). Assume these observations are free from mistakes and systematic errors. For convenience in analyzing the data, except for the first value, only the seconds' portions of the observations are tabulated. The data have been rearranged in column (1) so that entries begin with the smallest observed value and are listed in increasing size. If a certain value was obtained more than once, the number of times it occurred, or its *frequency*, is tabulated in column (2).

From Table 3.1, it can be seen that the *dispersion* (range in observations from smallest to largest) is $30.8 - 19.5 = 11.3$ sec. However, it is difficult to analyze the distribution pattern of the observations by simply scanning the tabular values; that is, beyond assessing the dispersion and noticing a general trend for observations toward the middle of the range to occur with greater frequency. To assist in studying the data, a *histogram* can be prepared. This is simply a bar graph showing the sizes of the observations (or their residuals) versus their frequency of occurrence. It gives an immediate visual impression of the distribution pattern of the observations (or their residuals).

For the data of Table 3.1, a histogram showing the frequency of occurrence of the residuals has been developed and is plotted in Figure 3.2. To plot a histogram of residuals, it is first necessary to compute the most probable value for the angle observation. This has been done with Equation (3.2). As shown at the bottom of Table 3.1, its value is $27°43'24.9''$. Then using Equation (3.3),

TABLE 3.1 ANGLE OBSERVATIONS FROM PRECISE TOTAL STATION INSTRUMENT

Observed Value (1)	No. (2)	Residual (Sec.) (3)	Observed Value (1 Cont.)	No. (2. Cont.)	Residual (Sec.) (3 Cont.)
27°43′19.5″	1	5.4	27°43′25.1″	3	−0.2
20.0	1	4.9	25.2	1	−0.3
20.5	1	4.4	25.4	1	−0.5
20.8	1	4.1	25.5	2	−0.6
21.2	1	3.7	25.7	3	−0.8
21.3	1	3.6	25.8	4	−0.9
21.5	1	3.4	25.9	2	−1.0
22.1	2	2.8	26.1	1	−1.2
22.3	1	2.6	26.2	2	−1.3
22.4	1	2.5	26.3	1	−1.4
22.5	2	2.4	26.5	1	−1.6
22.6	1	2.3	26.6	3	−1.7
22.8	2	2.1	26.7	1	−1.8
23.0	1	1.9	26.8	2	−1.9
23.1	2	1.8	26.9	1	−2.0
23.2	2	1.7	27.0	1	−2.1
23.3	3	1.6	27.1	3	−2.2
23.6	2	1.3	27.4	1	−2.5
23.7	2	1.2	27.5	2	−2.6
23.8	2	1.1	27.6	1	−2.7
23.9	3	1.0	27.7	2	−2.8
24.0	5	0.9	28.0	1	−3.1
24.1	3	0.8	28.6	2	−3.7
24.3	1	0.6	28.7	1	−3.8
24.5	2	0.4	29.0	1	−4.1
24.7	3	0.2	29.4	1	−4.5
24.8	3	0.1	29.7	1	−4.8
24.9	2	0.0	30.8	1	−5.9
25.0	2	−0.1	Σ = 2494.0	Σ = 100	

Mean = 2494.0/100 = 24.9″

Most Probable Value = 27°43′24.9″

52 THEORY OF ERRORS IN OBSERVATIONS

residuals for all observed values are computed. These are tabulated in column (3) of Table 3.1. The residuals vary from 5.4″ to −5.9″. (The sum of the absolute value of these two extremes is the dispersion, or 11.3″.)

To obtain a histogram with an appropriate number of bars for portraying the distribution of residuals adequately, the interval of residuals represented by each bar, or the *class interval*, was chosen as 0.7″. This produced 17 bars on the graph. The range of residuals covered by each interval, and the number of residuals that occur within each interval, are listed in Table 3.2. By plotting class intervals on the abscissa against the number (frequency of occurrence) of residuals in each interval on the ordinate, the histogram of Figure 3.2 was obtained.

If the adjacent top center points of the histogram bars are connected with straight lines, the so-called *frequency polygon* is obtained. The frequency polygon for the data of Table 3.1 is superimposed as a heavy dashed blue line in Figure 3.2. It graphically displays essentially the same information as the histogram.

If the number of observations being considered in this analysis were increased progressively, and accordingly the histogram's class interval taken smaller and smaller, ultimately the frequency polygon would approach a smooth continuous curve, symmetrical about its center like the one shown with the heavy solid blue line in Figure 3.2. For clarity, this curve is shown separately in Figure 3.3. The curve's "bell shape" is characteristic of a normally distributed group of errors, and thus it is often referred to as the *normal distribution curve*. Statisticians frequently call it the *normal density curve*, since it shows the densities of errors having various sizes. In surveying, normal or very nearly normal error distributions can be expected, and henceforth in this book that condition is assumed.

In practice, histograms and frequency polygons are seldom used to represent error distributions. Instead, normal distribution curves that approximate them are preferred. Note how closely the normal distribution curve superimposed on Figure 3.2 agrees with the histogram and the frequency polygon.

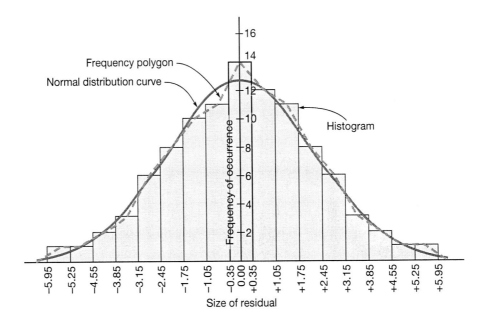

Figure 3.2
Histogram, frequency polygon, and normal distribution curve of residuals from angle measurements made with total station.

TABLE 3.2 RANGES OF CLASS INTERVALS AND NUMBER OF RESIDUALS IN EACH INTERVAL

Histogram Interval (Sec.)	Number of Residuals in Interval
−5.95 to −5.25	1
−5.25 to −4.55	1
−4.55 to −3.85	2
−3.85 to −3.15	3
−3.15 to −2.45	6
−2.45 to −1.75	8
−1.75 to −1.05	10
−1.05 to −0.35	11
−0.35 to +0.35	14
+0.35 to +1.05	12
+1.05 to +1.75	11
+1.75 to +2.45	8
+2.45 to +3.15	6
+3.15 to +3.85	3
+3.85 to +4.55	2
+4.55 to +5.25	1
+5.25 to +5.95	1
	$\Sigma = 100$

As demonstrated with the data of Table 3.1, the histogram for a set of observations shows the probability of occurrence of an error of a given size graphically by bar areas. For example, 14 of the 100 residuals (errors) in Figure 3.2 are between −0.35″ and +0.35″. This represents 14% of the errors, and the center histogram bar, which corresponds to this interval, is 14% of the total area of all bars. Likewise, the area between ordinates constructed at any two abscissas of a normal distribution curve represents the percent probability that an error of that size exists. Since the area sum of all bars of a histogram represents all errors, it therefore represents all probabilities, and thus its sum equals 1. Likewise, the total area beneath a normal distribution curve is also 1.

If the same observations of the preceding example had been taken using better equipment and more caution, smaller errors would be expected and the normal distribution curve would be similar to that in Figure 3.4(a). Compared to Figure 3.3, this curve is taller and narrower, showing that a greater percentage of values have smaller errors, and fewer observations contain big ones. For this comparison, the same ordinate and abscissa scales must be used for both curves. Thus,

54 THEORY OF ERRORS IN OBSERVATIONS

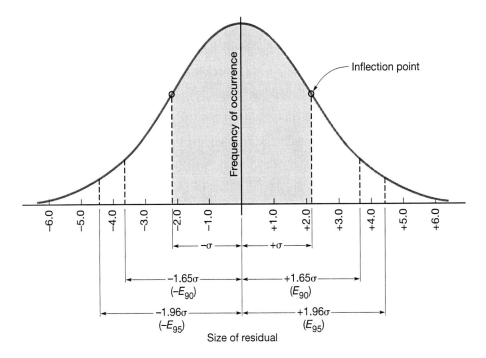

Figure 3.3
Normal distribution curve.

the observations of Figure 3.4(a) are more precise. For readings taken less precisely, the opposite effect is produced, as illustrated in Figure 3.4(b), which shows a shorter and wider curve. However, in all three cases the curve maintained its characteristic symmetric bell shape.

From these examples, it is seen that relative precisions of groups of observations become readily apparent by comparing their normal distribution curves. The normal distribution curve for a set of observations can be computed using parameters derived from the residuals, but the procedure is beyond the scope of this text. The reader should refer to the references at the end of this chapter for further exploration on this topic.

■ 3.13 GENERAL LAWS OF PROBABILITY

From an analysis of the data in the preceding section and the curves in Figures 3.2 through 3.4, some general laws of probability can be stated:

1. Small residuals (errors) occur more often than large ones; that is, they are more probable.
2. Large errors happen infrequently and are therefore less probable; for normally distributed errors, unusually large ones may be mistakes rather than random errors.
3. Positive and negative errors of the same size happen with equal frequency; that is, they are equally probable. [This enables an intuitive deduction of Equation (3.2) to be made: that is, the most probable value for a group of repeated observations, made with the same equipment and procedures, is the mean.]

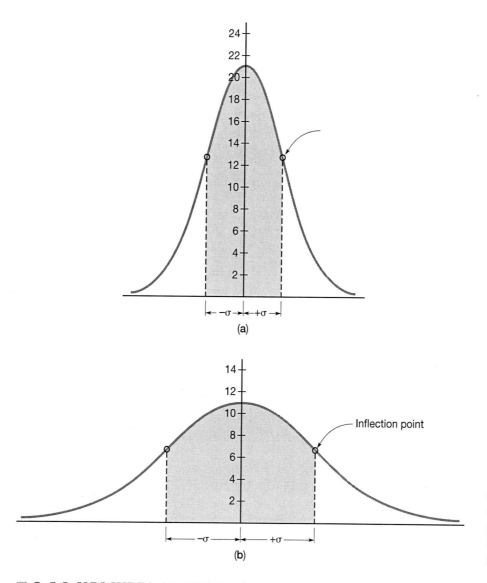

Figure 3.4
Normal distribution curves for:
(a) increased precision and,
(b) decreased precision.

3.14 MEASURES OF PRECISION

As shown in Figures 3.3 and 3.4, although the curves have similar shapes, there are significant differences in their dispersions; that is, their abscissa widths differ. The magnitude of dispersion is an indication of the relative precisions of the observations. Other statistical terms more commonly used to express precisions of groups of observations are *standard deviation* and *variance*. The equation for the standard deviation is

$$\sigma = \pm \sqrt{\frac{\Sigma v^2}{n-1}} \tag{3.4}$$

where σ is the standard deviation of a group of observations of the same quantity, v the residual of an individual observation, Σv^2 the sum of squares of the individual residuals, and n the number of observations. *Variance* is equal to σ^2, the square of the standard deviation.

Note that in Equation (3.4), the standard deviation has both plus and minus values. On the normal distribution curve, the numerical value of the standard deviation is the abscissa at the inflection points (locations where the curvature changes from concave downward to concave upward). In Figures 3.3 and 3.4, these inflection points are shown. Note the closer spacing between them for the more precise observations of Figure 3.4(a) as compared to Figure 3.4(b).

Figure 3.5 is a graph showing the percentage of the total area under a normal distribution curve that exists between ranges of residuals (errors) having equal positive and negative values. The abscissa scale is shown in multiples of the standard deviation. From this curve, the area between residuals of $+\sigma$ and $-\sigma$ equals approximately 68.3% of the total area under the normal distribution curve. Hence,

Figure 3.5 Relation between error and percentage of area under normal distribution curve.

it gives the range of residuals that can be expected to occur 68.3% of the time. This relation is shown more clearly on the curves in Figures 3.3 and 3.4, where the areas between $\pm \sigma$ are shown shaded. The percentages shown in Figure 3.5 apply to all normal distributions; regardless of curve shape or the numerical value of the standard deviation.

■ 3.15 INTERPRETATION OF STANDARD DEVIATION

It has been shown that the standard deviation establishes the limits within which observations are expected to fall 68.3% of the time. In other words, if an observation is repeated ten times, it will be expected that about seven of the results will fall within the limits established by the standard deviation, and conversely about three of them will fall anywhere outside these limits. Another interpretation is that one additional observation will have a 68.3% chance of falling within the limits set by the standard deviation.

When Equation (3.4) is applied to the data of Table 3.1, a standard deviation of ± 2.19 is obtained. In examining the residuals in the table, 70 of the 100 values, or 70%, are actually smaller than 2.19 seconds. This illustrates that the theory of probability closely approximates reality.

■ 3.16 THE 50%, 90%, AND 95% ERRORS

From the data given in Figure 3.5, the probability of an error of any percentage likelihood can be determined. The general equation is

$$E_P = C_P \sigma \tag{3.5}$$

where E_P is a certain percentage error and C_P the corresponding numerical factor taken from Figure 3.5.

By Equation (3.5), after extracting appropriate multipliers from Figure 3.5, the following are expressions for errors that have a 50%, 90%, and 95% chance of occurring:

$$E_{50} = 0.6745\ \sigma \tag{3.6}$$

$$E_{90} = 1.6449\ \sigma \tag{3.7}$$

$$E_{95} = 1.9599\ \sigma \tag{3.8}$$

The 50% error, or E_{50}, is the so-called *probable error*. It establishes limits within which the observations should fall 50% of the time. In other words, an observation has the same chance of coming within these limits as it has of falling outside of them.

The 90% and 95% errors are commonly used to specify precisions required on surveying (geomatics) projects. Of these, the 95% error, also frequently called the *two-sigma* (2σ) *error*, is most often specified. As an example, a particular project may call for the 95% error to be less than or equal to a certain value for the work to be acceptable. For the data of Table 3.1 with a standard deviation of $\pm 2.19''$, applying Equations (3.7) and (3.8), the 90% and 95% errors are $\pm 3.60''$ and $\pm 4.29''$, respectively. These errors are shown graphically in Figure 3.3.

58 THEORY OF ERRORS IN OBSERVATIONS

The so-called *three-sigma* (3σ) *error* is also often used as a criterion for rejecting individual observations from sets of data. From Figure 3.5, there is a 99.7% probability that an error will be less than this amount. Thus, within a group of observations, any value whose residual exceeds 3σ is considered to be a mistake, and either a new observation is taken or the computations based on one less value.

The *x*-axis is an asymptote of the normal distribution curve, so the 100% error cannot be evaluated. This means that no matter what size error is found, a larger one is theoretically possible.

■ Example 3.1

To clarify definitions and use the equations given in Sections 3.10 through 3.16, suppose that a line has been observed 10 times using the same equipment and procedures. The results are shown in column (1) of the following table. It is assumed that no mistakes exist, and that the observations have already been corrected for all systematic errors. Compute the most probable value for the line length, its standard deviation, and errors having 50%, 90%, and 95% probability.

Length (ft)(1)	Residual v (ft)(2)	v^2 (3)
538.57	−0.12	0.0144
538.39	+0.06	0.0036
538.37	+0.08	0.0064
538.39	+0.06	0.0036
538.48	−0.03	0.0009
538.49	−0.04	0.0016
538.33	+0.12	0.0144
538.46	−0.01	0.0001
538.47	−0.02	0.0004
538.55	−0.10	0.0100
$\Sigma = 5384.50$	$\Sigma = 0.00$	$\Sigma v^2 = 0.0554$

Solution

By Equation (3.2), $\overline{M} = \dfrac{5384.50}{10} = 538.45$ ft

By Equation (3.3), the residuals are calculated. These are tabulated in column (2) and their squares listed in column (3). Note that in column (2) the algebraic sum of residuals is zero. (For observations of equal reliability, except for round off, this column should always total zero and thus provide a computational check.)

By Equation (3.4), $\sigma = \pm\sqrt{\dfrac{\Sigma v^2}{n-1}} = \sqrt{\dfrac{0.0554}{9}} = \pm 0.078 = \pm 0.08$ ft.

By Equation (3.6), $E_{50} = \pm 0.6745\sigma = \pm 0.6745(0.078) = \pm 0.05$ ft.

By Equation (3.7), $E_{90} = \pm 1.6449(0.078) = \pm 0.13$ ft.

By Equation (3.8), $E_{95} = \pm 1.9599(0.078) = \pm 0.15$ ft.

The following conclusions can be drawn concerning this example.

1. The most probable line length is 538.45 ft.
2. The standard deviation of a single observation is ±0.08 ft. Accordingly, the normal expectation is that 68% of the time a recorded length will lie between 538.45 − 0.08 and 538.45 + 0.08 or between 538.37 and 538.53 ft; that is, about seven values should lie within these limits. (Actually, seven of them do.)
3. The probable error (E_{50}) is ±0.05 ft. Therefore, it can be anticipated that half, or five of the observations, will fall in the interval 538.40 to 538.50 ft. (Four values do.)
4. The 90% error is ±0.13 ft, and thus nine of the observed values can be expected to be within the range of 538.32 and 538.58 ft.
5. The 95% error is ±0.15 ft, so the length can be expected to lie between 538.30 and 538.60, 95% of the time. (Note that all observations indeed are within the limits of both the 90% and 95% errors.)

3.17 ERROR PROPAGATION

It was stated earlier that because all observations contain errors, any quantities computed from them will likewise contain errors. The process of evaluating errors in quantities computed from observed values that contain errors is called *error propagation*. The propagation of random errors in mathematical formulas can be computed using the general law of the propagation of variances. Typically in surveying (geomatics), this formula can be simplified since the observations are usually mathematically independent. For example, let $a, b, c, \ldots, n$ be observed values containing errors $E_a, E_b, E_c, \ldots, E_n$, respectively. Also let Z be a quantity derived by computation using these observed quantities in a function f, such that

$$Z = f(a, b, c, \ldots, n) \tag{3.9}$$

Then assuming that $a, b, c, \ldots, n$ are independent observations, the error in the computed quantity Z is

$$E_Z = \pm \sqrt{\left(\frac{\partial f}{\partial a}E_a\right)^2 + \left(\frac{\partial f}{\partial b}E_b\right)^2 + \left(\frac{\partial f}{\partial c}E_c\right)^2 + \cdots + \left(\frac{\partial f}{\partial n}E_n\right)^2} \tag{3.10}$$

where the terms $\partial f/\partial a, \partial f/\partial b, \partial f/\partial c, \ldots, \partial f/\partial n$ are the partial derivatives of the function f with respect to the variables $a, b, c, \ldots, n$. In the subsections that follow, specific cases of error propagation common in surveying are discussed, and examples are presented.

3.17.1 Error of a Sum

Assume the sum of independently observed observations $a, b, c, \ldots$ is Z. The formula for the computed quantity Z is

$$Z = a + b + c + \ldots$$

60 THEORY OF ERRORS IN OBSERVATIONS

The partial derivatives of Z with respect to each observed quantity are $\partial Z/\partial a = \partial Z/\partial b = \partial Z/\partial c = \ldots = 1$. Substituting these partial derivatives into Equation (3.10), the following formula is obtained, which gives the propagated error in the sum of quantities, each of which contains a different random error:

$$E_{Sum} = \pm\sqrt{E_a^2 + E_b^2 + E_c^2 + \ldots} \quad (3.11)$$

where E represents any specified percentage error (such as σ, E_{50}, E_{90}, or E_{95}), and a, b, and c are the separate, independent observations.

The error of a sum can be used to explain the rules for addition and subtraction using significant figures. Recall the addition of 46.7418, 1.03, and 375.0 from Example (a) from Section 2.4. Significant figures indicate that there is uncertainty in the last digit of each number. Thus, assume estimated errors of ± 0.0001, ± 0.01, and ± 0.1, respectively for each number. The error in the sum of these three numbers is $\sqrt{0.0001^2 + 0.01^2 + 0.1^2} = \pm 0.1$. The sum of three numbers is 422.7718, which was rounded, using the rules of significant figures, to 422.8. Its precision matches the estimated accuracy produced by the error in the sum of the three numbers. Note how the least accurate number controls the accuracy in the summation of the three values.

■ Example 3.2

Assume that a line is observed in three sections, with the individual parts equal to $(753.81, \pm 0.012)$, $(1238.40, \pm 0.028)$, and $(1062.95, \pm 0.020)$ ft, respectively. Determine the line's total length and its anticipated standard deviation.

Solution

Total length = 753.81 + 1238.40 + 1062.95 = 3055.16 ft.

By Equation (3.11), $E_{Sum} = \pm\sqrt{0.012^2 + 0.028^2 + 0.020^2} = \pm 0.036$ ft

3.17.2 Error of a Series

Sometimes a series of similar quantities, such as the angles within a closed polygon, are read with each observation being in error by about the same amount. The total error in the sum of all observed quantities of such a series is called the *error of the series*, designated as E_{Series}. If the same error E in each observation is assumed and Equation (3.11) applied, the series error is

$$E_{Series} = \pm\sqrt{E^2 + E^2 + E^2 + \ldots} = \pm\sqrt{nE^2} = \pm E\sqrt{n} \quad (3.12)$$

where E represents the error in each individual observation and n the number of observations.

This equation shows that when the same operation is repeated, random errors tend to balance out, and the resulting error of a series is proportional to the square root of the number of observations. This equation has extensive use—for instance, to determine the allowable misclosure error for angles of a traverse, as discussed in Section 9.7.

■ Example 3.3

Assume that each of the interior angles in a four-sided traverse has an estimate error of $\pm 3.5''$. Determine the error in the sum of the four interior angles.

Solution

By Equation (3.12), the error in the sum of the angles is

$$E_{Series} = \pm E\sqrt{n} = \pm 3.5''\sqrt{4} = \pm 7''$$

■ Example 3.4

A distance of 1000 ft is to be taped with an error of not more than 0.10 ft. Determine how accurately each 100-ft length must be observed to ensure that the error will not exceed the permissible limit.

Solution

Since by Equation (3.12), $E_{series} = \pm E\sqrt{n}$ and $n = 10$, the allowable error E in 100 ft is

$$E = \pm \frac{E_{Series}}{\sqrt{n}} = \pm \frac{0.10 \, \text{ft}}{\sqrt{10}} = \pm 0.03 \, \text{ft}$$

■ Example 3.5

Suppose it is required to tape a length of 2500 ft with an error of not more than ± 0.10 ft. How accurately must each tape length be observed?

Solution

Since 100 ft is again considered the unit length, $n = 25$, and by Equation (3.12), the allowable error E in 100 ft is

$$E = \pm \frac{0.10 \, \text{ft}}{\sqrt{25}} = \pm 0.02 \, \text{ft}$$

Analyzing Examples 3.4 and 3.5 shows that the larger the number of possibilities, the greater the chance for the errors to cancel out.

62 THEORY OF ERRORS IN OBSERVATIONS

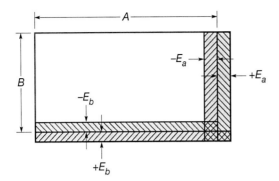

Figure 3.6
Error of area.

3.17.3 Error of a Product

The equation for propagation of errors in the product AB, where E_a and E_b, are the respective errors in A and B, is

$$E_{prod} = \pm\sqrt{A^2 E_b^2 + B^2 E_a^2} \qquad (3.13)$$

The physical significance of the error propagation formula for a product is illustrated in Figure 3.6, where A and B are shown to be observed sides of a rectangular parcel of land with errors E_a and E_b, respectively. The product AB is the parcel area. In Equation (3.13), $\sqrt{A^2 E_b^2} = AE_b$, represents area within either of the longer (horizontal) crosshatched bars and is the error caused by either $-E_b$ or $+E_b$ The term $\sqrt{B^2 E_a^2} = BE_a$ is represented by the area within the shorter (vertical) crosshatched bars, which is the error resulting from either $-E_a$ or $+E_a$

■ Example 3.6

For the rectangular lot illustrated in Figure 3.6, observations of sides A and B with their 95% errors are $(252.46, \pm 0.053)$ and $(605.08, \pm 0.072)$ ft, respectively. Calculate the parcel area and the estimated 95% error in the area.

Solution

Area $= 252.46 \times 605.08 = 152{,}760 \, \text{ft}^2$

By Equation (3.13),
$$E_{95} = \pm\sqrt{(252.46)^2(0.072)^2 + (605.08)^2(0.053)^2} = \pm 36.9 \, \text{ft}^2$$

Example 3.6 can also be used to demonstrate the validity of one of the rules of significant figures in computation. The computed area is actually $152{,}758.4968 \, \text{ft}^2$ However, the rule for significant figures in multiplication (see Section 2.4) states that there cannot be more significant figures in the answer than in any of the individual factors used. Accordingly, the area should be rounded off to 152,760 (five significant figures). From Equation (3.13), with an error of ± 36.9 ft^2 the answer could be $152{,}758.4968 \pm 36.9$, or from 152,721.6 to 152,795.4 ft^2. Thus, the fifth digit in the answer is seen to be questionable, and hence the number of significant figures specified by the rule is verified.

3.17.4 Error of the Mean

Equation (3.2) stated that the most probable value of a group of repeated observations of equal weight is the arithmetic mean. Since the mean is computed from individual observed values, each of which contains an error, the mean is also subject to error. By applying Equation (3.12), it is possible to find the error for the sum of a series of observations where each one has the same error. Since the sum divided by the number of observations gives the mean, the error of the mean is found by the relation

$$E_m = \frac{E_{series}}{n}$$

Substituting Equation (3.12) for E_{series}

$$E_m = \frac{E\sqrt{n}}{n} = \frac{E}{\sqrt{n}} \tag{3.14}$$

where E is the specified percentage error of a single observation, E_m the corresponding percentage error of the mean, and n the number of observations.

The error of the mean at any percentage probability can be determined and applied to all criteria that have been developed. For example, the standard deviation of the mean, $(E_{68})_m$ or σ_m is

$$(E_{68})_m = \sigma_m = \frac{\sigma}{\sqrt{n}} = \pm\sqrt{\frac{\sum v^2}{n(n-1)}} \tag{3.15a}$$

and the 90% and 95% errors of the mean are

$$(E_{90})_m = \frac{E_{90}}{\sqrt{n}} = \pm 1.6449 \sqrt{\frac{\sum v^2}{n(n-1)}} \tag{3.15b}$$

$$(E_{95})_m = \frac{E_{95}}{\sqrt{n}} = \pm 1.9599 \sqrt{\frac{\sum v^2}{n(n-1)}}. \tag{3.15c}$$

These equations show that *the error of the mean varies inversely as the square root of the number of repetitions*. Thus, to double the accuracy—that is, to reduce the error by one half—four times as many observations must be made.

Example 3.7

Calculate the standard deviation of the mean and the 90% error of the mean for the observations of Example 3.1.

Solution

By Equation (3.15a), $\sigma_m = \dfrac{\sigma}{\sqrt{n}} = \pm\dfrac{0.078}{\sqrt{10}} = \pm 0.025$ ft

Also, by Equation (3.15b), $(E_{90})_m = \pm 1.6449(0.025) = \pm 0.041$ ft

These values show the error limits of 68% and 90% probability for the line's length. It can be said that the true line length has a 68% chance of being within ± 0.025 of the mean, and a 90% likelihood of falling not farther than ± 0.041 ft from the mean.

3.18 APPLICATIONS

The preceding example problems show that the equations of error probability are applied in two ways:

1. To analyze observations already made, for comparison with other results or with specification requirements.
2. To establish procedures and specifications in order that the required results will be obtained.

The application of the various error probability equations must be tempered with judgment and caution. Recall that they are based on the assumption that the errors conform to a smooth and continuous normal distribution curve, which in turn is based on the assumption of a large number of observations. Frequently in surveying only a few observations—often from two to eight—are taken. If these conform to a normal distribution, then the answer obtained using probability equations will be reliable; if they do not, the conclusions could be misleading. In the absence of knowledge to the contrary, however, an assumption that the errors are normally distributed is still the best available.

3.19 CONDITIONAL ADJUSTMENT OF OBSERVATIONS

In Section 3.3, it was emphasized that the true value of any observed quantity is never known. However, in some types of problems, the sum of several observations must equal a fixed value; for example, the sum of the three angles in a plane triangle has to total 180°. In practice, therefore, the observed angles are adjusted to make them add to the required amount. Correspondingly, distances—either horizontal or vertical—must often be adjusted to meet certain conditional requirements. The methods used will be explained in later chapters, where the operations are taken up in detail.

3.20 WEIGHTS OF OBSERVATIONS

It is evident that some observations are more precise than others because of better equipment, improved techniques, and superior field conditions. In making adjustments, it is consequently desirable to assign *relative weights* to individual observations. It can logically be concluded that if an observation is very precise, it will have a small standard deviation or variance, and thus should be weighted more heavily (held closer to its observed value) in an adjustment than an observation of lower precision. From this reasoning, it is deduced that weights of observations should bear an inverse relationship to precision. In fact, it can be shown that relative weights are inversely proportional to variances, or

$$W_a \propto \frac{1}{\sigma_a^2} \tag{3.16}$$

where W_a is the weight of an observation a, which has a variance of σ_a^2. Thus, the higher the precision (the smaller the variance), the larger should be the relative weight of the observed value being adjusted. In some cases, variances are unknown originally, and weights must be assigned to observed values based on estimates of their relative precision. If a quantity is observed repeatedly and the individual observations have varying weights, the weighted mean can be computed from the expression

$$\overline{M}_W = \frac{\Sigma WM}{\Sigma W} \tag{3.17}$$

where $\overline{M}_W$ is the weighted mean, ΣWM the sum of the individual weights times their corresponding observations, and ΣW the sum of the weights.

Example 3.8

Suppose four observations of a distance are recorded as 482.16, 482.17, 482.20, and 482.18 and given weights of 1, 2, 2, and 4, respectively, by the surveyor. Determine the weighted mean.

Solution

By Equation (3.17)

$$\overline{M}_W = \frac{482.16 + 482.17(2) + 482.20(2) + 482.14(4)}{1 + 2 + 2 + 4} = 482.16 \text{ ft}$$

In computing adjustments involving unequally weighted observations, corrections applied to observed values should be made *inversely proportional to the relative weights*.

Example 3.9

Assume the observed angles of a certain plane triangle, and their relative weights, are $A = 49°51'15''$, $W_a = 1$; $B = 60°32'08''$, $W_b = 2$; and $C = 69°36'33''$, $W_c = 3$. Compute the weighted mean of the angles.

Solution

The sum of the three angles is computed first and found to be $4''$ less than the required geometrical condition of exactly 180°. The angles are therefore adjusted in inverse proportion to their relative weights, as illustrated in the accompanying tabulation. Angle C with the greatest weight (3) gets the smallest correction, $2x$; B receives $3x$; and A, $6x$.

	Observed Angle	Wt	Correction	Numerical Corr.	Rounded Corr.	Adjusted Angle
A	49°51'15''	1	6x	+2.18''	+2''	49°51'17''
B	60°32'08''	2	3x	+1.09''	+1''	60°32'09''
C	69°36'33''	3	2x	+0.73''	+1''	69°36'34''
Sum	179°59'56''	$\Sigma = 6$	11x	+4.00''	+4''	180°00'00''

$$11x = 4'' \text{ and } x = +0.36''$$

It must be emphasized again that adjustment computations based on the theory of probability are valid only if systematic errors and employing proper procedures, equipment, and calculations eliminates mistakes.

3.21 LEAST-SQUARES ADJUSTMENT

As explained in Section 3.19, most surveying observations must conform to certain geometrical conditions. The amounts by which they fail to meet these conditions are called misclosures, and they indicate the presence of random errors. In Example 3.9, for example, the misclosure was $4''$. Various procedures are used to distribute these misclosure errors to produce mathematically perfect geometrical conditions. Some simply apply corrections of the same size to all observed values, where each correction equals the total misclosure (with its algebraic sign changed), divided by the number of observations. Others introduce corrections in proportion to assigned weights. Still others employ rules of thumb, for example, the "compass rule" described in Chapter 10 for adjusting closed traverses.

Because random errors in surveying conform to the mathematical laws of probability and are "normally distributed," the most appropriate adjustment procedure should be based upon these laws. Least squares is such a method. It is not a new procedure, having been applied by the German mathematician Karl Gauss as early as the latter part of the 18th century. However, until the advent of computers, it was only used sparingly because of the lengthy calculations involved.

Least squares is suitable for adjusting any of the basic types of surveying observations described in Section 2.1, and is applicable to all of the commonly used surveying procedures. The method enforces the condition that *the sum of the weights of the observations times their corresponding squared residuals is minimized*. This fundamental condition, which is developed from the equation for the normal error distribution curve, provides most probable values for the adjusted quantities. In addition, it also (1) enables the computation of precisions of the adjusted values, (2) reveals the presence of mistakes so steps can be taken to eliminate them, and (3) makes possible the optimum design of survey procedures in the office before going to the field to take observations.

PROBLEMS

Asterisks (*) indicate problems that have answers given in Appendix G.

3.1 Discuss the differences between an error and a residual.
3.2 Give two examples of (a) direct and (b) indirect measurements.
3.3 Define the term *systematic error*, and give two surveying examples of a systematic error.
3.4 Define the term *random error*, and give two surveying examples of a random error.
3.5 Discuss the difference between accuracy and precision.
3.6 The observations of 156.53, 156.55, 156.51, and 165.52 are obtained when taping the length of a line. What should the observer consider doing before a mean length is determined from the set of observations?

A distance AB is observed repeatedly using the same equipment and procedures, and the results, in meters, are listed in Problems 3.7 through 3.10. Calculate (a) the line's most probable length, (b) the standard deviation, and **(c)** the standard deviation of the mean for each set of results.

3.7* 65.401, 65.400, 65.402, 65.396, 65.406, 65.401, 65.396, 65.401, 65.405, and 65.404
3.8 Same as Problem 3.7, but discard the 65.406 observation.
3.9 Same as Problem 3.7, but discard the 65.406 and 65.405 observations.
3.10 Same as Problem 3.7, but include two additional observations, 65.399 and 65.398.

In Problems 3.11 through 3.14, determine the range within which observations should fall (a) 90% of the time and (b) 95% of the time. List the percentage of values that actually fall within these ranges.

3.11* For the data of Problem 3.7.
3.12 For the data of Problem 3.8.
3.13 For the data of Problem 3.9.
3.14 For the data of Problem 3.10.

In Problems 3.15 through 3.17, an angle is observed repeatedly using the same equipment and procedures. Calculate (a) the angle's most probable value, (b) the standard deviation, and (c) the standard deviation of the mean.

3.15* 23°30′00″, 23°30′10″, 23°30′10″, and 23°29′55″.
3.16 Same as Problem 3.15, but with three additional observations, 23°29′56″, 23°30′02″ and 23°30′08″.
3.17 Same as Problem 3.15, but with two additional observations, 23°29′58″ and 23°30′02″.

3.18* A field party is capable of making taping observations with a standard deviation of ±0.005 ft per 100 ft tape length. What standard deviation would be expected in a distance of 300 ft taped by this party?

3.19 Repeat Problem 3.18, except that the standard deviation per 30-m tape length is ±5 mm and a distance of 120 m is taped. What is the expected 95% error in 120 m?

3.20 A distance of 200 ft must be taped in a manner to ensure a standard deviation smaller than ±0.04 ft. What must be the standard deviation per 100 ft tape length to achieve the desired precision?

3.21 Lines of levels were run requiring n instrument setups. If the rod reading for each backsight and foresight has a standard deviation σ, what is the standard deviation in each of the following level lines?
  (a) $n = 20, \sigma = \pm 0.01$ ft
  (b) $n = 44, \sigma = \pm 5$ mm

3.22 A line AC was observed in 2 sections AB and BC, with lengths and standard deviations listed below. What is the total length of AC, and its standard deviation?
  *(a) $AB = 60.00 \pm 0.015$ ft; $BC = 86.13 \pm 0.018$ ft
  (b) $AB = 30.000 \pm 0.003$ m; $BC = 13.150 \pm 0.001$ m

3.23 Line AD is observed in three sections, AB, BC, and CD, with lengths and standard deviations as listed below. What is the total length of AD and its standard deviation?
  (a) $AB = 302.97 \pm 0.017$ ft; $BC = 498.67 \pm 0.017$ ft; $CD = 784.65 \pm 0.017$ ft
  (b) $AB = 254.061 \pm 0.0052$ m; $BC = 486.578 \pm 0.0053$ m; $CD = 465.837 \pm 0.0058$ m

3.24 A difference in elevation between A and B was observed four times as 32.05, 32.03, 32.08, and 32.01 ft. The observations were given weights of 2, 3, 1, and 2, respectively, by the observer. *(a) Calculate the weighted mean for distance AB. (b) What difference results if later judgment revises the weights to 2, 3, 1, and 1, respectively?

3.25 Determine the weighted mean for the following angles:
  *(a) 222°12'48", wt 2; 222°12'42", wt 1; 222°12'50", wt 3
  (b) 96°14'50" ± 6"; 96°14'44" ± 4"; 96°14'36" ± 10"

3.26 Specifications for observing angles of an n-sided polygon limit the total angular misclosure to E. How accurately must each angle be observed for the following values of n and E?
  (a) $n = 4, E = \pm 8"$
  (b) $n = 8, E = \pm 10"$

3.27 What is the area of a rectangular field and its estimated error for the following recorded values:
  *(a) 243.89 ± 0.05 ft by 208.65 ± 0.04 ft
  (b) 220.45 ± 0.02 ft by 440.28 ± 0.04 ft
  (c) 344.092 ± 0.008 m by 180.403 ± 0.004 m

3.28 Adjust the angles of triangle ABC for the following angular values and weights:
  *(a) $A = 49°24'22"$, wt 2; $B = 39°02'16"$, wt 1; $C = 91°33'00"$, wt 3
  (b) $A = 79°23'50"$, wt 6; $B = 56°41'00"$, wt 2; $C = 43°55'32"$, wt 3

3.29 Determine relative weights and perform a weighted adjustment (to the nearest second) for angles A, B, and C of a plane triangle, given the following four observations for each angle:

Angle A	Angle B	Angle C
44°28'16"	65°56'13"	69°35'20"
44°28'12"	65°56'10"	69°35'24"
44°28'17"	65°56'06"	69°35'18"
44°28'11"	65°56'08"	69°35'24"

3.30 A line of levels was run from benchmarks A to B, B to C, and C to D. The elevation differences obtained between benchmarks, with their standard deviations, are listed below. What is the difference in elevation from bench mark A to D and the standard deviation of that elevation difference?

(a) BM A to BM B = +37.78 ± 0.12 ft; BM B to BM C = −73.50 ± 0.16 ft; and BM C to BM D = −84.09 ± 0.08 ft

(b) BM A to BM B = −60.821 ± 0.015 m; BM B to BM C = +94.378 ± 0.020 m; and BM C to BM D = +56.805 ± 0.015 m

BIBLIOGRAPHY

Alder, K. 2002. *The Measure of All Things—The Seven-Year Odyssey and Hidden Error That Transformed the World*. New York: The Free Press.

Cummock, M. and G. Wagstaff. 1999. "Part 1: Measurements—A Roll of the Dice." *Point of Beginning* 24 (No. 6): 34.

Foster. R. 2003. "Uncertainty about Positional Uncertainty." *Point of Beginning* 28 (No. 11): 40.

Ghilani, C. D. 2003. "Statistics and Adjustments Explained Part 1: Basic Concepts." *Surveying and Land Information Science* 63 (No. 2): 62.

____. 2003. "Statistics and Adjustments Explained Part 2: Sample Sets and Reliability." *Surveying and Land Information Science* 63 (No. 3): 141.

____. 2018. *Adjustment Computations: Spatial Data Analysis,* 6th Ed. John Wiley and Sons, Inc., Hoboken, NJ.

Hockert, C., et. al. 2006. *The International System of Units (SI) Conversion Factors for General Use. NIST Special Publication 1028*. Department of Commerce. Accessed at https://www.nist.gov/sites/default/files/documents/pml/wmd/metric/SP1038.pdf on February, 17, 2018.

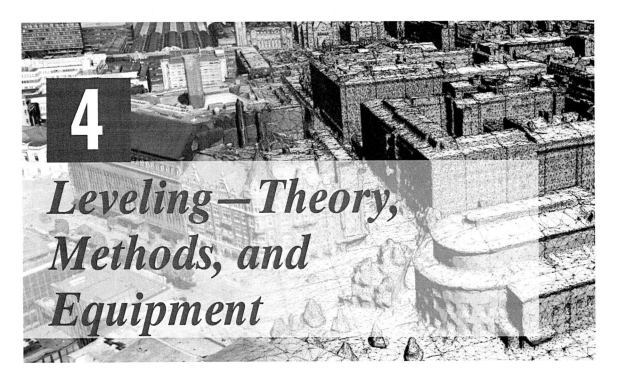

4 Leveling—Theory, Methods, and Equipment

PART I • LEVELING—THEORY AND METHODS

■ 4.1 INTRODUCTION

Leveling is the general term applied to any of the various processes by which elevations of points or differences in elevation are determined. It is a vital operation in producing necessary data for mapping, engineering design, and construction. Leveling results are used to (1) design highways, railroads, canals, sewers, water supply systems, and other facilities having grade lines that best conform to existing topography; (2) lay out construction projects according to planned elevations; (3) calculate volumes of earthwork and other materials; (4) investigate drainage characteristics of an area; (5) develop maps showing general ground configurations; and (6) study earth subsidence and crustal motion.

■ 4.2 DEFINITIONS

Basic terms in leveling are defined in this section, some of which are illustrated in Figure 4.1.

Vertical line. A line that follows the local direction of gravity as indicated by a plumb line.

Level surface. A curved surface that at every point is perpendicular to the local plumb line (the direction in which gravity acts). Level surfaces are approximately spheroidal in shape. A body of still water is the closest example of a level surface. Within local areas, level surfaces at different

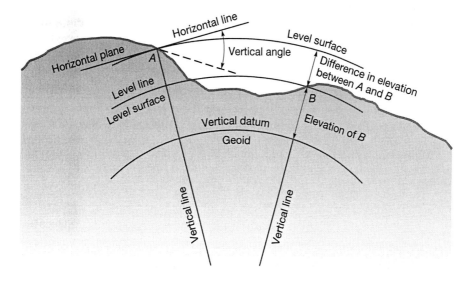

Figure 4.1
Leveling terms.

heights are considered to be concentric.[1] Level surfaces are also known as *equipotential surfaces* since, for a particular surface, the potential of gravity is equal at every point on the surface.

Level line. A line in a level surface—therefore, a curved line. A level line is also used to describe a unit of field work consisting of benchmarks and turning points, which are temporary benchmarks, connected by chains of differential leveling.

Horizontal plane. A plane perpendicular to the local direction of gravity. In plane surveying, it is a plane perpendicular to the local vertical line.

Horizontal line. A line in a horizontal plane. In plane surveying, it is a line perpendicular to the local vertical. This means that horizontal lines from two distant stations on opposite ends of the line are not parallel and may not even be of equal length.

Vertical datum. Any level surface to which elevations are referenced. This is the surface that is assigned an elevation of zero (see Section 19.6). This level surface is also known as a reference datum since points using this datum have heights relative to this surface.

Elevation. The distance measured along a vertical line from a vertical datum to a point or object. If the elevation of point *A* is 802.46 ft, *A* is 802.46 ft above the reference datum. The elevation of a point is also called its height above the datum and known as an *orthometric height*.

Geoid. A particular level surface that serves as a datum for all elevations and astronomical observations. As discussed in Section 4.3, the North American Vertical Datum of 1988 (NAVD88) was a hybrid geoid, which was based on a single benchmark called *Father Point* in Rimouski, which is on the mouth of the St. Lawrence Seaway in Quebec, Canada (see Section 4.3). In 2022, the National Geodetic Survey will be moving to

---

[1]Due to flattening of the Earth in the polar direction, level surfaces at different elevations and different latitudes are not truly concentric. This topic is discussed in more detail in Chapter 19.

a gravimetric geoid, which will use an agreed upon equipotential surface defined by 62,636,856.00 gal-m, (see Section 19.6) where gal is a unit of gravity. The unit gal-m is known as the *geopotential number* for an equipotential surface. Note, therefore, that after 2022 all equipotential surfaces including the geoid will be defined by the force of gravity and its height above this equipotential surface.

Mean sea level (MSL). This term is no longer applicable to benchmark elevations in NAVD 88. MSL was defined as the average height for the surface of the seas for all stages of tide over a 19-year period as determined by the National Geodetic Vertical Datum of 1929, further described in Section 4.3. It was derived from readings, usually taken at hourly intervals, at 26 gaging stations along the Atlantic and Pacific oceans and the Gulf of Mexico. The elevation of the sea differs from station to station depending on local influences of the tide; for example, at two points 0.5 mi apart on opposite sides of an island in the Florida Keys, it varies by 0.3 ft. MSL was accepted as the vertical datum for North America for many years. However, the current vertical datum uses a single benchmark as a reference (see Section 4.3).

Tidal datum. The vertical datum used in coastal areas for establishing property boundaries of lands bordering waters subject to tides. A tidal datum also provides the basis for locating fishing and oil drilling rights in tidal waters, and the limits of swamp and overflowed lands. Various definitions have been used in different areas for a tidal datum, but the one most commonly employed is the *mean high water* (MHW) line. Others applied include *mean higher high water* (MHHW), *mean low water* (MLW), and *mean lower low water* (MLLW).

Mean high-water is defined at the arithmetic mean of the high-water heights observed over a specific 19-year cycle. Mean low-water is defined at the arithmetic mean of the lower water heights of the tide observed over a specific 19-year cycle. Mean higher high-water is defined as the arithmetic mean of the higher high-water heights of the tide over the specific 19-year Metonic[2] cycle denoted as National Tidal Datum Epoch. This is a specific 19-year period that spans the longest periodic tidal variations resulting from astronomical tide-producing forces. Mean lower low-water is defined as the arithmetic mean of the lower low-water heights over the 19-year Metonic cycle.

Benchmark (BM). A relatively permanent object, natural or artificial, having a marked point whose elevation above or below a reference datum is known or assumed. Common examples are metal disks set in concrete (see Figure 19.8), reference marks chiseled on large rocks, unmovable parts of fire hydrants, curbs, etc.

Leveling. The process of finding elevations of points or their differences in elevation. Note that the leveled height difference between two points may not equal the orthometric height difference due to differences in gravity (see Section 19.15.4).

---

[2] Note that the Metonic cycle is named after the Greek astronomer Meton of Athens who observed from the surface of the Earth the time for astronomical objects, such as the planets, to appear in the same position was almost exactly 19 years.

Vertical control. A series of benchmarks or other points of known elevation established throughout an area, also termed *basic control* or *level control.* Vertical control for the United States was derived from first- and second-order leveling. Less precise third-order leveling has been used to fill gaps between second-order benchmarks, as well as for many other specific projects (see Section 19.14). Elevations of benchmarks, which are part of the National Spatial Reference System, can be obtained online from the National Geodetic Survey at http://www.ngs.noaa.gov. The data sheets for vertical control give the (1) approximate geodetic coordinates for the station, (2) adjusted NAVD88 elevation, (3) observed or modeled gravity reading at the station, and (4) a description of the station and its location among other things. Software plugins for an Internet browser exists that will plot these points in Google Earth to aid in the location of the monuments in the field.

4.3 NORTH AMERICAN VERTICAL DATUM

Precise leveling operations to establish a distributed system of reference benchmarks throughout the United States began in the 1850s. This work was initially concentrated along the eastern seaboard but in 1887 the U.S. Coast and Geodetic Survey[3] (USC&GS) began its first transcontinental leveling across the country's midsection. That project was completed in the early 1900s. By 1929, thousands of benchmarks had been set. In that year, the USC&GS began a general least-squares adjustment of all leveling completed in the United States and Canada. The adjustment involved over 100,000 km of leveling and incorporated long-term data from the 26 tidal gaging stations; hence it was related to mean sea level. In fact, that network of benchmarks with their resulting adjusted elevations defined the mean sea level datum. It was called the *National Geodetic Vertical Datum of 1929* (*NGVD 29*).

Through the years after 1929, the NGVD 29 deteriorated somewhat due to various causes including changes in sea level and shifting of the Earth's crust. Also, more than 625,000 km of additional leveling was completed. To account for these changes and incorporate the additional leveling, the National Geodetic Survey (NGS) performed a new general readjustment. Work on this adjustment, which included more than 1.3 million observed elevation differences, began in 1978. Although not finished until 1991, its planned completion date was 1988, and thus it has been named the *North American Vertical Datum of 1988* (*NAVD 88*). Besides the United States and Canada, Mexico was also included in this general readjustment. This adjustment shifted the position of the reference surface from the mean of the 26 tidal gage stations to a single tidal gage benchmark called *Father Point*, which is in Rimouski, Quebec, Canada, near the mouth of the St. Lawrence Seaway. Thus, elevations in NAVD 88 are no longer referenced to mean sea level. Benchmark elevations that were defined by the NGVD 29 datum have changed by relatively small, but nevertheless significant amounts in the eastern half of the continental United States (see Figure 19.7). However, the changes are much greater in the western part of the country and reach 1.5 m in the Rocky Mountain region. It is therefore

---

[3] The National Geodetic Survey was formerly known as the USC&GS, which was created by Thomas Jefferson and known as the U.S. Coast Survey.

imperative that surveyors positively identify the datum to which their elevations are referred. Listings of the new elevations are available from the NGS.[4] After the adjustment, adjusted elevations are properly known as orthometric heights.

With the introduction of NAVD 88, the NGS provided a "hybrid" geoid model, which serves as the official means for obtaining orthometric heights. At the time of this writing, the latest hybrid geoid model is known as GEOID18. The hybrid geoid model is found by fitting a gravimetric geoid model, which is based on gravity measurements and models of topographic variation, to NAVD 88 leveled benchmarks that have been observed with GNSS. In this manner, the hybrid geoid model provides an estimate of the separation between NAD 83 ellipsoid heights and NAVD 88 orthometric heights. Thus, it subsumes the systematic errors that exist between the two datums, such as any tilt and bias.

However, the National Geodetic Survey has decided that it is time to develop a four-dimensional national reference system with its geometric origin and orientation consistent with the global International Terrestrial Reference Frame (ITRF). In doing so, the National Geodetic Survey will also be creating a new geopotential (vertical) datum with its surface best fitting, in a least-squares sense, mean sea level. Current plans are to have these new georeferenced systems published in 2022. The vertical component of the reference system in the United States will be called the North American-Pacific Geopotential Datum of 2022 (NAPGD2022). This geopotential datum will contain all of the necessary information to provide mutually consistent orthometric heights, geoid undulations, gravity anomalies, surface gravity values, deflections of the vertical, and all other geodetic coordinates related to the gravity field. An important component of NAPGD2022 will be the time-dependent geoid model called GEOID2022, which will cover the entirety of North and Central America, Hawaii, Alaska, Greenland, and the Caribbean. Additional grids of GEOID2022 will also cover American Samoa and Guam and the Commonwealth of the Mariana Islands. GEOID2022 will be a high-accuracy, purely gravimetric geoid model based on the latest satellite, airborne, shipborne, and ground-based gravity observations.

4.4 CURVATURE AND REFRACTION

From the definitions of a level surface and a horizontal line, it is evident that the horizontal plane departs from a level surface because of curvature of the Earth. In Figure 4.2, the deviation DB from a horizontal line through point A is expressed approximately by the formulas

$$C_f = 0.667M^2 = 0.0239F^2 \tag{4.1a}$$

or

$$C_m = 0.0785K^2 \tag{4.1b}$$

---

[4]Descriptions and NAVD 88 elevations of benchmarks can be obtained from the National Geodetic Information Center at their website address http://www.ngs.noaa.gov/datasheet.html. Information can also be obtained by email at info_center@ngs.noaa.gov, or by writing to the National Geodetic Information Center, NOAA, National Geodetic Survey, 1315 East West Highway, Silver Spring, MD 20910; telephone: (301) 713-3242.

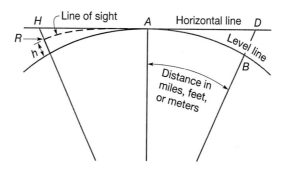

Figure 4.2
Curvature and refraction.

where the departure of a level surface from a horizontal line is C_f in feet or C_m in meters, M is the distance AB in miles, F the distance in *thousands of feet*, and K the distance in kilometers.

Since points A and B are on a level line, they have the same elevation. If a graduated rod was held vertically at B and a reading was taken on it by means of a telescope with its line of sight AD horizontal, the Earth's curvature would cause the reading to be read too high by length BD.

Light rays passing through the Earth's atmosphere are bent or refracted toward the Earth's surface, as shown in Figure 4.3. Thus, a theoretically horizontal line of sight, like AH in Figure 4.2, is bent to the curved form AR. Hence, the reading on a rod held at R is diminished by length RH.

The effects of refraction in making objects appear higher than they really are (and therefore rod readings too small) can be remembered by noting what happens when the Sun is on the horizon, as in Figure 4.3. At the moment when the Sun has just passed below the horizon, it is seen just above the horizon. The Sun's diameter of approximately 32 minutes is roughly equal to the average refraction on a horizontal sight. Since the red wavelength of light bends the least and thus clears the horizon last, it is not uncommon to see a red Sun in a clear sky at dusk and dawn.

Displacement resulting from refraction is variable. It depends on atmospheric conditions, length of line, and the angle a sight line makes with the vertical.

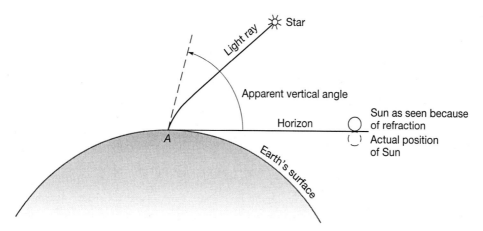

Figure 4.3
Refraction.

For a horizontal sight, refraction R_f in feet or R_m in meters is expressed approximately by the formulas

$$R_f = 0.093 M^2 = 0.0033 F^2 \qquad (4.2a)$$

or

$$R_m = 0.011 K^2 \qquad (4.2b)$$

This is about one-seventh the effect of curvature of the Earth, but in the opposite direction.

The combined effect of curvature and refraction, h in Figure 4.2, is approximately

$$h_f = 0.574 M^2 = 0.0206 F^2 \qquad (4.3a)$$

or

$$h_m = 0.0675 K^2 \qquad (4.3b)$$

where h_f is in feet and h_m is in meters.

For sights of 100, 200, and 300 ft, $h_f = 0.00021, 0.00082$, and 0.0019 ft, respectively, or 0.00068 m for a 100 m length. It will be explained in Section 5.4 that, although the combined effects of curvature and refraction produce rod readings that are slightly too large, proper field procedures in differential leveling can practically eliminate the error due to these causes. However, this is not true for trigonometric leveling (see Section 4.5.4) where this uncompensated systematic error can result in erroneous elevation determinations. This is one of several reasons why trigonometric leveling has never been used in precise geodetic surveys.

4.5 METHODS FOR DETERMINING DIFFERENCES IN ELEVATION

Differences in elevation have traditionally been determined by taping, differential leveling, barometric leveling, and indirectly by trigonometric leveling. A newer method involves measuring vertical distances electronically. Brief descriptions of these methods follow. Other new techniques, described in Chapters 13, 14, and 15, utilize satellite systems. Elevation differences can also be determined using photogrammetry, as discussed in Chapter 27.

4.5.1 Measuring Vertical Distances by Taping or Electronic Methods

Application of a tape to a vertical line between two points is sometimes possible. This method is used to measure depths of mine shafts, to determine floor elevations in condominium surveys, and in the layout and construction of multistory buildings, pipelines, etc. When water or sewer lines are being laid, a graduated pole or rod may replace the tape (see Section 23.4). In certain situations, especially on construction projects, reflectorless electronic distance measurement (EDM) devices (see Section 6.22) are replacing the tape for measuring vertical distances on construction sites. This concept is illustrated in Figure 4.4, 6.13(b), and 23.4.

Figure 4.4
Reflectorless EDMs are being used to measure elevation differences in construction applications. (Courtesy Leica Geosystems.)

4.5.2 Differential Leveling

In this most commonly employed method, a telescope with suitable magnification is used to read graduated rods held on fixed points. A horizontal line of sight within the telescope is established by means of a level vial or automatic compensator.

The basic procedure is illustrated in Figure 4.5. An instrument is set up approximately halfway between BM Rock and point X.[5] Assume the elevation of BM Rock is known to be 820.00 ft. After leveling the instrument, a plus sight taken on a rod held on the BM gives a reading of 8.42 ft. A *plus sight* (+S), also termed *backsight* (BS), is the reading on a rod held on a point of known or

Figure 4.5
Differential leveling.

---

[5] As noted in Section 4.4, the combination of Earth curvature and atmospheric refraction causes rod readings to be too large. However for any setup, if the backsight and foresight lengths are made equal (which is accomplished with the midpoint setup) the error from these sources is eliminated, as described in Section 5.4.

assumed elevation. This reading is used to compute the *height of instrument* (HI), defined as the vertical distance from datum to the instrument line of sight. Direction of the sight—whether forward, backward, or sideways—is not important. The term plus sight is preferable to backsight, but both are used. Adding the plus sight 8.42 ft to the elevation of BM Rock, 820.00, gives an HI of 828.42 ft.

If the telescope is then turned to bring into view a rod held on point X, a *minus sight* ($-S$), also called *foresight* (FS), is obtained. In this example, it is 1.20 ft. A minus sight is defined as the rod reading on a point whose elevation is desired. The term minus sight is preferable to foresight. Subtracting the minus sight, 1.20 ft, from the HI, 828.42, gives the elevation of point X as 827.22 ft.

Differential leveling theory and applications can thus be expressed by two equations, which are repeated over and over

$$HI = elev + BS \qquad (4.4)$$

and

$$elev = HI - FS \qquad (4.5)$$

Since differential leveling is by far the most commonly used method to determine differences in elevation, it is discussed in detail in Chapter 5.

4.5.3 Barometric Leveling

The barometer, an instrument that measures air pressure, can be used to find relative elevations of points on the Earth's surface since a change of approximately 1000 ft in elevation will correspond to a change of about 1 in. of mercury (Hg) in atmospheric pressure, which is approximately 1 torr in 120 m. Figure 4.6 shows a surveying altimeter. Calibration of the scale on different models is in multiples of 1 or 2 ft, 0.5 or 1 m. Air pressures are affected by circumstances other than difference in elevation, such as sudden shifts in temperature and changing weather conditions due to storms. Also, during each day a normal variation in barometric pressure amounting to perhaps a 100-ft difference in elevation occurs. This variation is known as the *diurnal range*.

In barometric leveling, various techniques can be used to obtain correct elevation differences in spite of pressure changes that result from weather variations. In one of these, a *control* barometer remains on a benchmark (base) while a *roving* instrument is taken to points whose elevations are desired. Readings are made on the base at stated intervals of time, perhaps every 10 minutes, and the elevations recorded along with temperature and time. Elevation, temperature, and time readings with the roving barometer are taken at critical points and adjusted later in accordance with changes observed at the control point. Methods of making field surveys using a barometer have been developed in which one, two, or three bases may be used. Other methods employ leapfrog or semi-leapfrog techniques. In stable weather conditions, and by using several barometers, elevations correct to within ± 2 to ± 3 ft are possible.

Figure 4.6
Surveying altimeter.

Barometers have been used in the past for work in rough country where extensive areas had to be covered but a high order of accuracy was not required. However, they are seldom used today having given way to other more modern and accurate equipment.

4.5.4 Trigonometric Leveling

The difference in elevation between two points can be determined by measuring (1) the inclined, which is known as a slant or slope distance, or horizontal distance between them and (2) the zenith angle or the altitude angle to one point from the other. (Zenith and altitude angles, described in more detail in Section 8.13, are measured in vertical planes. Zenith angles are observed downward from vertical, and altitude angles are observed up or down from horizontal.) Thus, in Figure 4.7 if slope distance S and zenith angle z or altitude angle α between C and D are observed, then V, the elevation difference between C and D, is

$$V = S \cos z \qquad (4.6)$$

or

$$V = S \sin \alpha \qquad (4.7)$$

Alternatively, if horizontal distance H between C and D is measured, then V is

$$V = H \cot z \qquad (4.8)$$

or

$$V = H \tan \alpha \qquad (4.9)$$

80 LEVELING—THEORY, METHODS, AND EQUIPMENT

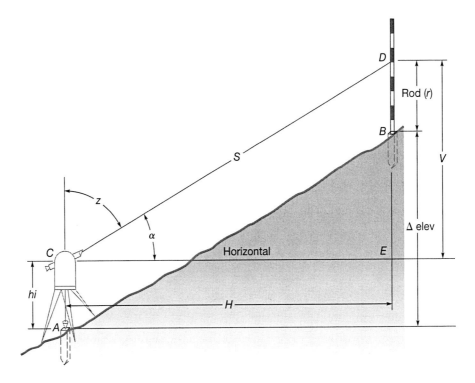

Figure 4.7 Trigonometric leveling—short lines.

The difference in elevation Δelev between points A and B in Figure 4.7 is given by

$$\Delta\text{elev} = hi + V - r \quad (4.10)$$

where hi is the height of the instrument above point A and r the reading on the rod held at B when zenith angle z or altitude angle α is read. If r is made equal to hi, then these two values cancel in Equation (4.10) and simplify the computations.

Note the distinction in this text between HI and hi. Although both are called height of instrument, the term *HI is the elevation of the instrument above datum*, as described in Section 4.5.2, while *hi is the height of the instrument above an occupied point*, as discussed here.

For short lines (up to about 1000 ft in length) elevation differences obtained in trigonometric leveling are appropriately depicted by Figure 4.7 and properly computed using Equations (4.6) through (4.10). However, for longer lines Earth curvature and refraction become factors that must be considered. Figure 4.8 illustrates the situation. Here an instrument is set up at C over point A. Sight D is made on a rod held at point B, and zenith angle z_m, or altitude angle α_m, is observed. The true difference in elevation (Δelev) between A and B is vertical distance HB between level lines through A and B, which is equal to $HG + GF + V - ED - r$. Since HG is the instrument height hi, GF is Earth curvature C [see Equations (4.1)], and ED is refraction R [see Equations (4.2)], the elevation difference can be written as

$$\Delta\text{elev} = hi + V + h_{CR} - r \quad (4.11)$$

4.5 Methods for Determining Differences in Elevation   81

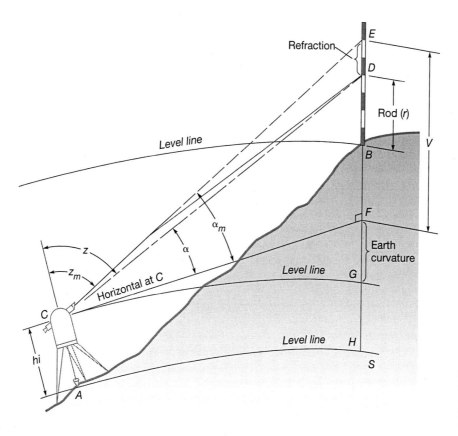

Figure 4.8
Trigonometric leveling—long lines.

The value of V in Equation (4.11) is obtained using one of Equations (4.6) through (4.9), depending on the quantities being observed. Again, if r is made equal to hi, these values cancel. Also, the term h_{CR} is given by Equations (4.3). Thus, except for the addition of the curvature and refraction correction, long and short sights may be treated the same in trigonometric leveling computations. Note that in developing Equation (4.11), angle F in triangle CFE was assumed to be 90°. Of course, as lines become extremely long, this assumption does not hold. However, for lengths within a practical range, errors caused by this assumption are negligible.

The hi used in Equation (4.11) can be obtained by simply observing the vertical distance from the occupied point up to the instrument's horizontal axis (axis about which the telescope rotates) using a graduated rule or rod. An alternate method can be used to determine the elevation of a point that produces accurate results and does not require measurement of the hi. In this procedure, which is especially convenient if a total station instrument is used, the instrument is set up at a location where it is approximately equidistant from a point of known elevation (benchmark) and the one whose elevation is to be determined. The slope distance and zenith (or vertical) angle are measured to each point. Because the distances from the two points are approximately equal, curvature and refraction errors cancel. Also, since the same instrument setup applies to both readings, the hi values cancel, and if the same rod reading r is sighted when making both angle readings, they cancel. Thus the elevation of the unknown point is simply the benchmark elevation, minus V calculated for the benchmark, plus V computed for the unknown point, where the V values are obtained using either Equation (4.6) or (4.7).

■ Example 4.1

The slope distance and zenith angle between points A and B were observed with a total station instrument as 9585.26 ft and 81°42′20″, respectively. The hi and rod reading r were equal. If the elevation of A is 1238.42 ft, compute the elevation of B.

Solution

By Equation (4.3a), the curvature and refraction correction is

$$h_f = 0.0206\left(\frac{9585.26 \sin 81°42'20''}{1000}\right)^2 = 1.85 \text{ ft}$$

(Theoretically, the horizontal distance should be used in computing curvature and refraction. In practice, multiplying the slope distance by the sine of the zenith angle approximates it.)

By Equations (4.6) and (4.11), the elevation difference is (note that hi and r cancel)

$$V = 9585.26 \cos 81°42'20'' = 1382.77 \text{ ft}$$

$$\Delta\text{elev} = 1382.77 + 1.85 = 1384.62 \text{ ft}$$

Finally, the elevation of B is

$$\text{elev}_B = 1238.42 + 1384.62 = 2623.04 \text{ ft}$$

Note that if curvature and refraction had been ignored, an error of 1.85 ft would have resulted in the elevation for B in this calculation. Although Equation (4.11) was derived for an uphill sight, it is also applicable to downhill sights. In that case, the algebraic sign of V obtained in Equations (4.6) through (4.9) will be negative because the vertical angles, α or z, will cause the trigonometric functions to return a negative value.

For uphill sights curvature and refraction is added to a positive V to increase the elevation difference. For downhill sights, it is again added, but to a negative V, which decreases the elevation difference. Therefore, if "reciprocal" zenith (or altitude) angles are read (simultaneously observing the angles from both ends of a line), and V is computed for each and averaged, the effects of curvature and refraction cancel. Alternatively, the curvature and refraction correction can be completely ignored if one calculation of V is made using the average of the reciprocal angles. This assumes atmospheric conditions remain constant, so that refraction is equal for both angles. Hence they should be observed within as short a time period as possible. This method is preferred to reading the zenith (or altitude) angle from one end of the line and correcting for curvature and refraction, as in Example 4.1. The reason is that Equations (4.3) assume a standard atmosphere, which may not actually exist at the time of observations.

■ Example 4.2

In Example 4.1, assume that at B the slope distance was observed again as 9585.26 ft and the zenith angle was read as 98°19′06″. The instrument height and r were equal. Compute (a) the elevation difference from this end of the line and (b) the elevation difference using the mean of reciprocal angles.

Solution

(a) By Equation (4.3a), $h_f = 1.85$ (the same as for Example 4.1).
By Equations (4.6) and (4.11) (note that hi and r cancel),

$$\Delta\text{elev} = 9585.26 \cos 98°19'06'' + 1.85 = -1384.88 \text{ ft}$$

Note that this disagrees with the value of Example 4.1 by 0.26 ft. (The sight from B to A was downhill, hence the negative sign.) The difference of 0.26 ft is most probably due partly to observational errors and partly to refraction changes that occurred during the time interval between vertical angle observations. The average elevation difference for observations made from the two ends is 1384.75 ft.

(b) The average zenith angle is $\dfrac{81°42'20'' + (180° - 98°19'06'')}{2} = 81°41'37''$

By Equation (4.10), $\Delta\text{elev} = 9585.26 \cos 81°41'37'' = 1384.75$ ft

Note that this checks the average value obtained using the curvature and refraction correction.

With the advent of total station instruments, trigonometric leveling has become an increasingly common method for rapid and convenient observation of elevation differences because slope distances and vertical angles are quickly and easily observed from a single setup. Trigonometric leveling is used for topographic mapping, construction stakeout, control surveys, and other tasks. It is particularly valuable in rugged terrain. In trigonometric leveling, accurate vertical angle observations are critical. For precise work, a 1″ to 3″ total station instrument is recommended and angles should be read direct and reversed from both ends of a line. Also, errors caused by uncertainties in refraction are mitigated if sight lengths are limited to about 1000 ft.

PART II • EQUIPMENT FOR DIFFERENTIAL LEVELING

■ 4.6 CATEGORIES OF LEVELS

Instruments used for differential leveling can be classified into four categories: *dumpy levels*, *tilting levels*, *automatic levels*, and *digital levels*. Although each differs somewhat in design, all have two common components: (1) a *telescope* to create a line of sight and enable a reading to be taken on a graduated rod and (2) a system to orient the line of sight in a horizontal plane. Dumpy and tilting levels use level vials to orient their lines of sight, while automatic levels employ *automatic compensators*. Digital levels also employ automatic compensators, but use bar-coded rods for automated digital readings. Automatic levels are the type most commonly employed today, although tilting levels are still used especially on projects requiring very precise work. Digital levels have gained prominence due to their ability to be interfaced with a survey controller (see Section 2.12) and their ease of use. These three types of levels are described in the sections that follow. Dumpy levels are rarely used today, having been replaced by these other newer types. *Hand levels*, although not commonly used for differential leveling, have many special uses where rough elevation differences over short distances are needed. They are also discussed in this chapter. Total station instruments can also be used for differential leveling. These instruments and their uses are described in Section 8.18.

Electronic laser levels that transmit beams of either visible laser or invisible infrared light are another category of leveling instruments. They are not commonly employed in differential leveling, but are used extensively for establishing elevations on construction projects. They are described in Chapter 23.

■ 4.7 TELESCOPES

The telescopes of leveling instruments define the line of sight and magnify the view of a graduated rod against a reference reticle, thereby enabling accurate readings to be obtained. The components of a telescope are mounted in a cylindrical tube. Its four main components are the objective lens, negative lens, reticle, and eyepiece. Two of these parts, the objective lens and eyepiece, are external to the instrument, and are shown on the automatic level illustrated in Figure 4.9.

Objective Lens. This compound lens, securely mounted in the tube's object end, has its optical axis reasonably concentric with the tube axis. Its main function is to gather incoming light rays and direct them toward the negative focusing lens.

Negative Lens. The negative lens is located between the objective lens and reticle, and mounted so its optical axis coincides with that of the objective lens. Its function is to focus rays of light that pass through the objective lens onto the reticle plane. During focusing, the negative lens slides back and forth along the axis of the tube.

Figure 4.9
Parts of an automatic level.
(Courtesy Leica Geosystems.)

Reticle. The reticle consists in a pair of perpendicular reference lines (usually called *cross hairs*[6]) mounted at the principal focus of the objective optical system. The point of intersection of the cross hairs, together with the optical center of the objective system, forms the so-called *line of sight*, also sometimes called the *line of collimation*. The cross hairs were originally created by stretching the hairs of a horse, which was readily available at the time, between two screws. Today they are fine lines etched on a thin round glass plate. The glass plate is held in place in the main cylindrical tube by two pairs of opposing screws, which are located at right angles to each other to facilitate adjusting the line of sight. Two additional lines parallel to and equidistant from the primary lines are commonly added to reticles for special purposes such as for *three-wire leveling* (see Section 5.8) and for *stadia* (see Section 5.4). The reticle is mounted within the main telescope tube with the lines placed in a horizontal-vertical orientation.

Eyepiece. The eyepiece is a microscope (usually with magnification from about 25 to 45 power) for viewing the image.

Focusing is an important required function when using a telescope. The process is governed by the fundamental principle of lenses stated in the following formula:

$$\frac{1}{f_1} + \frac{1}{f_2} = \frac{1}{f} \qquad (4.12)$$

where f_1 is the distance from the lens to the image at the reticle plane, f_2 the distance from the lens to the object, and f the lens *focal length*. The focal length of any lens is a function of the radii of the ground spherical surfaces of the lens, and of the index of refraction of the glass from which it is made. It is a constant for any particular single or compound lens. To focus for each varying f_2 distance, f_1 must be changed to maintain the equality of Equation (4.12).

Focusing the telescope of a level is a two-stage process. First the eyepiece lens must be focused. Since the position of the reticle in the telescope tube remains fixed, the distance between it and the eyepiece lens must be adjusted to suit the eye of an individual observer. This is done by bringing the cross hairs to a clear focus; that is, making them appear as black as possible when sighting at the sky or a distant, light-colored object. Once this has been accomplished, the adjustment need not be changed for the same observer, regardless of sight length, unless the eye fatigues.

The second stage of focusing occurs after the eyepiece has been adjusted. Objects at varying distances from the telescope are brought to sharp focus at the plane of the cross hairs by turning the focusing knob. This moves the negative focusing lens to change f_1 and create the equality in Equation (4.12) for varying f_2 distances.

After focusing, if the cross hairs appear to travel over the object sighted when the eye is shifted slightly in any direction, *parallax* exists. The objective lens, the eyepiece, or both must be refocused to eliminate this effect if accurate work is to be done.

---

[6]The term cross hair comes from the fact that very early surveying telescopes used the hairs from a horse in the reticle. These are now replaced by fine lines etched on glass plates.

4.8 LEVEL VIALS

Level vials are used to orient many different surveying instruments with respect to the direction of gravity. There are two basic types: the *tube* vial and the *circular* or so-called "bull's-eye" version. Tube vials are used on tilting levels (and also on the older dumpy levels) to precisely orient the line of sight horizontal prior to making rod readings. Circular vials are also used on tilting levels, and on automatic levels for quick, rough leveling, after which precise final leveling occurs. The principles of both types of vials are identical.

A tube level is a glass tube manufactured so that its upper inside surface precisely conforms to an arc of a given radius (see Figure 4.10). The tube is sealed at both ends, and except for a small air bubble, it is filled with a sensitive liquid. The liquid must be nonfreezing, quick acting, and maintain a bubble of relatively stable length for normal temperature variations. Purified synthetic alcohol is generally used. As the tube is tilted, the bubble moves, always to the highest point in the tube because air is lighter than the liquid. Uniformly spaced graduations etched on the tube's exterior surface, and spaced 2 mm apart, locate the bubble's relative position. The *axis of the level vial* is an imaginary longitudinal line tangent to the upper inside surface at its midpoint. When the bubble is centered in its run, the axis should be a horizontal line, as in Figure 4.10. For a leveling instrument that uses a level vial, if it is in proper adjustment, its line of sight is parallel to its level vial axis. Thus by centering the bubble, the line of sight is made horizontal.

Its radius of curvature, established in manufacture, determines the sensitivity of a level vial; the larger the radius, the more sensitive a bubble. A highly sensitive bubble, necessary for precise work, may be a handicap in rough surveys because more time is required to center it.

A properly designed level has a vial sensitivity correlated with the *resolving power* (resolution) of its telescope. A slight movement of the bubble should be accompanied by a small but discernible change in the observed rod reading at a distance of about 200 ft. Sensitivity of a level vial is expressed in two ways: (1) the angle, in seconds, subtended by one division on the scale and (2) the radius of the tube's curvature. If one division subtends an angle of 20″ at the center, it is called a 20″ bubble. A 20″ bubble on a vial with 2-mm division spacings has a radius of

Figure 4.10
Tube-type level vial.

Figure 4.11
Coincidence-type level vial correctly set in left view; twice the deviation of the bubble shown in the right view.

Figure 4.12
Bull's-eye level vial.

approximately 68 ft.[7] The sensitivity of level vials on most tilting levels (and the older dumpy levels) ranges from approximately 20″–40″.

Figure 4.11 illustrates the *coincidence-type tube level* vial used on precise equipment. A prism splits the image of the bubble and makes the two ends visible simultaneously. Bringing the two ends together to form a smooth curve centers the bubble. This arrangement enables bubble centering to be done more accurately.

Circular level vials are spherical in shape (see Figure 4.12), the inside surface of the sphere being precisely manufactured to a specific radius. Like the tube version, except for an air bubble, circular vials are filled with liquid. The vial may be graduated with concentric circles having 2-mm spacings. Its axis is actually a plane tangent to the radius point of the graduated concentric circles. When the bubble is centered in the smallest circle, the axis should be horizontal. Besides their use for rough leveling of tilting and automatic levels, circular vials are also used on total station instruments, tribrachs, rod levels, prism poles, and many other surveying instruments. Their sensitivity is much lower than that of tube vials—generally in the range from 2′ to 25′ per 2-mm division but they allow someone to quickly obtain an approximate level of the instrument.

---

[7]The relationship between sensitivity and radius is readily determined. In radian measure, an angle θ subtended by an arc whose radius and length are R and S, respectively, is given as

$$\theta = \frac{S}{R}$$

Thus, for a 20″ bubble with 2-mm vial divisions, by substitution

$$\frac{20''}{\frac{206{,}265''}{\text{rad}}} = \frac{2 \text{ mm}}{R}$$

Solving for R

$$R = \frac{2 \text{ mm} \left(\frac{206{,}265''}{\text{rad}} \right)}{20''} = 20{,}625 \text{ mm} = 20.6 \text{ m} = 68 \text{ ft (approx.)}$$

■ 4.9 TILTING LEVELS

Tilting levels were used for the most precise work. With these instruments, an example of which is shown in Figure 4.13, quick approximate leveling is achieved using a circular vial and the leveling screws. On some tilting levels, a ball-and-socket arrangement (with no leveling screws) permits the head to be tilted and quickly locked nearly level. Precise level in preparation for readings is then obtained by carefully centering a telescope bubble. This is done for each sight, after aiming at the rod, by tilting or rotating the telescope slightly in a vertical plane about a fulcrum at the vertical axis of the instrument. A micrometer screw under the eyepiece controls this movement.

The tilting feature saves time and increases accuracy, since only one screw needs to be manipulated to keep the line of sight horizontal as the telescope is turned about a vertical axis. The telescope bubble is viewed through a system of prisms from the observer's normal position behind the eyepiece. A prism arrangement splits the bubble image into two parts. Centering the bubble is accomplished by making the images of the two ends coincide, as in Figure 4.11.

The tilting level shown in Figure 4.13 has a three-screw leveling head, $42\times$ magnification, and sensitivity of the level vial equal to $10''/2\,\text{mm}$.

■ 4.10 AUTOMATIC LEVELS

Automatic levels of the type pictured in Figure 4.14 incorporate a self-leveling feature. Most of these instruments have a three-screw leveling head, which is used to quickly center a circular bubble, although some models have a ball-and-socket arrangement for this purpose. After the circular bubble is centered manually, an *automatic compensator* takes over, levels the line of sight, and keeps it level.

The operating principle of one type of automatic compensator used in automatic levels is shown schematically in Figure 4.15. The system consists of prisms suspended from wires to create a pendulum. The wire lengths, support locations,

Figure 4.13
Parts of a precise tilting level. (Courtesy Topcon Positioning Systems, Inc.)

Figure 4.14
Automatic level with micrometer. (Topcon Corp.)

When telescope tilts up, compensator swings backward.

Telescope horizontal

When telescope tilts down, compensator swings forward.

Figure 4.15
Compensator of self-leveling level. (Courtesy Keuffel & Esser Company.)

and nature of the prisms are such that only horizontal rays reach the intersection of cross hairs. Thus, a horizontal line of sight is achieved even though the telescope itself may be slightly tilted away from horizontal. Damping devices shorten the time for the pendulum to come to rest, so the operator does not have to wait.

Automatic levels have become popular for general use because of the ease and rapidity of their operation. Some are precise enough for high-order work if a *parallel-plate micrometer* is attached to the telescope front as an accessory, as with the instrument shown in Figure 4.14. When the micrometer plate is tilted, the line of sight is displaced parallel to itself, and decimal parts of rod graduations can be read by means of a graduated dial.

Under certain conditions, the damping devices of an automatic level compensator can stick. To check, with the instrument leveled and focused, read the rod held on a stable point, lightly tap the instrument, and after it vibrates, determine whether the same reading is obtained. Also, some unique compensator problems, such as residual stresses in the flexible links, can introduce systematic errors if not corrected by an appropriate observational routine on first-order work. Another problem is that some automatic compensators are affected by magnetic fields, which can result in systematic errors in rod readings. The sizes of the errors are azimuth-dependent, maximum for lines run north and south, and can exceed 1 mm/km. Thus, it is of concern for high-order control leveling only.

■ 4.11 DIGITAL LEVELS

The newest type of automatic level, the *digital level*, is pictured in Figure 4.16(a). It is classified in the automatic category because it uses a pendulum compensator to level itself, after an operator accomplishes rough leveling with a circular

(a)

(b)

Figure 4.16 (a) Digital level and (b) associated level rod. (Courtesy Topcon Corp.)

bubble. With its telescope and cross hairs, the instrument could be used to obtain readings manually, just like any of the automatic levels. However, this instrument is designed to operate by employing digital image processing. After leveling the instrument, its telescope is turned toward a special bar-coded rod [Figure 4.16(b)] and focused. At the press of a button, the image of bar codes in the telescope's field of view is captured and processed. This processing consists of an onboard computer comparing the captured image to the rod's entire pattern, which is stored in memory. When a match is found, the rod reading is displayed digitally. It can be recorded manually or automatically stored in a survey controller.

The length of rod appearing within the telescope's field of view is a function of the distance from the rod. Thus as a part of its image processing, the instrument is also able to automatically compute the sight length, a feature convenient for balancing backsight and foresight lengths (see Section 5.4). The instrument's maximum range is approximately 100 m, and its accuracy in rod readings is ± 0.5 mm. The bar-coded rods can be obtained with English or metric graduations on the side opposite the bar code. The graduated side of the rod can be used by the operator to manually read the rod in situations that prohibit the instrument from reading the bar codes such as when the rod is in heavy brush.

4.12 TRIPODS

Leveling instruments, whether tilting, automatic, or digital, are all mounted on tripods. A sturdy tripod in good condition is essential to obtain accurate results. Several types are available. The legs can be made of wood, fiberglass, or metal, may be fixed or adjustable in length, and solid or split. All models are shod with metallic conical points and hinged at the top, where they connect to a metal head. An adjustable-leg tripod is advantageous for setups in rough terrain or in a shop, but the type with a fixed-length leg may be slightly more rigid. The split-leg model is lighter than the solid type, but less rugged. (Adjustment of tripods is covered in Section 8.19.2.)

4.13 HAND LEVELS

The hand level (Figure 4.17) is a handheld instrument used on low-precision work, or to obtain quick checks on more precise work. It consists of a brass tube approximately 6 in long, having a plain glass objective and peep-sight eyepiece. A small level vial mounted above a slot in the tube is viewed through the eyepiece by means of a prism or 45° angle mirror. A horizontal line extends across the tube's center.

Figure 4.17
Hand level.
(Courtesy Topcon Corp.)

Figure 4.18
View of level rod through hand level.

As shown in Figure 4.18, the prism or mirror occupies only one half of the tube, and the other part is open to provide a clear sight through the objective lens. Thus, the rod being sighted, and the reflected image of the bubble, is visible beside each other with the horizontal cross line superimposed.

The instrument is held in one hand and leveled by raising or lowering the objective end until the cross line bisects the bubble. Resting the level against a rod or staff provides stability and increases accuracy. This instrument is especially valuable in quickly checking proposed locations for instrument setups in differential leveling.

4.14 LEVEL RODS

A variety of level rods are available, some of which are shown in Figure 4.19. They are made of wood, fiberglass, or metal and have graduations in feet and decimals, or meters and decimals.

A Philadelphia rod is shown in Figure 4.19(a) and (b). It consists of two sliding sections graduated in hundredths of a foot and joined by brass sleeves *a* and *b*. The rear section can be locked in position by a clamp screw *c* to provide any length from a *short rod* for readings of 7 ft or less to a *long rod* (*high rod*) for readings up to 13 ft. *When the high rod is needed, it must be extended fully, otherwise a serious mistake will result in its reading.* Graduations on the front faces of the two sections read continuously from zero at the base to 13 ft at the top for the high-rod setting. However, it should be realized by the user that in the extended position, the 7-foot mark is at the 13-foot position.

Rod graduations are accurately painted, alternate black and white spaces 0.01 ft wide. Spurs extending the black painting emphasize the 0.1- and 0.05-ft marks. Tenths are designated by black figures, and footmarks by red numbers, all straddling the proper graduation. Rodpersons should keep their hands off the painted markings, particularly in the 3- to 5-ft section, where a worn face will make the rod unfit for use. A Philadelphia rod can be read accurately with a level at distances up to about 250 ft.

A wide choice of patterns, colors, and graduations on single-piece, two-piece, three-section, and four-section leveling rods is available. The various types, usually named for cities or states, include the Philadelphia, New York, Boston, Troy, Chicago, San Francisco, and Florida rods.

Philadelphia rods can be equipped with targets [*d* in Figure 4.19(a) and (b)] for use on long sights. When employed, the rodperson sets the target at the instrument's line-of-sight height according to communications or hand signals from the instrument

Figure 4.19
(a) Philadelphia rod (front). (Courtesy Pearson Education.) (b) Philadelphia rod (rear). (Courtesy Pearson Education.) (c) Double-faced leveling rod with metric graduations. (Courtesy Leica, Inc.) (d) Lenker direct-reading rod. (Courtesy Lenker Manufacturing Company.)

operator. It is fixed using clamp *e*, then read and recorded by the rodperson. The *Vernier* scale at *f*, can be used to obtain readings to the nearest 0.001 ft if desired.

A Vernier scale is a short auxiliary scale set parallel to and beside a primary scale. It enables reading fractional parts of the smallest main-scale divisions without interpolation. Figure 4.20 shows a Vernier scale. The Vernier scale is constructed so that 10 of its divisions cover 9 divisions on the main scale. The difference between the length of one main-scale division and one Vernier scale division is therefore 0.1 of the main-scale division. This is the so-called *least count* of this Vernier scale. In general, the least count of a Vernier scale is given by

$$\text{least count} = d/n \qquad (4.14)$$

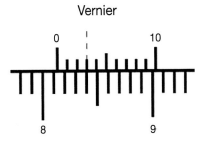

Figure 4.20
Reading a Vernier scale.

where d is the value of the smallest main-scale division, and n the number of Vernier scale divisions that span $(n - 1)$ main-scale units. By Equation (4.14), the least count of the Vernier scale of Figure 4.20 is $0.1/10 = 0.01$. This verifies the intuitive determination given above. *An observer cannot make readings using a Vernier scale without first determining its least count.* In Figure 4.20 the first two digits are read where using the last digit on the main scale across from the 0 mark on the Vernier scale; in this instance, the reading is 8.1. The final digit is the read from the Vernier scale where there is alignment with the main-scale division; in this instance, the 3 on the Vernier scale aligns with a division on the main scale. Thus Figure 4.20 is read as 8.13. Thus, a target on a Philadelphia rod yields a least-count reading of 0.001 ft since it is 0.01/10 ft.

The Chicago rod, consisting of independent sections (usually three) that fit together but can be disassembled, is widely used on construction surveys. The San Francisco model has separate sections that slide past each other to extend or compress its length, and is generally employed on control, land, and other surveys. Both are conveniently transported in vehicles.

The direct-reading Lenker level rod [Figure 4.19(d)] has numbers in reverse order on an endless graduated steel-band strip that can be revolved on the rod's end rollers. Figures run down the rod and can be brought to a desired reading—for example, the elevation of a benchmark. Rod readings are preset for the backsight, and then, due to the reverse order of numbers, foresight readings give elevations directly without manually adding backsights and subtracting foresights.

A rod consisting of a wooden, or fiberglass, frame and an Invar strip to eliminate the effects of humidity and temperature changes is used on precise work. The Invar strip, attached at its ends only is free to slide in grooves on each side of the wooden frame. Rods for precise work are usually graduated in meters and often have dual scales. Readings of both scales are compared to eliminate mistakes.

As described in Section 1.8, safety in traffic and near heavy equipment is an important consideration. The Quad-pod, an adjustable stand that clamps to any leveling rod, can help to reduce traffic hazards, and in some cases also lower labor costs.

■ 4.15 TURNING POINTS

Often when performing differential leveling, the distance between benchmarks is too far to accomplish in one setup. When this happens, turning points are established at appropriate locations between the two points to allow the surveyor to transfer heights between setups. This process is covered in Section 5.4. Of most importance is that temporary turning points be stable and also of such design to have only one high point. This can be typically accomplished with something that has a

Figure 4.21
A leveling plate also known as a turtle. (Courtesy SECO manufacturing.)

rounded headed on which to place the level rod. Figure 4.21 shows a leveling plate, which is commonly called a turtle. The turtle is placed on the ground and depressed into the earth. The rod is then placed on the rounded point at the center of the turtle. Other items that surveyors have used in these situations include railroad spikes and large marlinspikes. As last resort, a survey hub driven at an angle such that one corner of the hub is higher than the others has been used. However, when using any of these devices, it is important that the item provides a solid stable surface so that it does not change height between the foresight and backsight readings. If this is not accomplished, even the best leveling will yield poor leveling results.

■ 4.16 TESTING AND ADJUSTING LEVELS

Through normal use and wear, all leveling instruments will likely become maladjusted from time to time. The need for some adjustments may be noticed during use, for example, level vials on tilting levels. Others may not be so obvious, and therefore it is important that instruments be checked periodically to determine their state of adjustment. If the tests reveal conditions that should be adjusted, depending on the particular instrument, and the knowledge and experience of its operator, some or all of the adjustments can be made immediately in the field. However, if the parts needing adjustment are not readily accessible, or if the operator is inexperienced in making the adjustments, it is best to send the instruments away for adjustment by qualified technicians.

4.16.1 Requirements for Testing and Adjusting Instruments

Before testing and adjusting instruments, care should be exercised to ensure that any apparent lack of adjustment is actually caused by the instrument's condition and not by test deficiencies. To properly test and adjust leveling instruments in the field, the following rules should be followed:

1. Choose terrain that permits solid setups in a nearly level area enabling sights of at least 200 ft to be made in opposite directions.

2. Perform adjustments when good atmospheric conditions prevail, preferably on cloudy days free of heat waves. No sight line should pass through alternate sun and shadow, or be directed into the sun.
3. Place the instrument in shade, or shield it from direct rays of the sun.
4. Make sure the tripod shoes are tight and the instrument is screwed onto the tripod firmly. Spread the tripod legs well apart and position them so that the tripod plate is nearly level. Press the shoes into the ground firmly.

Standard methods and a prescribed order must be followed in adjusting surveying instruments. Loosening or tightening the proper adjusting nuts and screws with special tools and pins attains correct positioning of parts. Time is wasted if each adjustment is perfected on the first trial, since some adjustments affect others. The complete series of tests may have to be repeated several times if an instrument is badly off. A final check of all adjustments should be made to ensure that all have been completed satisfactorily.

Tools and adjusting pins that fit the capstans and screws should be used, and the capstans and screws handled with care to avoid damaging the soft metal. Adjustment screws are properly set when an instrument is shipped from the factory. Tightening them too much (or not enough) nullifies otherwise correct adjustment procedures and may leave the instrument in worse condition than it was before adjusting. Thus it is recommended that only qualified technicians should make adjustments on the instrument. However the procedure to determine if an instrument is out of adjustment is important for all to know.

4.16.2 Adjusting for Parallax

The parallax adjustment is extremely important and must always be kept in mind when using a leveling instrument, but especially during the testing and adjustment process. The adjustment is done by carefully focusing the objective lens and eyepiece so that the cross hairs appear clear and distinct, and so that the cross hairs do not appear to move against a background object when the eye is shifted slightly in position while viewing through the eyepiece.

4.16.3 Testing and Adjusting Level Vials

For leveling instruments that employ a level vial, the axis of the level vial should be perpendicular to the vertical axis of the instrument (axis about which the instrument rotates in azimuth). Then once the bubble is centered, the instrument can be turned about its vertical axis in any azimuth and the bubble will remain centered. Centering the bubble and revolving the telescope 180° about the vertical axis can quickly check this condition. The distance the bubble moves off the central position is twice the error. To correct any maladjustment, turn the capstan nuts at one end of the level vial to move the bubble *halfway back* to the centered position. Level the instrument using the leveling screws. Repeat the test until the bubble remains centered during a complete revolution of the telescope.

4.16.4 Preliminary Adjustment of the Horizontal Cross Hair

Although it is good practice to always sight an object at the center of the cross hairs, if this is not done and the horizontal cross hair is not truly horizontal when the instrument is leveled, an error will result. To test for this condition, sight a sharply defined point with one end of the horizontal cross hair. Turn the telescope slowly on its vertical axis so that the cross hair moves across the point. If the cross hair does not remain on the point for its full length, it is out of adjustment. To correct any maladjustment, loosen the four capstan screws holding the reticle. Rotate the reticle in the telescope tube until the horizontal hair remains on the point as the telescope is turned. The screws should then be carefully tightened in their final position.

4.16.5 Testing and Adjusting the Line of Sight

For tilting levels, described in Section 4.9, when the bubble of the level vial is centered, the line of sight should be horizontal. In other words, for this type of instrument to be in perfect adjustment, the axis of the level vial and the line of sight must be parallel. If they are not, a *collimation error* exists. For the automatic levels, described in Section 4.10, after rough leveling by centering the circular bubble, the automatic compensator must define a horizontal line of sight if it is in proper adjustment. If it does not, the compensator is out of adjustment, and again a collimation error exists. *The collimation error will not cause errors in differential leveling as long as the minus and plus sight distances are balanced.* However, it will cause errors when backsights and foresights are not balanced, which sometimes occurs in differential leveling, and cannot be avoided in profile leveling (see Section 5.9), and construction staking (see Chapter 23).

One method of testing a level for collimation error is to stake out four points spaced equally, each about 100 ft apart on approximately level ground as shown in Figure 4.22. The level is then set up at point 1, leveled, and rod readings (r_A) at A, and (R_B) at B are taken. Next the instrument is moved to point 2 and releveled. Readings R_A at A, and (r_B) at B are then taken. As illustrated in the figure, assume that a collimation error ε exists in the rod readings of the

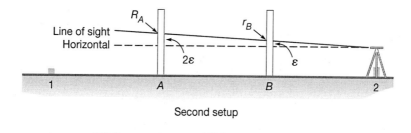

Figure 4.22 Horizontal collimation test.

two shorter sights. Then the error caused by this source would be 2ε in the longer sights because their length is double that of the shorter ones. Whether or not there is a collimation error, the difference between the rod readings at 1 should equal the difference of the two readings at 2. Expressing this equality, with the collimation error included, gives

$$(R_B - 2\varepsilon) - (r_A - \varepsilon) = (r_B - \varepsilon) - (R_A - 2\varepsilon) \tag{4.14}$$

Solving for ε in Equation (4.13) yields

$$\varepsilon = \frac{R_B - r_A - r_B + R_A}{2} \tag{4.15}$$

The corrected reading for the level rod at point A while the instrument is still setup at point 2 would be $R_A - 2\varepsilon$ and at point B would be $R_B - \varepsilon$. If an adjustment is necessary, it is done by loosening the top (or bottom) screw holding the reticle, and tightening the bottom (or top) screw to move the horizontal hair up or down until the required reading is obtained on the rod at A. This changes the orientation of the line of sight. Several trials may be necessary to achieve the exact setting. If the reticle is not accessible, or the operator is unqualified, the instrument should be serviced by a qualified technician.

As discussed in Section 19.14, it is recommended that the level instrument be tested before the observation process when performing precise differential leveling. A correction for the error in the line of sight is then applied to all field observations using the sight distances obtained by reading the stadia wires (see Section 5.4). The error in the line of sight is expressed in terms of ε per unit sight distance. For example, the collimation error C is unitless and expressed as 0.00005 ft/ft or 0.00005 m/m. Using the sight distances obtained in the leveling process this error can be mathematically eliminated. However, for most common leveling work, this error is removed by simply keeping minus and plus sight distances approximately equal between benchmarks.

■ Example 4.3

A horizontal collimation test is performed on an automatic level following the procedures just described. With the instrument setup at point 1, the rod reading at A was 5.630 ft, and to B was 5.900 ft. After moving and leveling the instrument at point 2, the rod reading to A was determined to be 5.310 ft and to B 5.560 ft. As shown in Figure 4.22, the distance between the points was 100 ft. What is the collimation error of the instrument, and the corrected reading to A from point 2?

Solution

Substituting the appropriate values into Equation (4.15), the collimation error is

$$\varepsilon = \frac{5.900 - 5.630 - 5.560 + 5.310}{2} = 0.010 \text{ ft}$$

Thus the corrected reading to A from point 2 is

$$R = 5.310 - 2(0.010) = 5.290 \text{ ft}$$

As noted above, if a collimation error exists but the instrument is not adjusted, accurate differential leveling can still be achieved when the plus sight and minus sight distances are balanced. In situations where these distances cannot be balanced, correct rod readings can still be obtained by applying collimation corrections to the rod readings. This procedure is described in Sections 5.12.1 and 19.14.

■ Example 4.4

The instrument in Example 4.3 was used in a survey between two benchmarks before the instrument was adjusted where the sight distance could not be balanced due to the physical conditions. The sum of the plus sights was 900 ft while the sum of the minus sights was 1300 ft between the two benchmarks. The observed elevation difference was 120.64 ft. What is the corrected elevation difference between the two benchmarks?

Solution

In Example 4.3, the error ε was determined to be 0.01 ft/100 ft. Thus, the collimation error C is -0.0001 ft/ft, and the corrected elevation difference is

$$\Delta \text{elev} = 120.64 - 0.0001(900 - 1300) = 120.68 \text{ ft}$$

PROBLEMS

Asterisks (*) indicate problems that have answers given in Appendix G.

4.1 Define the following leveling terms: (a) leveling, (b) level surface, and (c) geoid.

4.2* How far will a horizontal line depart from the Earth's surface in 1 km? 5 km? 10 km? (Apply both curvature and refraction.)

4.3 Visit the website of the National Geodetic Survey at http://www.ngs.noaa.gov, and obtain a data sheet description of a benchmark in your local area.

4.4 In your own words, define the term level surface.

4.5 What is meant by the term Metonic cycle? Why is this important in defining tidal datums?

4.6 Why are elevations today not referred to as mean sea-level heights?

4.7* On a large lake without waves, how far from shore is a sailboat when the top of its 30-ft mast disappears from the view of a person lying at the water's edge?

4.8 Similar to Problem 4.7, except for a 5-m mast and a person whose eye height is 0.5 m above the water's edge.

4.9 Readings on a line of differential levels are taken to the nearest 0.2 mm. For what maximum distance can the Earth's curvature and refraction be neglected?

4.10 Similar to Problem 4.9 except readings are to the 0.002 ft.

4.11 Describe how readings are determined in a digital level when using a bar-coded rod.

Successive plus and minus sights taken on a downhill line of levels are listed in Problems 4.12 and 4.13. The values represent the horizontal distances between the instrument and either the plus or minus sights. What error results from curvature and refraction?

4.12* 20, 225; 50, 195; 40, 135; 30, 250 ft.

4.13 20, 50; 15, 50; 10, 50; 10, 60 m.

4.14 What error results if the curvature and refraction correction is neglected in trigonometric leveling for sights: (a) 5000 ft long, (b) 1500 m long, and (c) 6500 ft long?

4.15* The slope distance and zenith angle observed from point P to point Q were 2406.787 m and 84°13'07" respectively. The instrument and rod target heights were equal. If the elevation of point P is 30.245 m, above datum, what is the elevation of point Q?

4.16 The slope distance and zenith angle observed from point X to point Y were 4889.45 ft and 83°15'17" The instrument and rod target heights were equal. If the elevation of point X is 847.89 ft above datum, what is the elevation of point Y?

4.17 Similar to Problem 4.15, except the slope distance was 1643.038 m, the zenith angle was 98°24'44".and the elevation of point P was 406.567 m above datum.

4.18 In trigonometric leveling from point A to point B, the slope distance and zenith angle measured at A were 5462.46 ft and 94°08'36" At B these measurements were 5462.58 ft and 85°51'32" respectively. If the instrument and rod target heights were equal, calculate the difference in elevation from A to B.

4.19 Describe how parallax in the viewing system of a level can be detected and removed.

4.20 What is the sensitivity of a level vial with 2-mm divisions for: (a) a radius of 41.4 m and (b) a radius of 20.6 m?

4.21* An observer fails to check the bubble, and it is off two divisions on a 500-ft sight. What error in elevation difference results with a 10-second bubble?

4.22 An observer fails to check the bubble, and it is off one division on a 100-m sight. What error results for a 30-second bubble?

4.23 Similar to Problem 4.21, except a 40-second bubble is off two divisions on a 300-ft sight.

4.24 With the bubble centered, a 200-m sight gives a reading of 1.352 m. After moving the bubble three divisions off center, the reading is 1.410 m. For 2-mm vial divisions, what is: (a) the vial radius of curvature in meters and (b) the angle in seconds subtended by one division?

4.25 Similar to Problem 4.24, except the sight length was 300 ft, the initial reading was 5.132 ft, and the final reading was 5.176 ft.

4.26 Sunshine on the forward end of a 30"/2mm level vial bubble draws it off 2 divisions, giving a plus sight reading of 6.36 ft on a 400-ft sight. Compute the correct reading.

4.27 Sunshine on the forward end of a 20"/2mm level vial draws the bubble off two divisions, giving a plus sight of 1.362 m on a 155-m observation. Compute the correct reading.

4.28* If a plus sight of 3.54 ft is taken on BM A, elevation 850.48 ft, and a minus sight of 7.84 ft is read on point X, calculate the HI and the elevation of point X.

4.29 If a plus sight of 1.097 m is taken on BM A, elevation 305.348 m, and a minus sight of 0.832 m is read on point X, calculate the HI and the elevation of point X.

4.30 Similar to Problem 4.28, except a plus sight of 5.65 ft is taken on BM A, elevation 103.65 ft, and a minus sight of 6.32 ft read on point X.

4.31 Describe the procedure used to test if the level vial is perpendicular to the vertical axis of the instrument.

4.32 A horizontal collimation test is performed on an automatic level following the procedures described in Section 4.15.5. With the instrument setup at point 1, the rod reading at A was 5.261 ft, and to B it was 5.453 ft. After moving and leveling the instrument at point 2, the rod reading to A was 4.940 ft and to B was 5.116 ft. The distance between the points in the test was 100 ft. What is the collimation error of the instrument and the corrected reading to A from point 2?

4.33 The instrument tested in Problem 4.32 was used in a survey immediately before the test where the observed elevation difference between two benchmarks was +44.65 ft. The sum of the plus sight distances between the benchmarks was 250 ft and the sum of the minus sight distances was 490 ft. What is the corrected elevation difference between the two benchmarks?

4.34 Similar to Problem 4.32 except that the rod readings are 1.677 and 1.722 m to A and B, respectively, from point 1, and 1.543 and 1.586 m to A and B, respectively, from point 2. The distance between the points in the test was 100 m.

4.35 The instrument tested in Problem 4.34 was used in a survey immediately before the test where the observed elevation difference between two benchmarks was −13.068 m. The sum of the plus sight distances between the benchmarks was 1065 m and the sum of the minus sight distances was 480 m. What is the corrected elevation difference between the two benchmarks?

4.36 Describe a field procedure which will compensate for the collimation errors described in Problems 4.33 and 4.35.

BIBLIOGRAPHY

Brazeal, R. 2014. "The Future of Determining Elevation." *XYHT* 1 (No. 1): 30.

Doyle, D. and D. Smith. 2012. "Definition and Densification of the Puerto Rico Vertical Datum of 2002." *Surveying and Land Information Science* 72 (No. 3): 109.

Naberezny, B. 2014. "National Surveyors Week 2014." *Professional Surveyor* 34 (No. 5): 8.

Penry, J. 2014. "The Earth Vertical Datums of the U.S. Geological Survey." *The American Surveyor* 11 (No. 5): 34.

Schowe, R. 2014. "New Record for Southeast Asia's Highest Peak." *The American Surveyor* 11 (No. 4): 10.

Smith, D. et al. 2015. "Recent Activities at the National Geodetic Survey: GRAV-D." *The American Surveyor* 12 (No. 2): 22.

Véronneau, M. et al. 2014. "Canada's New Vertical Datum: CGVD2013 — Part 1." *XYHT* 1 (No. 4): 43.

———. 2014. "Canada's New Vertical Datum: CGVD2013 — Part 2." *XYHT* 1 (No. 5): 41.

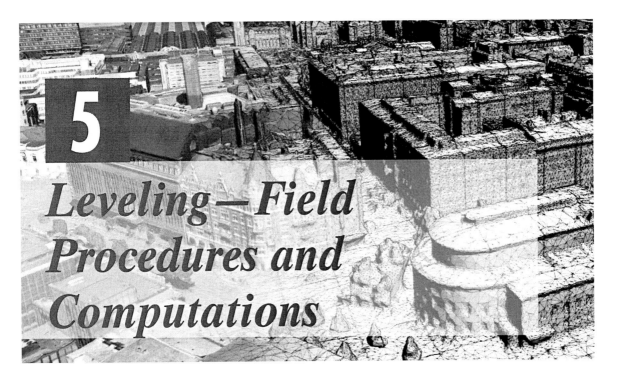

5
Leveling—Field Procedures and Computations

■ 5.1 INTRODUCTION

Chapter 4 covered the basic theory of leveling, briefly described the different procedures used in determining elevations, and showed examples of most types of leveling equipment. This chapter concentrates on differential leveling, and discusses handling the equipment, running and adjusting simple leveling loops, and performing some project surveys to obtain data for field and office use. Some special variations of differential leveling, useful or necessary in certain situations, are presented. Profile leveling, to determine the configuration of the ground surface along some established reference line, is described in Section 5.9. Finally errors in leveling are discussed. Leveling procedures for construction and other surveys, along with those of higher order to establish the nationwide vertical control network, will be covered in later chapters.

■ 5.2 CARRYING AND SETTING UP A LEVEL

The safest way to transport a leveling instrument in a vehicle is to leave it in its case. The case closes properly only when the instrument is set correctly in the padded supports. Whenever possible, a level should be removed from its container by lifting from the base, not by grasping the telescope. The head must be screwed snugly on the tripod. If the head is too loose, the instrument is unstable; if too tight, it may "freeze." Once the instrument is removed from the case, the case should be once again closed to prevent dirt and moisture from entering it.

The legs of a tripod must be tightened correctly. If each leg falls slowly of its own weight after being placed in a horizontal position, it is adjusted properly. Clamping them too tightly strains the plate and screws. If the legs are loose, unstable setups result.

Except for a few instruments that employ a ball-and-socket arrangement, all modern levels use a three-screw leveling head for initial rough leveling. Note that each of the levels illustrated in Chapter 4 (see Figures 4.9, 4.13, 4.14, and 4.16) has this type of arrangement. In leveling a three-screw head, the telescope is rotated until it is over two screws as in the direction *AB* of Figure 5.1. Using the thumb and first finger of each hand to adjust simultaneously the opposite screws approximately centers the bubble. This procedure is repeated with the telescope rotated 90° so that it is over *C*, the remaining single screw. Time is wasted by centering the bubble exactly on the first try, since it can be thrown off during the cross leveling process. Working with the same screws in succession about three times should complete the job. A simple but useful rule in centering a bubble, illustrated in Figure 5.1, is: *A bubble follows the left thumb when turning the screws*. A circular bubble is centered by alternately turning one screw and then the other two. The telescope need not be rotated during the process.

It is generally unnecessary to set up a level over any particular point. Therefore, it is inexcusable to have the base plate badly out of level before using the leveling screws. On sidehill setups, placing one leg on the uphill side and two on the downhill slope eases the problem. On very steep slopes, some instrument operators prefer two legs uphill and one downhill for stability. The most convenient height of setup is one that enables the observer to sight through the telescope without stooping or stretching.

Inexperienced instrument operators running levels up or down steep hillsides are likely to find, after completing the leveling process, that the telescope is too low for sighting the higher turning point (TP) or benchmark (BM). To avoid this, a hand level can be used to check for proper height of the setup before leveling the instrument precisely. As another alternative, the instrument can be quickly set up without attempting to level it carefully. Then the rod is sighted making sure

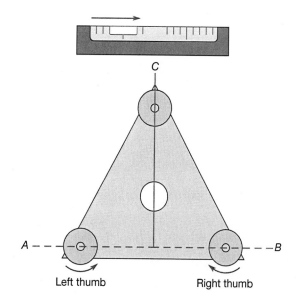

Figure 5.1
Use of leveling screws on a three-screw instrument.

the bubble is somewhat back of center. If it is visible for this placement, it obviously will also be seen when the instrument is leveled.

■ 5.3 DUTIES OF A ROD PERSON

The duties of a rod person are relatively simple. However, a careless rod person can nullify the best efforts of an observer by failing to follow a few basic rules.

A level rod must be held plumb on the correct monument or turning point to give the correct reading. In Figure 5.2, point A is below the line of sight by vertical distance AB. If the rod is tilted to position AD, an erroneous reading AE is obtained. It can be seen that the smallest reading possible, AB, is the correct one and is secured only when the rod is plumb.

A rod level of the type shown in Figure 5.3 ensures *fast* and *correct* rod plumbing. Its L-shape is designed to fit the rear and side faces of a rod, while the circular bubble is centered to plumb the rod in *both directions*. However, if a rod level is not available, one of the following procedures can be used to plumb the rod.

Waving the rod is one procedure used to ensure that the rod is plumb when a reading is taken. The process consists of *slowly* tilting the rod top, first perhaps a

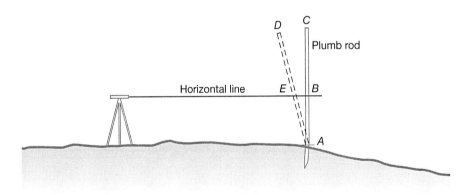

Figure 5.2
Plumbing a level rod.

Figure 5.3
Rod level.

foot or two toward the instrument and then just slightly away from it. The observer watches the readings increase and decrease alternately, and then selects the minimum value, the correct one. Beginners tend to swing the rod too fast or too slow and through too long an arc. It is also important that the rod swing an arc through the vertical. Small errors can be introduced in the process if the bottom of the rod is resting on a flat surface. A rounded-top monument, steel spike, or thin edge makes an excellent benchmark or intermediate point for leveling when this procedure is used.

On still days the rod can be plumbed by letting it balance of its own weight while lightly supported by the fingertips. An observer makes certain the rod is plumb in the lateral direction by checking its coincidence with the vertical wire and signals for any adjustment necessary. The rod person can save time by sighting along the side of the rod to line it up with a telephone pole, tree, or side of a building. Plumbing along the line toward the instrument is more difficult, but holding the rod against the toes, stomach, and nose can bring it close to a plumb position. A plumb bob suspended alongside the rod can also be used, and in this procedure the rod is adjusted in position until its edge is parallel with the string.

■ Example 5.1

In Figure 5.2, what error results if the rod is held in position AD, and if $AE = 10$ ft and $EB = 6$ in.?

Solution

Using the Pythagorean theorem, the vertical rod is

$$AB = \sqrt{10^2 - 0.5^2} = 9.987 \text{ ft}$$

Thus the error is $10.00 - 9.987 = 0.013$ ft, or 0.01 ft.

Errors of the magnitude of Example 5.1 are serious, whether the results are carried out to hundredths or thousandths. They make careful plumbing necessary, particularly for high-rod readings.

■ 5.4 DIFFERENTIAL LEVELING

Figure 5.4 illustrates the procedure followed in differential leveling. In the figure, the elevation of new BM Oak is to be determined by originating a leveling circuit at established BM Mil. In running this circuit, the first reading, a *plus sight*, is taken on the established benchmark. From it, the *height of instrument* (*HI*) can be computed using Equation (4.4). Then a *minus sight* is taken on the first intermediate point (called a *turning* point and labeled TP1 in the figure), and by Equation (4.5) its elevation is obtained. The process of taking a plus sight, followed by a minus sight, is repeated over and over until a circuit is completed.

As shown in the example of Figure 5.4, four instrument setups were required to complete half of the circuit (the run from BM Mil to BM Oak). Field notes for

106 LEVELING—FIELD PROCEDURES AND COMPUTATIONS

Figure 5.4
Differential leveling.

the example of Figure 5.4 are given in Figure 5.5. As illustrated in this figure, a tabular form of field notes is used for differential leveling, and the addition and subtraction to compute HIs and elevations is done directly in the notes at the time the data is collected, and should never be saved for later. These notes also show the data for the return run from BM Oak back to BM Mil to complete the circuit. It is important in differential leveling to run closed circuits so that the accuracy of the work can be checked, as will be discussed later.

As noted, the intermediate points upon which the rod is held in running a differential leveling circuit are called *turning points* (*TPs*). Two rod readings are taken on each, a minus sight followed by a plus sight. Turning points should be solid objects with a definite high point. Careful selection of stable turning points is essential to achieve accurate results. Steel turning pins, turtles, and railroad spikes driven into firm ground make excellent turning points when permanent objects are not conveniently available.

In differential leveling, horizontal lengths for the plus and minus sights should be made about equal. This can be done by pacing, by stadia measurements, by counting rail lengths or pavement joints if working along a track or roadway, or by any other convenient method. Stadia readings are the most precise of these methods and will be discussed in detail.

Stadia was once commonly used for mapping. The stadia method determines the horizontal distance to points through the use of readings on the upper and lower (stadia) wires in the reticle. The method is based on the principle that in similar triangles, corresponding sides are proportional. In Figure 5.6 which depicts a telescope with a simple lens, light rays from points *A* and *B* pass through the

DIFFERENTIAL LEVELS

Sta.	+B.S.	H.I.	−F.S.	Elev.	Adj. Elev.
BM Mil.	1.33			2053.18	2053.18
		2054.51		(−0.004)	
TP1	0.22		8.37	2046.14	2046.14
		2046.36	7.91	(−0.008)	
TP2	0.96		8.91	2038.45	2038.44
		2039.41		(−0.012)	
TP3	0.46		11.72	2027.69	2027.68
		2028.15		(−0.016)	
BM Oak	11.95		8.71	2019.44	2019.42
		2031.39		(−0.022)	
TP4	12.55		2.61	2028.78	2028.76
		2041.33		(−0.026)	
TP5	12.77		0.68	2040.65	2040.62
		2053.42		(−0.030)	
BM Mil.			0.21	2053.21	2053.18
Σ =	+40.24	Σ =	−40.21		
		Page Check:			
		2053.18			
		+ 40.24			
		2093.42			
		− 40.21			
		2053.21	Check		

GRAND LAKES UNIV. CAMPUS

BM Mil. to BM Oak

BM Mil. on GLU Campus 29 Sept. 2014
SW of Engineering Bldg. Clear, Warm 70° F
9.4 ft. north of sidewalk T.E. Henderson N
to instrument room and J.F. King ⌀
1.6 ft. from Bldg. Bronze D.R. Moore ⊤
disk in concrete flush Lietz Level #6
with ground, stamped "Mil"

BM Oak is a temporary
project bench mark located
at corner of Cherry and
Pine Sts., 14 ft. West of
computer laboratory. Twenty
penny spike in 18" Oak
tree, 1 ft. above ground.

Loop Misclosure = 2053.21 − 2053.18 = 0.03

Permissible Misclosure = $0.02\sqrt{n} = 0.02\sqrt{7}$
$\qquad\qquad\qquad\qquad\quad = 0.05$ ft.

Adjustment = $\dfrac{0.03}{7} = 0.004'$ per H.I.

J.E. Henderson

Figure 5.5
Differential leveling notes for Figure 5.4.

lens center and form a pair of similar triangles AmB and amb. Here $AB = I$ is the rod intercept (stadia interval), and $ab = i$ is the spacing between stadia wires.

Standard symbols used in stadia observations and their definitions are as follows (refer to Figure 5.6):

  $f = $ *focal length* of lens (a constant for any particular compound objective lens)
  $i = $ spacing between stadia wires (ab in Figure 5.6)
  $f/i = $ *stadia interval factor* usually 100 and denoted by K
  $I = $ rod intercept (AB in Figure 5.6), also called *stadia interval*
  $c = $ distance from instrument center (vertical axis) to objective lens center (varies slightly when focusing the objective lens for different sight lengths but is generally considered to be a constant)
  $C = $ stadia constant $= c + f$
  $d = $ distance from the focal point F in front of telescope to face of rod
  $D = $ distance from instrument center to rod face $= C + d$

108 LEVELING—FIELD PROCEDURES AND COMPUTATIONS

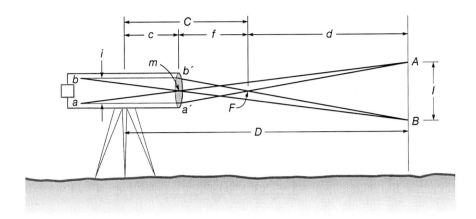

Figure 5.6
Principle of stadia.

From similar triangles of Figure 5.6

$$\frac{d}{f} = \frac{I}{i} \quad \text{or} \quad d = \frac{f}{i}I = KI$$

Thus

$$D = KI + C \tag{5.1}$$

The geometry illustrated in Figure 5.6 pertains to a simplified type of *external focusing telescope*. It has been used because an uncomplicated drawing correctly shows the relationships and aids in deriving the stadia equation. These telescopes are now obsolete in surveying instruments. The objective lens of an *internal focusing telescope* (the type now used in surveying instruments) remains fixed in position, while a movable negative-focusing lens between the objective lens and the plane of the cross hairs changes directions of the light rays. As a result, the stadia constant, (C), is so small that it can be assumed equal to zero and drops out of Equation (5.1). Thus the equation for a distance on a horizontal stadia sight reduces to

$$D = KI \tag{5.2}$$

Fixed stadia lines in theodolites, transits, levels, and alidades are generally spaced by instrument manufacturers to make the stadia interval factor $f/i = K$ equal to 100. It should be determined the first time an instrument is used or if the reticle or lenses are replaced or adjusted.

To determine the stadia interval factor K, rod intercept I for a horizontal sight of known distance D is read. Then in an alternate form of Equation (5.2), the stadia interval factor is $K = D/I$. As an example, at a measured distance of 300.0 ft, a rod interval of 3.01 was read. Then $K = 300.0/3.01 = 99.7$. Accuracy in determining K is increased by averaging values from several lines whose observed lengths vary from about 100 to 500 ft by 100 ft increments.

It should be realized by the reader that in differential leveling the actual sight distances to the rod are not important. All one needs to balance is the rod intervals on the plus and minus sights between benchmarks to ensure that the sight distances are balanced. However, as discussed in Section 19.14, in precise leveling the stadia readings are used to correct observations mathematically for instrument maladjustment (collimation error) and Earth curvature and refraction.

In most leveling procedures balancing plus and minus sight distances will eliminate errors due to instrument maladjustment (most important) and the combined effects of the Earth's curvature and refraction, as shown in Figure 5.6. Here e_1 and e_2 are the combined curvature and refraction errors for the plus and minus sights, respectively. If D_1 and D_2 are made equal, e_1 and e_2 are also equal. In calculations, e_1 is added and e_2 subtracted; thus they cancel each other. The procedure for reading all three wires of the instrument is known as three-wire leveling, which is discussed in Section 5.8.

Figure 5.7 can also be used to illustrate the importance of balancing sight lengths if a collimation error exists in the instrument's line of sight. This condition exists, if after leveling the instrument, its line of sight is not horizontal. For example, suppose in Figure 5.7 because the line of sight is systematically directed below horizontal, an error e_1 results in the plus sight. But if D_1 and D_2 are made equal, an error e_2 (equal to e_1) will result on the minus sight and the two will cancel, thus eliminating the effect of the instrumental error. On slopes it may be somewhat difficult to balance lengths of plus and minus sights, but following a zigzag path usually works. It should be remembered that Earth curvature, refraction, and collimation errors are systematic and will accumulate in long leveling lines if care is not taken to balance the plus and minus sight distances.

A benchmark is described in the field book the first time used, and thereafter by noting the page number on which it was recorded. Descriptions begin with the general location, and must include enough details to enable a person unfamiliar with the area to find the mark readily (see the field notes of Figures 5.5 and 5.12). A benchmark is usually named for some prominent object it is on or near, to aid in describing its location; one word is preferable. Examples are BM River, BM Tower, BM Corner, and BM Bridge. On extensive surveys, benchmarks are often numbered consecutively. Although advantageous in identifying relative positions along a line, this method is more subject to mistakes in field marking or recording. Digital images of the benchmark with one showing a close up of the monument and another showing the horizon of the benchmark with the leveling rod located on the monument can often help in later recovery of the monument.

Turning points are also numbered consecutively but not described in detail, since they are merely a means to an end and usually will not have to be relocated. However, if possible, it is advisable to select turning points that can be relocated,

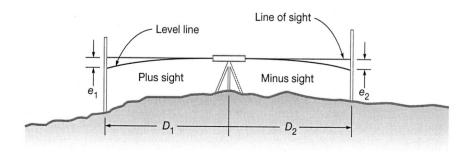

Figure 5.7
Balancing plus and minus sight distances to cancel errors caused by curvature and refraction.

so if reruns on long lines are necessary because of mistakes, fieldwork can be reduced. Before a party leaves the field, all possible note checks must be made to detect any mistakes in arithmetic and verify achievement of an acceptable closure. *The algebraic sum of the plus and minus sights applied to the first elevation should give the last elevation.* This computation checks the addition and subtraction for all HIs and turning points unless compensating mistakes have been made. When carried out for each left-hand page of tabulations, it is termed the *page check*. In Figure 5.5, for example, note that the page check is secured by adding the sum of plus sights, 40.24, to the starting elevation 2053.18, and then subtracting the sum of minus sights; 40.21, to obtain 2053.21, which checks the last elevation.

As previously noted, leveling should always be checked by running closed circuits or loops. This can be done either by returning to the starting benchmark, as demonstrated with the field notes in Figure 5.5, or by ending the circuit at another benchmark of equal or higher reliability. The final elevation should agree with the starting elevation if returning to the initial benchmark. The amount by which they differ is the *loop misclosure*. Note that in Figure 5.5, a loop misclosure of 0.03 ft was obtained.

If closure is made to another benchmark, the *section misclosure* is the difference between the closing benchmark's given elevation and its elevation obtained after leveling through the section. Specifications, or purpose of the survey, fix permissible misclosures (see Section 5.5). If the allowable misclosure is exceeded, one or more additional runs must be made. When acceptable misclosure is achieved, final elevations are obtained by making an adjustment (see Section 5.6).

Note that in running a level circuit between benchmarks, *a new instrument setup has to be made before starting the return run to get a complete check*. In Figure 5.5, for example, a minus sight of 8.71 was read on BM Oak to finish the run out, and a plus sight of 11.95 was recorded to start back, showing that a new setup had been made. Otherwise, an error in reading the final minus sight would be accepted for the first plus sight on the run back. An even better check is secured by tying the run to a different benchmark.

If the elevation above a particular vertical datum (i.e., NAVD 88) is available for the starting benchmark, elevations then determined for all intermediate points along the circuit will also be referenced to the same datum. However, if the starting benchmark's elevation above datum is not known, an assumed value may be used and all elevations converted to the datum later by applying a constant. Until a correct starting elevation is obtained, elevation differences between benchmarks can be obtained from the work. In some cases, this is all that is necessary.

A lake or pond undisturbed by wind, inflow, or outflow can serve as an extended turning point. Stakes driven flush with the water, or rocks whose high points are at this level should be used. However, this water level as a turning point should be used with caution since bodies of water generally flow to an outlet and thus may have differences in elevations along their surfaces.

Double-rodded lines of levels are sometimes used on important work. In this procedure, plus and minus sights are taken on two turning points, using two rods from each setup, and the readings carried in separate note form columns. A check on each instrument setup is obtained if the HI agrees for both lines. This same result can be accomplished using just one set of turning points, and reading both sides of a single rod that has two faces, that is, one side in feet and the other in meters. These rods are often used in precise leveling.

5.5 PRECISION

Precision in leveling is increased by repeating observations, making frequent ties to established benchmarks, using high-quality equipment, keeping it in good adjustment, and performing the measurement process carefully. However, no matter how carefully the work is executed, errors will exist and will be evident in the form of misclosures, as discussed in Section 5.4. To determine whether work is acceptable, misclosures are compared with permissible values on the basis of either number of setups or distance covered. Various organizations set precision standards based on their project requirements. For example, on a simple construction survey, an allowable misclosure of $C = 0.02\text{ ft }\sqrt{n}$ might be used, where n is the number of setups. Note that this criterion was applied for the level circuit in the field notes of 5.

The Federal Geodetic Control Subcommittee (FGCS) recommends the following formula to compute allowable misclosures:[1]

$$C = m\sqrt{K} \qquad (5.3)$$

where C is the allowable loop or section[2] misclosure, in millimeters; m is a constant; and K the total length leveled, in kilometers. For "loops" (circuits that begin and end on the same benchmark), K is the total perimeter distance, and the FGCS specifies constants of 4, 5, 6, 8, and 12 mm for the five classes of leveling, designated, respectively, as (1) first-order class I, (2) first-order class II, (3) second-order class I, (4) second-order class II, and (5) third-order. For "sections" the constants are the same, except that 3 mm applies for first-order class I and 4 mm applies to first-order class II. The order of accuracy recommended for a given type of project is discussed in Section 19.8.

■ Example 5.2

A differential leveling loop is run from an established BM A to a point 2 mi away and back, with a misclosure of 0.056 ft. What order leveling does this satisfy?

Solution

$$C = \frac{0.056 \text{ ft}}{0.00328 \text{ ft/mm}} = 17 \text{ mm}$$

$$K = (2 \text{ mi} + 2 \text{ mi}) \times 1.61 \text{ km/mi} = 6.4 \text{ km}$$

By a rearranged form of Equation (5.3), $m = \dfrac{C}{\sqrt{K}} = \dfrac{17}{\sqrt{6.4}} = 6.7$

---

[1] The FGCS was formerly the FGCC (Federal Geodetic Control Committee). Their complete specifications for leveling are available in a booklet entitled "Standards and Specifications for Geodetic Control Networks" (September 1984). Information on how to obtain this and other related publications can be obtained at the following website: http://www.ngs.noaa.gov. Inquiries can also be made by email at info_center@ngs.noaa.gov, or by writing to the National Geodetic Information Center, NOAA, National Geodetic Survey, 1315 East West Highway, Station 9202, Silver Spring, MD 20910; telephone: (301) 713-3242.

[2] A section consists of a line of levels that begins on one benchmark, and closes on another.

This leveling meets the allowable 8-mm tolerance level for second-order class II work, but does not quite meet the 6-mm level for second-order class I, and if that standard had been specified, the work would have to be repeated. It should be pointed out that even though this survey met the closure tolerance for a second-order class II as specified in the FGCS *Standards and Specifications for Geodetic Control Networks*, other requirements must be met before the survey can be certified to meet any level in the standards.

Since distance leveled is proportional to number of instrument setups, the misclosure criteria can be specified using that variable. As an example, if sights of 200 ft are taken, thereby spacing instrument setups at about 400 ft, approximately 8.2 setups/km will be made. For second-order class II leveling, the allowable misclosure will then be, again by Equation (5.1)

$$C = \frac{8}{\sqrt{8.2}}\sqrt{n} = 2.8\sqrt{n}$$

where C is the allowable misclosure, in millimeters; and n the number of times the instrument is set up.

It is important to point out that meeting FGCS misclosure criterion[3] alone does not guarantee that a certain order of accuracy has been met. Because of compensating errors, it is possible, for example, that crude instruments and low-order techniques can produce small misclosures, yet intermediate elevations along the circuit may contain large errors. To help ensure that a given level of accuracy has indeed been met, besides stating allowable misclosures, the FGCS also specifies equipment and procedures that must be used to achieve a given order of accuracy. These specifications identify calibration requirements for leveling instruments (including rods), and they also outline required field procedures that must be used. Then if the misclosure specified for a given order of accuracy has been met, while employing appropriate instruments and procedures, it can be reasonably expected that all intermediate elevations along the circuit are established to that order.

Field procedures specified by the FGCS include minimum ground clearances for the line of sight, allowable differences between the lengths of pairs of plus sight and minus sight distances, and maximum sight lengths. As examples, sight lengths of not more than 50 m are permitted for first-order class I, while lengths up to 90 m are allowed for third order. As noted in Sections 5.4 and 5.8, the stadia method is convenient for measuring the lengths of plus sights and minus sights to verify their acceptance. The reader should refer to the references listed at the end of this chapter for more information on the requirements specified in the standards.

5.6 ADJUSTMENTS OF SIMPLE LEVEL CIRCUITS

Since permissible misclosures are based on the lengths of lines leveled, or number of setups, it is logical to adjust elevations in proportion to these values. Observed elevation differences d and lengths of sections L are shown for a circuit

---

[3] A complete listing of the specifications for performing geodetic control leveling can be obtained at http://www.ngs.noaa.gov/fgcs/tech_pub/1984-stds-specs-geodetic-control-networks.htm.

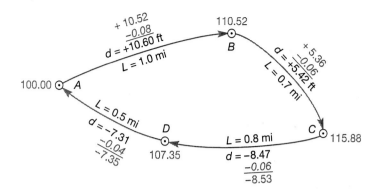

Figure 5.8
Adjustment of level circuit based on lengths of lines.

in Figure 5.8. The misclosure found by algebraic summation of the elevation differences is +0.10 ft. Adding lengths of the sections yields a total circuit length of 3.0 mi. Elevation adjustments are then (−0.10 ft/3.0) multiplied by the corresponding lengths, giving corrections of −0.03, −0.02, −0.03, and −0.02 ft (shown in the figure). The adjusted elevation differences (shown in black) are used to get the final elevations of benchmarks (also shown in black in the figure). Any misclosure that fails to meet tolerances may require reruns instead of adjustment. In Figure 5.5, adjustment for misclosure was made based on the number of instrument setups. Thus, after verifying that the misclosure of 0.03 ft was within tolerance, the correction per setup was −0.03/7 = −0.004 ft. Since errors in leveling accumulate, the first point receives a correction of 1(−0.004), the second 2(−0.004), and so on. The corrections are shown in parenthesis above each unadjusted elevation in Figure 5.5. However, the corrected elevations are rounded off to the nearest hundredth of a foot. Level circuits with different lengths and routes are sometimes run from scattered reference points to obtain the elevation of a given benchmark. The most probable value for a benchmark elevation can then be computed from a weighted mean of the observations, the weights varying inversely with line lengths.

In running level circuits, especially long ones, it is recommended that some turning points or benchmarks used in the first part of the circuit be included again on the return run. This creates a multiloop circuit, and if a blunder or large error exists, its location can be isolated to one of the smaller loops. This saves time because only the smaller loop containing the blunder or error needs to be rerun.

Although the least-squares method (see Section 16.6) is the best method for adjusting circuits that contain two or more loops, an approximate procedure can also be employed. In this method each loop is adjusted separately, beginning with the one farthest from the closing benchmark.

■ 5.7 RECIPROCAL LEVELING

Sometimes in leveling across topographic features such as rivers, lakes, and canyons, it is difficult or impossible to keep plus and minus sights short and equal. Reciprocal leveling may be utilized at such locations.

As shown in Figure 5.9, a level is set up on one side of a river at X, near A, and rod readings are taken on points A and B. Since XB is very long, several readings

Figure 5.9
Reciprocal leveling.

are taken for averaging. This is done by reading, turning the leveling screws to throw the instrument out of level, releveling, and reading again. The process is repeated two, three, four, or more times. Then the instrument is moved close to Y and the same procedure followed.

The two differences in elevation between A and B, determined with an instrument first at X and then at Y, will not agree normally because of curvature, refraction, and personal and instrumental errors. However, in the procedure just outlined, the long minus sight from X to B is balanced by the long plus sight from Y to A. Thus, the average of the two elevation differences cancels the effects of curvature, refraction, and instrumental errors, so the result is accepted as the correct value if the precision of the two differences appears satisfactory. Delays at X and Y should be minimized because refraction varies with changing atmospheric conditions.

5.8 THREE-WIRE LEVELING

As implied by its name, three-wire leveling consists in making rod readings on the upper, middle, and lower cross hairs. Formerly it was used mainly for precise work, but it can be used on projects requiring only ordinary precision. The method has the advantages of (1) providing checks against rod reading blunders, (2) producing greater accuracy because averages of three readings are available, and (3) furnishing stadia measurements of sight lengths to assist in balancing plus sight and minus sight distances. In the three-wire procedure, the difference between the upper and middle readings is compared with that between the middle and lower values. They must agree within one or two of the smallest units being recorded (usually 0.1 or 0.2 of the least count of the rod graduations); otherwise the readings are repeated. An average of the three readings is used as a computational check against the middle wire. As noted in Section 5.4, the difference between the upper and lower readings multiplied by the instrument *stadia interval factor* gives the sight distances. In leveling, the distances are often not important. What is important is that the sum of the plus sights is about equal to the sum of the minus sights between benchmarks, which minimizes errors due to curvature and refraction errors. Sections 5.12.1 and 19.14 discuss mathematical procedures to eliminate collimation error due to unbalanced sights distances.

A sample set of field notes for the three-wire method is presented in Figure 5.10. Plus sight readings on BM *A* of 0.718, 0.633, and 0.550 m taken on the upper, middle, and lower wires, respectively, give upper and lower differences (multiplied by 100) of 8.5 and 8.3 m, which agree within acceptable tolerance. Stadia measurement of the plus sight length (the sum of the upper and lower differences) is 16.8 m. The average of the three plus sight readings on BM *A*, 0.6337 m, agrees within 0.0007 m of the middle reading. The stadia minus sight length of 15.9 m at this setup is within 0.9 m of the plus sight length, and is satisfactory. The HI (104.4769 m)

THREE-WIRE LEVELING
TAYLOR LAKE ROAD

Sta.	+Sight	Stadia	−Sight	Stadia	Elev.
BM A					103.8432
	0.718		1.131		
	0.633	8.5	1.051	8.0	+0.6337
	0.550	8.3	0.972	7.9	104.4769
	3)1.901	16.8	3)3.154	15.9	−1.0513
	+0.6337		−1.0513		
TP1					103.4256
	1.151		1.041		
	1.082	6.9	0.969	7.2	+1.0820
	1.013	6.9	0.897	7.2	104.5076
	3)3.246	13.8	3)2.907	14.4	−0.9690
	+1.0820		−0.9690		
TP2					103.5386
	1.908		1.264		
	1.841	6.7	1.194	7.0	+1.8410
	1.774	6.7	1.123	7.1	105.3796
	3)5.523	13.4	3)3.581	14.1	−1.1937
	+1.8410		−1.1937		
BM B					104.1859
					Check
	+3.5567		−3.2140		
Page	Check:				
	103.8432	+3.5567	−3.2140	= 104.1859	

Figure 5.10
Sample field notes for three-wire leveling.

for the first setup is found by adding the plus sight reading to the elevation of BM A; subtracting the minus sight reading on TP1 gives its elevation (103.4256 m). This process is repeated for each setup.

5.9 PROFILE LEVELING

Before engineers can properly design linear facilities such as highways, railroads, transmission lines, aqueducts, canals, sewers, and water mains, they need accurate information about the topography along the proposed routes. Profile leveling, which yields elevations at definite points along a reference line, provides the needed data. The subsections that follow discuss topics pertinent to profile leveling and include staking and stationing the reference line, field procedures for profile leveling, and drawing and using the profile.

5.9.1 Staking and Stationing the Reference Line

Depending on the particular project, the reference line may be a single straight segment, as in the case of a short sewer line; a series of connected straight segments, which change direction at angle points, as with transmission lines; or straight segments joined by curves, which occur with highways and railroads. The required alignment for any proposed facility will normally have been selected as the result of a preliminary design, which is usually based on a study of existing maps and aerial photos. The reference alignment will most often be the proposed construction centerline, although frequently offset reference lines are used.

To stake the proposed reference line, key points such as the starting and ending points and angle points will be set first. Then intermediate stakes will be placed on line, usually at 100-ft intervals if the English system of units is used but sometimes at closer spacing. If the metric system is used, stakes are usually placed at 10-, 20-, 30-, or 40-m spacing, depending on conditions. Distances for staking can be taped or measured using the electronic distance measuring (EDM) component of a total station instrument operating in its tracking mode (see Sections 8.2 and 23.9).

In alignment surveys, a system called *stationing* is used to specify the relative horizontal position of any point along the reference line. The starting point is usually designated with some arbitrary value, for example in the English system of units, 10 + 00 or 100 + 00, although 0 + 00 can be used. If the beginning point was 10 + 00, a stake 100 ft along the line from it would be designated 11 + 00, the one 200 ft along the line 12 + 00, etc. The term *full station* is applied to each of these points set at 100-ft increments. This is the usual increment staked in rural areas. A point located between two full stations, say 84.90 ft beyond station 17 + 00, would be designated 17 + 84.90. Thus, locations of intermediate points are specified by their nearest preceding full station and their so-called *plus*. For station 17 + 84.90, the plus is 84.90. If the metric system is used, full stations are 1 km (1000 m) apart. The starting point of a reference line might be arbitrarily designated as 1 + 000 or 10 + 000, but again 0 + 000 could be used. In rural areas, intermediate points are normally set at 30- or 40-m increments along the line and are again designated by their pluses. If the beginning point was 1 + 000, and stakes were being set at 40-m intervals then 1 + 040, 1 + 080, 1 + 120, etc. would be set.

In rugged terrain and in urban situations, stakes are normally set closer together, for example at *half stations* (50-ft increments) or even *quarter stations* (25-ft increments) in the English system of units. In the metric system, 20-, 10-, or even 5-m increments may be staked.

Stationing not only provides a convenient unambiguous method for specifying positions of points along the reference line, it also gives the distances between points. For example, in the English system stations 24 + 18.3 and 17 + 84.9 are (2418.3 − 1784.9), or 633.4 ft, apart, and in the metric system stations 1 + 120 and 2 + 040 are 920 m apart.

5.9.2 Field Procedures for Profile Leveling

Profile leveling is differential leveling with the addition of intermediate minus sights (foresights) taken at required points along the reference line. Figure 5.11 illustrates an example of the field procedure, and the notes in Figure 5.12 relate to this example. Stationing for the example is in feet. As shown in the figure, the leveling instrument is initially set up at a convenient location and a plus sight of 10.15 ft taken on the benchmark. Adding this to the benchmark elevation yields a HI of 370.63 ft. Then intermediate minus sights are taken on points along the profile at stations as 0 + 00, 0 + 20, 1 + 00, etc. (If the reference line's beginning is far removed from the benchmark, differential levels running through several turning points may be necessary to get the instrument into position to begin taking intermediate minus sights on the profile line.) Notice that the note form for profile leveling contains all the same column headings as differential leveling but is modified to include another column labeled "Intermediate Sight."

When distances to intermediate sights become too long, or if terrain variations or vegetation obstructs rod readings ahead, the leveling instrument must be

Figure 5.11
Profile leveling.

moved. Establishing a turning point, as TP1 in Figure 5.11, does this. After reading a minus sight on the turning point, the instrument is moved ahead to a good vantage point both for reading the plus sight on the turning point, as well as to take additional rod readings along the profile line ahead. The instrument is leveled, the plus sight taken on TP1, the new HI computed, and further intermediate sights taken. This procedure is repeated until the profile is completed.

Whether the stationing is in feet or meters, intermediate sights are usually taken at all full stations. If stationing is in feet and the survey area is in rugged terrain or in an urban area, the specifications may require that readings also be taken at half- or even quarter-stations. If stationing is in meters, depending on conditions, intermediate sights may be taken at 40-, 30-, 20-, or 10-m increments. In any case, sights are also taken at high and low points along the alignment, as well as at changes in slope.

Intermediate sights should always be taken on "critical" points such as railroad tracks, highway centerlines, gutters, and drainage ditches. As presented in Figure 5.12, rod readings are normally only taken to the nearest 0.1 ft (English

PROFILE LEVELS

Station	+ Sight	HI	− Sight	Int. Sight	Elev.
BM Road	10.15	(370.62) 370.63			360.48
0+00				9.36	361.26
0+20				9.8	360.8
1+00				6.5	364.1
2+00				4.3	366.3
2+60				3.7	366.9
3+00				7.1	363.5
3+90				11.7	358.9
4+00				11.2	359.4
4+35		(366.48)		9.5	361.1
TP1	7.34	366.50	11.47		359.16
5+00				8.4	358.1
5+54				11.08	355.40
5+74				10.66	355.82
5+94				11.06	355.42
6+00				10.5	356.0
7+00		(363.77)		4.4	362.1
TP2	2.56	363.80	5.26		361.24
8+00				1.2	362.6
9+00				3.9	359.9
9+25.2				3.4	360.4
9+25.3				4.6	359.2
9+43.2				2.2	361.6
BM Store			0.76		363.04
Σ	20.05		17.49		(363.01)

BM ROAD to BM STORE

BM Road 3 miles SW of Mpls. 200 yds. N of Pine St. over pass 40 ft. E of ¢ Hwy. 169 Top of RW conc. post No. 268.
¢ Hwy. 169, painted "X"
West drainage ditch

Summit

Sag

Summit

E gutter, Maple St.
¢ Maple St.
W gutter, Maple St.

Summit   363.04−363.01=

Top of E curb, Elm St.
Bottom of E curb, Elm St.
¢ Elm St.
BM Store. NE corner Elm St. & 4th Ave. SE corner
Store foundation wall. 3˝ brass disc set in grout.
BM store elev. = 363.01

SW Minneapolis on Hwy 169
6 Oct. 2014
Cool, Sunny, 50° F
R.J. Hintz   N
N.R. Olson   φ
R.C. Perry   π
Wild Level #3

COPY

Page Check:
+20.05
−17.49
+ 2.56
360.48
363.04

Misclosure = 0.03

R.J. Hintz

Figure 5.12
Profile leveling notes for Figure 5.11.

system) or nearest cm (metric system) where the rod is held on the ground, but on critical points, and for all plus and minus sights taken on turning points and benchmarks, the readings are recorded to the nearest hundredth of a foot (English) or the nearest mm (metric).

In profile leveling, lengths of intermediate minus sights vary, and in general, they will not equal the plus sight length. Thus, errors due to an inclined line of sight and to curvature and refraction will occur. Because errors from these sources increase with increasing sight lengths, on important work the instrument's condition of adjustment should be checked (see Section 4.15), and excessively long intermediate minus sight distances should be avoided.

Instrument heights (HIs) and elevations of all turning points are computed immediately after each plus sight and minus sight. However, elevations for intermediate minus sights are not computed until after the circuit is closed on either the initial benchmark or another. Then the circuit misclosure is computed, and if acceptable, an adjustment is made, and elevations of intermediate points are then calculated. The procedure is described in the following subsection.

As in differential leveling, the page check should be made for each left-hand sheet. However, in profile leveling, intermediate minus sights play no part in this computation. As illustrated in Figure 5.12, the page check is made by adding the algebraic sum of the column of plus sights and the column of minus sights to the beginning elevation. This should equal the last elevation tabulated on the page for either a turning point or the ending benchmark if that is the case, as it is in the example of Figure 5.12.

5.9.3 Drawing and Using the Profile

Prior to drawing the profile, it is first necessary to compute elevations along the reference line from the field notes. However, this cannot be done until an adjustment has been made to distribute any misclosure in the level circuit. In the adjustment process, HIs are adjusted, because they will affect computed profile elevations. The adjustment is made progressively in proportion to the total number of HIs in the circuit. The procedure is illustrated in Figure 5.12, where the misclosure was 0.03 ft. Since there were three HIs, the correction applied to each is $-0.03/3 = -0.01$ ft per HI. Thus, a correction of 0.01 was applied to the first HI, -0.02 ft to the second, and -0.03 ft to the third. Adjusted HIs are shown in Figure 5.12 in parentheses above their unadjusted values. It is unnecessary to correct turning point elevations since they are of no consequence. After adjusting the HIs, profile elevations are computed by subtracting intermediate minus sights from their corresponding *adjusted* HIs. The profile is then drawn by plotting elevations on the ordinate versus their corresponding stations on the abscissa. By connecting adjacent plotted points, the profile is realized.

Until recently, profiles were manually plotted, usually on special paper like the type shown in Figure 5.13. Now with computer-aided drafting and design (CADD) systems (see Section 18.14), it is only necessary to enter the stations and elevations into the computer, and this special software will plot and display the profile on the screen. Hard copies, if desired, may be obtained from plotters interfaced with a computer. Often these profiles are generated automatically from

120 LEVELING—FIELD PROCEDURES AND COMPUTATIONS

Figure 5.13
Plot of profile.

the CADD software using only the alignment of the structure and an overlaying topographic map.

In drawing profiles, the vertical scale is generally exaggerated with respect to the horizontal scale to make differences in elevation more pronounced. A ratio of 10:1 is frequently used, but flatness or roughness of the terrain determines the desirable proportions. Thus, for a horizontal scale of 1 in. = 100 ft, the vertical scale might be 1 in. = 10 ft. The scale actually employed should be plainly marked. Plotted profiles are used for many purposes, such as (1) determining depth of cut or fill on proposed highways, railroads, and airports; (2) studying grade-crossing problems; and (3) investigating and selecting the most economical grade, location, and depth for sewers, pipelines, tunnels, irrigation ditches, and other projects.

The *rate of grade* (or *gradient* or *percent grade*) is the rise or fall in feet per 100 ft, or in meters per 100 m. Thus, a grade of 2.5% means a 2.5-ft difference in elevation per 100 ft horizontally. Ascending grades are plus; descending grades, minus. A gradeline of −0.15%, chosen to approximately equalize cuts and fills, is shown in Figure 5.13. Along this grade line, elevations drop at the rate of 0.15 ft per 100 ft. The grade begins at station 0 + 00 where it approximately meets existing ground at elevation 363.0 ft, and ends at station 9 + 43 and elevation 361.6 ft where again it approximately meets existing ground. The process of staking grades is described in Chapter 23.

The term *grade* is also used to denote the elevation of the finished surface on an engineering project.

■ 5.10 GRID, CROSS-SECTION, OR BORROW-PIT LEVELING

Grid leveling is a method for locating contours (see Section 17.9.3). It is accomplished by staking an area in squares of 10, 20, 50, 100, or more feet (or comparable meter lengths) and determining the corner elevations by differential leveling. Rectangular blocks, say 50 by 100 ft or 20 by 30 m, that have the longer sides roughly parallel with the direction of most contour lines may be preferable on steep slopes. The grid size chosen depends on the project extent, ground roughness, and accuracy required.

The same process, termed *borrow-pit leveling*, is employed on construction jobs to ascertain quantities of earth, gravel, rock, or other material to be excavated or filled. The procedure is covered in Section 26.10 and Plate B.2.

5.11 USE OF THE HAND LEVEL

A hand level can be used for some types of leveling when a low order of accuracy is sufficient. The instrument operator takes a plus and minus sight while standing in one position, and then moves ahead to repeat the process. A hand level is useful, for example, in cross-sectioning to obtain a few additional rod readings on sloping terrain where a turning point would otherwise be required.

5.12 SOURCES OF ERROR IN LEVELING

All leveling measurements are subject to three sources of error: (1) instrumental, (2) natural, and (3) personal. These are summarized in the subsections that follow.

5.12.1 Instrumental Errors

Line of Sight. As described in Section 4.15, a properly adjusted leveling instrument that employs a level vial should have its line of sight and level vial axis parallel. Then, with the bubble centered, a horizontal plane, rather than a conical surface, is generated as the telescope is revolved. In addition, if the compensators of automatic levels are operating properly, they should always produce a truly horizontal line of sight. If these conditions are not met, a *line of sight* (or *collimation*) error exists, and serious errors in rod readings can result. These errors are systematic, but they are canceled in differential leveling if the horizontal lengths of plus and minus sights are kept equal. The error may be serious in going up or down a steep hill where all plus sights are longer or shorter than all minus sights, unless care is taken to run a zigzag line. The size of the collimation error, ε, can be determined in a simple field procedure [see Equation (4.14) and Section 4.15.5]. If plus and minus sights cannot be balanced, a correction for this error can be made.

To apply the collimation correction, the value of ε from Equation (4.14) is divided by the length of the spaces between adjacent stakes in Figure 4.20. This yields the *collimation correction factor* in units of feet per foot, or meters per meter. Then for any plus or minus sight, the correction to be subtracted from the rod reading is obtained by multiplying the length of the sight by this correction factor. As an example, suppose that the distance between stakes in Example 4.3 was 100 ft. Then the collimation correction factor is $0.010/100 = 0.0001$ ft/ft. Suppose that a reading of 5.29 ft was obtained on a plus sight of 200 ft length with this instrument. The corrected rod reading would then be $5.29 - 200(0.00010) = 5.27$. As discussed in Section 19.13, when the three-wire leveling procedure is used the rod interval determined by the difference in the upper and lower wires can be used to determine the collimation correction factor.

Cross Hair Not Exactly Horizontal. Reading the rod near the center of the horizontal cross hair will eliminate or minimize this potential error.

Rod Not Correct Length. Inaccurate divisions on a rod cause errors in observed elevation differences similar to those resulting from incorrect markings on a measuring tape. Uniform wearing of the rod bottom makes HI values too large, but the effect is canceled when included in both plus and minus sights. Rod graduations should be checked by comparing them with those on a standardized tape.

Tripod Legs Loose. Tripod leg bolts that are too loose or too tight allow movement or strain that affects the instrument head. Loose metal tripod shoes cause unstable setups.

5.12.2 Natural Errors

Curvature of the Earth. As noted in Section 4.4, a level surface curves away from a horizontal plane at the rate of $0.667 M^2$ or $0.0785 K^2$, which is about 0.7 ft/mi or 8 cm/km. The effect of curvature of the Earth is to increase the rod reading. Equalizing lengths of plus and minus sights in differential leveling cancels the error due to this cause.

Refraction. Light rays coming from an object to the telescope are bent, making the line of sight a curve concave to the Earth's surface, which thereby decreases rod readings. Balancing the lengths of plus and minus sights usually eliminates errors due to refraction. However, large and sudden changes in atmospheric refraction may be important in precise work. Although, errors due to refraction tend to be random over a long period of time, they could be systematic on one day's run. Additionally due to the microclimate near surfaces, it is best to maintain a sight line that does not come within 1.5 ft or 0.5 m of any surface.

Temperature Variations. Heat causes leveling rods to expand but the effect is not important in ordinary leveling. If the level vial of a tilting level is heated, the liquid expands, and the bubble shortens. This does not produce an error (although it may be inconvenient), unless one end of the tube is warmed more than the other, and the bubble therefore moves. Other parts of the instrument warp because of uneven heating, and this distortion affects the adjustment. Shading the level by means of a cover when carrying it, and by an umbrella when it is set up, will reduce or eliminate heat effects. These precautions are followed in precise leveling.

Air boiling or heat waves near the ground surface or adjacent to heated objects make the rod appear to wave and prevent accurate sighting. Raising the line of sight by high tripod setups, taking shorter sights, avoiding any that pass close to heat sources (such as buildings and stacks), and using the lower magnification of a variable-power eyepiece reduce the effect.

Wind. Strong wind causes the instrument to vibrate and makes the rod unsteady. Precise leveling should not be attempted on excessively windy days.

Settlement of the Instrument. Settlement of the instrument during the time between a plus sight reading and a minus sight makes the latter too small and therefore the recorded elevation of the next point too high. The error is cumulative in a series of setups on soft material. Therefore setups on spongy ground, blacktop, or ice should be avoided if possible, but if they are necessary, unusual care is required to reduce the resulting errors. This can include taking readings in quick order, using two rods and two observers to preclude walking around the instrument, and alternating the order of taking plus and minus sights. Additionally, whenever possible, the instrument tripod's legs can be set on long hubs that are driven to refusal in the soft material.

Settlement of a Turning Point. This condition causes an error similar to that resulting from settlement of the instrument. It can be avoided by selecting firm, solid turning points or, if none are available, using a steel turning pin set firmly in the ground. A railroad spike can also be used in some situations.

5.12.3 Personal Errors

Bubble Not Centered. In working with levels that employ level vials, errors caused by the bubble not being exactly centered at the time of sighting are the most important of any, particularly on long sights. If the bubble runs between the plus and minus sights, *it must be recentered before the minus sight is taken.* Experienced observers develop the habit of checking the bubble before and after each sight, a procedure simplified with some instruments, which have a mirror–prism arrangement permitting a simultaneous view of the level vial and rod.

Parallax. Parallax caused by improper focusing of the objective or eyepiece lens results in incorrect rod readings. Careful focusing eliminates this problem.

Faulty Rod Readings. Incorrect rod readings result from parallax, poor weather conditions, long sights, improper target settings, and other causes, including mistakes such as those due to careless interpolation and transposition of figures. Short sights selected to accommodate weather and instrument conditions reduce the magnitude of reading errors. If a target is used, the rod person should read the rod, and the observer should check it independently.

Rod Handling. Using a rod level that is in adjustment, or holding the rod parallel to a plumb bob string eliminates serious errors caused by improper plumbing of the rod. Banging the rod on a turning point for the second (plus) sight may change the elevation of a point.

Target Setting. If a target is used, it may not be clamped at the exact place signaled by the observer because of slippage. A check sight should always be taken after the target is clamped.

5.13 MISTAKES

A few common mistakes in leveling are listed here.

Improper Use of a Long Rod. If the Vernier reading on the back of a damaged Philadelphia rod with English units is not exactly 6.500 ft or 7.000 ft for the short rod, the target must be set to read the same value before extending the rod.

Holding the Rod in Different Places for the Plus and Minus Sights on a Turning Point. The rod person can avoid such mistakes by using a well-defined point or by outlining the rod base with lumber crayon, keel, or chalk.

Reading a Foot Too High. This mistake usually occurs because the incorrect footmark is in the telescope's field of view near the cross line; for example, an observer may read 5.98 instead of 4.98. Noting the footmarks both above and below the horizontal cross line will prevent this mistake.

Waving a Flat Bottom Rod While Holding It on a Flat Surface. This action produces an incorrect rod reading because rotation is about the rod edges instead of the center or front face. In precise work, plumbing with a rod level, or other means, is preferable to waving. This procedure also saves time.

Recording Notes. Mistakes in recording, such as transposing figures, entering values in the wrong column, and making arithmetic mistakes, can be minimized by having the notekeeper repeat the value called out by an observer, and by making the standard field-book checks on rod sums and elevations. Digital levels that automatically take rod readings, store the values, and compute the level notes can eliminate these mistakes.

Touching Tripod or Instrument during the Reading Process. Beginners using instruments that employ level vials, may center the bubble, put one hand on the tripod or instrument while reading a rod, and then remove the hand while checking the bubble, which has now returned to center but was off during the observation. Of course, the instrument should not be touched when taking readings, but detrimental effects of this bad habit are practically eliminated when using automatic levels.

5.14 REDUCING ERRORS AND ELIMINATING MISTAKES

Errors in running levels are reduced (but never eliminated) by carefully adjusting and manipulating both instrument and rod (see Section 4.15 for procedures) and establishing standard field methods and routines. The following routines prevent most large errors or quickly disclose mistakes: (1) checking the bubble before and after each reading (if an automatic level is not being used), (2) using a rod level, (3) keeping the horizontal lengths of plus and minus sights equal, (4) running lines forward and backward, (5) making the usual field-book arithmetic checks, and (6) breaking long leveling circuits into smaller sections.

5.15 USING SOFTWARE

On the companion website for this book at http://www.pearsonhighered.com/ghilani is the software WOLFPACK. In this software is an option that takes the plus and minus readings from a simple leveling circuit to create a set of field notes and the file appropriate for a least-squares adjustment of the data (see Section 16.6). A sample file of the field notes from Figure 5.5 is depicted in Figure 5.14.

The software limits the length of the station identifiers to 10 characters. These characters must not include a space, comma, or tab since these are used as data delimiters in the file. All benchmark stations must start with the letters BM, while all turning points must start with the letters TP. This is used by the software to differentiate between a benchmark and a turning point in the data file.

While the format of the file is explained fully in the WOLFPACK help system, it will be presented here as an aid to the reader. The first line of the file shown in Figure 5.14 is a title line, which in this case is "Grand Lakes Univ. Campus Leveling Project." The second line contains starting and ending benchmark elevations. Since this line starts and ends on the same benchmark (BM_MIL), its elevation of 2053.18 need be listed only once. If a level circuit starts at one benchmark, but closes on another, then both the starting and ending elevations of the leveling circuit should be listed on this line in that order. The remainder of the file contains the plus and minus sights between each set of stations. Thus, each line contains the readings from one instrument setup. For example, a plus sight of 1.33 was made on BM_MIL and a minus sight of 8.37 was made on TP1, which is the first turning point. Each instrument setup is listed in order following the same procedure. Once the file is created and saved using the WOLFPACK editor, it can be read into the option *Reduction of differential leveling notes* as shown in Figure 5.15. The software then creates notes similar to those shown in Figure 5.5 adjusting the elevations, and demonstrating a page check.

For those who are interested in higher-level programming, the Mathcad® worksheet *C5.xmcd* is available on the companion website for this book. This worksheet reads a text file of observations that are obtained typically in differential leveling and creates and adjust the data placing the results in a format typically found in a field book. Additionally, the Excel® spreadsheet *C5.xls* demonstrates how a spreadsheet can be used to reduce the notes in Figure 5.5.

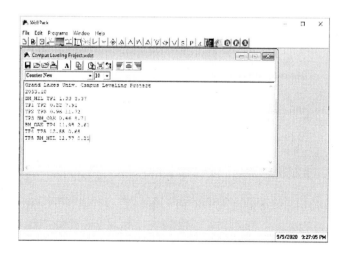

Figure 5.14
Sample data file for field notes in Figure 5.5. (Wolfpack software. Copyright © 2014 by Charles D. Ghilani. All Rights Reserved.)

126 LEVELING—FIELD PROCEDURES AND COMPUTATIONS

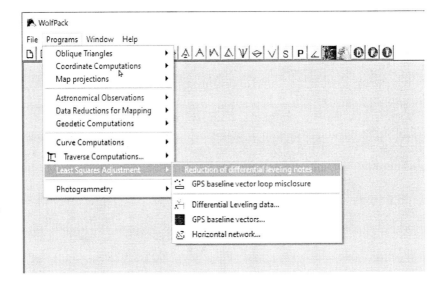

Figure 5.15
Option in WOLFPACK to reduce data file in Figure 5.14. (Wolfpack software. Copyright © 2014 by Charles D. Ghilani. All Rights Reserved.)

PROBLEMS

Asterisks (*) indicate problems that have partial answers given in Appendix G.

5.1 Discuss the duties of a rod person.

5.2 What properties make a good turning point?

5.3 Discuss how stadia can be used to determine the plus and minus sight distances in differential leveling.

5.4 What is the collimation error, and how can it be removed from the differential leveling process?

5.5 Discuss how errors due to Earth curvature and refraction can be eliminated from the differential leveling process.

5.6 When is it appropriate to use the reciprocal leveling procedure?

5.7 Discuss when double-rodded lines are used in leveling.

5.8* What error is created by a rod leaning 10 minutes from plumb at a 12.513-m reading on the leaning rod?

5.9 Similar to Problem 5.6, except the rod is leaned 20 minutes on a 7-m reading.

5.10 What error results on a 20-m sight with a level if the rod reading is 2.966 m but the top of the 4-m rod is 0.3 m out of plumb?

5.11 What error results on a 200-ft sight with a level if the rod reading is 11.307 ft but the top of the 12-ft rod is 0.3 ft out of plumb?

5.12 Prepare a set of level notes for the data listed. Perform a check and adjust the misclosure. Elevation of BM 7 is 562.458 m. If the total loop length is 900 m, what order of leveling is represented? (Assume all readings are in meters.)

POINT	+S (BS)	−S (FS)
BM 7	4.388	
TP1	6.907	4.538
TP2	4.680	8.800
BM 8	3.730	5.978
TP3	8.464	5.245
BM 7		3.598

5.13 Similar to Problem 5.12, except the elevation of BM 7 is 132.053 m and the loop length 1.560 km. (Assume all readings are in meters.)

5.14 A differential leveling loop began and closed on BM Tree (orthometric height 814.08 ft). The plus sight and minus sight distances were kept approximately equal. Readings (in feet) listed in the order taken are 6.18 (+S) on BM Tree, 7.76 (−S) and 4.44 (+S) on TP1, 5.05 (−S) and 5.63 (+S) on BM X, 2.64 (−S) and 7.35 (+S) on TP2, and 8.07 (−S) on BM Tree. Prepare, check, and adjust the notes.

5.15 A differential leveling circuit began on BM Hydrant (orthometric height 5869.31 ft) and closed on BM Rock (orthometric height 5880.03 ft). The plus sight and minus sight distances were kept approximately equal. Readings (in feet) given in the order taken are 3.88 (+S) on BM Hydrant, 5.59 (−S) and 10.83 (+S) on TP1, 7.66 (−S) and 11.81 (+S) on BM 1, 8.91 (−S) and 7.57 (+S) on BM 2, 5.14 (−S) and 9.01 (+S), on TP2, and 5.03 (−S) on BM Rock. Prepare, check, and adjust the notes.

5.16 A differential leveling loop began and closed on BM Bridge (orthometric height 12.063 m). The plus sight and minus sight distances were kept approximately equal. Readings (in meters) listed in the order taken are 1.374 (+S) on BM Bridge, 1.796 (−S) and 3.530 (+S) on TP1, 1.553 (−S) and 3.395 (+S) on BM X, 3.439 (−S) and 1.751 (+S) on TP2, and 3.250 (−S) on BM Bridge. Prepare, check, and adjust the notes.

5.17 A differential leveling circuit began on BM Rock (orthometric height 913.503 m) and closed on BM Manhole (orthometric height 916.783 m). The plus sight and minus sight distances were kept approximately equal. Readings (in meters) listed in the order taken are 2.695 (+S) on BM Rock, 2.399 (−S) and 3.623 (+S) on TP1, 1.664 (−S) and 3.086 (+S) on BM 1, 2.494 (−S) and 1.304 (+S) on BM 2, 2.255 (−S) and 2.743 (+S) on TP2, and 1.349 (−S) on BM Manhole. Prepare, check, and adjust the notes.

5.18 A differential leveling loop started and closed on BM Juno, orthometric height 1208.87 ft. The plus sight and minus sight distances were kept approximately equal. Readings (in feet) listed in the order taken are 4.92 (+S) on BM Juno, 2.88 (−S) and 8.34 (+S) on TP1, 4.80 (−S) and 5.92 (+S) on TP2, 7.83 (−S) and 3.72 (+S) on BM1, 9.12 (−S) and 7.35 (+S) on TP3, and 5.59 (−S) on BM Juno. Prepare, check, and adjust the notes.

5.19* A level setup midway between X and Y reads 6.29 ft on X and 7.91 ft on Y. When moved within a few feet of X, readings of 5.18 ft on X and 6.76 ft on Y are recorded. What is the true elevation difference, and the reading required on Y to adjust the instrument?

5.20 To test its line of sight adjustment, a level is setup near C (elev 257.064 m) and then near D. Rod readings listed in the order taken are C = 1.256 m, D = 1.115 m, D = 1.151 m, and C = 1.296 m. Compute the elevation of D, and the reading required on C to adjust the instrument.

5.21* The line of sight test shows that a level's line of sight is inclined downward 3 mm/50 m. What is the allowable difference between BS and FS distances at each setup (neglecting curvature and refraction) to keep elevations correct within 1 mm?

5.22 Reciprocal leveling gives the following readings in meters from a set up near A: on A, 1.110; on B, 0.987, 0.995, and 0.984. At the setup near B: on B, 1.549; on A, 1.673, 1.670, and 1.666. The elevation of A is 1045.343 m. Determine the misclosure and elevation of B.

5.23* Reciprocal leveling across a canyon provides the data listed (in meters). The elevation of Y is 2265.879 m. The elevation of X is required. Instrument at X: +S = 3.182; −S = 9.365, 9.370, and 9.368. Instrument at Y: +S = 10.223; −S = 4.037, 4.041, and 4.038.

5.24 Prepare a set of three-wire leveling notes for the data given and make the page check. The elevation of BM X is 83.097 m. Rod readings (in meters) are (U denotes upper cross-wire readings, M middle wire, and L lower wire): +S on BM X: U = 2.518, M = 1.681, L = 0.843; −S on TP1: U = 1.986, M = 1.715, L = 1.447; +S on TP1: U = 2.543, M = 2.263, L = 1.985; −S on BM Y: U = 1.548, M = 1.228, L = 0.909.

5.25 Similar to Problem 5.24, except the elevation of BM X is 876.98 ft, and rod readings (in feet) are: +S on BM X: U = 5.489, M = 4.739, L = 3.990; −S on TP1: U = 6.843, M = 5.918, L = 4.996; +S on TP1: U = 7.984, M = 6.969, L = 5.952; −S on BM Y: U = 4.872, M = 4.022, L = 3.173.

5.26 Assuming a stadia constant of 100.003, what is the distance leveled in Problem 5.24?

5.27 Assuming a stadia constant of 99.875, what is the distance leveled in Problem 5.25?

5.28 Prepare a set of profile leveling notes for the data listed and show the page check. All data is given in feet. The elevation of BM A is 653.72, and the elevation of BM B is 674.22. Rod readings are: +S on BM A, 5.98; intermediate foresight (IFS) on 11 + 00, 5.4; −S on TP1, 3.25; +S on TP1, 8.19; IFS on 12 + 00, 6.3 on 12 + 50, 5.5; on 13 + 00, 5.9; on 14 + 00, 6.1; −S on TP2, 3.00, +S on TP2, 8.78; IFS on 14 + 73, 5.4; on 15 + 00, 5.0; on 16 + 00, 5.5; −S on TP3, 3.11; +S on TP3, 9.88; −S on BM B, 3.05.

5.29 Same as Problem 5.28, except the elevation of BM A is 659.08 ft. and the elevation of BM B is 679.46 ft.

5.30 Plot the profile Problem 5.28 and design a grade line between stations 11 + 00 and 16 + 00 that balances cut and fill areas.

5.31* What is the percent grade between stations 11 + 00 and 16 + 00 in Problem 5.28?

5.32 Differential leveling between BMs A, B, C, D, and A gives elevation differences (in meters) of −15.632, +32.458, +38.214, and −55.025, and distances in km of 2.0, 4.0, 2.0, and 4.0, respectively. If the elevation of A is 634.597, compute the adjusted elevations of BMs B, C, and D, and the order of leveling.

5.33 Leveling from BM X to W, BM Y to W, and BM Z to W gives differences in elevation (in feet) of −30.26, +26.17, and +10.14, respectively. Distances between benchmarks are XW = 2000, YW = 2900, and ZW = 3550. True elevations of the benchmarks are X = 898.83, Y = 842.39, and Z = 858.41. What is the adjusted elevation of W? (Note: All data are given in feet.)

5.34 A 5-m level rod was calibrated, and its graduated scale was found to be uniformly contracted so that the distance between its 0 and 5.000 marks was actually 5.006 m. How will this affect elevations determined with this rod for (a) circuits run on relatively flat ground, (b) circuits run downhill, and (c) circuits run uphill?

5.35* A line of levels with 42 setups (84 rod readings) was run from BM Rock to BM Pond with readings taken to the nearest 3.0 mm; hence any observed value could have an error of ±1.5 mm. For reading errors only, what total error would be estimated in the elevation of BM Pond?

5.36 Same as Problem 5.35, except for 34 setups and readings to the nearest 0.01 ft with possible error of ±0.005 ft each.

5.37 Compute the permissible misclosure for the following lines of levels: (a) a 10-km loop of third-order levels, (b) a 20-km section of second-order class I levels, and (c) a 50-km loop of first-order class I levels.

BIBLIOGRAPHY

Crawford, W. G. 2008. "The One-Minute Peg Test." *Point of Beginning* 33 (No. 6): 52.
Federal Geodetic Control Subcommittee. 1984. *Standards and Specifications for Geodetic Control Surveys*. Silver Spring, MD: National Geodetic Information Branch, NOAA.
Reilly, J. P. 2004. "Tides and Their Relationship to Vertical Datums." *Point of Beginning* 29 (No. 4): 68.

6 Distance Measurement

PART I • METHODS FOR MEASURING DISTANCES

6.1 INTRODUCTION

Distance measurement is generally regarded as the most fundamental of all surveying observations. In traditional ground surveys, even though many angles may be read, the length of at least one line must be measured to supplement the angles in locating points. In plane surveying the distance between two points means the horizontal distance. If the points are at different elevations, the distance is the horizontal length between vertical lines at the points.

Lengths of lines may be specified in different units. In the United States, the foot, decimally divided, is usually used although the meter is becoming increasingly more common. Geodetic surveys and many highway surveys employ the meter. In architectural and machine work, and on some construction projects, the unit is a foot divided into inches and fractions of an inch. As discussed in Section 2.2, chains, varas, rods, and other units have been, and still are, utilized in some localities and for special purposes.

6.2 SUMMARY OF METHODS FOR MAKING LINEAR MEASUREMENTS

In surveying, linear measurements have been obtained by many different methods. These include (1) pacing, (2) odometer readings, (3) optical rangefinders, (4) tacheometry (stadia), (5) subtense bars, (6) taping, (7) electronic distance measurement (EDM), (8) satellite systems, and others. Of these, surveyors most commonly use taping, EDM, and satellite systems today. In particular, the satellite-supported *Global Navigation Satellite Systems* (GNSS) are rapidly replacing all other systems due to their many advantages, but most notably because of their range, accuracy, and efficiency. Methods (1) through (5) are discussed briefly in the following sections. Taping is discussed in Part II of this chapter, and EDM is described in Part III of this chapter. Satellite systems are described in Chapters 13, 14, and 15.

Triangulation is a method for determining positions of points from which horizontal distances can be computed (see Section 19.13.1). In this procedure, lengths of lines are computed trigonometrically from measured baselines and angles. *Photogrammetry* can also be used to obtain horizontal distances. This topic is covered in Chapter 27. Besides these methods, distances can be estimated, a technique useful in making field note sketches and checking observations for mistakes. With practice, estimating can be done quite accurately.

■ 6.3 PACING

Distances obtained by pacing are sufficiently accurate for many purposes in surveying, engineering, geology, agriculture, forestry, and military field sketching. Pacing is also used to detect blunders that may occur in making distance observations by more accurate methods.

Pacing consists of counting the number of steps, or paces, in a required distance. The length of an individual's pace must be determined first. This is best done by walking with natural steps back and forth over a level course at least 300 ft long, and dividing the known distance by the average number of steps. For short distances, the length of each pace is needed, but the number of steps taken per 100 ft is desirable for checking long lines.

It is possible to adjust one's pace to an even 3 ft, but a person of average height finds such a step tiring if maintained for very long. The length of an individual's pace varies when going uphill or downhill and changes with age. For long distances, a pocket instrument called a *pedometer* can be carried to register the number of paces, or a *passometer* attached to the body or leg counts the steps. Some surveyors prefer to count *strides*, a stride being two paces.

Pacing is one of the most valuable things learned in surveying since it has practical applications for everybody and requires no equipment. If the terrain is open and reasonably level, experienced pacers can measure distances of 100 ft or longer with an accuracy of 1/50 to 1/100 of the distance.

■ 6.4 ODOMETER READINGS

An odometer converts the number of revolutions of a wheel of known circumference to a distance. Lengths measured by an odometer on a vehicle are suitable for some preliminary surveys in route-location work. They also serve as rough checks on observations made by other methods. Other types of measuring wheels are available and useful for determining short distances, particularly on curved lines. Odometers give surface distances, which should be corrected to horizontal if the ground slopes severely (see Section 6.13). With odometers, an accuracy of approximately 1/200 of the distance is reasonable.

■ 6.5 OPTICAL RANGEFINDERS

These instruments operate on the same principle as rangefinders on single-lens reflex cameras. Basically, when focused, they solve for the object distance f_2 in

Equation (4.12), where focal length f and image distance f_1 are known. An operator looks through the lens and adjusts the focus until a distant object viewed is focused in coincidence, whereupon the distance to that object is obtained. These instruments are capable of accuracies of 1 part in 50 at distances up to 150 ft, but accuracy diminishes as the length increases. They are suitable for reconnaissance, sketching, or checking more accurate observations for mistakes.

6.6 TACHEOMETRY

Tacheometry (*stadia* is the more common term in the United States) is a surveying method used to quickly determine the horizontal distance to, and elevation of, a point. As discussed in Section 5.4, stadia observations are obtained by sighting through a telescope equipped with two or more horizontal cross hairs at a known spacing. The apparent intercepted length between the top and bottom wires is read on a graduated rod held vertically at the desired point. The distance from telescope to rod is found by proportional relationships in similar triangles. An accuracy of 1/500 of the distance is achieved with reasonable care.

6.7 SUBTENSE BAR

This indirect distance-measuring procedure involves using a theodolite to read the horizontal angle subtended by two targets precisely spaced at a fixed distance apart on a subtense bar. The unknown distance is computed from the known target spacing and the measured horizontal angle. Prior to observing the angle from one end of the line, the bar is centered over the point at the other end of the line, and oriented perpendicular to the line and in a horizontal plane. For sights of 500 ft (150 m) or shorter, and using a 1″ theodolite, an accuracy of 1 part in 3000 or better can be achieved. Accuracy diminishes with increased line length. Besides only being suitable for relatively short lines, this method of distance measurement is time consuming and is not used today, having been replaced by electronic distance measurement and GNSS surveys.

PART II • DISTANCE MEASUREMENTS BY TAPING

6.8 INTRODUCTION TO TAPING

With the accuracy and ease of use of electronic distance measuring (EDM) instruments discussed in Part III of this chapter, precise taping of lines over 100 ft is seldom, if ever, performed today. Similarly, tape corrections are seldom, if ever, made today. However, the proper use of a tape in measuring distances is still a required skill for the practicing surveyor. Part II of this chapter deals with the proper care and use of a tape when measuring distances. Since electronic distance measurement has become so accurate with today's instruments, actual tape corrections are seldom performed, and thus examples of tape correction have been moved to Appendix A of this book. However, with this stated, it is still important for surveyors to know the proper procedures for handling and using a tape.

Observation of horizontal distances by taping consists of applying the known length of a graduated tape directly to a line a number of times. Two types of problems arise: (1) observing an unknown distance between fixed points, such as between two stakes in the ground and (2) laying out a known or required distance with only the starting mark in place.

Taping is performed in six steps: (1) lining in, (2) applying tension, (3) plumbing, (4) marking tape lengths, (5) reading the tape, and (6) recording the distance. The application of these steps in taping on level and sloping ground is detailed in Sections 6.11 and 6.12.

6.9 TAPING EQUIPMENT AND ACCESSORIES

Over the years, various types of tapes and other related equipment have been used for taping in the United States. Tapes in current use are described here, as are other accessories used in taping.

Surveyor's and *engineer's tapes* are made of steel 1/4- to 3/8-in. wide and weigh 2 to 3 lbs/100 ft. Those graduated in feet are most commonly 100 ft in length, although they are also available in lengths of 200, 300, and 500 ft. They are marked in feet, tenths, and hundredths. Metric tapes have standard lengths of 30, 60, 100, and 150 m. All can either be wound on a reel [see Figure 6.1(a)] or done up in loops.

Invar tapes are made of a special nickel-steel alloy (35% nickel and 65% steel) to reduce length variations caused by differences in temperature. The thermal coefficient of expansion and contraction of this material is only about 1/30 to 1/60 that of an ordinary steel tape. However, the metal is soft and somewhat unstable. This weakness, along with the cost perhaps ten times that of steel tapes, made them suitable for precise geodetic work only and as a standard for comparison with working tapes. Another version, the *Lovar tape*, has properties and a cost between those of steel and Invar tapes.

Cloth (or *metallic*) *tapes* are actually made of high-grade linen, 5/8-in. wide with fine copper wires running lengthwise to give additional strength and prevent excessive elongation. Metallic tapes commonly used are 50, 100, and 200 ft in length and come on enclosed reels [see Figure 6.1(b)]. Although not suitable for precise work, metallic tapes are convenient and practical for many purposes.

Figure 6.1
Taping equipment for field party. (Courtesy W. & L.E. Gurley.)

Fiberglass tapes come in a variety of sizes and lengths and are usually wound on a reel. They can be employed for the same types of work as metallic tapes.

Chaining pins or *taping pins* are used to mark tape lengths. Most taping pins are made of number 12 steel wire, sharply pointed at one end, have a round loop at the other end, and are painted with alternate red and white bands [see Figure 6.1(c)]. Sets of 11 pins carried on a steel ring are standard. Since distances over one-hundred feet are typically observed using an EDM (see Part III), chaining pins are seldom used today.

The *hand level*, described in Section 4.13, is a simple instrument used to keep the tape ends at equal elevations when observing over rough terrain [see Figures 4.17 and Figure 6.1(d)].

Tension handles facilitate the application of a desired standard or known tension. A complete unit consists of a wire handle, a clip to fit the ring end of the tape, and a spring balance reading up to 30 lb in 1/2-lb graduations.

Clamp handles are used to apply tension by a positive, quick grip using a scissors-type action on any part of a steel tape. They do not damage the tape and prevent injury to hands and the tape.

A *pocket thermometer* permits reading data for making temperature corrections. It is about 5-in. long, graduated from perhaps 30° to 120° F in 1° or 2° divisions, and kept in a protective metal case.

Range poles (lining rods) made of wood, steel, or aluminum are about 1-in. thick and 6- to 10-ft long. They are round or hexagonal in cross section and marked with alternate 1-ft long red and white bands that can be used for rough measurements [see Figure 6.1(e)]. The main utility of range poles is to mark the line being measured so that the tape's alignment can be maintained.

Plumb bobs for taping [see Figure 6.1(f)] should weigh a minimum of 8 oz and have a fine point. However, most surveyors use 24-oz plumb bobs for stability reasons. At least 6 ft of good-quality string or cord, free of knots, is necessary for convenient work with a plumb bob. The points of most plumb bobs are removable, which facilitates replacement if they become dull or broken. The string can be wound on a spring-loaded reel that is useful for rough targeting. However, in taping, it is best to not use a reel.

■ 6.10 CARE OF TAPING EQUIPMENT

The following points are pertinent in the care of tapes and range poles:

1. Considering the cross-sectional area of the average surveyor's steel tape and its permissible stress, a pull of 100 lb will do no damage. But if the tape is kinked, a pull of less than 1 lb can break it. Therefore, always check to be certain that any loops and kinks are eliminated before tension is applied.
2. If a tape gets wet, wipe it first with a dry cloth, then with an oily one.
3. Tapes should be either kept on a reel or *thrown* into circular loops, but not handled both ways.
4. Each tape should have an individual number or tag to identify it.
5. Broken tapes can be mended by riveting or applying a sleeve device, but a mended tape should not be used on important work.
6. Range poles are made with the metal shoe and point in line with the section above. This alignment may be lost if the pole is used improperly.

6.11 TAPING ON LEVEL GROUND

The subsections that follow describe six steps in taping on level ground using a tape.

6.11.1 Lining In

Using range poles, the line to be measured should be marked at both ends, and at intermediate points where necessary, to ensure unobstructed sight lines. Taping requires a minimum of two people, a *forward tapeperson* and a *rear tapeperson*. The forward tapeperson is lined in by the rear tapeperson. Directions are given by vocal or hand signals.

6.11.2 Applying Tension

The rear tapeperson holding the 100-ft end of a tape over the first (rear) point lines in while the forward tapeperson, holding the zero end of the tape. For accurate results, the tape must be straight, and the two ends held at the same elevation. A specified tension, generally between 10 and 25 lb., is applied. To maintain a steady pull, tapepersons wrap the leather thong at the tape's end around one hand, keep forearms against their bodies, and face at right angles to the line. In this position, they are off the line of sight. Also, the body need only be tilted to hold, decrease, or increase the pull. Sustaining a constant tension with *outstretched* arms is difficult, if not impossible, for a pull of 15 lb. or more. Good communication between forward and rear tapepersons will avoid jerking the tape, save time, and produce better results.

6.11.3 Plumbing

Weeds, brush, obstacles, and surface irregularities may make it undesirable to lay a tape on the ground. In those cases, the tape is held above ground in a horizontal position. Placing the plumb-bob string over the proper tape graduation and securing it with one thumb, mark each end point on the tape. The rear tapeperson continues to hold a plumb bob over the fixed point, while the forward tapeperson marks the length. In measuring a distance shorter than a full tape length, the forward tapeperson moves the plumb-bob string to a point on the tape over the ground mark.

6.11.4 Marking Tape Lengths

When the tape has been lined in properly, tension has been applied, and the rear tapeperson is over the point, "stick" is called out. The forward tapeperson then places a pin exactly opposite the zero mark of the tape and calls "stuck." The marked point is checked by repeating the measurement until certainty of its correct location is assured.

After checking the measurement, the forward tapeperson signals that the point is OK, the rear tapeperson pulls up the rear pin, and they move ahead. The forward tapeperson drags the tape pacing roughly 100 ft and stops. The rear tapeperson calls "tape" to notify the forward tapeperson that they have gone 100 ft just before the 100-ft end reaches the pin that has been set. The process of measuring 100-ft lengths is repeated until a partial tape length is needed at the end of the line. Following this procedure, it is easy to see why the EDM has replaced taping for all but very short distances and in some construction work.

6.11.5 Reading the Tape

There are two common styles of graduations on 100-ft surveyor's tapes. *It is necessary to identify the type being used before starting work* to avoid making one-foot mistakes repeatedly.

The more common type of tape has a total graduated length of 101 ft. It is marked from 0 to 100 by full feet in one direction, and has an additional foot preceding the zero mark graduated from 0 to 1 ft in tenths, or in tenths and hundredths in the other direction. In measuring the last partial tape length of a line with this kind of tape, a full-foot graduation is held by the rear tapeperson at the last pin set [like the 87-ft mark in Figure 6.2(a)]. The actual footmark held is the one that causes the graduations on the extra foot between zero and the tape end to straddle the closing point. The forward tapeperson reads the additional length of 0.68 ft beyond the zero mark. In the case illustrated, to ensure correct recording, the rear tapeperson calls "87." The forward tapeperson repeats and adds the partial foot reading, calling "87.68." Since part of a foot has been added, this type of tape is known as an *add tape*.

The other kind of tape found in practice has a total graduated length of 100 ft. It is marked from 0 to 100 with full-foot increments, and the first foot at each end (from 0 to 1 and from 99 to 100) is graduated in tenths, or in tenths and hundredths. With this kind of tape, the last partial tape length is measured by holding a full-foot graduation at the last chaining pin set such that the graduated section of the tape between the zero mark and the 1-ft mark straddles the closing point. This is indicated in Figure 6.2(b), where the 88-ft mark is being held on the last chaining pin and the tack marking the end of the line is opposite 0.32 ft read from the zero end of the tape. The partial tape length is then $88.0 - 0.32 = 87.68$ ft. The quantity 0.32 ft is said to be *cut off*; hence this type of tape is called a *cut tape*. To ensure subtraction of a foot from the number at the full-foot graduation used, the following field procedure and calls are recommended: rear tapeperson calls "88"; forward tapeperson says "cut point three-two"; rear tapeperson answers "eighty seven point six eight"; forward tapeperson confirms the subtraction and replies "check" when satisfied it is correct.

An advantage of the add tape is that it is easier to use because no subtraction is needed when measuring decimal parts of a foot. Its disadvantage is that careless tapepersons will sometimes make measurements of 101.00 ft and record them as

Figure 6.2 Reading partial tape lengths.

100.00 ft. The cut tape practically eliminates this mistake. The advantage of a cut tape is in alignment surveys where the distance between stations is 100 ft or less. In this case, the 100-ft arc or chord lengths being laid out are often the same distance.

The same routine should be used throughout all taping by a party, and the results tested in every possible way. A single mistake in subtracting the partial foot when using a cut tape will destroy the precision of a hundred other good measurements. For this reason, the add tape is more foolproof. The greatest danger for mistakes in taping arises when changing from one style of tape to the other.

6.11.6 Recording the Distance

Accurate fieldwork may be canceled by careless recording. After the partial tape length is obtained at the end of a line, the rear tapeperson determines the number of full 100-ft tape lengths by counting the pins collected from the original set of 11. Since long distances are measured electronically today, tapes are never used for long distances today. Although taping procedures may appear to be relatively simple, high precision is difficult to achieve. Taping is a skill that can best be taught and learned by field demonstrations and practice.

■ 6.12 HORIZONTAL MEASUREMENTS ON SLOPING GROUND

In taping on uneven or sloping ground, it is standard practice to hold the tape horizontally and use a plumb bob at one or perhaps both ends. It is difficult to keep the plumb line steady for heights above the chest. Wind exaggerates this problem and may make accurate work impossible.

On steeper slopes, where a 100-ft length cannot be held horizontally without plumbing from above shoulder level, shorter distances are measured and accumulated to total a full tape length. This procedure, called *breaking tape*, is illustrated in Figure 6.3. As an example of this operation, assume that when taping down slope, the 100-ft end of the tape is held at the rear point, and the forward tapeperson can advance only 30 ft without being forced to plumb from above the chest. A pin is therefore set beneath the 70-ft mark, as in Figure 6.4. The rear tapeperson moves ahead to this pin and holds the 70-ft graduation there while another pin is set at, say, the 25-ft mark. Then, with the 25-ft graduation over the second pin, the full 100 ft distance is marked at the zero point. In this way, the partial tape lengths are added mechanically to make a full tape length by holding the proper graduations, and no mental arithmetic is required. The rear tapeperson returns the pins set at the intermediate points to the forward tapeperson to keep the tally clear on the number of full tape lengths established. To avoid kinking the tape, the full 100-ft length is pulled ahead by the forward tapeperson into position for measuring the next tape length. In all cases the tape is leveled by eye or hand level, with the tapepersons remembering the natural tendency to have the downhill end of a tape too low. Practice will improve the knack of holding a tape horizontally by keeping it perpendicular to the vertical plumb-bob string.

Taping downhill is preferable to measuring uphill for two reasons. First, in taping downhill, the rear point is held steady on a fixed object while the other end is plumbed. In taping uphill, the forward point must be set while the other end is

Figure 6.3
Breaking tape.

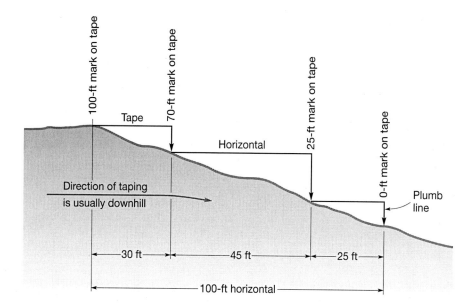

Figure 6.4
Procedure for breaking tape (when tape is not in box or on reel).

wavering somewhat. Second, if breaking tape is necessary, the head tapeperson can more conveniently use the hand level to proceed downhill a distance, which renders the tape horizontal when held comfortably at chest height.

■ 6.13 SLOPE MEASUREMENTS

In measuring the distance between two points on a steep slope, rather than break tape every few feet, it may be desirable to tape along the slope and compute the horizontal component. This requires measurement also of either the altitude angle α or the difference in elevation d (Figure 6.5). Breaking tape is more time

138 DISTANCE MEASUREMENT

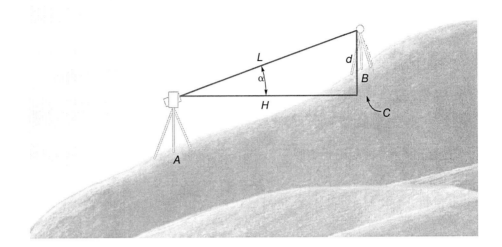

Figure 6.5
Slope measurement.

consuming and generally less accurate due to the accumulation of random errors from marking tape ends and keeping the tape level and aligned for many short sections.

In Figure 6.5, if altitude angle α is determined, the horizontal distance between points A and B can be computed from the following relation:

$$H = L \cos \alpha \tag{6.1a}$$

where H is the horizontal distance between points, L the slope length separating them, and α the altitude angle from horizontal, usually obtained with an *Abney hand level* and *clinometer* (hand device for measuring angles of inclination). If the difference in elevation d between the ends of the tape is measured, which is done by leveling (see Chapter 5), the horizontal distance can be computed using the following expression derived from the Pythagorean theorem:

$$H = \sqrt{L^2 - d^2} \tag{6.2a}$$

Another approximate formula, obtained from the first term of a binomial expansion of the Pythagorean theorem, may be used in lower-order surveys to reduce slope distances to horizontal:

$$H = L - \frac{d^2}{2L} (\text{approximate}) \tag{6.2b}$$

In Equation (6.2b), the term $d^2/2L$ equals C in Figure 6.5 and is a correction to be subtracted from the measured slope length to obtain the horizontal distance. The error in using the approximate formula for a 100-ft length grows with increasing slope. Equation (6.2b) is useful for making quick estimates, without a calculator, or error sizes produced for varying slope conditions. It should not be used as an alternate method of Equation (6.2a) when reducing slope distances.

■ 6.14 SOURCES OF ERROR IN TAPING

There are three fundamental sources of error in taping:

1. *Instrumental errors.* A tape may differ in actual length from its nominal graduated length because of a defect in manufacture or repair, or as a result of kinks.
2. *Natural errors.* The horizontal distance between end graduations of a tape varies because of the effects of temperature, wind, and weight of the tape itself.
3. *Personal errors.* Tapepersons setting pins, reading the tape, or manipulating the equipment.

With the precision of the EDM in today's total station, taping is seldom used for precise work; it has been relegated to use in areas where lower accuracy is required. However when a tape is used these sources of errors should be understood and avoided. For example, an offset measurement or tie measurements for a station taken with a tape should not be subjected to personal errors. The effects of personal and systematic error sources in taping are discussed in the subsections that follow. Due to the precision and accuracy of EDM instruments, precise taping, which required these corrections, is seldom performed today. Appendix A contains examples of tape corrections for systematic errors.

6.14.1 Incorrect Length of Tape

Incorrect length of a tape can be one of the most important errors. Tape manufacturers do not guarantee steel tapes to be exactly their graduated nominal length—for example, 100.00 ft—nor do they provide a standardization certificate unless requested and paid for as an extra. The true length is obtained by comparing it with a standard tape or distance. The National Institute of Standards and Technology (NIST)[1] of the U.S. Department of Commerce will make such a comparison and certify the exact distance between end graduations under given conditions of temperature, tension, and manner of support. A 100-ft steel tape usually is standardized for each of the two sets of conditions—for example, 68°F, a 12-lb pull, with the tape lying on a flat surface (fully supported throughout); and 68°F, a 20-lb pull, with the tape supported at the ends only.

An error, caused by incorrect length of a tape, occurs each time the tape is used. If the true length, known by standardization, is not exactly equal to its nominal value of 100.00 ft recorded for every full length, the correction can be determined as

$$C_L = \left(\frac{l - l'}{l'}\right)L \qquad (6.3)$$

where C_L is the correction to be applied to the measured (recorded) length of a line to obtain the true length, l the actual tape length, l' the nominal tape length,

---

[1] Information on tape calibration services of the National Institute of Standards and Technology can be obtained at the following website: http://www.nist.gov. Tapes can be sent for calibration to the National institute of Standards and Technology, Building 220, Room 113, 100 Bureau Dr., Gaithersburg, MD 20899; telephone: (301) 975-2465.

and L the measured (recorded) length of line. Units for the terms in Equation (6.3) can be in either feet or meters.

6.14.2 Temperature Other Than Standard

Steel tapes are standardized for 68°F (20°C) in the United States. A temperature higher or lower than this value causes a change in length that must be considered. The coefficient of thermal expansion and contraction of steel used in ordinary tapes is approximately 0.00000645 per unit length per degree Fahrenheit, and 0.0000116 per unit length per degree Celsius. For any tape, the correction for temperature can be computed as

$$C_T = k(T_1 - T)L \qquad (6.4)$$

where C_T is the correction in the length of a line caused by nonstandard temperature, k the coefficient of thermal expansion and contraction of the tape, T_1 the tape temperature at the time of measurement, T the tape temperature when it has standard length, and L the observed (recorded) length of line. The correction C_T will have the same units as L, which can be either feet or meters. Errors caused by temperature change may be practically eliminated by either (a) measuring temperature and making corrections according to Equation (6.4), or (b) using an Invar tape.

It should be pointed out that T_1 is the temperature of the tape and not the air temperature. Since the tape is often on the ground or on some solid surface, the temperature of the tape often does not match the air temperature. Tape thermometers are made to clamp onto the tape while in operation so that accurate tape temperatures can be obtained during precise taping.

Shop measurements made with steel scales and other devices likewise are subject to temperature effects. The precision required in fabricating a large airplane or ship can be lost by this one cause alone.

6.14.3 Inconsistent Pull

When a steel tape is pulled with a tension greater than its *standard pull* (the tension at which it was calibrated), the tape will stretch and become longer than its standard length. Conversely, if less than standard pull is used, the tape will be shorter than its standard length. The *modulus of elasticity* of the tape regulates the amount that it stretches. The correction for pull can be computed and applied using the following formula:

$$C_P = (P_1 - P)\frac{L}{AE} \qquad (6.5)$$

where C_P is the total elongation in tape length due to pull, in feet; P_1 the pull applied to the tape at the time of the observation, in pounds; P the standard pull for the tape in pounds; A the cross-sectional area of the tape in square inches; E the modulus of elasticity of steel in pounds per square inch; and L the observed (recorded) length of line. An average value of E is 29,000,000 lb/in.^{2} for the kind of steel typically used in tapes. In the metric system, to produce the correction C_P in meters, comparable units of P and P_1 are kilograms, L is meters, A is square centimeters, and E is kilograms per square centimeter. An average value of E for steel in these units is approximately 2,000,000 kg/cm^2. The cross-sectional area of a steel tape can be obtained from the manufacturer, by measuring its width and thickness with calipers, or by dividing the total tape weight by the product of its length (in feet) times the unit weight of steel (490 lb/ft²), and multiplying by 144 to convert square feet to square inches.

Errors resulting from incorrect tension can be eliminated by (a) using a spring balance to measure and maintain the standard pull, or (b) applying a pull other than standard and making corrections for the deviation from standard according to Equation (6.5).

Errors caused by incorrect pull may be either systematic or random. The pull applied by even an experienced tapeperson is sometimes greater or less than the desired value. An inexperienced person, particularly one who has not used a spring balance on a tape, is likely to apply less than the standard tension consistently.

6.14.4 Sag

A steel tape not supported along its entire length sags in the form of a *catenary*, a good example being the cable between two power poles. Because of sag, the horizontal distance (chord length) is less than the graduated distance between tape ends, as illustrated in Figure 6.6. Sag can be reduced by applying greater tension, but not eliminated unless the tape is supported throughout. The following formula is used to compute the sag correction:

$$C_S = -\frac{w^2 L_S^3}{24 P_1^2} \tag{6.6}$$

where in the English system C_S is the correction for sag (difference between length of curved tape and straight line from one support to the next), in feet; L_S the unsupported length of the tape, in feet; w the weight of the tape per foot of length, in pounds; and P_1 the pull on the tape, in pounds. Metric system units for Equation (6.6) are kg/m for w, kg for P_1, and meters for C_S and L_S.

The effects of errors caused by sag can be eliminated by (a) supporting the tape at short intervals or throughout, or (b) by computing a sag correction for each unsupported segment and applying the total to the recorded length according to Equation (6.6). It is important to recognize that Equation (6.6) is nonlinear, and thus must be applied to each unsupported section of the tape. It is incorrect to apply it to the overall length of a line unless the line was observed in one section.

As stated previously, when lines of unknown length are being measured, sag corrections are always negative, whereas positive corrections occur if the tension applied exceeds the standard pull. For any given tape, the so-called *normal tension*

Figure 6.6 Effect of sag.

needed to offset these two factors can be obtained by setting Equations (6.5) and (6.6) equal to each other and solving for P_1. Although applying the normal tension does eliminate the need to make corrections for both pull and sag, it is not commonly used because the required pull is often too great for convenient application.

6.14.5 Tape Not Horizontal and Tape Off-Line

Corrections for errors caused by a tape being inclined in the vertical plane are computed in the same manner as corrections for errors resulting from it being off-line in the horizontal plane. Corrected lengths can be determined by Equation (6.2), where in the vertical plane d is the difference in elevation between the tape ends, and in the horizontal plane, d is the amount where one end of the tape is off-line. In either case, L is the length of tape involved in the measurement.

Errors caused by the tape not being horizontal are systematic, and always make recorded lengths longer than true lengths. They are reduced by using a hand level to keep elevations of the tape ends equal, or by running differential levels (see Section 5.4) over the taping points and applying corrections for elevation differences. Errors from the tape being off-line are also systematic, and they too make recorded lengths longer than true lengths. This type of error can be eliminated by careful alignment.

6.14.6 Improper Plumbing

Practice and steady nerves are necessary to hold a plumb bob still long enough to mark a point. The plumb bob will sway, even in calm weather. On very gradual slopes and on smooth surfaces such as pavements, inexperienced tapepersons obtain better results by laying the tape on the ground instead of plumbing. Experienced tapepersons plumb most measurements.

Errors caused by improper plumbing are random since they may make distances either too long or too short. However, the errors would be systematic when taping directly against or in the direction of a strong wind. Heavier plumb bobs and touching the plumb bob on the ground, or steadying it with one foot, decreases its swing. Practice in plumbing will reduce errors.

6.14.7 Faulty Marking

Chaining pins should be set perpendicular to the taped line but inclined 45° to the ground. This position permits plumbing to the point where the pin enters the ground without interference from the loop.

Brush, stones, and grass or weeds deflect a chaining pin and may increase the effect of incorrect marking. Errors from these sources tend to be random and are kept small by carefully locating a point, then checking it.

When taping on solid surfaces such as pavement or sidewalks, pencil marks or scratches can be used to mark taped segments. Accuracy in taping on the ground can be increased by using tacks in stakes as markers rather than chaining pins.

6.14.8 Incorrect Reading or Interpolation

The process of reading to hundredths on tapes graduated only to tenths, or to thousandths on tapes graduated to hundredths, is called interpolation. Errors from this

TABLE 6.1 SUMMARY OF ERRORS

Error Type	Error Source*	Systematic (S) or Random (R)	Departure from Normal to Produce 0.01-ft Error for 100-ft Tape
Tape length	I	S	0.01 ft
Temperature	N	S or R	15°F
Pull	P	S or R	15 lb
Sag	N, P	S	0.6 ft at center for 100-ft tape standardized by support throughout
Alignment	P	S	1.4 ft at one end of 100-ft tape, or 0.7 ft at midpoint
Tape not level	P	S	1.4-ft elevation difference between ends of 100-ft tape
Plumbing	P	R	0.01 ft
Marking	P	R	0.01 ft
Interpolation	P	R	0.01 ft

*I, instrumental; N, natural; P, personal; R random.

source are random over the length of a line. They can be reduced by care in reading, employing a magnifying glass, or using a small scale to determine the last figure.

6.14.9 Summary of Effects of Taping Errors

An error of 0.01 ft is significant in many surveying measurements. Table 6.1 lists the nine types of taping errors; classifies them as instrumental (I), natural (N), or personal (P), and systematic (S) or random (R); and gives the departure from normal that produces an error of 0.01 ft in a 100 ft length.

PART III • ELECTRONIC DISTANCE MEASUREMENT

6.15 INTRODUCTION

A major advance in surveying instrumentation occurred in the mid-20th century with the development of electronic distance measuring (EDM) instruments. These devices measure lengths by indirectly determining the number of full and partial waves of transmitted electromagnetic energy required in traveling between the two ends of a line. In practice, the energy is transmitted from one end of the line to the other and returned to the starting point; thus, it travels the double path distance. Multiplying the total number of cycles by its wavelength and dividing by 2, yields the unknown distance.

The Swedish physicist Erik Bergstrand introduced the first EDM instrument in 1948. His device, called the *geodimeter* (an acronym for geodetic distance meter), resulted from attempts to improve methods for measuring the velocity of light.

The instrument transmitted visible light and was capable of accurately observing distances up to about 25 mi (40 km) at night. In 1957, a second EDM apparatus, the *tellurometer*, was introduced. Designed in South Africa by Dr. T. L. Wadley, this instrument transmitted microwaves, and was capable of observing distances up to 50 mi (80 km) or more, day or night.

The potential value of these early EDM models to the surveying profession was immediately recognized. However, they were expensive and not readily portable for field operations. Furthermore, observing procedures were lengthy, and mathematical reductions to obtain distances from observed values were difficult and time consuming. Continued research and development have overcome all of these deficiencies. Prior to the introduction of EDM instruments, taping made accurate distance measurements. Although seemingly a relatively simple procedure, precise taping is one of the most difficult and painstaking of all surveying tasks. In the great trigonometrical survey of India, a full year was spent observing the length of a single baseline. Now EDM instruments have made it possible to obtain accurate distance measurements rapidly and easily. Given a line of sight, long or short lengths can be measured over bodies of water, busy freeways, or terrain that is inaccessible for taping.

In the current generation, EDM instruments are combined with *digital theodolites* and *microprocessors* to produce *total station instruments* (see Figures 1.3 and 2.5). These devices can simultaneously and automatically observe both distances and angles. The microprocessor receives the measured slope length and zenith (or altitude) angle, calculates horizontal and vertical distance components, and displays them in real time. When equipped with *data collectors* (see Section 2.12), they can record field notes electronically for transmission to computers, plotters, and other office equipment for processing. These so-called *field-to-finish* systems are gaining worldwide acceptance and changing the practice of surveying substantially.

6.16 PROPAGATION OF ELECTROMAGNETIC ENERGY

Electronic distance measurement is based on the rate and manner that electromagnetic energy propagates through the atmosphere. The rate of propagation can be expressed with the following equation:

$$V = f\lambda \tag{6.7}$$

where V is the velocity of electromagnetic energy, in meters per second; f the modulated frequency of the energy, in hertz;[2] and λ the wavelength, in meters. The velocity of electromagnetic energy in a vacuum is 299,792,458 m/sec. Its speed is slowed somewhat in the atmosphere according to the following equation:

$$V = c/n \tag{6.8}$$

where c is the velocity of electromagnetic energy in a vacuum, and n the atmospheric *index of refraction*. The value of n varies from about 1.0001 to 1.0005,

---

[2]The hertz (Hz) is a unit of frequency equal to 1 cycle/sec. The kilohertz (KHz), megahertz (MHz), and gigahertz (GHz) are equal to 10^3, 10^6, and 10^9 Hz, respectively.

depending on pressure and temperature, but is approximately equal to 1.0003. Thus, accurate electronic distance measurement requires that atmospheric pressure and temperature be measured so that the appropriate value for n is known.

Temperature, atmospheric pressure, and relative humidity all influence the index of refraction. Because a light source emits light composed of many wavelengths, and since each wavelength has a different index of refraction, this group of waves has a *group index of refraction*. The value for the group refractivity N_g in *standard air*[3] for electronic distance measurement is

$$N_g = (n_g - 1)10^6 = 287.6155 + \frac{4.88660}{\lambda^2} + \frac{0.06800}{\lambda^4} \qquad (6.9)$$

where λ is the wavelength of the light expressed in micrometers (μm) and n_g is the group refractive index. The wavelengths of light sources commonly used in EDMs, are 0.6328 μm for red laser and 0.900 to 0.930 μm for infrared.

The actual group refractive index n_a for atmosphere at the time of observation due to variations in temperature, pressure, and humidity can be computed as

$$n_a = 1 + \left(\frac{273.15}{1013.25} \cdot \frac{N_g P}{t + 273.15} - \frac{11.27\, e}{t + 273.15} \right) 10^{-6} \qquad (6.10)$$

where e is the partial water vapor pressure in hectopascal[4] (hPa) as defined by the temperature and relative humidity at the time of the measurement, P the pressure in hPa, and t the dry bulb temperature in °C. The partial water vapor pressure, e, can be computed with sufficient accuracy for normal operating conditions as

$$e = E \cdot h / 100 \qquad (6.11)$$

where $E = 10^{[7.5t/(237.3+t)+0.7858]}$ and h is the relative humidity in percent.

■ Example 6.1

What is the actual wavelength and velocity of a near-infrared beam ($\lambda = 0.915\ \mu m$) of light modulated at a frequency of 320 MHz through an atmosphere with a (dry) temperature t of 34°C, relative humidity h of 56%, and an atmospheric pressure of 1041.25 hPa?

Solution

By Equation (6.9)

$$N_g = 287.6155 + \frac{4.88660}{(0.915)^2} + \frac{0.06800}{(0.915)^4} = 293.5491746$$

---

[3] A standard air is defined with the following conditions: 0.0375% carbon dioxide, temperature of 0°C, pressure of 760 mm of mercury, and 0% humidity.
[4] 1 Atmosphere = 101.325 kPa = 1013.25 hPa = 760 torr = 760 mm Hg

By Equation (6.11)

$$a = \frac{7.5\,(34)}{(237.3 + 34)} + 0.7858 = 1.7257$$

$$E = 10^a = 53.174$$

$$e = Eh = 53.174(56/100) = 29.7774$$

By Equation (6.10)

$$n_a = 1 + \left(\frac{273.15}{1013.25} \cdot \frac{293.5492 \times 1041.25}{34 + 273.15} - \frac{11.27 \times 29.7774}{34 + 273.15}\right)10^{-6}$$

$$= 1 + (268.268660 - 1.092597)\,10^{-6}$$

$$= 1.0002672$$

By Equation (6.8)

$$V = 299{,}792{,}458/1.0002672 = 299{,}712{,}382 \text{ m/sec}$$

Rearranging Equation (6.7) yields an actual wavelength of

$$\lambda = 299{,}712{,}382/320{,}000{,}000 = 0.9366012 \text{ m}$$

Note in the solution of Example 6.1 that the second parenthetical term in Equation (6.10) accounts for the effects of humidity in the atmosphere. In fact, if this term were ignored the actual index of refraction n_a would become 1.0002683 resulting in the same computed wavelength to five decimal places. This demonstrates why, in using EDM instruments that employ near-infrared light, the effects of humidity on the transmission of the wave can be ignored for all but the most precise work.

The manner by which electromagnetic energy propagates through the atmosphere can be represented conceptually by the sinusoidal curve illustrated in Figure 6.7. This figure shows one wavelength, or *cycle*. Portions of wavelengths or the positions of points along the wavelength are given by phase angles. Thus,

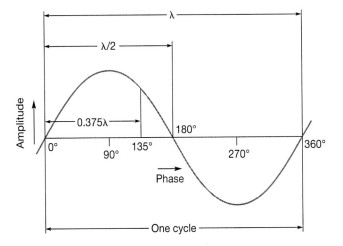

Figure 6.7
A wavelength of electromagnetic energy illustrating phase angles.

in Figure 6.7, a 360° *phase angle* represents a full cycle, or a point at the end of a wavelength, while 180° is a half wavelength, or the midpoint. An intermediate position along a wavelength having a phase angle of, say, 135° is 135/360, or 0.375 of a wavelength.

■ 6.17 PRINCIPLES OF ELECTRONIC DISTANCE MEASUREMENT

In Section 6.15, it was stated that distances are observed electronically by determining the number of full and partial waves of transmitted electromagnetic energy that are required in traveling the distance between the two ends of a line. In other words, this process involves determining the number of wavelengths in an unknown distance. Then, knowing the precise length of the wave, the distance can be determined. This is similar to relating an unknown distance to the calibrated length of a steel tape.

The procedure of measuring a distance electronically is depicted in Figure 6.8, where an EDM device has been centered over station A by means of a plumb bob or optical plumbing device. The instrument transmits a *carrier signal* of electromagnetic energy to station B. A *reference frequency* of a precisely regulated wavelength has been superimposed or *modulated* onto the carrier. A reflector at B returns the signal to the receiver, so its travel path is double the slope distance AB. In the figure, the modulated electromagnetic energy is represented by a series of sine waves, each having wavelength λ, The unit at A determines the number of wavelengths in the double path, multiplied by the wavelength in feet or meters, and divided by 2 to obtain distance AB.

Of course, it would be highly unusual if a measured distance was exactly an integral number of wavelengths, as illustrated in Figure 6.8. Rather, some fractional part of a wavelength would, in general, be expected; for example, the partial

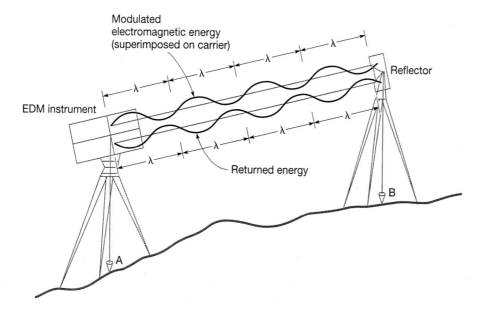

Figure 6.8
Generalized EDM procedure.

Figure 6.9
Phase difference measurement principle.

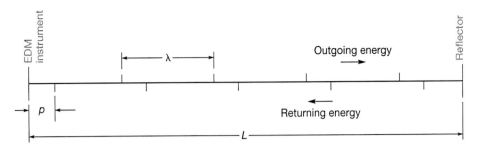

value p shown in Figure 6.9. In that figure, distance L between the EDM instrument and reflector would be expressed as

$$L = \frac{n\lambda + p}{2} \qquad (6.12)$$

where λ is the wavelength, n the number of full wavelengths, and p the length of the fractional part. The fractional length is determined by the EDM instrument from measurement of the *phase shift* (phase angle) of the returned signal. To illustrate, assume that the wavelength for the example of Figure 6.8 was precisely 20.000 m. Assume also that the phase angle of the returned signal was 115.7°, in which case length p would be $(115.7/360)20.000 = 6.428$ m. Then from the figure, since $n = 9$, by Equation (6.12), length L is

$$L = \frac{9(20.000) + 6.428}{2} = 93.214 \text{ m}$$

Considering the double path distance, the 20-m wavelength used in the example just given has an "effective wavelength" of 10 m. This is one of the fundamental wavelengths used in current EDM instruments. It is generated using a frequency of approximately 15 MHz.

EDM instruments cannot determine the number of full wavelengths in an unknown distance by transmitting only one frequency and wavelength. To resolve the ambiguity n, in Equation (6.12), they must transmit additional signals having longer wavelengths. This procedure is explained in the following section, which describes electro-optical EDM instruments.

6.18 ELECTRO-OPTICAL INSTRUMENTS

The majority of EDM instruments manufactured today are electro-optical and transmit infrared or laser light as a carrier signal. This is primarily because its intensity can be modulated directly, considerably simplifying the equipment. Earlier models used tungsten or mercury lamps. They were bulky, required a large power source, and had relatively short operating ranges, especially during the day because of excessive atmospheric scatter. EDM instruments using coherent light produced by gas lasers followed. These were smaller and more portable, and were capable of making observations of long distances in the daytime as well as at night.

Figure 6.10 is a generalized schematic diagram illustrating the basic method of operation of one particular type of electro-optical instrument. The transmitter uses a GaAs diode that emits *amplitude-modulated* (AM) infrared light. A crystal oscillator precisely controls the frequency of modulation. The modulation process may be thought of as similar to passing light through a stovepipe in which a damper plate is spinning at a precisely controlled rate or frequency. When the damper is closed, no light passes. As it begins to open, light intensity increases to a maximum at a phase angle of 90° with the plate completely open. Intensity reduces to zero again with the damper closed at a phase angle of 180°, and so on. This intensity variation or amplitude modulation is properly represented by sine waves such as those shown in Figures 6.7 and 6.8.

As shown in Figure 6.10, a beam splitter divides the light emitted from the diode into two separate signals: an *external* measurement beam and an *internal* reference beam. By means of a telescope mounted on the EDM instrument, the external beam is carefully aimed at a retroreflector that has been centered over the point at the line's other end. Figure 6.11 shows a triple corner cube retroreflector of the type used to return the external beam, coaxial, to the receiver.

The internal beam passes through a variable-density filter and is reduced in intensity to a level equal to that of the returned external signal, enabling a more accurate observation to be made. Both internal and external signals go through an interference filter, which eliminates undesirable energy such as sunlight. The internal and external beams then pass through components to convert them into electric energy while preserving the phase shift relationship resulting from their

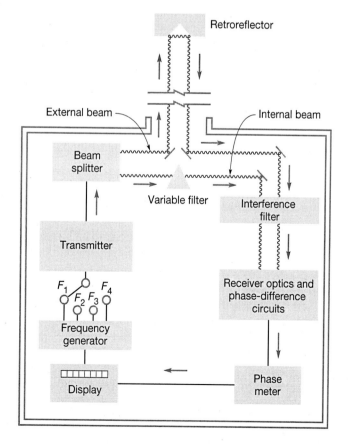

Figure 6.10
Generalized block diagram illustrating operation of electro-optical EDM instrument.

Figure 6.11
Triple retroreflector.
(Courtesy Topcon
America Corp.)

different travel path lengths. A phase meter converts this phase difference into direct current having a magnitude proportional to the differential phase. This current is connected to a *null meter* that is adjusted to null the current. The fractional wavelength is measured during the nulling process, converted to distance, and displayed.

To resolve the ambiguous number of full cycles a wave has undergone, EDM instruments transmit different modulation frequencies. The unit illustrated in the schematic of Figure 6.10 uses four frequencies: $F1$, $F2$, $F3$, and $F4$, as indicated. If modulation frequencies of 14.984 MHz, 1.4984 MHz, 149.84 KHz, and 14.984 KHz are used, and assuming the index of refraction is 1.0003, then their corresponding "effective" wavelengths are 10.000, 100.00, 1000.0, and 10,000 m, respectively. Assume that a distance of 3867.142 appears on the display as the result of measuring a line. The four rightmost digits, 7.142, are obtained from the phase shift measured while transmitting the 10.000-m wavelength at frequency $F1$. Frequency $F2$, having a 100.00-m wavelength, is then transmitted, yielding a fractional length of 67.14. This provides the digit 6 in the displayed distance. Frequency $F3$ gives a reading of 867.1, which provides the digit 8 in the answer, and finally, frequency $F4$ yields a reading of 3867, which supplies the digit 3, to complete the display. From this example, it should be evident that the high resolution of a measurement (nearest 0.001 m) is secured using the 10.000-m wavelength, and the others simply resolve the ambiguity of the number of these shorter wavelengths in the total distance.

With older instruments, changing of frequencies and nulling were done manually by setting dials and turning knobs. Now modern instruments incorporate microprocessors that control the entire measuring process. Once the instrument is aimed at the reflector and the measurement started, the final distance appears in the display almost instantaneously. Other changes in new instruments include improved electronics to control the amplitude modulation, and replacement of the null meter by an electronic phase detector. These changes have significantly improved the accuracy with which phase shifts can be determined, that in turn has reduced the number of different frequencies that need to be transmitted. Consequently, as few as two frequencies are now used on some instruments: one that produces a short wavelength to provide the high-resolution digits, and one with a long wavelength to provide the coarse numbers. To illustrate how this is possible, consider again the example measurement just described which used four frequencies. Recall that a reading of 7.142 was obtained with the 10.000-m wavelength, and that 3867 was read with the 10,000-m wavelength. Note the overlap of the common digit 7 in the two readings. Assuming that both phase shift measurements are reliably made to four significant figures, the leftmost digit of the first reading should indeed be the same as the rightmost one of the second reading. If these digits are the same in the measurement, this provides a check on the operation of the instrument. Modern instruments compare these overlapping digits, and will display an error message if they do not agree. If they do check, the displayed distance will take all four digits from the first (short wavelength) reading, and the first three digits from the second reading.

Manufacturers provide a full range of instruments with precisions that vary from $\pm(1 \text{ mm} + 1 \text{ ppm})$ to $\pm(10 \text{ mm} + 5 \text{ ppm})$.[5] Earlier versions were manufactured to stand alone on a tripod, and thus from any setup they could only measure distances. Now, as noted earlier, in most instances EDMs are combined with electronic digital theodolites to produce our modern and very versatile total station instruments. These are described in the following section.

6.19 TOTAL STATION INSTRUMENTS

Total station instruments combine an EDM instrument, an electronic digital theodolite, and a computer in one unit. These devices, described in more detail in Chapter 8, automatically observe horizontal and vertical angles, as well as distances, and transmit the results in real time to a built-in computer. The horizontal and vertical angle and slope distance can be displayed, and then upon keyboard commands, horizontal and vertical distance components can be instantaneously computed from these data and displayed. If the instrument is oriented in direction, and the coordinates of the occupied station are input to the system, the coordinates of any point sighted can be immediately obtained. This data can all be stored within the instrument, or in a survey controller, thereby eliminating manual recording.

Total station instruments are of tremendous value in all types of surveying, as will be discussed in later portions of this text. Besides automatically computing and displaying horizontal and vertical components of a slope distance and coordinates of points sighted, total station instruments can be operated in the *tracking mode*. In this mode, sometimes also called *stakeout*, a required distance (horizontal, vertical, or slope) can be entered by means of the control panel, and the instrument's telescope aimed in the proper direction. Then as the reflector is moved forward or back in position, the difference between the desired distance and that to the reflector is rapidly updated and displayed. When the display shows the difference

---

[5]Accuracies in electronic distance measurements are quoted in two parts; the first part is a constant, and the second is proportional to the distance measured. The abbreviation ppm = parts per million. One ppm equals 1 mm/km. In a distance 5000 ft long, a 5-ppm error equals $5000(5 \times 10^{-6}) = 0.025$ ft.

Figure 6.12
The LEICA TC1101 total station. (Courtesy Leica Geosystems, Inc.)

to be zero, the required distance has been established and a stake is set. This feature, extremely useful in construction stakeout, is described further in Section 23.9.

The total station instruments shown in Figures 2.5, 6.12, and 8.2 all have a distance range of approximately 3 km (using a single prism) with an accuracy of $\pm(1 \text{ mm} + 1.5 \text{ ppm})$ and read angles to the nearest $2''$.

■ 6.20 EDM INSTRUMENTS WITHOUT REFLECTORS

Some EDM instruments do not require reflectors for distance measurement. These devices use time-pulsed infrared laser signals, and in their *reflectorless* mode of operation, they can observe distances up to 200 m in length. The Leica Disto unit shown in Figure 6.13(a) is convenient for measuring lengths in a construction environment.

Some total station instruments, like that shown in Figure 6.12, utilize laser signals and can also observe distances up to 1000 m in the reflectorless mode. But as noted earlier, with prisms they can observe lengths greater than 3 km.

Using instruments in the reflectorless mode, observations can be made to inaccessible objects such as the features of a building as shown in Figure 6.13(b), faces of dams and retaining walls, structural members being assembled on bridges, and so on. These instruments can increase the speed and efficiency of surveys in any construction or fabrication project, especially when measuring to features that are inaccessible. It is this feature of distance measurement that is utilized in laser scanning as discussed in Section 17.9.6.

■ 6.21 COMPUTING HORIZONTAL LENGTHS FROM SLOPE DISTANCES

All EDM instruments measure the slope distance between two stations. As noted earlier, if the EDM unit is incorporated into a total station instrument, then it can reduce these distances to their horizontal components automatically using the vertical angle. With some of the earliest EDMs, this could not be done, and

Figure 6.13
(a) The LEICA DISTO handheld laser distance-measuring instrument, (b) using the LEICA DISTO to measure to an inaccessible point.
(Courtesy Leica Geosystems, Inc.)

reductions were carried out manually. The procedures used, whether performed internally by the microprocessor or done manually, follow those outlined in this section. It is presumed, of course, that slope distances are first corrected for instrumental and atmospheric conditions.

Reduction of slope distances to horizontal can be based on elevation differences, or on zenith (or vertical) angle. Because of Earth curvature, long lines must be treated differently in reduction than short ones and will be discussed in Chapter 19.

6.21.1 Reduction of Short Lines by Elevation Differences

If difference in elevation is used to reduce slope distances to horizontal, during field operations heights h_e of the EDM instrument, and h_r of the reflector above their respective stations are measured and recorded (see Figure 6.14). If elevations of stations A and B in the figure are known, Equation (6.2) will reduce the slope distance to horizontal, with the value of d (difference in elevation between EDM instrument and reflector) computed as follows:

$$d = (\text{elev}_A + h_e) - (\text{elev}_B + h_r) \tag{6.13}$$

■ Example 6.2

A slope distance of 165.360 m (corrected for meteorological conditions) was measured from A to B, whose elevations were 447.401 and 445.389 m above datum, respectively. Find the horizontal length of line AB if the heights of the EDM instrument and reflector were 1.417 and 1.615 m above their respective stations.

Solution

By Equation (6.13)

$$d = (447.401 + 1.417) - (445.389 + 1.615) = 1.814 \text{ m}$$

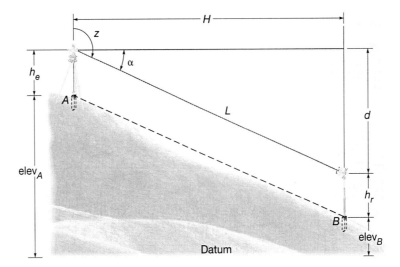

Figure 6.14
Reduction of EDM slope distance to horizontal.

By Equation (6.2)

$$H = \sqrt{(165.360)^2 - (1.814)^2} = 165.350 \text{ m}$$

6.21.2 Reduction of Short Lines by Vertical Angles

If zenith angle z (angle measured downward from the upward direction of the plumb line) is observed to the inclined path of the transmitted energy when measuring slope distance L (see Figure 6.14), then the following equation is applicable to reduce the slope length to its horizontal component:

$$H = L \sin(z) \tag{6.14}$$

If altitude angle α (angle between horizontal and the inclined energy path) is observed (see Figure 6.14), then Equation (6.1) is applicable for the reduction. For most precise work, especially on longer lines, the zenith (or altitude) angle should be observed in both the direct and reversed modes, and averaged (see Section 8.13). Also, as discussed in Section 19.14.2, the mean obtained from both ends of the line will compensate for curvature and refraction.

6.22 ERRORS IN ELECTRONIC DISTANCE MEASUREMENT

As noted earlier, accuracies of EDM instruments are quoted in two parts: a constant error, and a scalar error proportional to the distance observed. Specified errors vary for different instruments, but the constant portions range from 1 mm to 3 mm, and the proportion parts range from 1 ppm to 2 ppm. The constant error is most significant on short distances; for example, with an instrument having a constant error of ±2 mm, a measurement of 20 m is good to only $2/20,000 = 1/10,000$, or 100 ppm. For a long distance, say 2 km, the constant error becomes negligible and the proportional part more important.

The major error components in an observed distance are instrument and target miscentering, and the specified constant and scalar errors of the

EDM instrument. Using Equation (3.11), the error in an observed distance is computed as

$$E_d = \sqrt{E_i^2 + E_r^2 + E_c^2 + (ppm \times D)^2} \qquad (6.15)$$

where E_i is the estimated miscentering error in the instrument, E_r is the estimated miscentering error in the reflector, E_c the specified constant error for the EDM, ppm the specified scalar error for the EDM, and D the measured slope distance.

■ Example 6.3

A slope distance of 827.329 m was observed between two stations with an EDM instruments having specified errors of $\pm(2 \text{ mm} + 2 \text{ ppm})$. The instrument was centered with an estimated error of ± 1.5 mm. The estimated error in target miscentering was ± 3 mm. What is the estimated error in the observed distance?

Solution

By Equation (6.15)

$$E_d = \sqrt{1.5^2 + 3^2 + 2^2 + (2 \times 10^{-6} \times 827329)^2} = \pm 4.2 \text{ mm}$$

Note in the solution that the distance of 827.329 m was converted to millimeters to obtain unit consistency. This solution results in a distance precision of 4.2/827,329, or about 1:195,000.

From the foregoing, it is clear that except for very short distances, the order of accuracy possible with EDM instruments is very high. However, errors can seriously degrade the observations, and thus care should always be exercised to minimize their effects. Sources of error in EDM work may be personal, instrumental, or natural. The subsections that follow identify and describe errors from each of these sources.

The estimated error in the centering of the reflector or GNSS receiver that is atop a range pole follows the geometric equation of

$$S = Rf_d\mu \qquad (6.16)$$

where S is the displacement of the center of the top of the pole from its point at the bottom, R is the height of the rod, f_d is the fractional division the circular vial is miscentered, and μ is the sensitivity of the level vial on the pole. According to Franklin and Meyer (2016), the centering error of a reflector or GNSS receiver (see Chapters 13–16) that is placed on a pole follows a Rayleigh distribution. In this case, the estimated error in the centering can be determined as

$$E_r = \sqrt{(f_d\mu\sigma_R)^2 + (R\sigma_\theta)^2} \qquad (6.17)$$

where σ_R is the uncertainty in R and σ_θ is given by the Rayleigh distribution as

$$\sigma_\theta = f_d\mu\sqrt{0.5(4 - \pi)} \qquad (6.18)$$

Assume in Equation (6.18) that a calibrated 40′ circular vial is within one-half of a division of being centered, then σ_θ is computed as

$$\sigma_\theta = 0.5(0.011636)\sqrt{0.5(4-\pi)} = 0.003811 \text{ rad}$$

where 0.011636 is 40′ converted to radian units. If we assume that the range pole is 6 ft ± 0.01 ft, the estimated centering error in the reflector is given by Equation (6.17) as

$$E_r = \sqrt{\left[\frac{0.5(40)60}{206264.8}0.01\right]^2 + (6 \times 0.003811)^2}$$
$$= \sqrt{0.000058^2 + 0.02287^2}$$
$$= \pm 0.023 \text{ ft}$$

Note in this example that the error in the height of the rod has a minimal contribution to the overall error in centering when it is a reasonably accurate value. Thus, Equation (6.17) can be simplified as

$$E_r = R\sigma_\theta \tag{6.19}$$

where R and σ_θ are as previously defined.

6.22.1 Personal Errors

Personal errors include inaccurate setups of EDM instruments and reflectors over stations, faulty measurements of instrument and reflector heights [needed for computing horizontal lengths (see Section 6.23)], and errors in determining atmospheric pressures and temperatures. These errors are largely random. They can be minimized by exercising utmost care and by using good-quality barometers and thermometers.

Mistakes (not errors) in manually reading and recording displayed distances are common and costly. They can be eliminated with some instruments by obtaining the readings in both feet and meters and comparing them. Of course, data collectors (see Section 2.12) also circumvent this problem. Additionally, as

TABLE 6.2 ERROR IN OBSERVED DISTANCE DUE TO MISALIGNMENT OF THE PRISM

Misalignment in Degrees	0 mm Constant Prism Error (mm)	−30 mm Constant Prism Error (mm)
0	0.00	0.00
5	0.1	0.0
10	0.6	0.1
15	1.3	0.2
20	2.3	0.4
25	3.5	0.7
30	5.1	1.1

shown in Table 6.2, misalignment of the prism can cause significant errors when the reflector is set in its 0 mm constant position.

An example of a common mistake is failing to set the temperature and pressure in an EDM before obtaining an observation. Assume this occurred with the atmospheric conditions given in Example 6.1. The actual index of refraction was computed as 1.0002672. If the fundamental wavelength for a standard atmosphere was 10.000 m, then the actual wavelength produced by the EDM would be 10.000/1.0002672 = 9.9973 m. Using Equation (6.7) with an observed distance of 827.329 m, the error, e, in the observed distance would be

$$e = \left(\frac{9.9973 - 10.000}{10.000}\right)827.329 = -0.223 \text{ m}$$

The effect of failing to account for the actual atmospheric conditions produces a precision of only $|-0.223|/827.329$, or 1:3700. This is well below the computed precision of 1:195,000 in Example 6.3.

For each 1°C change in temperature, a 1 ppm error in the distance measurement will occur. As a rule, the current temperature and pressure should be set at the time of the measurement. However, it is often practical to set the temperature and pressure three or four times per day: morning, midmorning, noon, and midafternoon. At a minimum, the temperature and pressure should be set at least twice a day; once in the morning and at noon. However, a lower-accuracy survey will result. Table 6.3 depicts the distance error in millimeters versus the error in temperature entered into an EDM for various lengths of sight. Notice that an error of 1 mm can occur for all distances over 50 m when the temperature error is more than 9°C. This temperature difference can easily occur during certain times of the year between an early morning, midday, and late afternoon. Also note that for *a* this error will occur with only a 3°C temperature error for sight lengths that are greater than 300 m.

6.22.2 Instrumental Errors

If EDM equipment is carefully adjusted and precisely calibrated, instrumental errors should be extremely small. To assure their accuracy and reliability, EDM instruments

TABLE 6.3 ERROR IN OBSERVED DISTANCE IN MILLIMETERS VERSUS ERROR IN TEMPERATURE FOR VARIOUS SIGHT LENGTHS

Error in Temperature		Length of Sights (m)				
°C	°F	100	200	300	400	500
3	5.4	0.3	0.6	0.9	1.2	1.5
6	10.8	0.6	1.2	1.8	2.4	3
9	16.2	0.9	1.8	2.7	3.6	4.5
12	21.6	1.2	2.4	3.6	4.8	6
15	27.0	1.5	3	4.5	6	7.5
30	54.0	3	6	9	12	15

should be checked against a first-order baseline at regular time intervals. For this purpose, the National Geodetic Survey has established a number of accurate baselines in each state.[6] These are approximately a mile in length and placed in relatively flat areas. Monuments are set at the ends and at intermediate points along the baseline.

Although most EDM instruments are quite stable, occasionally they become maladjusted and generate erroneous frequencies. This results in faulty wavelengths that degrade distance measurements. Periodic checking of the equipment against a calibrated baseline will detect the existence of observational errors. It is especially important to make these checks if high-order surveys are being conducted.

The corner cube reflectors used with EDM instruments are another source of instrumental error. Since light travels at a lower velocity in glass than in air, the "effective center" of the reflector is actually behind the prism. Thus, it frequently does not coincide with the plummet, a condition that produces a systematic error in distances known as the *reflector constant offset*. This situation is shown in Figure 6.15. Notice that because the retroreflector is comprised of mutually perpendicular faces, the light always travels a total distance of $a + b + c = 2D$ in the prism. Additionally, given a refractive index for glass, which is greater than air, the velocity of light in the prism is reduced following Equation (6.8) to create an effective distance of nD where n is the index of refraction of the glass (approximately 1.517). The dashed line in Figure 6.15 shows the effective center thus created. The reflector constant, K in the figure, can be as large as 70 mm, and will vary with reflectors.

Once known, the *electrical center* of the EDM can be shifted forward to compensate for the reflector constant. However, if an EDM instrument is being used regularly with several unmatched reflectors, this shift is impractical. In this instance, the offset for each reflector should be subtracted from the observed distances to obtain corrected values.

With EDM instruments that are components of total stations and are controlled by microprocessors, this constant can be entered via the keyboard and included in the internally computed corrections. Equipment manufacturers also produce matching reflector sets for which the reflector constant is the same, thus allowing a single constant to be used for a set of reflectors with an instrument.

By comparing precisely known baseline lengths to observed distances, a so-called *system measurement constant* can be determined. This constant can then be applied to all subsequent observations for proper correction. Although calibration using a baseline is preferred, if one is not available, the constant can be obtained with the following procedure. Three stations, A, B, and C should be established in a straight line on flat ground, with stations A and C at a distance that is multiple units of the fundamental wavelength of the instrument apart. The fundamental wavelength of most instruments today is typically 10 m. Station B should be on the line between stations A and C and also at a multiple of the fundamental wavelength of the EDM. For example, the lengths AB and BC could be set at 40 m and 60 m, respectively, for an instrument with a fundamental wavelength of 10 m. The length of AC and the two components, AB and BC, should be observed several times with

---

[6]For locations of baselines in your area, contact the NGS National Geodetic Information Center by email at: info_center@ngs.noaa.gov; at their website address: http://www.ngs.noaa.gov/CBLINES/calibration.html; by telephone at (301)713-3242; or by writing to NOAA, National Geodetic Survey, Station 09202, 1315 East West Highway, Silver Spring, MD 20910.

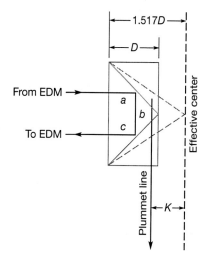

Figure 6.15
Schematic of retroreflector where D is the depth of the prism.

the instrument-reflector constant set to zero and the means of each length determined. From these observations, the following equation can be written:

$$AC + K = (AB + K) + (BC + K)$$

from which

$$K = AC - (AB + BC) \qquad (6.20)$$

where K is the system measurement constant to be added to correct the observed distances.

The procedure, including centering of the EDM instrument and reflector, should be repeated several times very carefully, and the average value of K adopted. Since different reflectors have varying offsets, the test should be performed with any reflector that will be used with the EDM, and the results marked on each to avoid confusion later. For the most precise calibration, lengths AB and BC should be carefully laid out as even multiples of the instrument's shortest measurement wavelength. Failure to do this can cause an incorrect value of K to be obtained. As shown in Figure 6.15, due to the construction of the reflector and the pole being located near the center of the reflector, the system measurement constant is typically negative.

While the above procedure provides method for determining a specific instrument-reflector constant, it is highly recommended that EDM instruments be calibrated using NGS calibration baselines. These baselines have been established throughout the country for use by surveyors. Their technical manual *Use of Calibration Base Lines*, which is listed in the bibliography at the end of the chapter, provides guidelines on the use of the baselines and reduction of the observations providing both the instrument-reflector offset constant and a scaling factor.

Another source of error can come from the level vial on the reflector rod being out of adjustment or not precisely centered at the time of the observation. As shown in Figure 6.16, a 4' to 6' error in leveling will cause an error in the centering of the reflector over the point of 0.01 ft or greater. This error is then propagated to the error in the observed horizontal distance by Equation (3.12). Also shown is that as the height of the reflector becomes greater, so does the amount of error in the observed horizontal distance. Thus, care should always be taken to ensure that the circular bubble on the rod is centered and calibrated. Additionally, this check of the centering of the level vial should be checked often or at regular intervals at a minimum.

6.22.3 Natural Errors

Natural errors in EDM operations stem primarily from atmospheric variations in temperature, pressure, and humidity, which affect the index of refraction, and modify the wavelength of electromagnetic energy. The values of these variables must be measured and used to correct observed distances. As demonstrated in Example 6.1, humidity can generally be neglected when using electro-optical instruments, but this variable was important when microwave instruments were employed.

The National Weather Service adjusts atmospheric pressure readings to sea level values. Since atmospheric pressure changes by approximately 1 in. of mercury (Hg) per 1000 ft of elevation, under no circumstances should radio broadcast values for atmospheric pressure be used to correct distances. Instead, atmospheric pressure should be measured by an aneroid barometer that is calibrated against a barometer not corrected to sea level. Many high school and college physics departments have these barometers.

EDM instruments within total stations have onboard microprocessors that use atmospheric variables, input through the keyboard, to compute corrected distances after making observations but before displaying them. For older instruments, varying the transmission frequency made corrections, or they could be computed manually after the observation. Equipment manufacturers provided tables and charts that assisted in this process. The magnitude of error in electronic distance measurement due to errors in observing atmospheric pressure and temperature is indicated in Figure 6.17. Note that a 10°C temperature error, or a pressure difference of 25 mm (1 in.) of mercury, each produce a distance error of about 10 ppm. Thus, if a radio broadcast atmospheric pressure is entered into an EDM in Denver, CO, the resulting distance error could be as great as 50 ppm and a 200 m distance could in error by as much as 1 cm.

Figure 6.16 Errors in EDM produced by temperature and pressure errors (based on atmospheric temperature and pressure of 15° and 760 mm of mercury).

A microclimate can exist in the layers of atmosphere immediately near a surface such as the ground. Field experiments prove that temperatures on or near the ground may be 10° to 25° higher or lower than those at shoulder height. Since this microclimate can substantially change the index of refraction, it is important to maintain a line of sight that is at least 0.5 m above the surface of the ground or any intervening object. On long lines of sight, the observer should be cognizant of intervening ridges or other objects that may exist between the instrument and reflector, which could cause problems in meeting this condition. If this condition cannot be met, the height of the reflector may be increased. Under certain conditions, it may be necessary to set an intermediate point on the encroaching surface to ensure that light from the EDM does not travel through these layers. This problem also exists near buildings and in parking lots where microclimates are created around the exterior of the buildings and vehicles. Again, it is prudent to keep the line of sight away from any intervening objects.

For the most precise work, on long lines, a sampling of the atmospheric conditions along the line of sight should be observed. In this case, it may be necessary to elevate the meteorological instruments. This can be difficult where the terrain becomes substantially lower than the sight line. In these cases, the atmospheric measurements for the ends of the line can be measured and averaged.

■ 6.23 USING SOFTWARE

On the companion website at http://www.pearsonhighered.com/ghilani is the Excel® spreadsheet *c6.xls*. This spreadsheet demonstrates the computations in Example 6.1 as well as tape corrections for systematic errors. For those wishing to see this programmed in a higher-level language, a Mathcad® worksheet *C6.xmcd* is available on the companion website also. This worksheet additionally demonstrates Example 6.2.

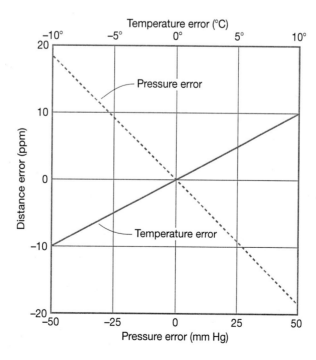

Figure 6.17

PROBLEMS

Asterisks (*) indicate problems that have partial answers given in Appendix G.

6.1 Describe a field procedure where pacing a distance is appropriate for distance measurement.

6.2* A student counted 92, 90, 92, 91, 93, and 91 paces in six trials of walking along a course of 200-ft known length on level ground. Then 85, 86, 86, and 84 paces were counted in walking four repetitions of an unknown distance AB. What is **(a)** the pace length and **(b)** the length of AB?

6.3 What difference in temperature from standard, if neglected in use of a steel tape, will cause an error of 1 part in 3000?

6.4 An add tape of 101 ft is incorrectly recorded as 100 ft resulting in a recorded 300-ft distance. What is the correct distance?

6.5* List five types of common errors in taping.

6.6 List the proper procedures taping a horizontal distance of about 84 ft down a 10% slope.

6.7 For the following data, compute the horizontal distance for a recorded slope distance AB,
  (a) $AB = 175.29$ ft, slope angle $= 8°26'32''$
  (b) $AB = 86.084$ m, difference in elevation A to $B = 6.849$ m.

6.8* When measuring a distance AB, the first taping pin was placed 0.45 ft to the right of line AB and the second pin was set 0.35 ft left of line AB. The recorded distance was 253.42 ft. Calculate the corrected distance. (Assume three taped segments, the first two 100 ft each.)

6.9 List the possible errors that can occur when measuring a distance with an EDM.

6.10 Discuss the advantages of making an EDM distance measurement compared to taping.

6.11 Discuss why the sight line for electronic distance measurement should be at least 0.5 m away from a parked vehicle near the line of sight.

6.12* Assume the speed of electromagnetic energy through the atmosphere is 299,836,182 m/sec for measurements with an EDM instrument. What time lag in the equipment will produce an error of 800 m in a measured distance?

6.13 What is the length of the partial wavelength for electromagnetic energy with a frequency of the 29.969 MHz and a phase shift of 245°?

6.14 What "actual" wavelength results from transmitting electromagnetic energy through an atmosphere having an index of refraction of 1.0043, if the frequency is:
  ***(a)** 29.988 MHz
  (b) 2.99989 MHz

6.15 What index of refraction in the atmosphere would produce the speed of light given in Problem 6.12?

6.16 To calibrate an EDM instrument, distances AC, AB, and BC along a straight line were observed as 98.567 m, 60.029 m, and 38.570 m, respectively. What is the system measurement constant for this equipment? Compute the length of each segment corrected for the constant.

6.17 Which causes a greater error in a line measured with an EDM instrument: **(a)** A disregarded 10°C temperature variation from standard or **(b)** a neglected atmospheric pressure difference from standard of 25 mm of mercury?

6.18* In Figure 6.14, h_e, h_r, z, $elev_A$, $elev_B$, and the measured slope length L were 5.56, 6.00, 603.45, 589.06, and 408.65 ft, respectively. Calculate the horizontal length between A and B.

6.19 Similar to Problem 6.18, except that the values were 1.676, 1.589, 987.623, 1006.324, and 435.687 m, respectively.

6.20 In Figure 6.14, h_e, h_r, z, and the measured slope length L were 5.55 ft, 5.50 ft, 96°18′56″, and 403.53 ft, respectively. Calculate the horizontal length between A and B if a total station measures the distance.

6.21* Similar to Problem 6.20, except that the values were 1.45 m, 1.55 m, 96°05′33″, and 1663.254 m, respectively.

6.22 What is the actual wavelength and velocity of a near-infrared beam ($\lambda = 0.901$ μm) of light modulated at a frequency of 318 MHz through an atmosphere with a dry bulb temperature, T, of 20°C; a relative humidity, h, of 88%; and an atmospheric pressure of 1034.56 hPa?

6.23 What is the actual wavelength and velocity of a near-infrared beam ($\lambda = 0.901$ μm) of light modulated at a frequency of 345 MHz through an atmosphere with a dry bulb temperature, T, of 32°C; a relative humidity, h, of 85%; and an atmospheric pressure of 1022 hPa?

6.24 The reflector rod with a 40′ circular level vial is being handheld with an estimated level error of 0.3 divisions. The reflector is at 6.00 ft ± 0.01 ft. What is the centering error of the reflector?

6.25* The standard deviation of taping a 30 m distance is ±5 mm. What should it be for a 90 m distance?

6.26 Determine the most probable length of a line AB, the standard deviation, and the 95% error of the measurement for the following series of taped observations made under the same conditions: 408.68, 408.68, 408.67, 408.68, 408.68, 408.65, 408.65, 408.68, 408.69, and 408.66 ft.

6.27 If an EDM instrument has a purported accuracy capability of ±(3.0 mm + 0.5 ppm), what error can be expected in a measured distance of **(a)** 20 m, **(b)** 804.98 ft, **(c)** 345.876 m? (Assume that the instrument and target miscentering errors are equal to zero.)

6.28 The estimated error for both instrument and target miscentering errors is ±1.2 mm. For the EDM in Problem 6.27, what is the estimated error in the observed distances?

6.29 If a certain EDM instrument has an accuracy capability of ±(2 mm + 2 ppm), what is the precision of measurements, in terms of parts per million, for line lengths of: **(a)** 50.000 m, **(b)** 500.000 m, **(c)** 5000.000 m? (Assume that the instrument and target miscentering errors are equal to zero.)

6.30 The estimated error for both instrument and target miscentering errors is ±1.5 mm. For the EDM and distances listed in Problem 6.29, what is the estimated error in each distance? What is the precision of the measurements in terms of parts per million?

6.31 Create a computational program that solves Problem 6.22.

BIBLIOGRAPHY

Anunciation, D. C., et al. 2012. "EDM Calibration and Finding Errors in the Baseline." *Surveying and Land Information Science* 72 (No. 4): 189.

Ernst, C. M. 2009. "Direct Reflex vs. Standard Prism Measurements." *The American Surveyor* 6 (No. 4): 48.

Fonczek, C. J. 1980. *Use of Calibration Base Lines*. NOAA Technical Memorandum NOS NGS-10.

Franklin, K. and T. Meyer. 2016. "Centering Error for Range Poles." *Surveying and Land Information Science* 75 (No. 2): 77.

GIA, 2001. "EDM PPM Settings." *Professional Surveyor* 21 (No. 6): 26.

_____. 2002. "EDM Calibration." *Professional Surveyor* 22 (No. 7): 50.

_____. 2003. "Phase Resolving EDMs." *Professional Surveyor* 23 (No. 10): 34.

Lemmens, Mathias. 2016. "EDM without Reflectors." *The American Surveyor* 13 (No. 8): 38.

Reilly, J. 2010. "Improving Geodetic Field Surveying Techniques." *2010 PSLS Surveyors' Conference*. Hershey, PA.

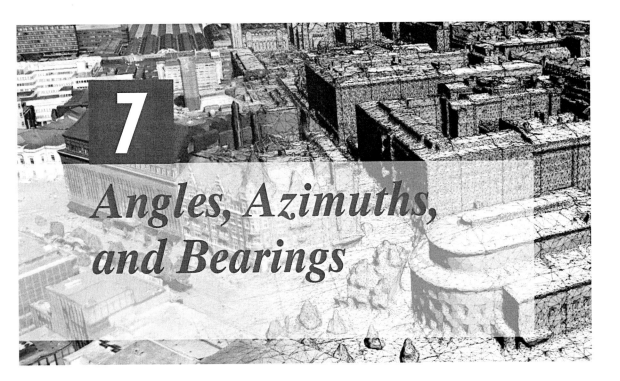

7
Angles, Azimuths, and Bearings

■ 7.1 INTRODUCTION

Determining the locations of points and orientations of lines frequently depends on the observation of angles and directions. In surveying, directions are given by *azimuths* and *bearings* (see Sections 7.5 and 7.6).

As described in Section 2.1, and illustrated in Figure 2.1, angles observed in surveying are classified as either *horizontal* or *vertical*, depending on the plane in which they are measured. Horizontal angles are the basic observations needed for determining bearings and azimuths. Vertical angles are used in trigonometric leveling and for the reduction of distances to horizontal (see Sections 6.23 and 19.14.2).

Angles are most often *directly* observed in the field with total station instruments, although in the past transits, theodolites, and compasses have been used. (The surveyor's compass is described in Section 7.10.) Three basic requirements determine an angle. As shown in Figure 7.1, they are (1) *reference* or *starting line*, (2) *direction of turning*, and (3) *angular distance* (value of the angle). Methods of computing bearings and azimuths described in this chapter are based on these three elements.

■ 7.2 UNITS OF ANGLE MEASUREMENT

A purely arbitrary unit defines the value of an angle. The *sexagesimal* system used in the United States and in many other countries is based on degrees, minutes, and seconds, with the last unit further divided decimally. In Europe the *grad* or *gon* is commonly used (see Section 2.2). Radians are more suitable in computer computations, but the sexagesimal system continues to be used in most U.S. surveys.

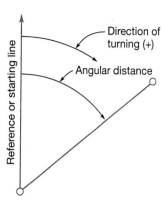

Figure 7.1
Basic requirements in determining an angle.

7.3 KINDS OF HORIZONTAL ANGLES

The kinds of horizontal angles most commonly observed in surveying are (1) *interior angles*, (2) *angles to the right*, and (3) *deflection angles*. Because they differ considerably, the kind used must be clearly indicated in field notes. Interior angles, shown in Figure 7.2, are observed on the inside of a *closed polygon*. Normally the angle at each apex within the polygon is measured. Then, as discussed in Section 9.7, a check can be made on their values because the sum of all interior angles in any polygon must equal $(n - 2)180°$, where n is the number of angles. Polygons are commonly used for boundary surveys and many other types of work since they provide a computational check on the quality of the work. Surveyors (geomatics engineers) normally refer to them as *closed traverses*.

Exterior angles, located outside a closed polygon, are explements of interior angles. The advantage to be gained by observing them is their use as another computational check, since the sum of the interior and exterior angles at any station must total 360°.

Angles to the right are measured *clockwise from the rear to the forward station*. Note: As a survey progresses, stations are commonly identified by consecutive alphabetical letters (as in Figure 7.2), or by increasing numbers. Thus, the interior angles of Figure 7.2(a) are also angles to the right. Most data collectors require that angles to the right be observed in the field. *Angles to the left*, turned

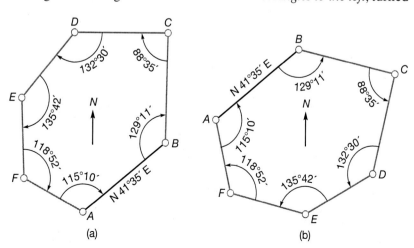

Figure 7.2
Closed polygon. (a) Clockwise interior angles (angles to the right). (b) Counterclockwise interior angles (angles to the left).

counterclockwise from the rear station, are illustrated in Figure 7.2(b). Note that the polygons of Figure 7.2 are "right" and "left"—that is, similar in shape but turned over like the right and left hands. Figure 7.2(b) is shown only to emphasize a serious mistake that occurs if counterclockwise angles are observed and recorded and then assumed to be clockwise later on. *To avoid this confusion, it is recommended that a uniform procedure of always observing angles to the right be adopted*, and the direction of turning noted in the field book with a sketch.

Angles to the right can be either interior or exterior angles of a closed-polygon traverse. Whether the angle is an interior or exterior angle depends on the direction the instrument proceeds around the traverse. If the direction around the traverse is counterclockwise, then the angles to the right will be interior angles. However, if the instrument proceeds clockwise around the traverse, then exterior angles will be observed. If this is the case, the sum of the exterior angles for a closed-polygon traverse will be $(n + 2)180°$. Analysis of a simple sketch should make these observations clear.

Deflection angles (Figure 7.3) are observed from an extension of the back line to the forward station. They are used principally on the long linear alignments of route surveys. As illustrated in the figure, deflection angles may be observed to the right (clockwise) or to the left (counterclockwise) depending on the direction of the route. Clockwise angles are considered plus, and counterclockwise ones minus, as shown in the figure. Deflection angles are always smaller than 180° and appending an R or L to the numerical value identifies the direction of turning. Thus, the angle at B in Figure 7.3 is (R), and that at C is (L). Deflection angles are the only exception where counterclockwise observation of angles should be made. In a closed polygon traverse, the sum of the deflection angles should be 360°.

■ 7.4 DIRECTION OF A LINE

The direction of a line is defined by the horizontal angle between the line and an arbitrarily chosen reference line called a *meridian*. Different meridians are used

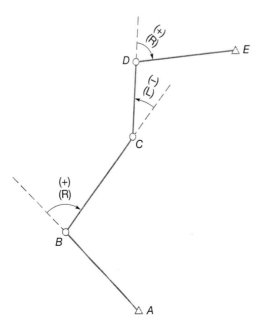

Figure 7.3
Deflection angles.

for specifying directions including (a) geodetic, (b) astronomic, (c) magnetic (also often called *true*), (d) grid, (e) record, and (f) assumed.

The *geodetic* meridian is the north–south reference line that passes through a mean position of the Earth's geographic poles. Since the poles wander, the positions of the poles are defined as their mean locations between the period of 1900.0 and 1905.0 (see Section 19.3).

Wobbling of the Earth's rotational axis, also discussed in Section 19.3, causes the position of the Earth's geographic poles to vary with time. At any point, the *astronomic* meridian is the north–south reference line that passes through the instantaneous position of the Earth's geographic poles. Astronomic meridians derive their name from the field operation to obtain them, which consists in making observations on the celestial objects, as described in Appendix C. Geodetic and astronomic meridians are very nearly the same, and the former can be computed from the latter by making small corrections (see Sections 19.3 and 19.5).

A *magnetic* meridian is defined by a freely suspended magnetic needle that is only influenced by the Earth's magnetic field. The Earth's magnetic poles move with time, and thus the positions of magnetic meridians are dependent on the time of the observation. Magnetic meridians are discussed in Section 7.10.

Surveys based on a state plane or other map projection coordinate systems employ a *grid* meridian for reference. Grid north is the direction of geodetic north for a selected *central meridian*, and held parallel to it over the entire area covered by a map projection coordinate system (see Chapter 20).

In boundary surveys, the term *record* meridian refers to directional references quoted in the recorded documents from a previous survey of a particular parcel of land. Another similar term, *deed* meridian, is used in the description of a parcel of land as recorded in a property deed. Chapters 21 and 22 discuss the use of record meridians and deed meridians in boundary retracement surveys.

An *assumed* meridian can be established by merely assigning any arbitrary direction—for example, taking a certain street line to be north. The directions of all other lines are then found in relation to it.

From the above definitions, it should be obvious that the terms north or due north, if used in a survey, must be defined, since they do not specify a unique line. This additional information such as all directions are based on geodetic north as defined by the inverse of two GNSS-established control points defined by the North American Datum of 1983 (2011) is extremely important for the interpretation and future use of the product.

7.5 AZIMUTHS

Azimuths are horizontal angles observed clockwise from any reference meridian. In plane surveying, azimuths are generally observed from north, but astronomers and the military have used south as the reference direction. The National Geodetic Survey (NGS) also used south as its reference for azimuths for NAD27, but north has been adopted for NAD83 (see Section 19.6). Examples of azimuths observed from north are shown in Figure 7.4. As illustrated, they can range from 0° to 360° in value. Thus, the azimuth of *OA* is 70°; of *OB*, 145°; of *OC*, 235°; and of *OD*, 330°. Azimuths may be *geodetic*, *astronomic*, *magnetic*, *grid*, *record*, or *assumed*, depending on the reference meridian used. To avoid any confusion, it is necessary

168 ANGLES, AZIMUTHS, AND BEARINGS

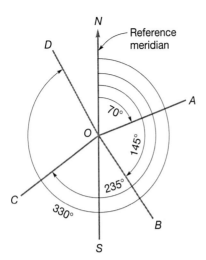

Figure 7.4
Azimuths.

to state in the field notes at the beginning of work, and on any documents or maps produced with this data, what reference meridian applies for azimuths, and whether they are observed from north or south.

A line's forward direction can be given by its *forward* azimuth, and its reverse direction by its *back* azimuth. In plane surveying, forward azimuths are converted to back azimuths, and vice versa, by adding or subtracting 180°. For example, if the azimuth of *OA* is 70°, the azimuth of *AO* is 70° + 180° = 250°. If the azimuth of *OC* is 235°, the azimuth of *CO* is 235° − 180° = 55°. However, as discussed in Sections 19.13.2 and 20.8.2, the convergence of the Earth's meridians must be taken into account for surveys covering large regions since forward and reverse geodetic azimuths typically differ by something other than 180°.

Azimuths can be read directly on the graduated circle of a total station instrument after the instrument has been oriented properly. As explained in Section 9.2.4, this can be done by sighting along a line of known azimuth with that value indexed on the circle, and then turning to the desired course. Azimuths are used advantageously in boundary, topographic, control, and other kinds of surveys, as well as in computations.

■ 7.6 BEARINGS

Bearings are another system for designating directions of lines. *The bearing of a line is defined as the acute horizontal angle between a reference meridian and the line.* The angle is observed from either the north or south toward the east or west, to give a reading smaller than 90°. The letter N or S preceding the angle, and E or W following it shows the proper quadrant. Thus, a properly expressed bearing includes quadrant letters and an angular value. An example is N80°E. In Figure 7.5, all bearings in quadrant *NOE* are measured clockwise from the meridian. Thus, the bearing of line *OA* is N70°E. All bearings in quadrant *SOE* are counterclockwise from the meridian, so *OB* is S35°E with the bearing angle computed as 180° − 145°. Similarly, the bearing of *OC* is S55°W, with the bearing angle computed as 235° − 180°, and that of *OD*, N30°W with the bearing angle

Figure 7.5
Bearing angles.

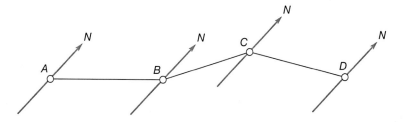

Figure 7.6
Forward and back bearings.

computed as 360° − 330°. When lines are in the cardinal directions, the bearings should be listed as "Due North," "Due East," "Due South," or "Due West."

Geodetic bearings are observed from the geodetic meridian, *astronomic bearings* from the local astronomic meridian, *magnetic bearings* from the local magnetic meridian, *grid bearings* from the appropriate grid meridian, and *assumed bearings* from an arbitrarily adopted meridian. The magnetic meridian can be obtained in the field by observing the needle of a compass, and used along with observed angles to get computed magnetic bearings. For this reason, many early boundary surveys were performed with a surveyor's compass, which is discussed in Section 7.10.

In Figure 7.6, assume that a compass is set up successively at points *A*, *B*, *C*, and *D* and bearings read on lines *AB*, *BA*, *BC*, *CB*, *CD*, and *DC*. As previously noted, bearings *AB*, *BC*, and *CD* are *forward bearings*; those of *BA*, *CB*, and *DC*, *back bearings*. Back bearings should have the same numerical values as forward bearings but opposite letters. Thus, if bearing *AB* is N44°E, bearing *BA* is S44°W.

■ 7.7 COMPARISON OF AZIMUTHS AND BEARINGS

Because bearings and azimuths are encountered in so many surveying operations, the comparative summary of their properties given in Table 7.1 should be helpful. Bearings are readily computed from azimuths by noting the quadrant in which the azimuth falls, then converting as shown in the table.

TABLE 7.1 COMPARISON OF AZIMUTHS AND BEARINGS

Azimuths	Bearings
Vary from 0 to 360°	Vary from 0 to 90°
Require only a numerical value	Require two letters and a numerical value
May be geodetic, astronomic, magnetic, grid, assumed, forward or back	Same as azimuths
Are measured clockwise only	Are measured clockwise and counterclockwise
Are measured either from north only, or from south only on a particular survey	Are measured from north and south

Quadrant	Formulas for computing bearing angles from azimuths
I (NE)	Bearing = Azimuth
II (SE)	Bearing = 180° − Azimuth
III (SW)	Bearing = Azimuth − 180°
IV (NW)	Bearing = 360° − Azimuth

Example directions for lines in the four quadrants (azimuths from north)

Azimuth	Bearing
54°	N54°E
112°	S68°E
231°	S51°W
345°	N15°W

■ Example 7.1

The azimuth of a boundary line is 128°13′46″. Convert this to a bearing.

Solution

The azimuth places the line in the southeast quadrant. Thus, the bearing angle is
$$180° - 128°13'46'' = 51°46'14''$$
and the equivalent bearing is S51°46′14″E.

■ Example 7.2

The first course of a boundary survey is written as N37°13′W. What is its equivalent azimuth?

Solution

Since the bearing is in the northwest quadrant, the azimuth is

$$360° - 37°13' = 322°47'.$$

■ 7.8 COMPUTING AZIMUTHS

Most types of surveys, but especially those that employ traversing, require computation of azimuths (or bearings). A traverse, as described in Chapter 9, is a series of connected lines whose lengths and angles at the junction points have been observed. Figures 7.2 and 7.3 illustrate examples. Traverses have many uses. To survey the boundary lines of a piece of property, for example, a "closed-polygon" type traverse like that of Figure 7.2(a) would normally be used. A highway survey from one city to another would usually involve a traverse like that of Figure 7.3. Regardless of the type used, it is necessary to compute the directions of its lines.

Many surveyors prefer azimuths to bearings for directions of lines because they are easier to work with, especially when calculating traverses with computers. Also sines and cosines of azimuth angles provide correct algebraic signs for departures and latitudes as discussed in Section 10.4.

Azimuth calculations are best made with the aid of a sketch. Figure 7.7 illustrates computations for azimuth *BC* in Figure 7.2(a). Azimuth *BA* is found by adding 180° to azimuth *AB*: 180° + 41°35' = 221°35' to yield its back azimuth. Then the angle to the right at *B*, 129°11', is added to azimuth *BA* to get azimuth

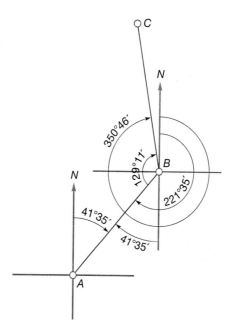

Figure 7.7
Computation of azimuth *BC* of Figure 7.2(a).

BC: 221°35′ + 129°11′ = 350°46′. This general process of adding (or subtracting) 180° to obtain the back azimuth and then adding the angle to the right is repeated for each line until the azimuth of the starting line is recomputed. If a computed azimuth exceeds 360°, then 360° is subtracted from it and the computations are continued. These calculations are conveniently handled in tabular form, as illustrated in Table 7.2. This table lists the calculations for all azimuths of Figure 7.2(a). Note that a check was secured by recalculating the beginning azimuth using the last angle. The procedures illustrated in Table 7.2 for computing azimuths are systematic and readily programmed for computer solution. The reader can view a Mathcad® worksheet *Azs.xmcd* on the companion website to review these computations.

Traverse angles must be adjusted to the proper geometric total before azimuths are computed. As noted earlier, in a closed-polygon traverse, the sum of interior angles equals $(n - 2)180°$, where n is the number of angles or sides. If the traverse angles fail to close by say 10″ and are not adjusted prior to computing azimuths, the original and computed check azimuth of *AB* will differ by the same 10″, assuming there are no other calculating errors. *The azimuth of any starting course should always be recomputed as a check using the last angle.* Any discrepancy shows that (a) an arithmetic error was made or (b) the angles were not properly adjusted prior to computing azimuths.

TABLE 7.2 Computation of Azimuths (from North) for Lines of Figure 7.2(a)

Angles to the Right [Figure 7.2(a)]

41°35′ = AB	211°51′ = DE
+ 180°00′	−180°00′
221°35′ = BA	31°51′ = ED
+ 129°11′	+ 129°11′
350°46′ = BC	167°33′ = EF
−180°00′	+ 180°00′
170°46′ = CB	347°33′ = FE
+ 88°35′	+ 118°52′
259°21′ = CD	*466°25′ − 360° = 106°25′ = FA
−180°00′	−180°00′
79°21′ = DC	286°25′ = AF
+ 132°30′	+ 115°10′
211°51′ = DE	*401°35′ − 360° = 41°35′ = AB ✓

*When a computed azimuth exceeds 360°, the correct azimuth is obtained by merely subtracting 360°.

■ 7.9 Computing Bearings

Drawing sketches similar to those in Figure 7.8 showing all data simplify computations for bearings of lines. In Figure 7.8(a), the bearing of line AB from Figure 7.2(a) is $N41°35'E$, and the angle at B turned clockwise (to the right) from known line BA is $129°11'$. Then the bearing angle of line BC is $180° - (41°35' + 129°11') = 9°14'$, and from the sketch the bearing of BC is $N9°14'W$.

In Figure 7.8(b), the clockwise angle at C from B to D was observed as $88°35'$. The bearing of CD is $88°35' - 9°14' = S79°21'W$. Continuing this technique, the bearings in Table 7.3 have been determined for all lines in Figure 7.2(a).

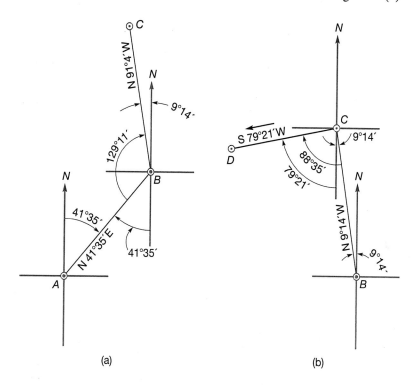

Figure 7.8
(a) Computation of bearing BC of Figure 7.2(a).
(b) Computation of bearing CD of Figure 7.2(a).

TABLE 7.3 BEARINGS OF LINES IN FIGURE 7.2(a)

Course	Bearing
AB	N41°35'E
BC	N9°14'W
CD	S79°21'W
DE	S31°51'W
EF	S12°27'E
FA	S73°35'E
AB	N41°35'E ✓

In Table 7.3, note that the last bearing computed is for *AB*, and it is obtained by employing the 115°10′ angle observed at *A*. It yields a bearing of N41°35′E, which agrees with the starting bearing. Students should compute each bearing of Figure 7.2(a) to verify the values given in Table 7.3.

An alternate method of computing bearings is to determine the azimuths as discussed in Section 7.8, and then convert the computed azimuths to bearings using the techniques discussed in Section 7.7. For example in Table 7.2, the azimuth of line *CD* is 259°21′. Using the procedure discussed in Section 7.7, the bearing angle is 259°21′ − 180° = 79°21′, and the bearing is S79°21′W.

Bearings, rather than azimuths, are used predominately in boundary surveying. This practice originated from the period of time when the magnetic bearings of parcel boundaries were determined directly using a surveyor's compass (see Section 7.10). Later, although other instruments (i.e., transits and theodolites) were used to observe the angles, and the astronomic meridian was more commonly used, the practice of using bearings for land surveys continued, and is still in common use today. Because boundary retracement surveyors *must follow the footsteps of the original surveyor* (see Chapter 21), they need to understand magnetic directions and their nuances. The following sections discuss magnetic directions, and explain how to convert directions from magnetic to other reference meridians, and vice versa.

■ 7.10 THE COMPASS AND THE EARTH'S MAGNETIC FIELD

Before transits, theodolites, and total station instruments were invented, directions of lines and angles were determined using compasses. Most of the early land-surveying work in the United States was done using these venerable instruments. Figure 7.9(a) shows the *surveyor's compass*. The instrument consists of a

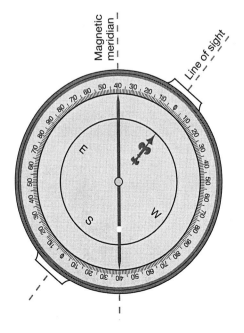

Figure 7.9
(a) Surveyor's compass. (Courtesy B Christopher / Alamy Stock Photo.)
(b) Compass box.

metal baseplate (A) with two sight vanes (B) at the ends. The compass box (C) and two small level vials (D) are mounted on the baseplate, the level vials being perpendicular to each other. When the compass was set up and the bubbles in the vials centered, the compass box was horizontal and ready for use.

A single leg called a Jacob staff supported early compasses. A ball-and-socket joint and a clamp were used to rotate the instrument and clamp it in its horizontal position. Later versions, such as that shown in Figure 7.9(a), were mounted on a tripod. This arrangement provided greater stability.

The compass box of the surveyor's compass was covered with glass to protect the magnetized steel needle inside. The needle was mounted on a pivot at the center of a circle that was graduated in degrees. A top view of a surveyor's compass box with its graduations is illustrated in Figure 7.9(b). In the figure, the zero graduations are at the north and south points of the compass and in line with the two sight-vane slits that comprise the line of sight. Graduations are numbered in multiples of 10° clockwise and counterclockwise from 0° at the north and south, to 90° at the east and west.

In using the compass, the sight vanes and compass box could be revolved to sight along a desired line, and then its magnetic bearing could be read directly. Note in Figure 7.9(b) for example, that the needle is pointing north and that the line of sight is directed in a northeast direction. The magnetic bearing of the line, read directly from the compass, is N 40° E. (Note that the letters *E* and *W* on the face of the compass box are reversed from their normal positions to provide the direct readings of bearings.)

Unless disturbed by *local attraction* (a local anomaly caused from such things as power lines, railroad tracks, metallic belt buckles, and so on that affect the direction a compass needle points at any location), a compass needle is free to spin and align itself with the Earth's magnetic field pointing in the direction of the magnetic meridian (toward the magnetic north pole in the northern hemisphere).[1]

The magnetic forces of the Earth not only align the compass needle, but they also pull or dip one end of it below the horizontal position. The *angle of dip* varies from 0° near the equator to 90° at the magnetic poles. In the northern hemisphere, the south end of the needle is weighted with a very small coil of wire to balance the dip effect and keep it horizontal. The position of the coil can be adjusted to conform to the latitude in which the compass is used. Note the coil (dark spot) on the south end of the needle of the compass in Figure 7.9(b).

The Earth's magnetic field resembles that of a huge dipole magnet located at the Earth's center, with the magnet offset from the Earth's rotational axis by about 13°. This field has been observed at about 200 magnetic observatories around the world, as well at many other temporary stations. At each observation point both the field's intensity and its direction are measured. Based upon many years of this data, models of the Earth's magnetic field have been developed. These models are used to compute the *magnetic declination* and *annual change* (see Sections 7.11 and 7.12), which are elements of importance to surveyors. The accuracy of the

---

[1]The locations of the north and south geomagnetic poles are continually changing, and in 2020, they were located at approximately 80.65° north latitude and 72.68° west longitude, and 80.65° south latitude and 107.32° east longitude, respectively.

models is affected by several items including the locations of the observations, the types of rocks at the surfaces together with the underlying geological structures in the areas, and local attractions. Today's models give magnetic declinations that are accurate to within about 30 min of arc, however, local anomalies of 3° to 4°, or more, can exist in some areas.

■ 7.11 MAGNETIC DECLINATION

Magnetic declination is the horizontal angle observed from the geodetic meridian to the magnetic meridian. Navigators call this angle *variation* of the compass; the armed forces use the term *deviation*. An east declination exists if the magnetic meridian is east of geodetic north; a west declination occurs if it is west of geodetic north. East declinations are considered positive and west declinations negative. The relationship between geodetic north, magnetic north, and magnetic declination is given by the expression

$$\text{Geodetic azimuth} = \text{magnetic azimuth} + \text{magnetic declination} \quad (7.1)$$

Because the magnetic pole positions are constantly changing, magnetic declinations at all locations also undergo continual changes. Establishing a meridian from astronomical or satellite (GNSS) observations and then reading a compass while sighting along the observed meridian can obtain the current declination at any location obtained baring any local attractions. Another way of determining the magnetic declination at a point is to interpolate it from an *isogonic chart*. An isogonic chart shows magnetic declinations in a certain region for a specific epoch of time. Lines on such maps connecting points that have the same declination are called *isogonic lines*. The isogonic line along which the declination is zero (where the magnetic needle defines geodetic north as well as magnetic north) is termed the *agonic line*. Figure 7.10 is an isogonic chart covering for the year 2015. On that chart, the agonic line cuts through the central part of the United States. It is gradually moving westward. Points to the west of the agonic line have east declinations and points to the east have west declinations. As a memory aid, the needle can be thought of as pointing toward the agonic line. Note there is about a 40° difference in declination between the northeast portion of Maine and the northwest part of Washington. This is a huge change if a pilot flies by compass between the two states!

The National Geophysical Data Center (NGDC) creates models and maps of the changes in the Earth's magnetic field every five years. Their maps such as the one in Figure 7.10 can be viewed at http://www.ngdc.noaa.gov/geomag/WMM/images.shtml. Of additional interest to surveyors is the secular (annual) changes (see Section 7.12) in magnetic declination in a one-year period. Maps showing secular variation can be viewed at the same website.

■ 7.12 VARIATIONS IN MAGNETIC DECLINATION

It has been stated that magnetic declinations at any point vary over time. These variations can be categorized as *secular, daily, annual,* and *irregular*, and are summarized as follows.

7.12 Variations in Magnetic Declination 177

Figure 7.10
Isogonic lines from World Magnetic Model for 2015. This image is from the NOAA National Data Centers, NGDC on the Internet at http://www.ngdc.noaa.gov/seg/geomag/declination.shtml

Secular Variation. Because of its magnitude, this is the most important of the variations. Unfortunately, no physical law has been found to enable precise long-term predictions of secular variation, and its past behavior can be described only by means of detailed tables and charts derived from observations. Records, which have been kept at London for four centuries, show a range in magnetic declination from 11°E in 1580, to 24°W in 1820, back to 3°W in 2000. Secular variation changed the magnetic declination at Baltimore, MD, from 5°11′W in 1640 to 0°35′W in 1800, 5°19′W in 1900, 7°25′W in 1950, 8°43′W in 1975, and 11°01′W in 2000.

In retracing old property lines run by compass or based on the magnetic meridian, it is necessary to allow for the difference in magnetic declination at the time of the original survey and at the present date. The difference is attributed mostly to secular variation.

Daily Variation. Daily variation of the magnetic needle's declination causes it to swing through an arc averaging approximately 8′ for the United States. The needle reaches its extreme easterly position at about 8:00 A.M. and its most westerly position at about 1:30 P.M. Mean declination occurs at around 10:30 A.M. and 8:00 P.M. These hours and the daily variation change with latitude and season of the year. Usually the daily variation is ignored since it is well within the range of error expected in compass readings.

Annual Variation. This periodic swing is less than 1 min of arc and can be neglected. It must not be confused with the annual change (the amount of secular-variation change in one year) shown on some isogonic maps.

Irregular Variations. Unpredictable magnetic disturbances and storms can cause short-term irregular variations of a degree or more.

7.13 SOFTWARE FOR DETERMINING MAGNETIC DECLINATION

As noted earlier, direct observations are only applicable for determining current magnetic declinations. In most situations, however, magnetic declinations that existed years ago, for example, on the date of an old property survey, are needed in order to perform retracement surveys. Until recently these old magnetic declinations had to be interpolated from isogonic charts for the approximate time desired, and the lines of annual change used to correct to the specific year required. Now software is available that can quickly provide the needed magnetic declination values. The software uses models that were developed from historical records of magnetic declination and annual change, which have been maintained for the many observation stations throughout the United States and the world.

The program WOLFPACK, which is on the companion website, contains an option for computing magnetic field elements. This program uses models that span five or more year time frames. Using the World Magnetic Model of 2015 (file: WMM2015.DAT), the declination and annual change for Portland, Oregon on January 1, 2019 are determined to be about 15.1°E[2] and 8′W per year, respectively

---

[2]The software indicates west declination as negative, and east declination as positive.

Figure 7.11 Magnetic declination data entry screen in WOLFPACK setup to compute magnetic field values for Portland, Maine.

TABLE 7.4 PREDICTED MAGNETIC DECLINATION AND ANNUAL CHANGE FOR VARIOUS LOCATIONS IN THE UNITED STATES ON JANUARY 1, 2019

City	Magnetic Declination	Annual Change
Boston, MA	14°35′W	4′E
Cleveland, OH	8°20′W	1′W
Chicago, IL	3°59′W	4′W
Denver, CO	8°09′E	6′W
San Francisco, CA	13°24′E	6′W
Seattle, WA	15°34′E	9′W

(see the input data in Figure 7.11). Using this same program, the declinations for various other cities in the United States were determined for January 1, 2019, and are shown in Table 7.4. It is important when using this software to select the appropriate model file for the desired date. Select the appropriate model from a drop-down list for the "Model File." The models are given by their source, and the year. The latitude, longitude, and elevation of the station must be entered in the appropriate data boxes and the time of the desired computation is selected from the drop-down list at the bottom of the box. After computing the magnetic field elements for the particular location and time, the results are displayed for printing. Similar computations to determine magnetic declination and rates of annual change can be made by using the NOAA National Geophysical Data Centers' (NGDC) online computation page at http://www.ngdc.noaa.gov/geomag/magfield.shtml. The location of any U.S. city can be found with the U.S. Gazetteer, which is linked to the software, or can be obtained at http://www.census.gov/cgi-bin/gazetteer on the Internet page of the U.S. Census Bureau. It should be noted that all of these models are only accurate to the nearest 30 min and should be used with caution.

7.14 LOCAL ATTRACTION

Metallic objects and direct-current electricity, both of which cause a local attraction, affect the main magnetic field. As an example, when set up beside an old-time streetcar with overhead power lines, the compass needle would swing toward the car as it approached, then follow it until it was out of effective range. If the source of an artificial disturbance is fixed, all bearings from a given station will be in error by the same amount. However, angles calculated from bearings taken at the station will be correct.

Local attraction is present if the forward and back bearings of a line differ by more than the normal observational errors. Consider the following compass bearings read on a series of lines:

AB	N24°15'W
BC	N76°40'W
CD	N60°00'E
DE	N88°35'E
BA	S24°10'E
CB	S76°40'E
DC	S61°15'W
ED	S87°25'W

Forward-bearing AB and back-bearing BA agree reasonably well, indicating that little or no local attraction exists at A, or B. The same is true for point C. However, the bearings at D differ from corresponding bearings taken at C and E by roughly 1°15' to the west of north. Local attraction therefore exists at point D and deflects the compass needle by approximately 1°15' to the west of north.

It is evident that to detect local attraction, successive stations on a compass traverse have to be occupied, and forward and back bearings read, even though the directions of all lines could be determined by setting up an instrument only on alternate stations.

■ 7.15 TYPICAL MAGNETIC DECLINATION PROBLEMS

Typical problems in boundary surveys require the conversion of geodetic bearings to magnetic bearings, magnetic bearings to geodetic bearings, and magnetic bearings to magnetic bearings for the declinations existing at different dates. The following examples illustrate two of these types of problems.

■ Example 7.3

Assume the magnetic bearing of a property line was recorded as S43°30'E in 1862. At that time the magnetic declination at the survey location was 3°15'W. What geodetic bearing is needed for a subdivision property plan?

Solution

A sketch similar to Figure 7.12 makes the relationship clear and should be used by beginners to avoid mistakes. Geodetic north is designated by a full-headed long arrow and magnetic north by a half-headed shorter arrow. The geodetic bearing is

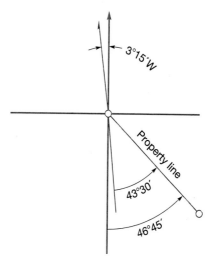

Figure 7.12
Computing geodetic bearings from magnetic bearings and declinations.

seen to be S43°30'E + 3°15' = S46°45'E. Using different colored pencils to show the direction of geodetic north, magnetic north, and lines on the ground helps clarify the sketch. Although this problem is done using bearings, Equation (7.1) could be applied by converting the bearings to azimuths. That is, the magnetic azimuth of the line is 136°30'. Applying Equation (7.1) using a negative declination angle results in a geodetic azimuth of 136°30' − 3°15' = 133°15', which correctly converts to the geodetic bearing of S46°45'E.

■ Example 7.4

Assume the magnetic bearing of line AB read in 1878 was N26°15'E. The declination at the time and place was 7°15'W. In 2000, the declination was 4°30'E. The magnetic bearing in 2000 is needed.

Solution

The declination angles are shown in Figure 7.13. The magnetic bearing of line AB is equal to the earlier date bearing minus the sum of the declination angles, or

$$N26°15'E - (7°15' + 4°30') = N14°30'E$$

Again, the problem can be computed using azimuths as 26°15' − (7°15' + 4°30') = 14°30', which converts to a bearing of N14°30'E.

182 ANGLES, AZIMUTHS, AND BEARINGS

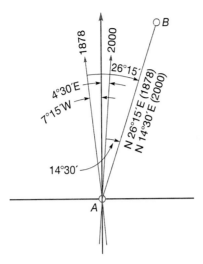

Figure 7.13
Computing magnetic bearing changes due to declination changes.

7.16 MISTAKES

Some mistakes made in using azimuths and bearings are:

1. Confusing magnetic and other reference bearings.
2. Mixing clockwise and counterclockwise angles.
3. Interchanging bearings for azimuths.
4. Listing bearings with angular values greater than 90°.
5. Failing to include both directional letters when listing a bearing.
6. Failing to change bearing letters when using the back bearing of a line.
7. Using an angle at the wrong end of a line in computing bearings—that is, using angle A instead of angle B when starting with line AB as a reference.
8. Not including the last angle to recompute the starting bearing or azimuth as a check—for example, angle A in traverse $ABCDEA$.
9. Subtracting 360°00′ as though it were 359°100′ instead of 359°60′, or using 90° instead of 180° in bearing computations.
10. Adopting an assumed reference line that is difficult to reproduce.
11. Reading degrees and decimals from a calculator as though they were degrees, minutes, and seconds.
12. Failing to adjust traverse angles before computing bearings or azimuths if there is a misclosure.

PROBLEMS

Asterisks (*) indicate problems that have partial answers given in Appendix G.

7.1 Describe the kinds and differences in horizontal angles.
7.2 List the three basic requirements in determining an angle.
7.3 Create a diagram with a four-sided polygon traverse that shows interior angles which are not angles to the right.
7.4 What is the relationship of a forward and back azimuth?
7.5 Convert: *(a) 203°26′48″ to grads, (b) 4.56783 radians to degrees, minutes, and seconds, (c) 316.35988 gons to degrees, minutes, and seconds.

In Problems 7.6 through 7.7, convert the azimuths from north to bearings, and compute the angles, smaller than 180° between successive azimuths.

7.6 47°35′00″, 126°40′12″, 205°17′17″, and 275°50′24″
7.7 82°46′51″, 166°36′28″, 193°15′06″, and 337°11′36″

Convert the bearings in Problems 7.8 through 7.9 to azimuths from north and compute the angle, smaller than 180°, between successive bearings.

7.8 N12°34′25″E, S44°59′06″E, S34°48′04″W, and N16°18′08″W
7.9 N66°26′24″E, S55°12′14″E, S68°14′48″W, and N64°03′16″W

Compute the azimuth from north of line CD in Problems 7.10 through 7.12. (Azimuths of AB are also from north.)

7.10* Azimuth $AB = 101°26′32″$ angles to the right $ABC = 50°54′26″, BCD = 38°36′38″$.
7.11 Bearing $AB = S29°15′54″W$; angles to the right $ABC = 15°13′56″$, $BCD = 154°47′08″$
7.12 Azimuth $AB = 336°17′58″$; angles to the right $ABC = 98°14′02″, BCD = 63°28′34″$.
7.13* For a bearing $DE = N8°53′56″W$ and angles to the right, compute the bearing of FG if angle $DEF = 88°12′29″$ and $EFG = 40°20′30″$.
7.14 Similar to Problem 7.13, except the azimuth of DE is 104°26′48″ and angles to the right DEF and EFG are 96°44′52″ and 102°15′14″, respectively.

Course AB of a five-sided traverse runs due north. From the given balanced interior angles to the right, compute and tabulate the bearings and azimuths from north for each side of the traverses in Problems 7.15 through 7.17.

7.15 $A = 122°22′35″, B = 106°25′58″, C = 108°45′22″, D = 95°16′28″$, and $E = 97°09′37″$
7.16* $A = 90°29′18″, B = 107°54′36″, C = 104°06′37″, D = 129°02′57″$, and $E = 108°26′32″$.
7.17 $A = 113°03′04″, B = 95°48′14″, C = 112°26′48″, D = 99°15′36″$, and $E = 119°26′18″$

In Problems 7.18 through 7.20, compute and tabulate the azimuths and bearings of the sides of a regular hexagon (polygon with six equal angles), given the starting direction of side AB and angles to the right.

7.18 Bearing of $AB = N65°26′41″E$
7.19 Azimuth of $AB = 289°16′22″$
7.20 Azimuth of $AB = 126°02′00″$

Compute and tabulate the azimuths and bearings of all lines for a closed traverse $ABCDA$ that has the following balanced angles to the right, using the directions listed in Problems 7.21 through 7.24. $DAB = 108°21′58″, ABC = 78°20′28″, BCD = 92°10′32″$, and $CDA = 81°07′02″$.

7.21 Bearing $AB = N15°37′10″W$.
7.22 Azimuth $AB = 88°15′59″$.
7.23 Bearing $AB = S68°46′08″W$.
7.24 Azimuth $AB = 344°48′32″$.
7.25 Geometrically show how the sum of the interior angles of a pentagon (five sides) can be computed using the formula $(n - 2)180°$.
7.26 Determine the predicted declinations on December 11, 2019, using the WMM-2015 model at the following locations.
  (a)* Latitude = 42°58′28″N, longitude = 77°12′36″W, elevation = 310.0 m;
  (b) Latitude = 37°56′44″N, longitude = 110°50′40″W, elevation = 1500.0 m;
  (c) Latitude = 41°18′15″N, longitude = 76°00′26″W, elevation = 240.0 m.

7.27 Using Table 7.4, what is the predicted total difference in magnetic declination between Boston, MA and Seattle, WA on January 1, 2019?

7.28 The magnetic declination at a certain place is 15°30′E. What is the magnetic bearing there: **(a)** of true north **(b)** of true south **(c)** of true east?

7.29 Same as Problem 7.28, except the magnetic declination at the place is 1°15′W.

For Problems 7.30 through 7.32 the observed magnetic bearing of line AB and its true magnetic bearing are given. Compute the amount and direction of local attraction at point A.

	Observed Magnetic Bearing	True Magnetic Bearing
7.30*	N32°30′E	N30°15′E
7.31	S56°45′W	S56°15′W
7.32	N13°40′W	N12°45′E

What magnetic bearing is needed to retrace a line for the conditions stated in Problems 7.33 through 7.36?

	1875 Magnetic Bearing	1875 Declination	Present Declination
7.33*	N32°45′E	8°12′W	2°30′E
7.34	S68°40′E	5°40′W	0°20′W
7.35	S24°23′E	12°30′W	16°15′W
7.36	N5°30′W	2°30′E	1°30′E

In Problems 7.37 through 7.38 calculate the magnetic declination in 1870 based on the following data from an old survey record.

	1870 Magnetic Bearing	Present Magnetic Bearing	Present Magnetic Declination
7.37	N14°20′E	N5°30′E	8°15′W
7.38	S40°30′E	S34°35′E	12°00′E

8 Total Station Instruments; Angle Observations

PART I • TOTAL STATION INSTRUMENTS

■ 8.1 INTRODUCTION

In the past, transits and theodolites were the most commonly used surveying instruments for making angle observations. These two devices were fundamentally equivalent and could accomplish basically the same tasks. Today, the total station instrument has replaced transits and theodolites. Total station instruments can accomplish all of the tasks that could be done with transits and theodolites and do them much more efficiently. In addition, they can also observe distances accurately and quickly and as discussed in Chapter 2 be connected to survey controllers. Furthermore, they can make computations with the angle and distance observations, and display the results in real time. These and many other significant advantages have made total stations the predominant instruments used in surveying practice today. They are used for all types of surveys including topographic, hydrographic, cadastral, and construction surveys. The use of total station instruments for specific types of surveys is discussed in later chapters. This chapter describes the general design and characteristics of total station instruments, and also concentrates on procedures for using them in observing angles.

■ 8.2 CHARACTERISTICS OF TOTAL STATION INSTRUMENTS

Total station instruments as shown in Figure 8.1, combine three basic components—an electronic distance measuring (EDM) instrument, an electronic angle measuring component, and a computer or microprocessor—into one integral unit. These devices can automatically observe horizontal and vertical angles, as well as slope distances from a single setup (see Chapter 6). From these data, they

186 TOTAL STATION INSTRUMENTS; ANGLE OBSERVATIONS

Figure 8.1
Parts of a total station instrument, with view of eyepiece end of telescope. (Courtesy Leica Geosystems, Inc.)

can compute horizontal and vertical distance components instantaneously, elevations and coordinates of points sighted, and display the results on a *liquid crystal display* (LCD). As discussed in Chapter 2, they can also store the data, either on board or in external data collectors connected to their communication ports.

The telescope is an important part of a total station instrument. It is mounted between the instrument's *standards* (see Figure 8.1), and after the instrument has been leveled, it can be revolved (or "plunged") so that its *axis of sight*[1] defines a vertical plane. The axis about which the telescope revolves is called the *horizontal axis*. The telescope can also be rotated in any azimuth about a vertical line called the *vertical axis*. Being able to both revolve and rotate the telescope in this manner makes it possible for an operator to aim the telescope in any azimuth, and along any slope, to sight points. This is essential in making angle observations, as described in Part II of this chapter. The three reference axes, the axis of sight, the horizontal axis, and the vertical axis, and their relationship to each other are discussed in Section 8.20.1.

---

[1]The axis of sight, also often called the "line of sight," is the reference line within the telescope which an observer uses for making pointings with the instrument. As defined in Section 4.7, it is the line connecting the optical center of the objective lens and the intersection of cross hairs in the reticle.

The EDM instruments that are integrated into total station instruments (described in Section 6.21), are relatively small, and as shown in Figure 8.1, are mounted with the telescope between the standards of the instrument. Although the EDM instruments are small, they still have distance ranges adequate for most work. Lengths up to about 4 km can be observed with a single prism, and even farther with a triple prism like the one shown in Figure 6.11.

Total station instruments are manufactured with two graduated circles, mounted in mutually perpendicular planes. Prior to observing angles, the instrument is leveled so that its *horizontal* circle is oriented in a horizontal plane, which automatically puts the vertical circle in a vertical plane. Horizontal and vertical angles can then be observed directly in their respective planes of reference. To increase the precision of the final horizontal angle, repeating instruments of the past had two vertical axes. This resulted in two horizontal motion screws. One set of motion screws allowed the instrument to be turned without changing the value on the horizontal circle. Today's total station instruments usually have only one vertical axis and are considered directional instruments. However, as discussed later, angles can be repeated on a total station by following the software procedures described in the instrument's manual. Most early versions of total station instruments employed level vials for orienting the circles in horizontal and vertical planes, but many newer ones now use automatic compensators, which are electronic tilt-sensing mechanisms.

The angle resolution of available total stations varies from as low as one-half second for precise instruments suitable for control surveys, up to 20″ for less expensive instruments made specifically for construction work. Formats used for displaying angles also vary with different instruments. For example, the displays of some actually show the degree, minute, and second symbols, but others use only a decimal point to separate the number of degrees from the minutes and seconds. Thus, 315.1743 is actually 315°17′43″. Most instruments allow a choice of units, such as the display of angular measurements in degrees, minutes, and seconds, or in grads (gons). Distances may be shown in either feet or meters. Also, certain instruments enable the choice of displaying vertical angles as either zenith or altitude angles. These choices are entered through the keyboard menu, and the microprocessor performs the necessary conversions accordingly. The keyboard, used for instrument control and data entry, is located just above the communication port, as shown in Figure 8.1.

Once the instrument has been set up and a sighting has been made through the telescope, the time required in displaying an angle and distance reading is approximately 2 to 4 sec when a total station instrument is being operated in the *normal mode*, and less than 0.5 sec when operated in the *tracking mode*. The normal mode, which is used in most types of surveys with the exception of construction layout, results in higher precision because multiple observations are made and averages taken. In the tracking mode, used primarily for construction layout, a prism is held on line near the anticipated final location of a stake. An observation is quickly taken to the prism, and the distance that it must be moved forward or back and left or right is instantly computed and displayed. The prism is moved ahead or back according to the results of the first observation, and another check of the distance is made. The process is quickly repeated as many times as necessary until the correct distance and direction are obtained, whereupon the stake is set. This procedure is discussed in more detail in Chapter 23.

Robotic total stations, which are further discussed in Section 8.6, have servomotors on both the horizontal and vertical axes that allow the instrument to perform a second pointing on a target or track a roving target without operator interaction. These instruments are often used in construction layout. In fact, robotic total stations are required in machine guidance and control on a construction site as discussed in Section 23.11. In machine guidance, the instrument guides a piece of construction equipment through the site preparation process, informing the construction equipment operator of the equipment's position on the job site and the amount of soil that needs to be removed or added at its location to match the project design. In machine control, the instrument sends data to a control unit on the machine that controls the equipment during the entire construction process.

■ 8.3 FUNCTIONS PERFORMED BY TOTAL STATION INSTRUMENTS

Total station instruments, with their microprocessors, can perform a variety of functions and computations, depending on how they are programmed. Most are capable of assisting an operator, step by step, through several different types of basic surveying operations. After selecting the type of survey from a menu, prompts will automatically appear on the display to guide the operator through each step. An example illustrating a topographic survey conducted using this procedure is given in Section 17.9.1.

In addition to providing guidance to the operator, microprocessors of total stations can perform many different types of computations. The capabilities vary with different instruments, but some standard computations include (1) averaging of multiple angle and distance observations, (2) correcting electronically observed distances for prism constants, atmospheric pressure, and temperature, (3) making approximate curvature and refraction corrections to vertical angles and elevations determined by trigonometric leveling, (4) reducing slope distances to their horizontal and vertical components, (5) calculating point elevations from the vertical distance components (supplemented with keyboard input of instrument and reflector heights), and (6) computing coordinates of surveyed points from horizontal angle and horizontal distance components (supplemented with keyboard input of coordinates for the occupied station), and a reference azimuth. The subject of coordinate computations is covered in Chapters 10 and 11.

Many total stations, but not all, are also capable of making corrections to observed horizontal and vertical angles for various instrumental errors. For example, by going through a simple calibration process, the *indexing error* of the vertical circle can be determined (see Section 8.13), stored in the microprocessor, and then a correction applied automatically each time a vertical angle is observed. A similar calibration and correction procedure applies to errors that exist in horizontal angles due to imperfections in the instrument (see Section 8.8). Some total stations are also able to correct for personal errors, such as imperfect leveling of the instrument. By means of electronic tilt-sensing mechanisms, they automatically measure the amount and direction of dislevelment, and then make corrections to the observed horizontal and vertical angles for this condition.

8.4 PARTS OF A TOTAL STATION INSTRUMENT

The upper part of the total station instrument, called the *alidade*, includes the telescope, graduated circles, and all other elements necessary for measuring angles and distances. The basic design and appearance of these instruments (see Figures 8.1 and 8.2) are:

1. The *telescopes* are short, have reticles with cross hairs etched on glass, and are equipped with rifle sights or collimators for rough pointing. Most telescopes have two focusing controls. The objective lens control is used to focus on the object being viewed. The eyepiece control is used to focus on the reticle. If the focusing of the two lenses is not coincident, a condition known as parallax will exist. Parallax is the apparent motion of an object caused by a movement in the position of the observer's eye. The existence of parallax can be observed by quickly shifting one's eye position slightly and watching for movement of the object in relation to the cross hairs. Careful adjustment of the eyepiece and objective lens will result in a sharp image of both the object and the cross hairs with no visible parallax. Since the eye tends to tire through use, the presence of parallax should be checked throughout the day. A common mistake of beginners is to have a colleague "check" their pointings. This is not recommended for many

Figure 8.2
Parts of a total station instrument with view of objective end of the telescope. (Courtesy Topcon America Corp.)

reasons including the personal focusing differences that exist between different individuals.

With newer instruments, objective lens auto focusing is available. This works in a manner similar to autofocusing for a camera, and increases the rate at which pointings can be made when objects are at variable distances from the instrument.

2. The *angle measurement system* functions by passing a beam of light through finely spaced graduations. The instrument shown in Figure 8.2 is representative of the way total stations operate and is briefly described here. For horizontal angle measurements, two glass circles within the alidade are mounted parallel, one on top of the other, with a slight spacing between them. After the instrument has been leveled, the circles should be in horizontal planes. The *rotor* (lower circle) contains a pattern of equally divided alternate dark lines and light spaces. The *stator* (upper circle) contains a slit-shaped pattern, which has the same pitch as that of the rotor circle. A *light-emitting diode* (LED) directs collimated light through the circles from below toward a photodetector cell above. A modern total station may have as many as 20,000 graduations!

When an angle is turned, the rotor moves with respect to the stator creating alternating variations of light intensity. Photodetectors sense these variations, convert them into electrical pulses, and pass them to a microprocessor for conversion into digital values. The digits are displayed using an LCD. Another separate system like that just described is also mounted within the alidade for measuring vertical angles. With the instrument leveled, this vertical circle system is aligned in a vertical plane. After making an observation, horizontal and vertical angles are both displayed, and can be manually read and recorded in field books, or alternatively, the instruments can be equipped with data collectors that eliminate manual reading and recording. (This helps eliminate mistakes!) Today's total stations can resolve angles to an accuracy of 1", 2", 3", or 5" typically.

3. The *vertical circle* of most total station instruments is precisely indexed with respect to the direction of gravity by an *automatic compensator*. These devices are similar to those used on automatic levels (see Section 4.10) and automatically align the vertical circle so that 0° is oriented precisely upward toward the *zenith* (opposite the direction of gravity). Thus, the vertical circle readings are *zenith angles*, that is, 0° occurs with the telescope pointing vertically upward, and either 90° or 270° is read when it is horizontal. Upon command, the microprocessor can convert zenith angles to *altitude angles* (i.e., values measured up or down from 0° at the horizontal). The *vertical motion*, which contains a *lock* and *tangent screw*, enables the telescope to be released so that it can be revolved about the horizontal axis, or locked (clamped) to prevent it from revolving. To sight a point, the lock can be opened, and the telescope tilted up or down about the horizontal axis as necessary to the approximate position needed to sight a point. The lock is then clamped, and fine pointing completed using the vertical tangent screw.

In servo-driven total stations (see Figure 8.7), the lock and tangent screw are replaced with a jog/shuttle mechanism. This device actuates an internal servo-drive motor that rotates the telescope about its horizontal axis. The speed at which the mechanism rotates determines the speed at which the telescope rotates.

4. Rotation of the telescope about the vertical axis occurs within a steel cylinder or on precision ball bearings, or a combination of both. The *horizontal motion*, which also contains a *lock* and *tangent screw*, controls this rotation. Clamping the lock can prevent rotation. To sight a point, the lock is released and the telescope rotated in azimuth to the approximate direction desired, and the lock clamped again. Then the horizontal tangent screw enables a fine adjustment to be made in the direction of pointing. (When sighting a point, both the vertical and horizontal locks are released so that the telescope can be simultaneously revolved and rotated. Then both are locked and fine pointing made using the two tangent screws.)

Similar to the vertical motion in servo-driven total stations, the horizontal lock and tangent screw is replaced with a jog/shuttle mechanism that actuates an internal servo-drive to rotate the instrument about its vertical axis. Again the speed at which the mechanism is rotated determines the speed at which the telescope rotates.

5. The *tribrach* consists of three screws or cams for leveling, a circular level, clamping device to secure the base of the total station or accessories (such as prisms and sighting targets), and threads to attach the tribrach to the head of a tripod. As shown in Figure 8.3, some tribrachs also have integral optical plummets (described next) to enable centering accessories over a point without the instrument.

6. The *bases* of total stations are often designed to permit interchange of the instrument with sighting targets and prisms in tribrachs without disturbing previously established centering over survey points. This can save a considerable amount of time. Most manufacturers use a standard "three-post" arrangement to enable interchangeability between different instruments and accessories.

7. An *optical plummet* built into either the tribrach or alidade of total station instruments, permits accurate centering over a point. Although either type enables accurate centering, best accuracy is achieved with those that are part of the alidade of the instrument. The optical plummet provides a line of sight that is directed downward, collinear with the vertical axis of the instrument but the

Figure 8.3 (a) Tribrach with optical plummet, (b) schematic of a tribrach optical plummet. [Figure (a), Courtesy Topcon America Corp.]

total station instrument or tribrach must be leveled first for the line of sight to be *vertical*. Figures 8.3(a) and (b) show a tribrach with optical plummet, and a schematic of the tribrach optical plummet, respectively. Due to the short length of the telescope in an optical plummet, it is extremely important to remove parallax before centering the instrument with this device.

In some instruments, laser plummets have replaced the optical plummet. This device produces a beam of collimated light that coincides with the vertical axis of the instrument. Since focusing of the objective and eyepiece lens is not required with a laser plummet, this option will increase both the speed and accuracy of setups. However, the laser mark may be difficult to see in bright sunlight. Shading the mark can help in these situations. Additionally, the defined laser point maybe larger than the mark the operator is trying to center over.

8. When being used, total station instruments stand on *tripods*. The tripods are the wide-frame type, and most have adjustable legs. Their primary composition may be wood, metal, or fiberglass.

9. The *microprocessor* provides several significant advantages to surveyors. As examples, (a) the circles can be zeroed instantaneously by simply pressing a button, or they can be initialized to any value by entry through the keyboard (valuable for setting the reference azimuth for a backsight); (b) angles can be observed with values increasing either left or right; and (c) angles observed by repetition (see Section 8.8), can be added to provide the total, even though 360° may have been passed one or more times. Other advantages include reduction of mistakes in making readings, and an increase in the overall speed of operation.

10. The *keyboard* and *display* (see Figure 8.2) provide the means of communicating with the microprocessor. Most total stations have a keyboard and display on both sides of the instrument, a feature that is especially convenient when operating the instrument in both the *direct* and *reverse* modes (see Section 8.8), as is usually done when observing angles to eliminate instrumental systematic errors. As shown in Figure 8.7, some robotic total stations (see Section 8.6) also have a keyboard and display mounted on a remote prism pole for "one-person" operations.

11. The *communication port* (see Figure 8.1) enables external data collectors to be connected to the instrument. Many instruments have internal data collection capabilities, and their communications ports permit them to be interfaced with a computer for direct downloading of data.

8.5 HANDLING AND SETTING UP A TOTAL STATION INSTRUMENT

A total station instrument should be carefully lifted from its carrying case by grasping the standards or handle, and the instrument securely fastened to the tripod by means of the tribrach. For most surveys, prior to observing distances and angles, the instrument must first be carefully set up over a specific point. The setup process using an instrument with an optical plummet, tribrach mount with circular bubble, and adjustable-leg tripod is accomplished most easily using the following steps: (1) extend the legs so that the scope of the instrument will be at an appropriate elevation for viewing, and then adjust the position of the tripod legs by lifting and moving the tripod as a whole until the point is roughly centered beneath the tripod head[2] (beginners can drop a stone from the center of the tripod head, or use a plumb bob to check nearness to the point); (2) firmly place the legs of the tripod in the ground and extend the legs so that the head of the tripod is approximately level; repeat step (1) if the tripod head is not roughly centered over the point; (3) roughly center the tribrach leveling screws on their posts; (4) mount the tribrach approximately in the middle of the tripod head to permit maximum translation in step (9) in any direction; (5) focus the plummet properly on the point, making sure to check and remove any parallax; (6) manipulate the leveling screws to aim the plummet's pointing device at the point below; (7) center the circular bubble by adjusting the lengths of the tripod extension legs; (8) level the instrument using the plate bubble and leveling screws; (9) if necessary, loosen the tribrach screw and translate the instrument (do not rotate it) to carefully center the plummet's pointing device on the point; and (10) repeat steps (8) and (9) until precise leveling and centering are accomplished. With total stations that have their plummets in the tribrach, the instrument can and should be left in the case until step (8).

To level a total station instrument that has a plate-level vial, the telescope is rotated to place the axis of the level vial parallel to the line through any two leveling screws, such as the line through A and B in Figure 8.4(a). The bubble is centered by turning these two screws, then the instrument is rotated 90°, as shown in Figure 8.4(b), and centered again using the third screw (C) only. This process is repeated in the initial two positions and carefully checked to ensure that the bubble remains centered. As illustrated in Figure 8.4, *the bubble moves in the direction of the left thumb when the foot screws are turned.* A solid tripod setup is essential, and the instrument must be shaded if set up in bright sunlight. Otherwise, the bubble will expand and run toward the warmer end as the liquid is heated.

Many instruments, such as that shown in Figure 8.1, do not have traditional level vials. Rather, they are equipped with an electronic, dual-axis leveling system as shown in Figure 8.5 in which four probes sense a liquid (horizontal) surface. After preliminary leveling is performed by means of the tribrach's circular bubble, signals from the probes are processed to form an image on the LCD, which guides an operator in performing rough leveling. The three leveling screws are used, but the instrument need not be turned about its vertical axis in the leveling process. After rough leveling, the amount and direction of any residual dislevelment is automatically and continuously received by the microprocessor, which corrects observed horizontal and vertical angles accordingly in real time.

As noted earlier, total stations are controlled with entries made either through their built-in keypads or through the keypads of handheld data collectors. Details for operating each individual total station vary somewhat and therefore

---

[2]Some prefer to place one leg firmly in the ground. The surveyor then looks through the optical plummet and moves the tripod with hands on the remaining two legs until the point is in view in the optical plummet. The remaining two legs are then set firmly in the ground.

194 TOTAL STATION INSTRUMENTS; ANGLE OBSERVATIONS

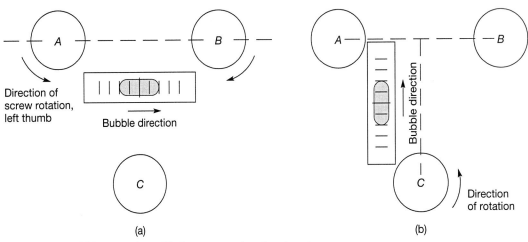

Figure 8.4 Bubble centering with three-screw leveling head.

Figure 8.5
A view of an electronic leveling system.
(Courtesy Leica Geosystems, Inc.)

are not described here. They are covered in the manuals provided with the purchase of instruments.

When moving between setups in the field, proper care should be taken. Before the total station is removed from the tripod, the foot screws should be returned to the midpoints of the posts. Many instruments have a line on the screw post that indicates the halfway position. The instrument should NEVER be transported on the tripod since this causes stress to tripod head, tribrach, and instrument base. Figure 8.6(a) depicts the proper procedure for carrying equipment in the field. With adjustable-leg tripods, retracting them to their shortest positions and lightly clamping them in position can avoid stress on the legs. Since the screws on the instrument are made of brass typically, overtightening screws on tripods and the instrument can cause serious harm to the instruments. Screws and locks should only be "finger" tight. Inexperienced users sometimes overtighten screws to the detriment of the equipment.

Figure 8.6
(a) A proper method of transporting a total station in the field.
(b) Total station in open case.
(Courtesy Leica Geosystems, Inc.)

(a) (b)

When returning the total station to its case, all locking mechanisms should be released. This procedure protects the threads and reduces wear when the instrument is jostled during transport and also prevents the threads from seizing during long periods of storage. If the instrument is wet, it should be wiped down and left in an open case until it is dry as shown in Figure 8.6(b). When storing tripods, it is important to loosen or lightly clamp all legs. This is especially true with wooden tripods where the wood tends to expand and contract with changes in humidity. Failure to loosen the clamping mechanism on wooden tripods can result in crushed wood fibers, which inhibit the ability of the clamp to hold the leg during future use. Additionally, during periods of high humidity, the clamps can seize on the legs, requiring a pliers or wrench to free them.

8.6 SERVO-DRIVEN AND REMOTELY OPERATED TOTAL STATION INSTRUMENTS

Manufacturers also produce "robotic" total station instruments equipped with servo-drive mechanisms that enable them to aim automatically at a point. The total station shown in Figure 8.7 is an example. When staking out points with these instruments, it is only necessary to identify the point's number with a keyboard entry. The computer retrieves the direction to the point from storage or computes it and activates a servomotor to turn the telescope to that direction within a few seconds. This feature is particularly useful for construction stakeout, but it is also convenient in control surveying when multiple observations are made in observing angles. In this instance, final precise pointing is done manually.

The survey controller shown in Figure 8.7, which is attached to a prism pole, has a built-in telemetry link for communication with the total station. The robotic instrument is equipped with an automatic search and aim function, as well as a link for communication with the survey controller. It has servomotors for automatic aiming at the prism both horizontally and vertically. Using the survey controller, the total station instrument can be controlled from a distance.

Figure 8.7
A robotic total station with its survey controller on the prism pole. (Courtesy Leica Geosystems, Inc.)

To operate the system, the robotic instrument must first be set up and oriented. This consists in entering the coordinates of the point where the total station is located and taking a backsight along a line of known azimuth. Once oriented, an operator carries the survey controller and prism to any convenient location and faces the robotic instrument. The instrument then scans for the prism both horizontally and vertically. Its horizontal servomotor then activates and swings around until it finds the prism. Once the total station has found the prism, which only takes a few seconds, and locks onto it, it will automatically follow its further movements. If lock is lost, the search routine is simply repeated.

With this and similar systems, the total station instrument is completely controlled through the keyboard of the remote unit at the prism pole. These systems enable one person to conduct a complete survey. They are exceptionally well suited for construction surveys and topographic surveys but can be used advantageously in other types as well. The system not only eliminates one person and speeds the work, but more importantly, it eliminates mistakes in identifying points that can occur when the prism is far from the total station and cannot be seen clearly.

PART II • ANGLE OBSERVATIONS

8.7 RELATIONSHIP OF ANGLES AND DISTANCES

Determining the relative positions of points often involves observing of both angles and distances. The best-quality surveys result when there is compatibility between the accuracies of these two different kinds of measurements. The formula for relating distances to angles is given by the geometric relationship

$$S = R\theta \tag{8.1}$$

In Equation (8.1), S is the arc length subtended at a distance R by an arc of θ in radians. To select instruments and survey procedures necessary for achieving consistency, and to evaluate the effects of errors due to various sources, it is helpful to consider the relationships between angles and distances given here and illustrated in Figure 8.8.

1' of arc = 0.03 ft at 100 ft, or 3 cm at 100 m (approx.)
1' of arc = 1 in. at 300 ft (approx.; actually 340 ft)
1" of arc = 1 ft at 40 mi, or 0.5 m at 100 km, or 1 mm at 200 m (approx.)
1" of arc = 0.000004848 radians (approx.)
1 radian = 206,264.8" of arc (approx.)

In accordance with the relationships listed, an error of approximately 1 min results in an observed angle if the line of sight is misdirected by 1 in. over a distance of 300 ft. This illustrates the importance of setting the instrument and targets over their respective points precisely, especially where short sights are involved. If an angle is expected to be accurate to within $\pm 5''$ for sights of 500 ft, then the distance must be correct to within $500(5'')0.000004848 = \pm 0.01$ ft for compatibility.

To appreciate the precision capabilities of a high-quality total station, an instrument reading to the nearest 0.5" is can measure the angle between two points approximately 1 cm apart and 4 km away theoretically! However, as discussed in Sections 8.19 through 8.21, errors from centering the instrument, sighting the point, reading the circle, and other sources, make it difficult, if not impossible, to actually accomplish this accuracy.

■ 8.8 OBSERVING HORIZONTAL ANGLES WITH TOTAL STATION INSTRUMENTS

As stated in Section 2.1, horizontal angles are observed in horizontal planes. After a total station instrument is set up and leveled, its horizontal circle is in a horizontal plane and thus in proper orientation for observing horizontal angles. To observe a horizontal angle, for example, angle *JIK* of Figure 8.9(a), the instrument is first set up and centered over station *I*, and leveled. Then a *backsight* is

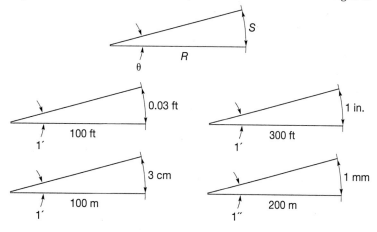

Figure 8.8
Angle and distance relationships.

taken on station *J*. This is accomplished by releasing the horizontal and vertical locks, turning the telescope in the approximate direction of *J*, and clamping both locks. A precise pointing is then made to place the vertical cross hair on the target using the horizontal and vertical tangent screws, and an initial value of 0°00′00″ is entered in the display. The horizontal motion is then unlocked, and the telescope turned clockwise toward point *K* to make the *foresight*. The vertical circle lock is also usually released to tilt the telescope for sighting point *K*. Again, the motions are clamped with the line of sight approximately on station *K*, and precise pointing is made as before using the horizontal and vertical tangent screws. When the foresight is completed, the value of the horizontal angle will automatically appear in the display.

To eliminate instrumental errors and increase precision, angle observations should be repeated an equal number of times in each of the *direct* and *reverse* modes, and the average taken. Built-in computers of total station instruments will perform the averaging automatically and display the final results. For instruments that have only a single keyboard and display, the instrument is in its direct mode when the eyepiece and keyboard are on the same side of the instrument. However, instruments do vary by manufacturer, and the operator should refer to the instrument's manual to determine the proper orientation of their instrument when in the direct mode. To get from the direct mode into the reverse mode, the vertical motion is unlocked the telescope is "plunged" (rotated 180° about the horizontal axis).

Procedures for repeating horizontal angle observations can differ with instruments and survey controllers of different manufacture, and operators must therefore become familiar with the features of their specific instrument by referring to its manual. The following is an example procedure that applies to some instruments. After making the first observation of angle *JIK*, as described earlier, the angular value in the display is held by pressing a button on the keyboard of the instrument. (Assume the first observation was in the direct mode.) To repeat the angle with the instrument in the same mode, a backsight is again taken on station *J*

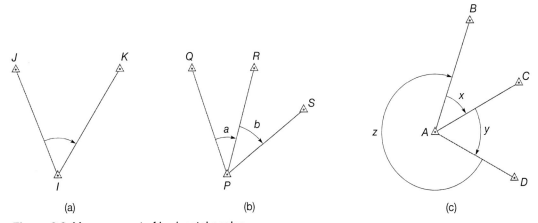

Figure 8.9 Measurement of horizontal angles.

using the horizontal lock and tangent screw. After the backsight is completed, with the next angle observation is taken by again pressing the appropriate button on the keyboard. Using the same procedures described earlier, a foresight is again taken on station *K*, after which the display will read the second angle. This procedure is repeated until the desired number of angles is observed in the direct mode. Then the telescope is plunged to place it in the reverse mode, and the angle repeated an equal number of times using the same procedure. In the end, the average of all angles turned, direct and reverse, will be displayed along with the individual observations and their residual errors. The operator can then accept the set of angles as observed or discard individual angles and repeat their observation.

The procedure just described for observing horizontal angles is often called the *repetition method*. As noted earlier, obtaining an average value from repeated observations increases precision, and by incorporating equal numbers of direct and reverse measurements, certain instrumental errors are eliminated (see Section 8.20).

An example set of field notes for observing the angle of Figure 8.9(a) by the repetition method is shown in Figure 8.10. In the example, four repetitions, two in each of the direct (Face I) and reverse (Face II) modes, were taken. In the notes, the identification of the angle being observed is recorded in column (1), the position of the instrument is placed in column (2) the values of the direct readings are tabulated in column (3), the values of the reverse readings are tabulated in column (4), and the mean of the four readings, which produces the final angle, is given in column (5). If these values agree within tolerable limits, the mean angle is accepted, if not the work is repeated.

Special capabilities are available with many total station instruments to enhance their accuracy and expedite operations. For example, most instruments have a dual-axis automatic compensator that senses any misorientation of the circles. This information is relayed to the built-in computer that corrects for any *indexing error* in the vertical circle (see Section 8.13), and any dislevelment of the horizontal circle, before displaying angular values. This real-time tilt sensing and correction feature makes it necessary to perform rough leveling of the instrument only, thus reducing setup time. In addition, some instruments observe angles by integration of electronic signals over the entire circle simultaneously; thus, errors due to graduations and eccentricities (see Section 8.20.1) are eliminated. Furthermore, the computer also corrects horizontal angles for instrumental

HORIZONTAL ANGLE MEASUREMENT

(1) Angle	(2) Face	(3) ° ′ ″	(4) ° ′ ″	(5) Mean ° ′ ″
JIK	I	66 37 40	66 37 42	
	II	66 37 40	66 37 48	66 37 40

Figure 8.10
Field notes for measuring the horizontal angle of Figure 8.9(a) by repetition.

errors if the axis of sight is not perpendicular to the horizontal axis, or if the horizontal axis is not perpendicular to the vertical axis. (These conditions are discussed in Sections 8.15 and 8.20.1, respectively.) This feature makes it possible to obtain angle observations free from instrumental errors without averaging equal numbers of direct and reverse readings. With these advantages, and more, it is obvious why these instruments have replaced the older optical instruments. However even with these advantages, it is best practice to keep your instrument in proper calibration but use it as if it is not, which means always reading and averaging the same number of direct and reverse angle observations.

■ 8.9 OBSERVING MULTIPLE HORIZONTAL ANGLES BY THE DIRECTION METHOD

As an alternative to observing a single horizontal angle by the repetition method described in the preceding section, total station instruments can be used to determine horizontal angles by the *direction method*. This procedure consists in observing *directions*, which are simply horizontal circle readings taken to successive stations sighted around the horizon. Then by taking the difference in directions between any two stations, the angle between them is determined. The procedure is particularly efficient when multiple angles are being observed at a station.

An example of this type of situation is illustrated in Figure 8.9(b), where angles a and b must both be observed at station P. Figure 8.11 shows a set of field notes for observing these angles by direction method. The method involves sighting the initial station Q in the direct mode (Face I) and zeroing the plates. Following this, all subsequent stations are sighted in the direct position and the readings written in column (3). After completing the readings in the direct mode, the telescope is plunged to its reverse (Face II) position, and all directions observed again [see the data entries in column (4)]. A set of readings in both the direct and reverse modes constitutes a so-called *position*.

The notes in Figure 8.11 are the results of four repetitions of direction measurements in each of the direct and reverse mode. In these notes, the repetition number is listed in column (1), the station sighted in column (2), direction readings taken in the direct and reverse modes in columns (3) and (4), respectively, the mean of direct and reverse readings in column (5), and the computed angles (obtained by subtracting the mean direction for station Q from that of station R, and subtracting R from S) in column (6).

Final values for the two angles are taken as the averages of the four angles in column (6). These are 37°30'28" and 36°43'14" for angles a and b, respectively. Note that in this procedure, as was the case with the repetition method, the multiple readings increase the precisions of the angles, and by taking equal numbers of direct and reverse readings, many instrumental errors are eliminated. As previously noted, this method of observing directions can significantly reduce the time at a station, especially when several angles with multiple repetitions are needed, for example, in triangulation.

The procedures for observing multiple angles with data collectors can vary by manufacturer. The reader should refer to their data collector manual to determine the proper procedures for their situation. One of the advantages of using a data collector to observe multiple angles is that they provide immediate

| \multicolumn{6}{c}{DIRECTIONS OBSERVED FROM STATION P} |
Repetition No.	Station Sighted	Reading Direct	Reading Reverse	Mean	Angle
(1)	(2)	(3)	(4)	(5)	(6)
		° ′ ″	° ′ ″	° ′ ″	° ′ ″
1	Q	0 00 00	0 00 00	0 00 00	
	R	37 30 27	37 30 21	37 30 24	37 30 24
	S	74 13 42	74 13 34	74 13 38	36 43 14
2	Q	0 00 00	0 00 00	0 00 00	
	R	37 30 32	37 30 28	37 30 30	37 30 30
	S	74 13 48	74 13 42	74 13 46	36 43 16
3	Q	0 00 00	0 00 00	0 00 00	
	R	37 30 26	37 30 26	37 30 26	37 30 26
	S	74 13 36	74 13 40	74 13 38	36 43 12
4	Q	0 00 00	0 00 00	0 00 00	
	R	37 30 34	37 30 30	37 30 32	37 30 32
	S	74 13 48	74 13 44	74 13 46	36 43 14

Figure 8.11 Field notes for measuring directions for Figure 8.9(b).

post-observation statistics. The residuals of each observation can be displayed after the observation process before accepting the average observations. The operator can view each residual and decide if any are too large to meet the job specifications, instrument specifications, and field conditions. If a single residual is deemed excessive, that observation can be removed and the observation repeated. If all the residuals are too large, the entire set of observations can be removed, and the entire angle observation process repeated.

■ 8.10 CLOSING THE HORIZON

Closing the horizon consists in using the direction method as described in the preceding section, but including *all* angles around a point. Suppose that in Figure 8.9(c) only angles x and y are needed. However, in closing the horizon angle z is also observed thereby providing for additional checks. An example set of field notes for this operation is shown in Figure 8.12. The angles are first turned around the horizon by making a pointing and direction reading at each station with the instrument in the direct mode [see the data entries in column (3) of Figure 8.12]. A final foresight pointing is made on the initial backsight station, and this provides a check because it should give the initial backsight reading (allowing for reasonable random errors). Any difference is the *horizon misclosure*, and if its value exceeds an allowable tolerance, that round of readings should be discarded, and the observations repeated. (Note that in the field notes of Figure 8.12, the maximum horizon misclosure is 4″.)

Figure 8.12
Field notes for closing the horizon at station A of Figure 8.9(c).

CLOSING THE HORIZON AT STATION A

Position No.	Station Sighted	Reading Direct	Reading Reverse	Mean	Angle
(1)	(2)	(3)	(4)	(5)	(6)
		° ′ ″	° ′ ″	° ′ ″	° ′ ″
1	B	0 00 00	0 00 00	0 00 00	
	C	42 12 12	42 12 14	42 12 13	42 12 13
	D	102 08 26	102 08 28	102 08 27	59 56 14
	B	0 00 02	0 00 02	0 00 02	257 51 35
				Sum	360 00 02
2	B	0 00 00	0 00 00	0 00 00	
	C	42 12 12	42 12 14	42 12 13	42 12 13
	D	102 08 28	102 08 28	102 08 28	59 56 15
	B	0 00 04	0 00 04	0 00 04	257 51 36
				Sum	360 00 04
3	B	0 00 00	0 00 00	0 00 00	
	C	42 12 14	42 12 12	42 12 13	42 12 13
	D	102 08 28	102 08 26	102 08 27	59 56 14
	B	0 00 04	0 00 00	0 00 02	257 51 35
				Sum	360 00 02
4	B	0 00 00	0 00 00	0 00 00	
	C	42 12 14	42 12 12	42 12 13	42 12 13
	D	102 08 32	102 08 28	102 08 30	59 56 17
	B	0 00 04	0 00 04	0 00 04	257 51 34
				Sum	360 00 04

The note-reduction process consists of calculating mean values of the direct and reverse directions to each station [see column (5)] and from them, the individual angles around the horizon are computed as discussed in Section 8.9 [see column (6)]. Finally, their sum is calculated, and checked against 360°. Any difference reveals a mistake or mistakes in computing the individual angles. Again, repeat values for each individual angle are obtained, and as another check on the work, these should be compared for their agreement.

As an alternative to closing the horizon by observing directions, each individual angle could be measured independently using the procedures outlined in Section 8.8. After observing all angles around the horizon, their sum could also be computed and compared against 360°. However, this procedure is not as efficient as closing the horizon using directions.

8.11 OBSERVING DEFLECTION ANGLES

A deflection angle is a horizontal angle observed from the prolongation of the preceding line, right or left, to the following line. In Figure 8.13(a) the deflection angle at F is 12°15′10″ to the right (12°15′10″R), and the deflection angle at G is 16°20′27″L.

A straight line between terminal points is theoretically the most economical route to build and maintain for highways, railroads, pipelines, canals, and transmission lines. Practically, obstacles and conditions of terrain and land-use require bends in the route, but deviations from a straight line are kept as small as possible. If an instrument is in perfect adjustment, which is unlikely, the deflection angle at F [see Figure 8.13(a)] is observed by setting the circle to zero and backsighting on point E with the telescope in the direct position. The telescope is then plunged to its reversed position, which places the line of sight on EF extended, as shown dashed in the figure. The horizontal lock is released for the foresight, point G sighted, the horizontal lock clamped, the vertical cross hair set on the mark carefully by means of the horizontal tangent screw, and the angle read.

Deflection angles are subject to serious errors if the instrument is not in adjustment, particularly if the line of sight is not perpendicular to the horizontal axis (see Section 8.15). If this condition exists, deflection angles may be read larger or smaller than their correct values, depending on whether the line of sight after plunging is to the right or left of the true prolongation [see Figure 8.13(b)]. To eliminate errors from this cause, angles are usually doubled or quadrupled by the following procedure: the first backsight is taken and the horizontal circle zeroed. The scope is then plunged to the reverse position and a foresight reading is taken and recorded. With the scope plunged, a second backsight is taken on the initial station and the horizontal circle zeroed. The scope is again plunged to the direct position and a second foresight is taken and recorded. The average of the two foresight readings is determined from which many instrumental errors are eliminated by cancellation. In outline fashion, the method is as follows:

1. Backsight with the telescope direct. Plunge to reversed mode and observe the angle. Record the reading.
2. Backsight with the telescope still reversed. Plunge again to direct mode and observe and record the angle.
3. Average the two angles.

Of course, making four, six, or eight repetitions and averaging can increase the precision in direction angle observation.

Figure 8.14 shows the left-hand page of field notes for observing the deflection angles at stations F and G of Figure 8.13(a). The procedure just outlined was followed. Four repetitions of each angle were taken with the instrument alternated from direct to reverse with each repetition. Readings were recorded only after the first, second, and fourth repetitions. The four angles observed should be checked for agreement. Any angle with a large discrepancy from the mean should be discarded and reobserved.

Figure 8.13
Deflection angles.

Figure 8.14
Field notes for measuring deflection angles.

Sta	BS/FS Sta	Reps	Circle Rdg ° ′ ″	Mean ° ′ ″	Right/Left
F	E	1	12 15 12		
		2	12 15 10		
		3	12 15 10		
	G	4	12 15 08	12 15 10	R
G	F	1	16 20 28		
		2	16 20 26		
		3	16 20 26		
	H	4	16 20 28	16 20 27	L

8.12 OBSERVING AZIMUTHS

Azimuths are observed from a reference direction, which itself must be determined from (a) a previous survey, (b) the magnetic needle, (c) a solar or star observation, (d) GNSS observations, (e) a north-seeking gyro, or (f) assumption. Suppose that in Figure 8.15 the azimuth of line AB is known to be 137°17′00″ from north. The azimuth of any other line that starts at A, such as AC in the figure, can be found directly using a total station instrument. In this process, with the instrument set up and centered over station A, and leveled, a backsight is first taken on point B. The azimuth of line AB (137°17′00″) is then set on the horizontal circle using the keyboard. The instrument is now "oriented," since the line of sight is in a known direction with the corresponding azimuth on the horizontal circle. If the circle were turned until it read 0°, the telescope would be pointing toward north (along the meridian). The next steps are to loosen the horizontal lock, turn the telescope clockwise to C and read the resultant direction, which is the azimuth of AC, and in this case is 83°38′00″.

In Figure 8.15, if the instrument is set up at point B instead of A, the azimuth of BA (317°17′00″) or the back azimuth of AB is put on the circle and point A sighted. The horizontal lock is released, and sights taken on points whose azimuths from B are desired. Again, if the instrument is turned until the circle reads zero, the telescope points north (or along the reference meridian). By following this procedure at each successive station of a traverse, for example, at $A, B, C, D, E,$ and F of the traverse of Figure 7.2(a), the azimuths of all traverse lines can be determined. With a closed polygon traverse like that of Figure 7.2(a), station A should be occupied a second time and the azimuth of AB determined again to serve as a check on the work. With an internal or external data collector, this procedure is accomplished by simply entering the point identifier for the backsight station.

8.13 OBSERVING VERTICAL ANGLES

A vertical angle is the difference in direction between two intersecting lines measured in a vertical plane. Vertical angles can be observed as either altitude or

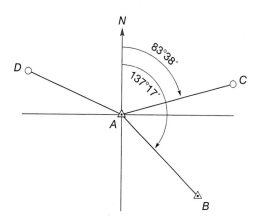

Figure 8.15
Orientation by azimuths.

zenith angles. An altitude angle is the angle above or below a horizontal plane through the point of observation. Angles above the horizontal plane are called *plus angles*, or *angles of elevation*. Those below it are *minus angles*, or *angles of depression*. Zenith angles are measured with zero on the vertical circle oriented toward the zenith of the instrument, and thus go from 0° to 360° in a clockwise circle about the horizontal axis of the instrument.

Most total station instruments are designed so that zenith angles are displayed rather than altitude angles. In equation form, the relationship between altitude angles and zenith angles is

$$\text{Direct mode} \quad \alpha = 90° - z \quad (8.2a)$$
$$\text{Reverse mode} \quad \alpha = z - 270° \quad (8.2b)$$

where z and α are the zenith and altitude angles, respectively. With a total station, therefore, a reading of 0° corresponds to the telescope pointing vertically upward. In the direct mode, with the telescope horizontal, the zenith reading is 90°, and if the telescope is elevated 30° above horizontal, the reading is 60°. In the reverse mode, the horizontal reading is 270°, and with the telescope raised 30° above the horizon it is 300°. Altitude angles and zenith angles are observed in trigonometric leveling, and in EDM work for reduction of observed slope distances to horizontal.

Observation of zenith angles with a total station instrument follows the same general procedures as those just described for horizontal angles, except that an automatic compensator orients the vertical circle. As with horizontal angles, instrumental errors in vertical angle observations are compensated for by computing the mean from an equal number of direct and reverse observations. With zenith angles, the mean is computed from

$$\bar{z}_D = \frac{\Sigma z_D}{n} + \frac{n(360°) - (\Sigma z_D + \Sigma z_R)}{2n} \quad (8.3)$$

where $\bar{z}_D$ is the mean value of the zenith angle [expressed according to its direct mode value], Σz_D the sum of direct zenith angles, Σz_R the sum of reverse angles, and

n the number of z_D and z_R pairs of zenith angles read. The latter part of Equation (8.3) accounts for the *indexing error* present in the instrument.

An indexing error exists if 0° on the vertical circle is not truly at the zenith with the instrument in the direct mode. This will cause all vertical angles read in this mode to be in error by a constant amount. For any instrument, an error of the same magnitude will also exist in the reverse mode, but it will be of opposite algebraic sign. The presence of an indexing error in an instrument can be detected by observing zenith angles to a well-defined point in both modes of the instrument. If the sum of the two values does not equal 360°, an indexing error exists. To eliminate the effect of the indexing error, equal numbers of direct and reverse angle observations should be made, and averaged. The averaging is normally done by the microprocessor of the total station instrument. Even though an indexing error may not exist, to be safe, *experienced surveyors always adopt field procedures that eliminate errors just in case the instrument is out of adjustment.*

With some total station instruments, indexing errors can be eliminated from zenith angles by computation, after going through a calibration procedure with the instrument. The computations are done by the microprocessor and applied to the angles before they are displayed. Procedures for performing this calibration vary with different manufacturers and are given in the manuals that accompany the equipment.

■ Example 8.1

A zenith angle was read twice direct giving values of 70°00′10″ and 70°00′12″, and twice reverse yielding readings of 289°59′44″ and 289°59′42″. What is the mean zenith angle?

Solution

Two pairs of zenith angles were read, thus $n = 2$. The sum of direct angles is 140°00′22″, and that of reverse values is 579°59′26″. Then by Equation (8.3)

$$\bar{z}_D = \frac{140°00'22''}{2} + \frac{2(360°) - (140°00'22'' + 579°59'26'')}{2(2)}$$

$$= 70°00'11'' + 0°00'03'' = 70°00'14''$$

Note that the value of 3″ from the latter part of Equation (8.3) is the indexing error.

■ 8.14 SIGHTS AND MARKS

Objects commonly used for sights when total station instruments are being used only for angle observations include prism poles, chaining pins, nails, pencils, plumb-bob strings, reflectors, and tripod-mounted targets. For short sights, a string is preferred to a prism pole because the small diameter permits more accurate sighting. Small red and white targets of thin plastic or cardboard placed on the string extend the length of observation possible. However, even small amounts of wind can cause the plumb bob to sway reducing its effectiveness as a target. Triangular marks placed on prisms as shown in Figure 8.16(a) provide excellent targets at both close and longer sight distances.

An error is introduced if the prism pole sighted is not plumb. The pole is kept vertical by means of a circular bubble. [The bubble should be regularly checked for adjustment, and adjusted if necessary (see Section 8.19.5).] The person holding the prism has to take special precautions in plumbing the pole, carefully watching the circular bubble on the pole. Bipods like the one shown in Figure 8.16(b) and tripods have been developed to hold the pole during multiple angle observation sessions. Use of pole supports can greatly increase both the precision and speed with which an angle can be observed.

The prism pole shown in Figure 8.16(b) has graduations for easy determination of the prism's height. The tripod mount shown in Figure 8.16(a) is centered over the point using the optical plummet of the tribrach. When sighting a prism pole, the vertical cross hair should bisect the pole just below the prism. Errors can result if the prism itself is sighted, especially on short lines since any misalignment of the face of the prism with the line of sight will cause and offset pointing on the prism.

In construction layout work, and in topographic mapping, permanent backsights and foresights may be established. These can be marks on structures such as walls, steeples, water tanks, and bridges, or they can be fixed artificial targets. They provide definite points on which the instrument operator can check orientation without the help of a rodperson.

The error in a horizontal angle due to miscentering of the line of sight on a target, or too large a target, can be determined with Equation (8.1). For instance, assume a prism pole that is 20 mm wide is used as a target on a direction of only 100 m. Assuming that the pointing will be within 1/2 of the width of the pole (10 mm), then according to Equation (8.1) the error in the direction would be $(0.01/100)206,264.8 = 21''$! For an angle where both sight distances are 100 m and

Figure 8.16 (a) Prism and sighting target with tribrach and tribrach adapter, and (b) pole and bipod, used when measuring distances and horizontal angles with total station instruments. (Courtesy Topcon America Corp.)

assuming that the pointings are truly random, the error would propagate according to Equation (3.12), and would result in an estimated error in the angle of $21''\sqrt{2}$, or approximately $30''$. From the angle–distance relationships of Section 8.7, it is easy to see why the selection of good targets that are appropriate for the sight distances in angle observations is so important.

■ 8.15 PROLONGING A STRAIGHT LINE

On route surveys, straight lines may be continued from one point through several others. To prolong a straight line from a backsight, the vertical cross wire is aligned on the back point by means of the lower motion, the telescope plunged, and a point, or points, set ahead on line. In plunging the telescope, a serious error can occur if the line of sight is not perpendicular to the horizontal axis. The effects of this error can be eliminated, however, by following proper field procedures. The procedure used is known as the *principle of reversion*. The method applied, actually double reversion, is termed *double centering*. Figure 8.17 shows a simple use of the principle in drawing a right angle with a defective triangle. Lines OX and OY are drawn with the triangle in direct and reverse positions. Angle XOY represents twice the error in the triangle at the $90°$ corner, and its bisector (shown dashed in the figure) establishes a line perpendicular to AB.

To prolong line AB of Figure 8.18 by double centering with a total station whose line of sight is not perpendicular to its horizontal axis, the instrument is set up at B. A backsight is taken on A with the telescope in the direct mode, and by plunging the telescope into the reverse position the first point C' is set. The horizontal circle lock is released, and the telescope turned in azimuth to take a second backsight on point A, this time with the telescope still plunged. The telescope is plunged again to its direct position and point C'' placed. Distance $C'C''$ is bisected to get point C, on line AB prolonged. In outline form, the procedure is as follows:

1. Backsight on point A with the telescope direct. Plunge to the reverse position and set point C'.
2. Backsight on point A with the telescope still reverse. Plunge to a direct position and set point C''.
3. Split the distance $C'C''$ to locate point C.

In the above procedure, each time the telescope is plunged, the instrument creates twice the total error in the instrument. Thus, at the end of the procedure, four times the error that exists in the instrument lies between points C' and C''. This procedure could be used to prolong a line in preparation for observing deflection angles. If this is done, the backsight for the deflection angle would be at C.

To adjust the instrument, the reticle must be shifted to bring the vertical cross wire one fourth of the distance back from C'' toward C'. For total station instruments that have exposed capstan screws for adjusting their reticles, an adjustment can be made in the field. Generally, however, it is best to leave this adjustment to qualified experts. If the adjustment is made in the field, it must be done very carefully! Figure 8.19 depicts the condition after the adjustment is completed. Since each cross hair has two sets of opposing capstan screws, *it is important to loosen one screw before tightening the opposing one by an equal*

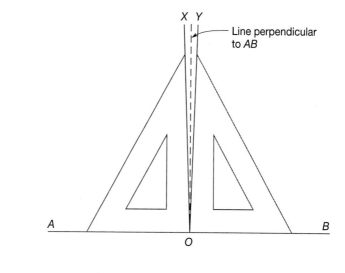

Figure 8.17 Principle of reversion.

Figure 8.18 Double centering.

amount. After the adjustment is completed, the procedure should be repeated to check the adjustment.

8.16 BALANCING-IN

Occasionally, it is necessary to set up an instrument on a line between two points already established but not intervisible—for example, A and B in Figure 8.20. This can be accomplished in a process called *balancing-in* or *wiggling-in*.

Location of a trial point C' on line is estimated and the instrument set over it. A sight is taken on point A from point C' and the telescope plunged. If the line of sight does not pass through B, the instrument is moved laterally a distance CC' estimated from the proportion $CC' = BB'(AC/AB)$, and the process repeated. Several trials may be required to locate point C exactly, or close enough for the purpose at hand. The shifting head of the instrument is used to make the final small adjustment. A method for getting a close first approximation of required point C takes two persons, X able to see point A and Y having point B visible, as shown in Figure 8.20. Each aligns the other in with the visible point in a series of adjustments, and two range poles are placed at least 20 ft apart on the course established. An instrument set at point C in line with the poles should be within a few tenths of a foot of the required location. From there the wiggling-in process can proceed more quickly.

8.17 RANDOM TRAVERSE

On many surveys it is necessary to run a line between two established points that are not intervisible because of obstructions. This situation arises repeatedly in

210 TOTAL STATION INSTRUMENTS; ANGLE OBSERVATIONS

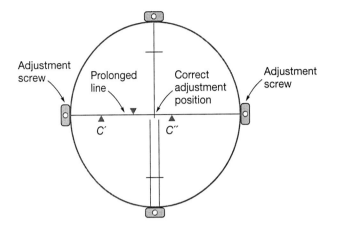

Figure 8.19
The cross-hair adjustment procedure.

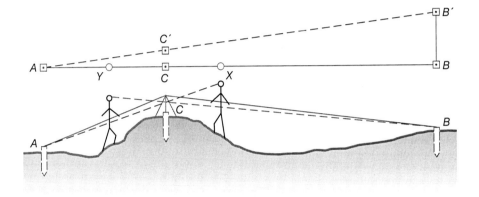

Figure 8.20
Balancing-in.

property surveys. To solve the problem, a *random traverse* is run from one point in the approximate direction of the other. Using coordinate computation procedures presented in Chapter 10, the coordinates of the stations along the random traverse are computed. Using these same computation procedures, coordinates of the points along the "true" line are computed, and observations necessary to stake out points on the line computed from the coordinates. With data collectors, the computed coordinates can be automatically determined in the field, and then staked out using the functions of the data collector.

As a specific example of a random traverse, consider the case shown in Figure 8.21 where it is necessary to run line *X–Y*. On the basis of a compass bearing, or information from maps or other sources, the general direction to proceed is estimated, and starting line *X-1* is given an assumed azimuth. Random traverse *X-1-2-3-Y* is then run, and coordinates of all points determined. Based upon these computations, coordinates are also computed for points *A* and *B*, which are on line *X-Y*. The distance and direction necessary for setting *A* with an instrument set up at point 1 are then computed using procedures discussed in Chapter 10. Similarly, the coordinates of *B* are determined and set from station 2. Using a data collector, these computations can be performed automatically. This procedure, known as *stake out*, is discussed in Chapter 23.

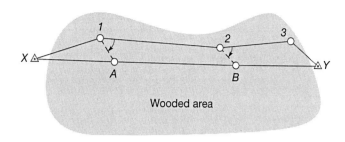

Figure 8.21
Random traverse
X-1-2-3-Y.

Once the angles and distances have been computed for staking points A and B, the actual stake out procedure is aided by operating the total station instrument in its *tracking mode* (see Section 6.21 and Chapter 23). If a robotic total station instrument is available, one person can perform the layout procedure. *This method of establishing points on a line is only practical when direct sighting along the line is not physically possible.*

■ 8.18 TOTAL STATIONS FOR DETERMINING ELEVATION DIFFERENCES

With a total station instrument, computed vertical distances between points can be obtained in real time from observed slope distances and zenith angles. In fact, this is the basis for *trigonometric leveling* (see Section 4.5.4). Several studies have compared the accuracies of elevation differences obtained by trigonometric leveling using modern total station instruments to those achieved by differential leveling as discussed in Chapters 4 and 5. Trigonometric leveling accuracies have always been limited by instrumental errors (discussed in Section 8.20) and the effects of earth curvature and refraction (see Section 4.4). Even with these problems, elevations derived from a total station survey are of sufficient accuracy for many applications such as for topographic mapping and other lower-order work.

However, studies have suggested that high-order results can be obtained in trigonometric leveling by following specific procedures. The suggested guidelines are: (1) place the instrument between two prisms so that sight distances are appropriate for the angular accuracy of the instrument, using Figure 8.22 as a guide;[3] (2) use target panels with the prisms; (3) keep rod heights equal so that their measurement is unnecessary; (4) observe the vertical distances between the prisms using two complete sets[4] of observations at a minimum; (5) keep sight distances approximately equal; and (6) apply all necessary atmospheric corrections and reflector constants as discussed in Chapter 6. This type of trigonometric leveling can be done faster than differential leveling, especially in rugged terrain where sight distances are limited due to rapid changes in elevation.

A set of notes from trigonometric leveling is shown in Figure 8.23. Column (a) lists the backsight and foresight station identifiers and the positions of the telescope [direct (D) and reverse (R)] for each observation; (b) tabulates the

---

[3]A description of ISO 17123-3 standard noted in Figure 8.22 is given in Section 8.21.
[4]One set of observations includes an elevation determination in both the direct and reverse positions.

Figure 8.22
Graph of appropriate sight distance versus angular accuracy.

backsight vertical distances, (BS+); (c) lists the backsight horizontal distances to the nearest decimeter; (d) gives the foresight vertical distances, (FS−); (e) lists the foresight horizontal distances to the nearest decimeter; and (f) tallies the elevation differences between the stations, computed as the difference of the FS vertical distances, minus the BS vertical distances. The observed elevation difference between stations A and E is −8.405 m.

■ 8.19 ADJUSTMENT OF TOTAL STATION INSTRUMENTS AND THEIR ACCESSORIES

The accuracy achieved with total station instruments is not merely a function of their ability to resolve angles and distances. It is also related to operator procedures and the condition of the total station instrument and other peripheral equipment being used with it. Operator procedure pertains to matters such as careful centering and leveling of the instrument, accurate pointing at targets, and observing proper field procedures such as taking averages of multiple angle observations made in both direct and reverse positions.

In Section 8.2, three reference axes of a total station instrument were defined: (a) the line of sight, (b) the horizontal axis, and (c) the vertical axis. These instruments also have a fourth reference axis, (d) the *axis of the plate-level vial* (see Section 4.8). For a properly adjusted instrument, the following relationships should exist between these axes: (1) the axis of the plate-level vial should be perpendicular to the vertical axis; (2) the horizontal axis should be perpendicular to the vertical axis, which makes the plate-level vial axis and horizontal axis parallel; and (3) the line of sight should be perpendicular to the horizontal axis. If these conditions do not exist, accurate observations can still be made by following proper procedures. However, it is more convenient if the instrument is in adjustment. Today, most total stations have calibration procedures that can electronically compensate for conditions (1) and (2) using sightings to well-defined targets with menu-defined procedures that can be performed in the field. However, if the operator is in doubt about the calibration procedures, a qualified technician should always be consulted.

TRIGONOMETRIC LEVELING NOTES

	(a) Sta/Pos	(b) BS(+)	(c) BD	(d) FS(−)	(e) FD	(f) ΔElev
A						
	D	1.211	98.12	1.403	86.34	
	D	1.210		1.403		
	R	1.211		1.404		
	R	1.211		1.403		
	Mean	1.2108		1.4033		−0.192
B						
	D	−5.238	101.543	−9.191	93.171	
	D	−5.236		−9.191		
	R	−5.238		−9.193		
	R	−5.237		−9.192		
	Mean	−5.2373		−9.1918		3.954
C						
	D	4.087	73.245	−3.849	97.392	
	D	4.088		−3.851		
	R	4.086		−3.849		
	R	4.087		−3.849		
	Mean	4.0870		−3.8495		7.936
D						
	D	3.214	89.87	6.507	97.392	
	D	3.214		6.507		
	R	3.214		6.508		
	R	3.215		6.507		
	Mean	3.2143		6.5072		−3.293
E						
					Sum	8.405

Figure 8.23
Trigonometric leveling field notes.

The adjustment for making the line of sight perpendicular to the horizontal axis was described in Section 8.15, and the procedure for making the axis of the plate bubble perpendicular to the vertical axis is given in Section 8.19.1. The test to determine if a total station's horizontal axis is perpendicular to its vertical axis is a simple one. With the instrument in the direct mode, it is set up a convenient distance away from a high vertical surface, say the wall of a two- or three-story building. After carefully leveling the instrument, sight a well-defined point, say A, high on the wall, at an altitude angle of at least 30°, and clamp the horizontal lock. Revolve (plunge) the telescope about its horizontal axis to set a point, B, on the wall below A just above ground level. Plunge the telescope to put it in reverse mode, turn the telescope 180° in azimuth, sight point A again, and clamp the horizontal lock. Plunge the telescope to set another point, C, at the same level as B. If B and C coincide, no adjustment is necessary. If the two points do not agree, then the horizontal axis is not perpendicular to the vertical axis. If an adjustment for this condition is necessary, the operator should refer to the manual that came with the instrument or send the instrument to a qualified technician.

Peripheral equipment that can affect accuracy includes tribrachs, plummets, prisms, and prism poles. Tribrachs must provide a snug fit without slippage. Plummets that are out of adjustment cause instruments to be miscentered over the point. Crooked prism poles or poles with circular bubbles that are out of adjustment also cause errors in placement of the prism over the point being observed. Prisms should be checked periodically to determine their constants (see Section 6.24.2), and their values stored for use in correcting distance observations. Surveyors should always heed the following axiom: *In practice, instruments should always be kept in good adjustment, but used as though they might not be.*

In the following subsections, procedures are described for making some relatively simple adjustments to equipment that can make observing more efficient and convenient, and also improve accuracy in the results.

8.19.1 Adjustment of Plate-Level Vials

As stated earlier, two types of leveling systems are used on total station instruments: (a) plate-level vials and (b) electronic leveling systems. These systems control the fine level of the instrument. If an instrument is equipped with a plate-level vial, it can easily be tested for its state of adjustment. To make the test, the instrument should first be leveled following the procedures outlined in Section 8.5. Then after carefully centering the bubble, the telescope should be rotated 180° from its first position. If the level vial is in adjustment, the bubble will remain centered. If the bubble deviates from center, the axis of the plate-level vial is not perpendicular to the vertical axis. The amount of bubble run indicates twice the error that exists. Level vials usually have a capstan adjusting screw for raising or lowering one end of the tube. If the level vial is out of adjustment, it can be adjusted by bringing the bubble *halfway back* to the centered position by turning the screw. Repeat the test until the bubble remains centered during a complete revolution of the telescope. If the instrument is equipped with an electronic level, follow the procedures outlined in the operator's manual to adjust the leveling mechanism.

If a plate bubble is out of adjustment, the instrument can be used without adjusting it and accurate results can still be obtained, but the specific procedures described in Section 8.20.1 must be followed.

8.19.2 Adjustment of Tripods

The nuts on the tripod legs must be tight to prevent slippage and rotation of the head. They are correctly adjusted if each tripod leg falls slowly of its own weight when placed in a horizontal position. If the nuts are overly tight, or if pressure is applied to the legs crosswise (which can break them) instead of lengthwise to fix them in the ground, the tripod is in a strained position. The result may be an unnoticed movement of the instrument head after the observational process has begun.

Tripod legs should be well spread to furnish stability and set so that the telescope is at a convenient height for the observer. Tripod shoes must be tight.

Proper field procedures can eliminate most instrument maladjustments, but there is no method that corrects a poor tripod with dried-out wooden legs, except to discard or repair it.

8.19.3 Adjustment of Tribrachs

The tribrach is an essential component of a secure and accurate setup. It consists of a minimum of three components, which are (1) a clamping mechanism, (2) leveling screws, and (3) a circular level bubble. As shown in Figure 8.3, some tribrachs also contain an optical plummet to center the tribrach over a station. The clamping mechanism consists of three slides that secure three posts that protrude from the base of the instrument or tribrach adapter. As the tribrach wears, the clamping mechanism may not sufficiently secure the instrument during observation procedures. When this happens, the instrument will move in the tribrach after it has been clamped, and the tribrach should be repaired or replaced.

8.19.4 Adjustment of Plummets

The line of sight in a plummet should coincide with the vertical axis of the instrument. Two different situations exist: (1) the plummet is enclosed in the alidade of the instrument and rotates with it when turned in azimuth, or (2) the plummet is part of the tribrach that is fastened to the tripod and does not turn in azimuth.

To adjust a plummet contained in the alidade, set the instrument over a fine point and aim the line of sight exactly at it by turning the leveling screws. Carefully adjust for any existing parallax. Rotate the instrument 180° in azimuth. If the plummet reticle moves off the point, bring it *halfway* back by means of the adjusting screws provided. These screws are similar to those shown in Figure 8.19. As with any adjustment, repeat the test to check the adjustment and correct if necessary.

For the second case where the optical plummet is part of the tribrach, carefully lay the instrument, with the tribrach attached, on its side (horizontally) on a stable, horizontal base such as a bench or desk, and clamp it securely. Fasten a sheet of paper on a vertical wall at least six feet away, such that it is in the field of view of the optical plummet's telescope. With the horizontal lock clamped, mark the position of the optical plummet's line of sight on the paper. Release the horizontal lock and rotate the tribrach 180°. If the reticle of the optical plummet moves off the point, bring it *halfway* back by means of the adjusting screws. Center the reticle on the point again with the leveling screws, and repeat the test.

8.19.5 Adjustment of Circular Level Bubbles

If a circular-level bubble on a total station does not remain centered when the instrument is rotated in azimuth, the bubble is out of adjustment. It should be corrected, although precise adjustment is unnecessary because it does not control fine leveling of the reference axes. To adjust the bubble, carefully level the instrument using the plate bubble and then center the circular bubble using its adjusting screws.

Circular bubbles used on prism poles and level rods must be in good adjustment for accurate work. To adjust them, carefully orient the rod or pole vertically by aligning it parallel to a long plumb line, and fasten it in that position using shims and C-clamps. Then center the bubble in the vial using the adjusting screws. Special adapters have been made to aid in the adjustment of the circular level bubble on rods or poles by some vendors.

For instruments such as automatic levels that do not have plate bubbles, use the following procedure. To adjust the bubble, carefully center it using the leveling screws and turn the instrument 180° in azimuth. *Half* of the bubble run is corrected by manipulating the vial-adjusting screws. Following the adjustment, the bubble should be centered using the leveling screws, and the test repeated.

8.20 SOURCES OF ERROR IN TOTAL STATION WORK

Errors in using total stations result from *instrumental*, *natural*, and *personal* sources. These are described in the subsections that follow.

8.20.1 Instrumental Errors

Figure 8.24 shows the fundamental reference axes of a total station. As discussed in Section 8.19, for a properly adjusted instrument, the four axes must bear specific relationships to each other. These are: (1) the vertical axis should be perpendicular to the axis of the plate-level vial, (2) the horizontal axis should be perpendicular to the vertical axis, and (3) the axis of sight should be perpendicular to the horizontal axis. If these relationships are not true, errors will result in measured angles unless

Figure 8.24
Reference axes of a total station instrument. (Courtesy Topcon America Corp.)

proper field procedures are observed. A discussion of errors caused by maladjustment of these axes, and of other sources of instrumental errors follows.

1. *Plate bubble out of adjustment.* If the axis of the plate bubble is not perpendicular to the vertical axis, the latter will not be truly vertical when the plate bubble is centered. *This condition causes errors in observed horizontal and vertical angles that cannot be eliminated by averaging direct and reverse readings.* The plate bubble is out of adjustment if after centering it runs when the instrument is rotated 180° in azimuth. The situation is illustrated in Figure 8.25. With the telescope initially pointing to the right and the bubble centered, the axis of the level vial is horizontal, as indicated by the solid line labeled *ALV*-1. Because the level vial is out of adjustment, it is not perpendicular to the vertical axis of the instrument, but instead makes an angle of $90° - \alpha$ with it. After turning the telescope 180°, it points left and the axis of the level vial is in the position indicated by the dashed line labeled *ALV*-2. The angle between the axis of the level vial and vertical axis is still $90° - \alpha$ but as shown in the figure, its indicated dislevelment, or bubble run, is E. From the figure's geometry, $E = 2\alpha$ is double the bubble's maladjustment. The vertical axis can be made truly vertical by bringing the bubble back *half of the bubble run*, using the foot screws. Then, even though it is not centered, the bubble should stay in the same position as the instrument is rotated in azimuth, and accurate angles can be observed. Although instruments can be used to obtain accurate results with their plate bubbles maladjusted, it is inconvenient and time consuming, so the required adjustment should be made as discussed in Section 8.19.1.

As noted earlier, some total stations are equipped with dual-axis compensators, which are able to sense the amount and direction of vertical axis tilt automatically. They can make corrections computationally in real time to both horizontal and vertical angles for this condition. Instruments equipped with single-axis compensators can only correct vertical angles. Procedures outlined in the manuals that accompany the instruments should be followed to properly remove any error.

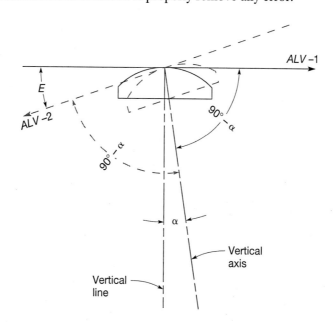

Figure 8.25
Plate bubble out of adjustment.

As was stated in Section 8.8, total station instruments with dual-axis compensators can apply a mathematical correction to horizontal angles, which accounts for any dislevelment of the horizontal and vertical axes. However, for this correction to be accurate, the instrument must be calibrated following the procedures outlined in step 8 of this section. In Figure 8.26, to sight on point S, the telescope is plunged upward. Because the instrument is misleveled, the line of sight scribes an inclined line SP' instead of the required vertical line SP. The angle between these two lines is α, the amount that the instrument is out of level. From this figure, it can be shown that the error in the horizontal direction, E_H, is

$$E_H = \alpha \tan(v) \tag{8.4}$$

In Equation (8.4), v is the altitude angle to point S. For the observation of any horizontal angle if the altitude angles for both the backsight and foresight are nearly the same, the resultant error in the horizontal angle is negligible. In flat terrain, this is approximately the case and the error due to dislevelment can be small. However, in mountainous terrain where the elevations of backsight and foresight pointings can vary by large amounts, this error can become substantial. For example, assume that an instrument that is 20″ out of level reads a backsight zenith angle as 93°, and the foresight zenith angle as 80°. The horizontal error in the backsight direction would be $20″\tan(-3°) = -1.0″$ and in the foresight is $20″\tan(10°) = 3.5″$ resulting in a cumulative error in the horizontal angle of $3.5″ - (-1″) = 4.5″$. This is a systematic error that becomes more serious as larger vertical angles are observed. It is critical in astronomical observations for azimuth as discussed in Appendix C.

Two things should be obvious from this discussion, it is important to check (1) the adjustment of the plate bubble often and (2) check the position of the bubble during the observation process.

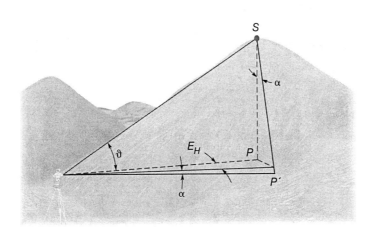

Figure 8.26
Geometry of instrument dislevelment.

2. *Horizontal axis not perpendicular to vertical axis.* This situation causes the axis of sight to define an inclined plane as the telescope is plunged and, therefore, if the backsight and foresight have differing angles of inclination, incorrect horizontal angles will result. Errors from this origin can be canceled by averaging an equal number of direct and reverse readings, or by double centering if prolonging a straight line. With total station instruments having dual-axis compensation, this error can be determined in a calibration process that consists of carefully pointing to the same target in both direct and reverse modes. From this operation the microprocessor can compute and store a correction factor. It is then automatically applied to all horizontal angles subsequently observed.

3. *Axis of sight not perpendicular to horizontal axis.* If this condition exists, as the telescope is plunged, the axis of sight generates a cone whose axis coincides with the horizontal axis of the instrument. The greatest error from this source occurs when plunging the telescope, as in prolonging a straight line or measuring deflection angles. Also, when the angle of inclination of the backsight is not equal to that of the foresight, observed horizontal angles will be incorrect. These errors are eliminated by double centering and by averaging equal numbers of direct and reverse readings.

4. *Vertical-circle indexing error.* As noted in Section 8.13, when the axis of sight is horizontal, an altitude angle of zero, or a zenith angle of either 90° or 270°, should be read; otherwise, an indexing error exists. The error can be eliminated by computing the mean from equal numbers of altitude (or zenith) angles read in the direct and reverse modes. With most total station instruments, the indexing error can be determined by carefully reading the same zenith angle both direct and reverse. The value of the indexing error is then computed, stored, and automatically applied to all observed zenith angles. However, the determination of the indexing error should be done carefully during calibration to ensure that an incorrect calibration is not applied to all subsequent angles observed with the instrument.

5. *Eccentricity of centers.* This condition exists if the geometric center of the graduated horizontal (or vertical) circle does not coincide with its center of rotation. Errors from this source are usually small. Total stations may also be equipped with systems that automatically average readings taken on opposite sides of the circles, thereby compensating for this error.

6. *Circle graduation errors.* If graduations around the circumference of a horizontal or vertical circle are nonuniform, errors in observed angles will result. These errors are generally very small. Some total stations always use readings taken from many locations around the circles for each observed horizontal and vertical angle, thus providing an elegant system for eliminating these errors.

7. *Errors caused by peripheral equipment.* Additional instrumental errors can result from worn tribrachs, plummets that are out of adjustment, unsteady

tripods, and sighting poles with maladjusted circular bubbles. This equipment should be regularly checked and kept in good condition or adjustment. Procedures for adjusting these items are outlined in Section 8.19.

8. *Adjustment of dual-axis compensators.* Most instruments today have dual-axis compensators. The correct adjustment of these instruments is critical to obtaining accurate horizontal and vertical readings from the instrument. Before performing this adjustment, read your instrument manual and understand the steps listed in the manual in performing this adjustment. Additionally, the adjustment should only be performed on a calm day with a sound tripod and a shaded instrument. However even with this adjustment recently performed, the instrument should always be used as if it was out of adjustment since one never knows if the adjustment is maintained during use.

To perform the adjustment, select an area where there is both a nearly level line of sight and one elevated at 10° or more above horizontal. The two targets used in the adjustment do not have to be aligned but they should provide well-defined horizontal and vertical sightings. The instrument must be truly leveled so if the instrument has a cross-axis level vial that is manually leveled, and make sure that the bubble remains centered in all directions before performing this adjustment. If it does not, follow the procedures in step 1 of this section to adjust the level vial; check and recheck the level ensuring that it stays centered in all positions about the horizon. After this step, you can also adjust the circular bubble on the instrument. If the total station has electronic levels, you will need to get the same tilt readings when the instrument is rotated 180° from its initial position. That is, the electronic compensators may not be centered but will be off center by the same amount in the first and rotated positions.

To start the adjustment, put the instrument into its calibration mode (see your manual) and sight a well-defined target that is near level in the face 1 position. Care must be taken in sighting both the horizontal and vertical positions on the target. Take an odd number of readings, such as three or five, in this position. Rotate the instrument 180° and sight the target in the face 2 position and take the same number of readings. Now in the face 2 position, sight and take readings on the target that is elevated. After obtaining these readings, rotate the instrument back into the face 1 position and repeat the procedure on the elevated target. At this point, most instruments will display the amount of adjustment in the x and y axes and its tilt while also correcting the electronic level vial display. You should now be able to sight and record readings in both faces and have these readings agree within the specifications of the instrument. You should also be able to set zero on the horizontal circle, rotate the telescope to 35° in elevation, and not have the horizontal circle reading vary by more than $5''.9$. Finally, as with any operation in surveying, if you are unsure about the procedures necessary to adjust and calibrate an instrument, you should leave these procedures to a qualified professional. However, checking and sometimes adjusting an instrument is an important step in obtaining accurate observations and in being a professional surveyor, and thus should be practiced on regular intervals with your equipment.

8.20.2 Natural Errors

1. *Wind.* Wind vibrates the tripod that the total station instrument rests on. On high setups, light wind can vibrate the instrument to the extent that precise

pointings become impossible. Shielding the instrument, or even suspending observations on precise work, may be necessary on windy days. An optical plummet is essential for making setups in this situation.

2. *Temperature effects.* Temperature differentials cause unequal expansion of various parts of total station instruments. This causes bubbles to run, which can produce erroneous observations. Shielding instruments from sources of extreme heat or cold reduces temperature effects.

3. *Refraction.* Unequal refraction bends the line of sight and may cause an apparent shimmering of the observed object. It is desirable to keep lines of sight well above the ground and avoid sights close to buildings, smokestacks, vehicles, and even large individual objects in generally open spaces. In some cases, observations may have to be postponed until atmospheric conditions have improved.

4. *Tripod settlement.* The weight of an instrument may cause the tripod to settle, particularly when set up on soft ground or asphalt highways. When a job involves crossing swampy terrain, stakes should be driven to support the tripod legs and work at a given station completed as quickly as possible. Stepping near a tripod leg or touching one while looking through the telescope will demonstrate the effect of settlement on the position of the bubble and cross wires. Most total station instruments have sensors that tell the operator when dislevelment has become too severe to continue the observation process.

8.20.3 Personal Errors

1. *Instrument not set up exactly over point.* Miscentering of the instrument over a point will result in an incorrect horizontal angle being observed. As shown in Figure 8.27, instrument miscentering will cause errors in both the backsight and foresight directions of an angle. The amount of error is dependent on the position of the instrument in relation to the point. For instance, in Figure 8.27(a), the miscentering that is depicted will have minimal effect on the observed angle since the error on the backsight to P_1 will partially cancel the error on the foresight to P_2. However, in Figures 8.27(b) and (c), the effect of the miscentering has a maximum effect on the observed angular values. Since the position of the instrument is random in relation to the station, it is important to carefully center the instrument over the station when observing angles. The position should be checked at intervals during the time a station is occupied to be certain it remains centered.

2. *Bubbles not centered perfectly.* The bubbles must be checked frequently but NEVER releveled between a backsight and a foresight—only *before* starting and *after* finishing an angular position.

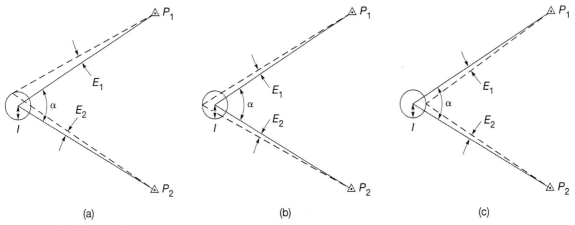

Figure 8.27 Effects of instrument miscentering on an angle.

3. *Improper use of clamps and tangent screws.* An observer must form good operational habits and be able to identify the various clamps and tangent screws by their touch without looking at them. Final setting of tangent screws is always made with a positive motion to avoid backlash. Clamps should be tightened just once and not checked again to be certain they are secure.

4. *Poor focusing.* Correct focusing of the eyepiece on the cross hairs, and of the objective lens on the target, is necessary to prevent parallax. Objects sighted should be placed as near the center of the field of view as possible. Focusing affects pointing, which is an important source of error. Some instruments like the one shown in Figure 8.24, automatic focusing of the objective lens is provided. These devices are similar to the modern photographic camera, and can increase the speed of the survey when sight distances to the targets vary.

5. *Overly careful sights.* Checking and double-checking the position of the crosshair setting on a target wastes time, and actually produces poorer results than one fast observation. The cross hair should be aligned quickly, and the next operation begun promptly. Beginners often want someone else to check their sights. This should never be done due to personal preferences, abilities, and physical limitations.

6. *Careless plumbing and placement of rod.* One of the most common errors results from careless plumbing of a rod when the instrument operator because of brush or other obstacles in the way can only see the top. Another is caused by placing a pole off-line behind a point to be sighted.

■ 8.21 PROPAGATION OF RANDOM ERRORS IN ANGLE OBSERVATIONS

Random errors are present in every horizontal angle observation. Whenever an instrument's circles are read, a small error is introduced into the final angle.

Similarly, each operator will have some miscentering on the target. These error sources are random. They may be small or large, depending on the instrument, the operator, and the conditions at the time of the angle observation. Increasing the number of angle repetitions can reduce the effects of reading and pointing.

With the introduction of total station instruments, standards were developed for estimating errors in angle observations caused by reading and pointing on a well-defined target. The original German standard, called DIN 18723, provides values for estimated errors in the mean of two direction observations, one each in the direct and reverse modes. It has since been replaced by the International Organization for Standards (ISO) 17123-3 standard. The instrument shown in Figure 8.1 has an ISO 17123-3 accuracy of $\pm 2''$, and the one in Figure 8.2 has an ISO 17123-3 accuracy of $\pm 5''$. Manufacturers often make a series of instruments that vary only in their ability to resolve angles. The DIN 18723 and ISO 17123-3 standards were created to allow individuals to determine the difference in the quality of the instruments. Neither standard is an accurate representation of one's ability to use and resolve an angle with the instrument. However, they do provide a means by which the uncertainty in angles can be estimated.

A set of angles observed with a total station will have an estimated error of

$$E = \frac{2E_{ISO}}{\sqrt{n}} \qquad (8.5)$$

where E is the estimated error in the angle due to pointing and reading, n is the total number of angles read in both direct and reverse modes, and E_{ISO} is the manufacturer's specified ISO 17123-3 or DIN 18723 error.

■ Example 8.2

Three sets of angles (3D and 3R) are measured with an instrument having an ISO 17123-3 specified accuracy of $\pm 2''$. What is the estimated error in the angle?

Solution

By Equation (8.5), the estimated error is

$$E = \frac{2(2'')}{\sqrt{6}} = \pm 1.6''$$

■ 8.22 MISTAKES

Some common mistakes in angle observation work are as follows:
1. Sighting on, or setting up over, the wrong point.
2. Calling out or recording an incorrect value.
3. Improper focusing of the eyepiece and objective lenses of the instrument.
4. Leaning on the tripod or placing a hand on the instrument when pointing or taking readings.

PROBLEMS

Asterisks (*) indicate problems that have partial answers given in Appendix G.

8.1 What are the defining characteristics of a total station?

8.2 What are the three axes of a total station?

8.3 What are the primary sources of random instrumental error in a total station instrument?

8.4 What are the defining characteristics of a robotic total station?

8.5 How does the angle measurement system work in a total station?

8.6 Why are the bases of total station instruments designed to be interchanged with other equipment?

8.7 Why is it important not to sight the EDM reflector when turning an angle?

8.8 Describe the process of setting up a total station instrument over a point.

8.9 Why should an instrument never be transported on the tripod?

8.10 Why should the EDM reflector never be sighted when turning an angle?

8.11 Name and briefly describe the natural errors that can affect the accuracy of an observed angle.

8.12 Determine the angles subtended for the following conditions:
   (a)* a 1-cm diameter pipe sighted by total station from 100 m.
   (b) a 1/8-in. tack sighted by total station from 300 ft.
   (c) a 3/16-in. diameter chaining pin observed by total station from 600 ft.

8.13 What is the error in an observed direction for the situations noted?
   (a) Setting a total station 5 mm to the side of a tack on a 50-m sight
   (b) Lining in the edge (instead of center) of an 3/16-in. diameter chaining pin at 500 ft
   (c) Sighting the edge (instead of center) of a 2.5-cm diameter range pole 100 m
   (d) Sighting a target at the top of a 2-m range pole that is 10′ off-level on a 200-m sight

8.14* Intervening terrain obstructs the line of sight so only the top of a 6-ft-long pole can be seen on a 250-ft sight. If the range pole is out of plumb and leaning sideways 0.025 ft per vertical foot, what maximum angular error results?

8.15 Same as Problem 8.14, except that it is a 2.0-m pole that is out of plumb and leaning sideways 1 cm per meter on a 50 m sight.

8.16 Discuss the advantages of a robotic total station instrument.

8.17 What instrumental errors are compensated by averaging an equal number of observations with the telescope direct and reversed?

8.18 Describe the process of placing a point on line between two existing stations when the two stations are not intervisible.

8.19 An interior angle x and its explement y were turned to close the horizon. Each angle was observed once direct and once reversed using the repetition method. Starting with an initial backsight setting of 0°00′00″ for each angle, the readings after the first and second turnings of angle x were 50°38′48″ and 50°38′40″ and the readings after the first and second turnings of angle y were 309°21′16″ and 309°21′10″. Calculate each angle and the horizon misclosure.

8.20* A zenith angle is measured as 83°13′56″ in the direct position. What is the equivalent zenith angle in the reverse position?

8.21 What is the average zenith angle given the following direct and reverse readings?
   Direct: 87°45′04″, 87°45′12″, 87°45′08″
   Reverse: 272°14′50″, 272°14′48″, 272°14′52″

In Figure 8.9(c), direct and reverse directions observed with a total station instrument from A to points B, C, and D are listed in Problems 8.22 and 8.23. Determine the values of the four angles, and the horizon misclosure.

8.22 Direct: 0°00′00″, 26°29′21″, 92°57′44″, 185°34′12″, 0°00′04″
Reverse: 0°00′00″, 26°29′17″, 92°57′46″, 185°34′14″, 0°00′02″

8.23 Direct: 0°00′00″, 106°52′06″, 191°38′43″, 267°42′44″, 359°59′58″
Reverse: 0°00′00″, 106°52′04″, 191°38′41″, 267°42′40″, 0°00′00″

8.24* The angles at point X were observed with a total station instrument. Based on four readings, the standard deviation of the angle was ±5.6″. If the same procedure is used in observing each angle within a six-sided polygon, what is the estimated standard deviation of closure at a 95% level of probability?

8.25 The line of sight of a total station is o of adjustment by ±10.0″.
  (a) In prolonging a line by plunging the telescope between backsight and foresight, but not double centering, what angular error is introduced?
  (b) What off-line linear error results on a foresight of 200.0 ft?

8.26 A line PQ is prolonged to point R by double centering. Two foresight points R' and R'' are set. What angular error would be introduced in a single plunging based on the following lengths of QR and $R'R''$, respectively?
  (a)* 650.50 ft and 0.35 ft.
  (b) 298.406 m and 22 mm.

8.27 Explain why the "principal of reversion" is important in angle measurement.

8.28* A total station with a 20″/div. level bubble is one division out of level on a point with an altitude angle of 38°50′44″ What is the error in the horizontal pointing?

8.29 What is the equivalent altitude angle for a zenith angle of 98°26′04″?

8.30 What is the equivalent altitude angle for a zenith angle of 85°14′30″?

8.31 What error in horizontal angles is consistent with the following linear precisions?
  (a) 1/5000, 1/10,000, 1/20,000, and 1/50,000
  (b) 1/4000, 1/8,000, 1/30,000, and 1/60,000

8.32 Why is it important to check for sloppiness in a tribrach?

8.33 Describe the procedure for testing a tripod.

8.34 List the procedures for prolonging a line of sight.

8.35 A total station has an ISO 17123-3 specified accuracy of ±3″. What is the estimated precision of an angle observed with eight repetitions?

8.36 Similar to Problem 8.35 except the instrument has an ISO 17123-3 specified accuracy of ±1″, and the angle is observed with two repetitions.

BIBLIOGRAPHY

Boyd, J. 2015. "Checking the 3 Axis Adjustment of an Electronic Total Station." *The Pennsylvania Surveyor* Winter: 17.

Crawford, W. 2001. "Calibration Field Tests of Any Angle Measuring Instrument." *Point of Beginning* 26 (No. 8): 54.

———. 2009. "Back to Basics: Quick Setup with a Laser Plummet." *Point of Beginning* 35 (No. 3): 38.

GIA. 2001. "Electronic Angle Measurement." *Professional Surveyor* 21 (No. 10): 47.

———. 2002. "2-axis Compensators." *Professional Surveyor* 22 (No. 9): 38.

———. 2002. "Basic Total Station Calibration." *Professional Surveyor* 22 (No. 5): 60.

———. 2005. "How Things Work: Modern Total Station and Theodolite Axes." *Professional Surveyor* 25 (No. 10): 42.

Stevens, K. 2003. "Locking in the Benefits." *Point of Beginning* 28 (No. 11): 16.

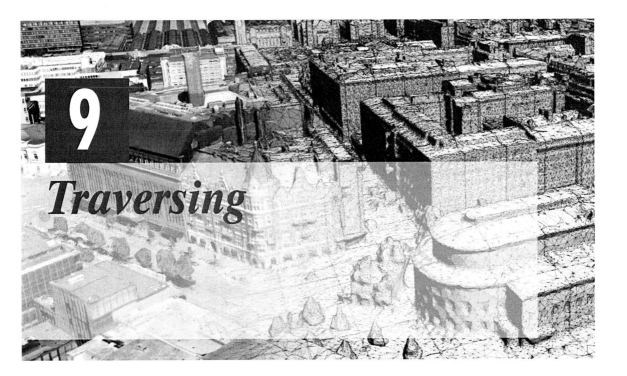

9
Traversing

■ 9.1 INTRODUCTION

A traverse is a series of consecutive lines whose ends have been marked in the field, and whose lengths and directions have been determined from observations. In traditional surveying by ground methods, *traversing*, the act of marking the lines, that is, establishing traverse stations and making the necessary observations, is one of the most basic and widely practiced means of determining the relative locations of points.

There are two kinds of traverses: *closed* and *open*. Two categories of closed traverses exist: *polygon* and *link*. In the polygon traverse, as shown in Figure 9.1(a), the lines return to the starting point, thus forming a closed figure that is both geometrically and mathematically closed. Link traverses finish upon another station that should have a positional accuracy equal to or greater than that of the starting point. The link type (geometrically open, mathematically closed), as illustrated in Figure 9.1(b), must have a closing reference direction, for example, line $E - AzMk_2$. Closed traverses provide checks on the observed angles and distances, which is an extremely important consideration. They are used extensively in control, construction, property, and topographic surveys.

If the distance between stations C and E in Figure 9.1(a) were observed, the resultant set of observations would become what is called a *network*. A network involves the interconnection of stations within the survey to create additional redundant observations. Networks offer more geometric checks than closed traverses. For instance in Figure 9.1(a), after computing coordinates on stations C and E using elementary procedures, the observed distance CE can be compared against a value obtained by inversing the coordinates (see Chapter 10 for discussion on computation of coordinates and inversing coordinates). Figure 9.7(b) shows another example where a network has been developed. Networks should be adjusted using the method of least squares as presented in Chapter 16.

9.1 Introduction   227

Figure 9.1
Examples of closed traverses.

An open traverse (geometrically and mathematically open) (Figure 9.2) consists of a series of lines that are connected but do not return to the starting point or close upon a point of equal or greater order accuracy. *Open traverses should be avoided because they offer no means of checking for observational errors and mistakes.* If they must be used, observations should be repeated carefully to guard against mistakes. The precise control-traversing techniques presented in Section 19.12.2 should be considered in these situations.

Hubs (wooden stakes with tacks to mark the points), steel stakes, or pipes are typically set at each traverse station A, B, C, etc., in Figures 9.1 and 9.2, where a change in direction occurs. Spikes, "P-K"[1] nails, and scratched crosses are used on blacktop pavement. Chiseled or painted marks are made on concrete. Traverse stations are sometimes interchangeably called *angle points* because an angle is observed at each one.

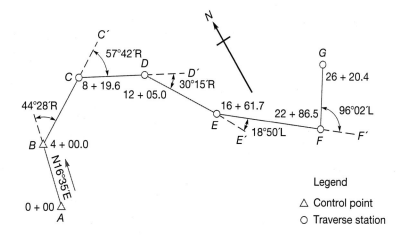

Figure 9.2
Open traverse.

---

[1]P-K is a trade name for concrete nails. The Parker–Kalon Company originally manufactured these nails. There is a small depression in the center of the nail that serves as a marker for the location of the station. Several companies now manufacture similar or better versions of this nail. Still the original name, P-K, is used to denote this type of nail.

9.2 OBSERVATION OF TRAVERSE ANGLES OR DIRECTIONS

The methods used in observing angles or directions of traverse lines vary and include (1) interior angles, (2) angles to the right, (3) deflection angles, and (4) azimuths. These are described in the following subsections.

9.2.1 Traversing by Interior Angles

Interior-angle traverses are used for many types of work, but they are especially convenient for property surveys. Although interior angles could be observed either clockwise or counterclockwise, *to reduce mistakes in reading, recording, and computing, they should always be turned clockwise* from the backsight station to the foresight station. The procedure is illustrated in Figure 9.1(a). In this text, except for left deflection angles, clockwise turning will always be assumed. Furthermore, when angles are designated by three station letters or numbers in this text, the backsight station will be given first, the occupied station second, and the foresight station third. Thus, angle *EAB* of Figure 9.1(a) is the angle observed at station *A*, with the backsight on station *E* and the foresight at station *B*.

Interior angles may be improved by averaging equal numbers of direct and reverse readings. As a check, exterior angles may also be observed to close the horizon (see Section 8.10). In the traverse of Figure 9.1(a), a reference line *A–Az MK* of known direction exists. Thus, the clockwise angle at *A* from *Az Mk* to *E* must also be observed to enable determining the directions of all other lines. This would not be necessary if the traverse contained a line of known direction, like *AB* of Figure 7.2, for example. Note that interior angles will be the same as angles to the right when stationing proceeds counterclockwise around a polygon traverse. However, with interior angles, the stationing can also proceed clockwise around a traverse also.

9.2.2 Traversing by Angles to the Right

Angles observed clockwise from a backsight on the "rearward" traverse station to a foresight on the "forward" traverse station [see Figures 9.1(a) and (b)] are called *angles to the right*. According to this definition, to avoid ambiguity in angle-to-the-right designations, the "sense" of the forward traverse direction must be established. This is normally done by consecutive numbering or lettering of traverse stations so that they increase in the forward direction. Depending on the direction of the traversing, angles to the right may be interior or exterior angles in a polygon traverse. If the direction of traversing is counterclockwise around the figure then clockwise interior angles will be observed. However, if the direction of traversing is clockwise, then exterior angles will be observed. Data collectors follow this convention when traversing. Thus, in Figure 9.1(b), for example, the direction from *A* to *B*, *B* to *C*, *C* to *D*, etc., is considered forward. By averaging equal numbers of direct and reverse readings, observed angles to the right can also be checked and their accuracy improved. From the foregoing definitions of interior angles and angles to the right, it is evident that in a polygon traverse the only difference between the two types of observational procedures may be ordering of the backsight and foresight stations since both procedures observe clockwise angles.

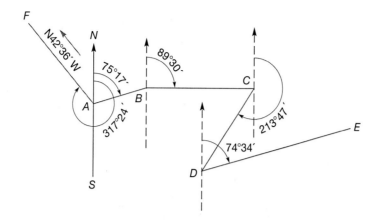

Figure 9.3
Azimuth traverse.

9.2.3 Traversing by Deflection Angles

Route surveys are commonly run by deflection angles observed to the right or left from the lines extended, as indicated in Figure 9.2. A deflection angle is not complete without a designation R or L, and, of course, it cannot exceed 180°. Each angle should be doubled or quadrupled, and an average value determined. The angles should be observed an equal number of times in face left and face right to reduce instrumental errors. Deflection angles can be obtained by subtracting 180° from angles to the right. Positive values so obtained denote right deflection angles; negative ones are left.

9.2.4 Traversing by Azimuths

With total station instruments, traverses can be run using azimuths. This process permits reading azimuths of all lines directly, and thus eliminates the need to calculate them. In Figure 9.3, azimuths are observed clockwise from the north end of the meridian through the angle points. The instrument is oriented at each setup by sighting on the previous station with either the back azimuth on the circle (if angles to the right are turned) or the azimuth (if deflection angles are turned), as described in Section 8.11. Then the forward station is sighted. The resulting reading on the horizontal circle will be the forward line's azimuth.

9.3 OBSERVATION OF TRAVERSE LENGTHS

The length of each traverse line (also called a course) must be observed, and this is usually done by the simplest and most economical method capable of satisfying the required precision of a given project. Their speed, convenience, and accuracy makes the EDM component of a total station instrument the most often used, although taping or other methods discussed in Chapter 6 could be employed. A distinct advantage of traversing with total station instruments is that both angles and distances can be observed with a single setup at each station. Averages of distances observed both forward and back will provide increased accuracy, and the repeat readings afford a check on the observations and are thus *redundant observations*. Additionally, as further discussed in Section 9.6, when backsighting a station, the distance can be observed and compared to the previous forward distance observation to check the instrument setup. Furthermore, as discussed in Chapter 16, these additional distance observations can be included in a least squares adjustment or used as an average length for the distance when performing traverse computations as discussed in Chapter 10. This additional check helps decrease potential errors in a survey and adds additional redundant observations to the survey.

Sometimes state statutes regulate the precision for a traverse to locate boundaries. On construction work, allowable limits of closure depend on the use and extent of the traverse and project type. Bridge locations and high-rise buildings, for example, demand a high degree of precision.

In closed traverses, each course is observed and recorded as a separate distance. On long, link traverses for highways and railroads, distances are carried along continuously from the starting point using stationing (see Section 5.9.1). In Figure 9.2, which uses stationing in feet, for example, beginning with station $0 + 00$ at point A, 100-ft stations $(1 + 00, 2 + 00,$ and $3 + 00)$ are marked until hub B at station $4 + 00$ is reached. Then stations $5 + 00, 6 + 00, 7 + 00, 8 + 00,$ and $8 + 19.60$ are set along course BC to C, etc. The length of a line in a stationed link traverse is the difference between stationing at its end points; thus, the length of line BC is $819.60 - 400.00 = 419.60$ ft.

9.4 SELECTION OF TRAVERSE STATIONS

Positions selected for setting traverse stations vary with the type of survey. In general, guidelines to consider in choosing them include accuracy, utility, and efficiency. Of course, intervisibility between adjacent stations, forward and back, must be maintained for angle and distance observations. The stations should also ideally be set in convenient locations that allow for easy access. Ordinarily, stations are placed to create lines that are as long as possible. This not only increases efficiency by reducing the number of instrument setups, but also increases accuracy in angle observations. However, utility may override using very long lines because intermediate hubs, or stations at strategic locations, may be needed to complete the survey's objectives. Seasonal variations may also improve sight lines. For example, lack of foliage may aid visibility between stations during the late fall, winter, and early spring.

Often the number of stations can be reduced, and the length of the sight lines increased by careful reconnaissance. It is always wise to "walk" the area being surveyed and find "ideal" locations for stations before the traverse stakes are set and the observation process undertaken.

Each different type of survey will have its unique requirements concerning traverse station placement. On property surveys, for example, traverse stations are placed at each corner if the actual boundary lines are not obstructed and can be occupied. If offset lines are necessary, a stake is located near each corner to simplify the observations and computations. Long lines and rolling terrain may necessitate extra stations.

On route surveys, stations are set at each angle point, and at other locations where necessary to obtain topographic data or extend the survey. Usually the centerline is run before construction begins, but it will likely be destroyed and need replacement one or more times during various phases of the project. An offset traverse can be used to avoid this problem.

A traverse run to provide control for topographic mapping serves as a framework to which map details such as roads, buildings, streams, and hills are referenced. Station locations must be selected to permit complete coverage of the area to be mapped. *Spurs* consisting of one or more lines may branch off as open (*stub*) traverses to reach vantage points. However, their use should be discouraged since a check on their positions cannot be made.

If a stub traverse must be executed, the surveyor should exercise extreme caution using multiple direct and reverse readings at each station to check their work. In some extreme cases, it may be advisable to repeat each setup at a later time. For example, setting over a station while both proceeding to the terminus of the stub and on the way back to the main traverse. It should be realized by the surveyor that observational errors can go undetected in an open traverse, thus additional observational checks such as closing the angular horizon and precise traversing techniques, which are discussed in Section 19.13.2, should be performed to ensure the correctness of the observations.

■ 9.5 REFERENCING TRAVERSE STATIONS

Traverse stations often must be found and reoccupied, months or even years after they are established. Also they may be destroyed through construction or other activity. Therefore, it is important that they be referenced by creating observational *ties* to them so that they can be relocated if obscured or reestablished, if destroyed.

Figure 9.4 presents a typical traverse tie. As illustrated, these ties consist of distance observations made to nearby fixed objects. Short lengths (less than 100 ft) are convenient if a steel tape is being used, but, of course, the distance to definite and unique points is a controlling factor. Two ties, preferably at about right angles to each other, are sufficient, but three should be used to allow for the possibility that one reference mark being destroyed. Ties to trees can be observed in hundredths of a foot if nails are driven into them. However, *permission must be obtained from the landowner before driving nails into trees*. It is always important to remember that the surveyor may be held legally responsible for any damages to property that may occur during the survey.

If natural or existing features such as trees, utility poles, or corners of buildings are not available, stakes may be driven and used as ties. Figure 9.5(a) shows an arrangement of *straddle hubs* well suited to tying in a point such as *H* on a

Figure 9.4
Referencing a point.

232 TRAVERSING

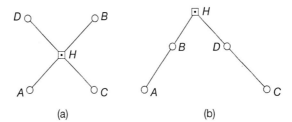

Figure 9.5
Hubs for ties.

highway centerline or elsewhere. Reference points A and B are carefully set on the line through H, as are C and D. Lines AB and CD should be roughly perpendicular, and the four points should be placed in safe locations, outside of areas likely to be disturbed. It is recommended that a third point be placed on each line to serve as an alternate in the event one point is destroyed. The intersection of the lines of sight of two total stations set up at A and C, and simultaneously aimed at B and D, respectively, will recover the point. The traverse hub H can also be found by intersecting strings stretched between diagonally opposite ties if the lengths are not too long. Hubs in the position illustrated by Figure 9.5(b) are sometimes used, but are not as desirable as straddle hubs for stringing.

■ 9.6 TRAVERSE FIELD NOTES

The importance of notekeeping was discussed in Chapter 2. Since a traverse is itself the end on a property survey and the basis for all other data in mapping, a single mistake or omission in recording is one too many. All possible field and office checks must therefore be made. A partial set of field notes for an interior-angle traverse run using a total station instrument is shown in Figure 9.6. Notice that details such as date, weather, instrument identifications, and party members and their duties are recorded on the right-hand page of the notes. Also a sketch with a north arrow is shown. The observed data are recorded on the left-hand page. First, each station that is occupied is identified, and the heights of the total station instrument and reflector that apply at that station are recorded. Then horizontal circle readings, zenith angles, horizontal distances, and elevation differences observed at each station are recorded. Notice that each horizontal angle is measured twice in the direct mode, and twice in the reverse mode. As noted earlier, this practice eliminates instrumental errors, and gives repeat angle values for checking. Zenith angles were also observed twice each direct and reverse. Although not needed for traversing, they are available for checking if larger than tolerable misclosures (see Chapter 10) should exist in the traverse. Details of making traverse observations with a total station instrument are described in Section 9.8.

An advantage of using a data collector in traversing is that each station setup can be checked immediately in the field by obtaining observations to the previous (backsight) station. Since data collectors compute station coordinates as the fieldwork progresses, it is possible to sight the previous station, obtain a distance observation and compare the forward coordinates of the station with those determined on the backsight. The typical procedure is to enter the previous station as the backsight in the controller software. The instrument is then zeroed while observing this backsight station, and a distance to this station is observed and recorded. Since the controller "knows" the azimuth to the previous station by coordinate inversing (see

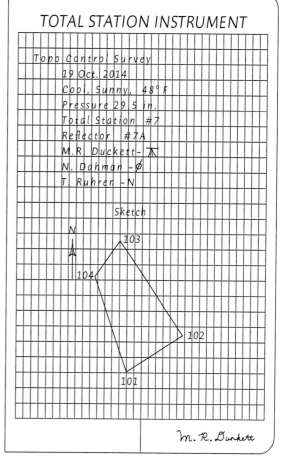

Figure 9.6
Example traverse field notes using a total station instrument.

Section 10.11), it can compute a second set of coordinates for the previous station. At this point, the difference in the two computed positions can be compared. If this difference is beyond tolerances, incorrect field procedures need to be addressed. Most controller software has the display of the differences in the coordinates shown automatically once this is done. A common field mistake uncovered in this procedure is failure to set the horizontal circle to zero on the backsight. Although it rarely happens, it is also possible to uncover observational distance errors.

9.7 ANGLE MISCLOSURE

The angular misclosure for an interior-angle traverse is the difference between the sum of the observed angles and the geometrically correct total for the polygon. The sum, Σ, of the interior angles of a closed polygon should be

$$\Sigma = (n - 2)180° \qquad (9.1)$$

where n is the number of sides, or angles, in the polygon. This formula is easily derived from known facts. The sum of the angles in a triangle is 180°; in a rectangle,

360°; and in a pentagon, 540°. Thus, each side added to the three required for a triangle increases the sum of the angles by 180°. As was mentioned in Section 7.3, if the direction about a traverse is clockwise when observing angles to the right, exterior angles will be observed. In this case, the sum of the exterior angles will be

$$\Sigma = (n + 2)180° \qquad (9.2)$$

Figure 9.1(a) shows a five-sided figure in which, if the sum of the observed interior angles equals 540°00′05″, the angular misclosure is 5″. Misclosures result from the accumulation of random errors in the angle observations. Permissible misclosure can be computed by the formula

$$c = K\sqrt{n} \qquad (9.3)$$

where n is the number of angles, and K a constant that depends on the level of accuracy specified for the survey. The Federal Geodetic Control Subcommittee (FGCS) recommends constants for five different orders of traverse accuracy: *first-order*, *second-order class I*, *second-order class II*, *third-order class I*, and *third-order class II*. Values of K for these orders, from highest to lowest, are 1.7″, 3″, 4.5″, 10″, and 12″, respectively. Thus, if the traverse of Figure 9.1(a) were being executed to second-order class II standards, its allowable misclosure error would be $4.5″\sqrt{5} = \pm 10″$.

The algebraic sum of the deflection angles in a closed-polygon traverse equals 360°, clockwise (right) deflections being considered plus and counterclockwise (left) deflections, minus. This rule applies if lines do not crisscross, or if they cross an even number of times. When lines in a traverse cross an odd number of times, the sum of right deflections equals the sum of left deflections.

A closed-polygon azimuth traverse is checked by setting up on the starting point a second time, after having occupied the successive stations around the traverse, and orienting by back azimuths. The azimuth of the first side is then obtained a second time and compared with its original value. Any difference is the misclosure. If the first point is not reoccupied, the interior angles computed from the azimuths will automatically check the proper geometric total, even though one or more of the azimuths may be incorrect.

Although angular misclosures cannot be directly computed for link traverses, the angles can still be checked. The direction of the first line may be determined from two intervisible stations with a known azimuth between them, or from a sun or Polaris observation, as described in Appendix C. Observed angles are then applied to calculate the azimuths of all traverse lines. The last line's computed azimuth is compared with its known value, or the result obtained from another Sun or Polaris observation. On long traverses, intermediate lines can be checked similarly. In using Sun or Polaris observations to check angles on traverses of long east-west extent, allowance must be made for *convergence of meridians*. This topic is discussed in Section 19.12.2.

■ 9.8 TRAVERSING WITH TOTAL STATION INSTRUMENTS

Total station instruments, with their combined electronic angle and distance measurement components, speed the process of traversing significantly because both the angles and distances can be observed from a single setup. The observing process is further aided because angles and distances are resolved automatically and displayed. Furthermore, the microprocessors of total stations can perform traverse computations, reduce slope distances to their horizontal and vertical components, and instantaneously calculate and store station coordinates and elevations. The reduction to obtain horizontal and vertical distance components was illustrated with the traverse notes of Figure 9.6.

To illustrate a method of traversing with a total station instrument, refer to the traverse of Figure 9.1(b). With the instrument set up and leveled at station A, a backsight is carefully taken on $Az\ MK_1$. The azimuth of line $A-Az\ MK_1$ is initialized on the horizontal circle by entering it in the unit using its keyboard. The coordinates and elevation of station A are also entered in memory. Next a foresight is made on station B. The azimuth of line AB will now appear on the display, and upon keyboard command, can be stored in the microprocessor's memory. Slope distance AB is then observed and reduced to its horizontal and vertical components by the microprocessor. Then the line's departure and latitude are computed and added to the coordinates of station A to yield the coordinates of station B. (Departures, latitudes, and coordinates are described in Chapter 10.) These procedures should be performed in both the direct and reverse modes, and the results averaged to account for instrumental errors.

The procedure outlined for station A is repeated at station B, except that the back-azimuth BA and coordinates of station B need not be entered; rather, they are recalled from the instrument's memory. From the setup at B, azimuth BC and coordinates of C are determined and stored. This procedure is continued until a station of known coordinates is reached, as E in Figure 9.1(b). Here, the known coordinates of E are entered in the unit's computer and compared to those obtained for E through the traverse observations. Their difference (or misclosure) is computed, displayed, and, if within allowable limits, distributed by the microprocessor to produce final coordinates of intermediate stations. (Procedures for distributing traverse misclosure errors are covered in Chapters 10 and 16.)

As discussed in Section 9.6, mistakes in orientation can be minimized when a data collector is used in combination with a total station. In this process, the coordinates of each backsight station are checked before proceeding with the angle and distance observations to the next foresight station. For example, in Figure 9.1(a), after the total station is leveled and oriented at station B, an observation is taken "back" on A. If the newly computed coordinates of A do not closely match their previously stored values, the instrument setup, leveling, and orientation should be rechecked, and the problem resolved before proceeding with any further measurements. This procedure often takes a minimal amount of time and typically identifies most field mistakes that occur during the observational process.

If desired, traverse station elevations can also be determined as a part of the procedure (usually the case for topographic surveys). Then entries hi (height of instrument) and hr (height of reflector) must be input (see Section 6.23). The microprocessor computes the vertical component of the slope distance, which includes a correction for curvature and refraction (see Section 4.5.4). The elevation difference is added to the occupied station's elevation to produce the next point's elevation. At the final station, any misclosure is determined by comparing the computed elevation with its known value, and if within tolerance, the misclosure is distributed to produce adjusted elevations of intermediate traverse stations.

236 TRAVERSING

All data from traversing with a total station instrument can be stored in a data collector for printing and transfer to the office for computing and plotting (see Sections 2.12 through 2.15). Alternatively, the traverse notes can be recorded manually as illustrated with Figure 9.6.

■ 9.9 RADIAL TRAVERSING

In certain situations, it may be most convenient to determine the relative positions of points by radial traversing. In this procedure, as illustrated in Figure 9.7(a), some point O, whose position is assumed known, is selected from which all points to be located can be seen. If a point such as O does not exist, it can be established. It is also assumed that a nearby azimuth mark, like Z in Figure 9.7(a), is available, and that reference azimuth OZ is known. With a total station instrument at point O, after backsighting on Z, horizontal angles to all stations A through F are observed. Azimuths of all radial lines from O (as OA, OB, OC, etc.) can then be calculated. The horizontal lengths of all radiating lines are also observed. By using the observed lengths and azimuths, coordinates for each point can be computed. (The subject of coordinate computations is discussed in Chapter 10.)

It should be clear that in the procedure just described, each point A through F has been surveyed independently of all others, and that no checks on their computed positions exist. To provide checks, lengths AB, BC, CD, etc., could be computed from the coordinates of points, and then these same lengths observed. This results in many extra setups and substantially more fieldwork, thus defeating one of the major benefits of radial traversing. To solve the problem of gaining checks with a minimum of extra fieldwork, the method presented in Figure 9.7(b) is recommended. Here, a second hub O' is selected from which all points can also be seen. The position of O' is determined by observations of the horizontal angle and distance from station O. This second hub O' is then occupied, and horizontal angles and distances to all stations A through F are observed as before. With the coordinates of both O and O' known, and by using the two independent sets

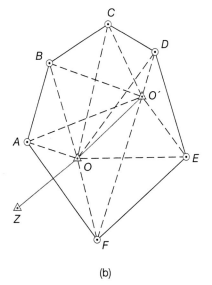

Figure 9.7
Radial traversing.
(a) From one occupied station.
(b) From two occupied stations.

of angles and distances, two sets of coordinates can be computed for each station, thus obtaining the checks. If the coordinates for each point agree within a reasonable tolerance, the average can be taken. However, a better adjustment is obtained using the method of least squares (see Section 3.21 and Chapter 16). Although radial traversing can provide coordinates of many points in an area rapidly, the method is not as rigorous as running closed traverses.

Radial traversing is ideal for quickly establishing many points in an area, especially when a total station instrument is employed. They not only enable the angle and distance observations to be made quickly, but they also perform the calculations for azimuth, horizontal distance, and station coordinates in real time. Radial methods are also very convenient for laying out planned construction projects with a total station instrument. In this application, the required coordinates of points to be staked are determined from the design, and the angles and distances that must be observed from a selected station of known position are computed. These are then laid out with a total station to set the stakes. The procedures are discussed in detail in Section 23.9.

9.10 SOURCES OF ERROR IN TRAVERSING

Some sources of error in running a traverse are:

1. Poor selection of stations, resulting in bad sighting conditions caused by (a) alternate sun and shadow, (b) visibility of only the rod's top, (c) line of sight passing too close to the ground, (d) lines that are too short, (e) line of sight passing close to an object like a vehicle, which causes a refracted line of sight, and (f) sighting into the sun.
2. Errors in observations of angles and distances.
3. Failure to observe angles an equal number of times direct and reversed.

9.11 MISTAKES IN TRAVERSING

Some mistakes in traversing are:

1. Occupying or sighting on the wrong station.
2. Incorrect orientation.
3. Confusing angles to the right and left.
4. Mistakes in note taking.
5. Misidentification of the sighted station.

PROBLEMS

Asterisks (*) indicate problems that have partial answers given in Appendix G.
- **9.1** How is angular closure achieved in a polygon traverse?
- **9.2** List the disadvantages of an open traverse.
- **9.3** How can an angular closure be obtained on a link traverse?
- **9.4** Discuss the difference between angles to the right and interior angles?
- **9.5** Draw two five-sided closed polygon traverses with station labels 1 to 5. The first traverse should show angles to the right that are interior angles, and the second should show angles to the right that are exterior angles.

9.6 List four pertinent considerations in selecting locations for traverse stations.

9.7 How should traverse stations be referenced?

9.8 Discuss the advantages and dangers of radial traversing.

9.9 What should be the sum of the interior angles for a closed-polygon traverse that has: (a)* 6 sides, (b) 10 sides, and (c) 24 sides?

9.10 What should be the sum of the exterior angles for a closed-polygon traverse that are listed in Problem 9.9?

9.11 Four interior angles of a five-sided polygon traverse were observed as $A = 106°47'59''$, $B = 112°06'12''$, $C = 118°33'32''$, and $D = 110°02'25''$. The angle at E was not observed. If all observed angles are assumed to be correct, what is the value of angle E?

9.12 Similar to Problem 9.11, except the traverse had six sides with observed angles of $A = 129°13'08''$, $B = 121°16'26''$, $C = 120°54'43''$, $D = 111°26'05''$, and $E = 133°38'28''$. Compute the angle at F, which was not observed.

9.13 What is the angular misclosure of a six-sided polygon traverse with observed angles of $A = 98°10'10''$, $B = 123°45'58''$, $C = 68°23'10''$, $D = 192°50'54''$, $E = 134°32'02''$, and $F = 102°17'40''$?

9.14 What FGCS standard would the angular misclosure in Problem 9.22 meet?

9.15* According to FGSC standards, what is the maximum acceptable angular misclosure for a second order, class I traverse having 20 angles?

9.16* What is the angular misclosure for a five-sided polygon traverse with observed exterior angles of $252°26'37''$, $255°55'13''$, $277°15'53''$, $266°35'02''$, and $207°47'05''$?

9.17 What is the angular misclosure for a five-sided polygon traverse with observed interior angles of $92°26'47''$, $109°55'03''$, $137°15'33''$, $106°35'22''$, and $93°47'20''$?

9.18 If the standard error of each angle observation in a traverse is $\pm 4''$, what is the estimated error in the closure of a six-sided traverse?

9.19* If the standard error for each measurement of a traverse angle is $\pm 5.4''$, what is the estimated standard error of the misclosure in the sum of the angles for an eight-sided traverse?

9.20 If the angles of a traverse are turned so that the 95% error of any angle is $\pm 3.4''$, what is the 95% error in a five-sided traverse?

9.21 What common mistake can be caught by observing the backsight station with a data collector?

9.22* The azimuth from station A of a link traverse to an azimuth mark is $212°12'36''$. The azimuth from the last station of the traverse to an azimuth mark is $192°12'16''$. Angles to the right are observed at each station: $A = 136°15'40''$, $B = 119°15'36''$, $C = 93°48'54''$, $D = 136°04'16''$, $E = 108°30'10''$, $F = 42°48'02''$, and $G = 63°17'16''$. What is the angular misclosure of this link traverse?

9.23 What FGCS order and class does the traverse in Problem 9.22 meet?

9.24* The interior angles in a five-sided closed-polygon traverse were observed as $A = 108°28'36''$, $B = 110°26'54''$, $C = 106°25'58''$, $D = 102°27'02''$, and $E = 112°11'15''$. Compute the angular misclosure. For what FGCS order and class is this survey adequate?

9.25 Similar to Problem 9.24, except for a six-sided traverse with observed exterior angles of $A = 244°28'36''$, $B = 238°26'54''$, $C = 246°25'58''$, $D = 234°27'02''$, $E = 235°08'55''$, and $F = 241°02'45''$.

9.26 In Figure 9.6, what is the average interior angle with the instrument at station 101?

9.27 Same as Problem 9.26 except at instrument station 103.

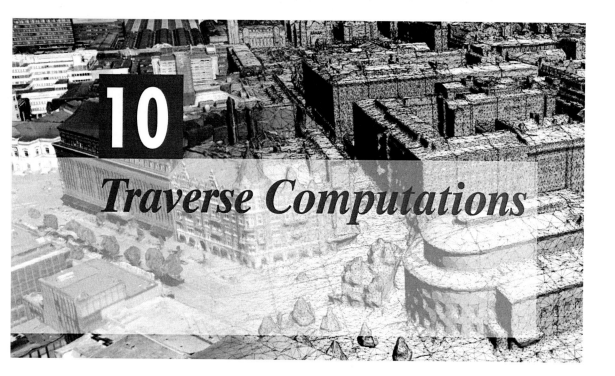

10 Traverse Computations

■ 10.1 INTRODUCTION

Measured angles or directions of closed traverses are readily investigated before leaving the field. Linear measurements, even though repeated, are more likely a source of error, and must also be checked. Although the calculations are lengthier than angle checks, with today's survey controllers they can also be done in the field to determine, before leaving, whether a traverse meets the required precision. If specifications have been satisfied, the traverse is then adjusted to create geometric "closure" or geometric consistency among angles and lengths; if not, field observations must be repeated until adequate results are obtained.

Investigation of precision, and acceptance or rejection of the field data is extremely important in surveying. Adjustment for geometric closure is also crucial. For example, in land surveying the law may require property descriptions to have exact geometric agreement.

Different procedures can be used for computing and adjusting traverses. These vary from elementary methods to more advanced techniques based on the method of least squares (see Chapter 16). This chapter concentrates on elementary procedures. The usual steps followed in making elementary traverse computations are (1) adjusting angles or directions to fixed geometric conditions, (2) determining preliminary azimuths (or bearings) of the traverse lines, (3) calculating departures and latitudes and adjusting them for misclosure, (4) computing rectangular coordinates of the traverse stations, and (5) calculating the lengths and azimuths (or bearings) of the traverse lines after adjustment. These procedures are all discussed in this chapter and are illustrated with several examples. Chapter 16 discusses traverse adjustment using the method of least squares.

■ 10.2 BALANCING ANGLES

In elementary methods of traverse adjustment, the first step is to balance (adjust) the angles to the proper geometric total. For closed traverses, angle balancing is done readily since the total error is known (see Section 9.7), although its exact distribution is not. Angles of a closed traverse can be adjusted to the correct geometric total by applying one of two methods:

1. Applying an average correction to each angle where observing conditions were approximately the same at all stations. The correction for each angle is found by dividing the total angular misclosure by the number of angles.
2. Making larger corrections to angles where poor observing conditions were present.

Of these two methods, the first is almost always applied.

■ Example 10.1

For the traverse of Figure 10.1, the observed interior angles are given in Table 10.1. Compute the adjusted angles using methods 1 and 2.

Solution

The computations are best arranged as shown in Table 10.1. The first part of the adjustment consists of summing the interior angles and determining the misclosure according to Equation (9.1), which in this instance, as shown beneath column 2, is +11″. The remaining calculations are tabulated, and the rationale for the procedures follows.

Figure 10.1
Traverse.

TABLE 10.1 ADJUSTMENT OF ANGLES

Method 1

Point (1)	Measured Interior Angle (2)	Multiples of Average Correction (3)	Correction Rounded To 1" (3)	Successive Differences (5)	Adjusted Angle (6)
A	100°45'37"	2.2"	2"	2"	100°45'35"
B	231°23'43"	4.4"	4"	2"	231°23'41"
C	17°12'59"	6.6"	7"	3"	17°12'56"
D	89°03'28"	8.8"	9"	2"	89°03'26"
E	101°34'24"	11.0"	11"	2"	101°34'22"
	$\Sigma = 540°00'11"$			$\Sigma = 11"$	$\Sigma = 540°00'00"$

Method 2

Point (1)	Measured Interior Angle (2)	Adjustment (7)	Adjusted Angle (8)
A	100°45'37"	2"	100°45'35"
B	231°23'43"	3"	231°23'40"
C	17°12'59"	3"	17°12'56"
D	89°03'28"	1"	89°03'27"
E	101°34'24"	2"	101°34'22"
	$\Sigma = 540°00'11"$	$\Sigma = 11"$	$\Sigma = 540°00'00"$

For work of ordinary precision, it is reasonable to adopt corrections that are even multiples of the smallest recorded digit or decimal place for the angle readings. Thus in this example, corrections to the nearest 1" will be made.

Method 1 consists of subtracting 11"/5 = 2.2" from each of the five angles. However, since the angles were read in multiples of 1", applying corrections to the nearest tenth of a second would give a false impression of their precision. Therefore it is desirable to establish a pattern of corrections to the nearest 1", as shown in Table 10.1. First multiples of the average correction of 2.2" are tabulated in column (3). In column (4), each of these multiples has been rounded off to the nearest 1". Then successive differences (adjustments for each angle) are found by subtracting the preceding value in column (4) from the one being considered. These are tabulated in column (5). Note that as a check, the sum of the corrections in this column must equal the angular misclosure of the traverse, which in this case is 11". The adjusted interior angles obtained by applying these corrections are listed in column (6). As another check, they must total exactly the true geometric value of $(n - 2)180°$, or 540°00'00" in this case.

In method 2, judgment is required because corrections are made to the angles estimated to contain the largest errors. In this example, 3″ is subtracted from the angles at B and C, since they have the shortest sights (along line BC), and 2″ is subtracted from the angles at A and E, because they have the next shortest sights (along line AE). A 1″ correction was applied to angle D because of its long sights. The sum of the corrections must equal the total misclosure. The adjustment made in this manner is shown in columns (7) and (8) of Table 10.1.

It should be noted that, although the adjusted angles by both methods satisfy the geometric condition of a closed figure, they may be no nearer to the true values than before adjustment. Unlike corrections for linear observations (described in Section 10.7), *adjustments applied to angles are independent of the size of the angle.*

10.3 COMPUTATION OF PRELIMINARY AZIMUTHS OR BEARINGS

After balancing the angles, the next step in traverse computation is calculation of either preliminary azimuths or preliminary bearings. This requires the direction of at least one course within the traverse to be either known or assumed. For some computational purposes an assumed direction is sufficient, and in that case the usual procedure is to simply assign north as the direction of one of the traverse lines. On certain traverse surveys, the magnetic bearing of one line can be determined and used as a reference for determining the other directions. However, in most instances, as in boundary surveys, true directions are needed. This requirement can be met by (1) incorporating within the traverse a line whose true direction was established through a previous survey; (2) including one end of a line of known direction as a station in the traverse [e.g., station A of line A-Az Mk of Figure 9.1(a)], and then observing an angle from that reference line to a traverse line; or (3) determining the true direction of one traverse line by astronomical observations (see Appendix C), or by GNSS surveys (see Section 14.6).

If a line of known direction exists within the traverse, computation of preliminary azimuths (or bearings) proceeds as discussed in Chapter 7. Angles adjusted to the proper geometric total must be used; otherwise the azimuth or bearing of the first line, when recomputed after using all angles and progressing around the traverse, will differ from its fixed value by the angular misclosure. Azimuths or bearings at this stage are called "preliminary" because they will change after the traverse is adjusted, as explained in Section 10.11. It should also be noted that since the azimuth of the courses will change, so will the angles, which were previously adjusted.

■ Example 10.2

Compute preliminary azimuths for the traverse courses of Figure 10.1, based on a fixed azimuth of 234°17′18″ for line AW, a measured angle to the right of 151°52′24″ for WAE, and the angle adjustment by method 1 of Table 10.1.

TABLE 10.2   COMPUTATION OF PRELIMINARY AZIMUTH USING THE TABULAR METHOD

126°55'17" = AB	+89°03'26" + D
+180°	284°35'20" = DE
306°55'17" = BA	−180°
+231°23'41" + B	104°35'20" = ED
538°18'58" − 360° = 178°18'58" = BC	+101°34'22" + E
−180°	206°09'42" = EA
358°18'58" = CB	−180°
+17°12'56" + C	26°09'42" = AE
375°31'54" − 360° = 15°31'54" = CD	+100°45'35" + A
−180°	126°55'17" = AB
195°31'54"	

Solution

Step 1: Compute the azimuth of course AB.

$$Az_{AB} = 234°17'18" + 151°52'24" + 100°45'35" - 360° = 126°55'17"$$

Step 2: Using the tabular method discussed in Section 7.8, compute preliminary azimuths for the remaining lines. The computations for this example are shown in Table 10.2. Figure 10.2 demonstrates the computations for line BC. Note that the azimuth of AB was recalculated as a check at the end of the table.

■ 10.4 DEPARTURES AND LATITUDES

After balancing the angles and calculating preliminary azimuths (or bearings), traverse closure is checked by computing the *departure* and *latitude* of each line. As illustrated in Figure 10.3, the departure of a course is its orthographic projection on the east-west axis of the survey and is equal to the length of the course multiplied by the sine of its azimuth (or bearing) angle. Departures are sometimes called *eastings* or *westings*.

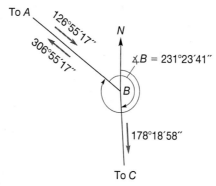

Figure 10.2
Computation of azimuth BC.

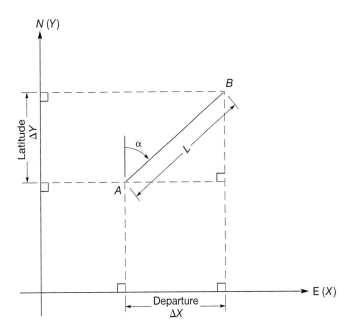

Figure 10.3
Departure and latitude of a line.

Also as shown in Figure 10.3, the latitude of a course is its orthographic projection on the north-south axis of the survey, and is equal to the course length multiplied by the cosine of its azimuth (or bearing) angle. Latitude is also called *northing* or *southing*.

In equation form, the departure and latitude of a line are

$$\text{departure} = L \sin \alpha \qquad (10.1)$$

$$\text{latitude} = L \cos \alpha \qquad (10.2)$$

where L is the horizontal length and α the azimuth of the course. Departures and latitudes are merely changes in the X and Y components of a line in a rectangular grid system, sometimes referred to as ΔX and ΔY. In traverse calculations, east departures and north latitudes are considered plus; west departures and south latitudes, minus. Azimuths (from north) used in computing departures and latitudes range from 0° to 360°, and the algebraic signs of sine and cosine functions automatically produce the proper algebraic signs of the departures and latitudes. Thus, a line with an azimuth of 126°55′17″ has a positive departure and negative latitude (the sine at the azimuth is plus and the cosine minus); a course of 284°35′20″ azimuth has a negative departure and positive latitude. In using bearings for computing departures and latitudes, the angles are always between 0° and 90°; hence their sines and cosines are invariably positive. Proper algebraic signs of departures and latitudes must therefore be assigned on the basis of the bearing angle directions, so a *NE* bearing has a plus departure and latitude, a *SE* bearing gets a plus departure and minus latitude, and so on. Because computers and calculators automatically affix correct algebraic signs to departures and latitudes using the azimuth angles' sines and cosines, it is more convenient to use azimuths than bearings for traverse computations.

■ 10.5 DEPARTURE AND LATITUDE CLOSURE CONDITIONS

For a closed-polygon traverse like that of Figure 10.1, it can be reasoned that if all angles and distances were measured perfectly, the algebraic sum of the departures of all courses in the traverse should equal zero. Likewise, the algebraic sum of all latitudes should equal zero. And for closed link-type traverses like that of Figure 9.1(b), the algebraic sum of departures should equal the total difference in departure (ΔX) between the starting and ending easting (X) coordinates. The same condition with the northing (Y) coordinates applies to latitudes (ΔY) in a link traverse. Because the observations are not perfect, and errors exist in the angles and distances, the conditions just stated rarely occur. The amounts by which they fail to be met are termed *departure misclosure* and *latitude misclosure*. Their values are computed by algebraically summing the departures and latitudes and comparing the totals to the required conditions.

The magnitudes of the departure and latitude misclosures for closed-polygon-type traverses give an "indication" of the precision that exists in the observed angles and distances. Large misclosures certainly indicate that either significant errors or even mistakes exist. Small misclosures usually mean the observed data are precise and free of mistakes, but it is not a guarantee that systematic or compensating errors do not exist.

■ 10.6 TRAVERSE LINEAR MISCLOSURE AND RELATIVE PRECISION

Because of errors in the observed traverse angles and distances, if one were to begin at point A of a closed-polygon traverse like that of Figure 10.1, and progressively follow each course for its observed distance along its preliminary bearing or azimuth, one would finally return not to point A, but to some other nearby point A'. Point A' would be removed from A in an east-west direction by the departure misclosure, and in a north-south direction by the latitude misclosure. The distance between A and A' is termed the linear misclosure of the traverse. It is calculated from the following formula:

$$\text{linear misclosure} = \sqrt{(\text{departure misclosure})^2 + (\text{latitude misclosure})^2} \quad (10.3)$$

The *relative precision* of a traverse is expressed by a fraction that has the linear misclosure as its numerator and the traverse perimeter or total length as its denominator, or

$$\text{relative precision} = \frac{\text{linear misclosure}}{\text{traverse length}} \quad (10.4)$$

The fraction that results from Equation (10.4) is then reduced to reciprocal form, and the denominator rounded to the same number of significant figures as the numerator. This is illustrated in the following example.

TABLE 10.3 COMPUTATION OF DEPARTURES AND LATITUDES

Station	Preliminary Azimuths	Length	Departure	Latitude
A	126°55′17″	647.25	517.451	−388.815
B	178°18′58″	203.03	5.966	−202.942
C	15°31′54″	720.35	192.889	694.045
D	284°35′20″	610.24	−590.565	153.708
E	206°09′42″	285.13	−125.715	−255.919
A		Σ = 2466.00	Σ = 0.026	Σ = 0.077

■ Example 10.3

Based on the preliminary azimuths from Table 10.2 and lengths shown in Figure 10.1, calculate the departures and latitudes, linear misclosure, and relative precision of the traverse.

Solution

In computing departures and latitudes, the data and results are usually listed in a standard tabular form, such as that shown in Table 10.3. The column headings and rulings save time and simplify checking.

In Table 10.3, taking the algebraic sum of east ($+$) and west ($-$) departures gives the misclosure, 0.026 ft. Also, summing north ($+$) and south ($-$) latitudes gives the misclosure in latitude, 0.077 ft. Linear misclosure is the hypotenuse of a small triangle with sides of 0.026 ft and 0.077 ft, and in this example its value is, by Equation (10.3)

$$\text{linear misclosure} = \sqrt{(0.026)^2 + (0.077)^2} = 0.081 \text{ ft}$$

The relative precision for this traverse, by Equation (10.4), is

$$\text{relative precision} = \frac{0.081}{2466.00} = \frac{1}{30,000}$$

■ 10.7 TRAVERSE ADJUSTMENT

For any closed traverse, the linear misclosure must be adjusted (or distributed) throughout the traverse to "close" or "balance" the figure. This is true even though the misclosure is negligible in plotting the traverse at map scale. There are several elementary methods available for traverse adjustment, but the one most used is the arbitrary method known as the *compass rule* (Bowditch method). As noted earlier, adjustment by least squares is a more advanced technique that can also be used. These two methods are discussed in the subsections that follow.

10.7.1 Compass (Bowditch) Rule

The compass, or Bowditch, rule adjusts the departures and latitudes of traverse courses in proportion to their lengths. Although not as rigorous as the least-squares method, it does result in a logical distribution of misclosures. Corrections by this method are made according to the following rules:

correction in departure for AB

$$= -\frac{(\text{total departure misclosure})}{\text{traverse perimeter}} \text{length of } AB \quad (10.5)$$

correction in latitude for AB

$$= -\frac{(\text{total latitude misclosure})}{\text{traverse perimeter}} \text{length of } AB \quad (10.6)$$

Note that the algebraic signs of the corrections are opposite those of the respective misclosures.

■ Example 10.4

Using the preliminary azimuths from Table 10.2 and lengths from Figure 10.1, compute departures and latitudes, linear misclosure, and relative precision. Balance the departures and latitudes using the compass rule.

Solution

A tabular solution, which is somewhat different than that used in Example 10.3, is employed for computing departures and latitudes (see Table 10.4). To compute departure and latitude corrections by the compass rule, Equations (10.5) and (10.6) are used as demonstrated. By Equation (10.5) the correction in departure for AB is

$$-\left(\frac{0.026}{2466}\right)647.25 = -0.007 \text{ ft}$$

And by Equation (10.6) the correction for the latitude of AB is

$$-\left(\frac{0.077}{2466}\right)647.25 = -0.020 \text{ ft}$$

The other corrections are likewise found by multiplying a constant—the ratio of misclosure in departure, and latitude, to the perimeter—by the successive course lengths.

In Table 10.4, the departure and latitude corrections are shown in parentheses above their unadjusted values. These corrections are added algebraically to

248 TRAVERSE COMPUTATIONS

TABLE 10.4 BALANCING DEPARTURES AND LATITUDES BY THE COMPASS (BOWDITCH) RULE

Station	Preliminary Azimuths	Length (ft)	Unadjusted Departure	Unadjusted Latitude	Balanced Departure	Balanced Latitude	X (ft) (Easting)	Y (ft) (Northing)
A							10,000.00	5000.00
	126°55'17"	647.25	(−0.007) 517.451	(−0.020) −388.815	517.444	−388.835		
B							10,517.44	4611.16
	178°18'58"	203.03	(−0.002) 5.966	(−0.006) −202.942	5.964	−202.948		
C							10,523.41	4408.22
	15°31'54"	720.35	(−0.008) 192.889	(−0.023) 694.045	192.881	694.022		
D							10,716.29	5102.24
	284°35'20"	610.24	(−0.006) −590.565	(−0.019) 153.708	−590.571	153.689		
E							10,125.72	5255.93
	206°09'42"	285.13	(−0.003) −125.715	(−0.009) −255.919	−125.718	−255.928		
A							10,000.00 ✓	5000.00 ✓
		Σ = 2466.00	Σ = 0.026	Σ = 0.077	Σ = 0.000	Σ = 0.000		

Linear precision = $\sqrt{(0.026)^2 + (-0.077)^2} = 0.081$ ft

Relative precision = $\dfrac{0.081}{2466} = \dfrac{1}{30{,}000}$

*Coordinates are rounded to same significance as observed lengths.

their respective unadjusted values, and the corrected quantities tabulated in the "balanced" departure and latitude columns. A check is made of the computational process by algebraically summing the balanced departure and latitude columns to verify that each is zero. In these columns, if rounding off causes a small excess or deficiency, revising one of the corrections to make the closure perfect eliminates this. However, if computations are carried out to one more decimal place than is justified, rounding seldom affects the final values.

10.7.2 Least-Squares Method

As noted in Section 3.21, the method of least squares is based on the theory of probability, which models the occurrence of random errors. This results in adjusted values having the highest probability. Thus, the least-squares method provides the best and most rigorous traverse adjustment, but until recently the method has not been widely used because of the lengthy computations required. The availability of computers has now made these calculations routine, and consequently the least-squares method has gained popularity.

In applying the least-squares method to traverses, angle and distance observations are adjusted simultaneously. Thus, no preliminary angle adjustment is made, as is done when using the compass rule. The least-squares method is valid for any type of traverse and has the advantage that observations of varying precisions can be weighted appropriately in the computations. Examples illustrating some elementary least-squares adjustments are presented in Chapter 16.

■ 10.8 RECTANGULAR COORDINATES

Rectangular X and Y coordinates of any point give its position with respect to an arbitrarily selected pair of mutually perpendicular reference axes. The X coordinate is the perpendicular distance, in feet or meters, from the point to the Y axis; the Y coordinate is the perpendicular distance to the X axis. Although the reference axes are discretionary in position, in surveying they are normally oriented so that the Y axis points north–south, with north the positive Y direction. The X axis runs east–west, with positive X being east. Given the rectangular coordinates of several points, their relative positions are uniquely defined.

Coordinates are useful in a variety of computations, including (1) determining lengths and directions of lines, and angles (see Section 10.11 and Chapter 11); (2) calculating areas of land parcels (see Section 12.5); (3) making certain curve calculations (see Sections 24.12 and 24.13); and (4) locating inaccessible points (see Section 11.9). Coordinates are also advantageous for plotting maps (see Section 18.8.1) and in developing geographic information systems (see Section 28.1).

In practice, *state plane coordinate systems*, as described in Chapter 20, are most frequently used as the basis for rectangular coordinates in plane surveys. However, for many calculations, any arbitrary system may be used. As an example, coordinates may be arbitrarily assigned to one traverse station. For example, to avoid negative values of X and Y an origin is assumed south and west of the traverse such that one hub has coordinates $X = 10,000.00$, $Y = 5,000.00$, or any other suitable values. In a closed traverse, assigning $Y = 0.00$ to the most southerly point and $X = 0.00$ to the most westerly station saves time in hand calculations.

Given the X and Y coordinates of any starting point A, the X coordinate of the next point B is obtained by adding the adjusted departure of course AB to X_A. Likewise, the Y coordinate of B is the adjusted latitude of AB added to Y_A. In equation form this is

$$X_B = X_A + \text{departure } AB$$
$$Y_B = Y_A + \text{latitude } AB \quad (10.7)$$

For closed polygons, the process is continued around the traverse, successively adding departures and latitudes until the coordinates of starting point A are recalculated. If these recalculated coordinates agree exactly with the starting ones, a check on the coordinates of all intermediate points is obtained (unless compensating mistakes have been made). For link traverses, after progressively computing coordinates for each station, if the calculated coordinates of the closing control point equal that point's control coordinates, a check is obtained.

■ Example 10.5

Using the balanced departures and latitudes obtained in Example 10.4 (see Table 10.4), and starting coordinates $X_A = 10,000.00$ and $Y_A = 5,000.00$, calculate coordinates of the other traverse points.

Solution

The process of successively adding balanced departures and latitudes to obtain coordinates is carried out in the two rightmost columns of Table 10.4. Note that the starting coordinates $X_A = 10,000.00$ and $Y_A = 5,000.00$ are recomputed at the end to provide a check. Note also that X and Y coordinates are frequently referred to as *eastings* and *northings*, respectively, as is indicated in Table 10.4.

■ 10.9 ALTERNATIVE METHODS FOR MAKING TRAVERSE COMPUTATIONS

Procedures for making traverse computations that vary somewhat from those described in preceding sections can be adopted. One alternative is to adjust azimuths or bearings rather than angles. Another is to apply compass rule corrections directly to coordinates. These procedures are described in the subsections that follow.

10.9.1 Balancing Angles by Adjusting Azimuths or Bearings

In this method, "unadjusted" azimuths or bearings are computed based on the observed angles. These azimuths or bearings are then adjusted to secure a geometric closure, and to obtain preliminary values for use in computing departures and latitudes. The method is equally applicable to closed-polygon traverses, like that of Figure 10.1, or to closed-link traverses, as shown in Figure 9.1(b) that begin on one control station and end on another. The procedure of making the adjustment for angular misclosure in this manner will be explained by an example.

■ Example 10.6

Table 10.5 lists observed angles to the right for the traverse of Figure 9.1(b). The azimuths of lines $A - AzMk_1$ and $E - AzMk_2$ have known values of 139°05′45″ and 86°20′47″ respectively. Compute unadjusted azimuths and balance them to obtain geometric closure.

Solution

From the observed angles of column (2) in Table 10.5, unadjusted azimuths have been calculated and are listed in column (3). Because of angular errors, the unadjusted azimuth of the final line $E - AzMk_2$ disagrees with its fixed value by 0°00′10″ This represents the angular misclosure, which is divided by 5, the number of observed angles, to yield a correction of −2″ per angle. The corrections to azimuths, which accumulate and increase by −2″ for each angle, are listed in

TABLE 10.5 BALANCING TRAVERSE AZIMUTHS

Station (1)	Measured Angle* (2)	Unadjusted Azimuth (3)	Azimuth Correction (4)	Preliminary Azimuth (5)
Az Mk1				
		319°05′45″		319°05′45″
A	283°50′10″			
		62°55′55″	−2″	62°55′53″
B	256°17′18″			
		139°13′13″	−4″	139°13′09″
C	98°12′36″			
		57°25′49″	−6″	57°25′43″
D	103°30′34″			
		340°56′23″	−8″	340°56′15″
E	285°24′34″			
		86°20′57″	−10″	86°20′47″
Az Mk₂				

$$
\begin{aligned}
&86°20′57″\\
&\underline{-86°20′47″}\\
&\text{misclosure} = 0°00′10″\\
&\text{correction per angle} = -10″/5 = -2″
\end{aligned}
$$

*Observed angles are angles to the right.

column (4). Thus, line AB, which is based on one observed angle, receives a $-2''$ correction; line BC, which uses two observed angles, gets a $-4''$ correction; and so on. The final azimuth, $E - AzMk_2$ receives a $-10''$ correction because all five observed angles have been included in its calculation. The corrected preliminary azimuths are listed in column 5.

10.9.2 Balancing Departures and Latitudes by Adjusting Coordinates

In this procedure, commencing with the known coordinates of a beginning station, unadjusted departures and latitudes for each course are successively added to obtain "preliminary" coordinates for all stations. For closed-polygon traverses, after progressing around the traverse, preliminary coordinates are recomputed for the beginning station. The difference between the computed preliminary X coordinate at this station and its known X coordinate is the departure misclosure. Similarly, the disagreement between the computed preliminary Y coordinate for the beginning station and its known value is the latitude misclosure. Corrections for these misclosures can be calculated using compass-rule Equations (10.5) and (10.6) and applied directly to the preliminary coordinates to obtain adjusted coordinates. The result is the same as if departures and latitudes were first adjusted and coordinates computed from them, as was done in Examples 10.4 and 10.5.

Closed traverses like the one shown in Figure 9.1(b) can be similarly adjusted. For this type of traverse, unadjusted departures and latitudes are also successively added to the beginning station's coordinates to obtain preliminary coordinates for all points, including the final closing station. Differences in preliminary X and Y coordinates, and the corresponding known values for the closing station, represent the departure and latitude misclosures, respectively. These misclosures are distributed directly to preliminary coordinates using the compass rule to obtain final adjusted coordinates. The procedure will be demonstrated by an example.

■ Example 10.7

Table 10.6 lists the preliminary azimuths (from Table 10.5) and observed lengths (in feet) for the traverse of Figure 9.1(b). The known coordinates of stations A and E are $X_A = 12{,}765.48$, $Y_A = 43{,}280$, $X_E = 14{,}797.12$, and $Y_E = 44{,}384.51$ ft. Adjust this traverse for departure and latitude misclosures by making corrections to preliminary coordinates.

Solution

From the lengths and azimuths listed in columns (2) and (3) of Table 10.6, departures and latitudes are computed and tabulated in columns (4) and (5). These unadjusted values are progressively added to the known coordinates of station A to obtain preliminary coordinates for all stations, including E, and are listed in columns (6) and (7). Comparing the preliminary X and Y coordinates of station E with its known values yields departure and latitude misclosures of $+0.179$ and -0.024 ft,

TABLE 10.6 TRAVERSE ADJUSTMENT BY COORDINATES

Station (1)	Length (ft) (2)	Preliminary Azimuth (3)	Departure (4)	Latitude (5)	Preliminary Coordinates (ft) X (6)	Preliminary Coordinates (ft) Y (7)	Corrections (ft) X (8)	Corrections (ft) Y (9)	Adjusted Coordinates* X (ft) (10)	Adjusted Coordinates* Y (ft) (11)
A					12,765.48	43,280.21			12,765.48	43,280.21
	1045.50	62°55'53"	930.978	475.762			−0.048 0.006			
B					13,696.458	43,755.972	(−0.048) (0.006)		13,696.41	43,755.98
	1007.38	139°13'09"	657.988	−762.802			−0.046 0.006			
C					14,354.446	42,993.170	(−0.094) (0.012)		14,354.35	42,993.18
	897.81	57°25'43"	756.604	483.336			−0.041 0.006			
D					15,111.050	43,476.506	(−0.135) (0.018)		15,110.92	43,476.52
	960.66	340°56'15"	−313.751	907.980			−0.044 0.006			
E					14,797.299	44,384.486	(−0.179) (0.024)		14,797.12✓	44,384.51✓
Σ = 3911.35					−14,797.12	−44,384.51				
			Misclosures		+0.179	−0.024				

Linear precision = $\sqrt{(0.179)^2 + (-0.024)^2} = 0.181$ ft

Relative precision = $\dfrac{0.181}{3911} = \dfrac{1}{21,000}$

*Adjusted coordinates are rounded to same significance as observed lengths.

respectively. From these values, the linear misclosure of 0.181 ft and relative precision of 1/21,000 are computed (see Table 10.6).

Compass-rule corrections for each course are computed and listed in columns (8) and (9). Their cumulative values obtained by progressively adding the corrections are given in parentheses in columns (8) and (9). Finally, by applying the cumulative corrections to the preliminary coordinates of columns 6 and 7, final adjusted coordinates (rounded to the nearest hundredth of a foot) listed in columns (10) and (11) are obtained.

10.10 INVERSING

If the departure and latitude of a line AB are known, its length and azimuth or bearing are readily obtained from the following relationships:

$$\tan \text{azimuth (or bearing)} \, AB = \frac{\text{departure} \, AB}{\text{latitude} \, AB} \tag{10.8}$$

$$\text{length} \, AB = \frac{\text{departure} \, AB}{\sin \text{azimuth (or bearing)} \, AB}$$

$$= \frac{\text{latitude} \, AB}{\cos \text{azimuth (or bearing)} \, AB}$$

$$= \sqrt{(\text{departure} \, AB)^2 + (\text{latitude} \, AB)^2} \tag{10.9}$$

Equations (10.7) can be written to express departures and latitudes in terms of coordinate differences ΔX and ΔY as follows:

$$\text{departure}_{AB} = X_B - X_A = \Delta X$$

$$\text{latitude}_{AB} = Y_B - Y_A = \Delta Y \tag{10.10}$$

Substituting Equations (10.10) into Equations (10.8) and (10.9)

$$\tan \text{azimuth (or bearing)} \, AB = \frac{X_B - X_A}{Y_B - Y_A} = \frac{\Delta X}{\Delta Y} \tag{10.11}$$

$$\text{length} \, AB = \frac{X_B - X_A \, (\text{or} \, \Delta X)}{\sin \text{azimuth (or bearing)} \, AB}$$

$$= \frac{Y_B - Y_A \, (\text{or} \, \Delta Y)}{\cos \text{azimuth (or bearing)} \, AB}$$

$$= \sqrt{(X_B - X_A)^2 + (Y_B - Y_A)^2}$$

$$= \sqrt{(\Delta X)^2 + (\Delta Y)^2} \tag{10.12}$$

Equations (10.8) through (10.12) can be applied to any line whose coordinates are known, whether or not it was actually observed in the survey. Note that X_B and Y_B must be listed first in Equations (10.11) and (10.12), so that ΔX and ΔY will have the correct algebraic signs. Computing lengths and directions of lines from departures and latitudes, or from coordinates, is called *inversing*.

10.11 COMPUTING FINAL ADJUSTED TRAVERSE LENGTHS AND DIRECTIONS

In traverse adjustments, as illustrated in Examples 10.4 and 10.7, corrections are applied to the computed departures and latitudes to obtain adjusted values. These in turn are used to calculate X and Y coordinates of the traverse stations. By changing departures and latitudes of lines in the adjustment process, their lengths and azimuths (or bearings) also change. In many types of surveys, it is necessary to compute the changed, or "final adjusted," lengths and directions. For example, if the purpose of the traverse was to describe the boundaries of a parcel of land, the final adjusted lengths and directions would be used in the recorded deed.

The equations developed in the preceding section permit computation of final values for lengths and directions of traverse lines based either on their adjusted departures and latitudes or on their final coordinates.

Example 10.8

Calculate the final adjusted lengths and azimuths of the traverse of Example 10.4 from the adjusted departures and latitudes listed in Table 10.4.

Solution

Equations (10.8) and (10.9) are applied to calculate the adjusted length and azimuth of line AB. All others were computed in the same manner. The results are listed in Table 10.7.

TABLE 10.7 FINAL ADJUSTED LENGTHS AND DIRECTIONS FOR TRAVERSE OF EXAMPLE 10.4

Line	Balanced Departure	Balanced Latitude	Length (ft)	Azimuth
AB	517.444	−388.835	647.26	126°55′23″
BC	5.964	−202.948	203.04	178°19′00″
CD	192.881	694.022	720.33	15°31′54″
E	−590.571	153.689	610.24	284°35′13″
EA	−125.718	−255.928	285.14	206°09′41″

By Equation (10.8)

$$\tan \text{azimuth}_{AB} = \frac{517.444}{-388.835} = -1.330755;$$

$$\text{azimuth}_{AB} = -53°04'37'' + 180° = 126°55'23''$$

By Equation (10.9)

$$\text{length}_{AB} = \sqrt{(517.444)^2 + (-388.835)^2} = 647.26 \text{ ft}$$

Comparing the observed lengths of Table 10.4 to the final adjusted values in Table 10.7, it can be seen that, as expected, the values have undergone small changes, some increasing, others decreasing, and length DE remaining the same because of compensating changes.

■ Example 10.9

Using coordinates, calculate adjusted lengths and azimuths for the traverse of Example 10.7 (see Table 10.6).

Solution

Equations (10.11) and (10.12) are used to demonstrate calculation of the adjusted length and azimuth of line AB. All others were computed in the same way. The results are listed in Table 10.8. Comparing the adjusted lengths and azimuths of this table with their unadjusted values of Table 10.6 reveals that all values have undergone changes of varying amounts.

$$X_B - X_A = 13{,}696.41 - 12{,}765.48 = 930.93 = \Delta X$$

$$Y_B - Y_A = 43{,}755.98 - 43{,}280.21 = 475.77 = \Delta Y$$

By Equation (10.11), $\tan \text{azimuth}_{AB} = 930.93/475.77 = 1.95668075$; $\text{azimuth}_{AB} = 62°55'47''$.

By Equation (10.12), $\text{length}_{AB} = \sqrt{(930.93)^2 + (475.77)^2} = 1045.46$ ft.

TABLE 10.8 FINAL ADJUSTED LENGTHS AND DIRECTIONS FOR TRAVERSE OF EXAMPLE 10.7

Line	Adjusted ΔX	Adjusted ΔY	Adjusted Length (ft)	Adjusted Azimuth
AB	930.93	475.77	1045.46	62°55'47"
BC	657.94	−762.80	1007.35	139°13'16"
CD	756.57	483.34	897.78	57°25'38"
DE	−313.80	907.99	960.68	340°56'06"

10.12 Coordinate Computations in Boundary Surveys

TABLE 10.9 FINAL ADJUSTED ANGLES FOR EXAMPLE 10.4

Angle	Foresight Azimuth	Backsight Azimuth	Adjusted Angle	Difference
A (EAB)	AB = 126°55′23″	AE = 26°09′41″	100°45′42″	7″
B (ABC)	BC = (178°19′00″ + 360°)	BA = 306°55′23″	231°23′37″	−4″
C (BCD)	CD = (15°31′54″ + 360°)	CB = (178°19′00″ + 180°)	17°12′54″	−2″
D (CDE)	DE = 284°35′13″	DC = (15°31′54″ + 180°)	89°03′19″	−7″
E (DEA)	EA = 206°09′41″	ED = (284°35′13″ − 180°)	101°34′28″	6″
			Σ = 540°00′00″	Σ = 0″

Because the final adjusted azimuths are different from their preliminary values, the preliminary adjusted angles have also changed. The backsight azimuth must be subtracted from the foresight azimuth to compute the final adjusted angles. A method of listing both the backsight and foresight stations for each angle helps in determining which azimuths should be subtracted. For example, the angle at A in Figure 10.1 is listed as EAB where E is the backsight station and B is the foresight station for the clockwise interior angle. As a pneumonic, angle A is computed as the difference in azimuths AB and AE where Az_{AB} is the foresight azimuth of angle A and Az_{AE} is the backsight azimuth. Thus, the angle at A is computed as

$$\angle EAB = Az_{AB} - Az_{AE}$$
$$= 126°55'23'' - (206°09'41'' - 180°)$$
$$= 100°45'42''$$

Notice in this example that the back azimuth of EA from Table 10.7 was needed for the backsight, and thus 180° was subtracted from azimuth EA. Also note that the final adjusted value for the angle at A differs from the preliminary adjusted value by 7″. The final adjusted angles for remainder of the traverse are shown in Table 10.9. For each angle the appropriate three-letter designator, which defines the clockwise interior angle, is shown in parentheses. Table 10.8 also shows the appropriate foresight and backsight azimuths and the final adjusted angle at each station. Notice that the sum of the angles again achieves geometric closure with a value of 540°. However, each angle differs from the value given in Table 10.1 by the amount shown in the last column.

10.12 COORDINATE COMPUTATIONS IN BOUNDARY SURVEYS

Computation of a bearing from the known coordinates of two points on a line is commonly done in boundary surveys. If the lengths and directions of lines from traverse points to the corners of a field are known, the coordinates of the corners can be determined and the lengths and bearings of all sides calculated.

258 TRAVERSE COMPUTATIONS

Figure 10.4
Plot of traverse for a boundary survey.

■ Example 10.10

In Figure 10.4, $APQDEA$ is a parcel of land that must be surveyed, but because of obstructions, traverse stations cannot be set at P and Q. Therefore offset stations B and C are set nearby, and closed traverse $ABCDE$ run. Lengths and azimuths of lines BP and CQ are observed as 42.50 ft, 354°50′00″ and 34.62 ft, 26°39′54″, respectively. Following procedures demonstrated in earlier examples, traverse $ABCEA$ was computed and adjusted, and coordinates were determined for all stations. They are given in the following table.

Point	X (ft)	Y (ft)
A	1000.00	1000.00
B	1290.65	1407.48
C	1527.36	1322.10
D	1585.70	1017.22
E	1464.01	688.25

Compute the length and bearing of property line PQ.

Solution

1. Using Equations (10.1) and (10.2), the departures and latitudes of lines BP and CQ are:

$$\text{Dep}_{BP} = 42.50 \sin(354°50′00″) = -3.83 \text{ ft}$$
$$\text{Dep}_{CQ} = 34.62 \sin(26°39′54″) = 15.54 \text{ ft}$$

$$\text{Lat}_{BP} = 42.50 \cos(354°50′00″) = 42.33 \text{ ft}$$
$$\text{Lat}_{CQ} = 34.62 \cos(26°39′54″) = 30.94 \text{ ft}$$

2. From the coordinates of stations B and C and the departures and latitudes just calculated, the following tabular solution yields X and Y coordinates for points P and Q:

	X	Y		X	Y
B	1290.65	1407.48	C	1527.36	1322.10
BP	−3.83	+42.33	CQ	+15.54	+30.94
P	1286.82	1449.81	Q	1542.90	1353.04

3. From the coordinates of P and Q, the length and bearing of line PQ are found in the following manner:

	X	Y
Q	1542.90	1353.04
P	−1286.88	−1449.81
PQ	$\Delta X = 256.02$	$\Delta Y = -96.77$

By Equation (10.11), $\tan \text{bearing}_{PQ} = 256.02/-96.77 = -2.64565$; $\text{bearing}_{PQ} = $ S69°17′40″E

By Equation (10.12), length $PQ = \sqrt{(-96.77)^2 + (256.02)^2} = 273.79$ ft

By using Equations (10.11) and (10.12), lengths and bearings of lines AP and QD can also be determined. As stated earlier, extreme caution must be used when employing this procedure, since no checks are obtained on the length and azimuth measurements of lines BP and CQ, nor are there any computational checks on the calculated lengths and bearings.

10.13 USE OF OPEN TRAVERSES

Although open traverses should be used with reluctance, sometimes there are situations where it is very helpful to run one and then compute the length and direction of the "closing line." In Figure 10.5, for example, suppose that improved horizontal alignment is planned for Taylor Lake and Atkins Roads, and a new construction line AE must be laid out. Because of dense woods, visibility between points A and E is not possible. A random line (see Section 8.17) could be run from A toward E and then corrected to the desired line, but that would be very difficult and time consuming due to tree density. One solution to this problem is to run open traverse $ABCDE$, which can be done quite easily along the cleared right-of-way of existing roads.

For this problem an assumed azimuth (e.g., due north) can be taken for line UA, and assumed coordinates (e.g., 10,000.00 and 10,000.00) can be assigned to station A. From observed lengths and angles, departures and latitudes of all lines, and

260 TRAVERSE COMPUTATIONS

Figure 10.5
Closing line of an open traverse.

coordinates of all points can be computed. From the resulting coordinates of stations A and E, the length and azimuth of closing line AE can be calculated. Finally, the deflection angle α needed to reach E from A can be computed and laid off.

In running open traverses, extreme caution must be exercised in all observations, because there is no check, and any errors or mistakes will result in an erroneous length and direction for the closing line. Procedures such as closing the horizon and observing the lengths of the lines from both ends of the lines should be practiced so that independent checks on all observations are obtained. Utmost care must also be exercised in the calculations, although carefully plotting the traverse and scaling the length of the closing line and the deflection angle can secure a rough check on them.

■ Example 10.11

Compute the length and azimuth of closing line AE and deflection angle α of Figure 10.5, given the following observed data:

Point	Length (ft)	Angle to the Right
A		115°18′25″E
	3305.78	
B		161°24′11″
	1862.40	
C		204°50′09″
	1910.22	
D		273°46′37″
	6001.83	
E		

Solution

Table 10.10 presents a tabular solution for computing azimuths, departures and latitudes, and coordinates.

From the coordinates of points A and E, the ΔX and ΔY values of line AE are

$$\Delta X = 7{,}004.05 - 10{,}000.00 = -2{,}995.95 \text{ ft}$$

$$\Delta Y = 17{,}527.05 - 10{,}000.00 = 7{,}527.05 \text{ ft}$$

TABLE 10.10 COMPUTATIONS FOR CLOSING LINE

Point	Azimuth	Departure	Latitude	X (ft)	Y (ft)
U					
	North (assumed)				
A				10,000.00	10,000.00
	295°18′25″	−2988.53	1413.11		
B				7011.47	11,413.11
	276°42′36″	−1849.64	217.61		
C				5161.83	11,630.72
	301°32′45″	−1627.93	999.39		
D				3533.90	12,630.11
	35°19′22″	3470.15	4896.94		
E				7004.05	17,527.05

Linear precision = $\sqrt{(0.026)^2 + (-0.077)^2} = 0.081$ ft

Relative precision = $\dfrac{0.081}{2466} = \dfrac{1}{30{,}000}$

*Coordinates are rounded to same significance as observed lengths.

$$\begin{aligned}&86°20′57″\\&\underline{-86°20′47″}\\\text{misclosure} =\ &0°00′10″\end{aligned}$$

correction per angle = −10″/5 = −2″

*Observed angles are angles to the right.

By Equation (10.12), the length of closing line AE is

$$\text{length}_{AE} = \sqrt{(-2995.95)^2 + (7527.05)^2} = 8101.37 \text{ ft}$$

By Equation (10.11), the azimuth of closing line AE is

$$\tan \text{azimuth}_{AE} = \frac{-2995.95}{7527.05} = -0.39802446; \text{azimuth}_{AE} = 338°17'46''$$

(Note that with a negative ΔX and positive ΔY the bearing of AE is northwest, hence the azimuth is $338°17'46''$.)

Finally, deflection angle α is the difference between the azimuths of lines AE and UA, or

$$-\alpha = 338°17'46'' - 360° = -21°42'14'' \text{ (left)}$$

With the emergence of GNSS, problems like that illustrated in Example 10.11 will no longer need to be solved using open traverses. Instead, receivers could be set at points $U, A,$ and E of Figure 10.5, and their coordinates determined. From these coordinates the azimuths of lines UA and AE can be calculated, as well as angle α.

■ 10.14 STATE PLANE COORDINATE SYSTEMS

Under ordinary circumstances, rectangular coordinate systems for plane surveys would be limited in size due to earth curvature. However, the National Geodetic Survey (NGS) developed statewide coordinate systems for each state in the United States, which retain an accuracy of 1 part in 10,000 or better while fitting curved geodetic distances to plane grid lengths. However, if reduction of observations is properly performed (see Section 20.8), little accuracy will be lost in the survey.

State plane coordinates are related mathematically to the geodetic coordinates of latitude and longitude, so control survey stations set by the NGS, as well as those set by others, can all be tied to the systems. As additional stations are set and their coordinates determined, they too become usable reference points in the state plane systems. These monumented control stations serve as starting points for local surveys and permit accurate restoration of obliterated or destroyed marks having known coordinates. If state plane coordinates of two intervisible stations are known, like A and $Az\ Mk$ of Figure 9.1(a), the direction of line A-$Az\ Mk$ can be computed and used to orient the total station instrument at A. In this way, azimuths and bearings of traverse lines are obtained without the necessity of making astronomical observations or resorting to other means.

Some cities and counties use their own local plane coordinate systems for locating street, sewer, property, and other lines. Because of their limited extent and the resultant discontinuity at city or county lines, such local systems are less desirable than a statewide system. Another plane coordinate system called the *Universal Transverse Mercator* (UTM) (see Section 20.11) is widely used to pinpoint the locations of objects by coordinates. The military and others use this system for a variety of purposes.

■ 10.15 TRAVERSE COMPUTATIONS USING COMPUTERS

Computers are particularly convenient for making traverse computations. Small programmable handheld units, survey controllers, and laptop computers are commonly taken into the field and used to verify data for acceptable misclosures before returning to the office. In the office, personal computers are widely used. A variety of software is available for use by surveyors. Some manufacturers supply standard programs, which include traverse computations, with the purchase of their equipment. Various software is also available for purchase from a number of suppliers. Spreadsheet software can also be conveniently used with personal computers to calculate and adjust traverses. Of course, surveying and engineering firms frequently write programs specifically for their own use. Standard programming languages employed include Fortran, Pascal, BASIC, C, and others.

A traverse computation program is provided in the software WOLFPACK on the companion website. It computes departures and latitudes, linear misclosure, and relative precision, and performs adjustments by the compass (Bowditch) rule. In addition, the program calculates coordinates of the traverse points and the area within polygon traverses using the coordinate method (discussed in Section 12.5). In Figure 10.6, the input and output files from WOLFPACK are shown for Example 10.4. For the data file of Figure 10.6, the information entered to the right of the numerical data is for explanation only and need not be included in the file. The format of any data file can be found in the accompanying help screen for the desired option.

Also on the companion website for this book, the Excel® file *C10.xls* demonstrates the traverse computations and for the data in Examples 10.4 and 10.6. For those interested in a higher-level programming language, Example 10.4 is computed in the Mathcad® worksheet *TRAV.XMCD*. This example is also demonstrated in the html file *Trav.html*.

Besides performing routine computations such as traverse solutions, personal computers have many other valuable applications in surveying and engineering offices. Two examples are their use with *computer-aided drafting* (CAD) software for plotting maps and drawing contours (see Section 18.14), and with increasing frequency they are also being employed to operate *geographic information system* (GIS) software (see Chapter 28).

■ 10.16 LOCATING BLUNDERS IN TRAVERSE OBSERVATIONS

A numerical or graphic analysis can often be used to determine the location of a mistake, and thereby save considerable field time in making necessary additional observations. For example, if the sum of the interior angles of a five-sided traverse gives a large misclosure—say 10'11"—it is likely that one mistake of 10' and several small errors accumulating to 11" have been made. Methods of graphically locating the station or line where the mistake occurred are illustrated in Figure 10.7. The procedure is shown for a five-sided traverse, but can be used for traverses having any number of sides.

DATA FILE

```
Figure 10.1, Example 10.4 //title line
5 1  //number of courses; 1 = angles to the right; -1 = clockwise direction
126 55 17 //azimuth of first course in traverse; degrees minutes seconds
647.25 100 45 37 //first distance and angle at control station
203.03 231 23 43 //distance and angle for second course and station, respectively
720.35 17 12 59 //and so on
610.24 89 03 28
285.13 101 34 24
10000.00 5000.00 //coordinates of first control station
```

OUTPUT FILE

~~~~~~~~~~~~~~~~~~~~~~~~~~~ Traverse Computation ~~~~~~~~~~~~~~~~~~~~~~~~~~~
Title: Figure 10.1, Example 10.4 //title line   Type: Polygon traverse

Angle Summary

Station	Unadj. Angle	Adj. Angle
1	100°45'37.0"	100°45'34.8"
2	231°23'43.0"	231°23'40.8"
3	17°12'59.0"	17°12'56.8"
4	89° 3'28.0"	89°03'25.8"
5	101°34'24.0"	101°34'21.8"

Angular misclosure (sec): 11"

Course	Length	Azimuth	Unbalanced Dep	Lat
1-2	647.25	126°55'17.0"	517.451	-388.815
2-3	203.03	178°18'57.8"	5.966	-202.942
3-4	720.35	15°31'54.6"	192.891	694.044
4-5	610.24	284°35'20.4"	-590.564	153.709
5-1	285.13	206°09'42.2"	-125.716	-255.919
Sum =	2,466.00		0.028	0.077

Balanced Dep	Lat	Point	Coordinates X	Y
517.443	-388.835	1	10,000.00	5,000.00
5.964	-202.949	2	10,517.44	4,611.16
192.883	694.022	3	10,523.41	4,408.22
-590.571	153.690	4	10,716.29	5,102.24
-125.719	-255.928	5	10,125.72	5,255.93

Linear misclosure = 0.082
Relative Precision = 1 in 30,200

Area: 272,600 sq. ft.
      6.258 acres {if distance units are feet}

Adjusted Observations

Course	Distance	Azimuth	Point	Angle
1-2	647.26	126°55'24"	1	100°45'42"
2-3	203.04	178°19'00"	2	231°23'37"
3-4	720.33	15°31'54"	3	17°12'54"
4-5	610.24	284°35'14"	4	89°03'20"
5-6	285.14	206°09'41"	5	101°34'28"

Figure 10.6 Data file and output file of traverse computations using WOLFPACK.

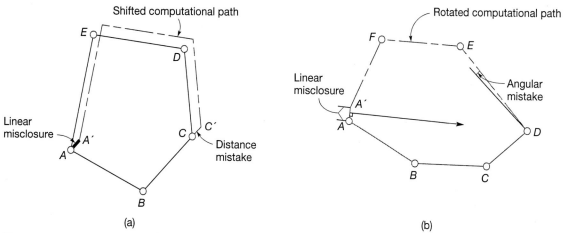

Figure 10.7 Locating a distance (a) or angle (b) blunder.

In Figure 10.7(a), a blunder in the distance BC has occurred. Notice that the mistake CC' shifts the computed coordinates of the remaining stations in such a manner that the azimuth of the linear misclosure line closely matches the azimuth of the course BC that contains the mistake. If no other errors, random or systematic, occurred in the traverse, there would be a perfect match in the directions of the two lines. However since random errors are inevitable, the direction of the course containing the mistake and that of the linear misclosure line never matches perfectly, but will be close.

As shown in Figure 10.7(b), a mistake in an angle (such as at D) will rotate the computed coordinates of the remaining stations. When this happens, the linear misclosure line AA' is a chord of a circle with radius AD. Thus, the perpendicular bisector of the linear misclosure line will point to the center of the circle, which is the station where the angular mistake occurred. Again, if no other errors occurred during the observational process, this perpendicular bisector would point directly to the station. Since other random errors are inevitable, it will most likely point very near the station.

Additional observations and careful field practice will help isolate mistakes. For instance, horizon closures often help isolate and eliminate mistakes in the field. A *cutoff line*, such as CE shown dashed in Figure 9.1(a), run between two stations on a traverse, produces smaller closed figures to aid in checking and isolating blunders. Additionally, the extra observations will increase the redundancy in the traverse, and hence the precision of the overall work. These additional observations can be used as checks when performing a compass rule adjustment or can be included in a least squares adjustment, which is discussed in Chapter 16.

For those wishing to program the computations presented in this chapter, the Mathcad worksheet *TRAV.XMCD*, which is available on the companion website for this book, demonstrates the examples presented in this chapter. Additionally, a traverse with a single angular blunder is used to demonstrate how the perpendicular bisector of the misclosure line seemingly points directly to the angle containing a 1-min blunder.

10.17 MISTAKES IN TRAVERSE COMPUTATIONS

Some of the more common mistakes made in traverse computations are:

1. Failing to adjust the angles before computing azimuths or bearings
2. Applying angle adjustments in the wrong direction and failing to check the angle sum for proper geometric total
3. Interchanging departures and latitudes, or their signs
4. Confusing the signs of coordinates

PROBLEMS

Asterisks (*) indicate problems that have partial answers given in Appendix G.

10.1 What are the usual steps followed in adjusting a closed traverse?

10.2* The sum of seven interior angles of a closed-polygon traverse each read to the nearest 3″ is 899°59′39″. What is the misclosure, and what correction would be applied to each angle in balancing them by method 1 of Section 10.2?

10.3 Similar to Problem 10.2, except the angles were read to the nearest 2″ and their sum was 720°00′12″ for a six-sided polygon traverse.

10.4 Similar to Problem 10.2, except the angles were read to the nearest 1″, and their sum for a nine-sided polygon traverse was 1259°59′51″.

10.5* Balance the angles in Problem 9.22. Compute the preliminary azimuths for each course.

10.5 Balance the following interior angles (angles-to-the-right) of a five-sided closed polygon traverse using method 1 of Section 10.2. If the azimuth of side AB is fixed at 138°16′27″ calculate the azimuths of the remaining sides. $A = 132°47′08″$; $B = 108°46′20″$; $C = 107°19′39″$; $D = 81°50′38″$; $E = 109°16′20″$.

10.6 Compute departures and latitudes, linear misclosure, and relative precision for the traverse of Problem 10.5 if the lengths of the sides (in feet) are as follows: $AB = 202.74$; $BC = 283.87$; $CD = 498.37$; $DE = 320.33$; and $EA = 380.78$.

10.7 Using the compass (Bowditch) rule, adjust the departures and latitudes of the traverse in Problem 10.6. If the coordinates of station A are $X = 20,000.00$ ft and $Y = 15,000.00$ ft calculate (1) coordinates for the other stations, (2) lengths and azimuths of lines BC and CD, and (3) the final adjusted angles at stations B and C.

10.8 Balance the following interior angles-to-the-right for a polygon traverse to the nearest 1′ using method 1 of Section 10.2. Compute the azimuths assuming a fixed azimuth of 259°34′07″ for line AB. $A = 57°00′38″$; $B = 88°24′44″$; $C = 126°37′24″$; $D = 46°03′50″$; $E = 221°53′34″$.

10.9 Determine departures and latitudes, linear misclosure, and relative precision for the traverse of Problem 10.8 if lengths of the sides (in meters) are as follows: $AB = 336.249$; $BC = 315.584$; $CD = 299.806$; $DE = 294.536$; and $EA = 234.374$.

10.10 Using the compass (Bowditch) rule adjust the departures and latitudes of the traverse in Problem 10.9. If the coordinates of station A are $X = 310,630.892$ m and $Y = 121,311.411$ m, calculate (1) coordinates for the other stations and, from them, (2) the lengths and bearings of lines BC and EA, and (3) the final adjusted angles at B and D.

10.11 Same as Problem 10.8, except assume line AB has a fixed azimuth of 125°09′32″.

10.12 Using the lengths from Problem 10.9 and azimuths from Problem 10.11, calculate departures and latitudes, linear misclosure, and relative precision of the traverse.

10.13 Adjust the departures and latitudes of Problem 10.12 using the compass (Bowditch) rule, and compute coordinates of all stations if the coordinates of station A are $X = 243,605.596$ m and $Y = 25,393.201$ m. Compute the length and azimuth of line AC.

10.14 Compute and tabulate for the following closed-polygon traverse: (1) preliminary azimuths, (2) unadjusted departures and latitudes, (3) linear misclosure, and (4) relative precision. (*Note:* line BC bears NW.)

Course	Azimuth	Length (m)	Interior Angle (Right)
AB	75°14′47″	519.674	A = 12°25′31″
BC		456.765	B = 48°19′25″
CD		433.926	C = 126°37′24″
DE		427.558	D = 46°03′36″
EA		107.934	E = 306°34′19″

10.15* In Problem 10.14, if one side and/or angle is responsible for most of the error of closure, which is it likely to be?

10.16 Adjust the traverse of Problem 10.14 using the compass rule. If the coordinates in meters of point A are 10,356.548 E and 98,761.391 N, determine the coordinates of all other points. Find the length and bearing of line AC.

For the closed-polygon traverses given in Problems 10.17 and 10.18 (lengths in feet), compute and tabulate: (1) unbalanced departures and latitudes, (2) linear misclosure, (3) relative precision, and (4) preliminary coordinates if $X_A = 10,000.00$ and $Y_A = 5000.00$. Balance the traverses by coordinates using the compass rule.

Course		AB	BC	CD	DA
10.17	Bearing	N 8°17′02″ E	N 87°02′05″ E	S 14°47′06″ W	N 68°43′20″ W
	Length	403.73	622.63	653.16	550.84
10.18	Azimuth	111°18′00″	25°03′12″	312°43′05″	205°05′04″
	Length	385.94	1016.88	403.50	1164.49

10.19 Compute the linear misclosure, relative precision, and adjusted lengths and azimuths for the sides after the departures and latitudes are balanced by the compass rule in the following closed-polygon traverse.

Course	Length (m)	Departure (m)	Latitude (m)
AB	412.516	−216.2394	−351.2975
BC	513.185	+512.9654	+15.0112
CA	448.495	−296.7364	+336.2964

10.20 The following data apply to a closed link traverse [like that of Figure 9.1(b)]. Compute preliminary azimuths, adjust them, and calculate departures and

268 TRAVERSE COMPUTATIONS

latitudes, misclosures in departure and latitude, and traverse relative precision. Balance the departures and latitudes using the compass rule, and calculate coordinates of points B and C. Compute the final lengths and azimuths of lines AB, BC, and CD.

Station	Measured Angle (to the Right)	Adjusted Azimuth	Measured Length (ft)	X (ft)	Y (ft)
AzMk$_1$					
		322°41′31″			
A	267°19′13″			2,517,347.31	395,025.36
			243.09		
B	248°59′20″				
			258.95		
C	110°29′24″				
			197.42		
D	92°03′00″			2,517,910.07	395,184.30
		321°32′20″			
AzMk$_2$					

10.21 Similar to Problem 10.20, except use the following data:

Station	Measured Angle (to the Right)	Adjusted Azimuth	Measured Length (m)	X (m)	Y (m)
AzMk$_1$					
		314°09′23″			
A	263°48′52″			194,325.090	25,353.988
			957.957		
B	263°45′24″				
			945.617		
C	98°55′04″				
			857.724		
D	110°21′40″			196,277.341	26,262.583
		331°00′27″			
AzMk$_2$					

The azimuths (from north of a polygon traverse are $AB = 38°17'02''$, $BC = 121°26'30''$, $CD = 224°56'59''$, and $DA = 308°26'56''$. If one observed distance contains a mistake, which course is most likely responsible for the closure conditions given in Problems 10.22 and 10.23? Is the course too long or too short?

10.22* Algebraic sum of departures = −4.24 ft latitudes = −4.28 ft.
10.23 Algebraic sum of departures = 2.044 m latitudes = 2.588 m.
10.24 Determine the lengths and bearings of the sides of a lot whose corners have the following X and Y coordinates (in meters): A (198.300, 617.109); B (330.087, 458.965); C (640.903, 718.007); D (508.146, 875.635).
10.25 Compute the lengths and azimuths of the sides of a closed-polygon traverse whose corners have the following X and Y coordinates (in feet): A (320.08, 155.40); B (384.13, 155.40); C (384.13, 281.29); D (320.01, 280.96).
10.26 In searching for a record of the length and true bearing of a certain boundary line which is straight between A and B, the following notes of an old random traverse were found (survey by compass and Gunter's chain, declination 4°45′ W). Compute the true bearing and length (in feet) of BA.

Course	A-1	1-2	2-3	3-B
Magnetic bearing	Due North	N20°00′E	Due East	S46°30′E
Distance (ch)	11.90	35.80	24.14	12.72

10.27 Describe how a blunder may be located in a traverse.

BIBLIOGRAPHY

Ghilani, C. D. 2018. *Adjustment Computations: Spatial Data Analysis*, 6th ed. New York, NY: Wiley.

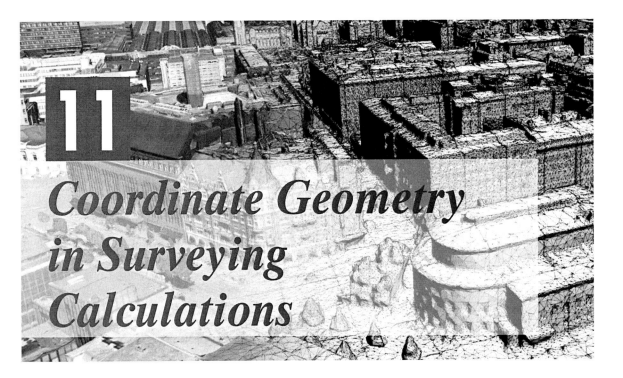

11
Coordinate Geometry in Surveying Calculations

■ 11.1 INTRODUCTION

Except for extensive geodetic control surveys, almost all other surveys are referenced to plane rectangular coordinate systems. State plane coordinates (see Chapter 20) are most frequently employed, although local arbitrary systems can be used. Advantages of referencing points in a rectangular coordinate system are (1) the relative positions of points are uniquely defined, (2) they can be conveniently plotted, (3) if lost in the field, they can readily be recovered from other available points referenced to the same system, and (4) computations are greatly facilitated.

Computations involving coordinates are performed in a variety of surveying problems. Two situations were introduced in Chapter 10, where it was shown that the length and direction (azimuth or bearing) of a line can be calculated from the coordinates of its end points. Area computation using coordinates is discussed in Chapter 12. Additional problems that are conveniently solved using coordinates are determining the point of intersection of (1) two lines, (2) a line and a circle, and (3) two circles. The solutions for these and other coordinate geometry problems are discussed in this chapter. It will be shown that the method employed to determine the intersection point of a line and a circle reduces to finding the intersection of a line of known azimuth and another line of known length. Also, the problem of finding the intersection of two circles consists of determining the intersection point of two lines having known lengths. These types of problems are regularly encountered in the horizontal alignment surveys where it is necessary to compute intersections of tangents and circular curves, and in boundary and subdivision work where parcels of land are often defined by straight lines and circular arcs.

The three types of intersection problems noted above are conveniently solved by forming a triangle between two stations of known position from which

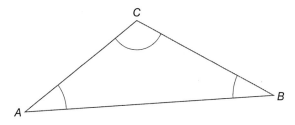

Figure 11.1
An oblique triangle.

the observations are made, and then solving for the parts of this triangle. Two important functions used in solving oblique triangles are (1) the law of sines, and (2) the law of cosines. The law of sines relates the lengths of the sides of a triangle to the sines of the opposite angles. For Figure 11.1, this law is

$$\frac{BC}{\sin A} = \frac{AC}{\sin B} = \frac{AB}{\sin C} \tag{11.1}$$

where AB, BC, and AC are the lengths of the three sides of the triangle ABC, and A, B, and C are the angles. The law of cosines relates two sides and the included angle of a triangle to the length of the side opposite the angle. In Figure 11.1, the following three equations can be written that express the law of cosines:

$$BC^2 = AC^2 + AB^2 - 2(AC)(AB)\cos A$$
$$AC^2 = BA^2 + BC^2 - 2(BA)(BC)\cos B \tag{11.2}$$
$$AB^2 = CB^2 + CA^2 - 2(CB)(CA)\cos C$$

In some coordinate geometry solutions, the use of the quadratic formula is used. Examples where this equation simplifies the solution are discussed in Sections 24.16.1 and 25.10. This formula, which gives the solution for x in any quadratic equation of form $ax^2 + bx + c = 0$, is

$$x = \frac{-b \pm \sqrt{b^2 - 4ac}}{2a} \tag{11.3}$$

In the remaining sections of this chapter, procedures using triangles and Equations (11.1) through (11.3) are presented for solving each type of standard coordinate geometry problem.

11.2 COORDINATE FORMS OF EQUATIONS FOR LINES AND CIRCLES

In Figure 11.2, straight line AB is referenced in a plane rectangular coordinate system. Coordinates of end points A and B are X_A, Y_A, and X_B, and Y_B. Length AB and azimuth Az_{AB} of this line in terms of these coordinates are

$$AB = \sqrt{(X_B - X_A)^2 + (Y_B - Y_A)^2} \tag{11.4}$$

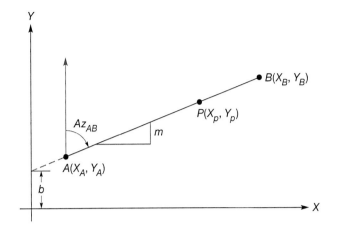

Figure 11.2
Geometry of a straight line in a plane coordinate system.

$$Az_{AB} = \tan^{-1}\left(\frac{\Delta X}{\Delta Y}\right) + C \tag{11.5a}$$

where ΔX is $X_B - X_A$, C is $0°$ if both ΔX and ΔY are greater than zero; C is $180°$ if ΔY is less than zero, and C is $360°$ if ΔX is less than zero, and ΔY is greater than zero. Another frequently used equation for determining the azimuth of a course in software is known as the atan2 function, which is computed as

$$Az_{AB} = \text{atan2}(\Delta Y, \Delta X) + D = 2\tan^{-1}\left(\frac{\Delta X}{\Delta Y + \sqrt{\Delta X^2 + \Delta Y^2}}\right) + D \tag{11.5b}$$

where D is the $0°$ if the results of the atan2 function are positive and $360°$ if the results of the function are negative. The general mathematical expression for a straight line is

$$Y_P = mX_P + b \tag{11.6}$$

where Y_P is the Y coordinate of any point P on the line whose X coordinate is X_P, m the slope of the line, and b the y-intercept of the line. Slope m can be expressed as

$$m = \frac{Y_B - Y_A}{X_B - X_A} = \cot Az_{AB} \tag{11.7}$$

From Equations (11.5a) and (11.7), it can be shown that

$$Az_{AB} = \tan^{-1}\left(\frac{1}{m}\right) + C \tag{11.8}$$

The mathematical expression for a circle in rectangular coordinates can be written as

$$R^2 = (X_P - X_O)^2 + (Y_P - Y_O)^2 \tag{11.9}$$

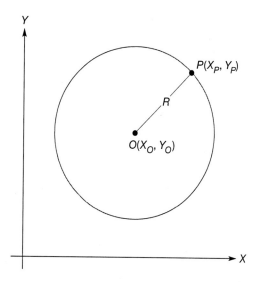

Figure 11.3
Geometry of a circle in a plane coordinate system.

In Equation (11.9), and with reference to Figure 11.3, R is the radius of the circle, X_O and Y_O are the coordinates of the radius point O, and X_P and Y_P the coordinates of any point P on the circle. The general form of the circle equation is

$$X_P^2 + Y_P^2 - 2X_O X_P - 2Y_O Y_P + f = 0 \qquad (11.10)$$

where the length of the radius of the circle is given as $R = \sqrt{X_O^2 + Y_O^2 - f}$ [Note: Equations (11.9) and (11.10) are not used in solving problems in this chapter, they are applied in later chapters.]

11.3 PERPENDICULAR DISTANCE FROM A POINT TO A LINE

A common problem encountered in boundary surveying is determining the perpendicular distance of a point from a line. This procedure can be used to check the alignment of survey markers on a block and is also useful in subdivision design. Assume in Figure 11.4 that points A and B are on the line defined by two block corners whose coordinates are known. Also, assume that the coordinates of point

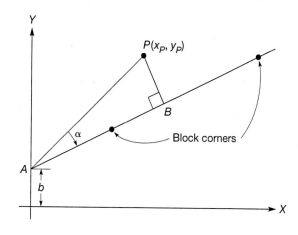

Figure 11.4
Perpendicular distance of a point from a line.

274 COORDINATE GEOMETRY IN SURVEYING CALCULATIONS

P are known. The slope, m, and y-intercept, b, of line AB are computed from the coordinates of the block corners. By assigning the X and Y coordinate axes as shown in the figure, the coordinates of point A are $X_A = 0$, and $Y_A = b$. Using Equations (11.4) and (11.5a), the length and azimuth of line AP can be determined from its coordinates. By Equation (11.8), the azimuth of line AB can be determined from the slope of the line AB. Now angle α can be computed as the difference in the azimuth AP and AB, which for the situation depicted in Figure 11.4 is

$$\alpha = Az_{AB} - Az_{AP} \qquad (11.11)$$

Recognizing that ABP is a right triangle, length BP is

$$BP = AP \sin \alpha \qquad (11.12)$$

where the length of AP is determined from the coordinates of points A and P using Equation (11.4).

■ Example 11.1

For Figure 11.4, assume that the XY coordinates of point P are (1123.82, 509.41) and that the coordinates of the block corners are (865.49, 416.73) and (1557.41, 669.09). What is the perpendicular distance of point P from line AB? (All units are in feet.)

Solution

Using the block corner coordinates and Equation (11.7), the slope of line AB is

$$m = \frac{669.09 - 416.73}{1557.41 - 865.49} = 0.364724245$$

Rearranging Equation (11.6), the y-intercept of line AB is

$$b = 416.73 - 0.364724245(865.49) = 101.065 \text{ ft}$$

By Equations (11.4) and (11.5a), the length and azimuth of line AP is

$$AP = \sqrt{(1123.82 - 0)^2 + (509.41 - 101.065)^2} = 1195.708 \text{ ft}$$

$$Az_{AP} = \tan^{-1}\left(\frac{1123.82 - 0}{509.41 - 101.07}\right) + 0° = 70°01'52.2''$$

By Equation (11.8), the azimuth of line AB is

$$Az_{AB} = \tan^{-1}\left(\frac{1}{0.364724245}\right) + 0° = 69°57'42.7''$$

Using Equation (11.11), angle α is

$$\alpha = 70°01'52.2'' - 69°57'42.7''$$

From Equation (11.12), the perpendicular distance from point P to line AB is

$$BP = 1195.708 \sin 0°04'09.5'' = 1.45 \text{ ft}$$

11.4 INTERSECTION OF TWO LINES, BOTH HAVING KNOWN DIRECTIONS

Figure 11.5 illustrates the intersection of two lines AP and BP. Each has known coordinates for one end point, and each has a known direction. Determining the point of intersection for this type of situation is often called the *direction-direction* problem. A simple method of computing the intersection point P is to solve for the parts of oblique triangle ABP. Since the coordinates of A and B are known, the length and azimuth of AB (shown dashed) can be determined using Equations (11.4) and (11.5a), respectively. Then, from the figure angle A is the difference in the azimuths of AB and AP, or

$$A = Az_{AP} - Az_{AB} \tag{11.13}$$

Similarly, angle B is the difference in the azimuths of BA and BP, or

$$B = Az_{BA} - Az_{BP} \tag{11.14}$$

With two angles of the triangle ABP computed, the remaining angle P is

$$P = 180° - A - B \tag{11.15}$$

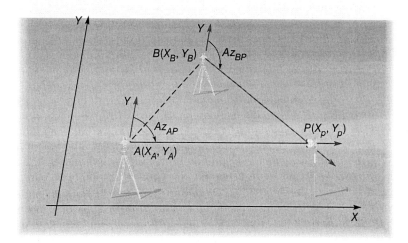

Figure 11.5
Intersection of two lines with known directions.

Substituting into Equation (11.1), and rearranging, the length of side AP is

$$AP = AB\frac{\sin B}{\sin P} \quad (11.16)$$

With both the length and azimuth of AP known, the coordinates of P are

$$X_P = X_A + AP \sin Az_{AP}$$
$$Y_P = Y_A + AP \cos Az_{AP} \quad (11.17)$$

A check on this solution can be obtained by solving for length BP, and using it together with the azimuth of BP to compute the coordinates of P. The two solutions should agree, except for round off error.

It should be noted that if the azimuths for lines AP and BP are equal, then the lines are parallel and have no intersection.

■ Example 11.2

In Figure 11.5, assuming the following information is known for two lines, compute coordinates X_P and Y_P of the intersection point. (Coordinates are in feet.)

$$X_A = 1425.07 \quad X_B = 7484.80 \quad Az_{AP} = 76°04'24''$$
$$Y_A = 1971.28 \quad Y_B = 5209.64 \quad Az_{BP} = 141°30'16''$$

Solution

By Equations (11.4) and (11.5a), the length and azimuth of side AB are

$$AB = \sqrt{(7484.80 - 1425.07)^2 + (5209.64 - 1971.28)^2} = 6870.757 \text{ ft}$$

$$Az_{AB} = \tan^{-1}\left(\frac{7484.80 - 1425.07}{5209.64 - 1971.28}\right) + 0° = 61°52'46.8''$$

By Equations (11.13) through (11.5), the three angles of triangle ABP are

$$A = 76°04'24'' - 61°52'46.8'' = 14°11'37.2''$$
$$B = (180° + 61°52'46.8'') - 141°30'16'' = 100°22'30.8''$$
$$P = 180° - (14°11'37.2'' + 100°22'30.8'') = 65°25'52.0''$$

By Equation (11.16), length AP is

$$AP = 6870.757\frac{\sin 100°22'30.8''}{\sin 65°25'52.0''} = 7431.224 \text{ ft}$$

By Equations (11.17), the coordinates of station P are

$$X_P = 1425.07 + 7431.224 \sin 76°04'24'' = 8637.85 \text{ ft}$$
$$Y_P = 1971.28 + 7431.224 \cos 76°04'24'' = 3759.83 \text{ ft}$$

Check:

$$BP = 6870.757 \frac{\sin 14°11'37.2''}{\sin 65°25'52''} = 1852.426 \, \text{ft}$$

$$X_P = 7484.80 + 1852.426 \sin 141°30'16'' = 8637.85 \checkmark$$

$$Y_P = 5209.64 + 1852.426 \cos 141°30'16'' = 3759.83 \checkmark$$

■ 11.5 INTERSECTION OF A LINE WITH A CIRCLE

Figure 11.6 illustrates the intersection of a line (AC) of known azimuth with a circle of known radius ($BP_1 = BP_2$). Finding the intersection for this situation reduces to finding the intersection of a line of known direction with another line of known length and is sometimes referred to as the *direction–distance* problem. As shown in the figure, notice that this problem has two different solutions, but as discussed later, the incorrect one can generally be detected and discarded.

The approach to solving this problem is like that employed in Section 11.4; that is, the answer is determined by solving an oblique triangle. This solution will demonstrate the use of the quadratic equation to obtain both solutions. In Figure 11.6, the coordinates of B (the radius point of the circle) are known. From the coordinates of points A and B, the length and azimuth of line AB (shown dashed) are determined by employing Equations (11.4) and (11.5a), respectively. Then angle A is computed from the azimuths of AB and AC as follows:

$$A = Az_{AP} - Az_{AB} \tag{11.18}$$

Substituting the known values of A, AB, and BP into the law of cosines [Equation (11.2)] yields

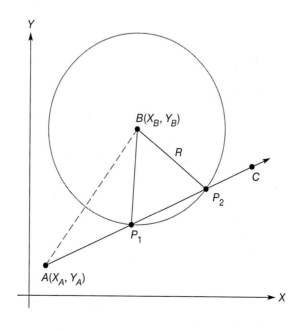

Figure 11.6
Intersection of a line and a circle.

$$BP^2 = AB^2 + AP^2 - 2(AB)(AP)\cos A \qquad (11.19)$$

In Equation (11.19), AP is an unknown quantity. Rearranging this equation gives

$$AP^2 - 2(AB)(AP)\cos A + (AB^2 - BP^2) = 0 \qquad (11.20)$$

Now Equation (11.20), which is a second-degree expression, can be solved using the quadratic formula [Equation (11.3)] as follows:

$$AP = \frac{2(AB)\cos A \pm \sqrt{[2(AB)\cos A]^2 - 4(AB^2 - BP^2)}}{2} \qquad (11.21)$$

In comparing Equation (11.21) to Equation (11.3), it can be seen that $a = 1, b = 2(AB)\cos A$ and $c = (AB^2 - BP^2)$. Because of the $\pm$ sign in the formula, there are two solutions for length AP. Once these two lengths are determined, the possible coordinates of station P are

$$\begin{aligned} X_{P1} &= X_A + AP_1 \sin Az_{AP} \text{ and } Y_{P1} = Y_A + AP_1 \cos Az_{AP} \\ X_{P2} &= X_A + AP_2 \sin Az_{AP} \text{ and } Y_{P2} = Y_A + AP_2 \cos Az_{AP} \end{aligned} \qquad (11.22)$$

If errors exist in the given data for the problem, or if an impossible design is attempted, the circle will not intersect the line. In this case, the terms under the radical in Equation (11.21) will be negative; that is, $[2(AB)\cos A]^2 - 4(AB^2 - BP^2) < 0$. It is therefore important when solving any of the coordinate geometry problems to be alert for these types of potential problems.

The sine law can also be used to solve this problem. However, care must be exercised when using the sine law since the two solutions will not be readily apparent. The procedure of solving this problem using the sine law is as follows:

1. Compute the length and azimuth of line AB from the coordinates using Equations (11.4) and (11.5a), respectively.
2. Compute the angle at A using Equation (11.18).
3. Using the sine law solve for the angles at P_1 as

$$\sin P = \frac{AB \sin A}{BP} \qquad (11.23)$$

4. Note that the sine function has the relationship $\sin x = \sin(180° - x)$. Thus, the solution for the angle at B is

$$\begin{aligned} B_1 &= 180° - (A + P) \\ B_2 &= P - A \end{aligned} \qquad (11.24)$$

5. Using the two solutions for angle B, determine the azimuth of line BP as

$$\begin{aligned} Az_{BP1} &= Az_{BA} - B_1 \\ Az_{BP2} &= Az_{BA} - B_2 \end{aligned} \qquad (11.25)$$

6. Finally using the two azimuths and the observed length of BP determine the two possible solutions for station P as

$$X_{P1} = X_B + BP \sin Az_{BP1} \text{ and } Y_{P1} = Y_B + BP \cos Az_{BP1}$$
$$X_{P2} = X_B + BP \sin Az_{BP2} \text{ and } Y_{P2} = Y_B + BP \cos Az_{BP2}$$
(11.26)

■ Example 11.3

In Figure 11.6, assume the (x, y) coordinates of point A are $(100.00, 130.00)$, and that the (x, y) coordinates point B are $(500.00, 600.00)$. If the azimuth of AP is $70°42'36''$, and the radius of the circle (length BP) is 350.00, what are the possible coordinates of point P? (Note: linear units are feet.)

Solution

By Equations (11.4) and (11.5a), the length and azimuth of AB are

$$AB = \sqrt{(500 - 100)^2 + (600 - 130)^2} = 617.171 \text{ ft}$$
$$Az_{AB} = \tan^{-1}\left(\frac{500 - 100}{600 - 130}\right) + 0° = 40°23'59.7''$$

By Equation (11.18), the angle at A is

$$A = 70°40'36'' - 40°23'59.7'' = 30°18'.3''$$

Substituting appropriate values according to Equation (11.20), the quadratic equation coefficients are

$$a = 1$$
$$b = -2(617.171)\cos 30°18'36.3'' = -1065.616$$
$$c = 617.171^2 - 350.00^2 = 258{,}400.043$$

Substituting these values into Equation (11.21) yields

$$AP = \frac{1065.616 \pm \sqrt{1065.616^2 - 4(258{,}400.043)}}{2}$$
$$= \frac{1065.616 \pm 319.276}{2}$$
$$= 373.170 \text{ or } 692.446$$

Using the azimuth and distances for AP, the two possible solutions for the coordinates of P are

$$X_{P1} = 100.00 + 373.170 \sin 70°42'36'' = 452.22 \text{ ft}$$
$$Y_{P1} = 130.00 + 373.170 \cos 70°42'36'' = 253.28 \text{ ft}$$

or

$$X_{P2} = 100.00 + 692.446 \sin 70°42'36'' = 753.57 \text{ ft}$$
$$Y_{P2} = 130.00 + 692.446 \cos 70°42'36'' = 358.75 \text{ ft}$$

In solving a quadratic equation, the decision to add or subtract the value from the radical can be made on the basis of experience, or by using a carefully constructed scaled diagram, which also provides a check on the computations. One answer will be unreasonable and should be discarded. An arithmetic check is possible by solving for the two possible angles at B to P in triangle ABP and determining the coordinates of P from station B, or by solving the problem using the second procedure. Readers should verify that the same solution can be obtained using Equations (11.23) through (11.26).

■ 11.6 INTERSECTION OF TWO CIRCLES

In Figure 11.7, the intersection of two circles is illustrated. Note that the circles are obtained by simply radiating two distances (their radius values R_A and R_B) about their radius points A and B. As shown, this geometry again results in two intersection points, P_1 and P_2. As with the two previous cases, these intersection points can again be located by solving for the parts of oblique triangle ABP. In this situation, two sides of the triangle are the known radii, and thus the problem is often called the *distance–distance* problem. The third side of the triangle, AB, can be computed from known coordinates of A and B, or the distance can be observed.

The first step in solving this problem is to compute the length and azimuth of line AB using Equations (11.4) and (11.5a). Then angle A can be determined using the law of cosines (Equation 11.2). As shown in Figure 11.7, the two solutions for P

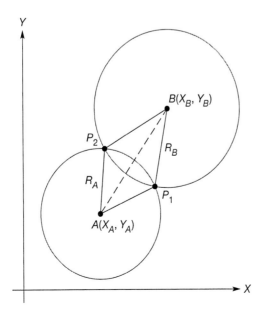

Figure 11.7
Intersection of two circles.

at either P_1 or P_2 are derived by either adding or subtracting angle A from the azimuth of line AB to obtain the direction of AP. By rearranging Equation 11.2, angle A is

$$A = \cos^{-1}\left[\frac{AB^2 + AP^2 - BP^2}{2(AB)(AP)}\right] \quad (11.27)$$

Thus, the azimuth of line AP is either

$$Az_{AP1} = Az_{AB} + A$$
$$Az_{AP2} = Az_{AB} - A \quad (11.28)$$

The possible coordinates of P are

$$X_{P1} = X_A + AP_1 \sin Az_{AP1} \text{ and } Y_{P1} = Y_A + AP_1 \cos Az_{AP1}$$
$$X_{P2} = X_A + AP_2 \sin Az_{AP2} \text{ and } Y_{P2} = Y_A + AP_2 \cos Az_{AP2} \quad (11.29)$$

The decision of whether to add or subtract angle A from the azimuth of line AB can be made on the basis of experience, or through the use of a carefully constructed scaled diagram. One answer will be unreasonable and should be discarded. As can be seen from Figure 11.7, there will be no solution if length of AB is greater than the sum of R_A and R_B.

■ Example 11.4

In Figure 11.7, assume the following data (in meters) are available:

$$X_A = 2851.28 \quad Y_A = 2851.28 \quad R_A = 2000.00$$
$$X_B = 3898.72 \quad Y_B = 2870.15 \quad R_B = 1500.00$$

Compute the X and Y coordinates of point P.

Solution

By Equations (11.4) and (11.5a), the length and azimuth of AB are

$$AB = \sqrt{(3898.72 - 2851.28)^2 + (2870.15 - 299.40)^2} = 2775.948 \text{ m}$$

$$Az_{AB} = \tan^{-1}\left[\frac{3898.72 - 2851.28}{2870.15 - 299.40}\right] + 0° = 22°10'6''$$

By Equation (11.27), A is

$$A = \cos^{-1}\left(\frac{2775.948^2 + 2000.00^2 - 1500.00^2}{2(2775.948)2000.00}\right) = 31°36'53.6''$$

By combining Equations (11.28) and (11.29), the possible solutions for P are

$$X_{P1} = 2851.28 + 2000.00 \sin(22°10'05.6'' + 31°36'53.6'') = 4464.85 \text{ m}$$
$$Y_{P1} = 299.40 + 2000.00 \cos(22°10'05.6'' + 31°36'53.6'') = 1481.09 \text{ m}$$

or

$$X_{P2} = 2851.28 + 2000.00 \sin(22°10'05.6'' - 31°36'53.6'') = 2523.02 \text{ m}$$
$$Y_{P2} = 299.40 + 2000.00 \cos(22°10'05.6'' - 31°36'53.6'') = 2272.28 \text{ m}$$

An arithmetic check on this solution can be obtained by determining the angle and coordinates of P from station B.

11.7 THREE-POINT RESECTION

This procedure locates a point of unknown position by observing horizontal angles from that point to three visible stations whose positions are known. The situation is illustrated in Figure 11.8, where a total station instrument occupies station P and angles x and y are observed. A summary of the method used to compute the coordinates of station P follows (refer to Figure 11.8):

1. From the known coordinates of A, B, and C calculate lengths a and c, and angle α at station B.
2. Subtract the sum of angles x, y, and α in figure $ABCP$ from $360°$ to obtain the sum of angles $A + C$

$$A + C = 360° - (\alpha + x + y) \tag{11.30}$$

3. Calculate angles A and C using the following:

$$A = \tan^{-1}\left[\frac{a \sin x \sin(A + C)}{c \sin y + a \sin x \cos(A + C)}\right] \tag{11.31}$$

$$C = \tan^{-1}\left[\frac{c \sin y \sin(A + C)}{a \sin x + c \sin y \cos(A + C)}\right] \tag{11.32}$$

4. From angle A and azimuth AB, calculate azimuth AP in triangle ABP. Then solve for length AP using the law of sines, where $\alpha_1 = 180° - A - x$. Calculate the departure and latitude of AP followed by the coordinates of P.
5. In the manner outlined in step 4, use triangle BCP to calculate the coordinates of P to obtain a check.

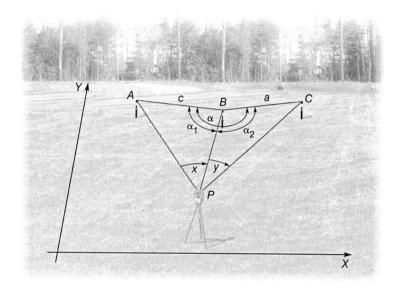

Figure 11.8
The resection problem.

■ Example 11.5

In Figure 11.8, angles x and y were measured as 48°53′12″ and 41°20′35″, respectively. Control points A, B, and C have coordinates (in feet) of $X_A = 5721.25$, $Y_A = 21{,}802.48$, $X_B = 13{,}542.99$, $Y_B = 22{,}497.95$, $X_C = 20{,}350.09$, and $Y_C = 24{,}861.22$. Calculate the coordinates of P.

Solution

1. By Equation (11.4)

$$a = \sqrt{(20{,}350.09 - 13{,}542.99)^2 + (24{,}861.22 - 22{,}497.95)^2} = 7205.67\,\text{ft}$$

$$c = \sqrt{(13{,}542.99 - 5721.25)^2 + (22{,}497.95 - 21{,}802.48)^2} = 7852.60\,\text{ft}$$

2. By Equation (11.5a)

$$Az_{AB} = \tan^{-1}\!\left(\frac{13{,}542.99 - 5721.25}{22{,}497.95 - 21{,}802.48}\right) + 0° = 84°55′08.1″$$

$$Az_{BC} = \tan^{-1}\!\left(\frac{20{,}350.09 - 13{,}542.99}{24{,}861.22 - 22{,}497.95}\right) + 0° = 70°51′15.0″$$

3. Calculate angle α

$$\alpha = 180° - (70°51′15.0″ - 84°55′08.1″) = 194°03′53.1″$$

4. By Equation (11.30)

$$A + C = 360° - 194°03′53.1″ - 48°53′12″ - 41°20′35″ = 75°42′19.9″$$

5. By Equation (11.31)

$$A = \tan^{-1}\left(\frac{7250.67 \sin 48°53'12'' \sin 75°42'19.9''}{7852.60 \sin 41°20'35'' + 7205.67 \sin 48°53'12'' \cos 75°42'19.9''}\right)$$

$$= 38°51'58.7''$$

6. By Equation (11.32)

$$C = \tan^{-1}\left(\frac{7852.60 \sin 41°20'35'' \sin 75°42'19.9''}{7205.67 \sin 48°53'12'' + 7852.60 \sin 41°20'35'' \cos 75°42'19.9''}\right)$$

$$= 36°50'21.2''$$

$(A + C = 38°51'58.7'' + 36°50'21.2'' = 75°42'19.9'' \checkmark)$

7. Calculate angle α_1

$$\alpha_1 = 180° - 38°51'58.7'' - 48°53'12'' = 92°14'49.3''$$

8. By the law of sines

$$AP = \frac{7852.60 \sin 92°14'49.3''}{\sin 48°53'12''} = 10,414.72 \text{ ft}$$

$$Az_{AP} = Az_{AB} + A = 84°55'1'' + 38°51'58.7'' = 123°47'06.8''$$

9. By Equations (10.1) and (10.2)

$$\text{Dep}_{AP} = 10,414.72 \sin 123°47'06.8'' = 8655.97 \text{ ft}$$

$$\text{Lat}_{AP} = 10,414.72 \cos 123°47'06.8'' = -5791.43 \text{ ft}$$

10. By Equation (10.7)

$$X_P = 5721.25 + 8655.97 = 14,377.22 \text{ ft}$$

$$Y_P = 21,802.48 - 5791.43 = 16,011.05 \text{ ft}$$

11. As a check, triangle BCP was solved to obtain the same results.

The three-point resection problem just described provides a unique solution for the unknown coordinates of point P, that is, there are no redundant observations, and thus no check can be made on the observations. This is actually a special case of the more general resection problem, which provides redundancy and enables a least-squares solution. In the general resection problem, in addition to observing the angles x and y, distances from P to one or more control stations could also have been observed. Other possible variations in resection that provide redundancy include observing (a) one angle and two distances to two control stations; (b) two angles and one, two, or three distances to three control points; or (c) the use of more than three control stations. Then all observations can be included in a least-squares solution to obtain the most probable coordinates of point *P*. Resection has become a popular method for quickly locating total station instruments, as discussed in Section 23.9. The procedure is convenient because these instruments can readily observe both angles and distances, and their onboard microprocessors can instantaneously provide the least-squares solution for the instrument's position.

It should be noted that the resection problem will not have a unique solution if points *A*, *B*, *C*, and *P* define a circle. Selecting points *B* and *P* so that they both lie on the same side of a line connecting points *A* and *C* avoids this problem. Additionally, the accuracy of the solution will decrease if the observed angles *x* and *y* become small. As a general guideline, the observed angles should be greater than 30° for best results.

■ 11.8 TWO-DIMENSIONAL CONFORMAL COORDINATE TRANSFORMATION

It is sometimes necessary to convert coordinates of points from one survey coordinate system to another. For example, this happens if a survey is performed in some local-assumed or arbitrary coordinate system, and later it is desired to convert it to state plane coordinates. The process of making these conversions is called *coordinate transformation*. If only planimetric coordinates (i.e., *X*s and *Y*s) are involved, and true shape is retained, it is called *two-dimensional (2D) conformal coordinate transformation*.

The geometry of a 2D conformal coordinate transformation is illustrated in Figure 11.9. In the figure, *X-Y* represents a local-assumed coordinate system,

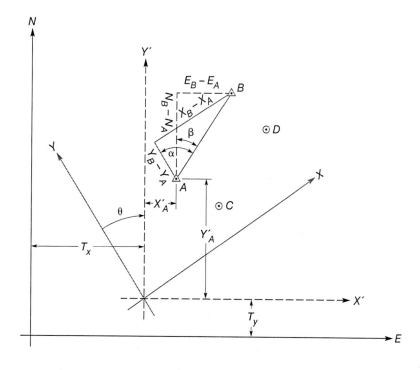

Figure 11.9
Geometry of the two-dimensional coordinate transformation.

and *E-N* a state plane coordinate system. Coordinates of points *A* through *D* are known in the *X-Y* system and those of *A* and *B* are also known in the *E-N* system. Points such as *A* and *B*, whose positions are known in both systems, are termed *control points*. At least two control points are required in order to determine *E-N* coordinates of other points such as *C* and *D*.

In general, three steps are involved in coordinate transformation: (1) rotation, (2) scaling, and (3) translation. As shown in Figure 11.9, rotation consists in determining coordinates of points in the rotated *X'-Y'* axis system (shown dashed). The *X'-Y'* axes are parallel with *E-N* but the origin of this system coincides with the origin of *X-Y*. In the figure, the rotation angle θ, between the *X-Y* and *X'-Y'* axis systems, is

$$\theta = \alpha - \beta \tag{11.33}$$

In Equation (11.33), azimuths, α and β, are calculated from the two sets of coordinates of control points *A* and *B* using Equation (11.5a) as follows:

$$\alpha = \tan^{-1}\left(\frac{X_B - X_A}{Y_B - Y_A}\right) + C$$

$$\beta = \tan^{-1}\left(\frac{E_B - E_A}{N_B - N_A}\right) + C$$

where as explained in Section 11.2, *C* places the azimuth in the proper quadrant.

In many cases, a scale factor must be incorporated in coordinate transformations. This would occur, for example, in transforming from a local arbitrary coordinate system into a state plane coordinate map projection. (Chapter 20 covers map projections) The scale factor relating any two coordinate systems can be computed according to the ratio of the length of a line between two control points obtained from *E-N* coordinates to that determined using *X-Y* coordinates. Thus,

$$s = \frac{\sqrt{(E_B - E_A)^2 + (N_B - N_A)^2}}{\sqrt{(X_B - X_A)^2 + (Y_B - Y_A)^2}} \tag{11.34}$$

(Note: If the scale factor is unity, the two surveys are of equal scale, and it can be ignored in the coordinate transformation.)

With θ and s known, scaled and rotated *X'* and *Y'* coordinates of any point, for example, *A*, can be calculated from

$$X'_A = sX_A \cos\theta - sY_A \sin\theta$$

$$Y'_A = sX_A \sin\theta + sY_A \cos\theta \tag{11.35}$$

Individual parts of the rotation formulas [right-hand sides of Equations (11.35)] are developed with reference to Figure 11.10. Translation consists of shifting the origin of the *X'-Y'* axes to that in the *E-N* system. This is achieved by

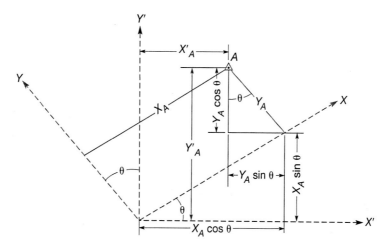

Figure 11.10
Detail of rotation formulas in two-dimensional conformal coordinate transformation.

adding translation factors T_X and T_Y (see Figure 11.9) to X' and Y' coordinates to obtain E and N coordinates. Thus, for point A

$$E_A = X'_A + T_X$$
$$N_A = Y'_A + T_Y \qquad (11.36)$$

Rearranging Equations (11.36) and using coordinates of one of the control points (such as A), numerical values for T_X and T_Y can be obtained as

$$T_X = E_A - X'_A$$
$$T_Y = N_A - Y'_A \qquad (11.37)$$

The other control point (i.e., point B) should also be used in Equations (11.37) to calculate T_X and T_Y and thus obtain a computational check.

Substituting Equations (11.35) into Equations (11.36) and dropping subscripts, the following equations are obtained for calculating E and N coordinates of noncontrol points (such as C and D) from their X and Y values:

$$E = sX \cos \theta - sY \sin \theta + T_X$$
$$N = sX \sin \theta + sY \cos \theta + T_Y \qquad (11.38)$$

In summary, the procedure for performing 2D conformal coordinate transformations consists of (1) calculating rotation angle θ using two control points, and Equations (11.5) and (11.33); (2) solving Equations (11.34), (11.35), and (11.37) using control points to obtain scale factor s, and translation factors T_X and T_Y; and (3) applying $\theta, s, T_X,$ and T_Y in Equations (11.38) to transform all noncontrol points. If more than two control points are available, an improved solution can be obtained using least squares. Coordinate transformation calculations require a significant amount of time if done by hand, but are easily performed when programmed for computer solution.

■ Example 11.6

In Figure 11.9, the following E-N and X-Y coordinates are known for points A through D. Compute E and N coordinates for points C and D.

288 COORDINATE GEOMETRY IN SURVEYING CALCULATIONS

Point	State Plane Coordinates (ft)		Arbitrary Coordinates (ft)	
	E	N	X	Y
A	194,683.50	99,760.22	2848.28	2319.94
B	196,412.80	102,367.61	5720.05	3561.68
C			3541.72	897.03
D			6160.31	1941.26

Solution

1. Determine α, β, and θ from Equations (11.5) and (11.33)

$$\alpha = \tan^{-1}\left(\frac{5720.05 - 2848.28}{3561.68 - 2319.94}\right) + 0° = 66°36'59.7''$$

$$\beta = \tan^{-1}\left(\frac{196{,}412.80 - 194{,}683.50}{102{,}367.61 - 99{,}760.22}\right) + 0° = 33°33'12.7''$$

$$\theta = 66°36'59.7'' - 33°33'12.7''$$

2. Compute the scale factor from Equation (11.34)

$$s = \frac{\sqrt{(196{,}412.80 - 194{,}683.50)^2 + (102{,}367.61 - 99{,}860.22)^2}}{\sqrt{(5720.05 - 2848.28)^2 + (3561.68 - 2319.94)^2}}$$

$$= \frac{3128.73}{3128.73}$$

$$= 1$$

(Since the scale factor is 1, it can be ignored.)

3. Determine T_X and T_Y from Equations (11.35) through (11.37) using point A

$$X'_A = 2848.28 \cos 33°03'47'' - 2319.94 \sin 33°03'47'' = 1121.39 \text{ ft}$$

$$Y'_A = 2848.28 \sin 33°03'47'' + 2319.94 \cos 33°03'47'' = 3498.18 \text{ ft}$$

$$T_X = 194{,}683.50 - 1121.39 = 193{,}562.11 \text{ ft}$$

$$T_Y = 99{,}760.22 - 3498.18 = 96{,}262.04 \text{ ft}$$

4. Check T_X and T_Y using point B

$$X'_B = 5720.05 \cos 33°03'47'' - 3561.68 \sin 33°03'47'' = 2850.69 \text{ ft}$$

$$Y'_B = 5720.05 \sin 33°03'47'' + 3561.68 \cos 33°03'47'' = 6105.58 \text{ ft}$$

$$T_X = 196{,}412.80 - 2850.69 = 193{,}562.11 \text{ ft} \checkmark$$

$$T_Y = 102{,}367.61 - 6105.68 = 96{,}262.03 \text{ ft} \checkmark$$

5. Solve Equations (11.38) for E and N coordinates of points C and D

$$E_C = 3541.72 \cos 33°03'47'' - 897.03 \sin 33°03'47'' + 193{,}562.11$$
$$= 196{,}040.93 \text{ ft}$$
$$N_C = 3541.72 \sin 33°03'47'' + 897.03 \cos 33°03'47'' + 96{,}262.04$$
$$= 98{,}946.04 \text{ ft}$$
$$E_D = 6160.31 \cos 33°03'47'' - 1941.26 \sin 33°03'47'' + 193{,}562.11$$
$$= 197{,}665.81 \text{ ft}$$
$$N_D = 6160.31 \sin 33°03'47'' - 1941.26 \cos 33°03'47'' + 96{,}262.04$$
$$= 101{,}249.78 \text{ ft}$$

With some simple modifications, Equations (11.38) can be rewritten in matrix form as

$$sR \begin{bmatrix} X \\ Y \end{bmatrix} + \begin{bmatrix} T_X \\ T_Y \end{bmatrix} = \begin{bmatrix} E \\ N \end{bmatrix} + \begin{bmatrix} v_E \\ v_N \end{bmatrix} \quad (11.39)$$

where the rotation matrix, R, is

$$R = \begin{bmatrix} \cos\theta & -\sin\theta \\ \sin\theta & \cos\theta \end{bmatrix} \quad (11.40)$$

Also v_E and v_N are residual errors, which must be included if more than two control points are available. Scaling the rotation matrix by s, and substituting a for $(s\cos\theta)$, b for $(s\sin\theta)$, c for T_X, and d for T_Y, Equation (11.39) can be rewritten as

$$\begin{bmatrix} a & -b \\ b & a \end{bmatrix} \begin{bmatrix} X \\ Y \end{bmatrix} + \begin{bmatrix} c \\ d \end{bmatrix} = \begin{bmatrix} E \\ N \end{bmatrix} + \begin{bmatrix} v_E \\ v_N \end{bmatrix} \quad (11.41)$$

With Equation (11.41), a least-squares adjustment (see Chapter 16) can be performed when more than two points are common in both coordinate systems. The program WOLFPACK, which is on the companion website, has this software option under the *coordinate computations* submenu. It will determine the unknown parameters for the 2D conformal coordinate transformation, and transform any additional points. The data file and the results of the adjustment for Example 11.6 are shown in 11.

Note that the transformed X and Y coordinates of points C and D obtained using the computer program agree (except for round off) with those computed in Example 11.6. Note also that in this solution with two control points, there are no redundancies and thus the residuals VX and VY are zeros.

```
DATA FILE
Example 11.6                                   {title line}
2                                              {number of control points}
A  194683.50  99760.22  2848.28  2319.94       {Point ID, SPCS E and N, arbitrary X and Y}
B  196412.80  102367.61  5720.05  3561.68
C  3561.68  897.03                             {Point ID, arbitrary system X and Y}
D  6160.31  1941.26
```

RESULTS OF ADJUSTMENT

```
Two Dimensional Conformal Coordinate Transformation of File: Example 11.6
--------------------------------------------------------------------------

ax - by + Tx = X + VX
bx + ay + Ty = Y + VY

Transformed Control Points
    POINT         X            Y           VX         VY
    -------------------------------------------------------
      A      194,683.50    99,760.22     -0.000     -0.000
      B      196,412.80   102,367.61      0.000      0.000

Transformation Parameters:
  a =   0.83807009
  b =   0.54556070
  Tx =  193562.110
  Ty =   96262.038

Rotation = 33°03'46.9"
  Scale = 1.00000

*******  Unique Solution Obtained !!  *******

    POINT       x           y            X            Y
    -------------------------------------------------------
      C      3,541.72     897.03    196,040.94   98,946.04
      D      6,160.31   1,941.26    197,665.81  101,249.77
```

Figure 11.11 Data file and results of adjustment for Example 11.6 using WOLFPACK.

11.9 INACCESSIBLE POINT PROBLEM

It is sometimes necessary to determine the elevation of a point that is inaccessible. This task can be accomplished by establishing a baseline such that the inaccessible point is visible from both ends. As an example, assume that the elevation of the chimney shown in Figure 11.12 is desired. Baseline AB is established, its length measured, and the elevations of its end points determined. Horizontal angles A and B, and altitude angles v_1, and v_2 are observed as shown in the figure. Points I_A and I_B are vertically beneath P. Using the observed values, the law of sines is applied to compute horizontal lengths AI_A and BI_B of triangle ABI as

$$AI_A = \frac{AB\sin(B)}{\sin[180° - (A + B)]} = \frac{AB\sin(B)}{\sin(A + B)} \quad (11.42)$$

$$BI_B = \frac{AB\sin(A)}{\sin(A + B)} \quad (11.43)$$

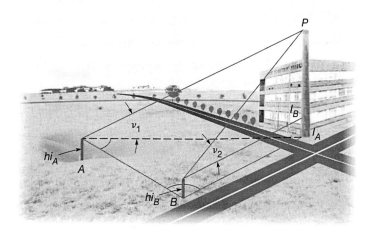

Figure 11.12
Geometry of the inaccessible point problem.

Length IP can be derived from either triangle AI_AP, or BI_BP as

$$I_AP = AI_A \tan(v_1) \quad (11.44)$$

$$I_BP = BI_B \tan(v_2) \quad (11.45)$$

The elevation of point P is computed as the average of the heights from the two triangles, which may differ because of random errors in the observation of v_1 and v_2, as

$$Elev_P = \frac{I_AP + Elev_A + hi_A + I_BP + Elev_B + hi_B}{2} \quad (11.46)$$

In Equation (11.46), hi_A and hi_B are the instrument heights at A and B, respectively.

■ Example 11.7

Stations A and B have elevations of 298.65 and 301.53 ft, respectively, and the instrument heights at A and B are $hi_A = 5.55$ and $hi_B = 5.48$ ft. The other field observations are

$$AB = 136.45 \text{ ft}$$
$$A = 44°12'34" \quad B = 39°26'56"$$
$$v_1 = 8°12'47" \quad v_2 = 5°50'10"$$

What is the elevation of the chimney stack?

Solution

By Equations (11.42) and (11.43), the lengths of AI_A and BI_B are

292 COORDINATE GEOMETRY IN SURVEYING CALCULATIONS

$$AI_A = \frac{136.45 \sin 39°26'56''}{\sin(44°12'34'' + 39°26'56'')} = 87.233 \text{ ft}$$

$$BI_B = \frac{136.45 \sin 44°12'34''}{\sin(44°12'34'' + 39°26'56'')} = 95.730 \text{ ft}$$

From Equation (11.44), length $I_A P$ is

$$I_A P = 87.233 \tan 8°12'47'' = 12.591 \text{ ft}$$

And from Equation (11.45), length $I_B P$ is

$$I_B P = 95.730 \tan 5°50'10'' = 9.785 \text{ ft}$$

Finally, by Equation (11.46), the elevation of point P is

$$Elev_P = \frac{12.591 + 298.65 + 5.55 + 9.785 + 301.53 + 5.48}{2} = 316.79 \text{ ft}$$

■ 11.10 THREE-DIMENSIONAL TWO-POINT RESECTION

The three-dimensional coordinates X_P, Y_P, and Z_P of a point such as P of Figure 11.13 can be determined based upon angle and distance observations made from that point to two other stations of known positions. This procedure is convenient for establishing coordinates of occupied stations on elevated structures, or in depressed areas such as in mines. In Figure 11.13, for example, assume that a total station instrument is placed at point P, whose X_P, Y_P, and Z_P coordinates are unknown, and that control points A and B are visible from P. Slope lengths PA and PB are observed along with horizontal angle γ and vertical angles v_1 and v_2. The computational process for determining X_P, Y_P, and Z_P is as follows.

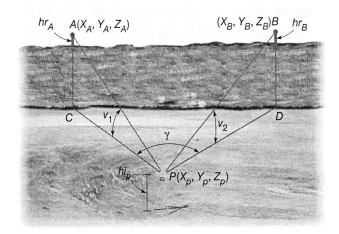

Figure 11.13
Geometry of the three-dimensional two-point resection problem.

1. Determine the length and azimuth of AB using Equations (11.4) and (11.5).
2. Compute horizontal distances PC and PD as

$$PC = PA \cos v_1$$
$$PD = PB \cos v_2 \qquad (11.47)$$

where C and D are vertically beneath A and B, respectively.

3. Using Equation (11.3), calculate horizontal angle DCP as

$$DCP = \cos^{-1}\left(\frac{AB^2 + PC^2 - PD^2}{2(AB)PC}\right) \qquad (11.48)$$

4. Determine the azimuth of line AP as

$$Az_{AP} = Az_{AB} + DCP \qquad (11.49)$$

5. Compute the planimetric (X-Y) coordinates of point P as

$$X_P = X_A + PC \sin Az_{AP}$$
$$Y_P = Y_A + PC \cos Az_{AP} \qquad (11.50)$$

6. Determine elevation differences AC and BD as

$$AC = PA \sin v_1$$
$$BD = PB \sin v_2 \qquad (11.51)$$

7. And finally calculate the elevation of P as

$$Elev_{P1} = Elev_A + hr_A - AC - hi_P$$
$$Elev_{P2} = Elev_B + hr_B - BD - hi_P \qquad (11.52)$$
$$Elev_P = \frac{Elev_{P1} + Elev_{P2}}{2}$$

In Equations (11.52), hi_P is the height of instrument above point P, and hr_A and hr_B are the reflector heights above stations A and B, respectively.

■ Example 11.8

For Figure 11.13, the X, Y, and Z coordinates (in meters) of station A are 7034.982, 5413.896, and 432.173, respectively, and those of B are 7843.745, 5807.242, and 428.795, respectively. Determine the three-dimensional position of a total station instrument at point P based upon the following observations:

$v_1 = 24°33'42''$ $PA = 667.413$ m $hr_A = 1.743$ m $\gamma = 77°48'08''$

$v_2 = 26°35'08''$ $PB = 612.354$ m $hr_B = 1.743$ m $hi_P = 1.685$ m

Solution

1. Using Equations (11.4) and (11.5), determine the length and azimuth of line AB.

$$AB = \sqrt{(7843.745 - 7034.982)^2 + (5807.242 - 5413.896)^2} = 899.3435 \text{ m}$$

$$Az_{AB} = \tan^{-1}\left(\frac{7843.745 - 7034.982}{5807.242 - 5413.896}\right) + 0° = 64°03'49.6''$$

2. By Equations (11.47), determine lengths PC and PD.

$$PC = 667.413 \cos 24°33'42'' = 607.0217 \text{ m}$$

$$PD = 612.354 \cos 26°35'08' = 547.6080 \text{ m}$$

3. From Equation (11.48), compute angle DCP.

$$DCP = \cos^{-1}\left(\frac{899.3435^2 + 607.0217^2 - 547.6080^2}{2(899.3435)607.0217}\right) = 36°31'24.2''$$

Note that this computed angle can be checked by using the law of sines, Equation (11.1), as

$$DCP = \sin^{-1}\left(\frac{547.6080 \sin 77°48'08''}{899.3435}\right) = 36°31'24.2'' \checkmark$$

4. Using Equation (11.49), find the azimuth of line AP.

$$Az_{AP} = 64°03'49.6'' + 36°31'24.2'' = 100°35'13.8''$$

5. From Equations (11.50), compute the X-Y coordinates of point P.

$$X_P = 7034.982 + 607.0217 \sin 100°35'13.8'' = 7631.670 \text{ m}$$

$$Y_P = 5413.896 = 607.0217 \cos 100°35'13.8'' = 5302.367 \text{ m}$$

6. By Equations (11.51), compute the vertical distances of AC and BD.

$$AC = 667.413 \sin 24°33'42'' = 277.425 \text{ m}$$

$$BD = 612.354 \sin 26°35'08'' = 274.049 \text{ m}$$

7. And finally, using Equations (11.52), compute and average the elevation of point P.

$$Elev_P = 432.173 + 1.743 - 277.425 - 1.685 = 154.806 \text{ m}$$

$$Elev_P = 428.795 + 1.743 - 274.049 - 1.685 = 154.804 \text{ m}$$

$$\text{Average Elevation} = 1554.805 \text{ m}$$

11.11 SOFTWARE

Coordinate geometry provides a convenient approach to solving problems in almost all types of modern surveys. Many problems that otherwise appear difficult can be greatly simplified and readily solved by working with coordinates. Although the calculations are sometimes rather lengthy, this has become inconsequential with the advent of computers and data collectors. Many software packages are available for performing coordinate geometry calculations. However, people involved in surveying (geomatics) must understand the basis for the computations, and they must exercise all possible checks to verify the accuracy of their results. While it is unfortunate, software has been known to yield incorrect results at times, and it is the engineers responsibility to check their software.

The Mathcad worksheet *C11.XMCD*, which is available on the companion website, demonstrates the programming of each example shown in this chapter. This software demonstrates the step-by-step approach in solving these problems. Programming of these problems in a higher-level programming language eliminates many of the mistakes that can occur when solving these problems by conventional methods. Figure 11.14 shows the coordinate geometry submenu from the WOLFPACK program, which is also available on the companion website. Also note in the figure, the menu options for a 2D conformal coordinate transformation, and a quadratic equation solver. The 2D conformal coordinate transformation requires a data file. The format for this file is discussed in the WOLFPACK help system, which is shown in Figure 11.15. This file can be created in a text editor. WOLFPACK contains an editor for this purpose. Its solution is also demonstrated in the Mathcad worksheet *C11-8.XMCD*, which is also available on the companion website for this book, demonstrates the least-squares solution of the example in Section 11.8.

Because of the nature of trigonometric functions, computations in some coordinate geometry problems will become numerically unstable when the angles

Figure 11.14 Coordinate geometry submenu from WOLFPACK program.

296 COORDINATE GEOMETRY IN SURVEYING CALCULATIONS

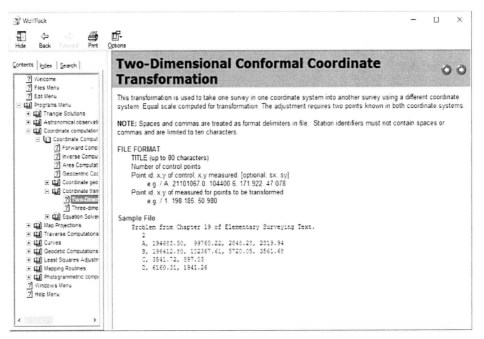

Figure 11.15 Help screen for two-dimensional conformal coordinate transformation from WOLFPACK program.

involved approach the cardinal directions of 0°, 90°, 180°, or 270°. Thus, if coordinate geometry is intended to be used to determine the locations of points, it is generally prudent to design the survey so that triangles used in the solution have angles between 30° and 60°. Also, it is important to observe good surveying practices in the field, such as taking the averages of equal numbers of direct and reverse angle observations, and exercising other checks and precautions.

As will be seen later, coordinate geometry plays an important role in computing highway alignments, in subdivision designs, and in the operation of geographic information systems.

PROBLEMS

Asterisks (*) indicate problems that have partial answers given in Appendix G.

11.1 The X and Y coordinates (in meters) of station Shore are 708.984 and 1524.346, respectively, and those for station Rock are 1132.945 and 973.482, respectively. What are the azimuth, bearing, and length of the line connecting station Shore to station Rock?

11.2 Same as Problem 11.1, except that the X and Y coordinates (in feet) of Shore are 654,127.06 and 394,087.52, respectively, and those for Rock are 652,531.62 and 392,134.06, respectively.

11.3* What are the slope, and y-intercept for the line in Problem 11.1?

11.4 What are the slope, and the y-intercept for the line in Problem 11.2?

11.5* If the slope (XY plane) of a line is 0.800946, what is the azimuth of the line to the nearest second of arc? (XY plane)

11.6 If the slope (XY plane) of a line is −1.260890, what is the azimuth of the line to the nearest second of arc? (XY plane)

11.7* What is the perpendicular distance of a point from the line in Problem 11.1, if the X and Y coordinates (in meters) of the point are 851.434 and 1341.594, respectively?

11.8 What is the perpendicular distance of a point from the line in Problem 11.2, if the X and Y coordinates (in feet) of the point are 653,595.01 and 393,436.37, respectively?

11.9* A line with an azimuth of 105°46′33″ from a station with X and Y coordinates of 5885.31 and 5164.15, respectively, is intersected with a line that has an azimuth of 200°31′24″ from a station with X and Y coordinates of 7337.08 and 5949.99, respectively. (All coordinates are in feet.) What are the coordinates of the intersection point?

11.10 A line with an azimuth of 56°23′02″ from a station with X and Y coordinates of 1234.87 and 898.56, respectively, is intersected with a line that has an azimuth of 321°53′47″ from a station with X and Y coordinates of 2034.79 and 962.48, respectively. (All coordinates are in feet.) What are the coordinates of the intersection point?

11.11 Same as Problem 11.9 except that the azimuth of the first line is 37°22′51″ and the azimuth of the second line is 342°18′35″.

11.12 In the accompanying figure, the X and Y coordinates (in meters) of station A are 1005.594 and 1868.720, respectively, and those of station B are 1564.865 and 644.491, respectively. Angle BAP was measured as 298°37′41″ and angle ABP was measured as 65°03′33″. What are the coordinates of station P?

Problems 11.12 through 11.16
Field conditions for intersections.

11.13* In the accompanying figure, the X and Y coordinates (in feet) of station A are 1248.16 and 3133.35, respectively, and those of station B are 1509.15 and 1101.89, respectively. The length of BP is 2657.45 ft, and the azimuth of line AP is 98°259000. What are the coordinates of station P?

11.14 In the accompanying figure, the X and Y coordinates (in feet) of station A are 912.76 and 1238.63, respectively, and those of station B are 1342.39 and 454.77, respectively. The length of AP is 544.61 ft, and angle ABP is 35°43′56″? What are the possible coordinates for station P?

11.15* A circle of radius 975.80 ft, centered at point A, intersects another circle of radius 963.09 ft, centered at point B. The X and Y coordinates (in feet) of A are 533.70 and 1157.86, respectively, and those of B are 1142.93 and 269.83, respectively. What are the coordinates of station P in the figure?

11.16 The same as Problem 11.15, except the radii from A and B are 945.62 ft and 989.63 ft, respectively.

11.17 For the subdivision in the accompanying figure, assume that lines AC, DF, GI, and JL are parallel, but that lines BK and CL are parallel to each other, but not parallel to AJ. What are the coordinates of each lot corner shown if the X and Y coordinates (in feet) of station A are (5000.00, 5000.00) and the remaining known values are as follows?

Course	Distance (ft)	Bearing
AB	150.00	S 88°44' E
BC	148.00	S 88°44' E
AD	80.00	N 10°27' E
DG	80.00	N 10°27' E
GJ	80.00	N 10°27' E
BK		N 10°12' E
CF		N 10°12' E
FI		N 10°12' E
IL		N 10°12' E

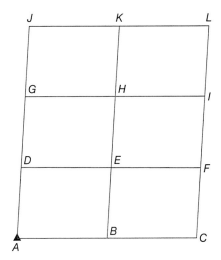

Problem 11.17 Subdivision.

11.18 If the X and Y coordinates (in feet) of station A are (1000.00, 1000.00), what are the coordinates of the remaining labeled corners in the accompanying figure given the following values?

Course	Distance (ft)	Bearing	Station(s)	Angle	Radius (ft)
AB	400.01	N 1°00' E	GBA	43°19'	
AJ	200.00	S 89°59' E	GCB	46°41'	
JK	30.00	S 89°59' E	G		30.00
KD		S 89°59' E			
BC	430.00	S 89°59' E			
DC		N 1°00' E			
JH		N 1°00' E			
KI		N 1°00' E			

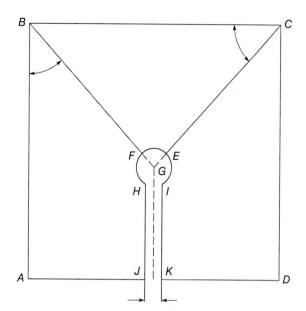

Problem 11.18 Subdivision.

11.20 In Figure 11.8, the X and Y coordinates (in feet) of A are 7322.70 and 9432.62, those of B are 7730.50 and 7588.65, and those of C are 9547.87 and 6453.90, respectively. Also angle x is 20°09′51″ and angle y is 25°51′20″. What are the coordinates of station P?

11.21 In Figure 11.9, the following EN and XY coordinates for points A through C are given. In a 2D conformal coordinate transformation, to convert the XY coordinates into the EN system, what are the

*(a) Scale factor?
 (b) Rotation angle?
 (c) Translations in X and Y?
 (d) Coordinates of points C in the EN coordinate system?

Point	State Plane Coordinates (m)		Arbitrary Coordinates (ft)	
	E	N	X	Y
A	719,542.829	111,493.468	4873.67	6609.04
B	719,899.341	111,844.860	6402.92	7207.45
C			5941.22	6337.23

11.22 Do Problem 11.21 with the following coordinates.

Point	State Plane Coordinates (m)		Arbitrary Coordinates (m)	
	E	N	X	Y
A	651,779.322	290,831.220	5504.32	3623.76
B	651169.151	290,542.891	3409.59	2906.42
C			3849.59	3857.64

11.23 In Figure 11.12, the elevations of stations A and B are 210.05 ft and 208.53 ft, respectively. Instrument heights hi_A and hi_B are both 5.50 ft. What is the average elevation of point P if the other field observations are

$AB = 136.52$ ft
$A = 50°11'22''$ $B = 51°06'08''$
$v_1 = 36°33'59''$ $v_2 = 37°26'46''$

11.24 In Problem 11.23, assume station P is to the left of the line AB, as viewed from station A. If the X and Y coordinates (in feet) of station A are 16,361.05 and 12,546.87, respectively, and the azimuth of line AB is $63°08'26''$, what are the X and Y coordinates of the inaccessible point?

11.25 In Figure 11.12, the elevations of stations A and B are 631.48 and 636.73 ft, respectively. Instrument heights hi_A and hi_B are 5.20 and 5.30 ft, respectively. What is the average elevation of point P if the other field observations are

$AB = 263.34$ ft
$A = 64°52'14''$ $B = 59°08'06''$
$v_1 = 4°33'44''$ $v_2 = 8°23'24''$

11.26 In Problem 11.25, assume station P is to the left of line AB as viewed from station A. If the X and Y coordinates (in feet) of station A are 1003.38 and 1240.45, respectively, and the azimuth of line AB is $152°30'13''$, what are the X and Y coordinates of the inaccessible point?

11.27 In Figure 11.13, the X, Y, and Z coordinates (in feet) of station A are 2897.37, 3406.73, and 234.56, respectively, and those of B are 3126.27, 3394.46, and 241.69, respectively. Determine the three-dimensional position of the occupied station P with the following observations:

$v_1 = 32°14'00''$ $PA = 243.67$ ft $hr_A = 6.53$ ft $\gamma = 64°39'51''$
$v_2 = 31°29'58''$ $PA = 260.10$ ft $hr_A = 5.33$ ft $hi_P = 5.50$ ft

11.28 Adapt Equations (11.43) and (11.47) so they are applicable for zenith angles.

11.29 In Figure 11.13, the X, Y, and Z coordinates (in meters) of station A are 2634.100, 3119.530, and 252.796, respectively, and those of B are 2540.210, 3277.25, and 245.809, respectively. Determine the three-dimensional position of occupied station P with the following observations:

$z_1 = 69°26'06''$ $PA = 179.439$ m $hr_A = 1.653$ m $\gamma = 57°51'53''$
$z_2 = 74°44'01''$ $PA = 212.851$ m $hr_A = 1.653$ m $hi_P = 1.600$ m

11.30 Use WOLFPACK to do Problem 11.9.
11.31 Use WOLFPACK to do Problem 11.10.
11.32 Use WOLFPACK to do Problem 11.12.
11.33 Use WOLFPACK to do Problem 11.13.
11.34 Use WOLFPACK to do Problem 11.15.
11.35 Use WOLFPACK to do Problem 11.16.
11.36 Use WOLFPACK to do Problem 11.17.

BIBLIOGRAPHY

Easa, S. M. 2007. "Direct Distance-Based Positioning without Redundancy—In Land Surveying." *Surveying and Land Information Science* 67 (No. 2): 69.

Ghilani, C. 2010. *Adjustment Computations: Spatial Data Analysis*, 5th ed. New York: Wiley.

Meyer, T. 2015. "A Simple Formula to Calculate Azimuth without a Two-Argument Arctangent Function." *Surveying and Land Information Science* 73 (No. 2): 91.

12
Area

12.1 INTRODUCTION

There are a number of important reasons for determining areas. One is to include the acreage of a parcel of land in the deed describing the property. Other purposes are to determine the acreage of fields, lakes, etc., or the number of square yards to be surfaced, paved, seeded, or sodded. Another important application is determining end areas for earthwork volume calculations (see Chapter 26).

In plane surveying, area is considered to be the orthogonal projection of the surface onto a horizontal plane. As noted in Chapter 2, in the United States the English system the most commonly used units for specifying small areas are the ft^2 and yd^2, and for large tracts the acre is most often used, where 1 acre = 43,560 ft^2 = 10 ch^2 (Gunter's). An acre lot, if square, would thus be 208.71^+ ft on a side. In the metric system, smaller areas are usually given in m^2, and for larger tracts *hectares* are commonly used, where 1 hectare is equivalent to a square having sides of 100 m, and thus equals 10,000 m^2. In converting areas between the English and metric systems, the conversion factors given in Table 12.1 are useful.

12.2 METHODS OF MEASURING AREA

Both field and map measurements are used to determine area. Field measurement methods are the more accurate and include (1) division of the tract into simple figures (triangles, rectangles, and trapezoids), (2) offsets from a straight line, (3) coordinates, and (4) double-meridian distances (DMDs). Each of these methods is described in sections that follow.

Methods of determining area from map measurements include (1) counting coordinate squares, (2) dividing the area into triangles, rectangles, or other regular geometric shapes, (3) digitizing coordinates, and (4) running a planimeter over

302 AREA

TABLE 12.1	APPROXIMATE AREA CONVERSION FACTORS	
To Convert from	**To**	**Multiply by**
ft^2	m^2	$(12/39.37)^2 \approx 0.09291$
m^2	ft^2	$(39.37/12)^2 \approx 10.76364$
yd^2	m^2	$(36/39.37)^2 \approx 0.83615$
m^2	yd^2	$(39.37/36)^2 \approx 1.19596$
acres	hectares	$(4.356 \times 12/39.37)^2 \approx 0.40470$
hectares	acres	$[39.37/(4.356 \times 12)]^2 \approx 2.47099$

the enclosing lines. These processes are described and illustrated in Section 12.9. Because maps themselves are derived from field observations, methods of area determination invariably depend on this basic source of data.

12.3 AREA BY DIVISION INTO SIMPLE FIGURES

A tract can usually be divided into simple geometric figures such as triangles, rectangles, or trapezoids. The sides and angles of these figures can be observed in the field and their individual areas calculated and totaled. An example of a parcel subdivided into triangles is shown in Figure 12.1.

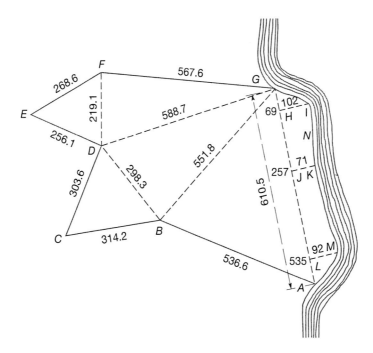

Figure 12.1
Area determination by triangles.

Formulas for computing areas of rectangles and trapezoids are well known. The area of a triangle whose lengths of sides are known can be computed by the formula

$$\text{area} = \sqrt{s(s-a)(s-b)(s-c)} \quad (12.1)$$

where a, b, and c are the lengths of sides of the triangle and $s = (a + b + c)/2$. Another formula for the area of a triangle is

$$\text{area} = \frac{1}{2}ab \sin C \quad (12.2)$$

where C is the angle included between sides a and b.

The choice of whether to use Equation (12.1) or (12.2) will depend on the triangle parts that are most conveniently determined; a decision ordinarily dictated by the nature of the area and the type of equipment available.

12.4 AREA BY OFFSETS FROM STRAIGHT LINES

Irregular tracts can be reduced to a series of trapezoids by observing right-angle offsets from points along a reference line. The reference line is usually marked by stationing (see Section 5.9.1), and positions where offsets are observed are given by their stations and pluses. The spacing between offsets may either be *regular* or *irregular*, depending on the conditions. These two cases are discussed in the subsections that follow.

12.4.1 Regularly Spaced Offsets

Offsets at regularly spaced intervals are shown in Figure 12.2. For this case, the area is found by the formula

$$\text{area} = b\left(\frac{h_0}{2} + h_1 + h_2 + \cdots + \frac{h_n}{2}\right) \quad (12.3)$$

where b is the length of a common interval between offsets, and $h_0, h_1, \ldots, h_n$ are the offsets. The regular interval for the example of Figure 12.2 is a half-station, or 50 ft.

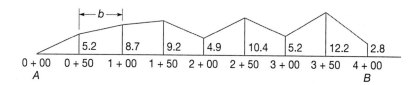

Figure 12.2 Area by offsets.

Example 12.1

Compute the area of the tract shown in Figure 12.2.

Solution

By Equation (12.3)

$$\text{area} = 50\left(\frac{0}{2} + 5.2 + 8.7 + 9.2 + 4.9 + 10.4 + 5.2 + 12.2 + \frac{2.8}{2}\right)$$

$$= 2860 \text{ ft}^2$$

In this example, a summation of offsets (terms within parentheses) can be secured by the *paper-strip method*, in which the area is plotted to scale and the mid-ordinate of each trapezoid is successively added by placing tick marks on a long strip of paper. The area is then obtained by making a single measurement between the first and last tick marks, multiplying by the scale to convert it to a field distance, and then multiplying by width b.

12.4.2 Irregularly Spaced Offsets

For irregularly curved boundaries like that in Figure 12.3, the spacing of offsets along the reference line varies. Spacing should be selected so that the curved boundary is accurately defined when adjacent offset points on it are connected by straight lines. A formula for calculating area for this case is

$$\text{area} = \frac{1}{2}[a(h_0 + h_1) + b(h_1 + h_2) + c(h_2 + h_3) + \cdots] \quad (12.4)$$

where $a, b, c, \ldots$ are the varying offset spaces, and $h_0, h_1, h_2, \ldots$ are the observed offsets.

Example 12.2

Compute the area of the tract shown in Figure 12.3.

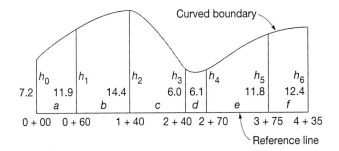

Figure 12.3 Area by offsets for a tract with a curved boundary.

Solution

By Equation (12.4)

$$\text{area} = \frac{1}{2}[60(7.2 + 11.9) + 80(11.9 + 14.4) + 100(14.4 + 6.0)$$
$$+ 30(6.0 + 6.1) + 105(6.1 + 11.8) + 60(11.8 + 12.4)]$$
$$= 4490 \text{ ft}^2$$

■ 12.5 AREA BY COORDINATES

Computation of area within a closed polygon is most frequently done by the coordinate method. In this procedure, coordinates of each angle point in the figure must be known. They are normally obtained by traversing, although any method that yields the coordinates of these points is appropriate. If traversing is used, coordinates of the stations are computed after adjustment of the departures and latitudes, as discussed and illustrated in Chapter 10. The coordinate method is also applicable and convenient for computing areas of figures whose coordinates have been digitized using an instrument like that shown in Figure 28.9. The coordinate method is easily visualized; it reduces to one simple equation that applies to all geometric configurations of closed polygons and is readily programmed for computer solution.

The procedure for computing areas by coordinates can be developed with reference to Figure 12.4. As shown in that figure, it is convenient (but not necessary) to adopt a reference coordinate system with the X and Y axes passing through the most southerly and the most westerly traverse stations, respectively. Lines BB', CC', DD', and EE' in the figure are constructed perpendicular to the Y axis. These lines create a series of trapezoids and triangles (shown by different color shadings). The area enclosed with traverse $ABCDEA$ can be expressed in terms of the areas of these individual trapezoids and triangles as

$$\text{area}_{ABCDEA} = E'EDD'E' + D'DCC'D' - AE'EA$$
$$- CC'B'BC - ABB'A \qquad (12.5)$$

The area of each trapezoid, for example $E'EDD'E'$ can be expressed in terms of lengths as

$$\text{area}_{E'EDD'E'} = \frac{E'E + DD'}{2} \times E'D'$$

In terms of coordinate values, this same area $E'EDD'E'$ is

$$\text{area}_{E'EDD'E'} = \frac{X_E + X_D}{2}(Y_E - Y_D)$$

306 AREA

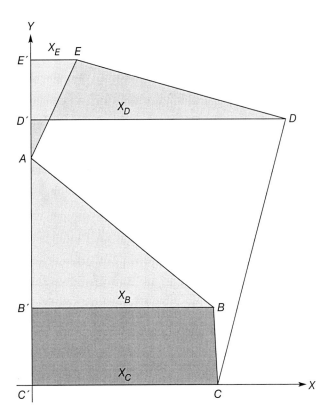

Figure 12.4
Area computation by the coordinate method.

Each of the trapezoids and triangles of Equation (12.5) can be expressed by coordinates in a similar manner. Substituting these coordinate expressions into Equation (12.5), multiplying by 2 to clear fractions, and rearranging

$$2(\text{area}) = + X_A Y_B + X_B Y_C + X_C Y_D + X_D Y_E + X_E Y_A \\ - X_B Y_A - X_C Y_B - X_D Y_C - X_E Y_D - X_A Y_E \quad (12.6)$$

Equation (12.6) can be reduced to an easily remembered form by listing the X and Y coordinates of each point in succession in two columns, as shown in Equation (12.7), *with coordinates of the starting point repeated at the end*. The products noted by diagonal arrows are ascertained with dashed arrows considered plus and solid ones minus. *The algebraic summation of all products is computed and its absolute value divided by 2 to get the area.*

$$
\begin{array}{cc}
X_A & Y_A \\
X_B & Y_B \\
X_C & Y_C \\
X_D & Y_D \\
X_E & Y_E \\
X_A & Y_A \\
\end{array}
\quad (12.7)
$$

The procedure indicated in Equation (12.7) is applicable to calculating any size traverse. The following formula, easily derived from Equation (12.6), is a variation that can also be used,

$$\text{area} = \frac{1}{2}[X_A(Y_E - Y_B) + X_B(Y_A - Y_C) + X_C(Y_B - Y_D)$$
$$+ X_D(Y_C - Y_E) + X_E(Y_D - Y_A)] \quad (12.8)$$

It was noted earlier that for convenience, an axis system can be adopted in which $X = 0$ for the most westerly traverse point, and $Y = 0$ for the most southerly station. Magnitudes of coordinates and products are thereby reduced, and the amount of work lessened, since four products will be zero. However, selection of a special origin like that just described is of little consequence if the problem has been programmed for computer solution. Then the coordinates obtained from traverse adjustment can be used directly in the solution. However, a word of caution applies when the coordinate values are extremely large as they would be normally when using state plane coordinate values (see Chapter 20). In those cases, to ensure sufficient precision and prevent serious round-off errors, double precision should be used. Or, as an alternative, the decimal place in each coordinate can arbitrarily be moved n places to the left, the area calculated, and then multiplied by 10^{2n}.

Either Equation (12.6) or Equation (12.8) can be readily programmed for solution by computer. The program WOLFPACK has this option under its *coordinate computations* menu. The format of the data file for this option is listed in its help screen. As was noted in Chapter 10, the "closed polygon traverse" option of WOLFPACK also computes areas using the coordinates of the adjusted traverse stations. A Mathcad® worksheet *C12.XMCD*, which is available on the companion website, demonstrates the computations in Sections 12.3 through 12.5.

■ Example 12.3

Figure 12.5 illustrates the same traverse as Figure 12.4. The computations in Table 10.4 apply to this traverse. Coordinate values shown in Figure 12.5, however, result from shifting the axes so that $X_A = 0.00$ (A is the most westerly station) and $Y_C = 0.00$ (C is the most southerly station). This was accomplished by subtracting 10,000.00 (the value of X_A) from all X coordinates and subtracting 4408.22 (the value of Y_C) from all Y coordinates. Compute the traverse area by the coordinate method. (Units are feet.)

Solution

These computations are best organized for tabular solution. Table 12.2 shows the procedure. Thus, the area contained within the traverse is

$$\text{area} = \frac{|1{,}044{,}861 - 499{,}684|}{2} = 272{,}588 \text{ ft}^2 \text{ (say } 272{,}600 \text{ ft}^2\text{)} = 6.258 \text{ acres}$$

308 AREA

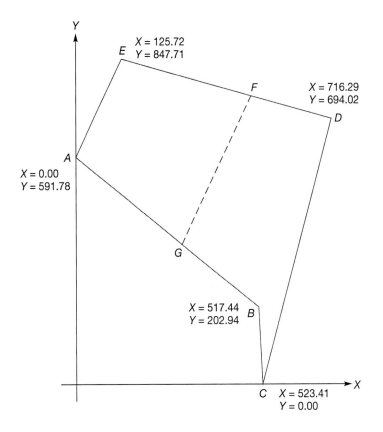

Figure 12.5 Traverse for computation of area by coordinates.

TABLE 12.2 COMPUTATION OF AREA BY COORDINATES

Point	X (ft)	Y (ft)	Double Area (ft²)	
			Plus (XY)	Minus (YX)
A	0.00	591.78		
B	517.44	202.94	0	306,211
C	523.41	0.00	0	106,221
D	716.29	694.02	363,257	0
E	125.72	847.71	607,206	87,252
A	0.00	591.78	74,398	0
			Σ = 1,044,861	Σ = 499,684
			−499,684	
			545,177	
			545,177 ÷ 2 = 272,588 ft² = 6.258 acres	

Notice that the precision of the computations was limited to four digits. This is due to the propagation of errors as discussed in Section 3.17.3. As an example, consider a square that has the same area as the parcel in Table 12.2. The length of its sides would be approximately 522.1 ft. Assuming that these coordinates have uncertainties of about ±0.1 ft, the error in the product as given by Equation (3.13) would be

$$E_{area} = \sqrt{(522.1 \times 0.1)^2 + (522.1 \times 0.1)^2} = \pm 74 \text{ ft}^2$$

Thus, rounding the computed area to the nearest hundred square feet is justified. As a *rule of thumb*, the accuracy of the area should not be stated any better than

$$E_{area} = S\sqrt{2}\sigma_S \qquad (12.9)$$

where S is the length of the side of a square having an area equivalent to the parcel being considered, and σ_S is the uncertainty in the coordinates of the points that bound the area in question.

Because of the effects of error propagation, it is important to remember that it is better to be conservative when expressing areas, and thus a phrase such as "6.258 acres more or less" is often adopted, especially when writing property descriptions (see Chapter 21).

12.6 AREA BY DOUBLE-MERIDIAN DISTANCE METHOD

The area within a closed figure can also be computed by the double-meridian distance (DMD) method. This procedure requires balanced departures and latitudes of the tract's boundary lines, which are normally obtained in traverse computations. The DMD method is not as commonly used as the coordinate method because it is not as convenient, but given the data from an adjusted traverse, it will yield the same answer. The DMD method is useful for checking answers obtained by the coordinate method when performing hand computations.

By definition, *the meridian distance of a traverse course is the perpendicular distance from the midpoint of the course to the reference meridian*. To ease the problem of signs, a reference meridian usually is placed through the most westerly traverse station.

In Figure 12.6, the meridian distances of courses AB, BC, CD, DE, and EA are MM', PP', QQ', RR', and TT', respectively. To express PP' in terms of convenient distances, MF and BG are drawn perpendicular to PP'. Then

$$PP' = P'F + FG + GP$$
$$= \text{meridian distance of } AB + \frac{1}{2} \text{ departure of } AB + \frac{1}{2} \text{ departure of } BC$$

310 AREA

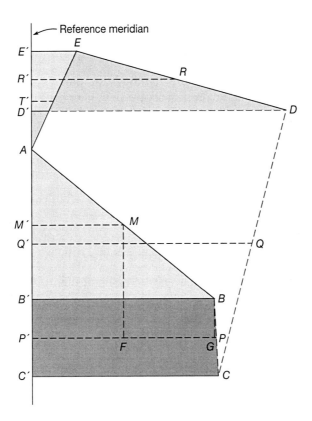

Figure 12.6
Meridian distances and traverse area computation by DMD method.

Thus, the meridian distance for any course of a traverse equals the meridian distance of the preceding course plus one half the departure of the preceding course plus half the departure of the course itself. It is simpler to employ full departures of courses. Therefore, DMDs equal to twice the meridian distances that are used, and a single division by 2 is made at the end of the computation.

Based on the considerations described, the following general rule can be applied in calculating DMDs: *The DMD for any traverse course is equal to the DMD of the preceding course, plus the departure of the preceding course, plus the departure of the course itself.* Signs of the departures must be considered. When the reference meridian is taken through the most westerly station of a closed traverse and calculations of the DMDs are started with a course through that station, *the DMD of the first course is its departure.* Applying these rules, for the traverse in Figure 12.6

DMD of AB = departure of AB
DMD of BC = DMD of AB + departure of AB + departure of BC

A check on all computations is obtained if the DMD of the last course, after computing around the traverse, is also equal to its departure but has the opposite sign. If there is a difference, the departures were not correctly adjusted before starting, or a mistake was made in the computations. With reference to

Figure 12.6, the area enclosed by traverse $ABCDEA$ may be expressed in terms of trapezoid areas (shown by different color shadings) as

$$\text{area} = E'EDD'E' + C'CDD'C' - (AB'BA + BB'C'CB + AEE'A) \tag{12.10}$$

The area of each figure equals the meridian distance of a course times its balanced latitude. For example, the area of trapezoid $C'CDD'C' = Q'Q \times C'D'$, where $Q'Q$ and $C'D'$ are the meridian distance and latitude, respectively, of line CD. The DMD of a course multiplied by its latitude equals double the area. Thus, the algebraic summation of all double areas gives *twice the area* inside the entire traverse. Signs of the products of DMDs and latitudes must be considered. If the reference line is passed through the most westerly station, all DMDs are positive. The products of DMDs and north latitudes are therefore plus, and those of DMDs and south latitudes are minus.

■ Example 12.4

Using the balanced departures and latitudes listed in Table 10.4 for the traverse of Figure 12.6, compute the DMDs of all courses.

Solution

The calculations done in tabular form following the general rule, are illustrated in Table 12.3.

■ Example 12.5

Using the DMDs determined in Example 12.4, calculate the area within the traverse.

TABLE 12.3 COMPUTATION OF DMDs

Departure of AB =	+517.444 = DMD of AB
Departure of AB =	+517.444
Departure of BC =	+5.964
	+1040.852 = DMD of BC
Departure of BC =	+5.964
Departure of CD =	+192.881
	+1239.697 = DMD of CD
Departure of CD =	+192.881
Departure of DE =	−590.571
	+842.007 = DMD of DE
Departure of DE =	−590.571
Departure of EA =	−125.718
	+125.718 = DMD of EA ✓

TABLE 12.4 COMPUTATION OF AREA BY DMDs

Course	Balanced Departure (ft)	Balanced Latitude (ft)	DMD (ft)	Double Areas (ft²)	
				Plus	Minus
AB	517.44	−388.84	517.44		201,201
BC	5.96	−202.95	1040.85		211,240
CD	192.88	694.02	1239.70	860,376	
DE	−590.57	153.69	842.01	129,408	
EA	−125.72	−255.93	125.72		32,176
Total	0.00	0.00		989,784	444,617
				−444,617	
				545,167	

$$\frac{545{,}167}{2} = 272{,}584 \text{ ft}^2 \text{ (say } 272{,}600 \text{ ft}^2) = 6.258 \text{ acres}$$

Solution

Computations for area by DMDs are generally arranged as in Table 12.4, although a combined form may be substituted. Sums of positive and negative double areas are obtained, and the absolute value of the smaller subtracted from that of the larger. The result is divided by 2 to get the area (272,600 ft²), and by 43,560 to obtain the number of acres (6.258). Note that the answer agrees with the one obtained using the coordinate method.

If the total of minus double areas is larger than the total of plus values, it signifies only that DMDs were computed by going around the traverse in a clockwise direction.

In modern surveying and engineering offices, area calculations are seldom done by hand; rather, they are programmed for computer solution. However, if an area is computed by hand, it should be checked by using different methods or by two persons who employ the same system. As an example, an individual working alone in an office could calculate areas by coordinates and check by DMDs. Those experienced in surveying (geomatics) have learned that a half-hour spent checking computations in the field and office can eliminate lengthy frustrations at a later time. The Mathcad worksheet *C12.XMCD*, which is available on the companion, demonstrates the programming of the coordinate method discussed in this book.

12.7 AREA OF PARCELS WITH CIRCULAR BOUNDARIES

The area of a tract that has a circular curve for one boundary, as in Figure 12.7, can be found by dividing the figure into two parts: polygon *ABCDEGFA* and *sector EGF*. The radius $R = EG = FG$ and either central angle $\theta = EGF$ or length EF

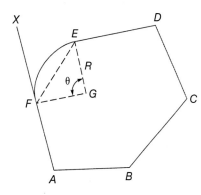

Figure 12.7
Area with circular curve as part of boundary.

must be known or computed to permit calculation of sector area *EGF*. If *R* and central angle θ are known, then the area of sector is

$$EGF = \pi R^2 \left(\frac{\theta}{360°}\right) \tag{12.11}$$

where θ is in decimal degrees. If chord length *EF* is known, angle $\theta = 2\sin^{-1}\left(\frac{EF}{2R}\right)$, and the preceding equation is used to calculate the sector area. To obtain the tract's total area, the sector area is added to area *ABCDEGFA* found by either the coordinate or DMD method.

Another method that can be used is to compute the area of the traverse *ABCDEFA*, and then add the area of the *segment*, which is the region between the arc and chord *EF*. The area of a segment is found as

$$\text{Area of segment} = 0.5R^2(\theta - \sin\theta) \tag{12.12}$$

where θ is expressed in radian units.

12.8 PARTITIONING OF LANDS

Calculations for purposes of partitioning land—that is, *cutting off* a portion of a tract for title transfer—can be aided significantly by using coordinates. For example, suppose the owner of the tract of land in Figure 12.5 wishes to subdivide the parcel with a line *GF*, parallel to *AE*, and have 3.000 acres in parcel *AEFG*. This problem can be approached by three different methods. The first involves trial and error and works quite well given today's computing capabilities. The second consists of writing equations for simple geometric figures such as triangles, rectangles, and trapezoids that enable a unique solution to be obtained for the coordinates of points *F* and *G*. The third approach involves setting up a series of coordinate geometry equations, together with an area equation, and then solving for the coordinates of *F* and *G*. The following subsections describe each of the above procedures.

12.8.1 Trial and Error Method

In this approach, estimated coordinates for the positions of stations *F* and *G* are determined, and the area of parcel *AEF'G'* is computed using Equation (12.6) where F' and G' are the estimated positions of F and G. This procedure is repeated until the area of the parcel equals 3.000 acres, or 130,680 ft^2.

Step 1: Using the final adjusted lengths and directions computed in Example 10.8 and coordinates of A and E from Example 12.3, and estimating the position of the cutoff line to be half the distance along line ED (i.e., $610.24/2 = 305.12$ ft), the coordinates of stations F' and G' in parcel $AEF'G'$ are computed as

Station F':

$$X = 125.72 + 305.12 \sin 104°35'13'' = 421.00$$
$$Y = 847.71 + 305.12 \cos 104°35'13'' = 770.87$$

Station G': is determined by direction-direction intersection using procedures discussed in Section 11.4. From WOLFPACK, the coordinates of Station G' are

$$X = 243.24 \text{ and } Y = 408.99$$

Creating a file for area computations, the area contained by these four stations is only 102,874 ft^2. Since 3.000 acres is equivalent to 130,680 ft^2, the estimated distance of 305.12 was short. It can now be increased and the process repeated.

Step 2: To estimate the needed increase to the distance, an assumption is made that the figure $F'FGG'$ is a rectangle, with $F'G'$ having a length of 403.18 ft, which is determined by coordinate inverse based on the coordinates of F' and G' from step 1. Thus, the amount to move the line $F'G'$ is determined as

$$(130{,}680 - 102{,}874)/403.18 = 68.97 \text{ ft}$$

For the second trial, the distance that F' is from E should be $305.12 + 68.97 = 374.09$ ft. Using the same procedure as in step 1, the area of $AEF'G'$ is 131,015 ft^2. The determined area is now too large, and can be reduced using the same assumption that was used at the beginning of this step. Thus, the distance EF' should be

$$EF' = 374.09 + (130{,}680 - 131{,}015)/(\text{length of } F'G')$$
$$= 374.09 - 0.78 = 373.31$$

This process is repeated until the final coordinates for F and G are determined. The next iteration yielded coordinates for F' of (487.00, 753.69) and for G' of (297.61, 368.14). Using these coordinates, the area of the parcel was computed to be 130,690 ft^2, or within 10 ft^2. The process is again repeated resulting in a reduction of the distance EF' of 0.02 ft, or $EF' = 373.29$ ft. The resulting area for $AEF'G'$ is 130,679 ft^2. Since this is within 1 ft^2 of the area, the coordinates are accepted as

$$F = (486.98, 753.70)$$
$$G = (297.59, 368.16)$$

Figure 12.8
Partitioning of lands by simple geometric figures.

The trial and error approach can be applied to solve many different types of land partitioning problems. Although the procedure may appear to involve a significant number of calculations, in many cases it provides the fastest and easiest solution when a computer program is available for doing the coordinate geometry calculations.

12.8.2 Use of Simple Geometric Figures

As can be seen in Figure 12.8, parcel $AEFG$ is a parallelogram. Thus, the formula for the area of a parallelogram $[A = 1/2(b_1 + b_2)h]$ can be employed, where b_1 is AE and b_2 is FG. In this procedure, a trigonometric relationship between the unknown length EF (denoted as d in Figure 12.8) and the missing parts h, FE', and $A'G$ must be determined. From the figure, angles α and β can be determined from azimuth differences, as

$$\alpha = Az_{EE'} - Az_{ED}$$
$$\beta = Az_{AB} - Az_{AA'}$$

Note in Table 10.7 that Az_{EA} is 206°09'41", and thus $Az_{AA'}$ and $Az_{EE'}$, which are perpendicular to line EA are 206°09'41" − 90° = 116°09'41". Also from

Table 10.7, AZ_{ED} and Az_{AB} are 104°35'13" and 126°55'23", respectively. Thus, the numerical values for α and β are:

$$\alpha = 116°09'41'' - 104°35'13'' = 11°34'28''$$
$$\beta = 126°55'23'' - 116°09'41'' = 10°45'42''$$

Now the parts h, FE', and $A'G$ can be expressed in terms of the unknown distance d as

$$h = d \cos \alpha$$
$$FE' = d \sin \alpha$$
$$A'G = h \tan \beta = d \cos \alpha \tan \beta \qquad (12.13)$$

The formula for the area of parallelogram $AEFG$ is

$$\frac{1}{2}(AE + FE' + AE + A'G)h = 130{,}680 \qquad (12.14)$$

Substituting Equations (12.13) into Equation (12.14), rearranging yields

$$(\cos^2 \alpha \tan \beta + \cos \alpha \sin \alpha)d^2 + [2(AE)\cos \alpha]d - 261{,}360 = 0 \qquad (12.15)$$

Expression (12.15) is a quadratic equation and can be solved using Equation (11.3). Substituting the appropriate values into Equation (12.15) and solving yields $d = EF = 373.29$ ft. This is the same answer as was derived in Section 12.8.1.

This approach of using the equations of simple geometric figures is convenient for solving a variety of land partitioning problems.

12.8.3 Coordinate Method

This method involves using Equations (10.11) and (12.8) to obtain four equations with the four unknowns X_F, Y_F, X_G, and Y_G, that can be solved uniquely. By Equation (10.11), the following three coordinate geometry equations can be written:

$$\frac{X_F - X_E}{Y_F - Y_E} = \frac{X_D - X_E}{Y_D - Y_E} \qquad (12.16)$$

$$\frac{X_G - X_A}{Y_G - Y_A} = \frac{X_B - X_A}{Y_B - Y_A} \qquad (12.17)$$

$$\frac{X_A - X_E}{Y_A - Y_E} = \frac{X_G - X_F}{Y_G - Y_F} \qquad (12.18)$$

Also by area Equation (12.8):

$$X_A(Y_G - Y_E) + X_E(Y_A - Y_F) + X_F(Y_E - Y_G)$$
$$+ X_G(Y_F - Y_A) = 2 \times \text{area} \qquad (12.19)$$

Substituting the known coordinates X_A, Y_A, X_B, Y_B, X_D, Y_D, X_E, and Y_E into Equations (12.16) through (12.19) yields four equations that can be solved for the four unknown coordinates. The four equations can be solved simultaneously, for example by using matrix methods, to determine the unknown coordinates for points F and G. (A program MATRIX is on the companion website, which can perform these matrix computations.)

Alternatively, the four equations can be solved by substitution. In this approach, Equations (12.16) and (12.17) are rewritten in terms of one of the unknowns, say X_F and X_G. These two new equations are then substituted into Equations (12.18) and (12.19). The resultant equations will now contain two unknowns Y_F and Y_G. The equation corresponding to Equation (12.18) can then be solved in terms of unknown, say Y_F, and this can be substituted into the equation corresponding to (12.19). The resultant expression will be a quadratic equation in terms of Y_G, which can be solved using Equation (11.3). This solution can then be substituted into the previous equations to derive the remaining three unknowns.

■ 12.9 AREA BY MEASUREMENTS FROM MAPS

To determine the area of a tract of land from map measurements its boundaries must first be identified on an existing map or a plot of the parcel drawn from survey data. Then one of several available methods can be used to determine its area. Accuracy in making area determinations from map measurements is directly related to the accuracy of the maps being used. Accuracy of maps, in turn, depends on the quality of the survey data from which they were produced, map scale, and the precision of the drafting process. Therefore, if existing maps are being used to determine areas, their quality should first be verified.

Even with good-quality maps, areas measured from them will not normally be as accurate as those computed directly from survey data. Map scale and the device used to extract map measurements are major factors affecting the resulting area accuracy. If, for example, a map is plotted to a scale of 1000 ft/1 in., and an engineer's scale is used, which produces measurements good to ± 0.02 in., distances or coordinates scaled from this map can be no better than about $(\pm 0.02 \times 1000) = \pm 20$ ft. This uncertainty can produce substantial errors in areas. Differential shrinkage or expansion of the material upon which maps are drafted is another source of error in determining areas from map measurements. Changes in dimensions of 2% to 3% are common for certain types of paper. (The subjects of maps and mapping are discussed in more detail in Chapters 17 and 18.)

Aerial photos can also be used as map substitutes to determine *approximate* areas if the parcel boundaries can be identified. The areas are approximate, as explained in Chapter 27, because except for flat areas the scale of an aerial photo is not uniform throughout. Aerial photos are particularly useful for determining areas of irregularly shaped tracts, such as lakes. Different procedures for determining areas from maps are described in the subsections that follow.

12.9.1 Area by Counting Coordinate Squares

A simple method for determining areas consists in overlaying the mapped parcel with a transparency having a superimposed grid. The number of grid squares included within the tract is then counted, with partial squares estimated and added to the total. Area is the product of the total number of squares times the area represented by each square. As an example, if the grids are 0.20 in. on a side, and a map at a scale of 200 ft/in. is overlaid, each square is equivalent to $(0.20 \times 200)^2 = 1600$ ft^2.

12.9.2 Area by Scaled Lengths

If the boundaries of a parcel are identified on a map, the tract can be divided into triangles, rectangles, or other regular figures, the sides measured, and the areas computed using standard formulas and totaled.

12.9.3 Area by Digitizing Coordinates

A mapped parcel can be placed on a digitizing table that is interfaced with a computer, and the coordinates of its corner points quickly and conveniently recorded. From the file of coordinates, the area can be computed using either Equation (12.6) or (12.8). It must be remembered, however, that even though coordinates may be digitized to the nearest 0.001 in., their actual accuracy can be no better than the map from which the data were extracted. Area determination by digitizing existing maps is now being practiced extensively in creating databases of geographic information systems. The area of a parcel on a map created in a computer-aided design and drafting (CADD) system can often be determined using this method by simply selecting the boundary of the parcel. This is the most common method employed today.

12.9.4 Area by Planimeter

A planimeter measures the area contained within any closed figure that is circumscribed by its tracer. There are two types of planimeters: mechanical and electronic. The major parts of the mechanical type are a scale bar, graduated drum and disk, Vernier, tracing point and guard, and anchor arm, weight, and point. The scale bar may be fixed or adjustable. For the standard fixed-arm planimeter, one revolution of the disk (dial) represents 100 in.^{2} and one turn of the drum (wheel) represents 10 in.^{2} The adjustable type can be set to read units of area directly for any particular map scale. The instrument touches the map at only three places: anchor point, drum, and tracing-point guard.

Because of its ease of use, the electronic planimeter (Figure 12.9) has replaced its mechanical counterpart. An electronic planimeter operates similarly to the mechanical type, except that the results are given in digital form on a display console. Areas can be measured in units of square inches or square centimeters,

Figure 12.9 Electronic planimeter. (Courtesy Topcon America Corporation.)

and by setting an appropriate scale factor, they can be obtained directly in acres or hectares. Some instruments feature multipliers that can automatically compute and display volumes.

As an example of using an adjustable type of mechanical planimeter, assume the area within the traverse of Figure 12.5 will be measured. The anchor point beneath the weight is set in a position outside the traverse (if inside, a polar constant must be added), and the tracing point brought over corner *A*. An initial reading of 7231 is taken, the 7 coming from the disk, 23 from the drum, and 1 from the Vernier. The tracing point is moved along the traverse lines from *A* to *B*, *C*, *D*, and *E*, and back to *A*. A triangle or a straightedge may guide the point, but normally it is steered freehand. A final reading of 8596 is made. The difference between the initial and final readings, 1365, is multiplied by the planimeter constant to obtain the area. To determine the planimeter constant, a square area is carefully laid out 5 in. on a side, with diagonals of 7.07 in., and its perimeter traced with the planimeter. If the difference between initial and final readings for the 5-in. square is, for example, 1250, then

$$5 \text{ in.} \times 5 \text{ in.} = 25 \text{ in.}^2 = 1250 \text{ units}$$

Thus, the planimeter constant is

$$1 \text{ unit} = \frac{25}{1250} = 0.020 \text{ in.}^2$$

Finally the area within the traverse is

$$\text{area} = 1365 \text{ units} \times 0.020 = 27.3 \text{ in.}^2$$

If the traverse is plotted at a map scale of 1 in. = 100 ft, then 1 in.^{2} = 10,000 ft^2 and the area measured is 273,000 ft^2.

As a check on planimeter operation, the outline may be traced in the opposite direction. The initial and final readings at point *A* should agree within a limit of perhaps two to five units.

The precision obtained in using a planimeter depends on operator skill, accuracy of the plotted map, type of paper, and other factors. Results correct to within 1/2% to 1% can be obtained by careful work.

A planimeter is most useful for irregular areas, such as that in Figure 12.3, and has many applications in surveying and engineering. The planimeter has been widely used in highway offices for determining areas of cross sections, and is also convenient for determining areas of drainage basins and lakes from measurements on aerial photos, checking computed areas in property surveys, etc.

12.10 SOFTWARE

As discussed in this chapter, there are several methods of determining the area of a parcel or figure. The method of area by coordinates is most commonly used in practice. However, other methods are sometimes used in unique situations that require a clever solution. Software typically uses the method of area by coordinates. For example, a CADD software package can use the coordinates of any irregularly shaped parcel to quickly determine their area by the coordinate method. WOLFPACK uses this method in determining the area enclosed by a figure from a listing of coordinates in sequential order. You may also enter the bounding coordinates of a parcel in a CADD package to determine the area enclosed by a parcel. For those wishing to see a higher-level programming of several of the examples discussed in this chapter, you are encouraged to explore the Mathcad worksheet *C12.XMCD*, which can be found on the companion website.

12.11 SOURCES OF ERROR IN DETERMINING AREAS

Some sources of error in area computations are:

1. Errors in the field data from which coordinates or maps are derived.
2. Making a poor selection of intervals and offsets to fit irregular boundaries.
3. Making errors in scaling from maps.
4. Shrinkage and expansion of maps.
5. Using coordinate squares that are too large and therefore make estimation of areas of partial blocks difficult.
6. Making an incorrect setting of the planimeter scale bar.
7. Running off and on the edge of the map sheet with the planimeter drum.
8. Using different types of paper for the map and planimeter calibration sheet.

12.12 MISTAKES IN DETERMINING AREAS

In computing areas, common mistakes include:

1. Forgetting to divide by 2 in the coordinate and DMD methods.
2. Confusing signs of coordinates, departures, latitudes, and DMDs.
3. Forgetting to repeat the coordinates of the first point in the area by coordinates method.
4. Failing to check an area computation by a different method.
5. Not drawing a sketch to scale or general proportion for a visual check.
6. Not verifying the planimeter scale constant by tracing a known area.

PROBLEMS

Asterisks (*) indicate problems that have answers given in Appendix G.

12.1* Compute the area enclosed within polygon *ABDFGA* of Figure 12.1 using triangles. Assume all distances are in units of feet.

12.2 Similar to Problem 12.1, except for polygon *EFGDE* of Figure 12.1.

12.3 Compute the area enclosed by *AGBA* and the shoreline of Figure 12.1 using the offset method.

12.4 By rule of thumb, what is the estimated uncertainty in a 437,853 ft^2 if the estimated error in the coordinates was ±0.1 ft?

12.5* Compute the area between a lake and the straight-line AG, from which offsets are taken at irregular intervals as follows (all distances in feet):

Offset Point	A	B	C	D	E	F	G
Stationing	0.00	0 + 54.80	1 + 32.54	2 + 13.02	2 + 98.74	3 + 45.68	4 + 50.17
Offset	12.3	34.2	56.5	85.4	69.1	68.9	23.9

12.6 Repeat Problem 12.5 with the following offset in meters.

Offset Point	A	B	C	D	E	F	G
Distance from A	0.000	30.000	78.940	128.003	163.654	193.973	227.608
Offset	3.15	6.32	10.04	19.57	15.43	9.42	3.65

12.7 Use the coordinate method to compute the area enclosed by the traverse of Problem 10.8.

12.8 Calculate by coordinates the area within the traverse of Problem 10.11.

12.9* Compute the area enclosed in the traverse of Problem 10.11 using DMDs.

12.10 Determine the area within the traverse of Problem 10.8 using DMDs.

12.11 By the DMD method, find the area enclosed by the traverse of Problem 10.20.

12.12 Compute the area within the traverse of Problem 10.17 using the coordinate method. Check by DMDs.

12.13 Calculate the area inside the traverse of Problem 10.18 by coordinates and check by DMDs.

12.14 Compute the area enclosed by the traverse of Problem 10.19 using the DMD method. Check by coordinates.

12.15 Find the area of the lot in Problem 10.25.

12.16* Determine the area of the lot in Problem 10.26.

12.17 Calculate the area of Lot 16 in Figure 21.2.

12.18 Plot the lot of Problem 10.25 to a scale of 1 in. = 100 ft.1 in. = 100 ft. Determine its surrounded area using a planimeter.

12.19 Similar to Problem 12.18, except for the traverse of Problem 10.26.

12.20 Plot the traverse of Problem 10.19 to a scale of 1 in. = 200 ft,1 in. = 200 ft, and find its enclosed area using a planimeter.

12.21 The (X,Y) coordinates (in feet) for a closed-polygon traverse $ABCDEFA$ follow. A (1000.00, 1000.00), B (1645.49, 1114.85), C (1675.95, 1696.05), D (1178.99, 1664.04), E (1162.62, 1337.78) and F (996.53, 1305.30). Calculate the area of the traverse by the method of coordinates.

12.22 Compute by DMDs the area in hectares within a closed-polygon traverse $ABCDEFA$ by placing the X and Y axes through the most southerly and most westerly stations, respectively. Departures and latitudes (in meters) follow. AB: E dep. 30, N lat. = 40; BC: E dep. = 70, N lat. = 10; CD = E dep. = 30, S lat. = 50; DE: W dep = 60, S lat. = 40; EF: W dep = 90, S lat = 30; FA: E dep. = 20, N lat = 70.

12.23 Calculate the area of a piece of property bounded by a traverse and circular arc with the following coordinates (in meters) at angle points: A (1479.838, 1803.583), B (1204.998, 1813.173), C (1204.728, 1447.473), D (1654.728, 1447.473) and a circular arc of radius CD starting at D and ending at A with the curve outside the course AD.

12.24 Calculate the area of a piece of property bounded by a traverse and circular arc with the following coordinates in feet at angle points: A (445.18, 913.88), B (453.67, 835.51), C (663.67, 835.51), D (663.67, 935.51), E (465.12, 936.04) with a circular arc of radius 20.00 ft starting at E, tangent to DE, and ending tangent at A.

12.25 Divide the area of the lot in Problem 12.23 into two equal parts by a line through point B. List in order the lengths and azimuths of all sides for each parcel.

12.26 Partition the lot of Problem 12.24 into two equal areas by means of a line parallel to BC. Tabulate in clockwise consecutive order the lengths and azimuths of all sides of each parcel.

12.27 Lot ABCD between two parallel street lines is 350.00 ft deep and has a 220.00 ft frontage (AB) on one street and a 260.00 ft frontage (CD) on the other. Interior angles at A and B are equal, as are those at C and D. What distances AE and BF should be laid off by a surveyor to divide the lot into two equal areas by means of a line EF parallel to AB?

12.28 Partition 1-acre parcel from the northern part of lot ABCDEFA in Problem 12.21 such that its southern line is parallel to the northern line.

13
Global Navigation Satellite Systems—Introduction and Principles of Operation

■ 13.1 INTRODUCTION

During the 1970s, the *global positioning system* (GPS) emerged. This system, which grew out of the space program, relies upon signals transmitted from satellites for its operation. It resulted from research and development paid for by the military to produce a system for global navigation and guidance. More recently, other countries are developing their own systems. Thus, the entire scope of satellite systems used in positioning is now referred to as *global navigation satellite systems* (GNSS). Receivers that use GPS satellites and another system such as GLONASS (see Section 13.10) are known as GNSS receivers. These systems provide precise timing and positioning information anywhere on the Earth with high reliability and low cost. The systems can be operated day or night, rain or shine, and do not require cleared lines of sight between survey stations. This represents a revolutionary departure from conventional surveying procedures, which rely on observed angles and distances for determining point positions. Since these systems all share similar features, GPS will be discussed in further detail herein.

Development of the first generation of satellite positioning systems began in 1958. This early system, known as the *Navy Navigation Satellite System* (NNSS), more commonly called the *TRANSIT* system, operated on the *Doppler* principle. In this system, *Doppler shifts* (changes in frequency) of signals transmitted from satellites were observed by receivers located on ground stations. The observed Doppler shifts are a function of the distances to the satellites and their directions of movement with respect to the receivers. The transmitting frequency was known and together with accurate satellite orbital position data and precise timing of observations, the positions of the receiving stations could be determined. The constellation of satellites in the TRANSIT system, which varied between five and seven in number, operated in polar orbits at altitudes of approximately 1100 km.

The objective of the TRANSIT system was to aid in the navigation of the U.S. Navy's Polaris submarine fleet. The first authorized civilian use of the system occurred in 1967, and the surveying community quickly adopted the new technology, finding it particularly useful for control surveying. Although these early instruments were bulky and expensive, the observation sessions lengthy, and the accuracy achieved moderate, the Doppler program was nevertheless an important breakthrough in satellite positioning in general, and in surveying in particular.

Because of the success of the Doppler program, the U.S. Department of Defense (DoD) began development of the NAVigation Satellite Timing and Ranging (NAVSTAR) GPS. The first satellite to support the development and testing of the system was placed in orbit in 1978. Since that date many additional satellites have been launched. The system, developed at a cost of approximately $12 billion, became fully operational in December of 1993. Like the earlier Doppler versions, the system is based on observations of signals transmitted from satellites whose positions within their orbits are precisely known. Also, the signals are picked up with *receivers* located at ground stations. However, the methods of determining distances from receivers to satellites, and of computing receiver positions, are different. These methods are described in later sections of this chapter. Current generation satellite receivers are illustrated in Figures 1.4 and 13.1. The size and cost of satellite surveying equipment have been substantially reduced from those of the Doppler program, and field and office procedures involved in surveys have been simplified so that now high accuracies can be achieved in real time.

13.2 OVERVIEW OF GPS

As noted in the preceding section, precise distances from the satellites to the receivers are determined from timing and signal information, enabling receiver positions to be computed. In satellite surveying, the satellites become the reference or *control* stations, and the *ranges* (distances) to these satellites, are used to compute the positions of the receiver. Conceptually, this is similar to the intersection of two circles in traditional horizontal surveying, as described in Section 11.6, except that distances are observed from moving control points to a ground station of unknown position in three dimensions.

The global positioning system can be arbitrarily broken into three parts: (1) the *space segment*, (2) the *control segment*, and (3) the *user segment*. The **space segment** consists nominally of 24 satellites operating in six orbital planes spaced at 60° intervals around the equator. Four additional satellites are held in reserve as spares. The orbital planes are inclined to the equator at a nominal angle of 55° [see Figure 13.2(b)]. This configuration provides 24-hour satellite coverage between the latitudes of 80°N and 80°S. The satellites travel in near-circular orbits that have a mean altitude of 20,200 km above the Earth and an orbital period of 12 sidereal hours.[1] The individual satellites are normally identified by their

---

[1]A sidereal day is approximately 4 minutes shorter than a solar day. See Appendix C for more information on sidereal years, and days.

(a) (b)

Figure 13.1 (a) The Trimble R8 and (b) the Sokkia GSR2700 receivers. (Courtesy of Trimble Navigation and Sokkia Corp.)

PseudoRandom Noise (PRN) number, (described below), but can also be identified by their satellite vehicle number (SVN), orbital position, or launch date.

Precise atomic clocks are used in the satellites to control the timing of the signals they transmit. These are extremely accurate clocks,[2] and extremely expensive as well. If the receivers used these same clocks, they would be cost prohibitive and would also require that users become trained in handling hazardous materials. Thus, the clocks in the receivers are controlled by the oscillations of a quartz crystal that, although also precise, are less accurate than atomic clocks. However, these relatively low-cost timing devices produce a receiver that is also relatively inexpensive.

The **control segment** consists of *monitoring stations*, which monitor the signals and track the positions of the satellites over time. The initial GPS monitoring stations are at Colorado Springs, and on the islands of Hawaii, Ascension, Diego Garcia, and Kwajalein.[3] The tracking information is relayed to the *master control station* in the Consolidated Space Operations Center (CSOC) located at Schriever Air Force base in Colorado Springs. The master control station uses this data to

---

[2]Atomic clocks are used, which employ either cesium or rubidium. The rubidium clocks may lose 1 second per 30,000 years, while the cesium type may lose 1 second only every 300,000 years. Hydrogen maser clocks, which may lose only 1 second every 30,000,000 years, have been proposed for future satellites. For comparison, quartz crystal clocks used in receivers may lose a second every 30 years.

[3]The number of tracking stations has been increased over the years to improve the accuracy of the broadcast ephemeris.

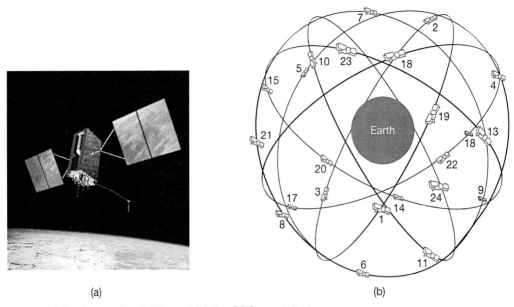

Figure 13.2 (a) A GPS satellite and (b) the GPS constellation.

make precise, near-future predictions of the satellite orbits, and their clock correction parameters. This information is uploaded to the satellites, and, in turn, transmitted by them as part of their *broadcast message* to be used by receivers to predict satellite positions and their clock *biases* (systematic errors).

The **user segment** in GPS consists of two categories of receivers that are classified by their access to two services that the system provides. These services are referred to as *Standard Position Service* (*SPS*) and the *Precise Positioning Service* (*PPS*). The SPS is provided on the L1 broadcast frequency and more recently the L2 (see Section 13.3) at no cost to the user. This service was initially intended to provide accuracies of 100 m in horizontal positions, and 156 m in vertical positions at the 95% error level. However, improvements in the system and the processing software have substantially reduced these error estimates. The PPS is broadcast on both the L1 and L2 frequencies and is only available to receivers having valid cryptographic keys, which are reserved for military and authorized users only. This message provides a published accuracy of 18 m in the horizontal, and 28 m in the vertical at the 95% error level.

■ 13.3 THE GPS SIGNAL

As the GPS satellites are orbiting, each continually broadcasts a unique signal on the two *carrier frequencies*. The carriers, which are transmitted in the L band of microwave radio frequencies, are identified as the L1 signal with a frequency of 1575.42 MHz and the L2 signal at a frequency of 1227.60 MHz. These frequencies are derived from a fundamental frequency, f_0, of the atomic clocks, which is 10.23 MHz. The L1 band has frequency of $154 \times f_0$, and the L2 band has a frequency of $120 \times f_0$.

Several different types of information (messages) are modulated upon these carrier waves using a phase modulation technique. Some of the information included in the broadcast message is the almanac, broadcast ephemeris, satellite clock correction coefficients, ionospheric correction coefficients, and satellite condition (also termed *satellite health*). These terms are defined later in this chapter.

In order for receivers to independently determine the ground positions of the stations, they occupy in real time, it was necessary to devise a system for accurate measurement of signal travel time from satellite to receiver. In GPS, this was accomplished by modulating the carriers with *pseudorandom noise* (PRN) codes. The PRN codes consist of unique sequences of binary values (zeros and ones) that appear to be random but, in fact, are generated according to a special mathematical algorithm using devices known as *tapped feedback shift registers*. Satellites transmit two or more different PRN codes. The L1 frequency is modulated with the *precise code*, or *P code*, and also with the *coarse/acquisition code*, or *C/A code*. This C/A code allows receivers to acquire the satellites as well as allowing receivers to determine their approximate positions. Until recently, the L2 frequency was modulated only with the P code.

The C/A and P codes are older technology. *Modernized satellites* are being equipped with new codes. The modernized satellites include a second civilian code on the L2 signal called the L2C. This code has both a civilian moderate (CM) and civilian long (CL) version. Additionally, the P code is being replaced by two new military codes, known as *M* codes. In 1999, the Interagency GPS Executive Board (IGEB) decided to add a third civilian signal known as the L5 to provide safety of life applications to GPS. L5 will be broadcast at a frequency of 1176.45 MHz, which is $115 \times f_0$. The L5 signal will carry both civilian codes along with a codeless component. This feature will greatly increase the strength of the signal due to the ability to use different processing techniques. Additionally as will be discussed in Section 13.6.2, these new codes will allow real-time ionospheric refraction corrections in code-based positioning. Both the L2C and L5 are added to the Block IIF and subsequent Block III satellites. The improvements in positioning due to these new codes will be discussed later in this chapter.

The C/A code has a frequency of 1.023 MHz and a wavelength of about 300 m. It is accessible to all users, and is a series of 1023 binary digits (*chips*) that are unique to each satellite. This chip pattern is repeated every millisecond in the C/A code. This code allows receivers to acquire the satellites and determine their approximate positions. The P code, with a frequency of 10.23 MHz and a wavelength of about 30 m, is 10 times more accurate for positioning than the C/A code. Additionally as discussed in Section 13.6.2, P-code users can make corrections for ionospheric refraction, which can be the largest error source in positioning. The P code has a chip pattern that takes 266.4 days to repeat. Each satellite is assigned a unique single-week segment of the pattern that is reinitialized at midnight every Saturday. Table 13.1 lists the GPS frequencies, and gives their factors of the fundamental frequency, f_0, of the P code.

To meet military requirements, the P code is encrypted with a W code to derive the Y code. This Y code can only be read with receivers that have the proper cryptographic keys. This encryption process is known as *anti-spoofing*

TABLE 13.1 FREQUENCIES TRANSMITTED BY GPS

Code Name	Frequency (MHz)	Factor of f_0
C/A	1.023	Divide by 10
P	10.23	1
L1	1575.42	Multiply by 154
L2	1227.60	Multiply by 120
L5	1176.45	Multiply by 115

(A-S). Its purpose is to deny access to the signal by potential enemies who could deliberately modify and retransmit it with the intention of "spoofing" unwary friendly users.

Because of its need for "one-way" communication, the satellite positioning systems depend on precise timing of the transmitted signal. To understand the concepts of the one-way system, consider the following. Imagine that the satellite transmits a series of audible beeps, and that the beeps are broadcast in a known irregular pattern. Now imagine that this same pattern is synchronously duplicated (but not transmitted) at the receiving station. Since the signal from the satellite transmitter must travel to the receiver, its reception will be delayed in relation to the signal generated by the receiver. This delay, which is approximately 0.07 seconds, can be measured, and converted to a time difference.

The process described above is like that used with GPS. In GPS, the chips of the PRN codes replace the beeps and the precise time of broadcast of the satellite code is placed into the broadcast message with a starting time indicated by the front edge of one of the chips. The receiver simultaneously generates a duplicate PRN code. Matching the incoming satellite signal with the identical receiver-generated signal derives the time it takes for the signal to travel from satellite to receiver. This yields the signal delay that is converted to travel time. From the travel time, and the known signal velocity, the distance to the satellite can be computed.

To aid in matching the codes, the broadcast message from each satellite contains a *Hand-Over Word* (HOW), which consists of some identification bits, flags, and a number. This number, times four, produces the *Time of Week* (TOW), which marks the leading edge of the next section of the message. The HOW and TOW assist the receiver in matching the signal received from the satellite to that generated by the receiver, so the delay can be quickly determined. This matching process is illustrated diagrammatically in Figure 13.3.

Figure 13.3 Determination of signal travel time by code matching.

13.4 REFERENCE COORDINATE SYSTEMS

In determining the positions of points on Earth from satellite observations, three different reference coordinate systems are important. First of all, satellite positions at the instant they are observed are specified in the "space-related" *satellite reference coordinate systems*. These are three-dimensional rectangular systems defined by the satellite orbits. Satellite positions are then transformed into a three-dimensional rectangular *geocentric coordinate system*, which is physically related to the Earth. As a result of satellite positioning observations, the positions of new points on Earth are determined in this coordinate system. Finally, the geocentric coordinates are transformed into the more commonly used and locally oriented *geodetic coordinate system*. The following subsections describe these three coordinate systems.

13.4.1 The Satellite Reference Coordinate System

Once a satellite is launched into orbit, its movement thereafter within that orbit is governed primarily by the Earth's gravitational force. However, there are a number of other lesser factors involved including the gravitational forces exerted by the Sun and moon, as well as forces due to solar radiation, which are sometimes called solar winds. Because of movements of the Earth, Sun, and moon with respect to each other, and because of variations in solar radiation, these forces are not uniform and hence satellite movements vary somewhat from their ideal paths. As shown in Figure 13.4, ignoring all forces except the Earth's gravitational pull, a satellite's idealized orbit is elliptical, and has one of its two foci at G, the Earth's mass center. The figure also illustrates a satellite reference coordinate system, X_S, Y_S, Z_S. *Perigee* and *apogee* are points where the satellite is closest to, and farthest away from G, respectively, in its orbit. The *line of apsides* joins these two points, passes through the two foci, and is the reference axis X_S. The origin of the X_S, Y_S, Z_S coordinate system is at G; the Y_S axis is in the mean orbital plane; and Z_S is perpendicular to this plane. Values of Z_S coordinates represent

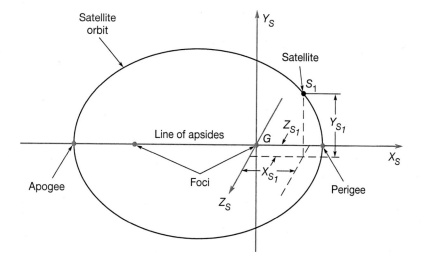

Figure 13.4
Satellite reference coordinate system.

orthogonal departures of the satellite from its mean orbital plane, and normally are very small. A satellite at position S_1 would have coordinates X_{S1}, Y_{S1}, and Z_{S1}, as shown in Figure 13.4. For any instant of time, the satellite's position in its orbit can be calculated from its orbital parameters, which are part of the broadcast ephemeris.

13.4.2 The Geocentric Coordinate System

Because the objective of satellite surveys is to locate points on the surface of the Earth, it is necessary to have a so-called *terrestrial* frame of reference, which enables relating points physically to the Earth. The frame of reference used for this is the geocentric coordinate system. Figure 13.5 illustrates a quadrant of a *reference ellipsoid*,[4] with a geocentric coordinate system (X_e, Y_e, Z_e) superimposed. This three-dimensional rectangular coordinate system has its origin at the mass center of the Earth. Its X_e axis passes through the Greenwich meridian in the plane

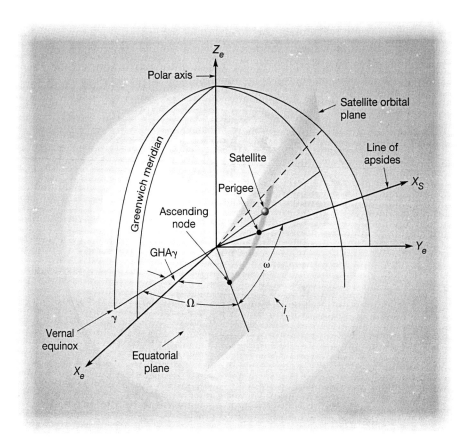

Figure 13.5 Parameters involved in transforming from the satellite reference coordinate system to the geocentric coordinate system.

---

[4]The reference ellipsoid used for most GPS work is the *World Geodetic System of 1984 (WGS84) ellipsoid*. As explained in Section 19.2, any ellipsoid can be defined by two parameters, for example the *semimajor axis (a)*, and the *flattening ratio (f)*. For the WGS84 ellipsoid these values are $a = 6{,}378{,}137$ m (exactly), and $f = 1/298.257223563$.

of the equator, and its Z_e axis coincides with the *Conventional Terrestrial Pole* (CTP) (see Section 19.3). Its Y_e axis lies in the plane of the equator and creates a right-handed coordinate system.

To make the conversion from the satellite reference coordinate system to the geocentric system, four angular parameters are required, which define the relationship between the satellite's orbital coordinate system and key reference planes and lines on the Earth. As shown in Figure 13.5, these parameters are (1) the *inclination* angle, i (angle between the orbital plane and the Earth's equatorial plane), (2) the *argument of perigee*, ω (angle in the orbital plane from the equator to the line of apsides), (3) the *right ascension of the ascending node*, Ω (angle in the plane of the Earth's equator from the vernal equinox to the line of intersection between the orbital and equatorial planes), and (4) the *Greenwich hour angle of the vernal equinox*, GHA_γ (angle in the equatorial plane from the Greenwich meridian to the vernal equinox). These parameters are known in real time for each satellite based upon predictive mathematical modeling of the orbits. Where higher accuracy is needed, satellite coordinates in the geocentric system for specific epochs of time are determined from observations at the tracking stations and distributed through precise ephemerides.

The equations for making conversions from satellite reference coordinate systems to the geocentric system are beyond the scope of this text. They are included in the software that accompanies the satellite positioning systems when they are purchased. However, an html file named satellite.html is available on the companion website, which demonstrates the transformation of satellite coordinates to the terrestrial coordinate system. Although the equations are not presented here, through this discussion students are apprised of the nature of satellite motion, and of the fact that there are definite mathematical relationships between orbiting satellites and the positions of points located on the Earth's surface.

13.4.3 The Geodetic Coordinate System

Although the positions of points in a satellite survey are computed in the geocentric coordinate system described in the preceding subsection, in that form they are inconvenient for use by surveyors (geomatics engineers). This is the case for three reasons: (1) with their origin at the Earth's center, geocentric coordinates are typically extremely large values, (2) with the *X-Y* plane in the plane of the equator, the axes are unrelated to the conventional directions of north-south or east-west on the surface of the Earth, and (3) geocentric coordinates give no indication about relative elevations between points. For these reasons, the *geocentric* coordinates are converted to *geodetic* coordinates of latitude (φ), longitude (λ), and height (h) so that reported point positions become more meaningful and convenient for users.

Figure 13.6 also illustrates a quadrant of the reference ellipsoid, and shows both the geocentric coordinate system (X, Y, Z), and the geodetic coordinate system (φ, λ, h). Conversions from geocentric to geodetic coordinates, and vice versa are readily made. From the figure, it can be shown that geocentric coordinates of point P can be computed from its geodetic coordinates using the following equations:

332 GLOBAL NAVIGATION SATELLITE SYSTEMS—INTRODUCTION AND PRINCIPLES OF OPERATION

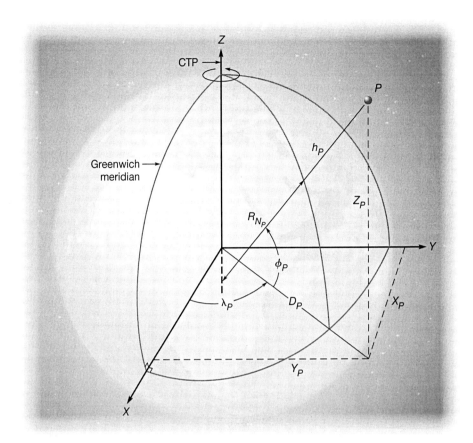

Figure 13.6
The geodetic and geocentric coordinate systems.

$$X_P = (R_{N_P} + h_P) \cos \varphi_P \cos \lambda_P$$
$$Y_P = (R_N + h_P) \cos \varphi_P \sin \lambda_P \quad \quad (13.1)$$
$$Z_P = [R_N (1 - e^2) + h_P] \sin \varphi_P$$

where

$$R_{N_P} = \frac{a}{\sqrt{1 - e^2 \sin^2 \varphi_P}} \quad \quad (13.2)$$

In Equations (13.1), X_P, Y_P, and Z_P are the geocentric coordinates of any point P, and the term e, which appears in both Equations (13.1) and (13.2), is the *eccentricity* of the WGS84 reference ellipsoid. Its value is 0.08181919084. In Equation (13.2), R_{N_P} is the *radius in the prime vertical*[5] of the ellipsoid at point P, and a, as noted earlier, is the semimajor axis of the ellipsoid. In Equations (13.1) and (13.2), north latitudes are considered positive and south latitudes negative. Similarly, east longitudes are considered positive and west longitudes negative. Additionally, the programming for the conversion of geodetic coordinates to

---

[5]The eccentricity and radius in the prime vertical are both described in Chapter 19.

geocentric coordinates and vice versa is demonstrated in Mathcad worksheet *C13.xcmd*, which is on the companion website for this book.

■ Example 13.1

The geodetic latitude, longitude, and height of a point A are 41°15′18.2106″ N, 75°00′58.6127″ W, and 312.391 m, respectively. Using WGS84 values, what are the geocentric coordinates of the point?

Solution

Substituting the appropriate values into Equations (13.1) and (13.2) yields

$$R_{N_A} = \frac{6{,}378{,}137}{\sqrt{1 - 0.0066943799 \sin^2 41°15''18.2106''}} = 6{,}387{,}440.3113 \text{ m}$$

$$X_A = (6{,}387{,}440.3113 + 312.391) \cos 41°15'18.2106'' \cos(-75°00'58.6127'')$$
$$= 1{,}241{,}581.343 \text{ m}$$

$$Y_A = (6{,}387{,}440.3113 + 312.391) \cos 41°15'18.2106'' \sin(-75°00'58.6127'')$$
$$= -4{,}638{,}917.074 \text{ m}$$

$$Z_A = [6{,}387{,}440.3113(1 - 0.00669437999) + 312.391] \sin(41°15'18.2106'')$$
$$= 4{,}183{,}965.568 \text{ m}$$

Conversion of geocentric coordinates of any point P to its geodetic values is accomplished using the following steps (refer again to Figure 13.6).

Step 1: Compute D_p as

$$D_P = \sqrt{X_P^2 + Y_P^2} \tag{13.3}$$

Step 1: Compute the longitude as[6]

$$\lambda_P = 2 \tan^{-1}\left(\frac{D_P - X_P}{Y_P}\right) \tag{13.4}$$

Step 3: Calculate approximate latitude, φ_0[7]

$$\varphi_0 = \tan^{-1}\left[\frac{Z_P}{D_P(1 - e^2)}\right] \tag{13.5}$$

Step 4: Calculate the approximate radius of the prime vertical, R_N, using φ_0 from step 3, and Equation (13.2).

Step 5: Calculate an improved value for the latitude from

$$\varphi = \tan^{-1}\left(\frac{Z_P + e^2 R_{N_P} \sin \varphi_0}{D_P}\right) \tag{13.6}$$

---

[6]This formula can conveniently be implemented in software with the function atan2(X_P, Y_P), which was show in Equation (11.5b)

[7]A Mathcad electronic book on the companion website for this book contains the routines to convert between geodetic and geocentric coordinates.

Step 6: Repeat the computations of steps 4 and 5 until the change in φ between iterations becomes negligible. This final value, φ_0, is the latitude of the station.

Step 7: Use the following formulas to compute the geodetic height of the station. For latitudes less than 45°, use

$$h_P = \frac{D_P}{\cos \varphi_P} - R_{N_P} \qquad (13.7a)$$

For latitudes greater than 45° use the formula

$$h_P = \left[\frac{Z_P}{\sin \varphi_P}\right] - R_{N_P}(1 - e^2) \qquad (13.7b)$$

It should be noted that the reason for Equations (13.7a) and (13.7b) are due to numerical stability of the trigonometric functions that they each employ.

■ Example 13.2

What are the geodetic coordinates of a point that has X, Y, Z geocentric coordinates of 1,241,581.343, −4,638,917.074, and 4,183,965.568, respectively? (Note: Units are meters.)

Solution

To visualize the solution, refer to Figure 13.6. Since the X-coordinate value is positive, the longitude of the point is between 0° and 90°. In addition, since the Y-coordinate value is negative, the point is in the western hemisphere. Similarly, since the Z-coordinate value is positive, the point is in the northern hemisphere. Substituting the appropriate values into Equations (13.3) through (13.7) yields

Step 1: $D = \sqrt{(1{,}241{,}581.343)^2 + (-4{,}638{,}917.074)^2} = 4{,}802{,}194.8993$ m

Step 2:
$$\lambda = 2 \tan^{-1}\left(\frac{4{,}802{,}194.8993 - 1{,}241{,}581.343}{-4{,}638{,}917.074}\right) = -75°00'58.6127'' \text{ (West)}$$

Step 3: $\varphi_0 = \tan^{-1}\left[\dfrac{4{,}183{,}965.568}{4{,}802{,}194.8993(1 - 0.00669437999)}\right] = 41°15'18.2443''$

Step 4: $R_N = \dfrac{6{,}378{,}137}{\sqrt{1 - 0.00669437999 \sin^2(41°15'18.2443'')}} = 6{,}387{,}440.3148$ m

Step 5: $\varphi_0 = \tan^{-1}\left[\dfrac{4{,}183{,}965.568 + e^2 6{,}387{,}440.3148 \sin 41°15'18.2443''}{4{,}802{,}194.8993}\right]$

$= 41°15'18.2107''$

Step 6: Repeat steps 4 and 5 until the latitude converges. The values for the next iteration are

$$R_N = 6,387,440,3113 \text{ m}$$
$$\varphi_0 = 41°15'18.2106'' \text{ N}$$

Repeating with the above values results in the same value for latitude to four decimal places, so the latitude of the station is 41°15'18.2106" N.

Step 7: Since the latitude is less than 45°, compute the geodetic height using Equation (13.7a) as

$$h = \frac{4,802,194.8993}{\cos 41°15'18.2106''} - 6,387,440.3113 = 312.391 \text{ m}$$

The geodetic coordinates of the station are latitude = 41°15'18.2106" N, longitude = 75°00'58.6127" W, and height = 312.391 m. Note that this example was the reverse computations of Example 13.1, and it reproduced the starting geodetic coordinate values for that example.

It is important to note that geodetic heights obtained with satellite surveys are measured with respect to the ellipsoid. That is, the geodetic height of a point is the vertical distance between the ellipsoid and the point as illustrated in Figure 13.7. As shown, these are not equivalent to *elevations* (more properly referred to as *orthometric heights*) given with respect to the geoid. Recall from Chapter 4 that the geoid is an equipotential gravitational reference surface that is used as a datum for elevations. To convert geodetic heights to elevations, the *geoid height* (vertical distance between ellipsoid and geoid) must be known. Then elevations can be expressed as

$$H = h - N \qquad (13.8)$$

where H is elevation above the geoid (orthometric height), h the geodetic height (determined from satellite surveys), and N the geoidal height. Figure 13.7 shows the correct relationship of the geoid and the WGS84 ellipsoid in the continental United States. Here the ellipsoid is above the geoid, and geoid height (*measured from the ellipsoid*) is negative. The geoid height at any point can be estimated with mathematical models developed by combining gravimetric data with distributed networks of points where geoidal height has been observed. One such model, *GEOID12B*, is a high-resolution model for the United States available from the National Geodetic Survey.[8] It uses latitude and longitude as arguments for determining geoid heights at any locations in the conterminous United States (CONUS), Hawaii, Puerto Rico, and the Virgin Islands.

---

[8] A disk containing GEOID12B can be obtained by writing to the National Geodetic Information Center, NOAA, National Geodetic Survey, N/CG17, SSMC3 Station 09535, 1315 East West Highway, Silver Spring, MD 20910, telephone (301) 713-3242, or it can be downloaded over the Internet at http://www.ngs.noaa.gov/PC_PROD/pc_prod.shtml.

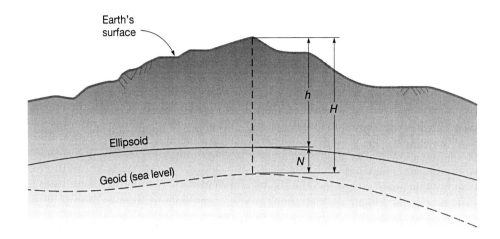

Figure 13.7 Relationships between elevation H, geodetic height h, and geoid undulation N.

■ Example 13.3

Compute the elevation (orthometric height) for a station whose geodetic height is 312.391 m, if the geoid undulation in the area is −33.000 m.

Solution

By Equation (13.8):

$$H = 312.391 - (-33.000) = 345.391 \text{ m}$$

Since the geoid height generally changes gradually in any region, a value that can be applied for it over a limited area can be determined. Including NAVD 88 benchmarks in the area in a GNSS survey can do this. Then with the ellipsoid heights and elevations known for these bench marks, the following rearranged form of Equation (13.8) is used to determine GPS observed geoidal heights:

$$N_{GPS} = h - H \qquad (13.9)$$

The value for N_{GPS} obtained in this manner should be compared with that derived from the model supplied by the National Geodetic Survey (NGS), and the difference should be computed as $\Delta N = N_{GPS} - N_{model}$. It is best to perform this procedure on several well-dispersed benchmarks in an area whenever possible. Then using an average ΔN for the survey area, the corrected orthometric height is

$$H = h - (N_{model} + \Delta N_{avg}) \qquad (13.10)$$

■ Example 13.4

The GNSS observed geodetic heights of benchmark stations *Red*, *White*, and *Blue* are 412.345, 408.617, and 386.945 m, respectively. The model geoidal heights for the stations are −29.894, −29.902, and −29.901 m, respectively, and their published elevations are 442.214, 438.490, and 416.822 m, respectively. What is the elevation of station *Brown*, which has an observed GNSS height of 397.519 m, if the model geoid height is published as −29.898 m?

Solution

By Equation (13.9), the observed geoid heights and ΔN's are

Station	N	ΔN
Red	412.345 − 442.214 = −29.869	−29.869 − (−29.894) = 0.025
White	408.617 − 438.490 = −29.873	−29.873 − (−29.902) = 0.029
Blue	386.945 − 416.822 = −29.877	−29.877 − (−29.901) = 0.024
		$\Delta N_{avg} = 0.026$

By Equation (13.10), the elevation of *Brown* is

$$Elev_{Brown} = 397.519 - (-29.898 + 0.026) = 427.391 \text{ m}$$

A word of caution should be added. Because the exact nature of the geoid is unknown, interpolated or extrapolated values of geoidal heights from an observed network of points, or those obtained from mathematical models, are not exact. Thus, orthometric heights obtained from ellipsoid heights will be close to their true values, but they may not be accurate enough to meet project requirements. Thus, for work that requires extremely accurate elevation differences, it is still best to obtain them by differential leveling from nearby benchmarks. The NGS is currently working on the *Gravity for Redefinition of the American Vertical Datum* (GRAV-D) project, which will result in the improvement of the geoid model for the United States to alleviate some of the error in conversion from geodetic to orthometric heights. The GRAV-D project (see Section 19.6.6) is expected to yield geoid heights to an accuracy of 1 cm throughout the United States and its territories. In fact, when the new datums are released in 2022, the NGS expects datum orthometric heights to come from satellite observations.

13.4.4 Evolution of WGS84 Reference Frame

It has always been the goal in surveying/geomatics to have one unifying coordinate system for the entire Earth. In 1987, the coordinates of the GPS tracking stations were realized by the coordinates of over 1000 terrestrial control station coordinates that were observed using TRANSIT. This became known as the WGS84 datum, which was considered to be coincident with the original NAD 83 (1986) horizontal datum.[9] However with the evolution of GPS, better fitting reference coordinate systems were realized for the Earth. The International Earth Rotation and Reference Systems Service (IERS), which consist of more than 200 worldwide agencies, has generated better fitting reference systems for the Earth based on

---

[9]The history of NAD83 and transformations between different reference coordinate systems is further discussed in Section 19.7

an expansive network of GNSS tracking stations, *very long baseline interferometry* (VLBI) stations, *satellite laser ranging* (SLR), and *Doppler ranging integrated on satellite* (DORIS) stations. These new coordinate systems were realized as the International Terrestrial Reference Frames (ITRF). The first was created in 1989 with ITRF89. Since then there have been the following reference coordinate systems: ITRF90, ITRF91, ITRF92, ITRF93, ITRF94, ITRF95, ITRF96, ITRF97, ITRF2000, ITRF2005, ITRF2008, and ITRF2014. All of these are known as *Earth-Centered, Earth-Fixed* (ECEF) coordinate systems since, as discussed in Section 13.4.2, they are based on the origin being at the mass center of the Earth and the axes defined by the Conventional Terrestrial Pole (CTP) and Greenwich meridian. All of these systems use the Geodetic Reference System of 1980 (GRS 80) ellipsoid.

Because of discrepancies between the original WGS84 reference frame and the better fitting ITRF coordinate systems, the Department of Defense began changing their control station coordinates to agree with the ITRF reference frames. For GPS these coordinate changes occurred during GPS weeks of 730, 873, 1150, 1674, and 1762. These new reference coordinate systems were designated as WGS84 (G730), WGS84 (G873), WGS84 (G1150), WGS84 (G1674), and WGS84 (G1762), respectively, where the "G" indicates that GPS measurements were used to establish the new datum, and the numeric value indicates the GPS week that the coordinate system was implemented. The most current WGS84 (G1762) epoch 2005.0 is in close agreement with the ITRF08 epoch 2005.0 reference frame but is significantly different from NAD83 (1986) or any of its later versions. These changes are made to account for the motions of the Earth's crustal plates and GNSS antenna calibration differences. Thus, each change in WGS84 indicates a different set of coordinate values for any station. Table 13.2 lists these changes in WGS84. However, it should be noted that the recent changes are under the measurement precisions of the typical GNSS survey.

When performing GNSS surveys or comparing coordinates from early GNSS surveys, it is always important to check the reference system for the station coordinates. Likewise, for future use, it is important to have the date and the reference system as part of the metadata to accompany the station coordinates. Since it is very possible that the position of stations given in coordinates can be in varying reference frames, several agencies such as IGS, NGS, and National Geospatial-Intelligence

TABLE 13.2 EVOLUTION OF WGS 84

Name	Implementation Date	Epoch	Accuracy*
WGS 84	1987		1–2 m
WGS 84 (G730)	Jun 29, 1994	1994.0	10 cm
WGS 84 (G873)	Jan 29, 1997	1997.0	5 cm
WGS 84 (G1150)	Jan 20, 2002	2001.0	1 cm
WGS 84 (G1674)	Feb 8, 2012	2005.0	<1 cm
WGS 84 (G1762)	Oct 16, 2013	2005.0	<1 cm

*Per component root mean square error.

Agency (NGA) along with private firms have created conversion software to transform coordinates between reference frames. The mathematics of these transformations is discussed in Section 19.7. *Horizontal Time Dependent Positioning* (HTDP) software, which is available from the NGS, allows users to transform coordinates between reference frames and dates. It is important for students early in their surveying careers to realize that coordinate systems will continue to evolve and change as we learn more about the Earth and the movement of its crustal plates. Thus, it is important to know not only the values of the coordinates for stations but also the defining reference coordinate system that is the basis for the coordinates, and the dates of the survey that established these coordinates.

13.5 FUNDAMENTALS OF SATELLITE POSITIONING

As discussed in Section 13.3, the precise travel time of the signal is necessary to determine the distance, or so-called *range*, to the satellite. Since the satellite is in an orbit approximately 20,200 km above the Earth, the travel time of the signal will be roughly 0.07 sec after the receiver generates the same signal. If this time delay between the two signals is multiplied by the signal velocity (speed of light in a vacuum) c, the range to the satellite can be determined from

$$r = c \times t \tag{13.11}$$

where r is the range to the satellite and t the elapsed time for the wave to travel from the satellite to the receiver.

Satellite receivers in determining distances to satellites employ two fundamental methods: *code ranging* and *carrier phase-shift measurements*. From distance observations made to multiple satellites, receiver positions can be calculated. Descriptions of the two methods, and their mathematical models, are presented in the subsections that follow. These mathematical models are presented to help students better understand the underlying principles of GPS operation. Computers that employ software provided by manufacturers of the equipment perform solutions of the equations.

13.5.1 Code Ranging

The code ranging method of determining the time it takes the signals to travel from satellites to receivers was the procedure briefly described in Section 13.3. With the travel times known, the corresponding distances to the satellites can then be calculated by applying Equation (13.11). With one range known, the receiver would lie on a sphere. If the range were determined from two satellites, the results would be two intersecting spheres. As shown in Figure 13.8(a), the intersection of two spheres is a circle. Thus, two ranges from two satellites would place the receiver somewhere on this circle. Now if the range for a third satellite is added, this range would add an additional sphere, which when intersected with one or both of the other two spheres would produce another circle of intersection. As shown in Figure 13.8(b), the intersection of two circles would leave only two possible locations for the position of the receiver. A "seed position" is used to quickly eliminate one of these two intersections.

Figure 13.8
(a) The intersection of two spheres and (b) the intersection of two circles.

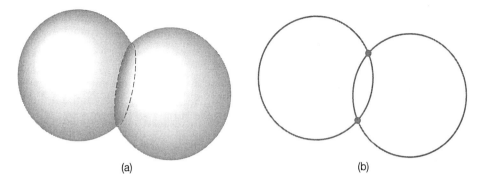

For observations taken on three satellites, the system of equations that could be used to determine the position of a receiver at station A is

$$\rho_A^1 = \sqrt{(X^1 - X_A)^2 + (Y^1 - Y_A)^2 + (Z^1 - Z_A)^2}$$
$$\rho_A^2 = \sqrt{(X^2 - X_A)^2 + (Y^2 - Y_A)^2 + (Z^2 - Z_A)^2} \qquad (13.12)$$
$$\rho_A^3 = \sqrt{(X^3 - X_A)^2 + (Y^3 - Y_A)^2 + (Z^3 - Z_A)^2}$$

where ρ_A^n are the *geometric ranges* for the three satellites to the receiver at station A, (X^n, Y^n, Z^n) are the geocentric coordinates of the satellites at the time of the signal transmission, and (X_A, Y_A, Z_A) are the geocentric coordinates of the receiver at transmission time. Note that the variable n pertains to superscripts and takes on values of satellites 1, 2, and 3.

However, in order to obtain a valid time observation, the systematic error (known as *bias*) in the clocks, and the refraction of the wave as it passes through the Earth's atmosphere, must also be considered. In this example, the receiver clock bias is the same for all three ranges since the same receiver is observing each range. With the introduction of a fourth satellite range, the receiver clock bias can be mathematically determined. This solution procedure allows the receiver to have a less accurate (and less expensive) clock. Algebraically, the system of equations used to solve for the position of the receiver and clock bias are:

$$R_A^1(t) = \rho_A^1(t) + c(\delta^1(t) - \delta_A(t))$$
$$R_A^2(t) = \rho_A^2(t) + c(\delta^2(t) - \delta_A(t))$$
$$R_A^3(t) = \rho_A^3(t) + c(\delta^3(t) - \delta_A(t)) \qquad (13.13)$$
$$R_A^4(t) = \rho_A^4(t) + c(\delta^4(t) - \delta_A(t))$$

where is $R_A^n(t)$ is the observed *range* (also called *pseudorange*) from receiver A to satellites 1 through 4 at epoch (time) t, $\rho_A^n(t)$ the geometric range as defined in Equation (13.12), c the speed of light in a vacuum, $\delta_A(t)$ the receiver clock bias, and $\delta^n(t)$ the satellite clock bias, which can be modeled using the coefficients supplied in the broadcast message. These four equations can be simultaneously solved yielding the position of the receiver (X_A, Y_A, Z_A), and the receiver clock bias $\delta_A(t)$. Equations (13.13) are known as the *point positioning equations* and as noted earlier they apply to code-based receivers.

As will be shown in Section 13.6, in addition to timing there are several additional sources of error that affect the satellite's signals. Because of the clock biases and other sources of error, the observed range from the satellite to receiver is not the true range, and thus it is called a *pseudorange*. Equations (13.13) are commonly called the *code pseudorange model*.

13.5.2 Carrier Phase-Shift Measurements

Better accuracy in measuring ranges to satellites can be obtained by observing *phase-shifts* of the satellite signals. In this approach, the phase-shift in the signal that occurs from the instant it is transmitted by the satellite, until it is received at the ground station, is observed. This procedure, which is similar to that used by EDM instruments (see Section 6.17), yields the fractional cycle of the signal from satellite to receiver.[10] However, it does not account for the number of full wavelengths or *cycles* that occurred as the signal traveled between the satellite and receiver. This number is called the *integer ambiguity* or simply *ambiguity*. Unlike EDM instruments, the satellites utilize one-way communication, but because the satellites are moving and thus their ranges are constantly changing, the ambiguity cannot be determined by simply transmitting additional frequencies. There are different techniques used to determine the ambiguity. All of these techniques require that additional observations be obtained. One such technique is discussed in Section 13.6. Once the ambiguity is determined, the mathematical model for carrier phase-shift, corrected for clock biases, is

$$\varphi_i^j(t) = \frac{1}{\lambda}\rho_i^j(t) + N_i^j + f^j[\delta^j(t) - \delta_i(t)] \tag{13.14}$$

where for any particular epoch in time, t, $\varphi_i^j(t)$ is the carrier phase-shift measurement between satellite j and receiver i, f^j the frequency of the broadcast signal generated by satellite j, $\delta^j(t)$ the clock bias for satellite j, λ the wavelength of the signal, $\rho_i^j(t)$ the range as defined in Equations (13.12) between receiver i and satellite j, N_i^j the integer ambiguity of the signal from satellite j to receiver i, and $\delta_i(t)$ the receiver clock bias.

13.6 ERRORS IN OBSERVATIONS

Electromagnetic waves can be affected by several sources of error during their transmission. Some of the larger errors include (1) satellite and receiver clock biases and (2) ionospheric and tropospheric refraction. Other errors in satellite surveying work stem from (a) ephemeris errors, (b) multipathing, (c) instrument miscentering, (d) antenna height measurements, (e) satellite geometry. These errors contribute to the total error of satellite-derived coordinates in the ground stations. These errors are discussed in the subsections that follow.

13.6.1 Clock Bias

Two errors already discussed in Section 13.5 were the satellite and receiver clock biases. The satellite clock bias can be modeled by applying coefficients that are part of the broadcast message using the polynomial

$$\delta^j(t) = a_0 + a_1(t - t_0) + a_2(t - t_0)^2 \tag{13.15}$$

---

[10]The phase-shift can be measured to approximately 1/100 of a cycle.

where $\delta^j(t)$ is the satellite clock bias for epoch t, t_0 the satellite clock reference epoch, and a_0, a_1, and a_2 the satellite clock offset, drift, and frequency drift, respectively, which are part of the broadcast message. As will be discussed in Section 13.9.1, when using *relative-positioning* techniques, and specifically *single differencing*, the satellite clock bias can be mathematically removed during post-processing.

As was shown in Section 13.5, the receiver clock bias can be treated as an unknown and computed using Equations (13.13) or (13.14). However as discussed in Section 13.9.2 when using relative positioning techniques, it can also be eliminated through *double differencing* during post-processing of the survey data.

13.6.2 Refraction

As discussed in Section 6.16, the velocities of electromagnetic waves change as they pass through media with different refractive indexes. The atmosphere is generally subdivided into regions. The subregions of the atmosphere that have similar composition and properties are known as *spheres*. The boundary layers between the spheres are called *pauses*. The two spheres that have the greatest effect on satellite signals are the *troposphere* and *ionosphere*. The troposphere is the lowest part of the atmosphere and is generally considered to exist up to 10–12 km in altitude. The *tropopause* separates the troposphere from the *stratosphere*. The stratosphere goes up to about 50 km. The combined refraction in the stratosphere, tropopause, and troposphere is known as *tropospheric refraction*.

There are several other layers of atmosphere above 50 km, but the one of most interest in satellite surveying is the *ionosphere* that extends from 50 to 1500 km above the Earth. As the satellite signals pass through the ionosphere and troposphere, they are refracted. This produces range errors similar to timing errors and is one of the reasons why observed ranges are referred to as pseudoranges.

The ionosphere is primarily composed of ions—positively charged atoms and molecules, and free negatively charged electrons. The free electrons affect the propagation of electromagnetic waves. The number of ions at any given time in the ionosphere is dependent on the Sun's ultraviolet radiation. Solar flare activity, which is part of space weather, can dramatically increase the number of ions in the ionosphere, and thus can be reason for concern when working with satellite surveying during periods of high sunspot activity. The activity of the Sun follows a periodic peak variation of 11 years. Since ionospheric refraction can be the single largest error in satellite positioning, it is important to check the space weather when performing surveys.

During periods of high solar activity, the errors due to ionospheric refraction can be large. Since the ionosphere will remain charged for extensive periods of time, there will be some days when a satellite survey simply should not be attempted. In periods of very high solar activity, radio signals from the satellites may be interrupted. Additionally, during these periods, radio communication between the base and roving receivers in an RTK survey (see Chapter 15) may be compromised. The National Oceanic and Atmospheric Administration (NOAA) Space Weather Prediction Center[11] provides forecasts for solar activity and its effect on radio communications.

---

[11] Space weather forecast are available at http://www.swpc.noaa.gov.

In particular, users should monitor geomagnetic storms, solar radiation storms, and radio blackouts. Geomagnetic storms may cause satellite orientation problems, increasing broadcast ephemeris errors, and satellite communication problems and can lead to problems in initialization. Solar radiation storms may also create problems with satellite operations, orientation, and communications, which can cause increased noise at the receiver resulting in degraded positioning. Radio blackouts can cause intermittent loss of satellite and radio communications, which can increase noise at the receiver degrading positional accuracy. These are identified on the NOAA website in five categories from mild to extreme. In general, a satellite survey should not be attempted when any of the three is rated in the range from strong to extreme. There is a global positioning dashboard on the space weather site that provides the GNSS user with a quick check on solar activity of the day and the predicted future.

Terms for both the ionospheric and tropospheric refraction can be incorporated into Equations (13.13) and (13.14) to account for those errors in the signal. Letting $\Delta\delta^j$ equal the difference between the clock bias for satellite j and the receiver at A for epoch t [i.e., $\Delta\delta^j = \delta^j(t) - \delta_A(t)$] then for any particular range listed in Equation (13.13) the incorporation of tropospheric and ionospheric refraction on the code pseudorange model yields

$$R^j_{L1}(t) = \rho^j(t) + c\Delta\delta^j + c[\delta^j_{f_{L1}}] + \delta^{Trop}(t)$$
$$R^j_{L2}(t) = \rho^j(t) + c\Delta\delta^j + c[\delta^j_{f_{L2}}] + \delta^{Trop}(t) \qquad (13.16)$$
$$R^j_{L5}(t) = \rho^j(t) + c\Delta\delta^j + c[\delta^j_{f_{L5}}] + \delta^{Trop}(t)$$

where $R^j_{L1}(t)$, $R^j_{L2}(t)$, and $R^j_{L5}(t)$ are the observed pseudoranges as computed with frequency $L1$, $L2$, and $L5$ (f_{L1}, f_{L2}, and f_{L5}) from satellite j to the receiver, $\rho^j(t)$ the geometric range as defined in Equation (13.12) from the satellite to the receiver, c the velocity of light in a vacuum, $\delta^{Trop}(t)$ the delay in the signal caused by the tropospheric refraction, and δ^{Iono} the ionospheric delay for the $L1$, $L2$, and $L5$ frequencies, respectively.

A similar expression can be developed for the carrier phase-shift model and is

$$\varphi^j_{L1}(t) = \frac{1}{\lambda_{L1}}\rho^j(t) + f_{L1}\Delta\delta^j + N_{L1} - f_{L1}\delta^{Iono} + f_{L1}\delta^{Trop}$$
$$\varphi^j_{L2}(t) = \frac{1}{\lambda_{L2}}\rho^j(t) + f_{L2}\Delta\delta^j + N_{L2} - f_{L2}\delta^{Iono} + f_{L2}\delta^{Trop} \qquad (13.17)$$
$$\varphi^j_{L5}(t) = \frac{1}{\lambda_{L5}}\rho^j(t) + f_{L5}\Delta\delta^j + N_{L5} - f_{L5}\delta^{Iono} + f_{L5}\delta^{Trop}$$

where φ^j_{L1}, φ^j_{L2}, and φ^j_{L5} are the carrier phase-shift observations from satellite j using frequencies $L1$, $L2$, and $L5$, respectively, at some epoch t, N_{L1}, N_{L2}, and N_{L5} are integer ambiguities for the frequencies $L1$, $L2$, and $L5$, and the other terms are as previously defined in Equations (13.14) and (13.16) for each frequency.

By taking observations on the three frequencies, and employing either Equations (13.16) or (13.17), the atmospheric refraction can be modeled and mathematically removed from the data. This is a major advantage of *dual-frequency receivers* (those that can observe both $L1$ and $L2$ signals) over their single-frequency counterparts and allows them to observe baselines up to 150 km accurately. The linear combination of the Li and Lj frequencies for the code pseudorange model, which is almost free of ionospheric refraction, is

$$R_{Li,Lj} = R_{Li} - \frac{(f_{Li})^2}{(f_{Lj})^2} R_{Lj} \qquad (13.18)$$

where $R_{Li,Lj}$ is the pseudorange observation for the combined Li and Lj signals and Li and Lj are a pair of the $L1$, $L2$, or $L5$ carrier frequencies. Until recently, only receivers capable of receiving the P code could perform the ionospheric refraction correction using code ranges. However with the addition of civilian codes on all three frequencies, civilian receivers will be able to process signals using Equation (13.18). This will result in much higher accuracies in code-based positioning due to that ability to nearly eliminate ionospheric refraction in real time.

The carrier-phase model, which is also almost free of ionospheric refraction, is

$$\varphi_{Li,Lj} = \varphi_{Li} - \frac{f_{Lj}}{f_{Li}} \varphi_{Lj} \qquad (13.19)$$

where $\varphi_{Li,Lj}$ is the phase observation of the linear combination of the Li and Lj waves and Li and Lj are replaced by a pair of $L1$, $L2$, or $L5$ carrier frequencies. By their very nature, single-frequency receivers cannot take advantage of the two separate signals, and thus they must use ionospheric modeling data that is part of the broadcast message. This limits their effective range to between 10 and 20 km, although, this limit is dependent on the space weather at the time of the survey.

The advantage in having the satellites at approximately 20,200 km above the Earth is that signals from one satellite going to two relatively close receivers pass through nearly the same atmosphere. Thus, the atmosphere has similar effects on the signals, and its affects can be practically eliminated using mathematical techniques as discussed in Sections 13.7 through 13.9. For long lines Equations (13.18) and (13.19) are typically used.

As can be seen in Figure 13.9, signals from satellites that are on the horizon of the observer must pass through considerably more atmosphere than signals coming from high above the horizon. Because of the difficulty in modeling the atmosphere at low altitudes, signals from satellites below a certain threshold angle, are typically omitted from the observations. The specific value for this angle (known as the satellite *mask angle*) is somewhat arbitrary. It can vary between 10° and 20° depending on the desired accuracy of the survey. Higher horizontal positioning accuracies will be obtained with satellites below 15°, and thus mask angles between 10° and 15° are typically used in surveying. This is discussed further in Chapter 14.

13.6.3 Other Error Sources

Several other error sources contribute to the positional errors of a receiver. These include (1) satellite ephemeris errors; (2) multipathing errors; (3) errors in centering the antenna over a point; (4) errors in measuring antenna height above the point; (5) errors due to satellite geometry, (6) errors in antenna calibration and (7) equipment noise.

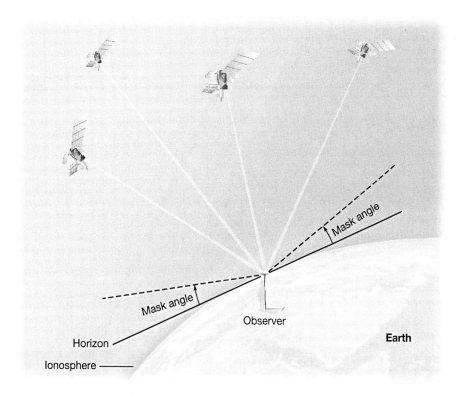

Figure 13.9
Relative positions of satellites, ionosphere, and receiver.

As noted earlier, the broadcast ephemeris predicts the positions of the satellites in the near future. However, because of fluctuations in gravity, solar radiation pressure, and other anomalies, these predicted orbital positions are always somewhat in error. In the code-matching method, these satellite-positioning errors are translated directly into the computed positions of ground stations. This problem can be reduced by updating the orbital data using information obtained later, which is based on the actual positions of the satellites determined by tracking stations. One disadvantage of this is the delay that occurs in obtaining the updated data. One of three updated postsurvey ephemerides is available: (1) ultra-rapid ephemeris, (2) *rapid ephemeris*, and (3) *precise ephemeris*. The ultra-rapid ephemeris is available four times a day; the rapid ephemeris is available within one to two days after the survey; the precise ephemeris (the most accurate of the three) is available two weeks after the survey. The ultrarapid or rapid ephemerides are sufficient for most surveying applications.

As shown in Figure 13.10(a), *multipathing* occurs when a satellite signal reflects from a surface and is directed toward the receiver. This causes multiple signals from a satellite to arrive at the receiver at slightly different times. Vertical structures such as vehicles, buildings, and chain link fences are examples of reflecting surfaces that can cause multipathing errors. Mathematical techniques have been developed to eliminate these undesirable reflections, but, in extreme cases, they can cause a receiver to *lose lock* on the satellite—loss of lock is essentially a situation where the receiver cannot use the signals from the satellite. This can be caused not only by multipathing, but also by obstructions, or high ionospheric activity. Multipathing can also cause incorrect resolution of the initial integer ambiguities with carrier phase-shift observations, which results in errors in positions throughout the project until the ambiguities are resolved a second time.

In satellite surveying, pseudoranges are observed to the receiver antennas. For precise work, the antennas are generally mounted on tripods, set up and carefully centered over a survey station, and leveled. Miscentering of the antenna over the point is another potential source of error. Set up and centering over a station should be carefully done following procedures like those described in Section 8.5. For any precise surveying work, including satellite surveys, it is essential to have a well-adjusted tripod, tribrach, level, and optical plummet. Any error in miscentering of the antenna over a point will translate directly into an equal-sized error in the computed position of that point.

Observing the height of the antenna above the occupied point is another source of error in satellite surveys. The ellipsoid height determined from satellite observations is determined at the phase center of the antenna. Therefore, to get the ellipsoid height of the survey station, it is necessary to measure carefully, and record the height of the *antenna's reference point* (ARP) above the occupied point, and account for it in the data reduction. The distance shown in Figure 13.10(b) is known as the *slant height* and can be observed. The observations are made to the *ground plane* (a plane at the base of the antenna, which protects it from multipath signals reflecting from the ground). The slant height should be observed at several locations around the ground plane, and if the observations do not agree, the instrument should be checked for level. Software within the system converts the slant height to the antenna's vertical distance above the station. Mistakes in identifying and observing heights to the ARP have caused errors as great as 10 cm in elevation.

Figure 13.10 (a) Multipathing and (b) Slant height measurements.

Since this error is avoidable with fixed-height tripods and poles, it is recommended that standard surveying tripods not be used in GNSS surveys. These fixed height devices provide a constant offset from the point to the ARP—typically set at 2 m.

Additionally, the *phase center*, which is the electronic center of the antenna, varies with the orientation of the antenna, elevation of the satellites, and frequency of the signals. In fact, the physical center of the antenna seldom matches the phase center of the antenna. This fact is accounted for by *phase center offsets*, which are translations necessary to make the phase center and physical center of the antenna match. This variation is corrected by using an antenna calibration report available from the NGS, which provides offsets from the electronic center of the antenna to the ARP for varying altitudes to the satellites. Thus it is important to correctly identify the antenna used in a survey during processing (see Section 14.5) to avoid large errors in positioning.

For older antennas, it is important to orient the antennas of multiple receivers in the same azimuth. This ensures the same orientation of the phase centers at all stations and eliminates a potential systematic error from the phase center, which is not precisely located at the geometric center of the antenna. The same antenna should always be used with a given receiver in a precise survey, but if other antennas are used, their phase center offsets must be accounted for during postprocessing. Newer antennas are directionally independent. They no longer require azimuthal alignment.

13.6.4 Geometry of Observed Satellites

An important additional error source in satellite surveying deals with the geometry of the visible satellite constellation at the time of observation. This is similar to the situation in traditional surveys, where the geometry of the network of observed ground stations affects the accuracies of computed positions. Figure 13.11 illustrates both weak and strong satellite geometry. As shown in Figure 13.11(a), small angles between incoming satellite signals at the receiver station produce weak geometry and generally result in larger errors in computed positions. Conversely, strong geometry, as shown in Figure 13.11(b), occurs when the angles between incoming satellite signals are large, and this usually provides an improved solution. Whether conducting a satellite survey or a traditional one, by employing least-squares adjustment in the solution, the effect of the geometry upon the expected accuracy of the results is determined. In general the larger the volume contained by the satellites and receiver is, the better the horizontal positioning. For better vertical positioning, at least one satellite also needs to be near the receiver's zenith.

Table 13.3 lists the various categories of errors that can occur in satellite positioning. For each category, the size of error that could occur in observed satellite ranges if no corrections or compensations were made are given, for example, ± 7.5 m could be expected as a result of ionospheric refraction during periods of high solar activity, etc. However, these error sizes assume ideal satellite geometry, that is, no further degradation of accuracy is included for weak satellite geometry. The anticipated size of these errors with the addition of the L2C and L5 signals is shown in the third column of Table 13.3. The L2C will be available to receivers as the satellites become available. The advantages of the L5 signal will not be

Figure 13.11
Weak and strong satellite geometry.

Weak geometry
(a)

Strong geometry
(b)

apparent to users until most of the satellite constellation has been upgraded. It is anticipated that the entire satellite constellation will be upgraded with these new signals by 2030. By comparing the current errors with those anticipated with the inclusion of newer coded signals, it is obvious why the decision was made to fund the newer satellites. Using Equation (3.11), the total *User Equivalent Range Error* (UERE) is currently approximately ±7.5 m. It is anticipated that this error will drop to approximately ±2.8 m with the L2C and L5 signals.

As noted above, by employing least squares in the solution, the effect of satellite geometry can be determined. In fact, before conducting a satellite survey, the number and positions of visible satellites at any particular time and location can be evaluated in a preliminary least-squares solution to determine their estimated effect upon the resulting accuracy of the solution. This analysis produces so-called *Dilution of Precision* (DOP) factors. The DOP factors are computed through error propagation (see Section 3.17). They are simply numbers, which when multiplied by the errors of Table 13.3, give the sizes of errors that could be expected if no corrections are applied to the ranges based upon the geometry of the observed constellation of satellites. For example, if the DOP factor is 2, then multiplying the sizes of errors listed in Table 13.3 by 2 would yield the estimated errors in the ranges for that time and location. Obviously, the lower the value for a DOP factor, the better the expected precision in computed positions of ground stations. If the preliminary least-squares analysis gives a higher DOP number than can be tolerated, the observations should be delayed until a more favorable satellite constellation is available.

The DOP factors that are of most concern to surveyors are PDOP (dilution of precision in position), HDOP (dilution of precision in horizontal position), and VDOP (dilution of precision in height). For the best possible constellation of satellites, the average value for HDOP is under 2 and under 5 for PDOP. Other DOP factors such as GDOP (dilution of precision in geometry) and TDOP (dilution of precision in time) can also be evaluated but are generally of less significance in

TABLE 13.3 ERROR SOURCES AND SIZES THAT CAN BE EXPECTED IN OBSERVED GPS RANGES

Error Source	Current Sizes of Errors (m)	Anticipated Sizes of Errors with Two or More Coded Signals (m)
Clock and ephemeris errors	±2.3 m	±2.3 m
Ionospheric Refraction	±7 m	±0.1 m
Tropospheric Refraction	±0.2 m	±0.2 m
Receiver Noise	±0.6 m	±0.6 m
Other (multipath, etc.)	±1.5 m	±1.5 m

TABLE 13.4 IMPORTANT CATEGORIES OF DILUTION OF PRECISION

Category of DOP	Stand. Dev. Terms	Equation	Acceptable Value (less than)*
PDOP, Positional DOP	σ in geocentric coordinates X, Y, Z	$\sqrt{\sigma_X^2 + \sigma_Y^2 + \sigma_Z^2}$	6
HDOP, Horizontal DOP	σ in local n, e coordinates	$\sqrt{\sigma_n^2 + \sigma_e^2}$	3
VDOP, Vertical DOP	σ in height, h	σ_h	5

*These recommended values are general guides for average types of GPS surveys, but individual project requirements may require other specific values.

surveying. Table 13.4 lists some important categories of DOP, explains their meanings in terms of standard deviations and equations, and gives maximum values that are generally considered acceptable for most surveys.

Multiplying the DOP factor by the UERE yields the positional error in code ranging using Equations (13.13). For example, the HDOP is typically about 1.5. Recall from Equation (3.8) that the 95% probable error is obtained using a multiplier of about 1.96. Using the error values from Table 13.3 and a HDOP of 1.5 the current 95% probable error in horizontal positioning is ±22.5 m (1.96 × 1.5 × 7.5). When the newer coded signals are available and used by receivers, the 95% horizontal positioning error will be approximately ±8.5 m if the ionospheric free models are implemented in the solution.

13.7 DIFFERENTIAL POSITIONING

As discussed in the two preceding sections, accuracies of observed pseudoranges are degraded by errors that stem from clock biases, atmospheric refraction, and other sources. Because of these errors, positions of points determined by point positioning techniques using a single code-based receiver can be in error by 20 m or more. While this order of accuracy is acceptable for certain uses, it is insufficient for most surveying applications. *Differential GPS* (DGPS) on the other hand, is a procedure that involves the simultaneous use of two or more code-based receivers. It can provide positional accuracies to within a one meter, and thus the method is suitable for certain types of lower-order surveying work.

In DGPS, one receiver occupies a so-called *base station* (point whose coordinates are precisely known from previous surveying), and the other receiver or receivers (known as the *rovers*) are set up at stations whose positions are unknown. By placing a receiver on a station of known position, the pseudorange errors in the signal can be determined using Equation (13.16). Since this base station receiver and the rover are relatively close to each other (often less than a kilometer but seldom farther than a few hundred kilometers), the pseudorange errors at both the base station and at the rovers will have approximately the same magnitudes. Thus after computing the corrections for each visible satellite at the base station, they can be applied to the roving receivers, thus substantially reducing or eliminating many errors listed in Table 13.3.

DGPS can be done in almost *real time* with a radio transmitter at the base station and compatible radio receivers at the rovers. This process is known as *real-time differential GPS* (RTDGPS). The radio transmissions to the rovers contain both *pseudorange corrections* (PRCs) for particular *epochs of time* (moments in time) and *range rate corrections* (RRCs)[12] so that they can interpolate corrections to signals between each epoch. Alternatively, the errors can be eliminated from coordinates determined for rover stations during postprocessing of the data.

To understand the mathematics in the procedure, a review of Equation (13.13) is necessary. Excluding multipathing, the various error sources presented in Section 13.6 cause the observed pseudorange $R_A^j(t_0)$ to be in error by a specific amount for any epoch, t_0. Letting this error at epoch t_0 be represented by $\Delta \rho_A^j(t_0)$, the *radial orbital error*, Equation (13.13) can be rewritten as

$$R_A^j(t_0) = \rho_A^j(t_0) + \Delta \rho_A^j(t_0) + c\delta^j(t_0) - c\delta_A(t_0) \quad (13.20)$$

where the other terms are as previously defined.

Because the coordinates of the base station are known, the geometric range $\rho_A^j(t_0)$ in Equation (13.20) can be computed using Equation (13.12). Also, since the pseudorange $R_A^j(t_0)$ is observed, the difference in these two values will yield the necessary correction for this particular pseudorange. Since the error conditions at each receiver are very similar, it can be assumed that the error in the pseudorange observed at the base station is the same as the error at the rovers. Obviously, this assumption will become closer to the truth with decreasing distance between the two receivers. This error at the base station is known as the code *pseudorange correction* (PRC) for satellite j at reference epoch t_0, and is represented as

$$\begin{aligned} \text{PRC}^j(t_0) &= -R_A^j(t_0) + \rho_A^j(t_0) \\ &= -\Delta \rho_A^j(t_0) - c[\delta^j(t_0) - \delta_A(t_0)] \end{aligned} \quad (13.21)$$

---

[12]Pseudorange corrections (PRCs) are differences between measured ranges and ranges that are computed based upon the known coordinates of both the occupied reference station and those of the satellite. Because the satellites are moving, measured ranges to them are constantly changing. The rates of these changes per unit of time are the range rate corrections (RRCs).

Because computation of the correction and transmission of the signal make it impossible to assign the PRC to the same epoch at the rovers, a RRC is approximated by numerical differentiation. This correction is used to extrapolate corrections for later epochs t. Thus, the pseudorange correction at any epoch t is given as

$$\text{PRC}^j(t) = \text{PRC}^j(t_0) + \text{RRC}^j(t_0)(t - t_0) \quad (13.22)$$

where $\text{RRC}^j(t_0)$ is the range rate correction for satellite j determined at epoch t_0.

Now this information can be used to correct the computed ranges at the roving receiver locations. For example, at a roving station B, the corrected pseudorange, $R_B^j(t)_{\text{corrected}}$, can be computed as

$$\begin{aligned} R_B^j(t)_{\text{corrected}} &= R_B^j(t) + \text{PRC}^j(t) \\ &= \rho_B^j(t) + [\Delta\rho_B^j(t) - \Delta\rho_A^j(t)] - c[\delta_B(t) - \delta_A(t)] \quad (13.23) \\ &= \rho_B^j(t) - c\Delta\delta_{AB}(t) \end{aligned}$$

where $\Delta\delta_{AB} = \delta_B(t) - \delta_A(t)$.

Notice that in the final form of Equation (13.23), it is assumed that the radial orbital errors at stations A and B, $\Delta\rho_A^j(t)$ and $\Delta\rho_B^j(t)$, respectively, are nearly the same, and thus are mathematically eliminated. Furthermore, the satellite clock bias terms will be eliminated. Finally, assuming the signals to the base and roving receivers pass through nearly the same atmosphere (which means they should be within a few hundred kilometers of each other), the ionospheric and tropospheric refraction terms are practically eliminated.

The U.S. Coast Guard maintains a system of beacon stations along the U.S. coast and waterways. Private agencies have developed additional stations. The correction signals described above are broadcast by modulation on a frequency between 285 and 325 kHz using the *Radio Technical Commission for Maritime Services Special Committee* 104 (RTCM SC-104) format. Among the data contained in this broadcast are C/A code differential corrections, delta differential corrections, reference station parameters, raw carrier phase measurements, raw code range measurements, carrier phase corrections, and code range corrections.

The *Wide Area Augmentation System* (WAAS) developed by the Federal Aviation Administration has a network of ground tracking base stations that collect GPS signals and determine range errors. These errors are transmitted to geosynchronous satellites that relay the corrections to rovers. GPS software typically allows users to access the WAAS system when performing RTK-GPS surveys (see Chapter 15). This option, sometimes called *RTK with infill*, accesses the WAAS corrections when base-station radio transmissions are lost. However, these corrections will provide significantly less accuracy than relative positioning techniques typically utilized by GPS receivers using carrier phase-shift measurements. In Europe, the *European Geostationary Navigation Overlay Service* (EGNOS) serves a similar role to WAAS. In Japan, the *Multifunctional Satellite Augmentation System* (MSAS) serves this purpose.

13.8 KINEMATIC METHODS

Methods similar to DGPS can also be employed with carrier phase-shift measurements to eliminate errors. The procedure, called *kinematic* surveying (see Chapter 15), again requires the simultaneous use of two or more receivers. All receivers must simultaneously collect signals from a least four of the same satellites through the entire observation process. Although single-frequency receivers can be used, kinematic surveying works best with dual-frequency receivers. The method yields positional accuracies to within a few centimeters, which makes it suitable for most surveying, mapping, and stakeout purposes.

As with DGPS, the fact that the base station's coordinates are known is exploited in kinematic surveys. Most manufacturers broadcast the observations at the base station to the rover. The roving receiver uses the relative positioning techniques discussed in Section 13.9 to determine the position of the roving receiver. However, it is also possible to compute and broadcast pseudorange corrections. Once the pseudorange corrections are determined, they are used at the roving receivers to correct their pseudoranges. Multiplying Equation (13.14) by λ, and including the radial orbital error term, the carrier phase pseudorange at base station A for satellites j at epoch t_0 is

$$\lambda \varphi_A^j(t_0) = \rho_A^j(t_0) + \Delta \rho_A^j(t_0) + \lambda N_A^j + c[\delta^j(t_0) - \delta_A(t_0)] \quad (13.24)$$

where N_A^j is the initially unknown ambiguity, and all other terms were previously defined in Equation (13.20). Recalling that the base station is a point with known coordinates, the pseudorange correction at epoch t_0 is given by

$$\begin{aligned} \text{PRC}^j(t_0) &= -\lambda \varphi_A^j(t_0) + \rho_A^j(t_0) \\ &= -\Delta \rho_A^j(t_0) - \lambda N_A^j - c[\delta^j(t_0) - \delta_A(t_0)] \end{aligned} \quad (13.25)$$

and the pseudorange correction at any epoch t is

$$\text{PRC}^j(t) = \text{PRC}^j(t_0) + \text{RRC}^j(t_0)(t - t_0) \quad (13.26)$$

Using the same procedure as was used with code pseudoranges, the corrected phase range at the roving receiver for epoch t is

$$\lambda \varphi_B^j(t)_{\text{corrected}} = \rho_B^j(t) + \lambda \Delta N_{AB}^j - c \Delta \delta_{AB}(t) \quad (13.27)$$

where $\Delta N_{AB}^j = N_B^j - N_A^j$ and $\Delta \delta_{AB}(t) = \delta_B(t) - \delta_A(t)$.

These equations can be solved as long as at least four satellites are continuously observed during the survey while the pseudorange corrections and the range rate corrections are transmitted to the receivers.

13.9 RELATIVE POSITIONING

The most precise positions are currently obtained using relative positioning techniques. Similar to both DGPS and kinematic surveying, this method removes most errors noted in Table 13.3 by utilizing the differences in either the code or carrier phase ranges. The objective of relative positioning is to obtain the coordinates of a point relative to another point. This can be mathematically expressed as

$$X_B = X_A + \Delta X$$
$$Y_B = Y_A + \Delta Y \qquad (13.28)$$
$$Z_B = Z_A + \Delta Z$$

where (X_A, Y_A, Z_A) are the geocentric coordinates at the base station A, (X_B, Y_B, Z_B) are the geocentric coordinates at the unknown station B, and $(\Delta X, \Delta Y, \Delta Z)$ are the computed *baseline vector components* (see Figure 13.12).

Relative positioning involves the use of two or more receivers simultaneously observing pseudoranges at the endpoints of lines. Simultaneity implies that the receivers are collecting observations at the same time from the same satellites. It is also important that the receivers collect data at the same epoch rate. This rate depends on the purpose of the survey and its final desired accuracy, but common intervals are 1, 5, or 15 seconds. Assuming that simultaneous observations have been collected, different linear combinations of the equations can be produced, and in the process certain errors can be eliminated. Figure 13.13 shows three linear combinations and

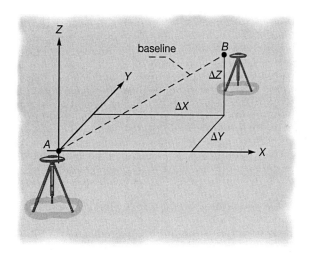

Figure 13.12
Computed baseline vector components.

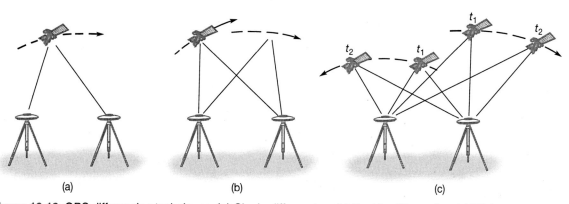

Figure 13.13 GPS differencing techniques. (a) Single differencing. (b) Double differencing. (c) Triple differencing.

the required receiver-satellite combinations for each. These are described in the subsections that follow, and only carrier-phase measurements are considered.

13.9.1 Single Differencing

As illustrated in Figure 13.13(a), single differencing involves subtracting two simultaneous phase observations made to one satellite from two points. This difference eliminates the satellite clock bias and much of the ionospheric and tropospheric refraction from the solution. Following Equation (13.14), the phase equations for the two points are

$$\varphi_A^j(t) - f\delta^j(t) = \frac{1}{\lambda}\rho_A^j(t) + N_A^j - f\delta_A(t)$$
$$\varphi_B^j(t) - f\delta^j(t) = \frac{1}{\lambda}\rho_B^j(t) + N_B^j - f\delta_B(t) \tag{13.29}$$

where the terms are as noted in Equation (13.14) for stations A and B. The difference in these two equations yields

$$\varphi_{AB}^j(t) = \frac{1}{\lambda}\rho_{AB}^j(t) + N_{AB}^j - f\delta_{AB}(t) \tag{13.30}$$

where the individual difference terms are

$$\varphi_{AB}^j(t) = \varphi_B^j(t) - \varphi_A^j(t),$$
$$\rho_{AB}^j(t) = \rho_B^j(t) - \rho_A^j(t),$$
$$N_{AB}^j = N_B^j - N_A^j, \text{and}$$
$$\delta_{AB}(t) = \delta_B(t) - \delta_A(t)$$

Note that in Equation (13.30), the satellite clock bias error, $f\delta^j(t)$ has been eliminated by this single-differencing procedure.

13.9.2 Double Differencing

As illustrated in Figure 13.13(b), double differencing involves taking the difference of two single differences obtained from two satellites j and k. The procedure eliminates the receiver clock bias. Assume the following two single differences:

$$\varphi_{AB}^j(t) = \frac{1}{\lambda}\rho_{AB}^j(t) + N_{AB}^j - f\delta_{AB}(t)$$
$$\varphi_{AB}^k(t) = \frac{1}{\lambda}\rho_{AB}^k(t) + N_{AB}^k - f\delta_{AB}(t) \tag{13.31}$$

Note that the receiver clock bias will be the same for each observation. Thus by taking the difference between these two single differences, the following double-difference equation is obtained, in which the receiver clock bias errors, $f\delta_{AB}^j(t)$ and $f\delta_{AB}^k(t)$ are eliminated.

$$\varphi_{AB}^{jk}(t) = \frac{1}{\lambda}\rho_{AB}^{jk}(t) + N_{AB}^{jk} \tag{13.32}$$

where the difference terms are

$$\varphi_{AB}^{jk}(t) = \varphi_{AB}^{k}(t) - \varphi_{AB}^{j}(t)$$
$$\rho_{AB}^{jk}(t) = \rho_{AB}^{k}(t) - \rho_{AB}^{j}(t)$$
$$N_{AB}^{jk} = N_{AB}^{k} - N_{AB}^{j}$$

13.9.3 Triple Differencing

The triple difference illustrated in Figure 13.13(c) involves taking the difference between two double differences obtained for two different epochs of time. This difference removes the integer ambiguity from Equation (13.32), leaving only the differences in the phase-shift observations and the geometric ranges. The two double-difference equations can be expressed as

$$\varphi_{AB}^{jk}(t_1) = \frac{1}{\lambda}\rho_{AB}^{jk}(t_1) + N_{AB}^{jk}$$
$$\varphi_{AB}^{jk}(t_2) = \frac{1}{\lambda}\rho_{AB}^{jk}(t_2) + N_{AB}^{jk}$$
(13.33)

The difference in these two double differences yields the following triple difference equation, in which the integer ambiguities have been removed. The triple difference equation is

$$\varphi_{AB}^{jk}(t_{12}) = \frac{1}{\lambda}\rho_{AB}^{jk}(t_{12})$$
(13.34)

In Equation (13.34) the two difference terms are:

$$\varphi_{AB}^{jk}(t_{12}) = \varphi_{AB}^{jk}(t_2) - \varphi_{AB}^{jk}(t_1), \text{and}$$

$$\rho_{AB}^{jk}(t_{12}) = \rho_{AB}^{jk}(t_2) - \rho_{AB}^{jk}(t_1).$$

The importance of employing the triple difference equation in the solution is that by removing the integer ambiguities, the solution becomes immune to *cycle slips*. Cycle slips are created when the receiver *loses lock* during an observation session. The three main sources of cycle slips are (1) obstructions, (2) low *signal to noise ratio* (SNR), and (3) incorrect signal processing. Signal obstructions can be minimized by careful selection of receiver stations. Low SNR can be caused by undesirable ionospheric conditions, multipathing, high receiver dynamics, or low satellite elevations. Malfunctioning satellite oscillators can also cause cycle slips, but this rarely occurs. It should be noted that today's processing software only uses triple differencing for detection of cycle slip detection since the integer ambiguities are resolved using more advanced on-the-fly techniques, which are discussed in Section 15.3.

■ 13.10 PRECISE POINT POSITIONING

While relative positioning works well in most environments, there are instances were physical limitations can hinder its results. These conditions exist in remote areas near Polar Regions and coastal surveys where the points with known coordinates are not near the area being surveyed. In these cases, *precise point positioning* (PPP) is another tool that can be used to postprocess both static and kinematic data. This technique can also be used in areas which have nearby electromagnetic interference (EMI) sources, do not have known coordinates for the base station, prohibit the use of radios in real-time surveys, or in areas without cellular coverage or outside the network's physical base stations for real-time network (RTN) surveys.

Precise point positioning involves the modeling of the errors sources as discussed in Section 13.6. The mathematical model for PPP is

$$\varphi_i = \rho_r^s + c(\delta t_r - \delta t^s) + \text{ISB} - \delta^{Iono} + \delta^{Trop} + N_r^s + m_p + \varepsilon \quad (13.35)$$

where φ_i is the observed phase shift for the i frequency in terms of length units, ρ_r^s the geometric range between receiver r and satellite s, c the speed of light in a vacuum, which is 299,792,458 m/s, δt_r and δt^s the receiver and satellite clock biases, respectively, ISB the intersystem biases, which account for differences hardware delays and residual time differences between systems, δ^{Iono} the ionospheric delay term, δ^{Trop} the tropospheric delay term, N_r^s the integer ambiguities between receiver r and satellite s, m_p the multipath error term, and ε the noise and unmodeled error terms caused by relativity, rotation of the Earth, tide loading, and other disturbing effects. Thus Equation (13.35) attempts to model all the errors that affect the results in positioning.

To understand how PPP works, it is necessary to understand how these terms are modeled. From Equation 13.12, we are trying to determine the geocentric coordinates (X, Y, Z) for the receiver. The satellite's geocentric coordinates at any epoch in time can be determined from a precise ephemeris. Table 13.5 list the accuracies and latency for the various precise ephemerides. It should be realized at this point that a 1 nanosecond (ns) error in time will result in approximately a 30 cm error in range. Additionally, a picosecond (ps) is 10^{-12} seconds, which

TABLE 13.5 AVAILABLE SATELLITE EPHEMERIDES WITH ACCOMPANYING ACCURACIES

Type		RMS Accuracy (Range Error)	Latency	Sample Interval
Broadcast	Orbit	~100 cm	Real time	Daily
	Sat. clock	~5 ns ± 2.5 ns (~150 cm)		
Ultra-Rapid (predicted)	Orbit	~5 cm	Real time	15 min
	Sat. clock	~3 ns ± 1.5 ns (~90 cm)		
Ultra-Rapid (observed)	Orbit	~3 cm	3–9 h	15 min
	Sat. clock	~150 ps ± 50 ps (~4.5 cm)		
Rapid	Orbit	~2.5 cm	17–41 h	15 min
	Sat. clock	~75 ps ± 25 ps (~2.2 cm)		5 min
Final	Orbit	~2.5 cm	12–18 days	15 min
	Sat. clock	~75 ps ± 20 ps (~2.2 cm)		30 s

results in about a 3 mm error in range. Thus, a 5 ns error in the broadcast ephemeris clock will result in nearly 1.5 m error in range whereas a 5 ps errors will result in a 1.5 mm error in the pseudorange. As one can see from Table 13.5, for precise point positioning a precise ephemeris is required.

The receiver clock bias δ_r^s in Equation (13.35) is estimated in a least squares Kalman filter solution as is the integer ambiguities N_r^s. The inter-system bias (*ISB*) is modeled as a constant error in the least squares solution. The ionospheric refraction term (δ^{Iono}) can be modeled using broadcast ephemeris terms or removed using an ionospheric free combination of two signals (dual frequencies) and is then mathematically constrained. Likewise, the tropospheric refraction term (δ^{Trop}) is modeled and constrained during the solution for position. The remaining unmodeled error term (ε) is estimated during the solution process.

It has been shown that PPP methods can produce valid positions with 2 or more hours of data. In the future, by streaming precise ephemeris products, it may be possible to achieve precise positions in real-time. Since this method of surveying only requires a single receiver and does not require a subscription service or cellular coverage, it is possible that this will be the dominant positioning method in the future.

13.11 OTHER SATELLITE NAVIGATION SYSTEMS

Satellite positioning affects all walks of life including transportation, agriculture, and data networks, cell phones, sporting events, and so on. In fact, the military and economic benefits of satellite positioning have been so great that other nations have or will be developing their own networks. This plethora of positioning satellites will greatly increase the utility and accuracy available from satellite positioning system. Other implemented or planned satellite positioning systems are discussed in the following subsections.

13.11.1 The GLONASS Constellation

GLONASS is the Russian navigation satellite system. The GLONASS constellation consists of 24 satellites equally spaced in three orbital planes making a 64.8° nominal inclination angle with the equatorial plane of the Earth. The satellites orbit at a nominal altitude of 19,100 km and have a period of approximately 11.25 hours. At least five are always visible to users. The system is free from selective availability but does have a restricted-access signal similar to the military codes in GPS. The current satellites broadcast at least two signals with frequencies that are unique using *frequency division multiple access* (FDMA) where satellites are assigned specific frequencies using the following algorithm:

$$f^j_{L1} = 1602.0000\,\text{MHz} + j \times 0.5625\,\text{MHz}$$
$$f^j_{L2} = 1246.0000\,\text{MHz} + j \times 0.4375\,\text{MHz} \qquad (13.36)$$

where j represents the channel number assigned to the specific satellite,[13] and varies from 1 to 24, and $L1$ and $L2$ represent the broadcast bands.

---

[13]Some antipodal satellites use the same frequencies.

However, to be fully compatible with the GPS, Galileo, and BeiDou systems, the modernized GLONASS-K series of satellites will also broadcast signals using *code division multiple access* (CDMA) techniques. When fully modernized, the system will broadcast civilian CDMA signals on three different frequencies. The initial GLONASS-K1 satellites broadcast the CDMA codes on the L3 band, which has a frequency of 1207.14 MHz. The GLONASS-K2 satellites broadcast CDMA codes on the L1 and L3 bands. The L1 band has frequency of 1575.42 MHz. The GLONASS-K3 satellites will broadcast CDMA codes on the L1, L3, and L5 bands. The L5 band has a frequency of 1176.45 MHz. The inclusion of these new CDMA codes means that the GLONASS navigation system will be fully compatible with the other systems and thus will provide easier access to the system.

As discussed in Section 13.3, GPS satellites broadcast their positions in every repetition of the broadcast message using the WGS84 reference system as the basis for coordinates. The GLONASS satellites only broadcast their positions every 30 min and use the PZ-90 reference ellipsoid as the basis for coordinates. Thus, GNSS receivers must extrapolate positions of the satellites for real-time reductions.

The time reference systems used in GPS and GLONASS are also different. At the request of the international community, the timing of GLONASS satellites has moved toward the international standard as set by the *Bureau Internationale de l'Heure* (International Bureau of Time). This standard is based on the frequency of the atom Cesium 133 in its ground state.[14] This standard differs from the orbital period of the Earth by approximately 1 sec every 6 months. To compensate, one leap second is periodically added to the atomic time (IAT) to create Universal Coordinated Time (UTC), which agrees with the solar day (see Section C.5). Currently, the GLONASS system clocks differ from Universal Coordinated Time by three hours. In contrast, the GPS system clocks never account for the leap second, and differ from IAT by a constant of 20 seconds. To account for the timing difference currently, two GLONASS satellites must be visible if GLONASS and GPS satellites are combined in a GNSS receiver. When the GLONASS system is fully modernized, this restriction will no longer occur.

13.11.2 The Galileo System

In 1998, the European Union decided to implement the Galileo satellite positioning system. The Galileo system will offer five levels of service with subscriptions required for some of the services. The five levels of service are (1) *open* service (OS), (2) *commercial* service (CS), (3) *safety-of-life* service (SOL), (4) *public-regulated* service (PR), and (5) *search and rescue* (SAR) service. Open service will provide free positioning down to 1 m. The commercial service is an encrypted, subscription service, which will provide positioning at the cm level. The safety-of-life service will be a free providing both guaranteed accuracy and integrity messages to warn of errors. The public-regulated service will be available only to government agencies, which is similar to the current P-code. The search-and-rescue service will receive distress beacon locations and be able to send feedback indicating that help is on the way.

---

[14]One second is defined as 9,192,631,770 periods of the radiation of the ground state of the cesium 133 atom.

The Galileo space segment will consist of 27 satellites plus 3 spares orbiting in three planes that are inclined to the equator at 56°. The satellites will have a nominal orbital altitude of 23,222 km above the Earth. The satellites will broadcast six navigation signals denoted as L1F, L1P, E6C, E6P, E5a, and E5b. The *European Space Agency* (ESA) is now launching operational satellites for the *In-Orbit Validation* (IOV) of the system with an anticipated completion date during 2020. Like the modernized GPS satellites, the strength of its signals should allow work in canopy situations. Like the GLONASS system, Galileo will be interoperable with GPS.

13.11.3 The BeiDou System

In 2006, China confirmed that it will create a fourth satellite positioning system. BeiDou[15] will contain 35 satellites. Five of these satellites will be geostationary Earth orbit (GEO) satellites with the remaining 30 satellites at about 21,000 km at approximately a 55° inclination angle. BeiDou will offer two levels of service—an open and commercial service with real-time positioning accuracy of 10 m. Its anticipated completion date is around 2020. Like GPS, it will broadcast using three different frequencies of B1, B2, and B3. Its B1 signal is transmitted with a frequency of 1589.472 MHz with a wavelength of approximately 18.8 mm. The B2 signal has a frequency of 1207.140 MHz with a wavelength of approximately 24.8 mm. The B3 signal has a frequency of 1268.520 MHz with a wavelength of approximately 23.6 mm.

13.11.4 Summary

Even though the Galileo and BeiDou satellite constellations of the systems may not be complete until 2020, manufacturers of satellite receiver technology are building receivers that will utilize all GPS, GLONASS, Galileo, and the BeiDou systems. The receivers shown in Figures 13.1 and 13.14 are currently capable of combining GPS, GLONASS, and Galileo satellite observations in their solutions. The obvious advantage of using multiple systems is that many more satellites are available for observation by receivers. In fact, in the near future it is possible that as many as 30 satellites may be available for positioning at any particular moment in time. By combining these systems, the surveyor can expect improvements in increased speed and accuracy. Furthermore, the combination of systems will provide a viable method of bringing satellite positioning to difficult areas such as canyons, deep surface mines, and urban areas surrounded by tall buildings, which are called *urban canyons*.

■ 13.12 THE FUTURE

The overall success of satellite positioning in the civilian sector is well documented by the number and variety of enterprises that are using the technology. This has led to increasing and improving GNSS constellations. By the end of this decade

---

[15]The Chinese name for their system is BeiDou, which means for North Dipper. In the past in some English text, this system has also been called the Compass System.

Figure 13.14
The GR5 receiver is capable of combining GPS, GLONASS, and Galileo signals into a combined solution.
(Courtesy Topcon Positioning Systems.)

improvements will occur in signal acquisition and positioning. For example, signals from all the satellite-positioning systems will be able to penetrate canopy situations and will provide satellite-positioning capabilities from within buildings. The additional signals from within each system will improve both ambiguity resolution and atmospheric corrections. For example, with the addition of the L2C and L5 signals, real-time ionospheric corrections to the code pseudoranges will become possible by implementation of Equations (13.18). Furthermore, the addition of the L2C and L5 signal will enhance our ability to determine the integer ambiguities correctly and quickly for phase-shift observations. In fact, in theory it will be possible to determine the ambiguities with a single epoch of data. It is anticipated that GPS and Galileo will provide 30 cm real-time point positioning accuracies with code solutions. Furthermore, it is anticipated that accuracies in relative positioning using the modernized systems will be reduced to the millimeter level. In fact, it is anticipated that differential code-based solutions will be available at the centimeter level. Additionally, the GPS week number is now represented as 13 bits, which translates into a rollover situation only once every 8192 weeks or once in 157 years. In its current configuration, GPS has experienced two rollover weeks, which is when the week number transmitted goes from 1027 back to 0.

The use of satellites in the surveying (geomatics) community has continued to increase as the costs of the systems have decreased. This technology has and will undoubtedly continue to have considerable impact on the way data is collected and processed. In fact, as the new satellite technologies are developed, the use of conventional surveying equipment will decrease. This will change much of how we survey today with more reliance on the speed and accuracy of the GNSS.

As is the current trend, less field time will be required for the surveyor (geomatics engineer), and more time will be used to analyze, manage, and manipulate the large volumes of data that this technology and others such as laser scanning (See Section 17.9.5) provide. Those engaged in surveying in the future will need to be knowledgeable in the areas of information management and computer science and will undoubtedly provide products to clients that currently do not exist.

PROBLEMS

Asterisks (*) indicate problems that have partial answers given in Appendix G.

- **13.1** What does the GNSS receiver provide the user?
- **13.2** What are the control in satellite positioning?
- **13.3** What is the frequency of the L5 signal and its relationship to the fundamental frequency of the satellite clock?
- **13.4*** Discuss the purpose of the pseudorandom noise codes.
- **13.5** What type of information is broadcast in the GPS signal?
- **13.6** What four parameters are required to convert satellite coordinates into the geocentric coordinate system?
- **13.7** What are the three geodetic coordinates?
- **13.8** Describe the satellite coordinate system.
- **13.9** Describe the geocentric coordinate system.
- **13.10** Define the terms "geodetic height," "geoid height," and "orthometric height." Include their relationship to each other.
- **13.11** Define PDOP, HDOP, and VDOP.
- **13.12** What reference ellipsoid is used in the broadcast message of GPS?
- **13.13** What is precise point positioning?
- **13.14** Why is it important to check the reference frame of previously determined coordinates for a station?
- **13.15** Approximately how far from a GPS satellite will a signal travel in ps?
- **13.16** What are the unknowns in Equation (13.13)?
- **13.17** What is integer ambiguity?
- **13.18** What is multipathing and why should it be avoided?
- **13.19*** What error can double differencing of phase-shift observations remove?
- **13.20** At a particular time of the day, all the satellites are near the mask angle in the southern portion of the sky. Will the horizontal positioning be more or less accurate than if there were satellites in positions near the mask angle but encircling the receiver's position?

For Problems 13.21 through 13.26 use the WGS84 ellipsoidal parameters.

- **13.21*** What are the geocentric coordinates in meters of a station in meters which has a latitude of 49°27′32.20144″ N, longitude of 122°46′53.56027″ W, and height of 303.436 m.
- **13.22** Same as Problem 13.21 except with geodetic coordinates of 32°16′02.34579″ N, longitude of 90°13′08.60792″ W and height of 108.593 m?
- **13.23** Same as Problem 13.21 except with geodetic coordinates of 41°18′15.02489″ N longitude of 69°24′15.64283″ W and height of 894.567 m?
- **13.24*** What are the geodetic coordinates in meters of a station with geocentric coordinates of (136,153.995, −4,859,278.535, 4,115,642.695)?

13.25 Same as Problem 13.24, except with geocentric coordinates in meters are (−1,059,346.748, −4,392,471.406, 4,488,270.465)?

13.26 Same as Problem 13.24, except with geocentric coordinates in meters are (690,658.287, −5,827,237.954, −2,491,164.434)?

13.27 The GNSS determined height of a station is 32.485 m. The geoid height at the point is −22.164 m. What is the orthometric of the point?

13.28* Same as Problem 13.27, except the height is 584.097 m and the geoid height is −31.543 m.

13.29 Same as Problem 13.27, except the height is 44.972 m and the geoid height is 28.968 m.

13.30* The orthometric height of a point is 124.886 m. The geoid height of the point is −29.998 m. What is the geodetic height of the point?

13.31 Same as Problem 13.30, except the orthometric height is 874.264 m, and the geoid height is −29.078 m.

13.32 The geodetic heights of two stations are 324.685 m and 309.879 m, and their orthometric heights are 356.496 m and 341.707 m, respectively. These stations have model-derived geoid heights of −31.827 and −31.835 respectively. What is the orthometric height of a station with a GNSS measured height of 305.645 m and a model-derived geoid height of −31.802 m?

13.33 Why are satellites at an elevation below 10° from the horizon eliminated from the positioning solution?

13.34 Why should NOAA's space weather center be consulted at the beginning of any day where a GNSS survey is planned?

BIBLIOGRAPHY

Altiner, Y., et al. 2014. "A Control Center for Real-Time GNSS Data." *Professional Surveyor* 34 (No. 2): 20.

Dodo, J. D., M. N. Kamarudin, and M. H. Yahya. 2008. "The Effect of Tropospheric Delay on GPS Height Differences along the Equator." *Surveying and Land Information Science* 68 (No. 3): 145.

Hofmann-Wellenhof, B., et al. 2008. *GNSS: Global Navigation Satellite Systems*. Springer-Wein, New York, NY.

Martin, D. J. 2003. "Around and Around with Orbits." *Professional Surveyor* 23 (No. 6): 50.

———. 2003. "Reaching New Heights in GPS, Part 3." *Professional Surveyor* 23 (No. 4): 42.

Petovello, M. 2014. "GNSS Solutions: Relative Positions." *Inside GNSS* 9 (No. 3): 38.

Reilly, J. 2003. "On Galileo, the European Satellite Navigation System." *Point of Beginning* 28 (No. 12): 46.

———. 2003. "On Geoid Models". *Point of Beginning* 29 (No. 12): 50.

Schrock, G. 2014. "GNSS: What Did Not Happen and Increases, Improvements, Innovations." *XYHT: Special Outlook Edition*: 16.

Tatarnikov, D., et al. 2015. "GNSS Antenna That Protect and Observe." *XYHT* 2 (No. 2): 34.

14
Global Navigation Satellite Systems— Static Surveys

■ 14.1 INTRODUCTION

Many factors can have a bearing on the ultimate success of a satellite survey, and there are many different approaches that can be taken in terms of equipment used and procedures followed. Because of these variables, satellite surveys should be carefully planned prior to going into the field. Small projects of lower-order accuracy may not require a great deal of preplanning beyond selecting receiver sites and making sure they are free from overhead obstructions and multipathing conditions. On the other hand, large projects that must be executed to a high order of accuracy will require extensive preplanning to increase the probability that the survey will be successful. As an example, a survey for the purpose of establishing control for an urban, rapid transit project will command the utmost care in selecting personnel, equipment, and receiver sites. It will also be necessary to make a presurvey site visit to locate existing control and identify possible overhead obstructions and multipath conditions that could interfere with incoming satellite signals at all proposed receiver sites. In addition, a careful preanalysis should be made to plan optimum *observation session*[1] times, the durations of the sessions, and to develop a plan for the orderly execution of the sessions. The project will probably require ground communications to coordinate survey activities, a transportation analysis to ensure reasonable itineraries for the execution of the survey, and installation of monuments to permanently mark the new points that will be located in the survey. Consideration of these factors, and others, in planning and executing global navigation satellite system (GNSS) projects are the subjects of this chapter.

---

[1]An observation session denotes the period of time during which all receivers being employed on a project have been set up on designated stations and are simultaneously engaged in receiving signals from the same satellites. When one session is completed, all receivers except one are generally moved to different stations and another observation session is conducted. The sessions are continued until all planned project observations have been completed.

Code-based receivers are used for positioning by people in all walks of life. They can be used by surveyors to gather details in situations not requiring typical survey accuracies. Examples are the approximate location of monuments, boundary, or otherwise, to aid in later relocation, the collection of data to update small-scale maps in a *geographic information system* (GIS—see Chapter 28), and the navigation to monuments that are part of the National Spatial Reference System (see Chapter 19). The use of code-based receivers in nonsurveying applications includes the tracking of vehicles in transportation. The shipping industry uses code-based receivers for navigation. Likewise, surveyors may use the navigation functions of a code-based receiver to locate control monuments or other features where geodetic coordinates are known. Since the use of code-based receivers is so extensive and reaches far beyond the realm of the surveying community, their uses will not be covered in detail in this book.

This chapter concentrates on the use of carrier phase-shift receivers, and the use of relative positioning techniques. This combination can provide the highest level of accuracy in determining the positions of points, and thus it is the preferred approach in some surveying applications. However, as noted in Chapter 13, the accuracy of a survey is also dependent on several additional variables. An important one is the type of carrier-phase receiver used on the survey. As noted in Chapter 13, there are several types: *GNSS* receivers, which can utilize the multiple signals available from several different satellite constellations; *multi-frequency receivers*, which can observe and process the multiple signals from the GPS constellation; *dual-frequency* receivers, which can observe and process the $L1$ and $L2$ frequencies from GPS; and *single-frequency receivers*, which can observe only the $L1$ band from GPS. In precise surveys, GNSS, multi-frequency, and dual-frequency receivers are preferred for several reasons: they can (a) collect the needed data faster; (b) observe longer baselines with greater accuracy; (c) almost eliminate certain errors, such as ionospheric refraction; and (d) process the newer signals when available, and therefore yield higher positional accuracies. Receivers also vary by the number of channels. This controls the number of satellites and signals that they can track simultaneously. As a minimum, carrier phase-shift receivers must have at least four channels, but some are capable of tracking as many as 30 satellites from the GPS, GLONASS, Galileo, and BeiDou constellations simultaneously using multiple frequency bands resulting in 200 or more channels. These receivers provide higher accuracies due to the increased number of satellites and increased strength in satellite geometry.

Some of these factors are beyond control of the surveyor (geomatics engineer), and therefore it is imperative that observational checks be made. These are made possible by the redundant observations. This chapter will discuss these checks.

The use of satellites for specific types of surveys, for example, construction surveys, land surveys, photogrammetric surveys, etc., are covered in later chapters in this text. A GNSS receiver being used on a construction site is shown in Figure 14.1. Use of satellite receivers for topographic surveys is covered in Section 17.9.4.

14.2 FIELD PROCEDURES IN STATIC GNSS SURVEYS

In practice, field procedures employed on surveys depend on the capabilities of the receivers and the type of survey. Some specific field procedures currently being

Figure 14.1
GNSS receiver being used on a construction site. (Courtesy Topcon Positioning Systems.)

used in surveying include the *static, rapid static, pseudokinematic,* and *kinematic* methods. These are described in the subsections that follow. All are based on carrier phase-shift measurements and employ *relative positioning* techniques (see Section 13.9); that is, two (or more) receivers, occupying different stations and simultaneously making observations to the same satellites. The vector (distance) between receivers is called a *baseline* as described in Section 13.9, and its ΔX, ΔY, and ΔZ coordinate difference components (in the geocentric coordinate system described in Section 13.4.2) are computed as a result of the observations.

14.2.1 Static Relative Positioning

Static surveying procedures produce the highest accuracies and are typically used in geodetic control surveys. In this procedure, two (or more) receivers are employed. Figure 14.2 depicts the typical equipment used in this and the following type of static surveys. The process begins with one receiver (called the *base receiver*) being located on an existing control station, while the remaining receivers (called the *roving receivers*) occupy stations with unknown coordinates. For the first observing session, simultaneous observations are made from all stations to four or more satellites for a time period of an hour or more depending on the baseline length (longer baselines require greater observing times). Except for one, all the receivers can be moved upon completion of the first session. The remaining receiver now serves as the base station for the next observation session. It can be selected from any of the receivers used in the first observation session. Upon completion of the second session, the process is repeated until all stations are occupied, and the observed baselines form geometrically closed figures. As discussed in Section 14.5, for checking purposes, some repeat baseline observations are made during the surveying process. Additionally, the actual coordinates of the base receiver do not

Figure 14.2
A GNSS receiver cabled to survey controller. (Courtesy Topcon Positioning Systems.)

need to be known at the time of the survey. However, they must be known when the data is post-processed (see Section 14.5.5). If sessions are sufficiently long, these coordinates can come from on-line processing software such as OPUS (discussed in Section 15.6), which will determine National Spatial Reference System coordinates in the United States. Similarly, other on-line processing software from various countries exists, which can determine coordinates of the base receiver in a system of international coordinates such as ITRF, which was discussed in Section 13.4.4.

The value for the *epoch rate*[2] in a static survey must be the same for all receivers during the survey. Typically, this rate is set to 15 sec to minimize the number of observations, and thus the data storage requirements. In longer sessions, 30 sec or 60 sec epoch rates are also employed. Most receivers either have internal memory capabilities or are connected to *controllers* that have internal memories for storing the observed data. After all observations are completed, data are transferred to a computer for post-processing.

Relative accuracies with static relative positioning are about ±(3 to 5 mm + 0.5 ppm) at a 68% level of probability. Typical durations for observing sessions using this technique, with both single- and dual-frequency receivers, are shown in Table 14.1.

14.2.2 Rapid Static Relative Positioning

This procedure is similar to static surveying with the exception of the maximum length of lines as well as data collection rates and times. Like a static survey, an observing session is conducted for each point, but the sessions and epoch rates

---

[2]GNSS satellites continually transmit signals, but if they were continuously collected by the receivers, the volume of data and hence storage requirements would become overwhelming. Thus, the receivers are set to collect samples of the data at a certain time interval which is called the epoch rate.

TABLE 14.1 TYPICAL SESSION LENGTHS FOR VARIOUS OBSERVATION METHODS

Method of Survey	Single Frequency	Dual Frequency
Static	30 min + 3 min/km	20 min + 2 min/km
Rapid Static	20 min + 2 min/km	10 min + 1 min/km

are shorter than for the static method. Table 14.1 shows the suggested session lengths for single- and dual-frequency receivers using the rapid static method. The rapid static method is suitable for observing baselines up to 20 km in length under good observation conditions. Rapid static relative positioning can also yield accuracies on the order of about $\pm(3 \text{ to } 5 \text{ mm} + 0.5 \text{ ppm})$ at a 68% level of probability. However, to achieve these accuracies, optimal satellite configurations (good PDOP), lack of multipathing, and favorable ionospheric conditions must exist. This method is ideal for small control surveys. As with static surveys, all receivers should be set to collect data at the same epoch rate. Typically, the epoch rate is set to 5 sec with this method, which is considerably shorter than the rate for a static session.

14.2.3 Pseudokinematic Surveys

This procedure is also known as the *intermittent* or *reoccupation* method, and like the other static methods requires a minimum of two receivers simultaneously observing the same satellites. In pseudokinematic surveying, the base receiver always stays on a control station, while the rover goes to each point of unknown position. Two relatively short observation sessions (around 5 min each in duration) are conducted with the rover on each station. The time lapse between the first session at a station and the repeat session should be about an hour. This produces an increase in the geometric strength of the observations due to the change in satellite geometry that occurs over the time period. Accuracies approaching those of static surveys can be achieved following reduction procedures similar to those described in Section 13.9.

A disadvantage of this method, compared to other static methods, is the need to revisit the stations. This procedure requires careful presurvey planning to ensure that sufficient time is available for site revisitation, and to achieve the most efficient travel plan. Pseudokinematic surveys are most appropriately used where the points to be surveyed are along a road, and rapid movement from one site to another can be readily accomplished. During the period of movement, the receiver can be turned off. Some projects for which pseudokinematic surveys may be appropriate include alignment surveys (see Chapters 24 and 25), photocontrol surveys (see Chapter 27), lower-order control surveys, and mining surveys. Given the speed and accuracy of kinematic surveys, however, this survey procedure is seldom used today in practice.

14.2.4 Establishing Initial Control

All the previously mentioned, survey methods require that the base receiver have known coordinates. This may seem impossible given passive control being sometimes unreliable and scarce. Additionally, it is advantageous to have the survey in some national or international reference system for current use in a GIS (see Chapter 28) or for future relocation. When occupation of preexisting control is not possible, the base receiver can be started with *autonomous* coordinates. Autonomous coordinates are obtained from the receiver using a code-based solution (see Section 13.5.1). While these coordinates are not sufficiently accurate for a typical survey, they allow the survey to progress and can be later modified by replacing them with a more accurate set of coordinates.

In most parts of the United States, more accurate coordinates can be obtained by connecting to the NOAA Continuously Operating Reference Station (CORS) Network (NCN) (see Section 14.3.5). With a sufficiently long session, which is dependent on the distance to the nearest NCN station, accurate coordinates in the national reference system can be obtained. These coordinates can then replace the autonomous set of coordinates for the base receiver so that its position and all positions determined by the roving receiver(s) can be updated.

However, in some remote parts of the country, such as in Alaska, this connection would require an extremely long observation session for the base receiver and may result in questionable coordinates. This condition can also occur along the seacoast such as in offshore mineral exploration, where all the NCN stations are at a distance and to one side of the base receiver. As mentioned in Section 13.10, precise point positioning (PPP) surveys can be performed in these as well as other areas. For precise coordinates to be obtained using PPP, the base observation session currently needs to be three hours or longer. The observational data can then be submitted to one of the free on-line PPP processing services for determination of the base receiver's position. As with coordinates obtained when using the NOAA CORS Network, they can be used to replace the autonomous position of the base receiver so that all surveyed points are corrected to the more accurate base receiver coordinates. There are several free on-line services which can process the base station's file and provide coordinates for the base station. These coordinates can then be used to process the rover's observation file. Table 14.2 list some of these services and their URLs.

Depending on the service used, the coordinates received can be in reference frames that differ from those required by the user. When this is the case, the coordinates may be transferred into the appropriate reference frame using online software such as the National Geodetic Survey's Horizontal Time-Dependent Positioning

TABLE 14.2 ONLINE PROCESSING SERVICES

Service	Owner	URL
Precise Point Positioning	Natural Resources of Canada (CSRS – PPP)	https://webapp.geod.nrcan.gc.ca/geod/tools-outils/ppp.php
	University of New Brunswick (GAPS)	http://gaps.gge.unb.ca/bestpractices.html
	Jet Propulsion Lab (APPS)	https://gipsy-oasis.jpl.nasa.gov/index.php?page=home
Relative Positioning	National Geodetic Survey (OPUS, OPUS-RS, and OPUS Projects)	http://www.ngs.noaa.gov/OPUS
	Australia Geosciences (AUSPOS)	http://www.ga.gov.au/scientific-topics/positioning-navigation/geodesy/auspos

(HTDP) software (see Section 14.3.5). Once the coordinates are in the appropriate reference frame, they may be entered into the user's software, and the remaining observation files from the rover can be processed to determine final project coordinates.

14.3 PLANNING SATELLITE SURVEYS

As noted earlier, small surveys generally do not require much in the way of project planning. However, for large projects, and for higher-accuracy surveys, project planning is a critical component in obtaining successful results. The subsections that follow discuss various aspects of project planning, with emphasis on control surveys.

14.3.1 Preliminary Considerations

All new high-accuracy survey projects that employ relative positioning techniques must be tied to nearby existing control points. Thus, one of the first things that must be done in planning a new project is to obtain information on the availability of existing control stations near the project area. For planning purposes, these should be plotted in their correct locations on an existing map or aerial photos of the area. Mapping products available on the Internet can provide excellent aerial coverage of an area with street-level views where locations can be entered by their geodetic coordinates.

Another important factor that must be addressed in the preliminary stages of planning for projects is the selection of the new station locations. Of course, they must be chosen so that they meet the overall project objective. In addition, terrain, vegetation, and other factors must also be considered in their selection. If possible, they should be reasonably accessible by either the land vehicles or aircraft that will be used to transport the survey hardware. The stations can be somewhat removed from vehicle access points since hardware components are relatively small and portable. Also, the receiver antenna is the only hardware component that must be accurately centered over the ground station. Once the preliminary station locations are selected, they should be plotted on the map or aerial photo of the area.

Another consideration in station selection is the assurance of an overhead view free of obstructions. This is known as *canopy restrictions*. Canopy restrictions may possibly block satellite signals, thus reducing observations and possibly adversely affecting satellite geometry. At a minimum, it is recommended that visibility be clear in all directions from a mask angle (altitude angle) of 10° to 20° above the horizon. However, to obtain the best horizontal positions, the mask angle should be at or below 15°. In some cases, careful station placement will enable this visibility criterion to be met without difficulty; in other situations, clearing around the stations may be necessary. Furthermore, as discussed in Section 13.6.3, potential sources that can cause interference and multipath errors should also be identified and avoided when visiting each site.

Selecting suitable *observation windows* is another important activity in planning surveys. This consists of determining the satellites that will be visible from a given ground station or project area during a proposed observation period. To aid in this activity, azimuth and elevation angles to each visible satellite can be predetermined using almanac data for times within the planned observation. Required computer input, in addition to observing date and time, include the station's approximate latitude and longitude, and a relatively current satellite almanac. Additionally, the space weather should be checked for possible solar storms during the periods

Figure 14.3 Satellite availability plot.

of occupation. Days where the solar radiation storm activity is rated from strong to extreme should be avoided. Section 15.2 discusses the effects of space weather on GNSS surveys in detail and is applicable to both static and kinematic surveys.

To aid in selecting suitable observation windows, a *satellite availability plot*, as shown in Figure 14.3, can be applied. The shaded portion of this diagram shows the number of satellites visible for a planned station occupation. The diagram is applicable for August 4, 2014 between the hours of 8:00 and 17:00 EDT. A mask angle of 15° has been used. In addition to showing the number of visible satellites, the lines running through the plot depict the predicted PDOP, HDOP, and VDOP (see Section 13.6.4) for this time period. It should be noted that, for the day shown in Figure 14.3, only two short time periods are unacceptable for data collection. DOP *spikes* occur between 8:02 and 8:12 when only four satellites are above the horizon mask angle, and between 13:45 and 14:00 when both VDOP and PDOP are unacceptable because of weak satellite geometry. However, during this latter period, the HDOP is acceptable, indicating that a horizontal control survey could still be executed. Note also that one of the better times for data collection is between 10:40 and 11:30 when PDOP is below 2, because 9 satellites are visible during that time. With the inclusion of GLONASS observations and as the satellite constellations of Galileo and BeiDou become mature, this type of situation will occur less frequently with multifrequency GNSS receivers.

Satellite visibility at any station is easily and quickly investigated using a *sky plot*. These provide a graphic representation of the azimuths and elevations to visible satellites from a given location. As illustrated in Figure 14.4(a) and (b), sky plots consist of a series of concentric circles. The circumference of the outer

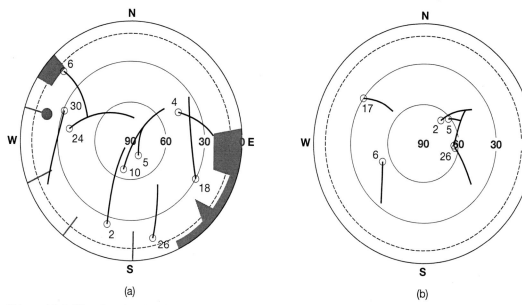

Figure 14.4 Sky plot showing (a) the obstructing objects around the station from 10 to 12. Plot (b) shows the weak satellite geometry from 13:30 to 14.

circle is graduated from 0 to 360° to represent azimuths to the satellites. Each successive concentric circle progressing toward the center represents an increment in the elevation angle, with the radius point corresponding to the zenith. Often an additional concentric circle is depicted indicating the mask angle from the horizon.

For each satellite, the PRN number is plotted beside its first data point, which is its location for the selected starting time of a survey. Then arcs connect successive plotted positions for the given time increments after the initial time. Thus, travel paths in the sky of visible satellites are shown. Sky plots are valuable in survey planning because they enable operators to quickly visualize not only the number of satellites available during a planned observation period, but also their geometric distribution in the sky.

When overhead obstructions are a concern, the elevations and azimuths of the obstructions can be overlaid on the sky plot to form *obstruction diagrams*. The diagram will then show whether crucial satellites are removed by the obstructions and indicate the best times to occupy the station to avoid the obstructions. As shown in Figure 14.4(a), a building will obscure satellite 6 briefly during the time period shown. Also note that signals from satellite 30 will experience a brief interruption caused by the presence of a nearby pole later in the session, and that satellite 4 will be lost near the end of the session because of a nearby building. None of these obstructions appear to be critical to the session.

The analysis of sky obstructions and satellite geometry are important for the highest accuracy in surveys. Recall from Section 13.6.4 that observations should be taken on groups of four or more widely spaced satellites that form a strong geometric intersection at the observing station. As stated in Section 13.6.4, better positioning will occur when larger volumes are contained by the satellites and receiver. This condition is illustrated in Figures 13.11(b) and 14.4. Weaker geometry, as shown in Figures 13.11(a) and 14.4(b), should be avoided if possible as it will yield lower coordinate accuracies. The PDOP and VDOP spikes shown in Figure 14.3 between the times from 13:45 and 14:00 are caused by the closely clustered distribution of the satellites as shown in Figure 14.4(b).

It is important to note that the optimal observation times will repeat 4 min earlier for each day following the planning session. That is, in Figure 14.3, the same satellite visibility chart will apply for the period from 7:56 to 16:56 on August 5, and from 7:52 to 16:52 on August 6, etc. Of course, the periods of poor PDOP will also shift by 4 min each day. This shift occurs because sidereal days are approximately 4 min shorter than solar days. Modern survey controllers can derive sky plots, compute PDOP values in the field, and display the results on their screens.

14.3.2 Selecting the Appropriate Survey Method

As discussed in Section 14.2, several different methods are available with which to accomplish a survey. Each method provides a unique set of procedural requirements for field personnel. In high-accuracy surveys that involve long baselines, the static surveying method with GNSS receivers is the best solution. However, in typical surveys limited to small areas, a single-frequency receiver using rapid static, pseudokinematic, or kinematic surveying methods (see Chapter 15) may be sufficient. Because of the variability in requirements, the selection of the appropriate survey method is dependent on the (1) desired level of accuracy in the final coordinates, (2) intended use of the survey, (3) type of equipment available for the survey, (4) size of the survey, (5) canopy and other local conditions for the survey, and (6) available software for reducing the data, there is seldom only one method for accomplishing the work. Since kinematic surveys are so much faster, static surveys are typically used where the highest achievable accuracy by a GNSS survey is required.

GNSS receivers will reduce the time required at each station in a static survey due to the increased number of visible satellites and the improved satellite geometry. Figure 14.5 is a schematic diagram that categorizes the various survey methods. The surveying community has used those shown on the left side of the diagram traditionally. However, a modernized GNSS will create the possibility of using differential GNSS surveys for lower-order surveys.

Figure 14.5
Schematic depicting GNSS positioning methods.

For mapping or inventory surveys where centimeter to submeter accuracy is sufficient, differential GNSS surveys or kinematic surveys (see Chapter 15) provide the most economical product. However, if the area to be mapped has several overhead obstructions, it may only be possible to use one of the static survey procedures to bring control into the region and do limited kinematic mapping in small areas where clear overhead views to the satellites are available. Recognizing this, many manufacturers have developed equipment that allows the surveyor to switch between a receiver and a total station instrument (see Chapter 8) using the same survey controller and project file. This capability is useful in areas where GNSS surveys are not practical. A more in-depth discussion on this topic is presented in Chapter 17. Restrictions in use of GNSS surveys due to canopy restrictions will be greatly reduced when modernized GNSS constellations are available.

The preferred approach in performing high-accuracy control surveys is the static method. Often a combination of the static methods will provide the most economical results for large projects. As an example, a static survey may be used to bring a sparse network of accurate control into a project area. This could be followed by a rapid static survey to densify the control. Finally, a kinematic survey could be used to establish control along smaller project areas in the region. In smaller areas with favorable viewing conditions, the best method for the survey may simply be the rapid static or kinematic methods. Available equipment, software, and experience often dictate the survey method of choice. In addition, it is a mistake to assume that one method of survey is always appropriate.

14.3.3 Field Reconnaissance

Once the existing nearby control points and proposed new stations have been located on paper, a reconnaissance trip to the field should be undertaken to check the selected observation sites for (1) overhead obstructions that rise above 10° to 15° from the horizon, (2) reflecting surfaces that can cause multipathing, (3) nearby electrical installations that can interfere with the satellite's signal, and (4) other potential problems. If reconnaissance reveals that any preliminary selected station locations are unsatisfactory, adjustments in their positions should be made. For the existing control stations that will be used in the survey, ties can be made to nearby permanent objects, and also photos or *rubbings*[3] of the monument caps should be created. These items will aid crews in locating the stations later during the survey, reduce the amount of time spent at each station, and minimize station misidentifications. Web mapping services can often be used to make preliminary decision about the suitability of a site for occupation by a GNSS receiver. However, a site visit is the only method to confirm its suitability since changes occur at all sites over time.

Once final sites have been selected for new stations, permanent monuments should be set and the positions of the stations documented with ties to nearby objects, photos, and rubbings. Also, at this time if required, an accurate horizon plot of any surrounding obstructions can be prepared and road directions and

---

[3]Monuments used to mark stations generally have metal caps (often brass) which give the name of the point and other information about the station. This information is stamped into the cap and can be obtained, by laying a piece of paper directly over the cap, and rubbing across the surface or obtaining a close image of the cap using a digital camera. Many survey controllers have an internal camera for capturing such notes.

approximate driving times between stations recorded. There are several web services that can be used to obtain driving times and directions between stations once their approximate positions are known. A valuable aid in identifying locations of stations is the use of a code-based receiver. These inexpensive devices will determine the geodetic coordinates of the stations with sufficient accuracy to enable plotting them on a map, navigation to the station, and project planning.

14.3.4 Developing an Observation Scheme

For surveying projects, especially those that employ relative positioning and are applied in control surveying, once existing nearby control points have been located and the new stations set, they and the observations that will be made comprise what is termed a *network*. Depending on the nature of the project and extent of the survey, the network can vary from only a few stations to very large and complicated configurations. Figure 14.6 illustrates a small network consisting of only two existing control points and four new stations.

After the stations are established, an observation scheme is developed for performing the work. The scheme consists of a planned sequence of observing sessions that accomplishes the objectives of the survey in the most efficient manner. As a minimum, it must ensure that every station in the network is connected to at least one other station by a *nontrivial* (also called *independent*) baseline as described later. However, the plan should also include some redundant observations (i.e., baseline observations between existing control stations, multiple occupations of stations, and repeat observations of certain baselines) to be used for checking purposes, and for improving the precision and reliability of the work. Desired accuracy is the principal factor governing the number and type of redundant observations. The *Federal Geodetic Control Subcommittee* (FGCS) has developed a set of standards and specifications for GPS relative positioning (see Section 14.5.1) that specify the number and types of redundant observations necessary for accuracy orders *AA*, *A*, *B*, and *C*. Generally, on larger, high-accuracy projects, these standards and specifications, or other similar ones, govern the conduct of the survey work and must be carefully followed.

In relative positioning, for any observing session the number of nontrivial baselines measured is one less than the number of receivers used in the session or

$$b = n - 1 \tag{14.1a}$$

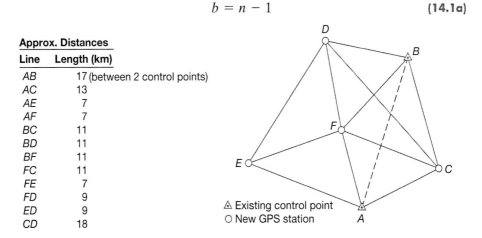

Figure 14.6 A GNSS network.

Approx. Distances

Line	Length (km)
AB	17 (between 2 control points)
AC	13
AE	7
AF	7
BC	11
BD	11
BF	11
FC	11
FE	7
FD	9
ED	9
CD	18

△ Existing control point
○ New GPS station

where b is the number of nontrivial baselines and n the number of receivers being employed in the session. Nontrivial baselines are a set of baselines that are mathematically independent of each other. When only two receivers are used in a session, only one baseline is observed, and it is nontrivial. If more than two receivers are used, both nontrivial and *trivial (mathematically dependent)* baselines will result. The total number of baselines can be computed as

$$T = \frac{n(n-1)}{2} \quad \text{(14.1b)}$$

where T is the total number of baselines possible. The number of trivial baselines for any session is

$$t = \frac{(n-1)(n-2)}{2} \quad \text{(14.1c)}$$

where t is the total number of trivial baselines. To differentiate between these two types of baselines, and to understand how trivial baselines occur, refer to Figure 14.7. In this figure, an observing session involving three receivers A, B, and C observing four satellites $1, 2, 3$, and 4 is shown. Pseudoranges $1A, 1B, 2A, 2B, 3A, 3B, 4A$, and $4B$ are employed to compute the baseline vector AB. Also, pseudoranges $1B, 1C, 2B, 2C, 3B, 3C, 4B$, and $4C$ are used in computing the baseline vector BC. Thus, all possible pseudoranges in this example have been used in calculating baselines AB and BC, and the computation of baseline CA would be redundant; that is, it would be based upon observations that have already been used. In this example, baselines AB and BC are considered nontrivial and CA trivial. This is reinforced with the fact that adding vectors AB and BC will result in vector CA, which demonstrates their mathematical dependence. However, the selection of the trivial baseline is arbitrary. That is, either AB, BC, or AC could be selected as

Figure 14.7
GNSS observation session using three receivers. In the case shown, *AB* and *BC* are considered nontrivial baselines. Thus, *AC* is a trivial baseline.

the trivial baseline, depending on which two baselines were selected as nontrivial. If four receivers are used in a session, six baselines will result: three nontrivial and three trivial. Students should verify this with a sketch. For meeting accuracy standards and obtaining valid statistics (see Chapter 16), only nontrivial baselines can be considered, and thus distinguishing between them is important. A simple rule of thumb to remember when selecting baselines for processing is that no single observation session should produce baselines that close a geometric figure.

When possible, at least one baseline should be observed between existing control monuments of higher accuracy for every receiver-pair used on a project to check field procedures, performance of software and equipment, and reliability of the control. Also, as noted above, some baselines should be observed more than once. These repeat baselines should ideally be observed at or near the beginning and at the end of the project observations to check the field procedures and equipment for repeatability. The analysis of these duplicate observations will be covered in Section 14.5.

For control surveys, the baselines from multiple observation sessions should form closed geometric figures since they are necessary to perform closure checks (see Section 14.5). The simple network of baselines, shown in Figure 14.6, will be used as an example to illustrate survey planning. Assume that the project is in the area of control station *PSU1* and that the survey dates will be as shown in the satellite availability plot and sky plots of Figures 14.3 and 14.4, respectively. Stations *A* and *B* in Figure 14.6 are existing control monuments and a baseline observation between them will be planned to (1) verify the accuracy of the existing control, (2) confirm that the equipment is in proper working condition, and (3) check field procedures.

In the example of Figure 14.6, it is assumed that two dual-frequency receivers are available for the survey and that the rapid static method will be used since all the baselines are less than 20 km and observation conditions are considered good. Following the minimum session lengths, as given in Table 14.1, of 10 min + 1 min/km baseline *AB* would require 10 + 1(17), or 27, minutes of observation time. The remaining baseline observation times are listed in Table 14.3 using the same computational techniques.

Two two-person crews, each working individually with separate vehicles, are assumed for conducting the survey. It is also assumed that setup and teardown times at each station are both approximately 15 min. By rounding each minimal observation session up to a nearest 5-min interval, the following observation sessions and times were planned for a two-day data collection project. It is important to remember that a session involves the simultaneous collection of satellite data. Thus, a session does not start until all receivers involved in the session are set up and running on their respective stations. Therefore, communication between the field crews is crucial to a successful survey.

The observation plan for this example is given in Table 14.4. The table gives the itinerary for both field crews, allowing time for setup and teardown of the equipment, travel between stations, and collecting sufficient observations. The sessions are identified by the day of the year and then by consecutive lettering starting at A. The plan includes all lines in the network as baselines. These include those for checking purposes, an observation of the control baseline *AB*, and repeat observations of *AF* and *BF*. Note that the two unfavorable times for collecting observations shown in Figure 14.3 are not scheduled as times for collecting data, but are used for other ancillary operations. In the event that operations should actually

TABLE 14.3 MINIMUM SESSION LENGTHS AND APPROXIMATE DRIVING TIMES BETWEEN STATIONS FOR BASELINES IN FIGURE 14.6

Line	Length (km)	Session Lengths (min)	Driving Time (min)
AB	17	27	15
AC	13	23	10
AE	7	17	8
AF	7	17	25
BC	11	21	15
BD	11	21	10
BF	11	21	20
FC	11	21	15
FE	7	17	15
FD	9	19	10
ED	9	19	15
CD	18	28	25

run ahead of or fall behind the planned schedule for some unforeseen reason, it is prudent to include a statement on the itinerary indicating that no baseline observations should be collected between the times of 8:00 to 8:15, and 13:40 to 14:00. Note that as indicated in the planned schedule, the crew with the stationary receiver should continue to collect data during the entire period of occupation of the station. This includes the periods of time when the other crew is moving between stations. The reason for this is that the data collected by the stationary crew can be used to connect to the National Spatial Reference System (NSRS) using the NCN (see Section 14.3.5). Optionally, the longer sessions can be processed using OPUS to create additional stronger ties to the national network. It is desirable to provide the field personnel with communication devices during the survey so that they can coordinate times for sessions and handle unforeseen logistical problems that will inevitably arise. The sessions are identified by the day of the year and then by consecutive lettering starting at A. Two two-person crews, each working individually with separate vehicles, are assumed for conducting the survey. It is also assumed that setup and teardown times at each station are both approximately 15 min. By rounding each minimal observation session up to a nearest 5-min interval, the following observation sessions and times were planned for a two-day data collection project. It is important to remember that a session involves the simultaneous collection of satellite data. Thus, a session does not start until all receivers involved in the session are set up and running on their respective stations. Therefore, communication between the field crews is crucial to a successful survey.

14.3.5 Availability of Reference Stations

As explained in Section 14.3.4, the availability of high-quality reference control stations is necessary to achieve the highest order of accuracy in positioning. To meet this need, individual states, in cooperation with the NGS, have developed *High Accuracy Reference Networks* (HARNs). The HARN is a network of

TABLE 14.4 Observation Itinerary for Figure 14.6

DAY 1 (August 4, 2014)

Time	Crew 1	Crew 2	Baseline	Session
8:00–8:45	Drive to Station A	Drive to Station C		
9:00–9:25	Collect data	Collect data	AC	216A
9:40–9:55	Collect data	Drive to Station F		
9:55–10:15	Collect data	Collect data	AF	216B
10:30–10:45	Collect data	Drive to Station E		
11:00–11:20	Collect data	Collect data	AE	216C
11:35–11:50	Drive to Station B	Drive to Station F		
12:05–12:30	Collect data	Collect data	BF	216D
12:45–1:00	Collect data	Drive to Station C		
13:15–13:40	Collect data	Collect data	BC	216E
13:55–14:05	Collect data	Drive to Station A		
14:20–14:50	Collect data	Collect data	AB	216F
15:05–15:15	Drive to Station D	Drive to Station C		
15:30–16:00	Collect data	Collect data	CD	216G
16:00–17:00	Return to office	Download data		

DAY 2 (August 5, 2014)

Time	Crew 1	Crew 2	Baseline	Session
8:00–9:00	Drive to Station A	Drive to Station F		
9:15–9:35	Collect data	Collect data	FA	217A
9:50–10:00	Drive to Station E	Collect data		
10:15–10:30	Collect data	Collect data	FE	217B
10:45–11:00	Drive to Station D	Collect data		
11:00–11:20	Collect data	Collect data	FD	217C
11:35–11:45	Drive to Station B	Collect data		
12:00–12:25	Collect data	Collect data	FB	217D
12:40–12:55	Drive to Station C	Collect data		
13:10–13:35	Collect data	Collect data	FC	217E
13:50–14:05	Drive to Station B	Drive to Station D		
14:20–14:45	Collect data	Collect data	BD	217E
15:00–15:15	Drive to Station D	Drive to Station E		
15:30–15:50	Collect data	Collect data	ED	217F
15:50–17:00	Return to office	Download data		

Note: No baseline observations should be made between 8:00–8:15 and 13:40–14:00 on August 4, and between 7:56–8:11 and 13:36–13:56 on August 5.

Figure 14.8
Locations that have continuously operating reference stations (CORS). (Courtesy National Geodetic Survey.)

control points that were precisely observed using GPS under the direction of the National Geodetic Survey (NGS). These HARN points are available to serve as reference stations for surveys in their vicinity.

Additionally, the NGS with cooperation from other public and private agencies has created a national system of CORS, called the NOAA CORS Network. The location of stations in the NCN as of 2020 is shown in Figure 14.8. These stations not only have their positions known to high accuracy, but they also are occupied by a receiver that continuously collects satellite data, which can be downloaded from the NGS website at http://www.ngs.noaa.gov/CORS/. This information can be used as base station data to support roving receivers operating in the vicinity of the NCN station. The data is stored in a format known as **R**eceiver **IN**dependent **EX**change (RINEX). This format is a standard that can be read by all postprocessing software. This Internet site also provides coordinates for the stations and the ultra-rapid, rapid, and precise ephemerides (see Section 13.6.3). The HARN and NCN make up what is known as the *National Spatial Reference Frame*.

Because of plate tectonics, velocity vectors accompany the coordinates for the stations. While motions in the eastern United States may be under 1 mm per year, these motions may be considerably more in the western United States. Thus, coordinates derived from control surveys should be accompanied by their reference frame (see Section 19.6.3) and epoch. The NGS has written software known as *horizontal time-dependent positioning* (HTDP) using 14 parameters to transform between various reference frames and epochs in time (see Section 19.6.6). The WGS84 (G1762) coordinates differ substantially from NAD83 coordinates. Since the broadcast ephemeris is determined in the WGS84 reference frame, this transformation is necessary to place coordinates derived by GNSS surveys into the NAD83 reference frame. This coordinate transformation is demonstrated in the Mathcad® file *C14.XMCD*, which is on the companion website for this book. The transformation can be avoided by performing a localization (see Section 19.6.6) on the GNSS survey, or by occupying a station with known NSRS coordinates.

The CORS data files are easily downloaded using the *User-Friendly CORS* (UFCORS) option on the NGS website. This utility provides the user with an interactive form which requests the (1) starting date, (2) time, (3) duration of the survey, (4) site selection, and (5) the collection interval among other things. The request is interpreted by the server and the appropriate data is sent, via the Internet, to the user within minutes.

Several factors can cause data not to be collected at a particular CORS site for short periods of time. These include local power outages, storm damage, and software and hardware failures. Thus, if a certain NCN station is planned for use on a particular survey, it is important to check the availability of the station prior to the beginning of the survey to ensure that it is functioning properly, and that the necessary data will be available following the survey. However, given the number of NCN stations available in the conterminous United States (CONUS), it is usually possible to select other NCN sites for processing when the desired station is unavailable.

■ 14.4 PERFORMING STATIC SURVEYS

During a control survey using either the static or rapid static method, the field crew initially locates the control station and erects the antenna over the station so that it is level. To minimize setup errors, 2-m fixed-height poles or tripods should be used. However, when using a fixed-height pole, be sure that the circular level vial sensitivity is less than 10′. As shown in Figure 14.2, cables should be connected to the antenna and controller. It should be wrapped around the pole to take up the extra cable and keep it safe from catching on neighboring objects accidentally. Procedures specified by the manufacturer are then followed to initiate data collection. Bluetooth connections are not recommended for use in a static survey since Bluetooth connections can be lost during lengthy observation sessions.

While the data is being logged, other ancillary information at the site can be collected and recorded. Typical ancillary data obtained during a survey includes (1) project and station names, (2) ties to the station, (3) a rubbing or photo of the monument cap, (4) panoramic photos of the setup showing identifying background features, (5) potential obstructive or reflective surfaces, (6) date and session number, (7) start and stop times, (8) name of observer, (9) receiver and antenna serial number, (10) meteorological data, (11) PDOP value at the start and end of the session, (12) antenna height, (13) orientation of antenna, (14) rate of data collection (epoch rate), and (15) notations on any problems experienced. This ancillary data is typically entered onto a site log sheet such as that shown in Figure 14.9.

Modern survey controllers can store the satellite observations internally, however, the receiver's internal memory should be used when available. In the latter case, a survey controller may not even be necessary to collect data since the control panel on the receiver can often perform this function. However, a survey controller is recommended since ancillary data such as sky plots, DOP values, antenna height, and so forth can be observed or entered during the survey. At least once a day, and preferably twice a day, the data should be transferred to a computer. The receiver or controller should not be used as a primary storage device since many manufacturers will automatically discard the older files when storage becomes limited. Downloading of the data can be done in the field with a computer, at the end of the day in the

Station Data Sheet

Station id: _____   Receiver s/n: _____

Location _____   City _____

PDOP Start _____ Midsession _____ End _____   **Time:** Start ___:___ End ___:___

Method of support: ☐ Fixed-height tripod ☐ Supported 2-m rod ☐ Tripod

Antenna height _____ ☐ ft ☐ m   **Type** ☐ Vertical ☐ Slant

Station Monument Inscription and Description _____

_____   _____
_____   _____
_____   _____

Visibility at start of session   Site Sketch (Image)
                                       Image number: _____

Approximate Position

Latitude _____ Longitude _____ Height _____ ☐ ft ☐ m

To Reach the Station _____

_____
_____
_____

Comments _____

_____   _____
_____   _____

By _____ Signature _____ Date ___/___/_____

Figure 14.9
Site log sheet for a static or rapid static survey.

office, or during the day with a data modem when cell coverage is available. File transfer is accomplished using software provided by the manufacturer or through the operating system of the controller. Some manufacturers now provide cloud service so that files can be immediately exchanged with office personnel.

14.5 DATA PROCESSING AND ANALYSIS

The first step in processing the data is to transfer the observation files from the receiver to the computer. A separate file is generally created for each session

Figure 14.10 Screen capture of GNSS baseline vector processing software. (Global Navigation Satellite System Software Defined Receiver http://gnss-sdr.org/ GNSS SDR.)

performed with the receiver. Typically, software provided by the manufacturer performs this function; however, it can also be performed using file transfer procedures in the operating system. In this process, the software will download all of the observation files into one subdirectory on the computer. This directory is part of the project's directory. As the observation files are downloaded, special attention should be given to checking station information that is read directly from the file with the site log sheets. Catching incorrectly entered items such as station identification, antenna heights, and antenna types at this point can greatly reduce later problems during processing. Files are batch processed typically during the reduction process. Figure 14.10 shows the postprocessing screen where the GNSS occupation data and observations are displayed at the bottom.

There are three types of processing software: (1) *single baseline*, (2) *session processing*, and (3) *multipoint solutions*. The single baseline solution is used most commonly and will be discussed here. The session processing software simultaneously computes all baseline vectors for any particular session. The user must then disable all trivial baselines from the reduced observations for each session. Typically, this step is best accomplished by creating an observation plan before the survey is started. The reduced observations can then be sorted by date and time of observation. This groups each session together so that the user can then easily disable the trivial baseline vectors for each session. It should be stated here that this step is not necessary if only two receivers are used during the observation sessions since no trivial baselines are generated in this configuration. However, most specifications for control surveys require three or more receivers, and thus according to Equation (14.1c) n number of trivial baseline vectors will be created for each observation session. Failure to remove trivial baseline vectors from the processing will result in false higher precisions.

Some users will disable baseline vectors based on the statistics of the vector. However, this method should be used with the understanding that often baselines that have poorer statistical values are typically also longer in length. Being longer in length also means that the baseline vector is expected to have higher uncertainties. Other reasons that can cause higher uncertainties in reduced baseline vectors are loss of lock problems at a particular station caused by obstructions, multipathing, or poor space weather conditions. In this situation, baselines connected to this station may have higher uncertainties than other baselines in the session. Of course, this condition is dependent on the length of the observed baseline vectors. However, this situation can be verified by observing the timeline for a session. Timelines that show discontinuity indicate loss of lock problems. If images are available for the setup, they may show the problem at the station. However, when no images are available, the station's geodetic coordinates can be entered into software, such as Google Earth® to view the setup conditions. Of course, in the latter case, it should be realized that the site conditions may have changed since the images were taken, and thus a site visit may be warranted.

When loss of lock problems are caused by obstructions, one technique that can be used is to remove portions of the satellite observations from the software's timeline. This is accomplished by viewing the timeline and disabling small portions of observations typically when the satellite is low in altitude. Since in a well-planned survey, obstructions typically affect only one or two satellites during the session, the data affected by the obstruction can be disabled. However, when these satellites are critical to the satellite geometry at the station, the baselines observed at this station may need to be reobserved at a time when the obstruction does not interfere with the satellite signals. However, this speaks to collecting more observations when occupying a site; the time and money lost in reobserving a session can often be avoided with a few more minutes of data collection in the field in the first place. Since drive, setup, and tear down times are typically longer than actual observation times, it is often wise to collect more than the minimum amount of data in any survey.

The initial processing can use the "broadcast" ephemeris, but it is recommended to use any of the "precise" ephemerides to achieve a higher level of accuracy since a precise ephemeris will remove most of the orbital errors from the processing (see Section 13.6.3). When NCN sites are used for a project, the most precise ephemeris available at the time of the download can be requested. Proper use of the "user-friendly" option, which is available on the NGS website, will automatically result in interpolated values for the data at the specified epoch rate to match the survey. Additionally, as mentioned in Section 13.6.3, NGS antenna calibration data should be used when determining the baseline vectors to account for changing vertical antenna offsets due to changes in satellite elevation and azimuth. Since the predicted ultra-rapid ephemeris is available within six hours, it is seldom necessary to perform this processing without a precise ephemeris. However, a broadcast ephemeris must still be loaded in the software when a precise ephemeris is used in the processing.

As shown in Figures 13.12 and 14.10, the baseline is computed as changes in X, Y, and Z between two stations. Thus, if station A has known coordinates, then the coordinates of station B can be computed using Equation (13.28). The single baseline processing software will (1) generate orbit files, (2) compute the best-fit point positions from the code pseudoranges, (3) compute baseline components (ΔX, ΔY, ΔZ) using the double difference Equation (13.32), and (4) compute statistical information for the baseline components.

After all baselines have been computed, their geometric closures can be analyzed. This step follows a series of procedures that check the data for internal consistency and eliminate possible blunders. No control points are needed for these analyses. Depending on the actual observations taken and the network geometry, these procedures may involve analyzing (1) differences between fixed and measured baseline components, (2) differences between repeated measurements of the same baseline components, and (3) loop closures. Procedures for making these analyses are described in the following subsections.

14.5.1 Specifications for Static Surveys

The *Federal Geodetic Control Subcommittee* (FGCS) has published the document entitled "Geometric Geodetic Accuracy Standards and Specifications for Using GPS Relative Positioning Techniques."[4,5] The document specifies seven different orders of accuracy for relative positioning and provides guidelines on instruments, field procedures, and office procedures to follow to achieve them. Table 14.5 lists these orders of accuracy.

The FGCS document also makes recommendations concerning categories of surveys for which the different orders of accuracy are appropriate. Some of these recommendations include: order AA for global and regional geodynamics and deformation measurements; order A for "primary" networks of the National Spatial Reference System (NSRS) and regional and local geodynamics; order B for "secondary" NSRS networks and high-precision engineering surveys; and the various classes of order C for mapping control surveys, property surveys, and engineering surveys. The allowable error ratios given in these standards imply

TABLE 14.5 GPS RELATIVE POSITIONING ORDERS OF ACCURACY

Order	Allowable Error Ratio	Parts Per Million (ppm)
AA	1:100,000,000	0.01
A	1:10,000,000	0.1
B	1:1,000,000	1.0
C-1	1:100,000	10
C-2-I	1:50,000	20
C-2-II	1:20,000	50
C-3	1:10,000	100

---

[4]Available from the National Geodetic Information Center, NOAA, National Geodetic Survey, N/CG17, SSMC3 Station 09535, 1315 East West Highway, Silver Spring, MD 20910, telephone (301) 713-3242. Their email address is info_center@ngs.noaa.gov, and their website address is http://www.ngs.noaa.gov.

[5]Information on solar activity can be on the Space Weather Prediction Center website at http://www.swpc.noaa.gov.

TABLE 14.6 NGS Geodetic Height Order and Class

Order	Class	Maximum Height Difference Accuracy (b)
First	I	0.5
First	II	0.7
Second	I	1.0
Second	II	1.3
Third	I	2.0
Third	II	3.0
Fourth	I	6.0
Fourth	II	15.0
Fifth	I	30.0
Fifth	II	60.0

the extremely high accuracies that are now possible with relative positioning techniques.

The National Geodetic Survey has created geodetic height accuracy standards that can be used for vertical surveys. These standards are based on the changes in geodetic heights between control stations. Following a correctly weighted, minimally constrained, least-squares baseline adjustment (see Section 14.5.5), the geodetic height order and class can be determined. This standard is based on the standard deviation (s) of the geodetic height difference (in millimeters) between two points obtained from the adjustment, and the distance (d) between the two control points (in kilometers). The ellipsoid height difference accuracy (b) is computed as

$$b = \frac{s}{\sqrt{d}} \quad (14.2)$$

Table 14.6 lists classifications versus value of b computed in Equation (14.2).

14.5.2 Analysis of Fixed Baseline Measurements

As noted earlier, job specifications often require that baseline observations be taken between fixed control stations. The benefit of making these observations is to verify the accuracy of both the observational process and the control being held fixed. Obviously, smaller discrepancies between observed and known baseline lengths mean better precisions. If the discrepancies are too large to be tolerated, the conditions causing them must be investigated before proceeding further. Table 14.7 shows the computed baseline vector data from a survey of the network of Figure 14.6. Note that one fixed baseline (between control points A and B) was observed.

Assuming the geocentric coordinates of station A are (402.3509, −4,652,995.3011, 4,349,760.7775) and those of station B are (8086.0318, −4,642,712.8474, 4,360,439.0833), the analysis of the fixed baseline is as follows.

TABLE 14.7 OBSERVED BASELINE VECTORS FOR FIGURE 14.6

Obs. Baseline	ΔX(m)	ΔY(m)	ΔZ(m)
AC	11644.2232	3601.2165	3399.2550
AE	−5321.7164	3634.0754	3173.6652
BC	3960.5442	−6681.2467	−7279.0148
BD	−11167.6076	−394.5204	−907.9593
DC	15128.1647	−6286.7054	−6371.0583
DE	−1837.7459	−6253.8534	−6596.6697
FA	−1116.4523	−4596.1610	−4355.9062
FC	10527.7852	−994.9377	−956.6246
FE	−6438.1364	−962.0694	−1182.2305
FD	−4600.3787	5291.7785	5414.4311
FB	6567.2311	5686.2926	6322.3917
BF	−6567.2310	−5686.3033	−6322.3807
AF	1116.4577	4596.1553	4355.9141
AB	7683.6883	10282.4550	10678.3008

1. Compute the coordinate differences between stations A and B as

$$\Delta X_{AB} = 8086.0318 - 402.3509 = 7683.6809$$
$$\Delta Y_{AB} = -4{,}642{,}712.8474 + 4{,}652{,}995.3011 = 10{,}282.4537$$
$$\Delta Z_{AB} = 4{,}360{,}439.0833 - 4{,}349{,}760.7775 = 10{,}678.3058$$

2. Compute the absolute values of differences between the observed and fixed baselines as

$$dX = |7683.6883 - 7683.6809| = 0.0074$$
$$dY = |10{,}282.4550 - 10{,}282.4537| = 0.0013$$
$$dZ = |10{,}678.3008 - 10{,}678.3058| = 0.0050$$

3. Using either the control coordinates or the baseline vector components, compute the length of the baseline as

$$AB = \sqrt{(7683.6809)^2 + (10{,}282.4537)^2 + (10{,}678.3058)^2} = 16{,}697.126 \text{ m}$$

4. Express the differences, as computed in Step 2, in parts per million (ppm) by dividing the difference by the length of the baseline computed in Step 3 as

$$\Delta X\text{-ppm} = (0.0074/16{,}697.126)\,1{,}000{,}000 = 0.44$$
$$\Delta Y\text{-ppm} = (0.0013/16{,}697.126)\,1{,}000{,}000 = 0.08$$
$$\Delta Z\text{-ppm} = (0.0050/16{,}697.126)\,1{,}000{,}000 = 0.30$$

5. Check the computed values for the ppm against a known standard. Typically, the FGCS standard given in Section 14.5.1 is used.

Another method that can be used to check any baseline or set of baselines is use the manufacturer's specifications for the receiver and survey methods along with estimated centering errors for the antennas as discussed in Section 6.22. For example, assume that the manufacturer specifications for a baseline using the static or rapid static survey method is 5 mm + 0.5 ppm. Also assume that an 8-minute circular level vial is assumed to be centered within 0.5 of a division of being centered. Using Equation (6.17), the centering error for a receiver-antenna on a 2-m fixed-height tripod, which is assumed to be within ±0.002 m, is approximately ±1.5 mm. Using Equation (3.11), the uncertainty in the observed fixed baseline is

$$\sigma_{AB} = \sqrt{2(1.5)^2 + 5^2 + \left[\frac{0.5(16,697,126)}{1,000,000}\right]^2} = \pm 10.0 \text{ mm}$$

This value would typically be increased in probability from 68% to some higher value such as 95%. Using Equation (3.8), the acceptable misclosure at the 95% probability level would be 1.9599(10.0) or about ±19.6 mm. The combined misclosure of the observation is $\sqrt{7.4^2 + 1.3^2 + 5.0^2}$, which equals 9.0 mm. Thus, the misclosure is well within its specifications and the check on the fixed baseline is assumed to be within tolerances.

14.5.3 Analysis of Repeat Baseline Measurements

Another procedure employed in evaluating the consistency of the observed data and in weeding out blunders is to make repeat observations of certain baselines. In fact, the FGCS guidelines require a minimum of 5% of the nontrivial baselines be repeated in the cardinal directions for each survey. These repeat measurements are taken in different observation sessions and the results compared. Significant differences in repeat baselines indicate problems with field procedures or hardware. For example, in the data of Table 14.7, baselines AF and BF were repeated. Table 14.8 gives comparisons of these observations using the same procedure that was used in Section 14.5.2. Column (1) lists the baseline vector components to be analyzed, columns (2) and (3) provide the repeat baseline vector components, column (4) lists the absolute values of the differences in these two observations,

TABLE 14.8 ANALYSIS OF REPEAT BASELINE OBSERVATIONS

(1) Component	(2) First Observation	(3) Second Observation	(4) Difference	(5) ppm
ΔX_{AF}	1116.4577	1116.4523	0.0054	0.84
ΔY_{AF}	4596.1553	4596.1610	0.0057	0.89
ΔZ_{AF}	4355.9141	4355.9062	0.0079	1.23
ΔX_{BF}	−6567.2310	−6567.2311	0.0001	0.01
ΔY_{BF}	−5686.3033	−5686.2926	0.0107	1.00
ΔZ_{BF}	−6322.3807	−6322.3917	0.0110	1.02

and column (5) lists the computed ppm values that are computed similar to the procedure given in Step 4 of Section 14.5.2.

Similar to fixed baselines, the repeat baselines can be checked against manufacturer specifications and estimated setup errors. For example, a check on repeat baselines AF using manufacturer specifications of 5 mm + 0.5 ppm and assuming a centering error of ±1.5 mm for each receiver would yield an estimated uncertainty for the observation of

$$\sigma_{AF} = \sqrt{2(1.5)^2 + 5^2 + \left[\frac{0.5(6{,}430{,}016)}{1{,}000{,}000}\right]^2} = \pm 6.3\,\text{mm}$$

Again, raising the probability to 95% will yield and estimated uncertainty in the observation of ±12.4 mm. However, a repeat baseline is the result of two observations, both of which contain errors. Thus, this 95% probable error of ±12.4 mm should be doubled to be ±24.8 mm. The actual difference in the two baselines is

$$\sqrt{5.4^2 + 5.7^2 + 7.9^2} = 11.1\,\text{mm}$$

Thus, this repeat baseline is within the errors estimated for the set of observations.

14.5.4 Analysis of Loop Closures

Static surveys consist of many interconnected closed loops typically. For example, in the network of Figure 14.6, points $ACBDEA$ form a closed loop. Similarly, $ACFA$, $CFBC$, $BDFB$, etc., are other closed loops. For each closed loop, the algebraic sum of the ΔX components should equal zero. The same condition should exist for the ΔY and ΔZ components. An unusually large closure within any loop will indicate that either a blunder or a large error exists in one (or more) of the baselines of the loop. It is important not to include any trivial baselines (see Section 14.3.4) in these computations, since they can yield false accuracies for the loop.

To compute loop closures, the baseline components are added algebraically for that loop. For example, the closure in X components for loop $ACBDEA$ is

$$cx = \Delta X_{AC} + \Delta X_{CB} + \Delta X_{BD} + \Delta X_{DE} + \Delta X_{EA} \qquad (14.3)$$

where cx is the loop closure in X coordinates. Similar equations apply for computing closures in Y and Z coordinates. Substituting numerical values into Equation (14.3), the closure in X coordinates for loop $ACBDEA$ is

$$cx = 11{,}644.2232 - 3960.5442 - 11{,}167.6076 - 1837.7459 + 5321.7164$$
$$= 0.0419\,\text{m}$$

Similarly, closures in Y and Z coordinates for that loop are

$$cy = 3601.2165 + 6681.2467 - 394.5204 - 6253.8534 - 3634.0754$$
$$= 0.0140\,\text{m}$$
$$cz = 3399.2550 + 7279.0148 - 907.9593 - 6596.6697 - 3173.6652$$
$$= -0.0244\,\text{m}$$

For evaluation purposes, loop closures are expressed in terms of the ratios of resultant closures to the total loop lengths. They are given in ppm. For any loop, the resultant closure is

$$lc = \sqrt{cx^2 + cy^2 + cz^2} \qquad (14.4)$$

where lc is the length of the misclosure in the loop.

Using the values previously determined for loop $ACBDEA$, the length of the misclosure is 0.0505 m. The total length of a loop is computed by summing its legs, each leg being computed from the square root of the sum of the squares of its observed $\Delta X, \Delta Y,$ and ΔZ components. For loop $ACBDEA$, the total loop length is 50,967 m and the closure ppm ratio is therefore $(0.0505/50,967)\, 1{,}000{,}000 = 0.99$ ppm. Again these ppm ratios can be compared against values given in the FGCS guidelines (Table 14.5) to determine if they are acceptable for the order or accuracy of the survey. As was the case with repeat baseline observations, the FGCS guidelines also specify other criteria that must be met in loop analyses besides the ppm values.

Again, the loop misclosure can be checked against manufacturer specifications and estimated setup errors. With this example, using manufacturer specifications of 5 mm + 0.5 ppm and assuming a centering error of ±1.5 mm for each receiver setup would yield an estimated uncertainty for the observation of

$$\sigma_{ACBDEA} = \sqrt{2(5)(1.5)^2 + 5(5^2) + \left[\frac{0.5(50{,}967{,}000)}{1{,}000{,}000}\right]^2} = \pm 26.0 \text{ mm}$$

Raising the probability to 95% will yield and estimated uncertainty in the observation of ±50.8 mm. Since the actual misclosure was 50.5 mm, which is under its estimated uncertainty at 95%, this loop is deemed as acceptable.

For any network, enough loop closures should be computed so that every baseline is included within at least one loop. Typically, all possible triangular loops are checked by the software at request. This should expose any large blunders that exist. If a blunder does exist, its location often can be determined through additional loop closure analyses. For example, assume that the misclosure of loop $ACDEA$ reveals the presence of a blunder. By also computing the closures of loops $AFCA, CFDC, DFED,$ and $EFAE$, the exact baseline containing the blunder can be detected. In this example, if a large misclosure was found in loop $DFED$ and all other loops appeared to be blunder free, the blunder would be in line DE, because that leg was also common to loop $ACDEA$, which contained a blunder as well. Note that in this analysis that vectors DF and FE in $DFED$ were checked in other loops and found to be acceptable. In most commercial software, the checking of loop closures is often automated and sometimes requires nothing more than the graphical selection of the baselines to analyze.

14.5.5 Baseline Network Adjustment

After the individual baselines are computed, a least-squares adjustment (see Section 16.8) of the observations is performed. This adjustment software is available from the receiver manufacturer and will provide final station coordinate values and their estimated uncertainties. If more than two receivers are used in a survey, trivial baselines will be computed during the single baseline reduction. These trivial baselines should be disabled or removed before the final network adjustment.

The observations used in the baseline network adjustment should be part of an interconnected network of baselines. Initially, a minimally constrained adjustment should be performed (see Section 16.11). The adjustment results should be analyzed both for mistakes and large errors. As an example, antenna height mistakes, which are not noticeable during a single baseline reduction, will be noticeable after the network adjustment. After the results of a minimally constrained adjustment are accepted, a fully constrained adjustment should be performed. During a fully constrained adjustment, all available control is added to the adjustment. At this time, any scaling problems between the control and the observations will become apparent by the appearance of overly large residuals. It is important not to adjust any check baselines between the control at this step since they will result in an indeterminate solution. Problems that are identified should be corrected and removed before the results are accepted.

Since these computations are performed in a geocentric coordinate system, the final adjusted values can be transformed into a geodetic coordinate system using procedures as outlined in Section 13.4.3, or into a plane coordinate system (see Chapter 20). These transformations typically occur by request in the software. As shown in Figure 14.10, the baseline vector components are given in (dN, dE, dHt) rather than in geocentric coordinate values at the user's request. Recall that geodetic elevations are measured from the ellipsoid and thus, as discussed in Section 13.4.3, the geoid height must be applied to these heights to derive orthometric elevations. In the United States and its territories, the GEOID18 software can be used to determine the geoid height. However, the choice of the geoid model should match the version of the North American datum that is being used for the geodetic coordinates. (see Section 19.6.5) This model is included in the software typically, and the user need only load the appropriate data file from the NGS for their region. Thus, the dHt value shown in Figure 14.10 may be a difference in orthometric heights or geodetic heights depending on whether an appropriate geoid model has been loaded into the software. Additionally, as can be seen from Equation (13.8), the accuracy of the orthometric height is not only dependent on the accuracy of the survey but also the accuracy of the geoid height model that is available.

Finally, the horizontal and vertical accuracy of the survey can be determined based on the FGCS or NGS horizontal and vertical accuracy standards (see Section 14.5.1).

14.5.6 The Survey Report

A final survey report is helpful in documenting the project for future analysis. At a minimum, the report should contain the following items.

1. The location of the survey and a description of the project area. An area map is recommended. This map can often be obtained using Internet mapping software.
2. The purpose of the survey and its intended specifications.
3. A description of the monumentation used including the tie sheets and photos or rubbings of the monuments.

4. A thorough description of the equipment used, including the serial numbers, antenna offsets, and the date the equipment was last calibrated.
5. A thorough description of the software used, including name and version number.
6. A listing for the datum of control coordinates such as NAD 83(2011), epoch 2005.0, and the time period of the survey. Recalling that the crustal plates are in motion, this provides a method to transform these coordinates into earlier or later reference frames. Additionally, it may help avoid users from combining coordinates from different reference frames without this transformation. This time-stamp data is known as metadata.
7. The observation scheme used, including the itinerary, the names of the field crew personnel, and any problems that were experienced during the observation phase.
8. The computation scheme used to analyze the observations and the results of this analysis.
9. A list of the problems encountered in performing the survey, or its analysis including high solar activity, potential multipathing problems, or other factors that can affect the results of the survey.
10. An appendix containing all written documentation, original observations, and analysis. Since the processing of the survey can produce volumes of printed material for a typical survey, only the most important files should be printed. However, all computer files should be copied onto some safe backup storage. An external hard disk drive (HDD) or cloud service provides excellent storage media that can be inserted into the back of the report for future reference.

■ 14.6 THINGS TO CONSIDER

With the introduction of GNSS to surveying, it is the first time that geodetic concepts and principles have been relevant to the traditional surveying community, which specializes in boundary and construction surveying. For example, while coordinate system reference frames have traditionally been thought of as absolute, it is now common to list velocity vectors with the coordinate values due to crustal plate movements. Surveyors must now be aware of the reference frame of the coordinates used in an adjustment, types of distances and azimuths displayed in a report, and quality of the adjustment. It is important that surveyors understand all aspects of the adjustment report. In Chapter 16, the process of least squares adjustments is discussed. An introduction to coordinate reference frames and the various types of distances and directions that can be computed between two points from baseline vectors are presented in this section and discussed further in Chapter 19.

As discussed in Section 13.4, the satellites in a GNSS survey broadcast their positions using orbital parameters. These positions are then converted from orbital coordinates to geocentric coordinates, which is the primary coordinate system used in GNSS computations. The results of the adjustment are then converted to the geodetic coordinate system as discussed in Section 13.4.3 providing geodetic latitude, longitude, and height for the user. Using a broadcast ephemeris, the geodetic coordinates will be in the reference frame of the broadcast ephemeris, which is the current WGS84 datum. If coordinates are needed in some local reference system such as one of the NAD83 reference systems discussed in Section 19.6, these positions can be converted using the localization procedures discussed in Sections 15.9 and 19.7, or by providing local reference system coordinates for the control in the network adjustment, which was discussed in Section 14.5.5. These coordinates can also be brought into the project using online processing services such as those listed in Table 14.2. If orthometric heights for the stations are desired, a geoid model can be applied to the geodetic heights to create orthometric heights, which are commonly used in leveling surveys. However, due to the limitations of GNSS and accuracy of the geoid models, these orthometric heights will seldom rival those obtained by traditional leveling.

It is important to recognize that single point solutions should only be used to establish a recognized national or global coordinate system for the project. The remaining receiver positions should be established during postprocessing as discussed in Section 14.5. The National Geodetic Survey realized that their products such as OPUS and OPUS-RS were being misused to establish the coordinates for all stations in an entire project. That is, users were not using the service to bring datum coordinates into the project, and then processing the remaining locations using network processing software but rather relying on independently determined OPUS solutions for all their stations in their project. To alleviate this problem, the National Geodetic Survey created OPUS Projects, which defines the project coordinate system and then adjusts the network baselines based on the base receiver's determined position.

One should never rely on the online services listed in Table 14.2, excluding OPUS Projects, to determine coordinates for all surveyed locations in their projects. Rather they should use these services to bring in either national or international coordinates into their project for their base receiver and then process the data as described in Section 14.5. Additionally, it is important to evaluate the reference frame of the coordinates determined by the online service. For example, most of the online services will provide international coordinates such as ITRFXX for their derived positions. If the project is to be in a national coordinate system such as NAD83 (2011), then the international coordinates will need to be transferred into the national system. In the United States, the National Geodetic Survey has created the Horizontal Time-Dependent Positioning (HTDP) service to perform this task. Other such transformation services exist where needed.

Sometimes surveyors wish to use GNSS surveys to establish control in an area where none exist so that coordinates in some recognized coordinate system can be provided to optical surveys. In these cases, surveyors need to understand the results of a GNSS adjustment, its accuracies, and implications on the optical survey. The results of the GNSS survey will often list the length and direction of a course between the endpoints of a baseline vector. It is here where the surveyor must use caution. The length and direction of the GNSS course are not readily reproducible by optical survey methods. For example, as discussed in Section 19.14, the length of course may be a geodetic distance, which is the arc distance between the two points on the ellipsoid, or it could be distance between the two marks on the surface of the Earth, which is known as the mark-to-mark distance of the line. While the mark-to-mark distance will be close to a slope distance measured at the height of the instrument and reflector by a optical instrument, as shown in Section 19.14 they are not the same. Additionally, if a map projection system, such as the 1983 state plane coordinate system discussed in Chapter 20, is provided in the GNSS software, the length provided may be a map projection or grid length. As discussed in Section 20.8, this grid length can be converted to a horizontal distance, which should closely match the optical survey length within the accuracy of the GNSS and the optical survey.

As covered in Chapter 7, the directions of a course provided by a GNSS survey can be either geodetic or grid. The relationship between a geodetic azimuth and grid azimuth are discussed in Section 20.8. If a localization is performed, it could also be in the local reference system, which could be arbitrary. If two points are established using a GNSS survey that are to provide direction for a optical survey, then the accuracy of a GNSS survey must also be considered. For example, the horizontal accuracies of a GNSS survey are often in the range of 1 to 2 cm, which is also dependent on external conditions that are out of the control of the surveyor. As discussed in Section 13.6, the accuracy of a GNSS survey is dependent on things such as the geometry of the satellites at the time of the survey, the solar activity at the time of the survey, multipathing of signals, and so on. Thus, the accuracy of the direction of the line is dependent on the accuracy of the coordinates at both endpoints of the line as well as the length of the line. As can be seen in Table 4.9, the accuracy of the azimuth of a line established by a GNSS survey can have large uncertainties when the length of the line is short. This accuracy has implications in map making and boundary retracement as discussed in Chapter 21. Figure 14.11 shows the accuracy of the azimuth of a line determined by a GNSS survey versus the length of the line. The solid line depicts the accuracy of the azimuth based on length of the line when the accuracy of the GNSS survey is ±1 cm. The dashed line shows the accuracy of the azimuth when the GNSS accuracy is ±2 cm. From this, it is obvious to see that while the relocation of a point established by a GNSS survey can be well within 0.1 ft, the azimuth determined from these coordinates can have large discrepancies from its true direction. Thus,

TABLE 14.9 ACCURACY OF A GNSS BASELINE AZIMUTH

Length of Line (m)	±1 cm	±2 cm
100	±41.3"	±82.5"
200	±20.6"	±41.3"
300	±13.8"	±27.5"
400	±10.3"	±20.6"
500	±8.3"	±16.5"
600	±6.9"	±13.8"
700	±5.9"	±11.8"
900	±4.6"	±9.2"
1300	±3.2"	±6.3"
1900	±2.2"	±4.3"
3700	±1.1"	±2.2"
7700	±0.5"	±1.1"

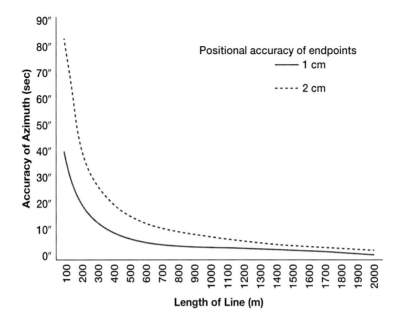

Figure 14.11 Accuracy of GNSS baseline vector azimuths versus length of line.

the surveyor must first determine the type of azimuth that is reported from the GNSS software, and then establish a baseline that has sufficient length to provide the desired accuracy for the survey.

When pulling adjusted data from a GNSS output, care must be taken. As an example, the adjustment file may list both azimuths and lengths of lines. The surveyor should explore the type of azimuth and distance that the adjustment is listing. The azimuth listed after an adjustment is typically a geodetic azimuth and the distance a mark-to-mark distance. (see Section 19.15) The type of azimuth and distance listed in the GNSS adjustment results can be established by performing a geodetic inverse on a pair of adjusted geodetic coordinates (see Section 19.16) or by using software such as that available from the National Geodetic Survey. For example, the Inverse3D software in the National Geodetic Survey Geodetic Tool Kit can be used to obtain geodetic and mark-to-mark distances of a line as well as the geodetic azimuth of the line. Simple inspection of the results listed in the GNSS adjustment results will identify if the azimuth and distance listed are of these types. If the azimuth of a line in the adjustment results is to be used to establish the direction of a line in a horizontal survey using state plane coordinates, it is important to ensure that the azimuth is a grid azimuth or that corrections are made to the geodetic azimuth. (see Sections 20.8 and 20.9)

14.7 A METHOD FOR OBTAINING ORTHOMETRIC HEIGHT DIFFERENCES USING GNSS

As has been previously stated, GNSS surveys result in geodetic heights and height differences. However, most projects require orthometric heights. In the past, the only method available to obtain reliable orthometric heights in a project area was through the use of differential leveling as discussed in Chapter 5. If the control benchmarks are not close to the project, differential leveling can be time and cost intensive. However, it is possible to obtain orthometric heights with accuracies at the 2-cm (0.07-ft) level using the static survey method. This requires the observation of a baseline where one end of the baseline is at a monumented benchmark with known orthometric height. Additionally, it is recommended that the project area be surrounded by four valid control benchmarks that are within 20 km of the project bench mark from which baselines are observed. Using a geoid model, the difference in orthometric heights between the control benchmarks and the project benchmark can be computed as

$$\Delta H = \Delta h - \Delta N \qquad (14.5)$$

where ΔH is the difference in orthometric height between the endpoints of the baselines, Δh the difference in geodetic height between the endpoints of the baselines, and ΔN the difference in the modeled geoid heights of the endpoints of the baselines. When this process is performed from multiple benchmarks surrounding the project, these varying orthometric heights for the project benchmark can be averaged or a least squares adjustment as discussed in Section 3.21 and illustrated in Section 16.6 can be performed to obtain an accuracy estimate of the project bench mark. It should be noted that, like differential leveling, using a GNSS survey still does not equal the accuracies obtainable from a carefully planned and executed precise-leveling survey as discussed in Section 19.14 due to the geometry of the intersection of the pseudoranges from the GNSS satellites. However, when the GRAV-D program is completed the goal is to have resulting gravimetric geoid model accurate to within 1–2 cm and, thus, accurate orthometric heights will be obtainable using a GNSS survey without the need for the control benchmarks by simply obtaining an accurate geodetic (ellipsoid) heights using a GNSS receiver.

■ Example 14.1

The nearest benchmark with an elevation of 235.642 m to a project is 10 km away. A baseline is observed using the static method and a geodetic height difference of 4.349 m is observed. The modeled geoid height at the benchmark is −31.672 m and at the project bench mark is −31.638 m. What is the orthometric height of the project benchmark?

Solution

Using Equation (14.5), the difference in orthometric height is

$$\Delta H = 4.349 - [-31.638 - (-31.672)] = 4.315 m$$

Thus, the orthometric height of the project benchmark is 235.642 + 4.315 = 239.957 m.

The National Geodetic Survey is moving to an active coordinate system with the release of the new four-dimensional datum in 2022. In this active system, benchmark elevations maybe determined from GNSS observations. The GNSS-derived orthometric heights and differential leveling elevation differences between the benchmarks will then be simultaneously adjusted (see Chapter 16). In this case, the benchmark elevations will be adjusted and not remain fixed as is typically done when working with passive monumentation. In order to achieve the best geodetic heights from GNSS observations and thus the best orthometric heights, it is important that some satellites be near the zenith of the receiver. In essence, at the time of the survey for the project benchmarks, there should be good VDOP (see Section 14.3.1).

14.8 SOURCES OF ERRORS IN SATELLITE SURVEYS

As is the case in any project, observations are subject to instrumental, natural, and personal errors. These are summarized in the following subsections.

14.8.1 Instrumental Errors

Clock Bias As mentioned in Section 13.6.1, both the receiver and satellite clocks are subject to errors. They can be mathematically removed using differencing techniques for all forms of relative positioning.

Setup Errors As with all work involving tripods, the equipment must be in good adjustment (see Section 8.19.2). Careful attention should be paid to maintaining tripods that provide solid setups and tribrachs with optical plummets that will center the antennas over the monuments. In GNSS work, tribrach adapters are often used that allow the rotation of the antenna without removing it from the tribrach. If these adapters are used, they should be inspected for looseness or "play" on a regular basis. Due to the many possible errors that can occur when using a standard tripod, special fixed-height tripods and rods should be used. The fixed-height rods can be set up using either a bipod or tripod with a rod on the point. They typically are set to a height of precisely two meters from the *antenna reference point* (ARP). These distances and the calibration of the circular level vials should be checked on a regular basis.

Nonparallelism of the Antennas Pseudoranges are observed from the phase center of the satellite antenna to the phase center of the receiver antenna. Whereas current antennas have their electronic centers coincident with the horizontal centers of the ARP, this may not be the case in older GPS antennas. As with EDM observations, the phase center of the antenna may not be the geometric center of the antenna. Each antenna must be calibrated to determine the phase center offsets for various frequencies. Antennas with horizontal phase center offsets should be aligned in the same direction. Generally, they are aligned according to local magnetic north using a compass.

Receiver Noise When working properly, the electronics of the receiver will operate within a specified tolerance. Within this tolerance, small variations occur in the generation and processing of the signals that can eventually translate into errors in the pseudorange and carrier-phase observations. Since these errors are not predictable, they are considered as part of the random errors in the system. However, periodic calibration checks and tests of receiver electronics should be made to verify that they are working within acceptable tolerances.

14.8.2 Natural Errors

Refraction Refraction due to the transit of the signal through the atmosphere plays a crucial role in delaying the signal from the satellites. The size of the error can vary from 0 to 7 m. Multifrequency receivers can mathematically model and remove this error using Equations (13.18) and (13.19). With single-frequency receivers, this error must be modeled. For surveys involving small areas using relative positioning methods, most of this error will be removed by differencing. Since high solar activity affects the amount of refraction in the ionosphere, it is best to avoid these periods. The Space Weather Prediction Center (Section 13.6.2) provides both current and predicted forecast on solar activity.

Relativity GNSS satellites orbit the Earth in approximately 12 h. The speed of the satellites causes their atomic clocks to slow down according to the theories of relativity. The master control station computes corrections for relativity and applies these to the satellite clocks.

Multipathing As discussed in Section 13.6.3, multipathing occurs when the signal emitted by the satellite arrives at the receiver after following more than one path. It is generally caused by reflective surfaces near the receiver. As discussed in Section 13.6.3, multipathing can become so great that it will cause the receiver to lose lock on the signal. Many manufacturers use signal filters to reduce the problems of multipathing. However, these filters will not eliminate all occurrences of multipathing and are susceptible to signals that have been reflected an even number of times. Thus, the best approach to reducing this problem is to avoid setups near reflective surfaces. Reflective surfaces include flat surfaces such as the sides of building, vehicles, water, chain link fences, and so on.

14.8.3 Personal Errors

Centering Error This error will directly affect the final accuracy of the coordinates. To minimize it, check the setup carefully before data collection begins, and again after it is completed. However, even with careful setups, centering errors can be significant due to the height of the receiver and the sensitivity of the circular bubble on the rod. For example, assume that a fixed-height tripod with an 8' circular bubble is used in a survey. If the bubble is off center by 0.1 of a division on a 2-m setup, the horizontal centering error determined using Equation (6.19) will be ±0.3 mm. However, if a fixed-height rod, such as that shown in Figure 14.2, is used, which typically has a 40' circular bubble, the same 0.1 of a division centering error results in a horizontal centering error of ±1.5 mm. Since a baseline involves two setups, one at each end of the line, this error occurs twice for each baseline observed. If the tip of the rod is placed in a well-defined monumented horizontal control point such as that shown in Figure 19.8, which has a depression with a 2-mm width, an additional 1-mm centering error can occur at each end of the line. Using Equation (3.11), the estimated centering error when using a fixed-height tripod is ±1.1 mm. However, for a fixed-height rod, the centering error will be ±1.8 mm. Thus, based on manufacturer specifications of 3 mm + 0.5 ppm for a 400-m baseline observed using the static method, the overall estimated error in the length of the baseline when using fixed-height tripods is computed as

$$E_{Sum} = \sqrt{2(0.3^2 + 1^2) + 3^2 + \left(\frac{400,000}{1,000,000}0.5\right)^2} = \pm 3.3 \text{ mm}$$

This error estimate is at 68%. Using Equation (3.8), we find the 95% error estimate for the length of the baseline to be

$$E_{95\%} = 1.9599(\pm 3.3) = \pm 6.6 \text{ mm}$$

Similarly, the estimated errors for a baseline observed using fixed-height rods is computed as ±3.7 mm at 68% and ±7.2 mm at 95%. Thus, it can be seen that to obtain the highest accuracies when performing a GNSS survey, it is important to use a fixed-height tripod with a calibrated 8′-circular level bubble since any calibration error in the level bubbles will result in even larger errors in the baseline length.

14.9 MISTAKES IN SATELITE SURVEYS

A few of the more common mistakes in GNSS work are listed here.

Misreading of the Antenna Height The height from the ground to the antenna ground plane or reference point should be read at several times. When measuring to a ground plane, it should be measured at several locations around the ground plane and the average recorded. To ensure that the tripod hasn't settled during the observation process and that the initial readings were correct, the slant height should be also measured at the end of the observation process. To avoid this problem, only fixed-height tripods or rods should be used in the most precise surveys. However, failure to properly lock these instruments into position can also result in incorrect heights. Additionally, field crews sometimes mistype or forget to change the height of the antenna into the survey controller. Another common mistake is to enter 2 ft when in reality it is 2 m. A separate site log sheet provides verification of the antenna height.

Incorrect Identification of the Station This mistake can cause hours of wasted time in processing of the data and will sometimes require that the survey be repeated. To limit this possibility, each station should be located from at least four readily visible permanent objects. Also, if possible, rubbings or photos of the monument caps should be obtained and photos taken of the area showing the location of the monument. During the observation phase of the survey, a second rubbing or photo of the monument and a photo of the setup showing the surrounding area should be taken. This data should be correlated before baseline processing to ensure that the correct names and monuments were used in the survey.

Processing of Trivial Baselines This mistake can only occur when more than two receivers are used in a survey. While this mistake will not generate false coordinates, it will generate false accuracies for the survey. Care should be taken to remove all trivial baselines before a network adjustment is attempted.

Misidentification of Antennas Since each antenna type has different phase center offsets, misidentifying an antenna will directly result in an error in the derived pseudorange. Using antennas from only one manufacturer can reduce this mistake or by correctly identifying and entering phase center offsets into the processing software.

PROBLEMS

Asterisks (*) indicate problems that have partial answers given in Appendix G.

14.1 For a 25-km baseline using a dual-frequency receiver, **(a)** what static surveying method should be used, **(b)** for what time period should the baseline be observed, and **(c)** what epoch rate should be used?

14.2* When using the static surveying method, what is the minimum recommended length of the session required to observe a baseline that is 30-km long with a dual-frequency receiver?

14.3 What would be the recommended epoch rate for the survey given in Problem 14.2?

14.4 Crew B starts their static session 3 min after crew A does. Both collect data for 25 min. What is the actual session time?

14.5 What variables affect the accuracy of a static survey?

14.6 What methods can be used to establish coordinates for the base receiver?

14.7 What site conditions are required for a good GNSS session?

14.8 Why is it recommended to use a precise ephemeris when processing a static survey?

14.9 What are the recommended rates of data collection in a
  (a) static survey?
  ***(b)** rapid static survey?

14.10 What restrictions apply to performing a rapid static survey?

14.11 *How many nontrivial baselines will be observed in one session with three receivers?

14.12 How many trivial baselines will be created in one session with five receivers?

14.13 A site has some overhead obstructions that are over 10° in altitude. What steps should occur in presurvey planning?

14.14 A baseline of approximately 12 mi is to be observed with a GNSS receiver.
  (a) What method of GNSS survey should be used?
  (b) How long should the observing session be?
  (c) What should the epoch rate be set to?

14.15 Describe what a trivial baseline is in a static survey.

14.16 What problems can overhead obstructions cause in a GNSS survey?

14.17 A planned survey site has several potential obstructions near its border. What tools can a surveyor use to find the optimal time for GNSS sessions?

14.18* When using four receivers, how many sessions will it take to independently observe all the baselines of a pentagon?

14.19 What determines the appropriate method to use in a GNSS survey?

14.20 What items should be checked during a field reconnaissance to ensure the site is suitable for a GNSS survey?

14.21 In preplanning it is noticed that for 20 min there is only one satellite in the NW quadrant of the sky plot. An obstruction will block this satellite for five min during this time.
  (a) What concerns should this raise about the survey at this site?
  (b) What can be done to ensure a successful survey at this site?

400 GLOBAL NAVIGATION SATELLITE SYSTEMS—STATIC SURVEYS

14.22 Why should the antenna type and serial number be listed on the site log sheet?

14.23 Why should repeat baselines be performed in a static survey?

14.24* What order of accuracy does a survey with a standard deviation in the geodetic height difference of 15 mm between two control stations that are 5 km apart meet?

14.25 Do Problem 14.24 when the standard deviation in the geodetic height difference is 5 mm for two control points 15 km apart.

14.26 An azimuth for a line with an accuracy of 10" is to be established by using the results of a GNSS survey at the endpoints of the line. What is the minimum length of the line?

14.27 What steps may be taken to identify an obstruction problem during the postprocessing of a GNSS static survey?

14.28 While data is being logged during a control survey, what ancillary data should be recorded on the session's site log sheet?

14.29 A 2-m rod with tripod support is used with a 40' circular vial in a control survey. It is estimated that the bubble is centered in the vial to within 0.3 div. and that height of the rod is within ±3 mm. What is the estimated centering error of the receiver?

14.30 Using loop *CBDF* from Figure 14.6, and the data from Table 14.7, what is the
  (a) misclosure in the X component?
  (b) misclosure in the Y component?
  (c) misclosure in the Z component?
  (d) length of the loop misclosure?
  (e) derived ppm for the loop?

14.31 Do Problem 14.30 with loop *DBFD*.

14.32 Do Problem 14.30 with loop *DEFD*.

14.33 A survey list the standard deviation in geodetic height of ±4.5 mm, which was derived from a baseline that is 5 km long. What NGS geodetic height order and class does this meet?

14.34 The observed baseline vector components in meters between two control stations is (2405.654, −5618.606, 1243.666). The geocentric coordinates of the control stations are (1,162,247.650, −4,882,012.315, 4,182,563.098) and (1,164,653.289, −4,887,630.930, 4,183,806.756). What are:
  ***(a)** ΔX ppm?
  (b) ΔY ppm?
  (c) ΔZ ppm?

14.35 Same as Problem 14.34 except the two control station have coordinates in meters of (1,130,295.165, − 5,498,572.893, 3,018,271.182) and (1,130,753.822, − 5,497,667.568, 3,019,353.825), and the baseline vector between them was (458.651, 905.318, 1082.649).

14.36 Baseline *EA* for Figure 14.6 is resurveyed at a later time as (5321.7200, − 3634.0709, − 3173.6573). What is
  ***(a)** ΔX ppm?
  (b) ΔY ppm?
  (c) ΔZ ppm?

14.37 Repeat Problem 14.36 for baseline *CF* of Figure 14.6, which had vector components of (−10,527.7651, 994.9519, 956.6093).

14.38 During the first three minutes of a session, the observer notices that the number of locked satellites varies between 7 and 8. What is the possible reason for this and how could it be checked in the field?

BIBLIOGRAPHY

Alsalman, A. S. A. 1999. "Evaluating the Accuracy of Differential, Trigonometric, and GPS Leveling." *Surveying and Land Information Systems* 59 (No. 1): 47.

Smith, D., et al. 2015. "Recent Activities at the National Geodetic Survey." *American Surveyor* 12 (No. 1) 18.

Steinberg, G. and G. Even-Tzur 2008. "Official GNSS-derived Vertical Orthometric Height Control Networks." *Surveying and Land Information Science* 68 (No. 1): 29.

Zilkowski, D., J. D'Onofrio, and S. Frakes. 1997. *NOAA Technical Memorandum NOS NGS-58: Guidelines for Establishing GPS-Derived Ellipsoid Heights (Standards: 2 CM and 5 CM) Version 4.3* accessed at http://www.ngs.noaa.gov/PUBS_LIB/NGS-58.html on January 28, 2016.

Zilkowski, D., E. Carlson, and C. Smith. 2008. *NOAA Technical Memorandum NOS NGS 59: Guidelines for Establishing GPS-Derived Orthometric Heights* accessed at http://www.ngs.noaa.gov/PUBS_LIB/NGS592008069FINAL2.pdf on January 28, 2016.

15
Global Navigation Satellite Systems— Kinematic Surveys

■ 15.1 INTRODUCTION

In most areas of surveying, speed and productivity are essential elements to financial success. In satellite surveying, the most productive form of surveying is kinematic surveying. It uses relative positioning techniques with carrier phase-shift observations as discussed in Sections 13.5.2 and 13.8. These surveys can provide immediate values to the coordinates of the points while the receiver is stationary or in motion. Its accuracy at $\pm(8-15 \text{ mm} + 1 \text{ ppm})$ is less than that obtainable with static surveys typically, but is adequate for many surveys. It has applications in several areas of surveying including mapping (Chapter 17), boundary (Chapters 21 and 22), construction (Chapters 23 and 24), and photogrammetry (Chapter 27). This chapter will look at both the post-processed and real-time kinematic surveying methods.

Kinematic surveying can provide immediate results in the field using a *real-time network* (RTN), *real-time kinematic* (RTK) mode, or in the office using the *post-process kinematic* (PPK) mode. Kinematic surveying provides positioning while the receiver is in motion. For example, kinematic surveys have been successfully used in positioning sounding vessels during hydrographic surveys (Section 17.13), mobile mapping units (Section 17.9.4), unmanned aerial vehicles (UAVs), and terrestrial and aerial cameras during photogrammetric surveys (Section 27.16). In large construction projects, it is used in machine guidance and control to guide earthwork operations. It is also used in nonsurveying applications such as high-precision agriculture.

It shares many commonalities with static surveys. For example, a kinematic survey requires two receivers collecting observations simultaneously from a pair of stations with one receiver, the base, occupying a station of known position and another, the rover, collecting data on points of interest. It also uses relative positioning computational procedures similar to those used in static surveys. Thus, it requires that the integer ambiguities (see Section 13.5.2) be resolved before the survey is started. The main difference between static and kinematic surveying techniques is the length of time per session. In a kinematic survey, observations from a single epoch may be all that are used to determine the position of the roving receiver. Establishment of control points using the static surveying method requires much longer sessions than are typically used in kinematic surveys.

As previously stated, the accuracy of kinematic surveys does not match that of static surveys. Some of the limiting factors are the use of the broadcast ephemeris, lack of repeated observations, and the length of the session. For example, a rapid static survey may use a 15-sec epoch rate to sample data over a 1-h session. This results in a total of 240 sets of observations per satellite. Additionally, during this observation session the satellite geometry changes creating different solution geometries. The combination of a large set of observations with varying satellite geometry results in a better solution for the receiver coordinates. In a kinematic survey, the receiver may collect 180 observations per satellite using a 1-sec epoch rate over a 3-min interval for a control point. Not only are the number of observations collected reduced, but since a real-time survey must use the broadcast ephemeris and the satellite geometry does not change significantly, the solution is often weaker than that obtained using static survey methods. Another accuracy degrading factor in RTK surveys includes the motion of the rover during data processing. Since observations from the base receiver must be transmitted, received, and processed at the rover, any motion by the rover during this time will cause errors in its computed position. The errors caused by this time difference, known as *data latency*, tend to be small, and are generally adequate for the lower-accuracy surveys previously cited. However, they can be significant in cases involving fast moving rovers such as in the positioning of a camera station during a photogrammetric mission. Other factors that limit the positioning accuracy of kinematic surveys are its susceptibility to errors such as DOP spikes, atmospheric and ionospheric refraction, multipathing, and obstructions to satellite signals. For example, if a kinematic survey collected data during the time of the PDOP spikes shown in Figure 14.3, the resulting positions during this time would be of considerably lower quality than those collected at other times of the day. Often the effect of these factors can be minimized with careful planning or by advanced processing techniques. An RTN survey mitigates some of these limiting factors by having corrections determined and broadcast to the rover's location using a network of physical base stations. However, most real-time networks are subscription services. Their use is also limited by cellular coverage and can still suffer from high PDOP, multipathing, canopy restrictions, and solar activity.

15.2 PLANNING OF KINEMATIC SURVEYS

All too often kinematic surveys are performed with little or no preplanning. Inevitably, when this is done an occasional survey or part of a survey will produce poor results. This happens due to several reasons including many of those discussed in Sections 13.6 and 14.3. However, kinematic surveys are particularly vulnerable to poor observation conditions due to the relatively low number of observations typically taken at any location and the lack of significant changes in satellite geometry. For example, if a site has canopy problems or is susceptible to multipathing, a weak survey with poor results will occur. However, since kinematic surveys are the most productive, they are used predominately in practice. Thus, the National Geodetic Survey has developed guidelines for performing kinematic control surveys.[1]

---

[1]*National Geodetic Survey User Guidelines for Classical Real Time GNSS Positioning* by William Henning is available at hhttp://www.ngs.noaa.gov/PUBS_LIB/NGSRealTimeUserGuidelines.v2.0.2.pdf.

A typical point located using kinematic surveying methods may have as few as 1 epoch to a few minutes of observational data. Thus, canopy restrictions, solar activity, multipathing, DOP spikes, and many other sources of error can have drastic effects on the determined locations of the receiver. For example, if a kinematic survey had been performed during the periods of DOP spikes shown in Figure 14.3, the resultant coordinates of these points would show larger errors when compared to others measured during periods of low PDOP. While a static survey is generally of sufficient length to "survive" the typical DOP spikes seen today, a kinematic survey is extremely vulnerable to them due to the short observation sessions. Additionally, canopy obstructions can cause high PDOP. Thus, it is important to be aware of the periods of DOP spikes and monitor PDOP while performing a kinematic survey.

In addition to the noise that multipathing can cause in a receiver, it can also cause problems in the resolution of the integer ambiguities. Multipathing is cyclical and can be modeled in the longer sessions typically present in static surveys by the postprocessing software. However, the short duration of the typical kinematic session prevents similar modeling in a kinematic survey. A receiver in multipathing conditions will continue to display precise results even though the opposite is true. Thus, the base station for a kinematic survey should never be placed in a location that is susceptible to multipathing, and the rover should similarly avoid these conditions. Tall buildings, trees, fencing, vehicles, poles, other similar reflective objects should be avoided. These objects can typically be located using offset procedures found in survey controllers.

Except when using real-time networks (see Section 15.8), the software used in kinematic surveys assumes that both base and roving receivers are in the same atmospheric conditions. Thus, baselines should be less than 20 km and surveys should be suspended when the base and roving receiver are not in similar conditions such as when a storm front passes through the project area.

As discussed in Section 13.6.3, refraction caused by the free electrons in the ionosphere and by weather conditions in the troposphere can adversely affect the positioning results in a kinematic survey. In periods of very high solar activity, radio signals from the satellites may be interrupted. During these periods, radio communication between the base and roving receivers in an RTK survey may be compromised. Additionally, solar storms may cause satellites to drift from their predicted paths resulting in larger errors in the broadcast ephemeris. It should be noted that in a post-processed kinematic (PPK) survey as with static surveys, the effects of geomagnetic storms creating ephemeral errors can be alleviated by using one of the available precise ephemerides during processing of the observations. Additionally, in a PPK survey, a radio link between the base and rover is not required, and thus local radio blackouts are not a problem. However, during these periods of high solar activity, it is possible that the signals from the satellites will be disrupted.

Of course, the equipment used in any survey should be calibrated. For example, fixed height poles and tripods should be checked for accuracy by measuring from the tip of the pole to the mounting plate of the receiver, which is also known as the antenna reference point (ARP). Poles should be checked for straightness and legs of tripods checked for tightness. Additionally, as discussed in Section 4.15.3, circular level vials should be regularly checked to ensure that they are in adjustment.

15.3 INITIALIZATION TECHNIQUES

To start a kinematic survey, the receivers must be *initialized*. This process includes determining the integer ambiguity (see Section 13.5.2) for each carrier-phase observation. Following any of the methods described next can yield initialization of the receivers.

One procedure for initializing the receivers uses a baseline whose ΔX, ΔY, and ΔZ components are known. A very short static observing session is conducted with base and roving receivers occupying two stations with known positions simultaneously. Because the baseline coordinate differences are known, differencing of the observations will yield the unknown integer ambiguities. These differencing computations are performed in the processing operation using the data from both receivers. If only one control station is available, a second one can be set using the rapid static surveying method described in Section 14.2.

An alternative initialization procedure, called *antenna swapping*, is also suitable if only one control station is available. Here, receiver A is placed on the control point and receiver B on a nearby, unknown point. For convenience, the unknown point can be within 30 ft (10 m) of the control station. After a few minutes of data collection with both receivers, their positions are interchanged while keeping the receivers running. In the interchange process, care must be exercised to make certain continuous tracking, or "lock" is maintained on at least four satellites. After a few more minutes of observations, the receivers are interchanged again, returning them to their starting positions. This procedure enables the baseline coordinate differences and the integer ambiguities to be determined, again by differencing techniques. However, most survey controller software does not support this method.

Most software today employs an initialization method known as *on-the-fly* (OTF) ambiguity resolution. These OTF methods require five usable satellites during the initialization process and dual-frequency receivers. OTF, which involves the solution of a sophisticated mathematical algorithm, has resolved ambiguities to the centimeter level in 2 min for a 20-km line. The LAMBDA method is the current OTF method used, which typically resolves the ambiguities in less than one minute. This happens while the surveyor is setting up the project in the controller software. However, longer sessions are sometimes necessary to resolve the ambiguities when less than ideal conditions are available.

Prudent surveyors will check the OTF solution for the ambiguities. To do this, a control baseline should be established immediately after the ambiguities have been resolved. This baseline is established using the initial ambiguity solution. The receiver is then "dumped," which is a process of losing lock on all the satellites by inverting the receiver. The OTF method is then allowed to resolve the ambiguities for each satellite again. Once this is accomplished, the baseline established previously is checked. The two solutions should match within the accuracy of the survey, which should be less than 3 cm typically. If the baseline does not check, a second dump of the receiver can be performed to check either of the previous solutions. If none of these attempts produce a suitable check on the baseline, other conditions such as canopy restrictions, multipathing, high solar activity, and so on should be investigated before the survey is attempted. Once satisfied with the solution, the surveyor can use this baseline to reestablish the ambiguities when necessary since the coordinate differences for this line have been established. This process ensures the same solution for the ambiguities during the entire project.

15.4 EQUIPMENT USED IN KINEMATIC SURVEYS

The operator's body can be an obstruction when performing a kinematic survey. Thus, as shown in Figure 15.1, the antenna is often mounted on a fixed-height rod that is 2 m in length to avoid operator obstructions. The base receiver should be mounted on a fixed-height tripod. In any case, the advantages of fixed-height rods and tripods in all GNSS surveys are that they minimize measurement errors in the height of the receiver and help avoid operator-caused obstructions.

Other equipment used in kinematic surveys includes traditional adjustable-height tripods and rods. However, the adjustments in these can often lead to errors in the measured heights to the antenna. Another factor to consider with traditional tripod equipment is the need of a tribrach for mounting of the antenna. As with conventional equipment such as total stations, when tribrachs are used, it is extremely important to check the adjustment of the optical plummets (see Section 8.19.4). Similarly, when using either fixed-height or adjustable rods, it is important to regularly check and adjust the circular level bubbles (see Section 8.19.5).

The sensitivity of the circular vial on a GNSS pole is another item that needs to be considered. Common sensitivities for circular bubbles used on GNSS poles include 8′, 20′, and 40′. If we assume that we can center the bubble to within half a division on a 2-m pole, it can be determined using Equation (6.17) that the

Figure 15.1
The roving receiver being used in a kinematic survey.
(Courtesy Topcon Positioning Systems, Inc.)

centering error caused by the different level vials are ±1.5 mm, ±3.8 mm, and ±7.6 mm, respectively. Obviously, it would be unwise to use a 40′ circular level vial on a job that is meant to establish control coordinates. However, this same pole would more than meet the requirements for a mapping survey.

For horizontal positioning, setup errors due to misleveling of the bubble can be minimized by using lower setups. The amount of setup error can be determined using Equation (6.19). For example, a fixed height tripod set at 2.000 m with a leveling error of 0.5 div on an 8′ circular vial will cause a horizontal positioning error of ±1.5 mm. At a height of 1.800 m, this same leveling error will result in horizontal positioning error of ±1.4 mm, and at a height of 1.500 m, this leveling error will result in a horizontal positioning error of ±1.1 mm. Since all of these errors are under the error achievable by satellite surveys, they are typically ignored. However, in kinematic surveys where the roving pole is often carried in a less careful manner, these errors can be significant. For example, a 2.000-m pole held within 10 min of level will result in a horizontal positioning error of 3.8 mm. This error is at the achievable horizontal precision in a satellite survey, and thus leads to higher positional errors in kinematic surveys when compared to static methods. Thus, many fixed-height tripods and some poles have several set positions for the mounting of the receiver. Fixed length extensions can be added to increase the height of the antenna, however as the height of the receiver increases, the amount of error in horizontal positioning will also increase.

As an aid in mitigating centering errors, some manufacturers have started to incorporate electronic levels in their receivers, which will interrupt the survey when the tilt of the rod is beyond set tolerances. Additionally, some modern GNSS receivers contain *inertial measurement units* (IMU), which allow the systems to mathematically correct for tilt of the unit thereby reducing centering errors caused by leveling errors. This is especially useful in topographic surveys where the surveyor may collect data along linear features while walking and thus finds it difficult to maintain level. However, tilting these units is only meant for mapping surveys. Typical IMUs provide a centering error of ±(10 mm + 0.7 mm per degree of tilt), which is above the accuracies of a static GNSS survey, and thus should not be relied on to mitigate leveling errors in a control survey but are more than adequate for lower-accuracy surveys. Another problem that is caused by tilting the receiver is the removal of some satellites from the antenna's field of view. However, this feature is useful for many kinematic surveys.

An analysis of error in misleveling on the derived height of a point can be determined as

$$e = r - r\cos\alpha \quad (15.1)$$

where e is the error in the height of the receiver, r the height of the receiver, and α is the amount of leveling error in the rod. For example, a 2.000-m pole that is held within 10 min of level will introduce a height error of only 0.009 mm $(2 - 2\cos 10')$. Whereas this error is insignificant in GNSS surveys, the method of carrying the rod during a kinematic survey can introduce vertical errors of 10 cm or more unless an IMU is incorporated with the receiver. However, even though these errors are greater than what is expected from a static survey, they are well below what is typically needed for mapping surveys where it is common to carry the rod off the ground.

In kinematic surveys, overhead obstructions should be avoided at the base and rover stations. Additionally, the base receiver location should be free of reflective objects such as buildings, fences, and vehicles. Manufacturers sell 100-ft cables to avoid these potential multipathing problems.

For most RTK surveys, the external radio antenna at the base receiver is often mounted on a nearby tripod. It is important to have the base radio antenna match the orientation of the antenna on the rover. Mounting the radio antenna high can increase the range of the base radio. However, repeater stations can also be used to extend the range of the base station radio in situations where necessary, as well as avoid obstructions.

Several factors may determine the "best" location for the base station in an RTK survey. Since the range of the radio can be increased with increasing height of the radio antenna, it is advantageous to locate the base station in a high location. When a suitable high location is not available on the project site, an external ultrahigh frequency (UHF) antenna can be used with a low impedance cable that allows the radio antenna to be mounted away from the receiver but well above the ground. As previously stated, the base station should be located in an area that is free of multipathing conditions and obstructions. Additionally, since the base station in an RTK survey requires the most equipment, it is also preferable to place the base station in an easily accessible location. The radios in an RTK survey are low power. Thus, it is wise to avoid sources of high electromagnetic activity such as power grid substations, high-tension power lines, or buildings containing large electric motors since these items generate substantial electromagnetic fields, which can disrupt radio transmissions. Furthermore, the radio signals can interfere with the receiver antenna. Thus, when possible, the radio antenna should be placed a few meters from the GNSS antenna.

Other equipment needed for RTK surveys include a radio and its power source. Typically, an external battery is used for the external base radio. When planning an RTK survey, it is important to provide some backup source of power for the inevitable event of a dead battery. A vehicle can serve as source for charging batteries and providing power for base station radios. Finally, a survey controller is required to control the collection of data from both the base and roving receivers.

■ 15.5 METHODS USED IN KINEMATIC SURVEYS

As the name implies, during kinematic surveys one receiver, the rover, can be in continuous motion. This is the most productive of the satellite survey methods, but also the least accurate. The accuracy of a kinematic survey is typically in the range of $\pm(8 \text{ to } 15 \text{ mm} + 1 \text{ ppm})$ at a 68% probability. This accuracy is sufficient for most types of surveys, and thus is the most common method used. Kinematic methods are applicable for any type of survey that requires many points to be located, which makes it very appropriate for most topographic and construction surveys. It is also excellent for dynamic surveying; that is, where the observation station is in motion. The range of a kinematic survey is typically limited to the broadcast range of the base radio. However, real-time networks have made kinematic surveys possible over large regions using data modems and the Internet.

After initialization has been completed using one of the techniques discussed in Section 15.3, the base receiver remains at the control station while the rover moves to either collect data on features or stakeout locations of points. Although the rover's positions can be determined at intervals as short as 0.2 sec, 1-sec epoch rates are typically used in kinematic surveys. During the survey, both receivers must maintain lock on at least four satellites. If lock is lost, the receivers must be reinitialized. Thus, care must be taken to avoid obstructing the rover's antenna by carrying it close to buildings, beneath trees or bridges, shielding it with the operator's body, and so on. During and at the end of the survey, the rover should be returned to the initial control station, or another, as a check. If less than four satellites are available during the survey, the receiver must reestablish the ambiguities. OTF methods often reset the ambiguities quickly once more than four satellites are visible. As mentioned in Section 15.3, it is often wise at this point to check the solution against a previously surveyed baseline.

Kinematic surveys generally follow two forms of data collection. In *true kinematic* mode, data are collected at a specific rate. This method is useful for collecting points along an alignment, or grade elevations for topographic surveys. An alternative to the true kinematic mode is to stop for several epochs of data at each point of interest, which are averaged. This method, known as *semikinematic* or the *stop-and-go* method, is useful for mapping and construction surveys where increased accuracy is desired for specific features and is always used to establish control. In the semikinematic mode, the antenna is positioned over a point of interest and an identifier, notes, mapping codes (see Section 17.13), etc., are entered into the survey controller. Since multiple epochs of data are typically recorded at each point, the accuracy of this mode is greater than that obtainable in the true kinematic mode. With both methods, the rate of data collection at the base station and rover is typically set to 1 sec.

As discussed in Section 15.2, in kinematic surveys the rover is never at a station long enough to survive a PDOP spike. For kinematic surveys, PDOP values should be less than four. Additionally, since high solar activity can cause significant ionospheric refraction errors, it is important to collect data only during periods of low solar activity. During periods of high solar activity, poor positioning results may be obtained and communications between the base receiver and rover can be disrupted. Both high PDOP values and high solar activity periods are avoided with careful project planning. Additionally, areas with potential multipathing sources should be avoided.

In PPK surveys, the computed coordinates should be stored in the survey controller, and the raw GNSS observations stored in the receiver until the fieldwork is completed. The data are then processed in the office using the same software and processing techniques used in static surveys. Data latency is not a problem in PPK surveys since the data is post-processed. Other advantages of PPK surveys are that (1) the precise ephemeris can be combined with the observational data to remove errors from the broadcast ephemeris and (2) the base station coordinates can be resolved after the fieldwork is complete. Thus, the base station's coordinates do not have to be known prior to the survey. The lack of data latency and use of a precise ephemeris result in PPK surveys having slightly higher accuracies than those obtainable from real-time surveys.

410 GLOBAL NAVIGATION SATELLITE SYSTEMS—KINEMATIC SURVEYS

As discussed in Section 13.8, RTK surveying, as implied by its name, enables positions of points to be determined instantaneously as the rover occupies a point. Like the PPK method, RTK surveying requires that two (or more) receivers be operated simultaneously. The unique aspect of this method is that a radio is used to transmit the base receiver coordinates and its raw observations to the rover. At the rover, the observations from both receivers are processed in real time by the unit's on-board computer to produce a nearly immediate determination of its location according to Equation (13.27). Like PPK surveys, the processing techniques are similar to those used in static surveys.

As shown in Figure 15.2, RTK surveys require compatible hardware at each end of the radio link. Normally, this equipment is purchased from one manufacturer. In North America and in other areas of the world, frequencies in the range of 150–174 MHz in the VHF radio spectrum, and from 450–470 MHz in the UHF radio spectrum can be used for RTK transmissions. Typically, the messages are updated at the rover every 0.5–2 sec. The data link for an RTK receiver requires a minimum of 2400 baud or higher for operation. However, it is typically much higher with baud rates of 38,400 or better.

Even with higher baud rates, the application of the base receiver data to the rover data is delayed because of delays in transmitting the base receiver's observations and position to the rover and the additional time required to compute the rover's position. Typically, the data latency is between 0.05 and 1.0 sec. Data latency can play a significant role in the final accuracies of derived positions.

The radio link used in a RTK survey can limit the distance between the base receiver and rover(s) to a maximum of 10 km, or about 6 mi. This distance can be increased with the use of repeater stations as shown in Figure 15.3. A repeater station receives the signal from a transmitter such as the base radio and retransmits it. Some transmitters require a Federal Communication Commission (FCC) license to broadcast the data. With low-powered radios, line of sight between the transmitter and receiver is required. An advantage of repeater stations is that they can be used to survey around obstructions and increase the range of the base radio. In areas where cellular coverage is available, data modems can also be used to broadcast data from the base receiver to the rover. However, it is important to note that as this distance between the base and rover increases, the error in positioning also increases.

The advantages of RTK surveys over PPK surveys are the reduction in office time and the ability to verify observations in the field. When using RTK, the data

Figure 15.2
A base receiver and rover with compatible internal radios used in RTK surveys.

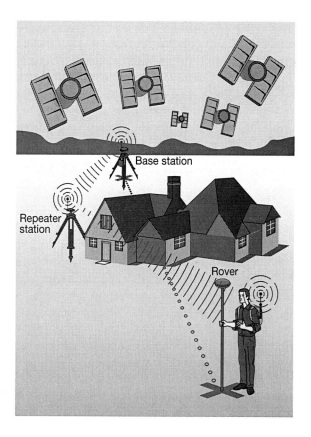

Figure 15.3
Use of a repeater radio to work around obstructions. (Based on Trimble Integrated Technologies.)

can also be downloaded immediately into a GIS (see Chapter 28) or an existing CAD surveying project. This increases the overall productivity of the survey.

15.6 PERFORMING POST-PROCESSED KINEMATIC SURVEYS

A PPK survey requires a base receiver that is collecting data at the same epoch rate as the rover. The base receiver is usually set over a reference station established from a prior survey. If a local NCN station is collecting data at the same rate, it can also be used as a base receiver. For example, if a local NCN station is gathering data at a 1- or 5-sec epoch rate and the rover is also set to a 1- or 5-sec epoch rate, the NCN station's observation files can be downloaded and used to reduce the rover's data. Some NCN stations have an epoch rate of 1 sec for this very purpose.

When no reference station coordinates are available for the base, it can be started using *autonomous* coordinates. These coordinates are simply the code-based solution for the position of the receiver, which is only good to several meters in accuracy. A single epoch[2] of data is all that is required to get the coordinates of the base, which allows the survey to start. Once in the office, the data from the base receiver can be used to correctly determine its position at the time of the survey using static surveying post-processing methods. The NGS provides the *Online*

---

[2]Due to the inaccuracy of the autonomous position of a GNSS receiver, it should never be used in RTK stakeout surveys.

Positioning User Service (OPUS)[3] for this use. Additionally, OPUS will provide the base station coordinates in the local NAD83 coordinate system. To achieve the maximum accuracy from the standard OPUS service, the base station must continuously collect data for a minimum of two hours. During this time, the surveyor can proceed with the PPK survey. If dual-frequency or multifrequency GNSS receivers are used in the survey, the NGS provides a rapid static OPUS (OPUS-RS) service that can determine the position of a receiver with as little as 15 min of data. During postprocessing, the autonomous coordinates for the base can be replaced by the OPUS or post-processed coordinates. The software then moves the rover coordinates to agree with the base station's new coordinates. As shown in Table 14.2, besides OPUS there are also other free online-processing software programs available such as AusPos from Australia, Automatic Precise Positioning Service (APPS) from Jet Propulsion Lab, GNSS Analysis and Positioning Software (GAPS) from the University of New Brunswick, and Canadian Spatial Reference System-Precise Point Positioning (CSRS-PPP) software from Natural Resources Canada. It should be remembered that to achieve the same accuracy using precise point positioning (PPP), the base station's observation file should be longer than three hours. This should not pose as a hindrance for the typical kinematic survey. Session lengths for the relative positioning services can vary to as short as 15 minutes for OPUS-RS. However, the two-hour minimum should be observed when it is practical when using precise point positioning.

PPK surveys are typically used to collect data for mapping surveys. They are especially useful for large surveys with minimal obstructions where the rover can be mounted on a vehicle. As mentioned in Section 15.5, features can be collected using the semikinematic or true kinematic modes. Controller software typically allows the surveyor to switch between these modes as necessary. In areas where canopy obstructions are a problem, the semikinematic mode can be used to establish temporary, higher-accuracy reference points for later occupation by a total station. A minimum of two points is required. One station is the reference station for the total station while the other is its backsight, which establishes the rotational position of its survey. To minimize the rotational uncertainty in the conventional survey, these two stations should be as far apart as the site will allow. Section 14.6 discusses the rotational error in an azimuth derived from a GNSS survey based on the length of the line. As the modernized satellites become available, the problem caused by canopy obstructions could disappear significantly, thus removing the need for a total station to gather data in canopy-restricted areas.

When collecting data using the true kinematic method, it is important to set a reasonable epoch rate based on the velocity of the rover. For example, if the rover is being hand carried, a 1-sec epoch rate would result in data being gathered every five or six feet. This is an excessive amount of data for the typical topographic survey. Furthermore, excessive data collection in lines would result in a weak triangulated irregular network (see Chapter 17). To avoid this, most survey controllers allow their users to set the data collection rate on a specified two- or three-dimensional distance. Section 17.8 discusses the importance of properly collecting data for topographic features. Section 17.13 discusses coding methods that are used to efficiently produce line work on maps.

---

[3]OPUS is available on the NGS website at http://www.ngs.noaa.gov/OPUS/.

As discussed previously, the receivers must be initialized before a kinematic survey can be started. Once this occurs, data collection can proceed as long as lock on four satellites is maintained. Thus, it is important to watch the number of visible satellites, PDOP, and RMS of the solution while surveying. If canopy restrictions obscure satellites that are crucial to an accurate solution, the displayed PDOP and the RMS of the coordinates will increase. In this event, the user needs to proceed to an area where the PDOP is sufficiently low and survey the area with high PDOP later using a total station or by using offset procedures. If the number of satellites drops below four, the rover must be reinitialized. The most common method of reinitialization is accomplished by moving the rover to a location where five satellites are visible. At this location, OTF will quickly reestablish lock on the satellites. OTF can reestablish lock on the satellites in less than 1 min in these situations. However, if this is not achievable, the user can move to a previously surveyed identifiable feature to reestablish lock on the satellites. As discussed earlier, a control baseline established at the start of the survey can be used to facilitate this solution and provide a check on the OTF ambiguity solutions. Since returning to a previously surveyed, identifiable feature can be time consuming, most users try to maintain lock on five or more satellites at all times and avoid situations where loss-of-lock problems can occur.

Since kinematic surveys use a small number of observations to establish the coordinates of points, a PDOP value that is less than four is recommended for most surveys. However, a value as high as six is acceptable for certain types of surveys—a mapping survey for example. The user can also watch the PDOP as the survey proceeds. When weak satellite geometry is present, the PDOP value will rise. No data should be collected if the PDOP value is greater than six. This limit can often be set in the survey controller. A sudden change in the PDOP value is usually caused by an obstruction that has removed a key satellite from the geometric solution of the coordinates. As mentioned previously, when this happens the user should proceed to a location where the PDOP is reduced and continue with the survey.

After the data is collected, it is loaded into the processing software. An advantage PPK surveys have over RTK surveys is that precise ephemeris can be used in the processing. As discussed in Section 13.6.3, this will result in a better solution for the positions of the surveyed points since it removes most ephemeral errors from the solution. The base station coordinates should be established or entered before the rover's observation file is downloaded. If the base station's coordinates are not known, the position of the base receiver should be computed in the processing software or obtained using online software such as OPUS. The base station's coordinates should be resolved and corrected before loading the rover's observation files to ensure that the vectors to the rover will radiate from a base station with the correct coordinates. However, when this is not feasible, entering the base station coordinates later will also shift the rover positions to their correct coordinates. The processing of the baseline vectors to the rover is then performed. Since this is a radial survey, no checks are available on the resultant coordinate values. However, for critical features, it is possible to resurvey these points from a second base station location. In this instance, the previously established control station can be used as the second base station. This procedure is similar to the radial traversing procedure discussed in Section 9.9 and should always be performed if precise control points are to be established during the survey.

As discussed in Section 13.4.3, the heights determined by satellite surveys are in the geodetic coordinate system. Typically, topographic maps are produced using a map coordinate system and orthometric heights. The conversion of geodetic coordinates to map coordinates is covered in Chapter 20. As shown in Equation (13.8), the geoid height at each point must be applied to the observed geodetic height to determine the orthometric height. If requested by the user and a geoid model is available, the software can determine the orthometric height of the points surveyed in real time. The current geoid model for the United States is GEOID18. This model has an accuracy of a few centimeters for most of the United States and its territories. Thus, the derived orthometric heights will be slightly less accurate. The software manufacturer usually supplies support files to upgrade both the controller and software to a specific geoid model.

15.7 COMMUNICATION IN REAL-TIME KINEMATIC SURVEYS

Roving receivers in RTK surveys require continuous communication with base receivers. These communications can be accomplished with radios, wireless Internet connections, or data modems. Using these devices, the base receiver transmits both corrections and raw data to the rover. The rover processes these data using procedures similar to those discussed in Section 13.9.

As shown in Figure 15.1, the most common form of communication between the base receiver and rover is by low-powered, 1-watt internal radios. The Federal Communications Commission (FCC) does not require a license for radios that broadcast in the range from 157–174 MHz, which is the typical frequency range for internal radios. However, all other frequencies given in Section 15.5 do require an FCC license. The external radio transmitters typically use the frequencies in the 450–470 MHz range. These frequencies require an FCC license. Since by FCC regulations voice communication takes precedence over data communication, radio transmitters generally come with as many as 10 or more preset frequencies or channels. The operator must first find a channel that is not in use already. Additionally, using an unlicensed channel is a violation of FCC regulations, which can result in stiff fines. Thus, it is wise to license several of the channels that are available on the transmitter. The maximum power of the radios is typically 35 watts. This form of communication will work in all areas of the world although additional licensing to use the frequencies may be required. When using radios, it is important to connect the antenna to the radio before powering the transmitter to avoid equipment failure.

Another option for communication between the base receiver and rover is data modems. These require cellular coverage in the area being surveyed. When cellular coverage is available, the data are transmitted via cellular technology to the rover. The cell provider charges a monthly service fee to use this option. Obviously, this form of communication is not available in areas that do not have cell coverage. Additionally, data latency with this form of communication will be greater than that experienced with radios.

In areas where wireless Internet connections are available, it is possible for the base receiver and rover to communicate over the Internet. This option requires that the base receiver and rover have Internet connections. Again, data latency will be greater than that experienced with radios using this form of communication.

Several problems can occur with communication equipment. Cables often develop breaks near connectors resulting in intermittent transmission problems. In severe cases, the cables fail and communication is impossible. Also, the power of the radio limits its range. When using receivers with internal radios, the range is often limited to small areas around the base station, which is less than 3 km typically. As discussed in Section 15.3, this range can be increased with repeater stations or by getting the base receiver on a higher location. With larger 35-watt external radios, the achievable range of the survey is maximized, but is generally limited to a maximum of 6 mi (10 km) in radius. Topography and vegetation can further limit this range. Again, longer ranges can be achieved with repeater stations. However, as these distances increase the accuracy of the determined positions will decrease.

■ 15.8 REAL-TIME NETWORKS

A base station requires additional receivers and personnel to perform a survey. If the base receiver could be used as a rover, the work could be performed in half the time. An option that eliminates the need for a base receiver in a RTK survey is known as a *Real-Time Network* (RTN). Both the private and public sectors are implementing this technology. The RTN is a network of base stations that are connected to a central processing computer using the Internet. Using the known positions of the base receivers and their observational data, the central processor models errors in the satellite ephemerides, range errors caused by ionospheric and tropospheric refraction, and the geometric integrity of the network stations. *Virtual reference station* (VRS) and *spatial correction parameter* (FKP)[4] are examples of two methods used in modeling these errors. Of course, these systems may not work reliably in areas that are cellularly challenged. Additionally, the uncertainty in the derived positions when using an RTN will dramatically increase if the survey goes outside the network's bounds.

Since the entire system involves communication from multiple base receivers to a central processor and finally to a rover, high traffic volume on the Internet, multiple connections between network servers to the central processor, and time of transmissions in the cellular world can create greater data latency than much simpler base-to-rover radio connections. Some manufacturers wait for the corrections from the central processor before processing the data at the rover. Others extrapolate the modeled errors to process the rover's observational data at the time of reception. The application software typically stops survey operations if the data latency becomes greater than a specified time interval. This value may be as great as four seconds! For this reason, real-time networks are not recommended in machine guidance and control operations (see Section 15.9).

As shown in Figure 15.4, in real-time kinematic surveys, the accuracy of the position degrades as the rover moves farther from the base station. This is principally due to differences in refraction of the signals between the base and the rover.

---

[4]FKP is an acronym for Flächenkorrekturparameter, which is German for spatial correction parameter.

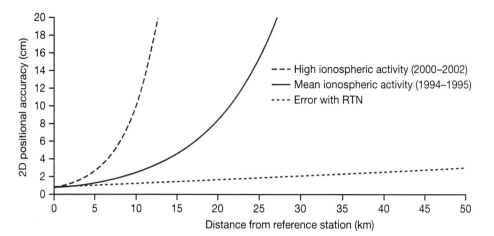

Figure 15.4 Comparison of horizontal errors during different periods of ionospheric activity. (Courtesy National Geodetic Survey.)

Notice that this distance changes with respect to solar activity and its effect on the ionosphere. This is particularly true in the vertical component where errors are traditionally 2 to 3 times greater than horizontal errors. In RTNs these errors are modeled and substantially reduced. However, as shown in Figure 15.4, these errors do increase as the rover moves farther from the network reference station. RTN providers[5] often have maps providing the locations of their base receivers. Thus, it is possible to check the distance from the nearest network station to the project site before the project begins to estimate the errors at the site. The typical accuracy for an RTN survey is $\pm(8 \text{ to } 15 \text{ mm} + 0.5 \text{ ppm})$ at a 68% probability. However, the scalar portion (ppm) of this error is based on the distance to the closest physical base in the network. Thus, the errors in a RTN survey can be larger than those in an RTK survey when the survey is far from any of the network base receivers.

When the rover connects to the RTN, the central processor interpolates the errors listed in Section 13.6 to a location at the survey site. A VRS is created that is used by the rover to determine its position. If the rover moves too far from the VRS, another virtual reference station is determined for the rover. When working with an RTN, the scalar error (ppm) in surveying is reduced resulting in better achievable accuracies generally than are present with a radio and single base receiver. However, this statement depends on the location of the survey in regards to the network base stations. The accuracy of positions determined using RTN is usually within 2 cm (<0.07 ft) anywhere within a distance of 30 km from a reference station. Another advantage of using a RTN is that the coordinates obtained from the network are referenced to a common datum,[6] and thus results from many surveys will fit together seamlessly.

These RTN systems are sold usually as a subscription service. Users of this service save costs since they do not need a base receiver or the additional personnel to monitor the base receiver while performing a survey. The system should

---

[5]Guidelines for real-time networks are available at http://www.ngs.noaa.gov/PUBS_LIB/NGS.RTN.Public.v2.0.pdf.

[6]Section 19.6 discusses the reference frames that are currently used on the North American continent.

be periodically calibrated by locating a known position in the RTN system with the rover. HARN stations can serve as good reference stations. Caution should be exercised when using an RTN outside the bounds of the network since errors increase rapidly when extrapolation of the corrections occurs beyond the limits of the base stations.

■ 15.9 PERFORMING REAL-TIME KINEMATIC SURVEYS

As previously stated, the main difference between RTK surveys and PPK surveys is the fact that RTK surveys provide immediate results in the field. Thus, RTK surveys are used primarily in construction stakeout. Since RTK surveys provide immediate results, some form of communication as discussed in Section 15.6 must be established and maintained during the entire RTK survey. Similar to PPK surveys, the receivers must be initialized before the survey is started and initialization must be maintained during the entire survey. However, the process of surveying is similar to the methods used in a PPK survey.

In fact, since observational data can be saved during an RTK survey, an RTK survey can become a PPK survey if this data is later post-processed. This may be helpful when problems are experienced in the field. However, this would serve no purpose on a stakeout survey. Thus, whether the survey is an RTK or PPK type of survey is often confused in conversation. Simply stated an RTK survey is required when positions are needed immediately in the field such as in a stakeout survey. A PPK survey is performed when locations are determined after postprocessing of the data, which would be appropriate for a mapping survey. The advantage of a PPK survey is that a radio connection between the base and rover is not necessary, which reduces the amount of battery power required to perform the survey. Additionally, a precise ephemeris can be used in the processing of a PPK survey resulting in better positioning since, as shown in Table 13.2, ephemeral errors can be one of the larger error sources in a GNSS survey. However, some manufacturers only support RTK surveys with the option on whether to save the raw observations or not. In this instance, the radio link is always required but not saving the raw observations prohibits post-processing the data later.

Stakeout surveys using RTK have some important distinctions from conventional surveys. One important difference is the reference frame (also called the datum). As discussed in Chapter 19, conventional surveys use some form of NAD 83 as their horizontal datum and NAVD 88 for their vertical datum in the United States. These reference frames are considered to be regional since they were developed using observations only on the North American continent. As discussed in Section 13.4.3, the broadcast ephemeris uses WGS 84, which is a worldwide reference frame. The current rendition of the WGS 84 reference frame (G1762) closely approximates ITRF 2008 at epoch 2005.0. The difference in the origins of the NAD 83 and ITRF 2008 data is about 2.2 m or about 7 ft. Thus, when performing a stakeout survey, the coordinates for stations produced by receivers can differ significantly from coordinates of the same stations in the regional reference frame that were used to perform the engineering design. Additionally, coordinates from surveys performed in priori

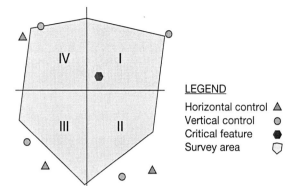

Figure 15.5 Proper locations for control when performing a localization of a GNSS survey.

reference frames will not match broadcast reference system when changes are made by the DoD. Thus, it is important to have the base station on a point with known regional coordinates when performing a stakeout survey. This concern is alleviated when using an RTN. However, if the coordinates used in the engineering design were determined in a previous realization of the regional datum, it will be important to set this reference frame in your survey controller before staking out the design.

As discussed in Section 19.7, the GNSS coordinates can be transformed into the current or a previous regional datum. A two-dimensional conformal coordinate transformation (see Equation 11.37) can be used to transform GNSS-derived horizontal coordinates. To do this, points having regional coordinate values must be established on the perimeter of the project area. A minimum of two horizontal control points and three vertical control points should exist. However, it is better to have four horizontal- and four vertical-control points with one in each quadrant of the survey near its exterior. As shown in Figure 15.5, it is important when performing these transformations to have control on the exterior or surrounding the project area to avoid extrapolation errors. It is also important to include control coordinates on key features. For example, in support of an alignment survey, benchmarks typically found on bridges in the alignment should be included in the transformation since they are often used in the design of the alignment. Additionally, the prudent surveyor will have additional local control, which will serve as checks once the transformation is completed.

Survey controllers have this transformation built into their software. Depending on the vendor, this transformation is known as *localization* or *site calibration* (see Section 19.7.2). This procedure should be performed at the beginning of each project that requires local or arbitrary coordinates. The procedure involves occupying the control stations with the GNSS receiver having the appropriate geoid model loaded into the software. It is important to include the geoid model when occupying control stations since the geoid model represents a systematic error in the vertical, which must be removed from the data. The controller then computes the transformation parameters and allows the user to view the errors. It is wise to perform this procedure whenever questions concerning the stability of the control arise to eliminate possible errors. However, once this transformation is performed on a project and its results accepted, it should not be performed again. With this in mind, an alignment survey should be planned with control points along the entire corridor to ensure their quick availability. Following this procedure, the stakeout of the design points can proceed.

15.10 MACHINE GUIDANCE AND CONTROL

Traditionally construction projects were executed by placing stakes at key locations in the project (see Chapter 23) to establish the levels of materials and grades of finished work. However, with RTK surveying methods, it is possible to load the project design, digital terrain models (see Section 18.14), and site calibration parameters into a computer that guides the vehicle during the construction process. This technology known as *machine guidance and control* allows the machine operator to see their position in a construction project, cut and fill elevations, and finished grades of the project, all in real time. As shown in Figures 15.6 and 15.7, this is achieved by placing RTK units on the construction equipment. One aspect of this technology is that the antenna must be calibrated with respect to the construction vehicle. For instance, the distance between the antenna reference point and the cutting edge of the blade on a machine must be observed and entered into the machine control system so that the height of the surface under the blade is accurately known.

To accurately achieve this level of automation, the surveyor must place sufficient horizontal and vertical control in and around the construction project area. A range of about 10 km (6 mi) from the base station is possible with a high-powered radio transmitter, but as stated earlier this distance is often less due to topography and vegetation. Additionally, in locations where obstructions may interfere with satellite transmissions, the surveyor must add sufficient control to

Figure 15.6 A dozer and grader using machine control to create an intersection of roads. (Courtesy Topcon Positioning Systems, Inc.)

Figure 15.7
GNSS antenna mounted on a grader blade. (Courtesy Topcon Positioning Systems, Inc.).

support the use of a robotic total station. In these areas, a robotic total station can guide the construction equipment past the obstructions. The surveyor must also create a digital terrain model (DTM—see Section 18.14) of the existing terrain prior to the start of the project and a proposed three-dimensional surface model of the finished project. These two items are loaded into the guidance system along with localization parameters. As discussed in Section 15.8, the localization parameters are needed to transform the GNSS coordinates into the project reference system. With an existing DTM, final digital surface, and localization parameters loaded into the guidance system, the operator of the construction vehicle is guided by the system in the project.

The accuracy in using RTK is about 1 cm in horizontal and 1.5 cm in vertical at a 68% probability. This accuracy is sufficient for excavation purposes. However, finished surfaces need to have accuracies under 0.02 ft (5 mm). This accuracy is achieved by augmenting the machine guidance system with laser levels as shown in Figure 23.2 or robotic total stations. One manufacturer has incorporated a laser level into their machine guidance systems to achieve millimeter accuracies in all three dimensions. When using this equipment, sufficient control must be placed at the perimeters of the project area to guide the construction vehicles through the project. For example, robotic total stations have a working radius of about 1000 ft from the construction vehicle and laser levels have a working radius of about 1500 ft from the construction vehicle. Thus, control must be placed within the

Figure 15.8
A large tractor pulling a land leveler. (Courtesy Topcon Positioning Systems, Inc.).

appropriate limits to provide sufficient support for the guidance system. Again, real-time networks should be used with caution in machine control since data latency can be large, which will lead to significant real-time errors during a construction project. In fact, manufacturers do not recommend the use of RTNs in machine guidance and control projects.

As shown in Figure 15.8, another area that is utilizing RTK is precision agriculture. This is an example of surveying technology reaching into nonsurveying fields. This area does not require a surveyor's expertise, but is interesting nonetheless. In precision agriculture, crop yields are monitored with respect to positions of the harvester on the field. Additionally, soil samples are located and tested for fertility, drainage, and so on to provide the farmer with a complete picture of yields versus growing conditions. In the following year, this information is fed into a control and guidance system that controls tillage equipment, planters, sprayers, and fertilizer spreaders so that appropriate tillage and chemicals are used as required for various locations on the field. The end results are an economy in fuel and chemicals that increases crop yields.

■ 15.11 ERRORS IN KINEMATIC SURVEYS

Kinematic surveys suffer from some same error sources that are found in conventional surveys. These include:

a. Setup errors at the base station and rover.
b. Errors in reading the height to the base station or rover antenna.
c. Leveling errors in setting base or holding rover.

Additionally, all satellite surveys have the following errors sources:

a. Ionospheric refraction
b. Tropospheric refraction
c. Errors in ephemerides
d. Base station coordinate errors
e. Weak satellite geometry

15.12 MISTAKES IN KINEMATIC SURVEYS

Some of the more common mistakes that can be made in kinematic surveys include:

a. Misidentification of stations
b. Incorrect identification of the station
c. Starting or proceeding with the survey before the integer ambiguities are resolved
d. Misidentification of the antenna
e. Surveying during high periods of solar activity
f. Incorrect settings for the radio or wireless connection
g. Surveying under overhead obstructions
h. Surveying near reflective surfaces
i. Failure to include a geoid model when performing a localization.

PROBLEMS

Asterisks (*) indicate problems that have partial answers given in Appendix G.

15.1* What are the three types of kinematic surveys?
15.2 List some areas of surveying where GNSS kinematic surveys can be employed.
15.3 What are commonalities between a static survey and a kinematic survey?
15.4 List six limiting factors in regards to the accuracies achievable in kinematic surveys.
15.5 How does an RTN survey mitigate some of a kinematics surveys limiting factors in accuracy?
15.6* How much error in horizontal position occurs if the GNSS antenna is mounted on a 2.000 m pole that is 10 min out of level?
15.7 Do Problem 15.6, but this time assume the level is 30 min out of level.
15.8 How much error in the vertical position occurs with the situation described in Problem 15.6?
15.9 How much error in vertical position occurs with the situation described in Problem 15.7?
15.10* Why should the radio antenna at the base station be mounted as high as possible?
15.11 A storm front passes over the base station, which is 4 km away from the rover. Why should the GNSS survey be stopped until the storm front passes over the rover?
15.12 Why should a control station be established at the start of the kinematic survey?

15.13 Why is a 2-m rod recommended for the roving receiver?
15.14 What are autonomous coordinates?
15.15 The manufacturer-specified accuracy for a PPK survey is ±(5 mm + 1 ppm) at 68% probability. What is the 95% error for a control baseline that is 4 km long, which is observed using the stop-and-go method?
15.16 What OTF method is currently employed to resolve the integer ambiguities?
15.17 Why should receivers with IMUs not be relied on to correct for leveling errors in a control survey?
15.18* What frequencies found in RTK radios require licensure?
15.19 Describe the difference between a true kinematic survey and a semikinematic survey.
15.20 How can an RTK survey become a PPK survey?
15.21 An RTN survey is being performed where the closest physical base station is 35 km away from the project area. It estimated horizontal precision is ±(10 mm + 0.5 ppm). What is the 95% horizontal precision using the stop-and-go method?
15.22 If the estimated horizontal positioning accuracy is ±(1 cm + 1 ppm), what is the 95% horizontal positioning error in an RTK with a base receiver 3.5 km away from the rover?
15.23 What three surveying elements are needed in machine guidance and control?
15.24 What errors are modeled in a VRS?
15.25 Why should a geoid model be included in the localization process?
15.26 What factors may determine the best location for a base station in an RTK survey?
15.27 Why is the use of an RTN discouraged in machine guidance and control applications?
15.28* How many total pseudorange observations will be observed using a 1-sec epoch rate for a total of 3 min with 12 visible satellites?
15.29 How many pseudorange observations will be observed using a 1-sec epoch rate for a total of 5 min with eight visible satellites?
15.30 How can precise point positioning (PPP) be used in a PPK survey where coordinates for the base receiver are unknown?
15.31 Why should periods of PDOP spikes be avoided in a kinematic survey?
15.32 Why should fixed height tripods or rods be used in a kinematic survey?

BIBLIOGRAPHY

Asher, R. 2009. "Crossing the RTK Bridge." *Professional Surveyor* 29 (No. 6): 18.
Barr, M. 2006. "Real-Time Connection." *Point of Beginning* 31 (No. 4): 22.
Crawford, W. 2006. "What Are Your Tolerances?" *Point of Beginning* 32 (No. 3): 46.
Henning, W. 2006. "The New RTK—Changing Techniques for GPS Surveying in the USA." *Surveying and Land Information Science* 66 (No. 2): 107.
Luccio, M. 2014. "Precision Ag: Satellite Signals Steer Farmers Straight." *XYHT* 1 (No. 5): 31.
Mosby, M. 2006. "Advancing with Machine Control." *Point of Beginning* 32 (No. 3): 32.
Pugh, N. 2007. "The Specifics on Managing Network RTK Integrity." *Point of Beginning* 33 (No. 1): 34.
Schrock, G. 2006. "RTN-101: An Introduction to Network Corrected Real-Time GPS/GNSS (Part 1) ." *The American Surveyor* 3 (No. 6): 28.
_____. 2006. "RTN-101: An Introduction to Network Corrected Real-Time GPS/GNSS (Part 2)." *The American Surveyor* 3 (No. 7): 38.

_____. 2006. "RTN-101: An Introduction to Network Corrected Real-Time GPS/GNSS (Part 3)." *The American Surveyor* 3 (No. 8): 38.

_____. 2006. "RTN-101: On-Grid – An Initiative in Support of RTN Development (Part 4)." *The American Surveyor* 3 (No. 9): 39.

Silver, M. 2014. "Sailing with the Pirate Surveyors." *American Surveyor* 11 (No. 1): 26.

Yüksel, A., et al. 2014. "A Control Center for Real-Time GNSS Data." *Professional Surveyor* 34 (No. 2): 20.

16
Adjustments by Least Squares

■ 16.1 INTRODUCTION

The subject of errors in measurements was introduced in Chapter 3 where the two types of errors, systematic and random, were defined. It was noted that systematic errors follow physical laws, and if conditions producing them are observed, corrections can be computed to eliminate them. However, random errors exist in all observed values. Additionally as discussed in Chapter 3, observations can contain mistakes (blunders). Examples of mistakes are setting an instrument on the wrong station, sighting the wrong station, transcription errors in recording observed values, and so on. Mistakes should be removed when possible before the adjustment process. As further explained in Chapter 3, experience has shown that random errors in surveying follow the mathematical laws of probability, and in any group of observations they are expected to conform to the laws of a *normal distribution*, as illustrated in Figure 3.3.

In surveying (geomatics), after eliminating mistakes and making corrections for systematic errors, the presence of the remaining random errors will be evident in the form of misclosures. Examples include sums of interior angles in closed polygons that do not total $(n-2)180°$, misclosures in closed leveling circuits, and traverse misclosures in departures and latitudes. To account for these misclosures, adjustments are applied to produce mathematically perfect geometric conditions. Although various techniques are used, the most rigorous adjustments are made by the method of least squares, which is founded on the laws of probability.

Although the theory of least squares was developed in the late 1700s, because of the lengthy calculations involved, the method was not used commonly prior to the availability of computers. Instead, arbitrary, or "rule of thumb," methods such as the compass (Bowditch) rule were applied. Now least-squares calculations are handled routinely, and adjusting by this method is rapidly becoming indispensable in modern surveying (geomatics). The method of least squares is currently being used to adjust all kinds of observations, including differences in elevation, horizontal distances, and horizontal and vertical angles. It has become essential in the adjustment of GNSS observations and is also widely used in adjusting photogrammetric data. Adjustments by the least-squares method have taken on added importance with the most recent surveying accuracy standards. These standards include the use of statistical quantities that result from least-squares adjustment. Thus in order to evaluate a survey for compliance with the standards, least-squares adjustments must first be performed.

Least-squares adjustments provide several advantages over other arbitrary methods. First of all, because the method is based upon the mathematical theory of probability, it is the most rigorous of adjustment procedures. It enables all observations to be included in an adjustment simultaneously, and each observation can be weighted according to its estimated precision. Furthermore, the least-squares method is applicable to any observational problem regardless of its nature or geometric configuration. In addition to these advantages, the least-squares method enables rigorous statistical analyses to be made on the results of the adjustment; that is, the precisions of all adjusted quantities can be estimated, and other factors investigated. The least-squares method even enables presurvey planning to be done to ensure that required precisions of adjusted quantities are obtained in the most economical manner.

A simple example can be used to illustrate the arbitrary nature of "rule of thumb" adjustments, as compared to least squares. Consider the horizontal survey network shown in Figure 16.1. If the compass rule was used to adjust the observations in the network, several solutions would be possible. To illustrate one variation, suppose that traverse $ABCDEFGA$ is adjusted first. Then holding the adjusted values of points G and E, traverse $GHKE$ is adjusted, and finally, holding the adjusted values on H and C, traverse HJC is adjusted. This obviously would yield a solution, but there are other possible approaches. In another variation, traverse $ABCDEFGA$ could be adjusted followed by $GHJC$, and then HKE.

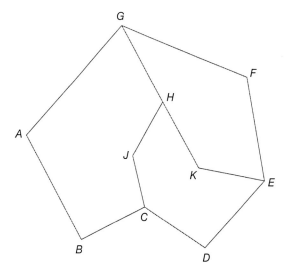

Figure 16.1
A horizontal network.

This sequence would result in another solution, but with different adjusted values for points H, J, and K. There are still other possible variations. This illustrates that the compass rule adjustment is properly referred to as an "arbitrary" method. On the other hand, the least-squares method adjusts all observations simultaneously, and for a given set of weights there is only one solution—that which yields the most probable values for the given set of observations.

In the sections that follow, the fundamental condition enforced in least squares is described, and elementary examples of least-squares adjustments are presented. Then systematic procedures for forming and solving least-squares equations using matrix methods are given and demonstrated with examples. The examples involving differential leveling, GNSS baselines, and horizontal networks are performed using the software WOLFPACK, which is on the companion website for this book at http://www.pearsonhighered.com/ghilani. For these examples, sample data files and the results of the adjustments are shown. A complete description of the data files is given in the *help* system provided with the software. For those wishing to see programming of these problems, Mathcad worksheets that demonstrate the differential leveling, GNSS baseline vector, and plane survey adjustments are available on the companion website for this book.

■ 16.2 FUNDAMENTAL CONDITION OF LEAST SQUARES

It was shown through the discussion in Section 3.12 and the normal distribution curves illustrated in Figures 3.2 and 3.3, that small errors (residuals) have a higher probability of occurrence than large ones in a group of normally distributed observations. Also discussed was the fact that in such a set of observations, there is a specific probability that an error (residual) of a certain size will exist within a group of errors. In other words, there is a direct relationship between probabilities and residual sizes in a normally distributed set of observations. The method of least-squares adjustment is derived from the equation for the normal distribution curve. *It produces that unique set of residuals for a group of observations that have the highest probability of occurrence.*

For a group of equally weighted observations, the fundamental condition enforced by the least-squares method is that the *sum of the squares of the residuals is a minimum*. Suppose a group of m observations of equal weight were taken having residuals $v_1, v_2, v_3, ..., v_m$. Then, in equation form, the fundamental condition of least squares is

$$\sum_{i=1}^{m} v_i^2 = v_1^2 + v_2^2 + v_3^2 + ... + v_m^2 \rightarrow \text{minimum}^1 \qquad (16.1)$$

For any group of observed values, weights may be assigned to individual observations according to *a priori* (before the adjustment) estimates of their relative worth or they may be obtained from the standard deviations of the

---

[1] Refer to Ghilani (2018) cited in the bibliography at the end of this chapter, for a derivation of this equation.

observations, if available. An equation expressing the relationship between standard deviations and weights, given in Section 3.20 and repeated here, is

$$w_i = \frac{1}{\sigma_i^2} \tag{16.2}$$

In Equation (16.2), w_i is the weight of the ith observed quantity and σ_i^2 the variance of the observation. *This equation states that weights are inversely proportional to variances.*

If observed values are to be weighted in least-squares adjustment then the fundamental condition to be enforced is that *the sum of the weights times their corresponding squared residuals is minimized* or, in equation form

$$\sum_i^m w_i v_i^2 = w_1 v_1^2 + w_2 v_2^2 + w_3 v_3^2 + \ldots + w_m v_m^2 \rightarrow \text{minimum} \tag{16.3}$$

Some basic assumptions underlying least-squares theory are that (1) mistakes and systematic errors have been eliminated, so only random errors remain in the observations; (2) the number of observations being adjusted is large; and (3) as stated earlier, the frequency distribution of the errors is normal. Although these basic assumptions are not always met, least-squares adjustments still provide the most rigorous error treatment available.

16.3 LEAST-SQUARES ADJUSTMENT BY THE OBSERVATION EQUATION METHOD

Two basic methods are employed in least-squares adjustments: (1) the *observation equation* method and (2) the *condition equation* method. The former is most common and is the one discussed herein. In this method, "observation equations" are written relating observed values to their residual errors and the unknown parameters. One observation equation is written for each observation. For a unique solution, the number of equations must equal the number of unknowns. If redundant observations are made, the least-squares method can be applied. In that case, an expression for each residual error is obtained from every observation equation. The residuals are squared and added to obtain the function expressed in either Equation (16.1) or Equation (16.3).

To minimize the function in accordance with either Equation (16.1) or Equation (16.3), partial derivatives of the expression are taken with respect to each unknown variable and set equal to zero. This yields a set of so-called *normal equations*, which are equal in number to the number of unknowns. The normal equations are solved to obtain most probable values for the unknowns. The following elementary examples illustrate the procedures.

■ Example 16.1

Using least squares, compute the most probable value for the equally weighted distance observations of Example 3.1.

Solution

1. For this problem, as was done in Example 3.1, let $\overline{M}$ be the most probable value of the observed length. Then write the following observation equations that define the residual for any observed quantity as the difference between the most probable value and any individual observation:

$$\overline{M} = 538.57 + v_1$$
$$\overline{M} = 538.39 + v_2$$
$$\overline{M} = 538.37 + v_3$$
$$\overline{M} = 538.39 + v_4$$
$$\overline{M} = 538.48 + v_5$$
$$\overline{M} = 538.49 + v_6$$
$$\overline{M} = 538.33 + v_7$$
$$\overline{M} = 538.46 + v_8$$
$$\overline{M} = 538.47 + v_9$$
$$\overline{M} = 538.55 + v_{10}$$

2. Solve for the residual in each observation equation and form the function $\sum v^2$ according to Equation (16.1)

$$\sum v^2 = (\overline{M} - 538.57)^2 + (\overline{M} - 538.39)^2 + (\overline{M} - 538.37)^2$$
$$+ (\overline{M} - 538.39)^2 + (\overline{M} - 538.48)^2 + (\overline{M} - 538.49)^2$$
$$+ (\overline{M} - 538.33)^2 + (\overline{M} - 538.46)^2 + (\overline{M} - 538.47)^2$$
$$+ (\overline{M} - 538.55)^2$$

3. Take the derivative of the function $\sum v^2$ with respect to $\overline{M}$, set it equal to zero (this minimizes the function)

$$\frac{\partial \sum v^2}{\partial \overline{M}} = 0 = 2(\overline{M} - 538.57) + 2(\overline{M} - 538.39) + 2(\overline{M} - 538.37)$$
$$+ 2(\overline{M} - 538.39) + 2(\overline{M} - 538.48) + 2(\overline{M} - 538.49)$$
$$+ 2(\overline{M} - 538.33) + 2(\overline{M} - 538.46) + 2(\overline{M} - 538.47)$$
$$+ 2(\overline{M} - 538.55)$$

4. Reduce and solve for $\overline{M}$

$$10\overline{M} = 5384.50$$
$$\overline{M} = \frac{5384.50}{10} = 538.45$$

Note that this answer agrees with the one given for Example 3.1. Note also that this procedure verifies the statement given earlier in Section 3.10 that *the most probable value for an unknown quantity, measured repeatedly using the same equipment and procedures, is simply the mean of the observations.*

Example 16.2

In Figure 8.9(c), the three horizontal angles observed around the horizon are $x = 42°12'13''$, $y = 59°56'15''$, and $z = 257°51'35''$. Adjust these angles by the least-squares method so that their sum equals the required geometric total of 360°. Note in this example only two of the three angles must be observed since the remaining angle could be computed. This means that there is one redundant angle measurement in this example.

Solution

1. Form the observation equations

$$x = 42°12'13'' + v_1 \qquad \text{(a)}$$

$$y = 59°56'15'' + v_2 \qquad \text{(b)}$$

$$z = 257°51'35'' + v_3 \qquad \text{(c)}$$

2. Write an expression that enforces the condition that the sum of the three adjusted angles total 360°.

$$x + y + z = 360° \qquad \text{(d)}$$

3. Substitute Equations (a), (b), and (c) into Equation (d), and solve for v_3

$$(42°12'13'' + v_1) + (59°56'15'' + v_2) + (257°51'35'' + v_3) = 360°$$
$$v_3 = -3'' - v_1 - v_2 \qquad \text{(e)}$$

(Because of the 360° condition, if v_1 and v_2 are fixed, v_3 is also fixed. Thus, there are only two independent residuals in the solution.)

4. Form the function Σv^2, which involves all three residuals but includes only the two independent variables v_1 and v_2

$$\Sigma v^2 = v_1^2 + v_2^2 + (-3'' - v_1 - v_2)^2 \qquad \text{(f)}$$

5. Take partial derivatives of Equation (f) with respect to the variables v_1 and v_2, and set them equal to zero.

$$\frac{\partial \Sigma v^2}{\partial v_1} = 0 = 2v_1 + 2(-3'' - v_1 - v_2)(-1); \qquad 4v_1 + 2v_2 = -6'' \qquad \text{(g)}$$

$$\frac{\partial \Sigma v^2}{\partial v_2} = 0 = 2v_2 + 2(-3'' - v_1 - v_2)(-1); \qquad 2v_1 + 4v_2 = -6'' \qquad \text{(h)}$$

6. Solve Equations (g) and (h) simultaneously

$$v_1 = -1'' \text{ and } v_2 = -1''$$

7. Substitute v_1 and v_2 into Equation (e) to compute v_3

$$v_3 = -3'' + 1'' + 1'' = -1''$$

8. Finally, substitute the residuals into Equations (a) through (c) to get the adjusted angles

$$x = 42°12'13'' - 1'' = 42°12'12''$$
$$y = 59°56'15'' - 1'' = 59°56'14''$$
$$z = 257°51'35'' - 1'' = \underline{257°51'34''}$$
$$\Sigma = 360°00'00'' \checkmark$$

Note that this result verifies another basic procedure frequently applied in surveying (geomatics) that *for equally weighted angles observed around the horizon, corrections of equal size are applied to each angle*. The same result occurs when equally weighted interior angles in a closed polygon traverse are adjusted by least squares. That is, each receives an equal-size correction.

Examples 16.1 and 16.2 are indeed simple, hardly the type for which least squares is best suited. However, they do supply the basis for some commonly applied simple adjustments, and illustrate procedures involved in making least-squares adjustments without complicating the mathematics. The following example illustrates least-squares adjustment of distance observations that are functionally related.

■ Example 16.3

Adjust the three equally weighted distance observations taken (in feet) between points A, B, and C of Figure 16.2.

Solution

1. Let the unknown distances AB and BC be x and y, respectively. These two unknowns are related through the observation equations as follows:

$$x + y = 393.65$$
$$x = 190.40 \qquad \text{(i)}$$
$$y = 203.16$$

2. Values for x and y could be obtained from any two of these equations so that the remaining equation is redundant. However, notice that values obtained for x and y will differ, depending on which two equations are solved. It is therefore apparent that the observations contain errors. Equations (i) may be rewritten as observation equations by including residual errors as follows:

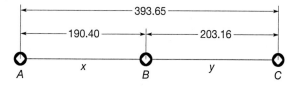

Figure 16.2
Equally weighted distance observations of Example 16.3.

$$x + y = 393.65 + v_1$$
$$x = 190.40 + v_2 \quad \text{(j)}$$
$$y = 203.16 + v_3$$

3. To obtain the least-squares solution, the observation equations (j) are rearranged to obtain expressions for the residuals. These are squared and added to form the function given in Equation (16.1) as follows:

$$\sum_{i=1}^{3} v_i^2 = (x + y - 393.65)^2 + (x - 190.40)^2 + (y - 203.16)^2 \quad \text{(k)}$$

4. Function (k) is minimized, enforcing the condition of least squares, by taking partial derivatives with respect to the unknowns x and y and setting them equal to zero. This yields the following two normal equations:

$$\frac{\partial \sum v^2}{\partial x} = 0 = 2(x + y - 393.65) + 2(x - 190.40)$$

$$\frac{\partial \sum v^2}{\partial y} = 0 = 2(x + y - 393.65) + 2(y - 203.16)$$

5. Reducing the normal equations and solving yields $x = 190.43$ ft and $y = 203.19$ ft. The residuals can now be calculated by substituting x and y into the original observation equations (j)

$$v_1 = 190.43 + 203.19 - 393.65 = -0.03 \text{ ft}$$
$$v_2 = 190.43 - 190.40 = +0.03 \text{ ft}$$
$$v_3 = 203.19 - 203.16 = +0.03 \text{ ft}$$

■ 16.4 MATRIX METHODS IN LEAST-SQUARES ADJUSTMENT[2]

It has been noted that least-squares computations are quite lengthy, and therefore generally performed on a computer. Their solution follows a systematic procedure that is conveniently adapted to matrix methods. In general, any group of observation equations may be represented in matrix form as

$$_m A^n {}_n X^1 = {}_m L^1 + {}_m V^1 \quad (16.4)$$

where A is the matrix of coefficients for the unknowns, X the matrix of unknowns, L the matrix of observations, and V the matrix of residuals. The detailed structures of these matrices are

---

[2]The balance of this chapter requires a basic understanding of matrix algebra. Students who do not have this background may consult Appendix E.

$$A = \begin{bmatrix} a_{11} & a_{12} & \cdots & a_{1n} \\ a_{21} & a_{22} & \cdots & a_{2n} \\ \vdots & \vdots & \vdots & \vdots \\ a_{m1} & a_{m2} & \cdots & a_{mn} \end{bmatrix} \quad X = \begin{bmatrix} x_1 \\ x_2 \\ \vdots \\ x_n \end{bmatrix} \quad L = \begin{bmatrix} l_1 \\ l_2 \\ \vdots \\ l_m \end{bmatrix} \quad V = \begin{bmatrix} v_1 \\ v_2 \\ \vdots \\ v_m \end{bmatrix}$$

The normal equations that result from a set of equally weighted observation equations [Equations (16.4)] are given in matrix form by

$$A^T A X = A^T L \tag{16.5}$$

In Equation (16.5), $A^T A$ is the matrix of normal equation coefficients for the unknowns. Premultiplying both sides of Equation (16.5) by $(A^T A)^{-1}$ and reducing yields

$$X = (A^T A)^{-1} A^T L \tag{16.6}$$

Equation (16.6) is the least-squares solution for equally weighted observations. The matrix X consists of most probable values for unknowns $x_1, x_2, x_3, \ldots, x_n$.

For a system of weighted observations, the following equation provides the X matrix:

$$X = (A^T W A)^{-1} A^T W L \tag{16.7}$$

In Equation (16.7), the matrices are identical to those of the equally weighted case, except that W is the inverse of the covariance matrix for the observations. However when the observations are independent of each other, which is typically found in surveying, the weight matrix is a diagonal matrix defined as[3]

$$W = \begin{bmatrix} w_1 & & & \\ & w_2 & & zeros \\ zeros & & \ddots & \\ & & & w_n \end{bmatrix}$$

If the observations in an adjustment are all equal weight, Equation (16.7) can still be used but the W matrix becomes an identity matrix. It therefore reduces exactly to Equation (16.6). Thus, Equation (16.7) is general and can be used for both the unweighted and weighted adjustments. This matrix solution is readily programmed for a computer.

---

[3] For a group of independent and *uncorrelated observations* (a case frequently encountered in surveying), the weight matrix is diagonal, that is, all off-diagonal elements are zeros. In certain cases, however, observations are correlated, that is, they are related to each other. An example occurs in GNSS baseline measurements, where the vector components result from least-squares adjustments and thus are correlated. As will be shown in Section 16.8, this yields off-diagonal elements in the W matrix.

Example 16.4

Solve Example 16.3 using matrix methods.

Solution

1. The observation equations of Example 16.3 can be expressed in matrix form as follows:

$$_3A^2\,_2X^1 = \,_3L^1 + \,_3V^1$$

where

$$A = \begin{bmatrix} 1 & 1 \\ 1 & 0 \\ 0 & 1 \end{bmatrix} \quad X = \begin{bmatrix} x \\ y \end{bmatrix} \quad L = \begin{bmatrix} 393.65 \\ 190.40 \\ 203.16 \end{bmatrix} \quad V = \begin{bmatrix} v_1 \\ v_2 \\ v_3 \end{bmatrix}$$

2. Solving matrix Equation (16.6)

$$A^T A = \begin{bmatrix} 1 & 1 & 0 \\ 1 & 0 & 1 \end{bmatrix} \begin{bmatrix} 1 & 1 \\ 1 & 0 \\ 0 & 1 \end{bmatrix} = \begin{bmatrix} 2 & 1 \\ 1 & 2 \end{bmatrix}$$

$$(A^T A)^{-1} = \frac{1}{3}\begin{bmatrix} 2 & -1 \\ -1 & 2 \end{bmatrix} \quad A^T L = \begin{bmatrix} 584.05 \\ 596.81 \end{bmatrix}$$

$$X = (A^T A)^{-1} A^T L = \frac{1}{3}\begin{bmatrix} 2 & -1 \\ -1 & 2 \end{bmatrix}\begin{bmatrix} 584.05 \\ 596.81 \end{bmatrix} = \begin{bmatrix} 190.43 \\ 203.19 \end{bmatrix}$$

Note that this solution yields $x = 190.43$ ft and $y = 203.19$ ft, which are exactly the same values obtained through the algebraic approach of Example 16.3.

16.5 MATRIX EQUATIONS FOR PRECISIONS OF ADJUSTED QUANTITIES

The matrix equation for calculating residuals after adjustment, whether the adjustment is weighted or not, is a rearrangement of Equation (16.4), or

$$V = AX - L \tag{16.8}$$

The standard deviation of unit weight for an unweighted adjustment is

$$\sigma_0 = \sqrt{\frac{V^T V}{r}} \tag{16.9}$$

The standard deviation of unit weight for a weighted adjustment is

$$\sigma_0 = \sqrt{\frac{V^T W V}{r}} \tag{16.10}$$

In Equations (16.9) and (16.10), r is the *number of degrees of freedom* in an adjustment, which usually equals the number of observations minus the number of unknowns, or $r = m - n$.

In the case of surveying, the number of degrees of freedom is the number of observations which are not required to solve a problem. In essence, these are redundant observations. For example, consider the traverse in Figure 9(a). There were six angles and five distances observed. In the figure, station A either has known or assumed coordinates, which fix the traverse in position. This leaves unknown coordinates for stations B, C, D, and E. Each station has two unknown coordinates for a total of eight unknowns. Thus, there are three (11 – 8) degrees of freedom in this traverse. These three additional observations are called *redundant* observations. It is these additional redundant observations that allow us to adjust the angles, latitudes, and departures of the traverse. As another example, as shown in Figure 5.8, there is one benchmark A with a known height of 100 ft and three benchmarks with unknown heights. The figure shows four differences in elevation observations. Thus, there is only one (4 – 3) degree of freedom in this adjustment. Finally, consider the GNSS network in Figure 14.6. This network has two stations with known geocentric coordinates (A and B). There are four stations with unknown geocentric coordinates or twelve (4 × 3) unknown coordinates in total. The check baseline AB between the control stations A and B cannot be adjusted since these stations have their coordinates fixed. Looking at Table 14.6, there is a total of 13 baseline observations excluding line AB. Thus, there are five (13 – 8) degrees of freedom in this network.

Standard deviations of the adjusted quantities are

$$\sigma_{x_i} = \sigma_0 \sqrt{q_{x_i x_i}} \qquad (16.11)$$

In Equation (16.11), σ_{x_i} is the standard deviation of the ith adjusted unknown x_i, that is, the value in the ith row of the X matrix; σ_0 the standard deviation of unit weight as calculated by Equation (16.9) or (16.10); and $q_{x_i x_i}$ the diagonal element in the ith row and ith column of matrix $(A^T A)^{-1}$ in the unweighted case, or matrix $(A^T W A)^{-1}$ in the weighted case. The matrices $(A^T A)^{-1}$ and $(A^T W A)^{-1}$ are the so-called *cofactor* matrices, and symbolized hereon by Q_{xx}, and $\sigma_0^2 Q_{xx}$ is the *covariance* matrix of the adjustment.

■ Example 16.5

Calculate the standard deviation of unit weight and the standard deviations of the adjusted quantities x and y for the unweighted problem of Example 16.4.

Solution

1. By Equation (16.8), the residuals are

$$V = \begin{bmatrix} 1 & 1 \\ 1 & 0 \\ 0 & 1 \end{bmatrix} \begin{bmatrix} 190.43 \\ 203.19 \end{bmatrix} - \begin{bmatrix} 393.65 \\ 190.40 \\ 203.26 \end{bmatrix} = \begin{bmatrix} -0.03 \\ 0.03 \\ 0.03 \end{bmatrix}$$

2. By Equation (16.9), the standard deviation of unit weight is

$$V^T V = \begin{bmatrix} -0.03 & 0.03 & 0.03 \end{bmatrix} \begin{bmatrix} -0.03 \\ 0.03 \\ 0.03 \end{bmatrix} = [0.0027]$$

$$\sigma_0 = \sqrt{\frac{0.0027}{3-2}} = \pm 0.052 \text{ ft}$$

3. Using Equation (16.11), the standard deviations of the adjusted values for x and y are

$$\sigma_x = \pm 0.052 \sqrt{\frac{2}{3}} = \pm 0.042 \text{ ft}$$

$$\sigma_y = \pm 0.052 \sqrt{\frac{2}{3}} = \pm 0.042 \text{ ft}$$

In part 3, the numbers 2/3 under the radicals are the elements in row 1, column 1, and row 2, column 2, respectively, of the $(A^T A)^{-1}$ matrix of Example 16.4. The interpretation of the standard deviations computed in step 3 of Example 16.5 is that a 68% probability exists that the adjusted values for x and y are within ± 0.042 ft of their true values. Note that for this simple example, the three residuals calculated in step 1 are equal, and the standard deviations of x and y are equal in step 3. This is caused by the symmetric nature of this particular problem (illustrated in Figure 16.2), but it is not the case generally with more complex problems.

■ 16.6 LEAST-SQUARES ADJUSTMENT OF LEVELING CIRCUITS

When control leveling is being done to establish new benchmarks, for example, benchmarks for a construction project, it is common practice to create a network like that illustrated in Figure 16.3. This enables each new benchmark to benefit from redundant observations and least-squares adjustment. In Figure 16.3, A and B are two new project benchmarks being established near a construction site. Each could be set by running a single loop, such as from BM 1 to A and back to establish A, and from BM 3 to B and back to set B. To build redundancy and checks into the survey, and to increase the precisions of the new benchmarks, additional lines from other nearby benchmarks can be run. Thus in Figure 16.3, five loops are run rather than the minimum of two needed to establish A and B. All observations within this leveling network can be adjusted simultaneously using the least-squares method to obtain most probable adjusted values for the two benchmarks.

In adjusting level networks, the observed difference in elevation for each course is treated as one observation containing a single random error. Observation equations are written that relate these observed elevation differences and their residual errors to the unknown elevations of the benchmarks involved. These can then be processed through the matrix equations given in Sections 16.4 and 16.5 to obtain adjusted values for the benchmarks and their standard deviations. The procedure is illustrated with the following example.

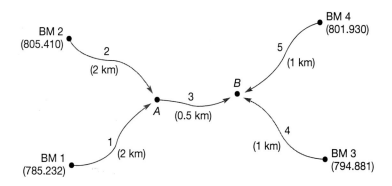

Figure 16.3
Level net for Example 16.6.

Example 16.6

Adjust the level net of Figure 16.3 by weighted least squares, and compute precisions of the adjusted benchmarks. In the figure, the benchmark elevations (in meters) and course lengths (in kilometers) are shown in parentheses. Observed elevation differences for courses 1 through 5 (given in order) are $+10.997, -9.169, +3.532, +4.858,$ and -2.202 m. Arrows on the courses in the figure indicate the direction of leveling. Thus for course 1 having a length of 2 km, leveling proceeded from BM 1 to A and the observed elevation difference was 10.997 m.

Solution

1. Observation equations are written relating each line's observed elevation difference to its residual error and the most probable values for unknown elevations A and B as follows:

$$
\begin{aligned}
A &= \text{BM 1} + 10.997 + v_1 \\
A &= \text{BM 2} - 9.169 + v_2 \\
B &= A + 3.532 + v_3 \\
B &= \text{BM 3} + 4.858 + v_4 \\
B &= \text{BM 4} - 2.202 + v_5
\end{aligned} \quad (\text{I})
$$

2. Substituting the elevations of BM 1, BM 2, BM 3, and BM 4 into Equations (I) and rearranging for the unknowns yields

$$
\begin{aligned}
A &= 796.229 + v_1 \\
A &= 796.241 + v_2 \\
-A + B &= 3.532 + v_3 \\
B &= 799.739 + v_4 \\
B &= 799.228 + v_5
\end{aligned}
$$

438 ADJUSTMENTS BY LEAST SQUARES

3. The A, X, L, and V matrices for this adjustment are

$$A = \begin{bmatrix} 1 & 0 \\ 1 & 0 \\ -1 & 1 \\ 0 & 1 \\ 0 & 1 \end{bmatrix} \quad X = \begin{bmatrix} A \\ B \end{bmatrix} \quad L = \begin{bmatrix} 796.229 \\ 796.241 \\ 3.532 \\ 799.739 \\ 799.728 \end{bmatrix} \quad V = \begin{bmatrix} v_1 \\ v_2 \\ v_3 \\ v_4 \\ v_5 \end{bmatrix}$$

4. As discussed in Section 5.6, weights in differential leveling are inversely proportional to course lengths. Thus, after inverting the lengths, the course weights are 0.5, 0.5, 2, 1, and 1, respectively, and the weight matrix is

$$W = \begin{bmatrix} 0.5 & 0 & 0 & 0 & 0 \\ 0 & 0.5 & 0 & 0 & 0 \\ 0 & 0 & 2 & 0 & 0 \\ 0 & 0 & 0 & 1 & 0 \\ 0 & 0 & 0 & 0 & 1 \end{bmatrix}$$

5. The weighted matrix solution for the most probable values according to Equation (16.7) is

$$A^T W = \begin{bmatrix} 1 & 1 & -1 & 0 & 0 \\ 0 & 0 & 1 & 1 & 1 \end{bmatrix} \begin{bmatrix} 0.5 & 0 & 0 & 0 & 0 \\ 0 & 0.5 & 0 & 0 & 0 \\ 0 & 0 & 2 & 0 & 0 \\ 0 & 0 & 0 & 1 & 0 \\ 0 & 0 & 0 & 0 & 1 \end{bmatrix} = \begin{bmatrix} 0.5 & 0.5 & -2.0 & 0.0 & 0.0 \\ 0.0 & 0.0 & 2.0 & 1.0 & 1.0 \end{bmatrix}$$

$$A^T W A = \begin{bmatrix} 0.5 & 0.5 & -2.0 & 0.0 & 0.0 \\ 0.0 & 0.0 & 2.0 & 1.0 & 1.0 \end{bmatrix} \begin{bmatrix} 1 & 0 \\ 1 & 0 \\ -1 & 1 \\ 0 & 1 \\ 0 & 1 \end{bmatrix} = \begin{bmatrix} 3 & -2 \\ -2 & 4 \end{bmatrix}$$

$$Q_{xx} = (A^T W A)^{-1} = \frac{1}{8}\begin{bmatrix} 4 & 2 \\ 2 & 3 \end{bmatrix}$$

$$A^T W L = \begin{bmatrix} 0.5 & 0.5 & -2.0 & 0.0 & 0.0 \\ 0.0 & 0.0 & 2.0 & 1.0 & 1.0 \end{bmatrix} \begin{bmatrix} 796.229 \\ 796.241 \\ 3.532 \\ 799.739 \\ 799.728 \end{bmatrix} = \begin{bmatrix} 789.171 \\ 1606.531 \end{bmatrix}$$

$$X = (A^T W A)^{-1} A^T W L = \frac{1}{8}\begin{bmatrix} 4 & 2 \\ 2 & 3 \end{bmatrix}\begin{bmatrix} 789.171 \\ 1606.531 \end{bmatrix} = \begin{bmatrix} 796.218 \\ 799.742 \end{bmatrix}$$

Thus, the adjusted benchmark elevations are $A = 796.218$ m and $B = 799.742$ m. Note in this example the term Q_{xx} was set equal to $(A^T W A)^{-1}$. Q_{xx} is known as the cofactor matrix for the unknowns and will be used in the next section to computer uncertainties in the computed unknowns.

6. The residuals by Equation (16.8) are

$$V = AX - L = \begin{bmatrix} 1 & 0 \\ 1 & 0 \\ -1 & 0 \\ 0 & 1 \\ 0 & 1 \end{bmatrix} \begin{bmatrix} 796.218 \\ 799.742 \end{bmatrix} - \begin{bmatrix} 796.229 \\ 796.241 \\ 3.532 \\ 799.739 \\ 799.728 \end{bmatrix} = \begin{bmatrix} -0.011 \\ -0.023 \\ -0.008 \\ 0.003 \\ 0.014 \end{bmatrix}$$

7. Utilizing Equation (16.10), the sum of the weighted residuals is

$$V^T W V = \begin{bmatrix} -0.011 & -0.023 & -0.008 & 0.003 & 0.014 \end{bmatrix}$$

$$\begin{bmatrix} 0.5 & 0 & 0 & 0 & 0 \\ 0 & 0.5 & 0 & 0 & 0 \\ 0 & 0 & 2 & 0 & 0 \\ 0 & 0 & 0 & 1 & 0 \\ 0 & 0 & 0 & 0 & 1 \end{bmatrix} \begin{bmatrix} -0.011 \\ -0.023 \\ -0.008 \\ 0.003 \\ 0.014 \end{bmatrix}$$

$$= [0.00066]$$

Thus, the standard deviation of unit weight is $\sigma_0 = \sqrt{\dfrac{0.00066}{5-2}} = \pm 0.015$ m

8. By Equation (16.11), the estimated standard deviations of adjusted benchmark elevations of A and B are

$$\sigma_A = \sigma_0 \sqrt{q_{AA}} = \pm 0.015 \sqrt{\dfrac{4}{8}} = \pm 0.010 \text{ m}$$

$$\sigma_B = \sigma_0 \sqrt{q_{BB}} = \pm 0.015 \sqrt{\dfrac{3}{8}} = \pm 0.009 \text{ m}$$

Note in these calculations that terms in the radicals are the diagonal elements, (1, 1) and (2, 2), of the cofactor matrix Q_{xx}. Note also that B has a lower standard deviation than A, indicating that its precision is better. A check of Figure 16.3 reveals that this should be expected, because the benchmarks closest to A are both 2 km away, and those closest to B are only 1 km away. Thus, the chance of introducing error into B is lower and the precision of its elevation higher than that of A.

The software WOLFPACK can be used to adjust the data of Example 16.6. The format for the data file is shown in Figure 16.4. Note that the lengths of the level lines, the number of setups, or the standard deviations of the observations can weight the observations. In this example, course lengths were used. The results of the adjustment are shown in Figure 16.5.

440 ADJUSTMENTS BY LEAST SQUARES

```
Differential leveling example of Section 16.6    {title line}
4 6 5                              {Number of benchmarks; number of stations; number of elev. diff.}
1 785.232                          {BM identifier and elevation}
2 805.410
3 794.881
4 801.930
1 A 10.997 2                       {from, to, elevation difference, [dist.|setups|σ}
2 A -9.169 2
A B 3.532 0.5
3 B 4.858 1
4 B -2.202 1
```

Figure 16.4 WOLFPACK data file for Example 16.6.

```
*******************************
Adjusted Elevation Differences
*******************************
      From         To      Elevation Difference      V         σ
===================================================================
        1          A           10.986             -0.011    0.0105
        2          A           -9.192             -0.023    0.0105
        A          B            3.524             -0.008    0.0091
        3          B            4.861              0.003    0.0091
        4          B           -2.188              0.014    0.0091


*******************
Adjusted Elevations
*******************
     Station       Elevation         σ
====================================================
        A           796.218        0.0105
        B           799.742        0.0091

Standard Deviation of Unit Weight: 0.015
```

Figure 16.5 Results from adjustment of data in Example 16.6 using WOLFPACK.

16.7 PROPAGATION OF ERRORS

In Section 3.17, the propagation of errors in functions using independent observations was discussed. At the completion of a least-squares adjustment, the unknowns are no longer independent as evidenced by the off-diagonal terms in the cofactor matrix Q_{xx}. When observations are not independent, errors propagate as

$$Q_{\ell\ell} = AQ_{xx}A^T \qquad (16.12)$$

where Q_{xx} equals the cofactor matrix $(A^TA)^{-1}$ in the unweighted case, or $(A^TWA)^{-1}$ in the weighted case.

The standard deviations in the computed observations are

$$\sigma_{\ell_i} = \sigma_0\sqrt{q_{\ell_i\ell_i}} \qquad (16.13)$$

In Equation (16.13), σ_{ℓ_i} is the standard deviation of the ith adjusted observation; σ_0 the standard deviation of unit weight as calculated by Equation (16.9) or (16.10); and $q_{\ell_i\ell_i}$ ith diagonal element of the cofactor matrix for the adjusted observations, $Q_{\ell\ell}$, in Equation (16.12).

■ Example 16.7

Compute the adjusted elevation differences and their standard deviations for Example 16.6.

Solution

By Equation (16.12) the cofactor matrix for the adjusted observations, $Q_{\ell\ell}$, is

$$Q_{\ell\ell} = \frac{1}{8}\begin{bmatrix} 1 & 0 \\ 1 & 0 \\ -1 & 1 \\ 0 & 1 \\ 0 & 1 \end{bmatrix}\begin{bmatrix} 4 & 2 \\ 2 & 3 \end{bmatrix}\begin{bmatrix} 1 & 1 & -1 & 0 & 0 \\ 0 & 0 & 1 & 1 & 1 \end{bmatrix} = \frac{1}{8}\begin{bmatrix} 4 & 4 & -2 & 2 & 2 \\ 4 & 4 & -2 & 2 & 2 \\ -2 & -2 & 3 & 1 & 1 \\ 2 & 2 & 1 & 3 & 3 \\ 2 & 2 & 1 & 3 & 3 \end{bmatrix}$$

A tabulation of the observations, their residuals, adjusted values, and standard deviations are

From	To	Obs. ΔElev	V	Adj. ΔElev	σ
BM 1	A	10.997	−0.011	10.986	±0.010
BM 2	A	−9.169	−0.023	−9.192	±0.010
A	B	3.532	−0.008	3.524	±0.009
BM 3	B	4.858	0.003	4.861	±0.009
BM 4	B	−2.202	0.014	−2.188	±0.009

The contents of the third line in the above table are explained to clarify the calculations. The adjusted elevation difference for the third observation, which consisted of leveling from A to B, is obtained by adding the observed elevation difference (see the data for Example Problem 16.6), and its residual (see step 6 of Example Problem 16.6) as

$$\text{Adj. } \Delta\text{Elev} = 3.532 - 0.008 = 3.524$$

Also the standard deviation, S, for the third observation is computed as

$$\sigma = \pm 0.015\sqrt{\frac{3}{8}} = \pm 0.009$$

where 3/8 is the third diagonal element of the $Q_{\ell\ell}$, and 0.015 is σ_0 as determined in Example 16.6. The adjusted elevations and standard deviations for the remaining observations are computed in similar fashion.

16.8 LEAST-SQUARES ADJUSTMENT OF GNSS BASELINE VECTORS

It was previously noted in Section 14.5.5 that the least-squares method is essential in adjusting GNSS observations. It is applied in this work in two different stages: (1) for adjusting the massive quantities of redundant data that result after several receivers have made repeated observations on multiple satellites over a period of time (this yields baseline components ΔX, ΔY, and ΔZ) and (2) it is applied in adjusting redundant observations of these baseline components to make them consistent in a static network. The reduction software, which is available when GNSS software is purchased, is programmed to perform the first stage of these two applications, and thus its development is not covered in this text. However, the second application is within the scope of this text and is illustrated with the adjustment of the network of Figure 14.10.

Since the baseline vector data are the results of the first least-squares adjustment noted earlier in the text, each baseline has its own 3×3 covariance matrix. This matrix not only contains terms along the diagonal, but it also has elements in the off-diagonal locations. The covariance terms depict the amount of correlation between the adjusted ΔX, ΔY, and ΔZ values. For each baseline, the carrier-phase reduction software will list the baseline components and their covariance terms. For example for baseline AC in Figure 14.10, the vector and its covariance terms were listed as

$$\begin{array}{llrrr} \Delta X & 11{,}644.2232 & 9.8E{-}4 & -9.6E{-}6 & 9.5E{-}6 \\ \Delta Y & 3601.2165 & & 9.4E{-}4 & -9.5E{-}6 \\ \Delta Z & 3399.2550 & & & 9.8E{-}4 \end{array}$$

Note that only the upper-triangular portion of the covariance matrix is shown to the right of the vector components. This is because the covariance matrix is always symmetric, and thus the elements of the lower-triangular portion mirror their upper-triangular values and need not be repeated. Since the baseline vectors are not independent, the weight matrix is computed as the inverse of the covariance matrix, or in matrix symbology is

$$W = \Sigma^{-1} \qquad (16.14)$$

where Σ is the covariance matrix for the baseline vectors, and W their weight matrix. It can be shown that this equation is also valid for independent observations, and thus is a general equation for weighting observations.

The baseline vector components (ΔX, ΔY, ΔZ), and their covariance terms for the survey of the network of Figure 14.10 are shown in Table 16.1. In the column labeled (1) of this table, the (1,1) element of the covariance matrix is listed; column (2) lists the (1,2) element of the covariance matrix; column (3) the (1,3) element; column (4) the (2,2) element; column (5) the (2,3) element; and column (6) the (3,3) element of the covariance matrix. In the survey, two HARN stations (see Section 14.3.5) were held fixed. These stations and their coordinates are listed in Table 16.2.

From the known X, Y, and Z coordinates of stations A and B, and the observed ΔX, ΔY, and ΔZ components, coordinates of new stations C, D, E, and F can be

TABLE 16.1 Observed Baseline Vectors and Their Covariance Matrix Values

Baseline	ΔX	ΔY	ΔZ	(1)	(2)	(3)	(4)	(5)	(6)
AC	11644.2232	3601.2165	3399.2550	9.8E−4	−9.6E−6	9.5E−6	9.4E−4	−9.5E−6	9.8E−4
DC	15128.1647	−6286.7054	−6371.0583	1.5E−4	−1.4E−6	1.3E−6	1.6E−4	−1.4E−6	1.3E−4
AE	−5321.7164	3634.0754	3173.6652	2.2E−4	−2.1E−6	2.2E−6	1.9E−4	−2.1E−6	2.0E−4
BC	3960.5442	−6681.2467	−7279.0148	2.3E−4	−2.2E−6	2.1E−6	2.5E−4	−2.2E−6	2.2E−4
BD	−11167.6076	−394.5204	−907.9593	2.7E−4	−2.8E−6	2.8E−6	2.7E−4	−2.7E−7	2.7E−4
DE	−1837.7459	−6253.8534	−6596.6697	1.2E−4	−1.2E−6	1.2E−6	1.3E−4	−1.2E−6	1.3E−4
FA	−11167.6076	−394.5204	−907.9593	7.5E−5	−7.9E−7	8.8E−7	6.6E−5	−8.1E−7	7.6E−5
FC	10527.7852	−994.9370	−956.6246	2.6E−4	−2.2E−6	2.4E−6	2.2E−4	−2.3E−6	2.4E−4
FE	−6438.1364	−962.0694	−1182.2305	9.4E−5	−9.2E−7	1.0E−6	1.0E−4	−8.9E−7	8.8E−5
FD	−4600.3787	5291.7785	5414.4311	9.3E−5	−9.9E−7	9.0E−7	9.9E−5	−9.9E−7	1.2E−4
FB	6567.2311	5686.2926	6322.3917	6.6E−5	−6.5E−7	6.9E−7	7.5E−5	−6.4E−7	6.0E−5
BF	−6567.2310	−5686.3033	−6322.3807	5.5E−5	−6.3E−7	6.1E−7	7.5E−5	−6.3E−7	6.6E−5
AF	1116.4577	4596.1553	4355.9141	6.6E−5	−8.0E−7	9.0E−7	8.1E−5	−8.2E−7	9.4E−5

TABLE 16.2 HARN Station Geocentric Coordinates

Station	X	Y	Z
A	402.3509	−4,652,995.3011	4,349,760.7775
B	8086.0318	−4,642,712.8474	4,360,439.0833

calculated. However, an adjustment is necessary because redundant observations exist. In applying least squares to this problem, observation equations are written that relate the unknown adjusted coordinates of new stations C, D, E, and F to the observed ΔX, ΔY, and ΔZ values and their residual errors. As shown in Figure 14.10, excluding the check observation of fixed baseline AB, there are 11 different baselines. However, two of these, AF and BF, were repeated giving a total of 13 baseline observations. The following observation equations are written for the first two baselines:

$$
\begin{aligned}
X_C &= X_A + \Delta X_{AC} + v_1 \\
Y_C &= Y_A + \Delta Y_{AC} + v_2 \\
Z_C &= Z_A + \Delta Z_{AC} + v_3 \\
X_C - X_D &= \Delta X_{DC} + v_4 \\
Y_C - Y_D &= \Delta Y_{DC} + v_5 \\
Z_C - Z_D &= \Delta Z_{DC} + v_6
\end{aligned}
\quad (16.15)
$$

Similar equations can be written for the other 11 baseline observations, giving a total of 39 observation equations. These observation equations can be expressed in matrix form according to Equation (16.4). To illustrate the contents of the matrices, the partial matrices that result from the observation equations of Equations (16.15) are

$$
{}_{39}A^{12} = \begin{bmatrix}
1 & 0 & 0 & 0 & 0 & 0 \\
0 & 1 & 0 & 0 & 0 & 0 \\
0 & 0 & 1 & 0 & 0 & 0 \\
1 & 0 & 0 & -1 & 0 & 0 & \cdots \\
0 & 1 & 0 & 0 & -1 & 0 \\
0 & 0 & 1 & 0 & 0 & -1 \\
& & & \vdots & &
\end{bmatrix}; \quad {}_{12}X^1 = \begin{bmatrix} X_C \\ Y_C \\ Z_C \\ X_D \\ Y_D \\ Z_D \\ \vdots \end{bmatrix};
$$

$$
{}_{39}L^1 = \begin{bmatrix}
X_A + \Delta X_{AC} \\
Y_A + \Delta Y_{AC} \\
Z_A + \Delta Z_{AC} \\
\Delta X_{DC} \\
\Delta Y_{DC} \\
\Delta Z_{DC} \\
\vdots
\end{bmatrix}; \quad {}_{39}V^1 = \begin{bmatrix} v_1 \\ v_2 \\ v_3 \\ v_4 \\ v_5 \\ v_6 \\ \vdots \end{bmatrix}
\quad (16.16)
$$

To complete the above A matrix, the coefficients for the remaining observation equations are entered expanding the dimensions of A matrix to 39 rows and

12 columns. Note that there are three unknowns for each of the four new ground stations. Thus, the X matrix has a total of 12 elements. The L and V matrices each have 39 elements, one for each observation equation.

The partial covariance matrix that results from the first two baseline observations is

$$_{39}\Sigma^{39} = \begin{bmatrix} 9.8E-4 & -9.6E-6 & 9.5E-6 & 0 & 0 & 0 & \\ -9.6E-6 & 9.4E-4 & -9.5E-6 & 0 & 0 & 0 & \\ 9.5E-6 & -9.5E-6 & 9.8E-4 & 0 & 0 & 0 & \\ 0 & 0 & 0 & 1.5E-E & -1.4E-6 & 1.3E-6 & \cdots \\ 0 & 0 & 0 & -1.4E-6 & 1.6E-4 & -1.4E-6 & \\ 0 & 0 & 0 & 1.3E-6 & -1.4E-6 & 1.3E-4 & \\ & & & \vdots & & & \ddots \end{bmatrix}$$

The Σ matrix, when completed, has 39 rows and 39 columns. Its formation follows the same procedure for each observed baseline as is demonstrated above. As can be seen, each observed baseline creates a 3×3 submatrix within Σ. Thus, the structure of this matrix is *block diagonal*, that is, each individual submatrix on the diagonal is 3×3, and all other elements are zeros. Thus, the first 3 rows and 3 columns are nonzero elements that pertain to the first baseline, the next three rows and 3 columns contain nonzero elements that pertain to the second baseline, and so on.

The weight matrix is obtained by inverting the Σ matrix according to Equation (16.14). Once all matrices are formed, the solution for the unknown station coordinates and their standard deviations can be determined using Equation (16.7). After the adjustment, the resulting geocentric coordinates can be used to determine geodetic coordinates using the procedures discussed in Section 13.4.3, and if requested map coordinates can be derived from the geodetic coordinates using procedures discussed in Chapter 20.

For this problem, the software program WOLFPACK was used to adjust the baselines using the least-squares method. The input file is shown in Figure 16.6, and the results of the adjustment are given in Figure 16.7. Note that a standard deviation

```
Example of Section 16.8                              {title line}
2 13                                                 {Number of control stations; Number of baselines}
A 402.3509 -4652995.3011 4349760.7775                {identifier, X, Y, Z coordinates of control station}
B 8086.0318 -4642712.8474 4360439.0833
A C 11644.2232 3601.2165 3399.2550 9.8E-4 -9.6E-6 9.5E-6 9.4E-4 -9.5E-6 9.8E-4    {baselines}
A E -5321.7164 3634.0754 3173.6652 2.2E-4 -2.1E-6 2.2E-6 1.9E-4 -2.1E-6 2.0E-4
B C 3960.5442 -6681.2467 -7279.0148 2.3E-4 -2.2E-6 2.1E-6 2.5E-4 -2.2E-6 2.2E-4
B D -11167.6076 -394.5204 -907.9593 2.7E-4 -2.8E-6 2.8E-6 2.7E-4 -2.7E-6 2.7E-4
D C 15128.1647 -6286.7054 -6371.0583 1.5E-4 -1.4E-6 1.3E-6 1.6E-4 -1.4E-6 1.3E-4
D E -1837.7459 -6253.8534 -6596.6697 1.2E-4 -1.2E-6 1.2E-6 1.3E-4 -1.2E-6 1.3E-4
F A -1116.4523 -4596.1610 -4355.9062 7.5E-5 -7.9E-7 8.8E-7 6.6E-5 -8.1E-7 7.6E-5
F C 10527.7852 -994.937 -956.6246 2.6E-4 -2.2E-6 2.4E-6 2.2E-4 -2.3E-6 2.4E-4
F E -6438.1364 -962.0694 -1182.2305 9.4E-5 -9.2E-7 1.0E-6 1.0E-4 -8.9E-7 8.8E-5
F D -4600.3787 5291.7785 5414.4311 9.3E-5 -9.9E-7 9.0E-7 9.9E-5 -9.9E-7 1.2E-4
F B 6567.2311 5686.2926 6322.3917 6.6E-5 -6.5E-7 6.9E-7 7.5E-5 -6.4E-7 6.0E-5
B F -6567.2310 -5686.3033 -6322.3807 5.5E-5 -6.3E-7 6.1E-7 7.5E-5 -6.3E-7 6.6E-5
A F 1116.4577 4596.1553 4355.9141 6.6E-5 -8.0E-7 9.00E-7 8.1E-5 -8.2E-7 9.4E-5
```

Figure 16.6 Input file for least-squares adjustment problem in Section 16.8.

446 ADJUSTMENTS BY LEAST SQUARES

```
                    Degrees of Freedom = 27
                    Reference Variance = 0.5010
              Standard Deviation of Unit Weight = 0.71
```

```
*************************
Adjusted Distance Vectors
*************************
```

From	To	dX	dY	dZ	Vx	Vy	Vz
A	C	11644.2232	3601.2165	3399.2550	0.00665	0.00219	0.03197
A	E	-5321.7164	3634.0754	3173.6652	0.02665	0.00579	0.01212
B	C	3960.5442	-6681.2467	-7279.0148	0.00475	0.01169	-0.00403
B	D	-11167.6076	-394.5204	-907.9593	-0.00730	-0.00137	-0.00065
D	C	15128.1647	-6286.7054	-6371.0583	-0.00084	-0.00783	-0.00058
D	E	-1837.7459	-6253.8534	-6596.6697	-0.00985	0.00267	0.00117
F	A	-1116.4523	-4596.1610	-4355.9062	0.00197	0.00527	-0.00773
F	C	10527.7852	-994.9370	-956.6246	-0.00568	-0.00004	-0.00236
F	E	-6438.1364	-962.0694	-1182.2305	-0.00368	-0.00514	-0.00611
F	D	-4600.3787	5291.7785	5414.4311	-0.00563	-0.00230	0.00083
F	B	6567.2311	5686.2926	6322.3917	-0.00053	0.00537	0.00017
B	F	-6567.2310	-5686.3033	-6322.3807	0.00043	0.00533	-0.01117
A	F	1116.4577	4596.1553	4355.9141	-0.00737	0.00043	-0.00017

```
**************************************************
Advanced Statistical Values
**************************************************
```

From	To	σ	Vector Length	Prec
A	C	0.0105	12,653.538	1,206,000
A	E	0.0091	7,183.255	794,000
B	C	0.0105	10,644.668	1,015,000
B	D	0.0087	11,211.408	1,282,000
D	C	0.0107	17,577.670	1,641,000
D	E	0.0097	9,273.836	960,000
F	A	0.0048	6,430.015	1,344,000
F	C	0.0104	10,617.871	1,019,000
F	E	0.0086	6,616.111	770,000
F	D	0.0083	8,859.035	1,066,000
F	B	0.0048	10,744.075	2,246,000
B	F	0.0048	10,744.075	2,246,000
A	F	0.0048	6,430.015	1,344,000

```
*******************
Adjusted Coordinates
*******************
```

Station	X	Y	Z	σx	σy	σz
A	402.3509	-4,652,995.3011	4,349,760.7775			
B	8,086.0318	-4,642,712.8474	4,360,439.0833			
C	12,046.5808	-4,649,394.0824	4,353,160.0645	0.0061	0.0061	0.0059
E	-4,919.3388	-4,649,361.2199	4,352,934.4548	0.0052	0.0053	0.0052
D	-3,081.5831	-4,643,107.3692	4,359,531.1234	0.0049	0.0051	0.0051
F	1,518.8012	-4,648,399.1454	4,354,116.6914	0.0027	0.0028	0.0028

Figure 16.7 Results of adjustment of data file in Figure 16.6 using WOLFPACK.

was computed for each baseline, and that the precision of each baseline was determined as the ratio of the standard deviation, σ, over the vector length times one million. For those wishing to explore the programming of this problem, the Mathcad worksheet *GPS.xmcd* demonstrates the programming of this network.

16.9 LEAST-SQUARES ADJUSTMENT OF CONVENTIONAL HORIZONTAL PLANE SURVEYS

Traversing (described in Chapters 9 and 10), and trilateration and triangulation (discussed in Chapter 19), are traditional conventional-surveying methods for conducting *horizontal surveys* [those surveys that establish X and Y coordinates, usually in some grid system such as state plane coordinates (see Chapter 20), or geodetic latitudes and longitudes of points (see Section 19.4)]. The basic observations that are made in traversing, trilateration, and triangulation are horizontal angles and horizontal distances. As with other types of surveys, they are most appropriately adjusted by the method of least squares.

To adjust horizontal surveys by the least-squares method, it is necessary to write observation equations for the horizontal distance and horizontal angle observations. These observation equations are nonlinear, and to facilitate solving them they are linearized using a first-order Taylor series expansion. The procedure is described in the following subsection.

16.9.1 Linearizing Nonlinear Equations

The general form of the first-order Taylor series expansion of a nonlinear equation is

$$F(x_1, x_2, \ldots, x_n) = F(x_{1_0}, x_{2_0}, \ldots, x_{n_0}) + \left(\frac{\partial F}{\partial x_1}\right)_0 dx_1$$

$$+ \left(\frac{\partial F}{\partial x_2}\right)_0 dx_2 + \cdots + \left(\frac{\partial F}{\partial x_n}\right)_0 dx_n + R \quad (16.17)$$

where $F(x_1, x_2, \ldots, x_n)$ is a nonlinear function in terms of the unknowns $x_1, x_2, \ldots, x_n$, which represents a measured quantity; $x_{1_0}, x_{2_0}, \ldots, x_{n_0}$ are approximate values for the unknowns $x_1, x_2, \ldots, x_n$; $(\partial F/\partial x_1)_0, (\partial F/\partial x_2)_0, \ldots, (\partial F/\partial x_n)_0$ are the partial derivatives of the function F with respect to $x_1, x_2, \ldots, x_n$ evaluated using the approximate values of $x_{1_0}, x_{2_0}, \ldots, x_{n_0}$; $dx_1, dx_2, \ldots, dx_n$ are corrections to the approximate values of $x_{1_0}, x_{2_0}, \ldots, x_{n_0}$, such that $x_1 = x_{1_0} + dx_1; x_2 = x_{2_0} + dx_2, \ldots, x_n = x_{n_0} + d_{x_n}$; and R is a remainder. In Equation (16.17), the only unknowns are $dx_1, dx_2, \ldots, dx_n$, and R. The term R is also nonlinear, but if the values assigned for $x_{1_0}, x_{2_0}, \ldots, x_{n_0}$ are close to the true values of the unknowns, then R is small and dropped, which linearizes Equation (16.17). However, this makes the equation an approximation, and thus the solution must be obtained *iteratively*, that is, the corrections $dx_1, dx_2, \ldots, dx_n$ are computed repetitively until their sizes become negligible.

After dropping R and rearranging Equation (16.17), the following general linear form of the equation is

$$\left(\frac{\partial F}{\partial x_1}\right)_0 dx_1 + \left(\frac{\partial F}{\partial x_2}\right)_0 dx_2 + \cdots + \left(\frac{F}{\partial x_n}\right)_0$$

$$= F(x_1, x_2, \ldots, x_n) - F(x_{1_0}, x_{2_0}, \ldots, x_{n_0}) \quad (16.18)$$

448 ADJUSTMENTS BY LEAST SQUARES

The subscripted zeros attached to the coefficients on the left side of Equation (16.18) indicate that these coefficients are simply numbers obtained by substituting the approximate values $x_{1_0}, x_{2_0}, \ldots, x_{n_0}$ into the partial derivative functions. Also, the right side of Equation (16.18) is the observed value, $[F(x_1, x_2, \ldots, x_n)]$, minus the computed value obtained by substituting the initial approximations into the original function, $[F(x_{1_0}, x_{2_0}, \ldots, x_{n_0})]$.

The process of solving a pair of nonlinear equations using the Taylor's series will be illustrated. Suppose that the following two nonlinear functions, $F(x, y)$ and $G(x, y)$ express the relationship between observed values 115 and 75, respectively, and the unknowns x and y

$$F(x, y) = x^2 + 3y = 115$$
$$G(x, y) = 5x + y^2 = 75 \qquad \text{(a)}$$

Partial derivatives of the functions with respect to the unknowns are

$$\frac{\partial F}{\partial x} = 2x; \quad \frac{\partial F}{\partial y} = 3; \quad \frac{\partial G}{\partial x} = 5; \quad \frac{\partial G}{\partial y} = 2y \qquad \text{(b)}$$

(A) First Iteration

Assume that through either estimation or preliminary calculations based upon one or more observations values of 9 and 4 are selected as initial estimates for the unknowns, x_0 and y_0. Then using the functions of Equations (a) and substituting these initial approximations and the partial derivatives from (b) into Equation (16.18), the following two linearized equations are obtained:

$$(2 \times 9)dx + 3dy = 115 - [9^2 + 3(4)]$$
$$5dx + (2 \times 4)dy = 75 - [5(9) + 4^2] \qquad \text{(c)}$$

Equations (c) are now in linear form and contain only two unknowns, dx and dy. The solution of this pair of equations yields the following corrections: $dx = 1.04$ and $dy = 1.10$. Using these corrections, new approximations x_0 and y_0 are computed as

$$x_0 = 9.00 + 1.04 = 10.04$$
$$y_0 = 4.00 + 1.10 = 5.10 \qquad \text{(d)}$$

These new approximations are now used to repeat the solution.

(B) Second Iteration

Substituting the approximate solution in Equation (d) into Equation (16.18) yields the following linearized equations:

$$(10.04)^2 + 3 \times 5.10 + 2(10.04)dx + 3dy = 115$$
$$5 \times 10.04 + (5.10)^2 + 5dx + 2(5.10)dy = 75 \qquad \text{(e)}$$

Solving Equations (e) for the unknown parameters yields: $dx = -0.08$
These corrections are used to get updated values for the coordinates as

$$x_0 = 10.04 - 0.08 = 9.96$$
$$y_0 = 5.10 - 0.08 = 5.02 \tag{f}$$

(C) Third Iteration

New linearized equations are formed by substituting the initial approximations in Equation (f) into Equations (16.18) as follows

$$(9.96)^2 + 3 \times 5.02 + 2(9.96)dx + 3dy = 115$$
$$5 \times 9.96 + (5.02)^2 + 5dx + 2(5.02)dy = 75 \tag{g}$$

Solving Equations (g) for the unknown corrections gives $dx = 0.04$ and $dy = -0.02$. The updated values for x and y are therefore

$$x_0 = 9.96 + 0.04 = 10.00$$
$$y_0 = 5.02 - 0.02 = 5.00$$

A fourth iteration (not shown) yields zeros for both dx and dy, and thus the solution has converged. The final answers are $x = 10.00$ and $y = 5.00$.

Although only two unknowns existed in the above example, the first-order Taylor's series expansion is applicable to linearizing and solving nonlinear equations with any number of unknowns. All that is necessary is to select an initial approximation for each unknown and take partial derivatives of the function with respect to each unknown, as indicated in Equation (16.18). The initial approximations for the unknown station coordinates in a horizontal survey are easily solved using methods discussed previously in Chapters 10 and 11. As is discussed in the subsections that follow, in least-squares adjustments of horizontal surveys, up to four unknowns can appear in distance observation equations and up to six unknowns can appear in angle observation equations. In more advanced types of geodesy and photogrammetry problems, many more unknowns can appear in the nonlinear equations used.

16.9.2 The Distance Observation Equation

The observation equation for an observed distance is expressed in terms of the X and Y coordinates of its end points, and includes its residual error. Referring to Figure 16.8, the following distance observation equation can be written for the line whose end points are identified by I and J:

$$L_{IJ} + v_{IJ} = \sqrt{(X_J - X_I)^2 + (Y_J - Y_I)^2} \tag{16.19}$$

In Equation (16.19), L_{IJ} is the observed length of the line IJ; v_{IJ} the residual error in the observation; X_I, Y_I, X_J, and Y_J the most probable values for stations I

450 ADJUSTMENTS BY LEAST SQUARES

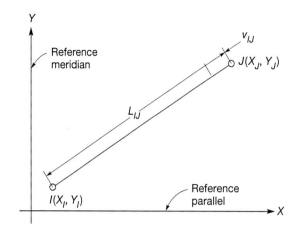

Figure 16.8
Observed distance expressed in terms of coordinates.

and J, respectively. By applying the first-order Taylor's series expansion [Equation (16.18)] to this nonlinear distance equation, the following linearized form of the equation results:[4]

$$\left(\frac{X_I - X_J}{IJ}\right)_0 dX_I + \left(\frac{Y_I - Y_J}{IJ}\right)_0 dY_I + \left(\frac{X_J - X_I}{IJ}\right)_0 dX_J$$
$$+ \left(\frac{Y_J - Y_I}{IJ}\right)_0 dY_J = k_{IJ} + v_{IJ} \quad (16.20)$$

In Equation (16.20), $k_{ij} = L_{ij} - (IJ)_0$, where L_{ij} is the observed length of the line, and $(IJ)_0$ the length based upon initial approximations for the coordinates of points I and J. It is computed as

$$(IJ)_0 = \sqrt{(X_{J_0} - X_{I_0})^2 + (Y_{J_0} - Y_{I_0})^2} \quad (16.21)$$

For a specific distance such as AB of Figure 16.1, Equation (16.21) would be written as

$$\left(\frac{X_A - X_B}{AB}\right)_0 dX_A + \left(\frac{Y_A - Y_B}{AB}\right)_0 dY_A + \left(\frac{X_B - X_A}{AB}\right)_0 dX_B$$
$$+ \left(\frac{Y_B - Y_A}{AB}\right)_0 dY_B = k_{AB} + v_{AB} \quad (16.22)$$

In Equation (16.22), each coefficient of the unknown dx and dy corrections is evaluated using initial approximations selected for the unknown coordinates of stations A and B; k_{AB} is $L_{AB} - (AB)_0$, where L_{AB} is the observed distance, and $(AB)_0$ the distance computed using Equation (16.21) and the approximate coordinates; and v_{AB} the residual error in the distance.

---

[4]The complete development of this linearized equation, and the linearized equations for azimuths and horizontal angles, which are given in the following subsections, are presented in Ghilani (2018) cited in the bibliography at the end of this chapter.

Example 16.8

Write the linearized observation equation for the distance AB whose observed length is 132.823 m. Assume the approximate coordinates for station A and B are (1023.151, 873.018) and (1094.310, 985.163), respectively.

Solution

Step 1: Compute the appropriate coordinate differences

$$(X_B - X_A)_0 = 1094.310 - 1023.151 = 71.159 \text{ m}$$
$$(Y_B - Y_A)_0 = 985.163 - 873.018 = 112.145 \text{ m}$$

Step 2: Compute $(AB)_0$

$$(AB)_0 = \sqrt{71.159^2 + 112.145^2} = 132.816 \text{ m}$$

Step 3: Substitute the appropriate values into Equation (16.22) to develop the linearized observation equation as

$$\left(\frac{-71.159}{132.816}\right) dX_A + \left(\frac{-112.145}{132.816}\right) dY_A + \left(\frac{71.159}{132.816}\right) dX_B$$

$$+ \left(\frac{112.145}{132.816}\right) dY_B = (132.823 - 132.816) + v_{AB}$$

Reducing:

$$-0.53577 \, dX_A - 0.84436 \, dY_A + 0.53577 \, dX_B + 0.84436 \, dY_B = 0.007 + v_{AB}$$

16.9.3 The Azimuth Observation Equation

Equation (10.11) expressed the azimuth of a line in terms of the X and Y coordinates of the line's end points. That expression is written here in observation equation form, and to make it general, the line designation has been changed from AB to IJ, and thus subscripts I and J have replaced A and B:

$$Az_{IJ} + v_{IJ} = \tan^{-1}\left(\frac{X_J - X_I}{Y_J - Y_I}\right) + C \tag{16.23}$$

In Equation (16.23), Az_{IJ} is the observed azimuth of line IJ, v_{IJ} the residual error in the observation and the coordinates in the expression on the right side of the equation are most probable values of the line's end points. The value of the constant C depends on the direction of the line as discussed in Section 11.2. If the azimuth of the line is between 0° and 90°, the value of C is 0°. If the azimuth of the line is between 90° and 270°, C is 180° and if the line's azimuth is between 270° and 360°, C is 360°. Equation (16.23) is nonlinear, but again by applying Equation (16.18) the following linearized form of this equation is obtained:

$$\rho\left(\frac{Y_I - Y_J}{IJ^2}\right)_0 dX_I + \rho\left(\frac{X_J - X_I}{IJ^2}\right)_0 dY_I + \rho\left(\frac{Y_J - Y_I}{IJ^2}\right)_0 dX_J$$

$$+ \rho\left(\frac{X_I - Y_J}{IJ^2}\right)_0 dY_J = k_{IJ} + v_{IJ} \quad (16.24)$$

As with the linearized distance observation equation, the coefficients of the unknown dx and dy terms in Equation (16.24) are obtained by using initial approximations for the coordinates of the end points of the line. The IJ^2 terms in the denominators of the coefficients are simply the squares of the line lengths as computed using initial approximations for the coordinates in Equation (16.21). The right side of Equation (16.24), which includes the constant term k_{IJ} and the residual, is given in seconds. Thus, to make the units consistent on both sides of the equation, the coefficients on the left are multiplied by rho (ρ), which is 206,265 sec/rad. Alternatively, k_{IJ} and v_{IJ} could be converted to radians to make the units consistent. The constant term k_{IJ} is computed as follows:

$$k_{IJ} = Az_{IJ} - \tan^{-1}\left(\frac{X_{J_0} - X_{I_0}}{Y_{J_0} - X_{I_0}}\right) + C \quad (16.25)$$

In Equation (16.25), Az_{IJ} is the observed azimuth, the arc tangent function is the computed azimuth based upon initial approximations for the coordinates, and C is the constant, as previously described.

16.9.4 The Angle Observation Equation

As illustrated in Figure 16.9, an angle can be expressed as the difference between the azimuths of two lines. Thus, angle BIF in the figure is simply azimuth IF minus azimuth IB. The nonlinear observation equation for angle, BIF, is therefore

$$\theta_{BIF} + v_{BIF} = \tan^{-1}\left(\frac{X_F - X_I}{Y_F - Y_I}\right) - \tan^{-1}\left(\frac{X_B - X_I}{Y_B - Y_I}\right) + D \quad (16.26)$$

In Equation (16.26), θ_{BIF} is the observed value for angle BIF and v_{BIF} the residual error in the observation. The right side of this equation is simply the

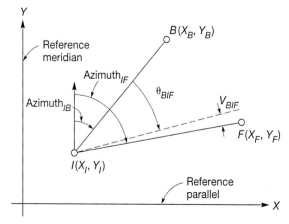

Figure 16.9 Measured angle expressed in terms of coordinates. (Note that an angle is simply the difference between the two azimuths.)

difference in the azimuths IF and IB, where these azimuths are expressed in the manner of Equation (16.23). The constant term $D = C_{IF} - C_{IB}$, where C was defined for Equation (16.23), and C_{IF} and C_{IB} apply to the azimuths of IF and IB, respectively. By applying Equation (16.18), the following linearized form of Equation (16.26) is obtained

$$\rho\left(\frac{Y_I - Y_B}{IB^2}\right)_0 dX_B + \rho\left(\frac{X_B - X_I}{IB^2}\right)_0 dY_B + \rho\left(\frac{Y_B - Y_I}{IB^2} - \frac{Y_F - Y_I}{IF^2}\right)_0 dX_I$$
$$+ \rho\left(\frac{X_I - X_B}{IB^2} - \frac{X_I - X_F}{IF^2}\right)_0 dY_I + \rho\left(\frac{Y_F - Y_I}{IF^2}\right)_0 dX_F \quad (16.27)$$
$$+ \rho\left(\frac{X_I - X_F}{IF^2}\right)_0 dY_F = k_{BIF} + v_{BIF}$$

In Equation (16.27), the term k_{BIF} is

$$k_{BIF} = \theta_{BIF} - \left[\tan^{-1}\left(\frac{Y_F - Y_I}{X_F - X_I}\right)_0 - \tan^{-1}\left(\frac{X_B - X_I}{Y_B - Y_I}\right)_0 + D\right]$$

where θ_{BIF} is the observed value for the angle. As with the distance and azimuth equations, the coefficients of the unknown corrections dX and dY in Equation (16.27) are developed using approximate coordinates for stations B, I, and F. In this equation, B represents the backsight station, I the instrument station, and F the foresight station for clockwise angle BIF. In Equation (16.27), it should again be noted that the coefficients on the left side of the equation are multiplied by $\rho(206,265''/\text{rad})$, so that the units on both sides are seconds. For a specific angle, such as angle GAB of Figure 16.1, after substitution of corresponding subscripts into Equation (16.27), the following linearized angle observation equation results:

$$\rho\left(\frac{Y_A - Y_G}{AG^2}\right)_0 dX_G + \rho\left(\frac{X_G - X_A}{AG^2}\right)_0 dY_G + \left(\frac{Y_G - Y_A}{AG^2} - \frac{Y_B - Y_A}{AB^2}\right)_0 dX_A$$
$$+ \left(\frac{X_A - X_G}{AG^2} - \frac{X_A - X_B}{AB^2}\right)_0 dY_A + \rho\left(\frac{Y_B - Y_A}{AB^2}\right)_0 dX_B \quad (16.28)$$
$$+ \rho\left(\frac{X_A - X_B}{AB^2}\right)_0 dY_B = k_{GAB} + v_{GAB}$$

■ Example 16.9

Angle GAB was observed as $107°29'40''$. The backsight station G, instrument station A, and foresight station B had the following approximate X and Y coordinates, respectively: (578.741, 1103.826); (415.273, 929.868); and (507.934, 764.652). (Note that all coordinate values are given in units of meters.) Write the linearized observation equation for this angle.

Solution

Step 1: Compute the appropriate coordinate differences for substitution into Equation (16.28).

$$(X_G - X_A)_0 = 578.741 - 415.273 = -163.468 \text{ m}$$
$$(Y_G - Y_A)_0 = 1103.826 - 929.868 = 173.958 \text{ m}$$
$$(X_B - X_A)_0 = 507.934 - 415.273 = 92.661 \text{ m}$$
$$(Y_B - Y_A)_0 = 764.652 - 929.868 = -165.216 \text{ m}$$

Step 2: Compute the distances AG and AB, and angle GAB_0

$$(AG)_0 = \sqrt{(163.468)^2 + (173.958)^2} = 238.711 \text{ m}$$
$$(AB)_0 = \sqrt{(92.661)^2 + (-165.216)^2} = 189.726 \text{ m}$$
$$GAB_0 = \tan^{-1}\left(\frac{92.661}{-165.216}\right) - \tan^{-1}\left(\frac{163.468}{173.958}\right) + 180° = 107°29'42''$$

Step 3: Substitute the appropriate values into Equation (16.28).

$$\rho\left(\frac{-173.958}{238.711^2}\right) dX_G + \rho\left(\frac{163.468}{238.711^2}\right) dY_G$$
$$+ \rho\left(\frac{173.958}{238.711^2} - \frac{-165.216}{189.726^2}\right) dX_A$$
$$+ \rho\left(\frac{-16.468}{238.711^2} - \frac{-92.661}{189.726^2}\right) dY_A + \rho\left(\frac{-165.216}{189.726^2}\right) dX_B$$
$$+ \rho\left(\frac{-92.661}{189.726^2}\right) dY_B$$
$$= (107°29'40'' - 107°29'42'') + v_{GAB}$$

Reducing:

$$-629.684'' \, dX_G + 591.713'' \, dY_G + 1579.405'' \, dX_A$$
$$- 59.065'' \, dY_A - 949.721'' \, dX_B - 532.649'' \, dY_B = -2'' + v_{GAB}$$

16.9.5 A Traverse Example Using WOLFPACK

Traverse adjustments by the least-squares method involve distance and angle observation equations, and sometimes include azimuth observation equations as well. Because of the lengthy calculations involved in forming and solving the observation equations, and because the solution is iterative, which requires repetitive computations, the least-squares adjustment for even small traverses should be done by computers. To perform a least-squares adjustment by computer, a data file must be prepared in which all observations and their identity (given by the end stations of lines for distances and azimuths, and by backsight, instrument and foresight stations for angles), must be entered. The computer can be programmed to calculate initial approximations for the coordinates of the unknown stations by using some limited amount of the observed data. In a traverse like that of Figure 11.1(a) for example, the angles at stations A, B, C, and D, and distances AB, BC, CD, and DE could be used to compute coordinates of the four unknown stations B, C, D, and E. Once the initial approximations have been calculated, a computer can easily determine the coefficients of the unknowns in the observation equations, as well as the constant terms. By employing Equation (16.2), relative weights can be determined from the standard deviations of the observed quantities. Thus, the computer is able to prepare the A, W, and L matrices, so that Equation (16.7) can be solved.

The WOLFPACK software, which is available on the companion website, has been used to adjust the traverse network of Figure 16.1. For this adjustment, station A with coordinates of $x = 415.273$ m and $y = 929.868$ m, was held fixed. Also azimuth AB was held fixed at $150°42'51''$ by applying a large weight, which was obtained by assigning it a standard deviation of $\pm 0.001''$. The coordinates of station A fix the survey in position while the azimuth AB fixes the survey in rotation. These two elements are required for any horizontal adjustment. The observed data for the traverse network, which includes distance observations and their standard deviations, and angle observations and their standard deviations, are listed in Tables 16.3 and 16.4, respectively.

Figure 16.10 shows the format and order of preparing the data file for input to the least-squares adjustment program of the WOLFPACK software. A printout giving the results of the adjustment is shown in Figure 16.11. Note that this latter table lists the adjusted coordinates of all new stations, as well as adjusted values for all distance and angle observations (obtained by adding the residuals to their corresponding observed values). The adjusted azimuth of line AB is also listed, which is the same as its input value. This is expected since that azimuth was held in the adjustment by assigning it a large weight.

TABLE 16.3 DISTANCE OBSERVATIONS FOR NETWORK SHOWN IN FIGURE 16.1

From	To	Distance (m)	σ (m)
A	B	189.436	0.007
B	C	122.050	0.007
C	D	121.901	0.007
D	E	145.256	0.007
E	F	168.180	0.007
F	G	231.021	0.007
G	A	238.714	0.007
G	H	143.780	0.007
H	K	119.631	0.007
K	E	114.695	0.007
H	J	96.036	0.007
J	C	85.908	0.007

TABLE 16.4 ANGLE OBSERVATIONS FOR NETWORK SHOWN IN FIGURE 16.1

Backsight Station	Instrument Station	Foresight Station	Angle	σ
G	A	B	107°29′40″	8.9″
A	B	C	94°44′24″	11.7″
B	C	D	235°09′26″	13.7″
C	D	E	104°08′40″	12.7″
D	E	F	124°27′36″	11.2″
E	F	G	121°37′08″	9.5″
F	G	A	112°23′00″	8.3″
F	G	H	38°25′46″	9.9″
G	H	J	243°15′20″	14.6″
H	J	C	135°08′30″	18.0″
J	H	G	116°44′44″	14.6″
J	H	K	296°44′38″	15.0″
H	K	E	131°16′30″	14.3″
K	E	F	68°40′36″	12.3″

These output results contain some of the quantities that are needed to meet newer surveying accuracy standards as was noted in Section 16.1. For those who wish to see this problem programmed, it is also demonstrated in the Mathcad® worksheet *hlsq.xmcd*, which is on the companion website for this book.

■ 16.10 THE ERROR ELLIPSE

Error ellipses depict a two-dimensional representation of the uncertainties of the adjusted coordinates of points as determined in a least-squares adjustment. They can be plotted at enlarged scales directly on scaled diagrams showing the points in the horizontal survey. When plotted in this manner, their sizes and orientations enable a quick visual analysis to be made of the overall relative precisions for all adjusted points. As discussed later in this section, this is useful in planning surveys and in analyzing the results of surveys for acceptance or rejection.

On the output listing of Figure 16.11, the adjusted coordinates for the stations in Figure 16.1 are listed, and to their right are columns titled σ_U, σ_V, and t. Respectively, these contain the semimajor axes, semiminor axes, and clockwise rotation angle from the Y-axis to the semimajor axis of the ellipse computed at each station. To compute these three terms, values from the Q_{xx} matrix (see Section 16.5), and the standard deviation of unit weight [see Equations (16.9) and (16.10)] are used with the following formulas.[5]

---

[5] For the derivations of these equations, see Ghilani (2010), which is cited in the bibliography at the end of this chapter.

```
Example 16.9.5 data for least squares adjustment    {Title line}
12 14 1 1 10              {Number of distances, angles, azimuths, control and total stations}
A 415.273 929.868         {Control station: Identification, X, Y}
B 507.934 764.652         {Unknown station: Identification, X, Y}
C 618.952 815.353
D 723.852 753.287
E 826.128 856.438
F 794.659 1021.655
G 578.741 1103.826
H 652.221 980.245
J 600.595 899.272
K 713.362 877.418
A B 189.436 0.007         {Distance obs: Stations: From, To, Measured Distance, and Std Dev}
B C 122.050 0.007
C D 121.901 0.007
D E 145.256 0.007
E F 168.180 0.007
F G 231.021 0.007
G A 238.714 0.007
G H 143.780 0.007
H K 119.631 0.007
K E 114.695 0.007
H J 96.036 0.007
J C 85.908 0.007
G A B 107 29 40 8.9       {Angle obs: Stations: B, I, F, Measured Angle, and Std Dev}
A B C 94 44 24 11.7
B C D 235 09 26 13.4
C D E 104 08 40 12.7
D E F 124 27 36 11.2
E F G 121 37 08 9.5
F G A 112 23 00 8.3
F G H 38 25 46 9.9
G H J 243 15 20 14.6
H J C 135 08 30 18
J H G 116 44 44 14.6
J H K 296 44 38 15.0
H K E 131 16 30 14.3
K E F 68 40 36 12.3
A B 150 42 51 0.001       {Azimuth obs: Stations: I, F, Observed Azimuth, and Standard Deviation}
```

Figure 16.10 Data file for adjustment of Figure 16.1.

1. Rotation angle, t

$$\tan(2t) = \frac{2q_{xy}}{q_{yy} - q_{xx}} \quad (16.29)$$

In Equation (16.29), the values of q_{xx} and q_{yy} are the diagonal elements from the Q_{xx} matrix, and q_{xy} is the off-diagonal element in the Q_{xx} matrix for a particular station. When computing t, it is important to establish its quadrant before dividing by 2. Failure to do this will result in a semimajor axis that is shorter than the semiminor axis of the ellipse. Obviously, this is an indication that something is wrong.

2. *Semimajor axis*:

$$\sigma_U = \sigma_0 \sqrt{q_{xx} \sin^2(t) + 2q_{xy} \cos(t) \sin(t) + q_{yy} \cos^2(t)} \quad (16.30)$$

3. *Semiminor axis*:

$$\sigma_V = \sigma_0 \sqrt{q_{xx} \cos^2(t) - 2q_{xy} \cos(t) \sin(t) + q_{yy} \sin^2(t)} \quad (16.31)$$

458 ADJUSTMENTS BY LEAST SQUARES

```
*****************
Adjusted Stations
*****************
 Station      N           E         σn       σe       σu       σv       t
===========================================================================
    B       764.645    507.938    0.0038   0.0021   0.0044   0.0000   150.7°
    C       815.350    618.955    0.0049   0.0046   0.0050   0.0045   159.3°
    D       753.285    723.867    0.0069   0.0064   0.0074   0.0058    36.6°
    E       856.441    826.133    0.0092   0.0053   0.0093   0.0052     7.5°
    F      1021.654    794.661    0.0086   0.0058   0.0091   0.0049   156.3°
    G      1103.827    578.746    0.0045   0.0058   0.0060   0.0042   111.7°
    H       980.245    652.226    0.0061   0.0049   0.0063   0.0047   157.6°
    J       899.270    600.599    0.0058   0.0050   0.0058   0.0050   176.2°
    K       877.418    713.370    0.0073   0.0056   0.0073   0.0056   176.8°

******************************
Adjusted Distance Observations
******************************
 Station   Station
Occupied   Sighted   Distance       V          σ
=================================================
    A        B       189.434    -0.0016    0.0044
    B        C       122.048    -0.0022    0.0045
    C        D       121.895    -0.0056    0.0044
    D        E       145.257     0.0006    0.0044
    E        F       168.184     0.0040    0.0042
    F        G       231.024     0.0028    0.0042

***************************
Adjusted Angle Observations
***************************
  Station      Station     Station
Backsighted   Occupied    Foresighted     Angle         V        σ"
====================================================================
     G           A            B        107°29'39"    -0.8"     5.0
     A           B            C         94°44'17"    -6.6"     6.4
     B           C            D        235°09'20"    -5.6"     7.8
     C           D            E        104°08'39"    -1.2"     7.1
     D           E            F        124°27'46"     9.6"     5.9
     E           F            G        121°37'16"     7.9"     5.1
     F           G            A        112°23'03"     2.7"     4.4

*****************************
Adjusted Azimuth Observations
*****************************
 Station   Station
Occupied   Sighted      Azimuth         V           σ"
=======================================================
    A        B       150°42'51"        0.0"        0.0"

Reference s₀ = ±0.6983
```

Figure 16.11 Abbreviated results from adjustment of data file in Figure 16.10 from WOLFPACK.

To demonstrate these calculations, part of the Q_{xx} and X matrices that were generated in the least-squares adjustment of the horizontal network in Section 16.9.4 are listed next. (Note: These were not shown in the abbreviated output listing of Figure 16.11.) Only those parts of the matrices that pertain to stations B and C are shown. The elements in the rows of the X matrix indicate the order of the unknowns and identify the elements of Q_{xx} that apply in computing error ellipses. The upper left 2×2 submatrix of Q_{xx} (shown bold), contains the elements that apply to station B. Because X_B is in row one of the X matrix, q_{xx} for point B occupies the row one column one, or one, one position. The Y_B coordinate is in row two of the X matrix and thus q_{yy} is located in the two, two position of X. Also, the one, two element (or the two, one element which is the same because of symmetry) contains q_{xy}. The following computations yield the error ellipse data for station B:

$$_{18}[Q_{xx}]^{18} = \begin{bmatrix} \mathbf{0.00009440} & \mathbf{-0.00001683} & 0.00000739 & -0.00001304 & \cdots \\ \mathbf{-0.00001683} & \mathbf{0.00003001} & -0.00001318 & 0.00002325 & \cdots \\ 0.00000739 & -0.00001318 & 0.00004327 & -0.00000288 & \cdots \\ 0.00001304 & 0.00002325 & -0.00000288 & 0.00004981 & \cdots \\ \vdots & \vdots & \vdots & \vdots & \ddots \end{bmatrix}$$

$$_{18}X^1 = \begin{bmatrix} X_B \\ Y_B \\ X_C \\ Y_C \\ \vdots \end{bmatrix}$$

Step 1: By Equation (16.29), compute rotation angle, t.

$$\tan(2t) = \frac{2(-0.00001683)}{0.00003001 - 0.00000944} = -1.59261$$

Since the numerator is negative and the denominator is positive, angle $2t$ is in the fourth quadrant,[6] and thus angle t is

$$t = \frac{1}{2}[\tan^{-1}(-1.59261) + 360°] = \frac{1}{2}[302.1247164°] = 151°03'44''$$

Step 2: Compute the semimajor axis using Equation (16.30). In these computations, the standard deviation of unit weight, σ_0, is taken from the bottom of the printout given in Figure 16.11, and has been rounded to 0.70.

$$\sigma_U = 0.70\sqrt{0.00000944 \sin^2(t) + 2(-0.00001683)\cos(t)\sin(t) + 0.00003001\cos^2(t)}$$
$$= 0.70\sqrt{0.000039447}$$
$$= 0.004 \text{ m}$$

Step 3: Compute the semiminor axis using Equation (16.31).

$$\sigma_V = 0.70\sqrt{0.00000944\cos^2(t) - 2(-0.00001683)\cos(t)\sin(t) + 0.00003001\sin^2(t)}$$
$$= 0.70\sqrt{0.000000002}$$
$$= 0.000 \text{ m}$$

---

[6]In Equation (16.29), $\tan 2t = \sin 2t/\cos 2t$. Thus, the sines and cosines enable determining the quadrant of $2t$. If the sine and cosine are both plus, that is, the numerator and denominator of Equation (16.29) are both plus, then $2t$ is in the first quadrant (between 0° and 90°). Similarly, if the numerator is plus and the denominator minus, $2t$ is in the second quadrant; and if the numerator and denominator are both minus, $2t$ is in the third quadrant.

Note again that only the upper left 2 × 2 submatrix (in bold) of the full Q_{xx} matrix was necessary for the computations. To compute the error ellipse for Station C, only the (3,3), (3,4), and (4,4) elements of Q_{xx} are needed. This pattern is continued for each station.

In the computed error ellipse for station B of this example, the semiminor axis is nearly zero. Also note that the rotation angle of the semimajor axis closely matches that of the azimuth of the line AB. This could be predicted since the azimuth AB was held fixed during the adjustment by giving it a large weight. This shows both the power and the danger of weights. In this example, the large weight is necessary to fix the horizontal network rotationally in azimuth. If the azimuth is not fixed by weighting, the network would be free to rotate about Station A and no solution would be found. However, inappropriate application of weights can cause unwanted corrections in the adjustment. *In least-squares adjustments, it is very important to weight the observations according to their estimated uncertainties.*

As noted previously, when an adjustment is completed, it is often informative to view a graphic of the error ellipses. The error ellipses for this example, shown in Figure 16.12, have had their semiaxes magnified 200 times to make their relatively small values easily visible on the plot. If an error ellipse approximates a circle, this would indicate that point is of approximately equal precision in all directions. Long and slender error ellipses indicate low precisions in their long directions (along their semimajor axes) and high precisions in their narrow directions (along their semiminor axes). In Figure 16.12, it can be seen that the directions of the semimajor axes for stations D, E, F, and G are aligned on circles with their centers approximately at A. This shows the rotational instability in the observational system, that is, the held azimuth of line AB alone fixes the rotation in the network. This instability could be improved by observing an azimuth on another line such as EF, or by connecting either Stations E or F to another neighboring control point (assuming one is available). In Figure 16.12, it can also be seen that the largest uncertainties exist on the stations that are farthest from the control station. This is the usual manner that errors propagate in observational systems—that is, the

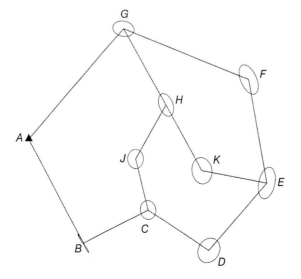

Figure 16.12
Error ellipses plotted at 200 times their actual sizes.

farther the unknown stations are from the control, the larger the estimated errors in their coordinates. With this in mind, the sizes of the errors at all stations, but especially *E* and *F*, could be reduced by connecting either of these stations to a neighboring control station (assuming one is available).

The analysis of plotted error ellipses is very useful in scrutinizing results of adjusted horizontal surveys, as illustrated by the simple analyses given in the previous paragraph. As demonstrated, both the sizes and shapes of error ellipses give an immediate visual impression about the relative accuracies of the points in a network and enable the most efficient plan to be developed for making additional observations to strengthen the observational scheme.

16.11 ADJUSTMENT PROCEDURES

Regardless of the nature of the specific adjustment problem, certain procedures should be followed. For example, before the adjustment is undertaken, all data must be carefully analyzed for blunders. Mistakes such as station misidentifications, transcription mistakes, reading blunders, and others must be identified and corrected. Failure to remove them will result in either an unsatisfactory adjustment or no adjustment at all. In several types of surveys, performing loop closures on the data can identify blunders. This is true in leveling, GNSS networks, and horizontal surveys, including traversing. Also in traverses, the methods discussed in Section 10.16 can be employed to detect blunders.

The minimum amount of control required for adjusting observations varies with the type of problem. In differential leveling only one benchmark is needed, and in a network of GNSS baseline observations, only one station with known coordinate values is necessary. For horizontal surveys such as traverses and networks, one station with known coordinates and one course with known direction must be available. If more than the minimum amount of control is present, the adjustment should be performed in two stages as a further means of detecting blunders. The first adjustment, called a *minimally constrained adjustment*, should contain only the minimum amount of control necessary to fix the observations in space. This should be followed by a *constrained adjustment*, in which all available control is used.

The minimally constrained adjustment provides checks on the geometric closures and the consistency of the observations. After the adjustment is completed, the residuals should be analyzed. Any unusually large residuals, or a preponderance of minus or plus residuals[7] will be keys to the existence of blunders in the data. However, even though a minimally constrained adjustment will check the data for *internal consistency*, it may fail to identify the presence of systematic errors and mistakes. For instance, assume during the observation of the leveling circuit shown in Figure 16.13 that a mistake of +1 m occurred during the observation of line 1 and that a −1-m error occurred during the observation of line 7. Even though every observed benchmark would have a 1m error, the loop misclosure

---

[7]According to normal distribution theory, a group of random errors should contain approximately an equal number of positive and negative residuals.

462 ADJUSTMENTS BY LEAST SQUARES

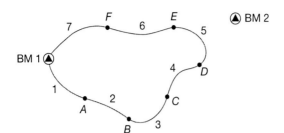

Figure 16.13
Differential leveling loop.

would not uncover this mistake and the observations would close geometrically. However, in that figure, if an additional line of differential levels is observed from BM 2 to D, the presence of these mistakes would become apparent.

Upon acceptance of the minimally constrained adjustment, all additional available control should be added to the data and the constrained adjustment performed. This will aid in identifying compensating or systematic errors in the data and blunders in control. For instance, suppose that the coordinate values for control station A in Figure 16.12 are state plane coordinate values, but the distances are not properly reduced to the state plane grid (see Chapter 20). In this situation, the geometric closures of the adjustment would appear to be fine but because of the distance scaling error, the computed coordinates for the unknown stations would all be incorrect. In this case, the minimally constrained adjustment would fail to catch the scaling error. However, if observations were connected to a second control station with known state plane coordinates, the scaling error would become apparent in the constrained adjustment. Similarly, a blunder in the control coordinates or benchmarks will not become apparent until the constrained adjustment is performed. Thus, it is important to perform a minimally constrained adjustment to obtain geometric checks on the data, and a constrained adjustment to find possible compensating errors, systematic errors, and control blunders. It follows logically that every survey should have redundant observations to provide geometric checks and isolate mistakes caused by careless work.

Upon acceptance of the constrained adjustment, the post-adjustment statistics, as given in Sections 16.5, 16.7, and 16.10, should be computed. When possible, these statistical values should be compared against published accuracy standards. If the survey fails to meet the required standards, additional observations should be taken or the work repeated using more precise equipment.

■ 16.12 OTHER MEASURES OF PRECISION FOR HORIZONTAL STATIONS

Since error ellipses are part of a bivariate distribution, their probability level is between 35% and 39%. Generally, surveyors prefer to state their results at a much higher level of confidence. For the semiminor and semimajor axes of the error ellipse, this is accomplished by using a multiplier that is based on critical values taken from an F distribution. This distribution is a function of the number of *degrees of freedom* (number of redundant observations) that existed in the adjustment. Some of the critical values from the F distribution are shown in

TABLE 16.5 $F_{\alpha, 2, \text{degrees of freedom}}$ CRITICAL STATISTIC VALUES FOR SELECTED PROBABILITIES

	Probability		
Degrees of Freedom	90%	95%	99%
1	49.50	199.50	4999.50
2	9.00	19.00	99.00
3	5.46	9.55	30.82
4	4.32	6.94	18.00
5	3.78	5.79	13.27
9	3.01	4.26	8.02
10	2.92	4.10	7.56
15	2.70	3.68	6.36
20	2.59	3.49	5.85
30	2.49	3.32	5.39

Table 16.5.[8] The multipliers for the semimajor and semiminor axes of an error ellipse expressed at other probabilities levels are determined from

$$c = \sqrt{2(F_{\alpha, 2, \text{degrees of freedom}})} \tag{16.32}$$

where the semimajor and semiminor axes would be computed as

$$\sigma_{U\%} = c\sigma_u \tag{16.33}$$
$$\sigma_{V\%} = c\sigma_v \tag{16.34}$$

Other measures of precision that can be used include the *circular error probable* (CEP), which is a function of the computed standard deviations for a horizontal station, or

$$\text{CEP} = 0.5887(\sigma_X + \sigma_Y) \tag{16.35}$$

The 90% region of the CEP is called the *circular map accuracy standard* (CMAS) and is computed as

$$\text{CMAS} = 1.8227 \, \text{CEP} \tag{16.36}$$

The *distance root mean square* (DRMS) error is another measure of precision. It can be computed as

$$\text{DRMS} = \sqrt{\sigma_X^2 + \sigma_Y^2} \tag{16.37}$$

---

[8]Critical values for the F distribution can often be obtained in spreadsheets like Excel. To obtain the F distribution critical value for Example 16.10 in Excel enter "=FINV(0.05,2,9)" where $0.05 = 1 - 0.95$, 2 represents the 2 unknowns at the station, and 9 represents the number of degrees of freedom in the adjustment into a cell in the spreadsheet.

■ Example 16.10

What are the 95% probability values for the semimajor and semiminor axes of station B in Section 16.10?

Solution

The adjustment of Figure 16.1 had 12 distance observations, 14 angle observations, and 1 azimuth observation for a total of 27 observations. There were 9 stations having 2 unknowns each for a total of 18 unknowns. Thus, the number of degrees of freedom are 27 − 18, or 9. From Table 16.5, the appropriate F value for nine degrees of freedom is 4.26. By Equation (16.32), the c-multiplier is

$$c = \sqrt{2\,(4.26)} = 2.92$$

From Section 16.10, the values for σ_U and σ_V were 0.004 and 0.000 m, respectively. Thus by Equations (16.33) and (16.34), the 95% values for the semimajor and semiminor axes are

$$\sigma_{U95\%} = 2.92(0.004) = \pm 0.012 \text{ m}$$
$$\sigma_{V95\%} = 2.92(0.000) = \pm 0.000 \text{ m}$$

The Federal Geographic Data Committee (FGDC) introduced the error circle, which produces a single uncertainty for a point based upon the radius of a circle in which the true location of the point falls 95% of the time. This error circle is derived from the lengths of the semiaxes of the standard error ellipse and a multiplier from the bivariate normal distribution curve. According to Leenhouts (1985), the 95% multiplier can be approximated as

$$k_p = 1.9600790 + 0.004071C + 0.114276C^2 + 0.371625C^3 \quad (16.38)$$

where k_p is the 95% multiplier and $C = S_v/S_u$. The radius of the 95% error circle is then computed as

$$r_{95\%} = k_p S_u \quad (16.39)$$

■ Example 16.11

Compute the radius of the 95% error circle for station B in Section 16.10.

Solution

In this example, we will need more than the three decimal places provided in Section 16.10. The value of S_u is ±0.0043965 and the value for S_v is ±0.0000313. Thus,

$$C = \frac{00.0000313}{0.0043964} = 0.00712$$

By Equation (16.39) the 95% multiplier is

$$k_p = 1.9600790 + 0.004071(0.00712) + 0.114276(0.00712)^2$$
$$+ 0.371625(0.00712)^3 = 1.96011$$

And thus, the radius of the 95% error circle is

$$r_{95\%} = 1.96011(0.0043964) = \pm 0.009 \text{ ft}$$

Note that this value is less than the semimajor axis of the 95% error ellipse computed in Example 16.9.

16.13 SOFTWARE

In a typical surveying office, software is used to perform the adjustments discussed in this chapter. On the companion website is the WOLFPACK program, which incorporates some of the adjustments discussed in this chapter. The help file, which accompanies the software, describes the formats for each type of data file supported. For those wishing to program these adjustments in a higher-level programming language, several Mathcad worksheets are available on the companion website also. In particular, the worksheet *LLSQ.XMCD* demonstrates a differential leveling least-squares adjustment, *HLSQ.XMCD* demonstrates a horizontal survey least-squares adjustment, and *GPS.XMCD* demonstrates a GNSS network least-squares adjustment. Additionally, the Excel spreadsheet *C16.XLS* on the companion website demonstrates the computations in Examples 16.4, 16.5, 16.6, and the solution of nonlinear equations presented in Section 16.9.

16.14 CONCLUSIONS

As discussed in Chapter 3, the presence of errors in observations is inevitable. However if the method of least squares is employed, the sizes of the errors can be evaluated and if they are within acceptable limits, the observations can be adjusted to determine the most probable values for the unknowns. If some of the observations contain unacceptable errors, these observations must be reobserved or removed from the final adjustment. The advantages of the least-squares method over other adjustment techniques are many. Some benefits of a least-squares adjustment are that it: (1) conforms to the laws of probability, (2) provides the most probable solution for a given set of observations, (3) allows individual weighting of observations, (4) forces geometric closures on the observations, (5) adjusts all observations simultaneously, (6) provides a single unique solution for a set of data, and (7) yields estimated precisions of adjusted quantities. The least-squares method is readily programmed for computer solution and the data is easily prepared for making adjustments. Because of these advantages and the fact that data from least-squares adjustments are now necessary for assessing compliance of surveys with modern standards such as the Federal Geodetic Control Subcommittee (FGCS) standards and specifications for GNSS Relative Positioning (see Section 14.5.1) and the ALTA-ACSM Land Title Survey Standards (see Section 21.10), all surveying offices should employ the least-squares method of adjustment.

PROBLEMS

Asterisks (*) indicate problems that have partial answers given in Appendix G.

16.1 List at least three checks on observational data that indicate the presence of errors in a closed polygon traverse.

16.2 What is true about the set of residuals produced by a least-squares adjustment?

16.3 Define the term redundant observations.

16.4* What is the most probable value for the following set of 10 distance observations in meters? 532.688, 532.682, 532.682, 532.684, 532.689, 532.686, 532.690, 532.684, 532.686, 532.686.

16.5 What is the standard deviation of the adjusted value in Problem 16.4?

16.6 Three horizontal angles were observed around the horizon of station A. Their values are 35°18'16", 85°19'27", and 239°22'26". Assuming equal weighting, what are the most probable values for the three angles?

16.7 What are the standard deviations of the adjusted values in Problem 16.6?

16.8 In Problem 16.6, the standard deviations of the three angles are ±2.5", ±1.0", and ±2.0", respectively. What are the most probable values for the three angles?

16.9* Determine the most probable values for the x and y distances of Figure 16.2, if the observed lengths of AC, AB, and BC (in meters) are 294.081, 135.467, and 158.607, respectively.

16.10* What are the standard deviations of the adjusted values in Problem 16.9?

16.11 A network of differential levels is run from existing benchmark Juniper through new stations A and B to existing benchmarks Red and Rock as shown in the accompanying figure. The elevations of Juniper, Red, and Rock are 987.56, 856.84, and 899.41 ft, respectively. Develop the observation equations for adjusting this network by least squares, using the following elevation differences.

From	To	Elev. Diff. (m)	σ (m)
Juniper	A	−45.41	0.016
A	B	−40.51	0.020
B	Red	−44.86	0.023
B	Rock	−2.22	0.025

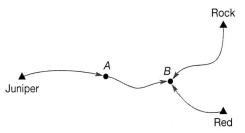

Problem 16.11

16.12 For Problem 16.11, following steps outlined in Example 16.6 perform a weighted least-squares adjustment of the network. Determine the weights based upon the given standard deviations. What are the
  (a)* most probable values for the elevations of A and B?
  (b) standard deviation of unit weight?
  (c) standard deviations of the adjusted elevations?
  (d) adjusted elevation differences and their residuals?
  (e) standard deviations of the adjusted elevation differences?

16.13 Repeat Problem 16.12 using distances for weighting. Assume the following course lengths for the problem.

From	To	Dist (ft)
Juniper	A	500
A	B	800
B	Red	1000
B	Rock	1300

16.14 Use WOLFPACK to do Problems 16.12 and 16.13 and compare the solutions for A and B.

16.15 Repeat Problem 16.12 using the following data.

From	To	Elev. Diff. (m)	σ (m)
Juniper	A	−45.42	0.020
A	B	−40.56	0.028
B	Red	−44.78	0.025
B	Rock	−2.25	0.025

16.16 A network of differential levels is shown in the accompanying figure. The elevations of benchmarks A and G are 835.24 ft and 865.64 ft, respectively. The observed elevation differences and the distances between stations are shown in the following table. Using WOLFPACK, determine the
  (a) most probable values for the elevations of new benchmarks $B, C, D, E, F,$ and H.
  (b) standard deviation of unit weight.
  (c) standard deviations of the adjusted elevations.
  (d) adjusted elevation differences and their residuals.
  (e) standard deviations of the adjusted elevation differences.

From	To	Elev. Diff. (m)	σ (ft)
A	B	30.55	0.022
B	C	−45.22	0.025
C	D	24.34	0.022
D	E	10.38	0.016
E	F	−15.16	0.013
F	A	−4.83	0.011
G	F	−25.59	0.008
G	H	−7.66	0.010
H	D	−13.10	0.009
G	B	0.14	0.010
G	E	−10.42	0.011

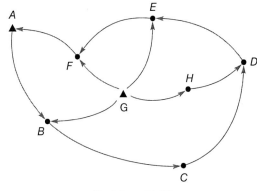

Problem 16.16

16.17 Develop the observation equations for lines *AB* and *BC* in Problem 16.16.

16.18 A network of GNSS observations shown in the accompanying figure was made with two receivers using the static method. Known coordinates of the two control stations are in the geocentric system. Develop the observation equations for the following baseline vector components.

Station	*X* (m)	*Y* (m)	*Z* (m)
Jim	−2,451,759.570	4,651,768.078	3,602,460.828
Al	−2,456,094.078	4.643.564.164	3,609,692.401

Jim to Troy

1420.351 9.13E−6 3.38E−8 1.12E−7
−822.733 9.33E−6 1.29E−7
2507.389 9.11E−6

Al to Troy

5754.864 1.77E−5 7.44E−7 1.28E−7
7381.183 1.78E−5 6.32E−7
−4724.192 1.82E−5

16.19 For Problem 16.18, construct the *A* and *L* matrices.
16.20 For Problem 16.18, construct the covariance matrix.
16.21 Use WOLFPACK to adjust the baselines of Problem 16.18.

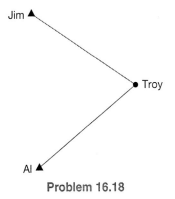

Problem 16.18

16.22 Convert the geocentric coordinates obtained for station Troy in Problem 16.21 to geodetic coordinates using the WGS84 ellipsoidal parameters. (*Hint*: See Section 13.4.3.)

16.23 A network of GNSS observations shown in the accompanying figure was made with two receivers using the static method. Use WOLFPACK to adjust the network, given the following data.

Station	X (m)	Y (m)	Z (m)
Bonnie	−214,600.461	4,500,508.196	4,499,662.395
Tom	−205,255.076	4,500,955.860	4,499,655.435

Bonnie to Ray
4527.069 5.90E−6 −4.58E−8 1.84E−7
6053.237   5.91E−6   1.57E−8
−5764.238   6.04E−6

Bonnie to Herb
5509.160 5.86E−6 1.05E−7 6.85E−8
−2088.296   5.99E−6 1.48E−7
2372.089   5.53E−6

Tom to Ray
−4818.311 5.89E−6 −5.54E−8 −1.62E−7
5605.572   5.81E−6   3.22E−7
−5757.281   6.11E−6

Tom to Herb
−3836.219 5.96E−6 −6.15E−8 1.15E−7
−2535.957   7.68E−6   1.22E−7
2379.050   5.96E−6

Bonnie to Tom (Fixed line—Don't use in adjustment.)
9345.386 6.05E−6 −1.87E−7 −3.94E−8
447.666   5.96E−6   −2.62E−7
−6.959   5.98E−6

16.24 For Problem 16.23, write the observation equations for the baselines "Bonnie to Ray" and "Tom to Herb."

16.25 For Problem 16.23, construct the A, X, and L matrices for the observations.

16.26 For Problem 16.23, construct the covariance matrix.

16.27* After completing Problem 16.23, convert the geocentric coordinates for station Ray and Herb to geodetic coordinates using the WGS84 ellipsoidal parameters. (*Hint*: See Section 13.4.3.)

16.28 Following the procedures discussed in Section 14.5.2, analyze the fixed baseline from station Bonnie to Tom.

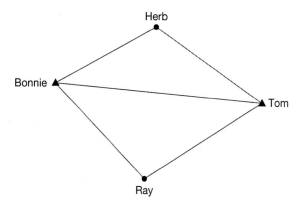

Problem 16.23 through 16.28

16.29 For the horizontal survey of the accompanying figure, determine initial approximations for the unknown stations. The observations for the survey are

Station	X (ft)	Y (ft)	From	To	Azimuth	σ
Dave	2340.12	3363.45	Dave	Wes	13°30'07"	±6.9"

From	To	Distance (ft)	σ (ft)
Dave	Steve	464.39	0.022
Steve	Frank	550.97	0.022
Frank	Wes	443.73	0.022
Wes	Dave	521.53	0.022
Dave	Frank	604.61	0.022
Steve	Wes	788.37	0.022

Backsight Station	Instrument Station	Foresight Station	Angle	σ''
Wes	Dave	Frank	45°40'37"	±8.4
Frank	Dave	Steve	60°22'14"	±9.0
Dave	Steve	Wes	39°28'32"	±8.3
Wes	Steve	Frank	33°02'59"	±7.5
Steve	Frank	Dave	47°06'14"	±8.2
Dave	Frank	Wes	57°13'20"	±9.2
Frank	Wes	Steve	42°37'28"	±8.6
Steve	Wes	Dave	34°28'48"	±7.7

16.30* Using the data in Problem 16.29, write the linearized observation equation for the distance from Steve to Frank.

16.31 Using the data in Problem 16.29, write the linearized observation equation for the angle Wes–Dave–Frank.

16.32 Assuming a standard deviation of ±6.9" for the azimuth line Dave–Wes, use WOLFPACK to adjust the data in Problem 16.29.

16.33* Given the following inverse matrix and a standard deviation of unit weight of 1.23, determine the parameters of the error ellipse.

$$(A^T W A)^{-1} = \begin{bmatrix} q_{xx} & q_{xy} \\ q_{xy} & q_{yy} \end{bmatrix} = \begin{bmatrix} 0.00016159 & -0.00001827 \\ -0.00001827 & 0.00028020 \end{bmatrix}$$

16.34 Compute S_x and S_y in Problem 16.33.

16.35* Compute the 95% radial error for the station in Problem 16.33.

16.36 Given the following inverse matrix and a standard deviation of unit weight of 1.45, determine the parameters of the error ellipse.

$$(A^T A W)^{-1} = \begin{bmatrix} q_{xx} & q_{xy} \\ q_{xy} & q_{yy} \end{bmatrix} = \begin{bmatrix} 0.0005043 & 0.000092 \\ 0.000092 & 0.0002734 \end{bmatrix}$$

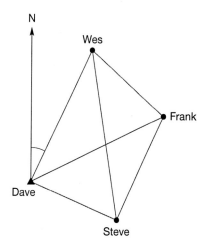

Problem 16.29 through 16.32

16.37 Compute S_x and S_y in Problem 16.36.

16.38 Compute the 95% radial error for the station in Problem 16.36.

16.39 The well-known observation equation for a line is $mx + b = y + v_y$. What is the slope and y-intercept of the best fit line for a set of points with coordinates of (1446.81, 2950.79), (2329.79, 2432.66), (3345.74, 1837.13), (478.72, 3517.64), (4382.98, 1229.16)?

16.40 Use WOLFPACK and the following standard deviations for each observation to do a least-squares adjustment of Example 10.4 and describe any differences in the solution. What advantages are there to using the least-squares method in adjusting this traverse?

Stations	Angle ± σ	Stations	Distance ± σ
E-A-B	100°45′37″ ± 16.7″	AB	647.25 ± 0.027
A-B-C	231°23′43″ ± 22.1″	BC	203.03 ± 0.026
B-C-D	17°12′59″ ± 21.8″	CD	720.35 ± 0.027
C-D-E	89°03′28″ ± 10.2″	DE	610.24 ± 0.027
D-E-A	101°34′24″ ± 16.9″	EA	285.13 ± 0.026
AZIMUTH AB	126°55′17″ ± 0.001″		

BIBLIOGRAPHY

Ghilani, C. D. 2003. "Statistics and Adjustments Explained—Part 1: Basic Concepts." *Surveying and Land Information Science* 63 (No. 2): 73.

———. 2003. "Statistics and Adjustments Explained—Part 2: Sample Sets and Reliability." *Surveying and Land Information Science* 63 (No. 3): 141.

———. 2004. "Statistics and Adjustments Explained—Part 3: Error Propagation." *Surveying and Land Information Science* 64 (No. 1): 23.

———. 2010. *Adjustment Computations: Spatial Data Analysis*. New York, NY: Wiley.

Leenhouts, P. 1985. *On the Computation of Bi-Normal Radial Error*. Navigation 32 (No. 1): 16–28.

Schwarz, C. R. 2005. "The Effects of Unestimated Parameters." *Surveying and Land Information Science* 65 (No. 2): 87.

Tan, W. 2002. "In What Sense a Free Net Adjustment?" *Surveying and Land Information Science* 62 (No. 4): 251.

17
Mapping Surveys

■ 17.1 INTRODUCTION

Mapping surveys are made to determine the locations of *natural* and *cultural* features on the Earth's surface and to define the configuration (*relief*) of that surface. Once located, these features can be represented on maps. Natural features normally shown on maps include vegetation, rivers, lakes, oceans, etc. Cultural (*artificial*) features are the products of people, and include roads, railroads, buildings, bridges, canals, boundary lines, etc. The relief of the Earth includes its hills, valleys, plains, and other surface irregularities. Lines and symbols are used to depict features displayed on maps. Names and legends are added to identify the different objects shown.

Two different types of maps, *planimetric* and *topographic*, are prepared as a result of mapping surveys. The former depicts natural and cultural features in the plan (*X-Y*) view only. Objects displayed are called *planimetric features*. Topographic maps also include planimetric features, but in addition they show the configuration of the Earth's surface. Both types of maps have many applications. They are used by engineers and planners to determine the most desirable and economical locations of highways, railroads, canals, pipelines, transmission lines, reservoirs, and other facilities; by geologists to investigate mineral, oil, water, and other resources; by foresters to locate access- or haul-roads, fire-control routes, and observation towers; by architects in housing and landscape design; by agriculturists in soil conservation work; and by archeologists, geographers, and scientists in numerous fields. Maps are used extensively in geographic information system (GIS) applications (see Chapter 28). Conducting the surveys necessary for preparing maps and the production of the maps from the survey data are the mainstay of many surveying businesses.

Relief is displayed on maps by using various conventions and procedures. For topographic maps, *contours* are commonly used and are preferred by surveyors

Figure 17.1
Total station capturing image of area to be surveyed. (Courtesy Topcon Positioning Systems.)

and engineers. *Digital elevation models* (DEMs) and *three-dimensional perspective models* are methods for depicting relief, made possible by computers. *Color, hachures, shading*, and *tinting* can also be used to show relief, but these methods are not quantitative enough and thus are generally unsuitable for surveying and engineering work. Contours, DEMs, and three-dimensional perspective models are discussed in later sections of this chapter and in Chapter 18.

Traditionally, maps were prepared using manual drafting methods. Now however, as described in Chapter 18, most maps are produced using computers, computer-aided drafting (CAD) software, and survey controllers. Today's survey controllers include drafting software so that field personnel can map their data in the field to check for mistakes and missing elements. Furthermore, modern survey controllers allow field personnel to add photo notes to help identify particular difficult features to describe. The image station shown in Figure 17.1 is capable of capturing images of the area that is to be surveyed to aid in the identification of features on the map. This chapter discusses procedures for collecting planimetric and topographic mapping data.

■ 17.2 BASIC METHODS FOR PERFORMING MAPPING SURVEYS

Mapping surveys are conducted by one of two basic methods: *aerial* (photogrammetric) or *ground* (field) techniques, but often a combination of both is employed. Refined equipment and procedures available today have made photogrammetry and airborne laser scanning (LiDAR), which is discussed in Section 27.18, very accurate and economical. Additionally, as discussed in Section 27.19, small unmanned aerial vehicles (sUAVs) have made it possible for surveyors to perform aerial surveys in house without contracting with an external photogrammetric firm. Hence, almost all mapping projects covering large areas now employ these methods. Ground surveys are still commonly used in preparing large-scale maps of smaller areas. Even when photogrammetry or airborne laser mapping is utilized, ground surveys are necessary to establish control and to field-check mapped features for accuracy. This chapter concentrates on ground methods including conventional instruments and terrestrial laser scanning, and describes several field procedures for locating topographic features, both horizontally and vertically. Photogrammetric methods, sUAVs, and LiDAR are discussed in Chapter 27.

■ 17.3 MAP SCALE

Map scale is the ratio of the length of a feature on a map to the true length of the feature. Map scales are given in three ways: (1) by *ratio* or *representative fraction*, such as 1:2000 or 1/2000; (2) by an *equivalence*, for example, 1 in. = 200 ft; and (3) by graphically using either a bar scale or labeled grid lines spaced throughout the map at uniform distances apart. Graphic scales permit accurate measurements to be made on maps, even though the paper upon which the map is printed may change dimensions.

An equivalence scale of 1 in./100 ft indicates that 1 in. on the map is equivalent to 100 ft on the object. In giving scale by ratio or representative fraction, the same units are used for the map distance and the corresponding object distance, and thus 1:1200 could mean 1 in. on the map is equivalent to 1200 in. on the object, but any other units would also apply. Obviously, it is possible to convert from an equivalence scale to a ratio, and vice versa. As an example, 1 in. = 100 ft is converted to a ratio by multiplying 100 ft by 12, which converts it to inches and gives a ratio of 1:1200. Those engaged in surveying (geomatics) and engineering generally prefer an equivalence scale and grid lines on their maps, while geographers often utilize a representative fraction and bar scale.

Choice of scale depends on the purpose, size, and required precision of the finished map. Dimensions of a standard map sheet, number and type of topographic symbols used, and accuracy requirements for scaling distances from the map are some additional considerations. Maps produced using the English system of units usually have their scales selected to be compatible with one of the standard graduations on engineer's scales. These standard graduations have 10, 20, 30, 40, 50, or 60 units per inch. Thus, scales of 1 in. = 100 ft and 1 in. = 1000 ft are compatible with the 10 scale; 1 in. = 200 ft and 1 in. = 2000 ft are consistent with the 20 scale, and so on. In the metric system, ratios or representative fractions such as 1:1000, 1:2000, 1:5000, and so on are usually employed.

Map scales may be classified as *large*, *medium*, and *small*. Their respective scale ranges are as follows:

Large scale, 1 in. = 200 ft (1:2400) or larger

Medium scale, 1 in. = 200 ft to 1 in. = 1000 ft (1:2400 to 1:12,000)

Small scale, 1 in. = 1000 ft (1:12,000) or smaller

Large-scale maps are applied where relatively high accuracy is needed over limited areas; for example, in subdivision design and the design of engineering projects like roads, dams, airports, and water and sewage systems. Medium scales are often used for applications such as general preliminary planning where larger areas are covered but only moderate accuracy is needed. Applications include mapping the general layout of potential construction sites, proposed transportation systems, and existing facilities. Small-scale maps are commonly used for mapping large areas where a lower order of accuracy will suffice. They are suitable for general topographic coverage, applications in site-suitability analysis, preliminary layout of expansive proposed construction projects, and for special applications in forestry, geology, environmental impact and management, etc.

Maps in graphic form can have their scales enlarged or reduced photographically or digitally using computer applications. The enlargement ratios possible by either of these methods are virtually unlimited. *However, enlargements must be produced with caution since any errors in the original maps or digital data are also magnified, and the enlarged product may not meet required accuracy standards.*

The scale at which a map will be plotted directly affects the choice of instruments and procedures used in performing the mapping survey. This is because the accuracy with which the position of an object is depicted on a map is related to the map's scale, which in turn dictates the accuracy with which features must be surveyed. Consider, for example, a map plotted at a scale of 1 in. = 20 ft. If distances and locations can be scaled from the map to within say 1/50th in., this represents a scaling error of $(1/50)20 = \pm 0.4$ ft. To ensure that the accuracy of the surveyed data does not limit the accuracy with which information can be scaled from a map, features must be located on the map to an accuracy better than ± 0.4 ft for this map. As a safety factor, many surveying and mapping agencies apply a rule of thumb in which they require features to be located in the field to at least twice the scaling accuracy, which in this instance would require accuracy to within ± 0.2 ft or better. Following this same reasoning, if map scale is 1 in. = 200 ft, then ground features should be located to an accuracy of ± 2 ft so as not to limit the accuracy of the map. Another consideration regarding map scale that affects surveying accuracy is the thicknesses of lines used to plot features. Assume, for example, that line widths on a map with a scale of 1 in. = 20 ft are 0.3 mm. This means that each line represents $0.3(240) = 72 mm = 0.24$ ft on the map. Therefore, to accurately depict an object on a map with this line width, the survey needs to be accurate to at least half the line width, or ± 0.1 ft. Obviously, the equipment and procedures used for the mapping work must be selected so that these accuracies are met.

While scaling factors such as those discussed earlier must be taken into account for each specific mapping project, it is important to also consider the possible future use of the map data being collected. Thus, even though the first map produced for a particular project may be a small-scale reconnaissance map, it is possible that as the project progresses medium-scale planning maps and large-scale design maps will be needed, and that some or all of the data collected could also be used for these maps. Thus, even though relaxed accuracies may suffice for the reconnaissance map, for efficiency, the data should be collected to accuracy suitable for other maps that may follow.

■ 17.4 CONTROL FOR MAPPING SURVEYS

Whether the mapping is done by ground or aerial methods, the first requirement for any project is good control. As discussed in Chapter 19, control is classified as either horizontal or vertical.

Horizontal control for a mapping survey is provided by two or more points on the ground, permanently or semipermanently monumented, and precisely fixed in position horizontally by distance and direction or coordinates. It is the basis for locating map features. Horizontal control can be established by the traditional ground surveying methods of *traversing*, *triangulation*, or *trilateration* (see Section 19.12), or by using *GNSS* surveys (see Chapters 14 and 15). For large areas, a sparse network of horizontal (and vertical) control can be densified using *photogrammetry* (see Chapter 27) or GNSS surveys. For small areas, horizontal control for mapping surveys is generally established by traversing or GNSS surveys. Monuments whose positions have been established through higher-order control surveys and referenced in the state plane coordinate systems (see Chapter 20) are used to initiate surveys of all types, but unfortunately more are needed in most locations. When local control is nearby traversing may be the method of choice. The advantage of a GNSS survey is that it can bring datum coordinates into any region where satellite visibility is available and additionally can be used to map the area.

Vertical control is provided by benchmarks in or near the tract to be surveyed and becomes the foundation for correctly portraying relief on a topographic map. Vertical control is usually established by running lines of differential levels starting from and closing on established benchmarks (see Chapters 4 and 5). Project benchmarks are established throughout the mapping area in strategic locations and their elevations determined by including them as turning points in the differential leveling lines. In rugged areas, trigonometric leveling with total station instruments is practical and frequently used to establish vertical control for mapping. GNSS surveys are also suitable for establishing vertical control for topographic mapping but the geodetic heights derived from GNSS surveys must first be converted to orthometric heights using Equation (13.8). This conversion is performed automatically when the appropriate geoid model is loaded into the software. The latter two methods are of sufficient accuracy to support most mapping surveys. However, the local accuracy of the geoid model should be checked to ensure that the orthometric heights derived from GNSS surveys are sufficiently accurate for the purposes of the map. As discussed in Section 13.4.3 once the GRAV-D project progresses and is completed, the accuracy of orthometric heights derived from a GNSS survey may be sufficient for most applications.

Regardless of the methods used in conducting the control surveys for mapping projects, specified maximum allowable closure errors for both horizontal and vertical control should be determined prior to the fieldwork then used to guide it. The locations of the features, which comprise the map (often also called *map details*), are based upon the framework of control points whose positions and elevations are established. Thus, any errors in the surveyed positions or elevations of the control points will result in erroneous locations of the details on the map. Therefore, it is advisable to run, check, and adjust the horizontal and vertical control surveys before beginning to locate map details, rather than carry on both processes simultaneously. The method selected for locating map details will govern the speed, cost, and efficiency of the survey. In later sections of this chapter, the different basic field procedures and the varying equipment that can be used are described.

17.5 CONTOURS

As stated earlier, surveyors and engineers most often use contours to depict relief. The reason is that they provide an accurate quantitative representation of the terrain. Because planimetric features and contours are located simultaneously in most field topographic surveys, it is important to understand contours and their characteristics before discussing the various field procedures used to position them.

A *contour is a line connecting points of equal elevation.* Since water assumes a level surface, the shoreline of a lake is a visible contour, but in general, contours cannot be seen in nature. On maps, contours represent the planimetric locations of the traces of level surfaces for different elevations (see the plan view of Figure 17.2). Contours are drawn on maps by interpolating between points whose positions and elevations have been observed and plotted. As noted earlier, computerized mapping and contouring systems are replacing manual plotting methods, but the principles of plotting terrain points and of interpolating contours is still basically the same in either method.

The vertical distance between consecutive level surfaces forming the contours on a map (the elevation difference represented between adjacent contours) is called the *contour interval.* For the small-scale U.S. Geological Survey *quadrangle maps* (plotted at 1:24,000 scale), depending on the nature of the terrain one of the following contour intervals is used: 5, 10, 20, 40, or 80 ft. For larger-scale maps used in engineering design, in the English system of units contour intervals of 1, 2, 5, or 10 ft are commonly used. In the metric system, a contour interval of 0.5, 1, 2, 5, or 10 m is generally selected. Figure 17.3 is a topographic map having 10-ft contours.

The contour interval selected depends on a map's purpose and scale, and upon the diversity of relief in the area. As examples, on a map to be used for designing the streets and water and sewer systems for a subdivision, a contour interval of 1 or 2 ft would perhaps be necessary, whereas a 10- or 20-ft contour interval may be suitable for mapping a large ravine to determine the reservoir capacity that would result from constructing a dam. Also, a smaller contour interval will normally be necessary to adequately depict gently rolling terrain with only moderate elevation differences, while rugged areas with large elevation differences

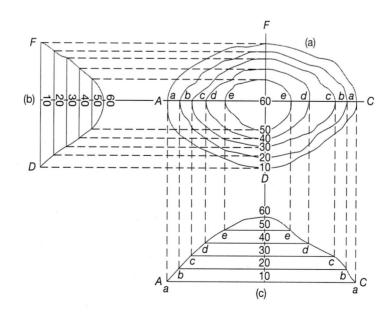

Figure 17.2
(a) Plan view of contour lines, (b) and (c) profile views.

Figure 17.3 Part of U.S.G.S. Lone Butte quadrangle map. (Courtesy U.S. Geological Society.)

normally require a larger contour interval so that the contours do not become too congested on the map. In general, reducing the contour interval requires more costly and precise fieldwork. In regions where both flat coastal areas and mountainous terrain are included in a map, supplementary contours, at one half or one fourth the basic contour interval, are often drawn (and shown with dashed lines).

Spot elevations are used on maps to mark unique or critical points such as peaks, potholes, valleys, streams, and highway crossings. They may also be used in lieu of contours for defining elevations on relatively flat terrain that extends over a large area.

Topographic mapping convention calls for drawing only those contours that are evenly divisible by the contour interval. Thus, for the 10-ft contour interval on the map in Figure 17.3, contours such as the 1100, 1110, 1120, and 1130 are shown. Elevations are shown in breaks in the contour lines, and to avoid confusion, at least every fifth contour is labeled. To aid in reading topographic maps, every fifth contour (each that is evenly divisible by five times the contour interval) is drawn using a heavier line. Thus, in Figure 17.3, the 1100, 1150, 1200, and so on contours are drawn more heavily.

17.6 CHARACTERISTICS OF CONTOURS

Although each contour line in nature has a unique shape, all contours adhere to a set of general characteristics. Important ones, fundamental to their proper field location and correct plotting, are listed.

1. Contour lines must close on themselves, either on or off a map. They cannot dead end.
2. Contours are perpendicular to the direction of maximum slope.
3. The slope between adjacent contour lines is assumed to be uniform. (Thus, it is necessary that breaks [changes] in grade be located in topographic surveys.)
4. The distance between contours indicates the steepness of a slope. Wide separation denotes gentle slopes; close spacing, steep slopes; even and parallel spacing, uniform slope.
5. Irregular contours signify rough, rugged country. Smooth lines imply more uniformly rolling terrain.
6. Concentric closed contours that increase in elevation represent hills. A contour forming a closed loop around lower ground is called a depression contour. (Spot elevations and hachures inside the lowest contour and pointing to the bottom of a hole or sink with no outlet make map reading easier.)
7. Contours of different elevations never meet except on a vertical surface such as a wall, cliff, or natural bridge. They cross only in the rare case of a cave or overhanging shelf. Knife-edge conditions are never found in natural formations.
8. A contour cannot branch into two contours of the same elevation.
9. Contour lines crossing a stream point upstream and form V's; they point down the ridge and form U's when crossing a ridge crest.
10. Contour lines go in pairs up valleys and along the sides of ridge tops.
11. A single contour of a given elevation cannot exist between two equal-height contours of higher or lower elevation. For example, an 820-ft contour cannot exist alone between two 810- or two 830-ft contours.

12. Cuts and fills for earth dams, levees, highways, railroads, canals, etc., produce straight or geometrically curved contour lines with uniform, or uniformly graduated spacing. Contours cross sloping or crowned streets in typical V- or U-shaped lines.

Keeping these characteristics in mind will (1) make it easier to visualize contours when looking at an area, (2) assist in selecting the best array of points to locate in the field when conducting a topographic survey, and (3) prevent serious mistakes in drawing contours.

17.7 METHOD OF LOCATING CONTOURS

Contours can be established by either the direct method (*trace-contour method*) or the indirect method (*controlling-point method*). The controlling-point method is generally more convenient and faster, and therefore it is most often selected. It is also the most frequent choice when data is entered into a computer for automated contouring. Since the direct method is seldom, if ever, used in practice today only the indirect method of establishing contours will be discussed. Previous early editions of this book discuss the direct method.

In the indirect method, the rod is set on "controlling points" that are critical to the proper definition of the topography. They include high and low points on the terrain, and locations where changes in ground slope occur, such as B, C, D, E, F, and G in Figure 17.4. Channels of drainage features and ridgelines must be included. Elevations are determined on these points using a total station instrument employing trigonometric leveling (see Section 4.5.4) or by using GNSS receivers with a geoid model incorporated into the survey. Horizontal distances and azimuths are also observed to locate the points. The positions of controlling points are then plotted and contours interpolated between elevations of adjacent points.

Figure 17.4(a) illustrates a set of controlling points labeled A through N that have been plotted according to their surveyed horizontal positions. Observed elevations (to the nearest foot) of the points are given in parentheses. Contours having a 10-ft interval have been sketched freehand between adjacent points by interpolation. It is improper to interpolate along lines that cross controlling features such as gullies, streams, rivers, ridge lines, roads, etc. Thus, to properly draw the contours of Figure 17.4(a) with the stream located on the map, elevations were first interpolated along its thread between surveyed points E, G, I, and J. Interpolations were then made from the stream to points on either side. As an example, it would have been incorrect to interpolate across the stream between points D and F. Rather the elevation of the stream on the line between D and F (approximately 9 ft) was used to interpolate both ways from the stream to points D and F. Note in Figure 17.4(a) that the gently curved contours tend to duplicate the naturally rolling topography of nature. Note also that contours crossing the stream form Vs pointing upstream.

Numerous controlling points may be needed to locate a contour in certain types of terrain. For example, in the unusual case of a nearly level field that is at or close to a contour elevation, the exact location of that contour would be time consuming or perhaps impossible to determine. In these situations, a uniform distribution of *spot elevations* can be determined in the field and plotted on the map to convey the area's relief.

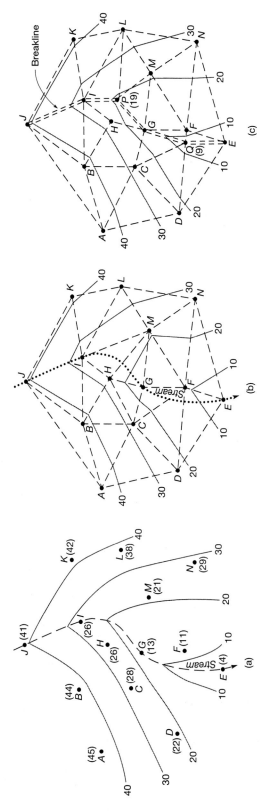

Figure 17.4 Direct method of locating contours.
(a) Contours compiled by hand from controlling points *A* through *N*. (b) TIN model (dashed lines) constructed from data of (a), and contours (solid lines) derived from TIN model. The stream is shown with a dotted line. Note the striking differences between the 10- and 20-ft contours of (a) and (b). (c) TIN model (dashed lines) constructed from data of (a) but with the addition of two points, *P* and *Q*, and the designation of lines *EQ*, *QG*, *GP*, *PI*, and *IJ* as breaklines. Contours shown with solid lines were derived from this TIN model. Note the agreement of these contours with those of (a).

17.8 DIGITAL ELEVATION MODELS AND AUTOMATED CONTOURING SYSTEMS

Data for use in automated contouring systems are collected in arrays of points whose horizontal positions are given by their X and Y coordinates and whose elevations are given as Z coordinates. Such three-dimensional arrays provide a *digital* representation of the continuous variation of relief over an area and are known as *digital elevation models* (DEMs). Alternatively, the term *digital terrain model* (DTM) is sometimes used.

Two basic geometric configurations are normally used in the field for collecting DEM data: the *grid method*, and the *irregular method*, although often a combination of the two methods is employed. In the grid method, elevations are determined on points that conform to a regular square or rectangular grid. The procedure is described in Section 17.9.3 and sample field notes are given in Plate B.2 of Appendix B. From the array of grid data, the computer interpolates between points along the grid lines to locate contour points and then draws the contour lines. The major disadvantage of this method is that critical high and low points and changes in slope do not generally occur at the grid intersections, and thus they are missed in the data collection process, which results in inaccurate relief portrayal.

The irregular method is simply the controlling-point method, but additional information (to be described later) is included. As previously noted, the controlling-point method involves determining the elevations of all high and low points and points where slopes change. Of course, this produces a DEM with an irregularly spaced configuration of surveyed points.

The first step taken by computerized contouring systems that utilize irregularly spaced spot elevations is to create a so-called *triangulated irregular network* (TIN) model of the terrain from the spot elevations. It is very important to understand the TIN model concept to ensure that an appropriate array of controlling points is selected and surveyed in the field if an automated contouring system is to be used. A TIN model is constructed by connecting points in the array to create a network of adjoining triangles. The dashed lines in Figure 17.4(b) show a TIN model created for the data in Figure 17.4(a). Various criteria can be used in the development of TIN models from an array of surveyed points, but one commonly used standard creates the "most equilateral network" of triangles.

Automated contouring systems generally make two assumptions concerning TIN models: (1) all triangle sides have a constant slope and (2) the surface area of any triangle is a plane. Based on these assumptions, elevations of contour crossings are interpolated along triangle edges, and contours are constructed such that they change direction only at triangle boundaries. Contours derived in this manner from the TIN model of Figure 17.4(b) are shown in the figure as solid lines. Note the disparities between the hand-drawn contours of Figure 17.4(a) and those derived from the TIN model of Figure 17.4(b). Differences are particularly obvious between the 10- and 20-ft contours. These occur because (1) the computer did not interpret the curved thread of the stream [shown as a dotted line in Figure 17.4(b)], and (2) in creating the network of triangles, several sides were constructed that cross the stream, resulting in improper interpolation across the stream.

From this example it is apparent, as noted earlier, that additional information must be provided for computer-driven systems to depict contours accurately. That important added information is the identification of controlling features, also more often called *breaklines*, or *fault lines* in modern computer mapping terminology. Breaklines are linear topographic features that delineate the intersection of two surfaces that have uniform slopes, and thus define changes in grade. *Automated mapping algorithms use these lines to define sides of the triangles that form the TIN model, and thus elevations are interpolated along them.* Streams, lake shores, roads, railroads, ditches, ridgelines, etc., are examples of controlling features or breaklines. Curved breaklines such as streams must have enough data points so that when adjacent ones are connected with straight lines, they adequately define the feature's alignment.

The dashed lines of Figure 17.4(c) represent the TIN model constructed from the same data set as in Figure 17.4(b), except that the stream (shown with a double dashed line) has now been identified as a breakline, and two additional points, P and Q, have been added to better approximate the curvature of the stream. In this figure, contours derived from the TIN model are shown. Note that these, now very nearly, duplicate the hand-drawn contours.

The important lesson of the foregoing is that if an automated contouring system is used, field points must be selected carefully, breaklines identified, and the data properly input to meet the system's assumptions. As indicated by this example, a few more controlling points may have to be surveyed, but the benefits of automated contouring systems make it worthwhile.

In order to avoid missing significant data during topographic surveys, it is usually best to collect features in groups. That is, data should be gathered first for (1) planimetric features, followed by (2) breaklines, (3) significant controlling points of elevation, and finally (4) sufficient *grade points* (those remaining points surveyed only to enable accurate depictions to be made of slopes and grades between the other types of points). Grade points are often most efficiently collected in a grid pattern throughout the entire area to be mapped. This grid should be sufficiently dense to avoid triangles in the TIN that are geometrically weak; that is, long and slender figures with one small angle. Varying grid sizes can be used, with larger spacing applied in areas of gradual slopes, and more dense patterns employed as the terrain becomes more undulating.

■ 17.9 BASIC FIELD METHODS FOR LOCATING TOPOGRAPHIC DETAILS

Objects to be located in a mapping survey can range from single points or lines to meandering streams and complicated geological formations. The process of tying mapping details to the control net is called *detailing*. Regardless of their shape, all objects can be located by considering them as composed of a series of connected straight lines, with each line being determined by two points. Irregular or curved lines can be assumed straight between points sufficiently close together; thus, detailing becomes a process of locating points.

Varying instruments may be involved in a mapping survey. These include total stations, laser scanners, GNSS receivers, manned or unmanned space-borne and aerial systems, or a combination of these. Location of planimetric features and contours can be accomplished by one of the following field procedures: (1) radiation by total station instrument, (2) coordinate squares or "grid" method, (3) offsets from a reference line, (4) use of portable GNSS units, (5) use of laser scanners, or (6) a combination of these methods. An explanation of each system and a discussion on their uses, advantages, and disadvantages follow.

17.9.1 Radiation by Total Station

In the radiation method, illustrated in Figure 17.5, with a total station instrument set up on a control point, the zenith angle, slope distance, and direction are observed to each desired item of mapping detail. From the zenith angle and slope distance, the elevation of the point can be determined, and by incorporating the direction, its horizontal position can be computed. These computations are often performed by the internal computer in a total station or by the survey controller. As shown in the figure, the sights to details radiate from the occupied station, hence the name for the procedure. This method is especially efficient if a survey controller (see Sections 2.12 through 2.15) is used to record the point identities and their associated descriptions, vertical distances, horizontal distances, and directions. The survey controller permits downloading the observations directly into a computer for processing through an automated mapping system. The field procedure of radiation with a total station can be made most efficient if the instrument is placed at a good vantage point (on a hill or ridge) that overlooks a large part or all of the area to be surveyed. This permits more and longer radial lines and reduces the number of setups required.

Table 17.1 is an example illustrating the use of a total station with survey controller for topographic mapping. The example relates to Figure 17.5. In Figure 17.5, a total station instrument was set up at control station 1 (John) and oriented in azimuth with a backsight at control station 2 (Bill). Observations of horizontal angle, slope distance, and zenith angle, respectively, were then taken to points 3, 4, and 5, which

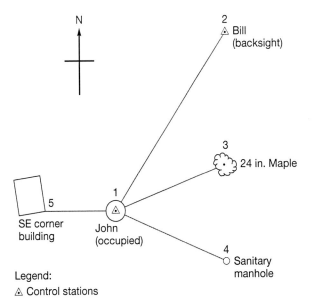

Figure 17.5
Radiation survey from control traverse.

are *sideshots* to mapping details. From these sideshots, two- or three-dimensional coordinates can be computed that are used to locate the points on a map sheet.

When using a survey controller in this process, initial data for the setup are first entered in the unit via the keyboard. These include the X and Y coordinates of stations John and Bill, the elevation of John, and the heights of both the total station instrument and reflector. This raw data file is written in LandXML. In the file each set of observations to points 3, 4, and 5 are bracketed by the command *RawObservation*. Note that each shot was taken using only one position of the instrument. Each raw observation contains the horizontal angle, slope distance, and zenith angle to the target, which then may have a description and feature code. As described in Section 17.13, these additional items add value to the file as well allowing field-to-finish software to automatically draw symbology for the point on the map.

As shown in Figure 17.6, details that have width such as trees are located with two separate observations. The first observation locates the azimuth to the object by observing an angle from a reference line to the front center of the object. The second shot measures the distance to the side center of the object. Using the azimuth of the first shot and the distance from the second shot, coordinates near the center of the object can be determined. Survey controllers have various names

TABLE 17.1 EXCERPT OF SURVEY CONTROLLER FIELD NOTES OF A RADIAL SURVEY FOR TOPOGRAPHIC DETAILS

Snippet of LandXML code from raw data file

```
<RawObservation setupID="pos1" purpose="sideshot" targetHeight="5.600" horizAngle="16.374400"
    slopeDistance="321.450" zenithAngle="90.25500">
    <TargetPoint name="4" desc="Maple"/>
    <Feature code="TREE" desc="Tree">
        <Property label="Trunk" value="24 in."/>
    </Feature>
</RawObservation>
<RawObservation setupID="pos1" purpose="sideshot" targetHeight="5.600" horizAngle="70.042600"
    slopeDistance="332.630" zenithAngle="91.154800">
    <TargetPoint name="4" desc="Sanitary manhole"/>
    <Feature code="SMH" desc="Sewer MH">
    </Feature>
</RawObservation>
<RawObservation setupID="pos1" purpose="sideshot" targetHeight="5.600"
      horizAngle="225.162300" slopeDistance="192.490" zenithAngle="88.234500">
    <TargetPoint name="5" desc="SE Building Corner "/>
    <Feature code="BLA" desc="Building Area" string="1"/>
</RawObservation>
```

486 MAPPING SURVEYS

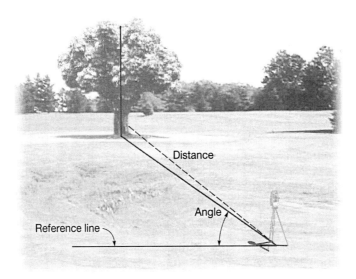

Figure 17.6
Proper location of objects such as trees.

for this offset data collection routine. This procedure should only be used when the diameter of the object is sufficiently large to cause a plotting error on the map. For smaller objects where the diameter will not noticeably displace the center of the object on the map, this procedure is unnecessary. Thus, the use of this method is dependent on the scale of the map and size of the object.

17.9.2 Coordinate Squares or "Grid" Method

The method of coordinate squares (grid method) is better adapted to locating contours than planimetric features, but can be used for both. The area to be surveyed is staked in squares 10, 20, 50, or 100 ft (5, 10, 20, or 40 m) on a side, the size depending on the terrain and accuracy required. A total station instrument can be used to lay out control lines at right angles to each other, such as AD and $D3$ in Figure 17.7. Grid lengths are marked and the other corners staked and identified by the number and letter of intersecting lines.

Elevations of the corners can be obtained by differential or trigonometric leveling. Contours are interpolated between the corner elevations (along the sides of the blocks) by estimation or by calculated proportional distances. Elevations obtained by interpolation along the diagonals will generally not agree with those from interpolation along the four sides because the ground's surface is not a plane. Except for plotting contours, this is the same procedure as that used in the borrow-pit problem in Section 26.10. In plotting contours by the grid method, a widely spaced grid can be used for gently sloping areas, but it must be made denser for areas where the relief is rolling or rugged.

A drawback of the method is that no matter how dense the grid, critical points (high and low spots and slope changes) will not generally occur at grid locations, thus degrading the accuracy of the resulting contour map. However, this method can be enhanced by collecting the critical points and breaklines that are not at grid intersections.

Figure 17.7 Coordinate squares.

17.9.3 Offsets from a Reference Line

This procedure is most often selected for mapping long linear features, and for performing surveys necessary for route locations. After the reference line or centerline has been staked and stationed, planimetric details are located by observing perpendicular offsets from it and noting the stations from which the offsets were taken. Features such as streams, trails, fences, buildings, utilities, trees, etc., can be located in this manner. Elevations for determining contour locations can also be determined by *cross-sectioning* (observing ground profiles along lines perpendicular to the reference line) as described in Section 26.3.

Shorter offset distances are generally most easily and quickly observed by taping, but longer lengths may be more efficiently obtained by electronic distance measurement using a total station instrument. Where steep slopes run transverse to the reference line, better accuracy and efficiency can often be obtained using a total station instrument. Perpendiculars to a reference line can be quickly established using a *pentagonal prism* (see Figure 17.8). This device works well when offset distances are being taped. While standing on the reference line at a location where a perpendicular is desired, the operator holds the instrument upright and looks through it sighting either ahead or back on the reference line. By means of the prisms, perpendicular views both left and right can be simultaneously seen. Alternatively, the total station instrument can be set up on the reference line at the location of a desired offset, oriented by sighting ahead or back on the line, and then used to turn a 90° angle.

Figures 17.9(a) and (b) illustrate examples of using offsets for mapping. In Figure 17.9(a), the location of Crooked Creek was determined by observing distances to the edge of the creek at intervals from the reference line. The offsets can be taken at regular intervals as illustrated in the figure, or they can be spaced at

488 MAPPING SURVEYS

Figure 17.8
Double pentagonal prism (on side) for establishing perpendiculars to a reference line. (Courtesy Leica Geosystems, Inc.)

Figure 17.9
(a) Location of creek by perpendicular offsets from a reference line.
(b) Locating objects by offsets from a reference line.

distances that permit the curved nature of the stream to be considered as a series of straight segments between successive offsets. Figure 17.9(b) illustrates an example of locating planimetric features along a road right-of-way. This type of survey would be useful to locate buildings, utilities, trees, and other features along a road for highway design, or for excavation to install an underground utility. After locating at least two corners of a building by observing their offset distances, for example, x_1 and x_2, in the figure, and their plusses on the reference line, its remaining dimensions can usually be quickly obtained by taping, for example, a and b in the figure.

In both of these examples, it would be convenient to include a sketch in the field book, and to record the observations directly on the sketch. Since data collected by this method are based purely on distances, it is difficult to merge them with data collected by the radiation method and is seldom used when employing computer drafting techniques.

17.9.4 Topographic Detailing with GNSS

GNSS receivers for topographic work are specially designed, small, and portable. The receivers are interfaced with a survey controller for system control and entry of codes to identify features surveyed. The unit shown in Figure 14.1 is suitable for topographic work. These receivers can determine (in real time) the coordinates of locations where the receiver antenna is placed and can store data for the points in files. The files are then directly downloaded to a computer for further processing, which could include automatic map drafting. These systems make topographic data collection a simple and very fast one-person operation. The kinematic surveying methods discussed in Chapter 15 are most often used for mapping surveys, however, static surveys (Chapter 14) are sometimes used to establish control in the project area. In areas where canopy restrictions or obstructions restrict the use of a GNSS survey, these control points can be used to orient a total station to complete the survey. The use of a real-time network (RTN), which is discussed in Section 15.8, can provide an accurate one-person real-time GNSS mapping survey without the need for establishing local control.

The stop-and-go method (semikinematic) has the advantage over the true kinematic method in that the operator can stop and collect multiple epochs of data for a point to increase positional accuracies. The semikinematic method generally results in a file size smaller than the file from the true kinematic method. Additionally, the stop-and-go method can be used to locate features or establish control points for occupation by conventional instruments such as total stations.

True-kinematic surveys collect data points at an operator-selected epoch, which can be set by time (usually 1–5 sec interval), horizontal distance, or slope distance, and thus this method can be used to quickly obtain cross-sections and locate linear features such as breaklines, curbs, roads, and streams. However, since the operator has no control over the actual instance of data collection, this method does not provide a convenient means for surveying key topographical and planimetric features. To overcome this most manufacturers have provided user-specified data collection options when performing true kinematic surveys so that the user can suspend the true kinematic collection and collect specific features with the stop-and-go method. Once the feature is collected the user can resume collecting data on the linear feature using the true kinematic method.

During a mapping survey, it is possible to switch from a true kinematic to semikinematic, or static survey (see Chapter 14). Therefore, if the conditions within a mapping project change, the operator can select the survey method that is most appropriate for the task at hand. Regardless of the method selected, it is essential that the antennas have clear visibility to the satellites. Thus, GNSS surveys are generally not be suitable for direct location of large trees, buildings, or other objects that could obscure the view of the satellites or create multipathing conditions. In these situations, a procedure known as offset location can be used. When locating an object by offsets, two points are established using the GNSS receiver where clear view to the satellites is possible. These two locations need to create a line that points to the object being located; this establishes the azimuth of the line. A distance is then observed from one of the two points using a tape and manually entered into the survey controller. Using the azimuth of the line and the distance, the coordinates of the object are determined in the collector. If the area contains several objects that are not accessible to GNSS receivers, two nearby temporary control stations can be established using the stop-and-go method and conventional data collection with total stations used to locate these map details by occupying one point and backsighting the other. For maximum accuracy in the azimuth, it is important that these two points be as far apart as is visually and physically possible on the site. Due to overhead obstructions and multipathing conditions, GNSS receivers should not be placed at the corners of buildings. In this instance, successive corners of the building can be located by offsets. Some survey controller software provides an option to tape the remaining sides of the building to create footprint of the building. However, when the modernized GNSS constellations are fully implemented, this limitation is expected to decrease due to the increased signal strength, processing advantages, and the availability of satellites.

As discussed in Chapter 2, most survey controllers work with GNSS receivers as well as total station instruments. However, since orthometric heights are normally required for the vertical component of a survey, it is important to transform GNSS-derived ellipsoid heights to their orthometric values using procedures discussed in Sections 13.4.3 and 15.9. Typically, survey controller software provides an option to perform this conversion using an appropriate geoid model.

Code-based GNSS receivers can also be applied in topographic mapping, but their use is generally limited to lower-order work such as those found in GIS applications. These receivers are very affordable and have postprocessed accuracies that are less than a few meters. Even though this is a relatively high positional uncertainty, these units can be used to do surveys for maps with scales smaller than 1:20,000 since the plotting errors become negligibly small. As additional channels are added to the GPS satellites and other satellite systems become mature, it is anticipated that the accuracies of code-based GNSS will be reduced to the meter level and possibly decimeter level in the differential positioning mode allowing code-based receivers to be used where total stations or carrier phase-shift receivers were previously required.

17.9.5 Unmanned Systems

As shown in Figure 17.10, a new player in capturing mapping detail on small projects is the unmanned aerial system (UAS). The quadcopter shown in Figure 17.10(a) captured the image mosaic shown in Figure 17.10(b). UAS systems consist of the unmanned air vehicle (UAV), navigation units, which can consist of GNSS receivers and inertial measurement units (IMUs), sensors such as digital cameras, command and control datalinks, operator station, and ground support equipment required for launch, recovery, and maintenance operations. Unmanned aerial vehicles can vary in size and payloads from one that can fit in the palm of your hand to full-size aircraft. However, a sUAS is defined as an aircraft weighing less than 55 lbs. including its payload. UASs are capable of providing high-resolution imagery, which can be used to create large-scale, three-dimensional point clouds. These system provide the users with the capability of capturing data in areas that would be costly or dangerous to gather using other methods or of such size that they are too large for other methods of data capture and too small for a full-scale photogrammetric campaign as discussed in Chapter 27.

Figure 17.10 (a) Unmanned quadcopter with digital camera. (Courtesy of Maria Dryfhout/Shutterstock.) (b) Image mosaic of open-pit mine captured using quadcopter shown in (a). (Courtesy of Angyalosi Beata/123RF.)

Similar to a full-scale photogrammetric campaign, the flight plan (see Section 27.17) of the vehicle must be designed and then loaded into the sUAS. This flight plan must contain the waypoints where the images will be captured, the altitude of the flight, and the extents of the project area. Once the images are captured, they must be reviewed for possible rejection. Reasons for rejection include images captured during take-off or landing, images that are blurry, over- or underexposed, and images that are outside the area of interest. From the remaining images a three-dimensional image can be produced during data processing and object extraction. The uses for these systems are rapidly expanding including uses such as three-dimensional mapping, cadastral mapping, urban planning, construction and mining projects, archeological reconnaissance and mapping, and precision agriculture. In a similar effort, unmanned water vehicles are being used to perform hydrographic surveys (see Section 17.14) in harbors and bays as well as explore the depths of the oceans. Most assuredly, the use of unmanned vehicles in surveying will expand rapidly in the upcoming years.

17.9.6 Laser Scanning

Laser scanners automate digital angle observation with reflectorless, laser-EDM technologies. They can quickly produce grids of three-dimensional coordinates for user-specified scenes. A combination of rotating mirrors allows the instruments to observe distances and orthogonal angles in precise grid patterns. These instruments vary in capabilities to match the varying requirements of jobs. Characteristics of an instrument are defined by the *number of observations* they can observe per second, method of distance observation by either *time of flight* (TOF) or *phase based*, observable distance from the instrument known as its *range*, minimum spacing between observations or its *resolution*, *accuracy*, and *field of view*. In general, the higher the number of observations per second, the faster the data acquisition. Instruments vary from less than a hundred observations per second to 500,000 observations per second. The range of instruments can vary from several meters to several kilometers. Instrument resolutions can vary from a few millimeters to several centimeters. Many ground-based instruments can produce finer resolutions on targets. However, it must be remembered that the higher the resolution, the larger the data files. Generally, ground-based laser scanners have distance measurement accuracies of a few centimeters. The laser scanner shown in Figure 1.5 has a range accuracy of ±6 mm at 50 m and angular accuracies of ±60 μrad.[1] The field of view dictates the area that a laser scanner can observe in a single setup. Some instruments can rotate 360° in horizontal and 270° in vertical allowing them to survey an entire scene surrounding the instrument. In general, users can specify the field of view to match the area of interest. Several manufacturers have added scanning capabilities to selected robotic total stations. Thus, the surveyor has the option of performing a traditional radiation survey or scanning the object. This is especially useful in rugged areas with large vertical relief and in industrial settings.

When ranging is done using TOF technology, the instrument sends out pulses of light. The time it takes for the pulse to reach its target and return to the instrument is timed. Thus, the distance to the target is computed as

$$D = v \frac{TOF}{2}$$

where D is the distance to the reflecting object, v is the velocity of pulse, and TOF is the time it takes the pulse to travel from the instrument and return. The advantage of these types of scanners is that they are typically capable of longer ranges.

The phase-based technology works similar to an EDM where the phase shift is used to determine the distance to the reflecting object. The advantage of this type of instrument is that it typically has a higher accuracy and higher data-capture rate, and thus collects more points per second than the time-of-flight instruments. Its disadvantage is that it can only do this for shorter ranges to the target. For example, the phase-based instruments can typically capture more than 1,000,000 points per second but only to distances of about 100 m whereas the TOF instruments may only capture 50,000 points per second but can collect points farther from the instrument and also provide photographic images of the scene. Both types of instruments provide a grid of points over the scene.

Another difference in scanners is the field of view. Most scanners can rotate 360° in the horizontal plane. However, there can be limitations in the vertical plane. Some scanners provide 270° in the vertical plane allowing the scanner to capture points below horizontal as well as directly overhead.

The resulting grid of scanned, three-dimensional points can be so dense that a visual image of the scene is formed. This so-called "point cloud" differs from a photographic image in that every point has a three-dimensional coordinate assigned to it. These coordinates can be used to obtain dimensions between any two observed points in the scene. In Figure 17.11, a point-cloud image of piping at a refinery provided engineers with the information needed to design a new pipe addition shown in white. The detail in this image would be difficult if not impossible to recreate using other surveying processes. Because of the high density of the observed points in a scene, it is sometimes referred to as "high-definition" surveying. Some instruments also capture a digital image of the scene. The digital image can be integrated with the scanned points to create a three-dimensional image having color and texture. This process was used for the bridge shown in

---

[1] A microradian equals 0.000001 radians, which is about 0.2″.

17.9 Basic Field Methods for Locating Topographic Details **493**

Figure 17.11 Laser-scanned image of refinery showing designed pipe alignment in white. (Courtesy Christopher Gibbons, Leica Geosystems, Inc.)

Figure 17.12 Point cloud from a mobile mapping system. (Courtesy Topcon Positioning Systems.)

Figure 23.15. Laser scanning can play a significant role in as-built surveys, archeology, and mapping of artifacts. Section 27.18 discusses the use of airborne laser mapping known as LiDAR. Some states have mapped their state using this technology. Similarly, it is now possible to view 3D images of cities produced from LiDAR data.

Figure 17.12 depicts a scene that was captured by the *mobile mapping system* shown in Figure 1.4. A mobile mapping system incorporates multiple LiDAR scanners, GNSS receiver, inertial measurement unit, high-quality digital camera, and odometry from the host vehicle data to provide geo-referenced coordinates for the point cloud. This system is capable of scanning objects within 100 m of the vehicle as it moves at highway speeds. It collects up to 1.3 million points per second. In Figure 17.12, the dots running down the street corridor are the GNSS position fixes of the system. The inertial measurement unit provides positioning for the system when canopy prohibits the GNSS fixes. The larger sphere in the center of the figure is a window into the high-resolution digital image of the scene. This digital image overlays the point cloud allowing the viewer to see a picture quality image of the scene. From this point, items can be identified and their positions determined or distances between objects computed. Because the scene is geo-referenced, additional surveying and ground control are not required. However, higher accuracies can be obtained by post-processing the GNSS fixes against a GNSS base station (see Chapter 14).

17.10 PLANNING A LASER-SCANNING SURVEY

The original point cloud established by a laser-scanning survey can be determined in an arbitrary three-dimensional coordinate system. If it is necessary to have coordinates in a project-based coordinate system, a traditional survey can be used to establish coordinates on targets in the scene or of the scanner's setup locations. Control must be strategically located at the edges of each scene. A minimum of three control points per scene are required. However, additional control is often used to provide redundancy. Multiple scenes can be connected using common targets. After determining the project coordinates of the control, a three-dimensional conformal coordinate transformation discussed in Section 17.11 can be performed to transform points from the arbitrary coordinate system to the project coordinate system.

Another method of establishing project coordinates is to traverse around the object that is to be scanned establishing three-dimensional project coordinates around the object. Figure 17.13 depicts such a traverse around a building. This set of coordinated points can be established using conventional instruments such as a total station or by a GNSS survey. Once the coordinates of the stations are established, the scanner can be placed at each station and a backsight can be taken on another station. When the coordinates of the stations are entered into the scanner's project, the resulting point cloud coordinates will be based on the coordinates of the stations that the scanner occupies and backsights. Alternatively,

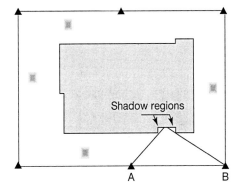

Figure 17.13
A possible traverse for capturing details on a building using laser scanning.

the three-dimensional coordinates of the scanner station and its backsight can be entered during post-processing to provide the system with sufficient information to transform the coordinates of the point cloud.

Part of the planning for a laser scanning survey is to determine the ideal locations to set up the scanner. For instance, when scanning the building with the footprint shown in Figure 17.13, the scanner must be set so that shadows created in one scan can be filled in by a second location's scan. This can be performed over the entire scene but typically it is performed on the area of interest only reducing the amount of overlap and redundant points in the scene. Notice in Figure 17.13 that the scanner at point A is unable to collect points in the shadow region on the left side of the entry to the building but the scanner set at point B will be able to collect data in that region. The same condition occurs at station B where the scanner is unable to collect data on the right side of the entry way. In this scenario, the region to the right side of the entry way is covered by the scanner at A and the region on the left side of the entry way is covered by the scanner at B. To avoid collecting too many scanned points on the building, the second scan at B can be limited to the shadow region on the left side of the entry way. Intervening obstructions such as tree shown in Figure 17.13 will also cast shadow regions on the structure. Again, enough scanning locations must be planned to capture data behind these obstructions.

A second method that can be used to provide sufficient information to join multiple scans is to place three or more reflective targets on each side of the building's face so that resection procedures (Section 11.10) can be performed to extend the coordinates of the survey around each face of the building. In this instance, the project coordinates will be based on the initial setup of the scanner, which is typically defaulted to three-dimensional coordinates of $(0,0,0)$, and its initial orientation, which establishes an arbitrary azimuth of zero. In this instance, the reflective targets shown in Figure 17.14 must be located in positions that will not fall into a shadow region for the next scanner's location. Again, the locations of the scanner must be planned before the targets are placed on the buildings and the first scan is ever taken. After the data is captured, the coordinates of the point clouds can be transformed into project coordinates by changing the coordinates of the scanner's location.

Depending on the complexity of the object being scanned and the needs of the project, the resolution of the scan must also be planned. The resolution of the scan is established by setting the horizontal and vertical intervals of data collection in the scanner, which is typically set in terms of angular advances by the instrument in both the horizontal and vertical planes. However as shown in Figure 17.15 since the object being scanned is not at a uniform distance from the scanner typically, the resolution of the scan will vary with the varying distance to the object. The density of the scan in the outlying regions of the point cloud can be densified if necessary by additional overlapping scans.

Suppose the positions and dimensions of the windows, doors, and even the door handles are desired from the survey. Further assume that the door handles have a width of 0.07 ft. Then the scan may need a resolution of 0.03 ft at the distance between the scanner and the door in order to guarantee that this particular object is captured in the scan. If the scanner is set 200 ft from the door handles, then the angular resolution of the scan can be determined using Equation (8.1)

Figure 17.14
Reflective targets for laser scanning. Targets come in various sizes with adhesive backing to allow placement on surfaces or with magnetic backing material that allows placement of target over a traverse station.

Figure 17.15
Changing resolutions caused by object being different distances from scanner.

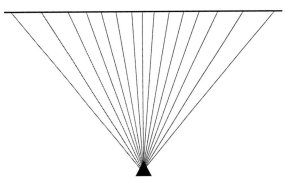

as approximately 30″. If this is the only location on the building that needs this dense of a scan, then two scans can be taken from the same station where one scan captures the bulk of the object and another is limited to the area requiring a higher density scan. In fact, it is this technique that is used by the scanner to accurately determine the center of the targets shown in Figure 17.14.

17.11 THREE-DIMENSIONAL CONFORMAL COORDINATE TRANSFORMATION

The three-dimensional conformal coordinate transformation transfers points from one three-dimensional coordinate system (xyz) into another (XYZ). This transformation is similar to the two-dimensional coordinate transformation covered in Section 11.8. However, the three-dimensional conformal coordinate transformation involves seven unknown parameters (three rotation angles, three translation factors, and one scale factor). The development of the rotations is covered in Section 19.17. The mathematical model for the transformation is

$$X = S(m_{11}x + m_{21}y + m_{31}z) + T_X \tag{17.1a}$$

$$Y = S(m_{12}x + m_{22}y + m_{32}z) + T_Y \tag{17.1b}$$

$$Z = S(m_{13}x + m_{23}y + m_{33}z) + T_Z \tag{17.1c}$$

17.11 Three-Dimensional Conformal Coordinate Transformation

where S is a scale factor, T_X, T_Y, and T_Z are the translations in x, y, and z, respectively, and m_{11} through m_{33} are elements of the rotation matrix.

$$m_{11} = \cos(\theta_y) \cos(\theta_z)$$
$$m_{12} = \sin(\theta_x) \sin(\theta_y) \cos(\theta_z) + \cos(\theta_x) \sin(\theta_z)$$
$$m_{13} = -\cos(\theta_x) \sin(\theta_y) \cos(\theta_z) + \sin(\theta_x) \sin(\theta_z)$$
$$m_{21} = -\cos(\theta_y) \sin(\theta_z)$$
$$m_{22} = -\sin(\theta_x) \sin(\theta_y) \sin(\theta_z) + \cos(\theta_x) \cos(\theta_z)$$
$$m_{23} = \cos(\theta_x) \sin(\theta_y) \sin(\theta_z) + \sin(\theta_x) \cos(\theta_z)$$
$$m_{31} = \sin(\theta_x)$$
$$m_{32} = -\sin(\theta_x) \cos(\theta_y)$$
$$m_{32} = \cos(\theta_x) \cos(\theta_y)$$

θ_x, θ_y, and θ_z are the counterclockwise rotation angles about the X, Y, and Z axes, respectively, as viewed from their positive ends. To solve this set of nonlinear equations, methods like those discussed in Section 16.9.1 are employed. The linearized set of equations for this transformation given in matrix form is

$$\begin{bmatrix} \left(\dfrac{\partial X}{\partial S}\right)_0 & 0 & \left(\dfrac{\partial X}{\partial \theta_y}\right)_0 & \left(\dfrac{\partial X}{\partial \theta_z}\right)_0 & 1 & 0 & 0 \\ \left(\dfrac{\partial Y}{\partial S}\right)_0 & \left(\dfrac{\partial Y}{\partial \theta_x}\right)_0 & \left(\dfrac{\partial Y}{\partial \theta_y}\right)_0 & \left(\dfrac{\partial Y}{\partial \theta_z}\right)_0 & 0 & 1 & 0 \\ \left(\dfrac{\partial Z}{\partial S}\right)_0 & \left(\dfrac{\partial Z}{\partial \theta_x}\right)_0 & \left(\dfrac{\partial Z}{\partial \theta_y}\right)_0 & \left(\dfrac{\partial Z}{\partial \theta_z}\right)_0 & 0 & 0 & 1 \end{bmatrix} \begin{bmatrix} dS \\ d\theta_x \\ d\theta_y \\ d\theta_z \\ dT_X \\ dT_Y \\ dT_Z \end{bmatrix} = \begin{bmatrix} X - X_0 \\ Y - Y_0 \\ Z - Z_0 \end{bmatrix} \quad (17.2)$$

where X_0, Y_0, and Z_0 are determined using Equations (17.1a) through (17.1c), respectively, with approximations for the unknown parameters. The coefficients from the linearized equations are

$$\frac{\partial X}{\partial S} = m_{11}x + m_{21}y + m_{31}z$$

$$\frac{\partial X}{\partial \theta_y} = S(-x \sin\theta_y \cos\theta_z + y \sin\theta_y \sin\theta_z + z \cos\theta_y)$$

$$\frac{\partial X}{\partial \theta_z} = S(m_{21}x - m_{11}y)$$

$$\frac{\partial Y}{\partial S} = m_{12}x - m_{22}y + m_{33}z$$

$$\frac{\partial Y}{\partial \theta_x} = -S(m_{13}x + m_{23}y + m_{33}z)$$

$$\frac{\partial Y}{\partial \theta_y} = S(x \sin \theta_x \cos \theta_y \cos \theta_z - y \sin \theta_x \cos \theta_y \sin \theta_z + z \sin \theta_x \sin \theta_y)$$

$$\frac{\partial Y}{\partial \theta_z} = S(m_{22}x - m_{12}y)$$

$$\frac{\partial Z}{\partial S} = m_{13}x + m_{23}y + m_{33}z$$

$$\frac{\partial Z}{\partial \theta_x} = S(m_{12}x + m_{22}y + m_{32}z)$$

$$\frac{\partial Z}{\partial \theta_y} = S(-x \cos \theta_x \cos \theta_y \cos \theta_z + y \cos \theta_x \cos \theta_y \sin \theta_z - z \cos \theta_x \sin \theta_y)$$

$$\frac{\partial Z}{\partial \theta_z} = S(m_{23}x - m_{13}y)$$

As outlined in Section 16.9.1, Equation (17.2) is formed using approximate values for the unknowns and iterated until the corrections to the unknown parameters become negligibly small. This process is demonstrated in a Mathcad worksheet *3DC.xmcd* on the companion website for the book. The complete mathematical development of the transformation is covered in several books listed in the bibliography at the end of this chapter.

17.12 SELECTION OF FIELD METHOD

Selection of the field method to be used on any topographic survey depends on many factors, including (1) purpose of the survey, (2) map use (accuracy required), (3) map scale, (4) contour interval, (5) size and type of terrain involved, (6) costs, (7) equipment and time available, and (8) experience of the survey personnel.

Items (1) to (5) are interdependent. The cost, of course, will be a minimum if the most suitable method is selected for a project. On large projects, personnel costs rather than equipment investment will usually govern. However, the equipment owned may govern the method chosen by a private surveyor making a topographic survey of 50 or 100 acres. In many instances, the larger surveys are performed photogrammetrically using techniques discussed in Chapter 27.

17.13 WORKING WITH SURVEY CONTROLLERS AND FIELD-TO-FINISH SOFTWARE

Surveying instruments equipped with survey controllers can record and store field notes for electronic transmission to computers, plotters, and other office equipment for processing. Such systems, called "field to finish," rely on sophisticated software for their operation. Their use can increase productivity tremendously in surveying and mapping.

It is important that field personnel understand the field-to-finish system so that field data are designated properly. Additionally, collecting more information than is necessary to create a map makes the field data value added, which can be mined later for special-use situations such as in a GIS (see Chapter 28). For instance, it is not sufficient to simply collect the features without adding additional information about the features such the type of sign, tree, road pavement material, or owner of the building and whether it is a commercial or private dwelling.

In using field-to-finish systems for topographic surveys, the survey controller will store a point identifier and the *NEH* coordinates for each point located. However, in addition, ancillary descriptive information can accompany each surveyed point. For instance, in Figure 17.16, points 1, 2, 14, and 15 are the corners of a building; points 5, 6, 7, and 8 mark the corners of a sidewalk; points 4, 9, 10, 11, and 12 are points along a property line, with points 10 and 11 being on a curve; and points 3 and 13 are deciduous trees. It is possible to add and store this descriptive information in the survey controller through the use of *notes*. If the notes are entered in a manner that is understandable by the field-to-finish software and the points are collected in a manner that is consistent with and supports the drafting system, the software will use appropriate symbols for plotting each feature, draw and close polygons, and create a complete and finished drawing. To accomplish the finished drawing, some software systems require that different features be placed on different drafting layers to control the drawing of the scene. No matter the method employed, correctly entering the field data at the time of collection will greatly reduce the time in generating the final map product. However, in order for the system to operate properly, the field personnel collecting the data must understand the requirements of the plotting software and implement these requirements during data collection.

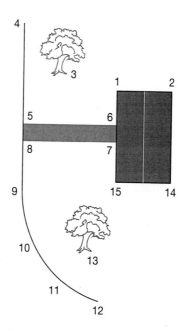

Figure 17.16
Example survey showing line work for planimetric features.

While the field-to-finish software of the various vendors use somewhat different techniques in reducing field data to a finished map, the basic concepts of these systems can be discussed. Typically drawing designators are used to construct the line work from the field data. By entering appropriate designators in the notes for each point, as it is located, the field personnel instruct office personnel and field-to-finish software on how to draw the lines. Specific *codes* for point symbols in the drawing can also be entered in the notes in the field as work progresses. For example, the identifier DTREE could indicate the deciduous trees shown in Figure 17.16. Typically, there are three different types of drawing codes. There are codes for points, lines, and areas. DTREE is an example of a point code. BL may indicate a building line and BLA may indicate a building area or its footprint. Since there may be several buildings on a map, the line and area codes typically require a number associated with them as shown in Figure 17.17. Thus, BL1 and BL2 will not be connected as the same line in the drawing but rather be treated as two different buildings. Another advantage of using these codes in the field is that the map will appear on the screen as the data are being collected.

Even though different features can be collected in varying order, it is important that successive shots on any object be collected in successive order. For example, if the field personnel had collected the building corners in the order of 1, 2, 15, and 14, the line work for the building would cross, creating an hourglass shape. Similarly, it is important to collect point 4 before point 5 so that the line work for the right-of-way is drawn in a linear fashion. Since GNSS surveys are frequently used to collect data for mapping surveys, survey controllers often have offset routines that allow the user to tape to features such as trees and buildings that would typically create multipathing or overhead obstruction problems. Some survey controllers even contain features to allow the user to tape the entire perimeter of a building once two successive corners have been located.

Since survey controllers viewing screens are typically small, many allow users to place field shots on different drafting layers. For example, trees may all be placed on a tree layer, buildings on a building layer, roads and walkways on a cartway layer, and so on. By doing this the field crew can turn on and off layers to ensure that all the features are appropriately defined at the time of the survey.

Figure 17.17 Typical data entry screen for entering a line code. (TopSURV Reference Manual, page 5-39 Figure 5-50 (39 of 110). All contents in this manual are copyrighted by Topcon Positioning Systems. Copyright © 2009, Topcon America Corporation, All Rights Reserved. Used by permission.)

This feature also allows field personnel to readily identify their progress on collecting data for various features in their project site. For example, grade shots and breaklines, which are points located simply for later contouring, can be placed on a single layer. Again, by doing so, the field personnel can identify topographic features that are missing or may require additional data for proper definition.

With the complexities of collecting data, selecting the locations of points, and properly noting the features, it is easy to see that some orderly plan for data collection should be developed before the instruments are taken from their cases. Also, there must be coordination between the field and drafting personnel. While each organization may develop their own procedures, some guidelines for collecting data suggest collecting planimetric feature data first, paying special attention to the sequence in which the data is collected. It is often most efficient to collect data for one feature type before beginning another; that is, locating all buildings, then the roadways, then vegetation, and so on in an orderly fashion at each instrument setup. Again, if these various features are placed on separate layers, the field personnel can more easily identify planimetric features that are missing or require more points for proper location. For topographic surveys, all controlling points may be collected, followed by break lines, and finally sufficient grade shots to allow accurate interpolation of the contours. (See Chapter 18 for a discussion on interpolation of contours.)

From the foregoing, it should be obvious that it is essential for field personnel to understand not only what data to collect, but also the order and manner in which to collect it so that proper plotting will occur. It should be noted here that different coding requirements are necessary for the various software used in practice. However, all field-to-finish systems have some drawing conventions that must be understood by both the field and office personnel. Correctly performed, a field-to-finish survey will greatly reduce the time it takes to create a finished and correct map.

■ 17.14 HYDROGRAPHIC SURVEYS

Hydrographic surveys determine depths and terrain configurations of the bottoms of water bodies. Usually the survey data are used to prepare hydrographic maps. In navigation and dredging, they may be recorded in electronic formats for real-time analysis. Bodies of water surveyed include rivers, reservoirs, harbors, lakes, and oceans.

Hydrographic surveys and maps are used in a variety of ways. As examples, engineers employ them for planning and monitoring harbor and river dredging operations, and to ascertain reservoir capacities for flood control and water supply systems; petroleum engineers use them to position offshore drilling facilities and locate underwater pipelines; navigators need them to chart safe passageways and avoid reefs, bars, and other underwater hazards; biologists and conservationists find them helpful in their study and management of aquatic life; and anglers use them to locate structures where fish will likely be located.

Field procedures for hydrographic surveys are similar to those for topographic work; hence, the subject is discussed in this chapter. There are some basic differences in procedures used by surveyors since the land area being mapped cannot be seen, and the depth measurements must be made in water.

Two basic tasks involved in traditional hydrographic surveys are (1) making *soundings* (measuring depths) from the water surface to bottom, and (2) locating the positions where soundings were made. Techniques used to perform these tasks vary depending on the water body's size, accuracy required, type of equipment to be used, and number of personnel available. The subsections that follow briefly describe procedures for mapping small to moderate-sized water bodies.

17.14.1 Equipment for Making Soundings

The size of a water body and its depth control the type of equipment used to measure depths. For shallow areas of limited size, a *sounding pole* can be used. This is usually a wooden or fiberglass staff resembling a level rod. It is perhaps 15 ft long, graduated in feet or tenths of feet, with a metal shoe on the bottom. Direct depth measurements are made by lowering the pole vertically into the water until it hits the bottom, and then reading the graduation at the surface. Today shallow coastline mapping is also performed using LiDAR using a green light, which can penetrate water and reflect back to the receptors.

Lead lines can be used where depths are greater than can be reached with a sounding pole. These consist of a suitable length of stretch-resistant cord or other material, to which a heavy lead weight (perhaps 5–15 lb.) is attached. The cord is marked with foot graduations, and these should be checked frequently against a steel tape for their accuracy. In use, the weight is lowered into the water, being careful to keep the cord vertical. The graduation at the surface is read when the weight hits the bottom.

In deep water, or for hydrographic surveys of appreciable extent, electronically operated sonic depth recorders called *echo sounders* are used to measure depths. These devices, an example of which is shown in Figure 17.18, transmit an acoustic pulse vertically downward and measure the elapsed time for the signal

Figure 17.18 Depth sounder used in small lake hydrographic survey. (Sea Floor Systems, Inc.)

to travel to the bottom, be reflected, and return. The travel time is converted to depth and displayed in either digital or graphic form. A graphic profile of the depths, such as that shown in Figure 17.19 can be displayed on a computer screen. This graphic plot can be referred to repeatedly for plotting and checking.

Sounding poles and lead lines yield spot depths and are restricted for use in relatively shallow water. However, electronic depth sounders provide continuous profiles of the surface beneath the boat's path and can be used in water of virtually any depth. For example, in the profile of Figure 17.19, the chart's vertical range was set to 40 ft and profile depths shown vary from 10 to 24 ft.

The reference plane from which depth soundings are measured is the water surface. Because of surface fluctuations, its elevation or *stage* at the time of survey must be determined with respect to a fixed datum—for example, orthometric height in NAVD88. Running a level circuit to the water from a nearby benchmark can do this. In situations where soundings are repeated at regular intervals, a graduated staff can be permanently installed in the water so that its stage, in feet above the datum, can be read directly each time soundings are repeated.

Remote-controlled survey vessels, as shown in Figure 17.20, are also being used to perform hydrographic surveys. These vehicles are light enough to be carried by hand and can perform hydrographic surveys in mines, sewage treatment plants, contaminated lakes, harbors, and rivers. The systems consist of GNSS receiver(s) to record the boat's position along with either depth sounders as shown in Figure 17.20 or multibeam (swath) echosounders. Combined, these systems can measure the location and depth of underwater physical and biological objects.

Another option for mapping shallow waters includes LiDAR, (Section 27.18) which can penetrate water by using light with a wavelength of 532 nm^2 (green light). Thus, unmanned aerial systems can collect point data on the water's

Figure 17.19 Bottom profile produced by electronic depth sounder.

504 MAPPING SURVEYS

Figure 17.20
HyDrone-RCV Portable, Remote Control Survey Boat. (Courtesy Seafloor Systems, Inc.)

floor or on intervening objects. In optimal environmental conditions such as clear water, green LiDAR may reach a depth of 50 m with a horizontal accuracy of ±2.5 m and a vertical accuracy of ±0.25 m. These technologies are growing fast and provide additional tools to map a water body's floor.

17.14.2 Locating Soundings

Any of the traditional ground-surveying procedures can be used to locate positions where soundings are taken. In addition to these techniques, other methods have also been applied in hydrographic surveys, for example, GNSS receivers. If ground-surveying techniques are used, some horizontal control must first be established on shore. Ideal locations for control stations are on peninsulas or in open areas that afford a wide unobstructed view of the water body for tracking a sounding boat. The coordinate positions of the control points can be established by traverse, but triangulation and trilateration are also well suited for this work.

Among the various boat-positioning methods, *radiation* and *angle intersection* are usually selected if total station instruments are used. Radiation is particularly efficient, especially if a total station instrument is used, because only one person on shore is needed to track the boat. After setting up on one control station and back sighting another, angles and distances are measured to locate each boat position. Robotic total station instruments and specialized reflectors, which are also used in machine control, are manufactured for this work to facilitate sighting and observing distances electronically to a moving target. From the angles and distances, which are automatically read, the total station's computer determines the boat's coordinates. These can either be stored in a survey controller for later office use in mapping or transmitted by radio to the boat if real-time positioning is required, as in controlling ongoing dredging.

---

[2]1 nanometer (nm) is equal to 1×10^{-9} meters.

Figure 17.21 illustrates the use of angle intersections in the hydrographic survey of a lake. Here, the boat travels back and forth along *range lines* while the depth sounder continuously records bottom profiles. At regular intervals, *fixes* are taken by observing angles to the boat from shore stations. Two angles establish the boat's position, but three or more provide redundancy and a check. For example, in Figure 17.21, angle observations e, g, and h, for fix number 50 (indicated by dashed lines) have been made from shore stations E, G, and H, respectively. Prior to observing angles, the total stations or theodolites were oriented by backsighting on another visible control station, as at station G from E.

Flag or radio signals are given from the boat to coordinate fixes and ensure that angles from all shore stations are observed simultaneously. At the precise moment of any fix, the profile is also marked and the fix number noted. For example, in Figure 17.21, fixes 48 through 52 are identified and marked on the profile of Figure 17.19. This correlates bottom depths with specific locations in the water body—a necessity for mapping.

If the boat is driven back and forth along parallel range lines to cover the area of interest, and then the area traversed again with perpendicular courses, a grid of profiles results from which contours can be drawn. In larger bodies of water, a compass is valuable to assist in keeping the range lines parallel. Required accuracy dictates the spacing between range lines, with closer spacing yielding more accurate results. Various other boat-positioning systems can be used, depending on circumstances. One that works well for hydrographic surveys of rivers or other relatively narrow water bodies consist in laying out uniformly spaced reference lines, which cross the water. Placing tall painted stakes on the bank on either side marks the lines. Then, fixes can be taken as the sounding boat navigates along the reference lines. However, to position each fix along the lines,

Figure 17.21
Angle intersection procedure for locating boat fixes along range lines.

either a distance must be observed from one reference point, or an azimuth to the boat from an independent control point. When the boat is moving perpendicular to the marked lines, its passage across projections between stakes locates fixes, but again a distance or angle is needed to complete the fix position.

Kinematic surveying methods (see Chapter 15) are ideal for establishing sounding locations for hydrographic surveys and also for guiding the boat along planned range lines on larger water bodies. With its many advantages, it has replaced other hydrographic positioning techniques when overhead obstructions are not a limiting factor. In this case, the transducer is located by measuring offset from the antenna reference point. Often the antenna is mounted on the sounding pole well above the top of the boat and directly over the transducer. In this case, only the distance from the antenna reference point to the transducer needs to be measured.

17.14.3 Hydrographic Mapping

Procedures for preparing hydrographic maps do not differ appreciably from those used in topographic mapping discussed in Chapter 18. Basically, depths are plotted in their surveyed positions and contours drawn. If an echo sounder is used, depths are interpolated from the profiles and plotted between fix locations. In addition to depth contours, the shoreline and other prominent features are also located on hydrographic maps. This is especially important for navigation and fishing maps, as the features are the means by which users line in and locate themselves on the water body. Planimetric features are most often located photogrammetrically (see Chapter 27), but the techniques for topographic mapping discussed in this chapter can also be used.

Modern hydrographic surveying systems utilize sophisticated electronic positioning and depth-recording devices. These, coupled with computers interfaced with plotters, enable rapid automated production of hydrographic maps in near real time. But the basic principles discussed here still apply.

■ 17.15 SOURCES OF ERROR IN MAPPING SURVEYS

Some sources of error in planimetric and topographic surveys are:

1. Instrumental errors, especially an index error that affects vertical angles.
2. Errors in reading instruments.
3. Control not established, checked, and adjusted before beginning to collect details.
4. Control points too far apart and poorly selected for proper coverage of an area.
5. Sights taken on detail points which are too far away.
6. Poor selection of points for contour delineation.

■ 17.16 MISTAKES IN MAPPING SURVEYS

Some typical mistakes in planimetric and topographic surveys are:

1. Unsatisfactory equipment or field method for survey and terrain conditions.
2. Mistakes in instrument reading and data recording.

3. Failure to periodically check azimuth orientation when many detail points are located from one instrument station.
4. Too few (or too many) contour points taken.
5. Failure to collect some mapping details.
6. Mistakes in entering point identifiers, drawing designators, and symbols when using field-to-finish surveying and mapping systems.

PROBLEMS

Asterisks (*) indicate problems that have answers given in Appendix G.

17.1 List four items that maybe cultural features on a map.

17.2 What is the difference between a planimetric and topographic map?

17.3 Describe different methods of displaying relief on a map.

17.4 What different methods can be used to show a map's scale?

17.5 What determines the scale used for a map?

17.6* On a map sheet having a scale of 1 in. = 360 ft, what is the smallest distance (in feet) that can be plotted with an engineer's scale? (Minimum scale graduations are 1/60th in.)

17.7 The same map is plotted at two different scale. The first is plotted at a scale of 1:1000. The second map is plotted at a scale of 1:2000. Which map has the larger scale?

17.8 A topographic map has a contour interval of 1 ft and a scale of 1:4800. If two adjacent contours are 0.05 in. apart, what is the average slope of the ground between the contours?

17.9 What are the ratio scales for the equivalent scales of 1 in. = 50 ft and 1 cm = 10 m?

17.10* On a map whose scale is 1 in. = 40 ft, how far apart (in inches) would 1-ft contours be on a uniform slope (grade) of 2%?

17.11 Define the term *small-map scale*?

17.12 On a map drawn to a scale of 1:1200, contour lines are 0.5 inches apart at a center location. The contour interval is 1 ft. What is the ground slope in percent, between adjacent contours?

17.13 The line width on a map of scale 1:2000 is 0.3 mm. What should the accuracy of the survey be to draw this map?

17.14 Why is it important to consider the future use of a map for a specific mapping project?

17.15* What conditions in the field need to exist when using kinematic satellite survey?

17.16 What is a digital elevation model?

17.17 What is a contour? What is a contour interval?

17.18 Why are spot elevations placed on maps?

17.19 Between the elevations of 5233 ft and 5333 ft at what elevations will contours be drawn for a 20-ft contour interval?

17.20 Using the rules of contours, list the contouring mistakes that are shown in the accompanying figure.

17.21 What is a triangulated irregular network?

17.22 Assuming *xy* coordinates for the instrument station of (5000, 5000), a backsight azimuth of 175°22′48″, an HI of 362.18 ft, and a HR of 6.00 ft, determine the coordinates and elevations for points 3, 4, and 5 in Table 17.1.

17.23 What assumptions are made in an automatic contouring system?

17.24 Discuss how a survey controller with a total station instrument can be combined with GNSS methods to collect data for a topographic map.

Problem 17.20

17.25 What does the term "point cloud" describe in laser-scanning?

17.26 What factors must be considered when planning a laser scanning survey?

For Problems 17.27 through 17.29, calculate the X, Y, and Z coordinates of point B for radial readings taken to B from occupied station A, if the backsight azimuth at A is 93°21′03″, the elevation of A = 810.96 ft, and hi = 5.60 ft. Assume the XY coordinates of A are (5000.000, 5000.000).

17.27* Clockwise horizontal angle = 23°37′42″, zenith angle 86°29′35″, slope distance = 125.64 ft, hr = 5.50 ft.

17.28 Clockwise horizontal angle = 212°38′08″, zenith angle = 93°28′16″, slope distance = 98.64 ft, hr = 5.50 ft.

17.29 Clockwise horizontal angle = 275°24′08″, zenith angle = 88°26′37″, slope distance = 168.59 ft, hr = 5.50 ft.

17.30 What are breaklines?

17.31 List the various methods that can be used to collect mapping data.

17.32 What is the advantage of coding data points when combined with field-to-finish software?

17.33 What instruments maybe involved in a mapping survey?

17.34* On a map having a scale of 200 ft/in. the distance between plotted fixes 49 and 50 of Figure 17.20 is 3.15 in. From measurements on the profile of Figure 17.19, determine how far from fix 50 the 20-ft contour (existing between fixes 49 and 50) should be plotted on the map.

17.35 Name three considerations that must be considered when planning a laser scanning survey.

17.36 What is a hydrographic survey?

BIBLIOGRAPHY

Andelin, E. 2009. "On the Right Track." *Point of Beginning* 35 (No. 3): 12.

ASPRS. 1987. *Large Scale Mapping Guidelines*. Bethesda, MD: American Society for Photogrammetry and Remote Sensing.

Battista, T. and C. Taylor. 2015. "Finding the Fish Hot Spots." *XYHT* 2 (No. 3): 14.

Barnes, G. and W. Volkmann. 2015. "High-Resolution Mapping with Unmanned Aerial Systems." *Surveying and Land Information Science* 74 (No. 1): 5.

Barton, J. 2014. "Digital Heritage: When Is Heritage Worth Digitizing?" *The American Surveyor* 11 (No. 3): 33.

Bernardini, F. 2015. "Power Scanning." *XYHT* 2 (No. 3): 33.
Cooper, P. R. 2014. "Hydrography." *XYHT* Special Outlook Edition (Outlook '15): 4.
Duffy, L. 2014. "Going Below the Surface." *Point of Beginning* 39 (No. 6): 14.
Ethen, T. 2015. "Mapping the USS Arizona." *XYHT* 2 (No. 3): 37.
Graham, L. 2014. "Accuracy with Small UAS Mapping." *XYHT* 1 (No. 1): 48.
Jacobs, G. 2009. "3D Scanning: Accuracy of Scan Points." *Professional Surveyor* 29 (No. 8): 24.
_____. 2009. "3D Scanning: Laser Scanner Versatility Factors, Part 1." *Professional Surveyor* 29 (No. 10): 34.
_____. 2010. "3D Scanning: Laser Scanner Versatility Factors, Part 2." *Professional Surveyor* 30 (No. 1): 32.
Luccio, M. 2014. "From Shore to Floor." *XYHT* 1 (No. 4): 16.
Manyoky, M., P. Theiler, D. Steudler, and H. Eisenbeiss. 2011. "Unmanned Aerial Vehicle in Cadastral Applications." In *Proceedings of ISPRS and UAV-g*, Zurich, Switzerland. [http://www.geometh.ethz.ch/uav_g/proceedings/manyoky].
Neitzel, F. and J. Klonowski. 2011. "Mobile 3D Mapping with a Low-Cost UAV System." In *Proceedings of ISPRS and UAV-g*, Zurich, Switzerland. [http://www.geometh.ethz.ch/uav_g/proceedings/neitzel].
Silver, M. "The Dinosaur Surveyors." *The American Surveyor* 11 (No. 8): 18.
Spring, A. and C. Peters. 2014. "3D Imaging Shines New Light on Tristan Stone." *XYHT* 1 (No. 5): 38.
Stocking, A. 2014. "The Promise of Scanning." *Point of Beginning* 39 (No. 9): 8.
Trojak, L. 2014. "Road Scholars." *The American Surveyor* 11 (No. 3): 42.
Velde, G. 2014. "Out of Africa." *The American Surveyor* 11 (No. 2): 16.
Wagner, M. Jo. 2009. "Scanning the Horizon." *Point of Beginning* 35 (No. 2): 24.

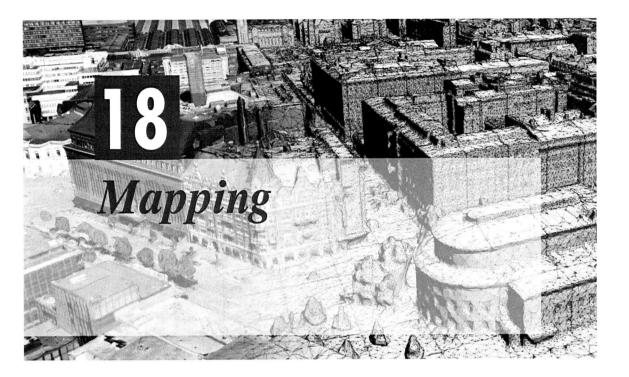

18
Mapping

18.1 INTRODUCTION

Maps are visual expressions of portions of the Earth's surface. Features are depicted using various combinations of points, lines, and standard symbols. Maps have traditionally been produced in *graphic*, or "hardcopy," form; that is, printed on paper or a stable-base plastic material such as Mylar. However, today most mapping data are collected in digital form and are then processed using *Computer Aided Drafting and Design* (CADD) systems to develop "softcopy" maps. Softcopy maps are stored within a computer, can be analyzed, modified, enlarged, or reduced in scale, and have their contour intervals changed while being viewed on the monitors of CADD systems. Different types or "layers" of information can also be extracted from digital maps to be represented and analyzed separately, and softcopy maps can be transferred instantaneously to other offices or remote locations electronically. Of course, they can also be printed in hardcopy form if desired. Softcopy maps are indispensable in the development and operation of modern *Land Information Systems* (LISs) and *Geographic Information Systems* (GISs) (see Chapter 28).

Throughout the ages, maps have had a profound impact on human activities, and today the demand for them is perhaps greater than ever. They are important in engineering, resource management, urban and regional planning, management of the environment, construction, conservation, geology, agriculture, navigation, and many other fields. Maps show various features—for example, topography, property boundaries, transportation routes, soil types, vegetation, land ownership, and mineral and resource locations. Maps are especially important in engineering for planning project locations, designing facilities, and estimating contract quantities.

As noted in the preceding text, maps are essential in the development and operation of LISs and GISs. These systems for spatial data analysis and management use the computer to store, retrieve, manipulate, merge, analyze, display, and disseminate information by means of digital maps (see Chapter 28). GISs have applications in virtually every field of endeavor. Spatial databases to support the systems are generally developed either by digitizing existing graphic maps or by generating new digital maps in the computer based upon digitized ground survey or photogrammetric data. Maps of various types needed to create spatial databases for LISs and GISs include topographic maps, which display the natural and cultural features and relief in an area; cadastral maps, which give boundaries of land ownership; natural resource maps, which provide the location and distribution of forest and water resources, wetlands, soil types, etc.; facilities maps, which show existing transportation networks, water and sewer mains, and distribution lines for electric power; and land-use maps, which show the various activities of humans related to the land. Applications of GIS technology have been expanding at a substantial rate and these activities will impose a heavy demand for high-quality maps of various types and scales in the future.

Cartography, the term applied to the overall process of map production, includes map design, preparing or compiling manuscripts, final drafting, and reproduction. These processes, which apply whether the maps are graphic or digital, are described in this chapter.

18.2 AVAILABILITY OF MAPS AND RELATED INFORMATION

Maps for a variety of different purposes, prepared at scales varying from large to small, and in both graphic and digital form, are prepared by private surveying and engineering companies, industries, public utilities, cities, counties, states, and agencies of the federal government. Unfortunately, with such a wide range in organizations and agencies involved, some duplication of effort has occurred because mapping activities generally have not been coordinated. Also, the existence of available maps and related information is often unknown to potential users. However steps have been taken to improve this situation. The U.S. Geological Survey (USGS) now coordinates all mapping activities at the federal level. They offer nationwide information and sales service for map products and Earth science publications. The USGS provides information about topographic, land use, geologic and hydrologic maps, books and reports; Earth science and map data in digital format and related applications software; aerial, satellite, and radar images and related products; and geodetic data.[1]

Several states have established state cartographers' offices. One of their functions is the dissemination of local maps and related products and information to surveyors, engineers, cartographers, and the general public. Additionally many county offices have mapping departments, which provide maps to the public at the local level.

18.3 NATIONAL MAPPING PROGRAM

The *National Mapping Program* was established to provide maps and other cartographic products needed by the citizens of the United States. This is the responsibility of the *National Mapping Division* of the USGS. The USGS began publishing

---

[1]The U.S. Geological Survey can be reached by telephone at (888) ASK-USGS, [(888)275-8747]. Information can also be obtained, and selected maps and other products can be downloaded at the following website: http://www.usgs.gov/pubprod/. Contact by mail can be made to the U.S. Department of the Interior, U.S. Geological Survey, 12201 Sunrise Valley Drive, Reston, VA 20192.

topographic maps in 1886 as an aid to scientific studies. It now produces a variety of topographic maps at differing scales; however, its standard series has a scale of 1:24,000. In this series, individual sheets cover *quadrangles* of 7 1/2-min in both latitude and longitude. Each quadrangle map is named, usually according to the most prominent feature within its bounds. Except for Alaska, the entire United States is covered at the 1:24,000 scale and over 57,000 maps are involved. [Maps covering 15' quadrangles at a scale of 1:63,360 (1 in./mi) are standard for Alaska.] On the quadrangle maps, cultural features are shown in black, contours in brown, water features in blue, urban regions in red, and woodland areas in green. Topographic coverage of the United States is also available at scales of 1:50,000 (county maps), 1:62,500 (the older 15' quadrangles produced until about 1950), 1:100,000, and 1:250,000. The USGS has also published a state map series. Most are at a scale of 1:500,000, but a few are at 1:1,000,000 or other scales.

As noted in Section 18.1, requirements for digital cartographic data are growing rapidly to support LISs and GISs. To meet these needs, the USGS has developed two very useful types of digital data: (1) *digital line graphs* (DLGs) and (2) *digital elevation models* (DEMs). The DLGs contain only linear features or planimetry in an area. Included are political boundaries, hydrography, transportation networks, and the subdivision lines of the U.S. Public Land Survey System (see Chapter 22). The DEMs are arrays of elevation values, produced in grids of varying dimensions, depending on the source of the information. The horizontal positions of points in the DEMs are X and Y coordinates referenced to the Universal Transverse Mercator coordinate system (see Section 20.11). A grid of 30 m is used for DEMs generated from 7 1/2' quadrangles, with larger intervals being used for those generated from smaller scale maps.

In addition to topographic maps, DLGs, and DEMs, a variety of other special maps and related products are published as a part of the National Mapping Program. As noted in the preceding section, information on all of their available maps and other related products is obtained through the USGS.

18.4 ACCURACY STANDARDS FOR MAPPING

To provide a set of uniform standards for guiding the production of maps, and to protect consumers of maps, the United States Bureau of the Budget developed the *National Map Accuracy Standards* (NMAS). These standards, first published in 1941 and revised in 1947, provide specifications governing both the horizontal and vertical accuracy with which features are depicted on maps. Published maps meeting these accuracy requirements may have the following note in their legends: "This map complies with National Map Accuracy Standards," thereby providing assurance that the map meets these specified accuracy levels.

To meet the NMAS horizontal position specification, for maps produced at scales larger than 1:20,000, not more than 10% of well-defined points tested shall be in error by more than 1/30 in. (0.8 mm). Accordingly, on a map plotted to a scale of 1 in. = 100 ft, point positions would have to be correctly portrayed to within ±3.3 ft to meet this specification. On smaller scale maps, the limit of horizontal error is 1/50 in. (0.5 mm), or approximately ±40 ft on the ground at a map scale of 1:24,000. These limits of accuracy apply to positions of well-defined points only, such as monuments, benchmarks, highway intersections, and building corners.

The NMAS vertical accuracy requirements specify that not more than 10% of elevations tested shall be in error by more than one half the contour interval, and none can exceed the interval. To meet this requirement, contours may be shifted by distances up to the horizontal positional tolerance (discussed in the preceding text), if necessary.

The accuracy of any map can be tested by comparing the positions of points whose locations or elevations are shown on it with corresponding positions determined by surveys of a higher order of accuracy. Plotted horizontal positions of objects are checked by running an independent traverse or other survey to points selected by the person or organization for which the map was made. To check vertical accuracy, elevations obtained from field profile surveys are compared with elevations taken from profiles made from plotted contours. These procedures provide a check on both fieldwork and map drafting.

When the NMAS were developed, maps were being produced in hardcopy form. But as noted in Section 18.1, softcopy maps are now most common. To accommodate this change, the *Federal Geographic Data Committee* (FGDC) drafted a more current set of accuracy standards called the *Geospatial Positioning Accuracy Standards*.[2] The FGDC is composed of representatives from 19 Federal agencies and was established to coordinate policies, standards, and procedures for producing and sharing geographic information. The new Geospatial Positioning Accuracy Standards are completed, and like NMAS, the document specifies accuracies in separate horizontal and vertical components. But unlike NMAS, accuracies are specified in terms of coordinates of points, ground distances, and elevations at the 95% confidence level. Thus, these new standards are applicable to all types and scales of maps, including those in digital form. The test for maps intended to meet this standard involves checking a set of at least 20 or more well-defined points against information obtained from an independent source of higher accuracy. Root-mean-square errors (RMSE) are computed and converted to the 95% confidence level by using appropriate multipliers. A digital planimetric map that passes at the 1-m level, for example, would contain the statement "Tested 1-meter horizontal accuracy at 95% confidence level" in its legend. A similar statement can be included that applies to a map's vertical accuracy.

The USGS has developed its own standards to govern the production of the maps and other products it provides through the National Mapping Program. Standards have been developed not only for their hardcopy maps, but also for their digital products including digital elevation models, digital line graphs, digital orthophotos (see Section 27.15), and others.

The *American Society for Photogrammetry and Remote Sensing* (ASPRS) has also adopted its own standards[3] to govern digital orthoimagery, digital planimetric data, and digital elevation data. It specifies standards for varying types

---

[2]Information on the status of the **Geospatial Positioning Accuracy Standards** can be obtained from the Federal Geographic Data Committee by telephone at (703)648-5514, or at the following website: http://www.fgdc.gov. Copies of the current standards can also be downloaded at this website. Contact can also be made by mail at The Federal Geodetic Data Committee, U.S. Geological Survey, 590 National Center, Reston, VA 20192.

[3]The 2014 *ASPRS Accuracy Standards for Digital Geospatial Data* can be found at http://www.asprs.org/wp-content/uploads/2015/01/ASPRS_Positional_Accuracy_Standards_Edition1_Version100_November2014.pdf.

of ground coverage including vegetated and nonvegetated areas. All positional accuracies are based on a RMSE at the 95% level of confidence. It also has procedures to compare older standards, such as NMAS, into these newer standards and has accuracy standards for LiDAR data within a swath of coverage and at seams between the swaths. These standards recommend a minimum of 20 clearly defined checkpoints for any map covering less than 500 km² (50,000 ha) but requires more checkpoints for maps covering larger regions.

The *American Society of Civil Engineers* (ASCE) has also developed a set of standards for topographic mapping that are aimed primarily at large-scale engineering maps. In addition to suggesting accuracies for various map scales, they also provide standards for contouring, map symbols, abbreviations, lettering, and other factors important in mapping.

The importance of classifying the accuracy of maps with statistical values is that all forms of maps whether hardcopy or digital can have their accuracy stated so that users understand the product and it potential uses. The RMSE is defined as the square root of the average of squared discrepancies for points tested. Discrepancies are the differences between coordinates and elevations of points taken from the map, and their values as determined by check surveys. Mathematically, it is denoted as

$$\text{RMSE} = \sqrt{\sum_{i=1}^{n} \frac{[f(x_i) - x_i]^2}{n}} \quad (18.1)$$

where n is the number of tested samples from the map, $f(x_i)$ is the position obtained from the map, and x_i is the position obtained from the check survey. For example, consider the data shown in Table 18.1. The map-derived values for coordinates of the points are shown in columns (1)–(3). The surveyed checkpoint coordinate values are shown in columns (4)–(6). The discrepancies shown in columns (7)–(9) are computed as the difference between the map and checkpoint coordinate values. Since the checkpoints must be surveyed at a higher accuracy than the map, these discrepancies can be thought of as residuals. Thus, the residual for the x coordinate of point 1 is computed as 624,891.153 − 624,891.114 = 0.039. Similarly, all other coordinate differences are determined. Using Equation (18.1), the RMSE in x (RMSE_x), y (RMSE_y), and z (RMSE_z) are computed from their corresponding residuals as

$$\text{RMSE}_x = \sqrt{\frac{0.039^2 + (-0.019)^2 + 0.035^2 + (-0.018)^2 + (-0.057)^2}{5}} = 0.037 \text{ m}$$

$$\text{RMSE}_y = \sqrt{\frac{0.031^2 + 0.055^2 + (-0.008)^2 + 0.045^2 + (-0.064)^2}{5}} = 0.045 \text{ m}$$

$$\text{RMSE}_z = \sqrt{\frac{(-0.115)^2 + 0.079^2 + 0.089^2 + (-0.036)^2 + (-0.054)^2}{5}} = 0.079 \text{ m}$$

From the RMSE_x and RMSE_y, a radial value for the horizontal positional accuracy can be determined as

$$\text{RMSE}_r = \sqrt{\text{RMSE}_x^2 + \text{RMSE}_y^2} = \sqrt{0.037^2 + 0.045^2} = 0.058 \text{ m}$$

TABLE 18.1 Map Coordinates versus Surveyed Checkpoint Coordinates

	Map Coordinates			Surveyed Checkpoint Coordinates			Discrepancies/Residuals		
Point	(1) x (m)	(2) y (m)	(3) z (m)	(4) X (m)	(5) Y (m)	(6) H (m)	(7) Δx	(8) Δy	(9) Δz
1	624,891.153	365,874.351	124.589	624,891.114	365,874.320	124.704	0.039	0.045	−0.115
2	623,480.214	365,789.318	138.674	623,480.233	365,789.263	138.595	−0.019	0.055	0.079
3	638,924.086	366,182.489	156.483	638,924.051	366,182.497	156.394	0.035	−0.008	0.089
4	639,716.158	358,315.168	105.641	639,716.176	358,315.123	105.677	−0.018	0.045	−0.036
5	625,314.604	373,648.084	143.891	625,314.661	373,648.148	143.945	−0.057	−0.064	−0.054
				Number of Checkpoints			5	5	5
						RMSE (m)	0.037	0.045	0.079
						RMSEr	0.058		
						95% RMSE	0.100		0.155

To obtain the 95% positional horizontal accuracy for the map, the $RMSE_r$ is multiplied by 1.7308, which in this case results in a 95% positional horizontal accuracy of 0.100 m for this map. To compute the 95% positional vertical accuracy for the map ($RMSE_z$) is multiplied by 1.9600, which in this case yields 0.155 m. In order for RMSEs to be credible, the check survey must be performed at a higher level of accuracy than the map product. For example, assume a laser scanner (see Section 17.9.6) is used to develop a map of a structure. In this instance, a conventional survey with a reflectorless total station will typically provide checkpoints at a higher level of accuracy and could be used to establish coordinates on clearly defined points, which are also visible and measured in the scan of the site. If a GNSS survey is to be used, special procedures as outlined in the NGS' *Guidelines for Establishing GPS-Derived Ellipsoid Heights* (Zilkoski, 1997) must be used to establish geodetic heights on clearly defined points as discussed in Section 14.7. Additionally, the most recent geoid model must be applied to the geodetic heights to derive orthometric heights for the points using Equation (13.8). However, if this method is used realize that the geoid model for the area of interest may contain more error than is allowable by the standards. If sufficient benchmarks surround the area of interest, a method of establishing orthometric heights (elevations) of stations as discussed in Section 13.4.3 can be used to alleviate any problems in the geoid model. Alternately, a GNSS survey could be used to establish horizontal positioning while differential leveling or trigonometric leveling could be used to establish orthometric heights on points.

■ 18.5 MANUAL AND COMPUTER-AIDED DRAFTING PROCEDURES

As previously stated, maps may be drafted manually, or produced with CADD systems. Manual procedures utilize standard drafting tools such as scales, protractors, compasses, triangles, and T-squares. CADD systems employ computers programmed with special software and interfaced with electronic plotting devices. With either approach, after deciding on scale and other factors that control overall map design, a manuscript is prepared. When completed, final drafting is performed.

In manual drafting, the manuscript is usually compiled in pencil. It should be prepared carefully to locate all features and contours as accurately as possible and be complete in every detail, including placement of symbols and letters. Lettering on the manuscript need not be done with extreme care, for its major purpose at this stage is to ensure good overall map design and proper placement. A well-prepared manuscript goes a long way toward achieving a good-quality final map.

The completed version of the manually compiled manuscript is drafted in ink or *scribed*. If inked, the manuscript is placed on a light-table, and features are traced on a stable-base transparent overlay material. Lettering is usually done first; then planimetric features and contours are made. Scribing is performed on sheets of transparent stable-base material coated with an opaque emulsion. Manuscript lines are transferred to the coating in a laboratory process. Lines representing features and contours are then made by cutting and scraping to remove the coating. Special scribing tools are used to vary line weights and make standard symbols. Scribing is generally easier and faster than inking. Reproductions are made from the finished inked or scribed product.

In drafting maps with CADD, a softcopy manuscript is compiled in the computer and displayed on its screen as work progresses. CADD software provides instructions to the computer, which basically duplicate manual drafting functions. An operator interactively designs and compiles the map by entering commands into the computer's keyboard or using a mouse to activate functions on a menu. Points, lines of various types, and a variety of symbols are available to the operator. Letters of differing sizes and styles can also be selected. When the manuscript is completely finished, simply activating the electronic plotter will draw the final map. Map production using CADD systems has many advantages over manual methods, and therefore, it is the method of choice by mapping firms. However, it is still important to learn the basics of manual map drafting techniques, since these are often duplicated in the CADD processes. CADD systems are described in more detail in Section 18.14.

■ 18.6 MAP DESIGN

Before beginning the design of a map, the following two basic questions should be answered: (1) What is the purpose of the map? and (2) who is the map intended to serve? All maps have a purpose, which in turn dictates the information that the map must convey. Once the purpose of the map is fixed, emphasis should be placed on achieving the design that best meets its objectives and conveys the necessary information clearly to its users.

Maps typically depict many different types and classes of details in portraying natural and cultural features and, if properly designed, they can convey an enormous amount of information. On the other hand, maps that are carelessly designed can be confusing, difficult to read, understand, or interpret. To achieve maximum effectiveness in map design, the following elements or factors should be considered: (1) clarity, (2) order, (3) balance, (4) contrast, (5) unity, and (6) harmony. Definitions of these six elements in relation to map design, and explanations of their interdependence are discussed in the following text.

1. *Clarity* relates to the ability of a map to convey its intended information completely and unambiguously. It can only be achieved after fully examining the objectives of the map and then emphasizing the features necessary to carry out those objectives. Maps should not be overloaded with details, as this can cause congestion and confusion. If considerable detail must be included on a map, the information can be placed in a table. Other alternatives consist of preparing larger-scale *inset maps* of areas that contain dense detail, or creating an overlay to display some of the detail. The proper use of textual elements is very important in achieving clarity. (The subject of map lettering is covered in Section 18.11.)

2. *Order* refers to the logic of a map, and relates to the path that a user's eye would follow when looking at one. A design should be adopted that first draws the user's attention to the subject area of the map, then the map title, and then to any notes. Never let auxiliary elements such as bar scales and directional arrows dominate the map. A common mistake made by beginners is to make bar scales and north arrows so large and bold that they attract attention away from the subject of the map.

3. All elements on a map have weight, and they should be distributed uniformly around the "visual center" of the map to create good overall *balance*. The visual center is slightly above the geometrical center of the map sheet. In general, the weight of an element is affected by factors such as size, color, font, position, and line width. Map elements that appear at the center have less weight than those on the edges. Elements in the top or right half of the map will appear to have more weight than those in the bottom or left half of the map. Also map elements identified with thicker line widths will appear to have heavier weights than their slimmer counterparts. Colors such as red appear heavier than blue or yellow. Two examples illustrating balance are shown in Figure 18.1. In Figure 18.1(a), the map appears to be too heavily weighted to the left, and thus has poor overall balance. A redistribution of the map features, as shown in Figure 18.1(b), produces a more visually balanced product. The use of thumbnail sketches can often help to achieve a balanced layout for a map. It is important to place highest weights on those elements that enhance the purpose of the map.
4. *Contrast* relates primarily to the use of different line weights, and fonts of varying sizes. Contrast can be used to enhance balance, order, and clarity. For example, the title of the map should be displayed in a larger font than the other textual elements. This will attract the viewer's attention, thereby enhancing the order and clarity of the map. Various fonts can also be used to provide balance with other elements on the map. Another example where contrast supports the clarity of a map is in contouring. Here, *index contours* (every fifth contour) should be drawn with a heavier line than the other contours. This enhances the map's clarity and facilitates the determination of elevations.
5. *Unity* refers to the interrelationships between the backgrounds, shading, and colors on a map. Again these items can enhance clarity, balance, and contrast. They can also detract from these same items. For example, yellow lettering on a white background is difficult to see and often overlooked by

Figure 18.1
(a) The layout of a poorly balanced map sheet. (b) A better layout for the same map.

(a)

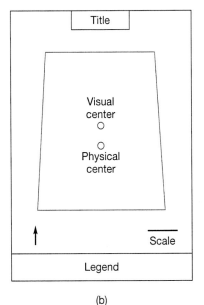

(b)

the reader. However, this same yellow lettering on a black background will stand out and appear emphasized. A map with good unity is visualized as a unit, and not as an assemblage of individual elements.

6. *Harmony* relates to the interrelationships between all elements on the map. If a map has good harmony, the elements work together. Common errors are the use of too many fonts, a north arrow that is too fancy or large, or a bar scale that is too large. It should be remembered when placing a north arrow and graphic scale on a map that they are not the most important features of the map and should not distract from the subject of the map.

In designing maps, it is important to remember that different audiences may require different maps. For example, it would be difficult for a layperson to read and understand a map produced for an engineering project. Accordingly, maps that are developed for design professionals are not generally suitable for public hearings. In fact because laypeople often have no training in map reading, it may be best to develop specialized three-dimensional maps or models that depict relief, boundaries, proposed buildings, landscaping, and so on.

Sometimes when designing a map, certain elements of map design will be in conflict. When this happens, priorities must be established that provide a reasonable solution to the conflict. A perfect map rarely, if ever, exists, and there are generally several equally acceptable designs that could be adopted. Often there are design conflicts that cannot be resolved and a compromising solution may have to be accepted. Map creation is often subjective and the production of a well-designed map requires a combination of skill, art, and patience.

18.7 MAP LAYOUT

In general, the subject area of the map should be plotted at the largest scale that will enable it to fit neatly within its borders without producing overcrowding. It should also be centered on the map sheet and, if possible, should be aligned so that the edges of the map sheet coincide with the cardinal directions. If this is not done, users may experience some confusion when viewing the map. Accordingly, the size and shape of the map sheet, the size and shape of the area to be mapped, the orientation of the subject area on the map sheet, and map scale, must be jointly considered in map layout.

To illustrate, consider the example of Figure 18.2, which is a simple traverse from a planimetric survey. Before any plotting is done, the proper scale for a sheet of given size must be selected. Assume in this example that an 18- by 24-in. sheet will be used, with a 1-in. border on the left (for possible binding) and 1/2-in. borders on the other three sides. A borderline somewhat heavier than all other lines can be drawn to outline this area. If the most westerly station (A in the example) has been chosen as the origin of coordinates, then divide the total departure to the most easterly point C by the number of inches available for plotting in the east–west direction. The maximum scale possible in Figure 18.2 is 774.25 divided by 22.5, or 1 in. = 34 ft. The nearest standard scale that will fit is 1 in. = 40 ft.

This scale must be checked in the Y direction by dividing the total difference in Y coordinates, 225.60 + 405.57 = 631.17 by 40 ft, giving 15.8 in. required in the

520 MAPPING

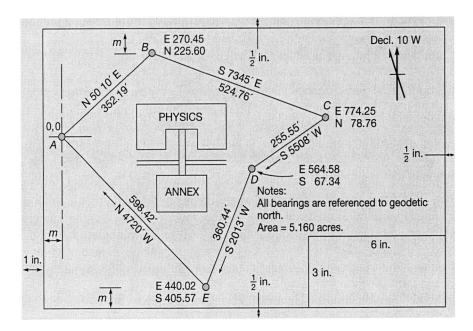

Figure 18.2
Map layout.

north–south direction. Since 17 in. are usable, a scale of 1 in. = 40 ft is satisfactory, although a smaller scale would yield a larger border margin. If a scale of 1 in. = 40 ft is not suitable for the map's purpose, a sheet of different size should be selected, or alternatively more than one sheet employed to map the required area.

In Figure 18.2, the traverse is centered between the borderlines in the Y direction by making each distance m equal to $1/2(17 - 631.17/40)$, or 0.61 in. The same 0.61 in. can be used for the left side. Weights of the title, notes, and north arrow compensate for the traverse being to the left of the sheet center and leave ample space for including the necessary auxiliary elements[4] of the map.

If a compromising choice must be made between map scale and the sizes of auxiliary elements, it is better to maximize the map's scale and minimize the size of the auxiliary elements. Beginners should avoid using oversized auxiliary elements to use up available or leftover space since doing so detracts from the map's order and balance.

■ 18.8 BASIC MAP PLOTTING PROCEDURES

Map plotting may be done either manually, or by employing automated CADD systems. Regardless of which method is used, the procedure consists fundamentally of plotting individual points. Lines are then drawn from point to point to portray features. Although this process may seem simple in principle, accurate work requires skill, patience, and care. While points could be laid out using angles and distances, or lines scaled and plotted directly, the most convenient method for laying out points and drafting maps involves plotting points by coordinates. Plotting by this procedure is also consistent with today's modern data collection

---

[4]Auxiliary elements include the map title block, notes, legend, bar scale, and north arrow. These elements are described in Section 18.12.

systems; that is, total station instruments and portable GNSS units, because those devices provide coordinates directly. The following subsections describe manual and CADD coordinate plotting procedures.

18.8.1 Plotting Manually by Coordinates

To plot points by coordinates, the map sheet is first laid out precisely in a grid pattern with unit squares of appropriate size. Squares of 2, 4, or 5 in. are commonly used, and depending on map scale, they may represent 100, 200, 400, 500, or 1000 ft, or 50, 100, 200, or 500 m. The grids are constructed using a sharp, hard pencil, and are checked by carefully measuring diagonals. The grid lines are labeled with coordinate values, making sure that the range of coordinates covered on the map will accommodate the most extreme X and Y coordinates to be plotted.

Initially, coordinates for all features to be mapped must be determined. These points are then plotted by laying off their X and Y coordinates from the ruled grid lines. Mistakes in plotting can be detected by comparing scaled lengths (and directions) of lines with their measured or computed values. Since each point is plotted independently, a mistake in one will not affect the others, and that point can simply be corrected.

Many mapping elements such as bar scales, legends, north arrows, etc., are prepared on separate map sheets. These items along with the map are then cut to their dimensions and manually located for optimal presentation. After the desired arrangement is achieved, the entire layout is copied onto a more stable medium as discussed in Section 18.5.

18.8.2 Plotting Using CADD

Fundamentally, CADD systems plot points and lines in a manner similar to manual drafting techniques. However compared to manual map drafting, computer-assisted mapping offers advantages of increased accuracy, speed, flexibility, and reduced cost. Computers are capable of quickly performing many drafting chores that are tedious and time consuming if done by manual methods, for example, drawing complicated line types and symbols, and performing lettering. With CADD systems, lettering reduces to simply choosing letter sizes and styles and selecting and monitoring placement. Placement of text can be changed as the map is created. Since these systems can often read files of coordinates, such as those from data collectors, the plotting process can become almost totally automated (see Section 17.12). For example, many common features of a map such as bar scale, north arrow, legend, and title block can be created as *blocks* and imported into any map with varied scales. This process simplifies the entire map production process and creates a standardized look for a mapping agency or company. Additionally, the digital environment of a CADD system allows for the easy arrangement of the mapping elements, which simplifies the process of map design and enables colors to be readily selected and changed. Often standard symbology is imported as blocks that either come with the software or are created in-house.

In preparing the topographic data for computer drafting, it is advantageous to develop files that group similar categories of features in separate layers. As an example, individual layers can be created for buildings, vegetation, transportation routes, utilities, hydrology, contours, and so on. By developing the data structure in this manner, various types of maps at different scales can be produced from the same original topographic data file. This is particularly advantageous in mapping for LISs and GISs. This feature also enables specialized products to be created such as *ephemeral maps*—that is, those produced only for current conditions and then redrawn as the conditions change. Examples of these types of maps are those used in guidance, where a map is generated showing the location of a vehicle at a particular moment in time, and as the vehicle moves, the map is instantaneously redrawn on the monitor to replace its earlier version.

In order that digital map files developed during CADD drafting processes are suitable for importing into GISs, it is important that all closed features are actually "physically" closed in the mapping files. As shown in Figure 18.3, a frequent mistake in CADD mapping is the failure to close polygons that appear on the screen to be "visually" closed, but which in fact are not. Since GIS software packages use polygons to represent features, the visually closed but physically open features could appear as simply a series of random lines, or even be viewed as errors in the drawing when imported into the GIS package.

18.9 CONTOUR INTERVAL

As noted in Section 17.5, the choice of contour interval to be used on a topographic map depends on the map's intended use, required accuracy, type of terrain, and scale. If, according to National Map Accuracy Standards (see Section 18.4), elevations can be interpolated from a map to within one half the contour interval, and then if elevations taken from the map must be accurate to within ±1 ft, a 2-ft maximum interval is necessary. However, if only 10-ft accuracy is required, a 20-ft contour interval will suffice.

Terrain type and map scale combine to regulate the contour interval needed to produce a suitable density (spacing) of contours. Rugged terrain requires a larger contour interval than gently rolling country and flat ground mandates a relatively small one to portray the surface adequately. Also if the map scale is reduced, the contour interval must be increased; otherwise, lines are crowded, confuse the user, and possibly obscure other important details.

Figure 18.3
A visually closed feature versus a physically closed feature.

Open entity

Visually closed polygon

Physically closed polygon

For average terrain, the following large and medium map scales and contour interval relationships generally provide suitable spacing:

English System		Metric System	
Scale (ft/in.)	Contour Interval (ft)	Scale	Contour Interval (m)
50	1	1:500	0.5
100	2	1:1000	1
200	5	1:2000	2
500	10	1:5000	5
1000	20	1:10,000	10

■ 18.10 PLOTTING CONTOURS

In plotting contours, points used in locating them are first plotted on the map following techniques described in Section 18.8. Contours found by the *indirect method* (see Section 17.7) are interpolated between plotted points.

Interpolation to find contour locations between points of known elevation can be done in several ways:

1. Estimating.
2. Scaling the distance between points of known elevation and locating the contour points by proportion.
3. Using special devices called *variable scales*, which contain a graduated spring. The spring may be stretched to make suitable marks fall on the known elevations.
4. Using a triangle and scale, as indicated in Figure 18.4. To interpolate for the 420-ft contour between point A at elevation 415.2 and point B at elevation 423.6, first set the 152 mark on any of the engineer's scales opposite A. Then, with one side of the triangle against the scale and the 90° corner at 236, the scale and triangle are pivoted together around A until the perpendicular

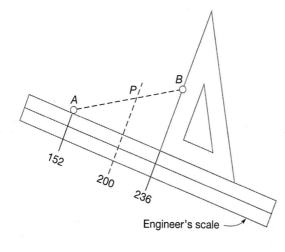

Figure 18.4
Interpolating using engineer's scale and triangle.

edge of the triangle passes through point *B*. The triangle is then slid to the 200 mark and a dash drawn to intersect the line from *A* to *B*. This is the interpolated contour point *P*.

Contours are drawn only for elevations evenly divisible by the contour interval. Thus for a 20-ft interval, elevations of 800, 820, and 840 are shown, but 810, 830, and 850 are not. To improve legibility, every fifth line (those evenly divisible by five times the contour interval) is made heavier. So for a 20-ft interval, the 800, 900, and 1000 lines would be heavier.

■ 18.11 LETTERING

An important part of the contents of any map is its textual information. The title and all feature names, coordinate values, contour elevations, and other items must be clearly identified. To produce a professional looking drawing and one that clearly conveys the intended information, a suitable style of lettering must be selected. That style should be used consistently throughout the map, but the size varied in accordance with the importance of each particular item identified. Lettering that is too big or bold should not be used, but the letters must be large enough to be readable without difficulty.

Lettering should be carefully placed so that it is clearly associated with the item it identifies, and so that letters do not interfere with other features being portrayed. Typically, the best balance results if names are centered in the objects being identified. Also both appearance and clarity are generally improved by aligning letters parallel with linear objects that run obliquely, as has been done with the traverse lengths and bearings of Figure 18.2. For ease in map reading, letters should be placed so that the map can be read from either the bottom or its right side.

Text should take precedence over line work. If necessary, lines should be broken where text is placed, as this improves clarity. An example of this is in the labeling of contours, where the lines are preferably broken, and the contour elevation inserted in the break. It is best to select straight, or nearly straight, sections of contours for labeling. Contours should not be labeled around tight turns since this will remove valuable topographic information expressed by the contours. When manually drafting a map from a manuscript, the text should be lettered before the line work, and then during drafting the lines can be broken where text is encountered. When using automated drafting techniques, manuscripts must be carefully examined to make sure that the text and lines do not overwrite each other, and any observed overwrites corrected.

Because of the importance of lettering to a map's overall appearance and utility, even when drafting is done manually, the text is seldom hand-lettered. Rather, mechanical lettering devices that produce uniform sizes and styles of inked letters, or special machines that print letters on adhesive-backed transparent tape are usually used. With the latter device, a variety of fonts and sizes are available. After the letters are printed, they are pasted onto the map, but can be lifted and moved later if necessary.

In computer-assisted mapping, lettering is greatly simplified. A wide choice of fonts and sizes are available, and letters can be easily placed, aligned, rotated, and moved. However to ensure a good-quality final product, the same rules stated above for manual lettering should be followed when using CADD. Common mistakes by some in CADD are to use too many fonts and label contours on tight turns.

▪ 18.12 CARTOGRAPHIC MAP ELEMENTS

Notes, legends, bar scales, meridian arrows, and title blocks are essential cartographic elements included on maps. *Notes* cover special features pertaining specifically to a particular map. The following are examples:

- All bearings are geodetic (or magnetic or grid or record/deed number).
- Coordinates are based on the NAD 83 (2011) Pennsylvania North Zone State Plane Coordinates.
- Datum for elevations is the NAVD 88.
- Area by calculation is X acres (or hectares).

Notes must be in a prominent place where they are certain to be seen upon even a cursory examination of the map. The best location is near the title block. This is suggested because the user of a map will find and identify a specific plot by its title, and then presumably also check any special notes beside the title before examining the drawing.

Cartographic symbols and different line types are commonly used to represent and portray different topographic features on maps, and *legends* are employed to explain the meaning of those symbols and lines. Figure 18.5 gives some of the hundreds of symbols and line types employed in topographic mapping. The symbols shown in the legend should be replicas of those used on the map. In a CADD environment, it is usually expedient to copy the element to the legend from the map and reduce its scale to match the size of the font. Often legend symbols are created as blocks in CADD and recalled for later use in both the map and legend. Any symbol that is not self-explanatory should appear in the legend. Often the legend can be used to balance other map elements. Sometimes, especially if there are an unusually large number of elements in the legend, it is best to create a separate legend sheet for the map.

The *scale* of the map should preferably be presented as both a representative fraction and a graphical element. A few typical examples of graphical scales are shown in Figure 18.6. Note that the units are associated with each scale. If a map sheet is enlarged or reduced in a reproduction process, the graphical scale will change accordingly, and thus the original scale of the map will be preserved on the reproduction. When designing a bar scale, it is important to maintain a narrow bar. If the bar becomes too wide, it will draw undue attention. A graphic scale element should be placed in the CADD drawing such that when plotted, the graphic element will change indicating the unintended use of the product.

Every map must display a *meridian arrow* for orientation purposes. However, the arrow should not be so large or elaborate that it becomes the focal point of a sheet. Geodetic, grid, or magnetic north (or all three) may be shown. Often the true-meridian arrow is identified by a full head and full feather; and a grid and/or magnetic arrow by a half-head and half-feather. The half-head and half-feather are put on the side away from the true north arrow to avoid touching it. The identity of

Figure 18.5 Topographic symbols.

Figure 18.6
Typical graphical scales in mapping.

the reference meridian used should be noted above or below the arrow in text to define the reference system. When magnetic directions are shown, the declination at the time of the survey should be indicated on the map. If a grid meridian is used, the grid system should be referenced. If an assumed meridian is used, information that will allow the reader to recreate the meridian in the field should be provided.

The *title block* should state the type of map, name of property or project and its owner or user, location or area, date completed, scale, contour interval, horizontal and vertical reference systems (datum) used, and for property surveys, the name of the surveyor with his or her license number. Additional data may be required on special-purpose maps. The title block may be placed wherever it will best balance the sheet, but it always should be kept outside of the subject area. Searching for a particular map in a bound set of maps, or loose pile of drawings, is facilitated if all titles are in the same location. Since sheets are normally filed flat, bound along the left border, or hung from the top, the lower right-hand corner is the most convenient position. Lettering within the title block should be simple in style rather than ornate and conform in size with the individual map sheet. Emphasis should be placed on the most important parts of the title block by increasing letter size or using uppercase (capital) letters for them. Perfect symmetry of outline about a vertical centerline is necessary since the eye tends to exaggerate any defection. An example title block is given in Figure 18.7. No part of a map better portrays the artistic ability of the compiler than a neat, well-arranged title block. Today many companies and government agencies use sheets with preprinted title forms to be filled in with individual job data, or with CADD systems standard title blocks are stored, retrieved, and modified as appropriate for each new project.

18.13 DRAFTING MATERIALS

Polyester film and tracing and drawing papers are the materials commonly used for preparing maps in surveying and engineering offices. Polyesters such as Mylar are by far most frequently employed because they are dimensionally stable, are also strong, durable, and waterproof. In addition, they take pencil, ink, and stickup items, and withstand erasing, so they are ideal for manual drafting. Tracing papers are available in a variety of grades, and good ones also are stable, take pencil, ink,

The Pennsylvania State University
Surveying Program

SURVEY OF CAMPUS

Scale: 1:480   Date: 16 Sept. 2014
Survey by: P. Dills and J. Tills   Map by: S. Smith

Figure 18.7
Title arrangement.

and stickups, and endure some erasing. Both Mylar and tracing paper are transparent, so blueprints can be made from them.

Papers of different types and grades are used for printing maps made with CADD systems. When CADD is used, the paper quality can be relaxed somewhat because erasing, stickup lettering, etc., will not be necessary. *If accurate measurements are to be extracted from the maps, then material with good dimensional stability must be used.*

■ 18.14 AUTOMATED MAPPING AND COMPUTER-AIDED DRAFTING SYSTEMS

Digital computers have had a profound impact in all areas of life and surveying and mapping is certainly no exception. *Automated mapping* (AM) and *computer-aided drafting and design* (CADD) systems have now become commonplace in surveying and engineering offices throughout the world. Generic CADD systems developed for general drafting and engineering work are widely used for map drafting. In addition, special AM systems have been designed specifically for surveying, mapping, and GIS work.

The hardware necessary for AM and CADD systems varies, but as a minimum it will include a computer and a high-resolution monitor; an input device such as a digitizer and/or mouse; and a plotting device. The most important component of any CADD system is its software. This enables an operator to interact with the computer and activate the system's various functions.

CADD systems enable operators to design and draw manuscript maps in real time using the computer. A visual display of the manuscript can be examined on the monitor as it is being compiled, and any additions, deletions, or changes can be made as needed. Lines can be added, deleted, or their styles altered; placement of symbols and lettering modified; and lettering sizes and styles varied. Parts of the drawing may be "picked up" and moved to other areas to resemble a "cut and paste" operation that is handy for subdivision design or placement of frequently occurring symbols. A zoom feature allows more complicated or crowded parts of the manuscript to be magnified for better viewing. In the end, the map can be checked for completeness and accuracy, and when the operator is satisfied that all requirements are met and the design is optimum, the final product plotted.

Required input to a computer for automated mapping includes a set of specific mapping instructions and a file of point locations and elevations. The instructions will include map scale, contour interval, line styles, lettering sizes and styles, symbols, and other items of information. Point locations are usually entered as a file of X, Y, Z coordinates, but angle and distance data can be entered, and the coordinates computed. Special CADD systems for mapping with data collected by total station instruments and GNSS receivers have been developed.

As explained in Section 17.8, most automated mapping systems draw contours after constructing a *triangulated irregular network* (TIN) model. These are networks of nonoverlapping triangles, which represent the individual facets of the terrain. The computer interpolates contour crossings along the edges of the triangles and then draws the contours. A portion of a TIN model from a mapping project is illustrated in Figure 18.8(a), and the contours constructed from it are shown in Figure 18.8(b).

To understand the process of automatically constructing a contour line using a TIN, imagine contouring the TIN shown in Figure 18.9. To begin the procedure, a line on the edge of the TIN is randomly selected for contouring; in this instance, line 1 has been chosen. Assume that the *X, Y, Z* coordinates of the end points of line 1 are (5401.08, 4369.79, 865.40) and (5434.90, 4456.90, 868.30), respectively, and that the 868-contour line is to be drawn. From the *X* and *Y* coordinate values, and applying Equations (10.11) and (10.12), the length and azimuth of line 1 are 93.45 ft and 21°13′06″, respectively. Also, from the *Z* coordinates, the elevation difference is $(868.30 - 865.50) = 2.9$ ft. The elevation difference from the first end point of line 1 to the 868-contour line is $(868.00 - 865.40) = 2.6$ ft. Now the following equivalent ratios can be formed:

$$\frac{2.6}{2.9} = \frac{\Delta L}{93.45}$$

$$\Delta L = 93.45 \left(\frac{2.6}{2.9}\right) = 83.78 \text{ ft}$$

This computation indicates that the distance (ΔL) from the first end point of the line to the 868-contour line is 83.78 ft. Using the previously derived azimuth for line 1, and Equations (10.7), the *X, Y, Z* coordinate values for the intersection of line 1 and the 868-contour are (5431.40, 4447.89, 868.0), respectively. This is the initial point of the 868-ft contour. Now a search algorithm checks the elevations of the end points of lines 2 and 3 (the other two sides of triangle *A*) for the continuation of the 868-contour. Once it determines that line 2 contains the continuation of the line, it again uses the same linear interpolation procedure to determine the coordinates of the intersection of the 868-contour with line 2 and draws the contour line from line 1 to 2. The contour is now ready to enter triangle *B*. It proceeds to check the elevations of the endpoints of lines 4 and 5 and determines that the 868-contour intersects with line 5. Then it again uses linear interpolation to determine the coordinate values for the intersection and continues with the drawing of the 868-contour to line 5. It continues to draw straight-line

(a)

(b)

Figure 18.8
(a) Triangulated irregular network (TIN) model derived from digital-elevation model. (b) Contours derived by automated mapping system from TIN model of (a). Note the roadway edges were defined by breaklines in (a). (Courtesy Wisconsin Department of Transportation.)

530 MAPPING

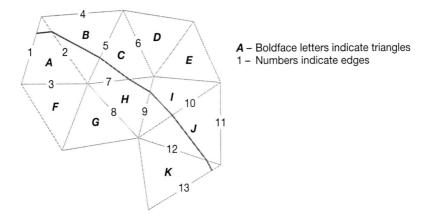

Figure 18.9
Automated contouring of TIN example.

segments for the 868-contour as it passes through each triangle until it finally exits triangle *K* on line 13. From the preceding, it can be seen that this algorithm is repetitive and ideally suited for computer solution.

Contours compiled automatically should be carefully edited for correctness. In certain areas, *breaklines* (see Section 17.8) may be added or changed to obtain the proper terrain representation. Incorrect interpolation often occurs along the outer edges of automatically contoured areas, so these areas require special processing and additional field data. Thus, it is good practice to carry the field survey somewhat beyond the area of interest and "trim" the edges of the map.

Once all the contours have been drawn as straight-line segments, the software then uses a *smoothing* algorithm to round the corners created at each intersection. The operator can usually control the amount of smoothing with a single entry called a *smoothing factor*. The higher the smoothing factor selected by the user, the smoother the intersections become, but the farther the contours depart from their computed values. Thus, the operator must choose a value for the smoothing factor carefully to ensure that the lines do not depart excessively from their original positions.

With terrain information stored in the computer in the form of TIN models, profiles and cross sections along selected lines can be derived automatically and plotted if desired. By including grade lines and design templates, earthwork computations can be made and stakeout information automatically derived for projects such as highways, railroads, and canals.

A TIN is a *digital terrain model* (DTM), which is also commonly called a *digital elevation model* (DEM), of the Earth's surface. A DTM shows only topographic features of the Earth and is devoid of any vegetation, or structures that lie on the surface. A DTM can be created with a TIN or by locating spot elevations in a grid, although, the former is more prevalent in practice. A DTM supports projects in machine control (see Section 15.9), flood modeling, highway design (see Chapters 24 and 25), as well as other construction projects (see Chapter 23). Figure 18.10 shows a DTM of a highway alignment in a three-dimensional perspective view.

The *three-dimensional perspective grid* is an alternative form of terrain representation. It can also be produced by the computer from TIN models. The example shown in Figure 18.10 illustrates its important advantage—it gives a very vivid impression of relief.

Figure 18.10
Three-dimensional perspective grid. (Courtesy Ashtech Corp.)

Figure 18.11 is a portion of a topographic map for an engineering design project created with a CADD system, and Figure 18.12 is a subdivision plat, also designed and drawn using CADD. The condominium plat shown in Figure 21.5 is another example of a product designed and drafted with CADD.

There are numerous advantages derived from using CADD systems in map design and drafting. A major one is increased speed in completing projects. Others include reduction or elimination of errors, increased accuracy, and preparation of a consistently more uniform final product. With completed maps stored in digital form, copies can be quickly reproduced at any time and revisions easily made.

Map data compiled using CADD systems can be stored in a data bank, with different numerical codes for each of the various kinds of features. They can be retrieved later for plotting in total, or in so-called *layers*, or parts, for special-purpose maps. For example, a city engineer may only be interested in a map showing the roads and utilities, while the assessor may want only property boundaries and buildings. This concept of layered maps is fundamental to land and geographic information systems (see Chapter 28).

Another significant advantage of producing maps in digital form is that they can be transmitted electronically from one office to others at remote locations. As an example, the Wisconsin Department of Transportation produces digital maps photogrammetrically for roadway design at its central office in Madison. These maps are then transmitted instantaneously to any of nine district offices located throughout the state, where they are immediately available to engineers for additional computer-aided design or production of hard copies. In spite of the many improvements made in automated mapping systems, there is still a possibility that mistakes can occur. For this reason it is good practice to have the field party chief, who is familiar with the area, review the completed maps. The different AM and CADD systems all have varying individual capabilities. Books and brochures available from the manufacturers provide detailed descriptions of these products.

Figure 18.11 Engineering design map prepared using CADD system. (Courtesy Wisconsin Department of Transportation.)

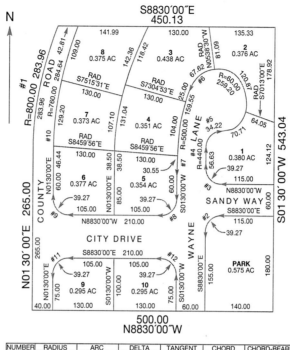

Figure 18.12 Subdivision map compiled automatically by computer-driven plotter. (Courtesy Technical Advisors, Inc.)

18.15 MIGRATING MAPS BETWEEN SOFTWARE PACKAGES

The surveyor often creates maps to exchange with other professionals. Since vendors of software use proprietary file formats, the exchange of drawings can be burdensome. To alleviate this problem, a consortium of people with interest in geospatial technologies has developed a common map exchange format known as *landxml*. As shown in Figure 18.13, there is much more information given about the survey than simply a set of coordinates. For example, the <CgPoint> field shown in Figure 18.13 has a total of 26 attributes associated with it. These attributes allow various packages to share not only the coordinates of the points

```
<CgPoints>
    <CgPoint> name="5000" code="EP" state=" existing"> 55.51243735 37.61100014 491.75260475</CgPoint>
    <CgPoint> name="5001" code="EP" state="existing"> 55.51244858 37.61100010 491.75334163</CgPoint>
    <CgPoint> name="5002" code="EP" state="existing"> 55.51245991 37.61100021 491.74966512</CgPoint>
    <CgPoint> name="5003" code="EP" state="existing"> 55.51247103 37.61100007 491.75405488</CgPoint>
    <CgPoint> name="5004" code="EP" state="existing"> 55.51248222 37.61100028 491.76003092</CgPoint>
    <CgPoint> name="5005" code="EP" state="existing"> 55.51249359 37.61100008 491.75272352</CgPoint>
    <CgPoint> name="100" desc="Stop" code="SIGN" state="existing"> 55.51260359 37.61100015 491.75348631</CgPoint>
    <CgPoint> name="5006" code="EP" state="existing"> 55.51266198 37.61100011 491.75532883</CgPoint>
```

Figure 18.13 A portion of a landxml file from a data collector field-to-finish map.

but also their relationships with each other. A complete description of the various fields and their attributes is available on the LandXML website.[5] Additionally, when using field-to-finish data collection methods, the landxml format allows the user to take the map as drawn in the field and bring it into a CAD software environment thus greatly reducing the amount of errors and time necessary to achieve a finished product. It should also be pointed out here that many survey controller software packages can produce shape files of mapping data, which can be imported into GIS mapping software.

18.16 IMPACTS OF MODERN LAND AND GEOGRAPHIC INFORMATION SYSTEMS ON MAPPING

Land Information Systems (LISs), Geographic Information Systems (GISs), and automated mapping and facilities management (AM/FM) systems all require enormous quantities of position-related land data. From this information, maps and other special-purpose graphic displays can be made and analyzed. For example, a typical LIS or GIS may include attribute data such as political boundaries, land ownership, topography, land use, soil types, natural resources, transportation routes, utilities, and many others. From the stored information, a user can display a map of each attribute category (or *layer*) on a screen, or several layers of data can be merged to produce combination maps. This merging, or *overlaying*, concept, discussed in Chapter 28, greatly facilitates data analysis and aids significantly in management and decision making. If printed maps of any selected layers or combinations are desired, they can be produced rapidly using automated drafting equipment.

The position-related land attribute data needed for LISs and GISs can be collected from a variety of sources and entered in the computer. These sources include field surveys (see Chapter 17), aerial photographs (see Chapter 27), and existing maps. Modern surveying instruments such as the total station instruments, laser scanners, GNSS receivers, and digital photogrammetric plotters can produce huge quantities of digital terrain data in *X, Y, Z* coordinate form rapidly and economically. *Raster scanners* (see Section 28.7.4) are able to systematically scan existing maps and other printed documents line-by-line and convert the information to numerical form. The processes of collecting and digitizing data to support

---

[5]Information on the LandXML organization and the landxml file format can be found at http://www.landxml.org.

LISs and GISs are expected to place a heavy workload on surveyors (geomatics engineers) for many years to come.

18.17 THE IMPORTANCE OF METADATA

Metadata is information that describes the content, quality, condition, origin, and other information or data about the data contained in a product. Metadata for a map should include the when, where, how, and by whom the data were collected including items such as scale at the time of production. As discussed in Chapters 4, 7, 13, 19, and 20 there have been multiple horizontal and vertical datums in the United States and the world over the years. Thus, it is important to clearly define the basis for the coordinates, directions and elevations denoted on maps and other surveying documents. For example, simply stating the map was constructed using state plane coordinates from the North American Datum of 1983 (NAD 83) is not sufficiently clear to the user of the product since there is the original definition of NAD 83, which is now known as NAD 83 (1986), followed by NAD 83 (HARN), NAD 83 (CORS 93), NAD 83 (CORS 94), NAD 83 (CORS 96), NAD 83 (2007), and NAD 83 (2011).[6] This has similarly happened with benchmark elevations, which have defining datums of NGVD 29, IGLD 85, PRVD 02, and NAVD 88 as well as many national, regional, and local municipality defined datums. Thus, simply stating an elevation or identifying a contour elevation will not sufficiently let the user know the basis for the elevation. It is extremely important to list the defining basis for observations represented on a map including the basis for the directions of the lines, which can be magnetic (Chapter 7), astronomic or true (Appendix C), grid (Chapter 20), or geodetic (Chapter 19). In these cases, defining even the type of direction is not sufficient since all are related to a specific period in time and/or a specific defining datum. Failure to list metadata in any type of surveying product such as maps and deeds can make the product difficult to interpret or even useless to future generations of users.

18.18 SOURCES OF ERROR IN MAPPING

Sources of error in mapping include:

1. Errors in the data used in plotting.
2. Errors in the scales used for laying out lengths and coordinate values.
3. Errors in laying out grids for plotting by coordinates.
4. Using a soft pencil, or one with a blunt point, for plotting.
5. Variations in the dimensions of map sheets due to temperature and moisture.

18.19 MISTAKES IN MAPPING

Some common mistakes in mapping are:

1. Selecting an inappropriate scale or contour interval for the map.

---

[6]See Section 19.6 for a discussion on the evolution of NAD 83.

2. Failing to check grids by measuring diagonals, and not checking points plotted from coordinates by measuring distances between them.
3. Using the wrong edge of an engineer's scale.
4. Making the north arrow too large or too complex.
5. Neglecting to identify the meridian of reference, that is, geodetic, grid, magnetic, etc.
6. Omitting the scale or necessary notes.
7. Failing to balance the sheet by making a preliminary sketch.
8. Drafting the map on a poor-quality medium.
9. Failing to realize that errors are also magnified when maps are enlarged electronically or photographically.
10. Operating AM and CADD systems without sufficient prior training.
11. Failure to list all metadata on the map.

PROBLEMS

Asterisks (*) indicate problems that have partial answers given in Appendix G.

18.1 List some of the features shown on map.
18.2* On a map drawn to a scale of 1:4800, a point has a plotting error of 1/30 in. What is the equivalent ground error in units of inches?
18.3 What operations can a GIS perform with maps?
18.4 What federal agency in the United States is responsible for coordinating all mapping activities at the federal level?
18.5 What is contained in a digital line graph?
18.6 What is contained in a digital elevation model?
18.7 Why did the Federal Geographic Data Committee accuracy standards to replace the National Map Accuracy Standards?
18.8 What is the 95% horizontal positional accuracy for the following set of data?

	Map Coordinates		Check Coordinates	
Point	x (m)	y (m)	X (m)	Y (m)
1	556,987.118	196,408.166	556,987.063	196,408.136
2	555,783.409	198,183.218	555,783.444	198,183.119
3	554,986.093	197,205.495	554,986.105	197,205.557
4	557,063.173	194,318.148	557,063.150	194,318.174

18.9 What is the 95% positional vertical accuracy for the following data?

	Map Coordinates	Check Coordinates
Point	z (m)	H (m)
1	348.995	349.001
2	339.523	339.560
3	355.502	355.491
4	368.885	368.864

18.10* For a 20-ft contour interval, what is the greatest error in elevation expected of any definite point read from a map if it complies with National Map Accuracy Standards?

18.11 An area that varies in elevation from 1035 to 1096 ft is being mapped. What contour intervals will be drawn if a 10-ft interval is used? Which lines are emphasized?

18.12 Similar to Problem 18.11, except elevations vary from 213 to 248 m and a 5-m interval is used.

18.13 If a map is to have a 10-ft contour interval, which contours are labeled between the elevations of 5089 and 5230 ft?

18.14 Often the north arrow on a surveying map is large, which draws attention to it. What basic principle is being violated when this is done?

18.15* What is the largest acceptable error in position for 90% of the well-defined points on a map with a 1:24,000 scale that meets national map accuracy standards?

18.16 What questions must be answered before designing a map?

18.17 Discuss why insets are sometimes used on maps.

18.18* If a map is to have a 1-in. border on all sides, what is the largest nominal scale that may be used for a subject area with dimensions of 604 and 980 ft on a paper of dimensions 24 by 36 in?

18.19 Similar to Problem 18.18, except the dimensions of the subject area are 826 ft and 1034 ft.

18.20 If a map is to have 1-1/2 in. borders on the top and left sides and 1/2 in. borders on the bottom and right sides, what is the largest nominal scale that may be used for a subject area with dimensions of 786 and 1435 ft on a paper of dimensions 24 by 36 in?

18.21 If 90 percent of all elevations on a map must be interpolated to the nearest ±0.5 ft, what contour interval is necessary according to the National Map Accuracy Standards? Explain.

18.22 If an area having an average slope of 10% is mapped using a scale of 1:1000 and contour interval of 0.5 m, how far apart will contours be on the map?

18.23 Similar to Problem 18.22, except average slope is 3.8%, map scale is 500 ft/in., and contour interval is 5 ft.

18.24* Similar to Problem 18.22, except average slope is 4%, map scale is 1:1000, and contour interval is 0.5 m.

18.25* The three-dimensional (X, Y, Z) coordinates in meters of vertexes A, B, and C in the accompanying figure are (5412.456, 4480.621, 248.147), (5463.427, 4459.660, 253.121), and (5456.081, 4514.382, 236.193), respectively. What are the coordinates of the intersection of the 250-m contour with side AB? With side BC?

18.26 The three-dimensional (X, Y, Z) coordinates in meters for vertices $A, B,$ and C in the accompanying figure are (8649.220, 6703.670, 214.685), (8762.040, 6649.770, 218.345), and (8752.640, 6770.200, 220.534), respectively. What are the coordinates of the intersections of the 215-m contour as it passes through the sides of the triangle?

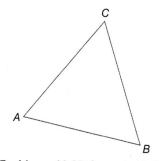

Problems 18.25 through 18.27

18.27 Similar to Problem 18.26, except compute the coordinates of the intersection of the 248-m contour.

18.28 Discuss how balance is achieved on a map.

The following table gives elevations at the corners of 50-ft coordinate squares, and they apply to Problems 18.29 and 18.31.

74	70	66	67	66	69
76	73	71	68	63	70
77	69	68	66	69	68

18.29 At a horizontal scale of 1 in. = 50 ft draw 2-ft contours for the area.

18.30 Similar to Problem 18.29, except at the bottom of the table add a fourth line of elevations: 79, 78, 70, 66, 61, and 65 (from left to right).

18.31 If a map is drawn with 1-ft contour intervals, what contours between 243 and 273 ft are drawn with heavier line weight?

18.32 Why is it important to physically close lines in a CADD drafting system?

18.33 What important item should be included with any meridian arrow?

18.34 What is the purpose of a smoothing factor in a CADD environment?

18.35 Define metadata.

BIBLIOGRAPHY

American Society for Photogrammetry and Remote Sensing. 2014. "ASPRS Positional Accuracy Standards for Digital Geospatial Data." *Photogrammetric Engineering and Remote Sensing* 81 (No. 3): A1.

———. 1990. "ASPRS Standards for Large-Scale Maps." *Photogrammetric Engineering and Remote Sensing* 56 (No. 7): 1068.

———. 1987. "*Large Scale Mapping Guidelines*." Bethesda, MD: American Society for Photogrammetry and Remote Sensing.

Crawford, K. A. 2009. "Model Behavior: The How-to Guide to Successful Surface Modeling, Part 1." *The American Surveyor* 6 (No. 6): 63.

———. 2009. "Model Behavior: The How-to Guide to Successful Surface Modeling, Part 2." *The American Surveyor* 6 (No. 8): 44.

———. 2009. "Model Behavior: The How-to Guide to Successful Surface Modeling, Part 3." *The American Surveyor* 6 (No. 10): 64.

———. 2010. "Model Behavior: The How-to Guide to Successful Surface Modeling, Part 4." *The American Surveyor* 7 (No. 5): 50.

Davis, T. G. 2009. "USGS Quadrangles in Google Earth." *The American Surveyor* 6 (No. 9): 28.

Schrock, G. 2015. "Go Wide." *XYHT* 2 (No. 3): 20.

Zilkoski, D., J. D'Onofrio, and S. Frakes. 1997. "NOAA Technical Memorandum NOS NGS-58, V. 4.3: Guidelines for Establishing GPS-Derived Heights (Standards: 2 cm and 5 cm)." URL: *http://www.ngs.noaa.gov/PUBS_LIB/NGS-58.html* (accessed on March 4, 2015).

Zilkoski, D., E. Carlson, and C. Smith. 2008. "NOAA Technical Memorandum NOS NGS 59, Guidelines for Establishing GPS-Derived Orthometric Heights." URL: *http://www.ngs.noaa.gov/PUBS_LIB/NGS592008069FINAL2.pdf* (accessed on March 4, 2015).

19
Control Surveys and Geodetic Reductions

■ 19.1 INTRODUCTION

Control surveys establish precise horizontal and vertical positions of reference monuments. These serve as the basis for originating or checking subordinate surveys for projects such as topographic and hydrographic mapping; property boundary delineation, route and construction planning, design, and layout, and control for aerial imagery. They are also essential as a reference framework for giving locations of data entered into Land Information Systems (LISs) and Geographic Information Systems (GISs).

Traditionally, there have been two general types of control surveys: *horizontal* and *vertical*. Horizontal surveys establish *geodetic latitudes* and *longitudes* (see Section 19.4) of stations over large areas. From these values, plane rectangular coordinates, usually in a state plane or Universal Transverse Mercator (UTM) coordinate system (see Chapter 20) can be computed. On control surveys in smaller areas, plane rectangular coordinates may be determined directly without obtaining geodetic latitudes and longitudes.

Field procedures used in horizontal control surveys have traditionally been the ground methods of *triangulation, precise traversing, trilateration*, and combinations of these basic approaches (see Section 19.13). In addition, astronomical observations (see Appendix C) were made to determine azimuths, latitudes, and longitudes. Rigorous photogrammetric techniques (see Chapter 27) have also been used to densify the control in areas.

During the 1970s, *inertial surveying systems* (ISSs) were introduced. Their operating principle consisted fundamentally in making measurements of accelerations over time. This was done while the instrument was carried from point to point in a land vehicle or helicopter. The acceleration and time observations were taken independently in three mutually orthogonal planes, which were oriented north–south, east–west, and in the direction of gravity. Orientation was achieved by means of gyroscopes. From the acceleration and time data, components of the instrument's movement in each of three reference planes could be computed, and hence relative positions of points determined. Inertial surveying systems were used in a variety of surveying applications, one of the most important being control surveying. Drawbacks of the systems were their high initial cost, equipment that was bulky, and an overall accuracy less than that attainable with *global navigation satellite system* (GNSS) receivers. As a result, ISSs are no longer used for control surveys. However, these systems have been replaced by more modern *inertial measurement units* (IMUs) that are used in *unmanned aerial* systems (UASs) and mobile mapping units. The marriage of an IMU and GNSS in mobile mapping units (see Section 17.9.5) allow them to carry coordinates in areas where canopy conditions obstruct GNSS satellite signals.

Satellite surveying (see Chapters 13, 14, and 15) has been employed with increasing frequency, especially in control surveys. GNSS surveys are rapidly replacing the other methods because of several advantages including its ease of use, speed, and extremely high accuracy capabilities over long distances, which do not require an unobstructed line of sight. However, in small areas, traditional methods of establishing control are still being used.

Vertical-control surveys establish elevations for a network of reference monuments called *benchmarks*. Depending on accuracy requirements, they have traditionally been run by either *differential leveling* or *trigonometric leveling* (see Chapters 4 and 5). GNSS survey can also establish vertical control but are limited by the need for a precise geoid model (see Section 19.2). Thus, the most accurate and widely applied method is still precise differential leveling (see Section 19.14). However, the National Geodetic Survey (NGS) is currently working on the Gravity for Redefinition of the Vertical Datum (GRAV-D) project. As discussed in Section 19.6.6, at the completion of this project the NGS will create a new four-dimensional datum based on an equipotential surface defined by this project.

This chapter will define elements of geodetic reference systems used for control surveys, describe the *National Spatial Reference System* (NSRS), discuss some of the traditional ground methods used in control surveying, and explain some basic computational methods used in making geodetic reductions on conventional observations.

19.2 THE ELLIPSOID AND GEOID

It was noted in Section 19.1 that horizontal control surveys generally determine geodetic latitudes and longitudes of points. To explain geodetic latitude and longitude, it is necessary to first define the *geoid*. The geoid is an equipotential gravitational surface, which is everywhere perpendicular to the direction of gravity. Because of variations in the Earth's mass distribution and the rotation of the Earth, the geoid has an irregular shape.

The ellipsoid is a mathematical surface obtained by revolving an ellipse about the Earth's polar axis. The dimensions of the ellipse are selected to give the best mathematical fit, in a least-squares sense, of the ellipsoid to the geoid over a large area, and are based upon surveys made in the area. With the advent of satellites, current-day ellipsoids provide a best fit for the Earth.

A two-dimensional view, which illustrates conceptually the geoid and ellipsoid, is shown in Figure 19.1. As illustrated, the geoid contains nonuniform undulations (which are exaggerated in the figure for clarity), and is therefore not readily defined mathematically. Ellipsoids, which approximate the geoid and can be defined mathematically, are therefore used to compute positions of widely spaced points that are located through control surveys. The *Clarke Ellipsoid of 1866, which was developed to fit the African continent*, approximates the geoid in North America very well, and from 1879 until the 1980s it was the ellipsoid used as a reference surface for specifying geodetic positions of points in the United States, Canada, and Mexico. Currently, the *Geodetic Reference System of 1980* (GRS 80) and *World Geodetic System of 1984* (WGS 84) ellipsoids are commonly used in the United States because they provide a good worldwide fit to the geoid. This is important because of the global surveying capabilities of GNSS.

Sizes and shapes of ellipsoids can be defined by two parameters. Table 19.1 lists the parameters for the three ellipsoids noted above. For the Clarke 1866 ellipsoid, the defining parameters are the semiaxes a and b. For GRS 80 and WGS84, the defining parameters are the semimajor axis a and flattening factor f. The relationship between these three parameters is

$$f = 1 - \frac{b}{a} \tag{19.1}$$

Other quantities commonly used in ellipsoidal computations are the *first eccentricity*, e, and the second eccentricity, e' of the ellipse, where

$$e = \frac{\sqrt{a^2 - b^2}}{a} = \sqrt{2f - f^2} \tag{19.2a}$$

$$e' = \frac{\sqrt{a^2 - b^2}}{b} = \frac{e}{\sqrt{1 - e^2}} \tag{19.2b}$$

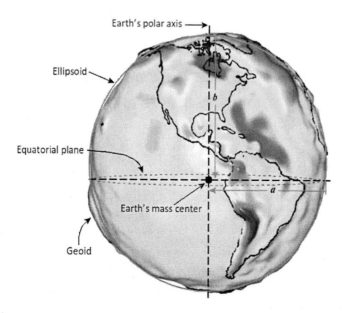

Figure 19.1
Ellipsoid and geoid.

TABLE 19.1 DEFINING ELLIPSOIDAL PARAMETERS

Ellipsoid	Semiaxis a (m)	Semiaxis b (m)	Flattening f
Clarke, 1866	6,378,206.4*	6,356,583.8*	1/294.978698214
GRS80	6,378,137.0*	6,356,752.3	1/298.257222101*
WGS84	6,378,137.0*	6,356,752.3	1/298.257223563*
PZ-90.11	6,378,136.0*	6,356,751.4	1/298.257839303*

*Defining parameters for the ellipsoids.

Often the term eccentricity is understood to mean the first eccentricity, and this book will follow that convention. For each ellipsoid, the polar semiaxis is only about 21 km (13 mi) shorter than the equatorial semiaxis b. This means the ellipsoid is nearly a sphere and hence for some calculations involving moderate lengths (usually up to about 50 km) this assumption can be made.[1]

■ Example 19.1

Using the defining parameters, what are the first eccentricities of the Clarke 1866 and GRS80 ellipsoids?

Solution

For the Clarke 1866 ellipsoid, Equation (19.2a) yields

$$e = \frac{\sqrt{6378,206.4^2 - 6,365,583.8^2}}{6,378,206.4} = 0.082271854$$

For the GRS80 ellipsoid, Equation (19.2a) yields

$$e = \sqrt{\frac{2}{298.257222101} - \frac{1}{298.257222101^2}} = 0.081819191$$

■ 19.3 THE CONVENTIONAL TERRESTRIAL POLE

As discussed in the preceding section, an ellipsoid is defined on the basis of the size of an ellipse that is rotated about the polar axis of the Earth. In reality, since the principal axis of inertia of the Earth does not coincide with the rotational axis of the Earth, the polar axis at any moment in time is not fixed in position. Rather, as illustrated in Figure 19.2, it rotates with respect to the inertial system. This motion is generally divided into two major categories called *precession* and *nutation*. Precession is the greater of the two and is the wander of the polar axis over a long

---

[1] In computations if the ellipsoid is assumed a sphere, its radius is usually taken such that its volume is the same as the reference ellipsoid. It is computed from $r = \sqrt[3]{a^2 b}$. For the GRS 80 ellipsoid, its rounded value is 6,371,000 m.

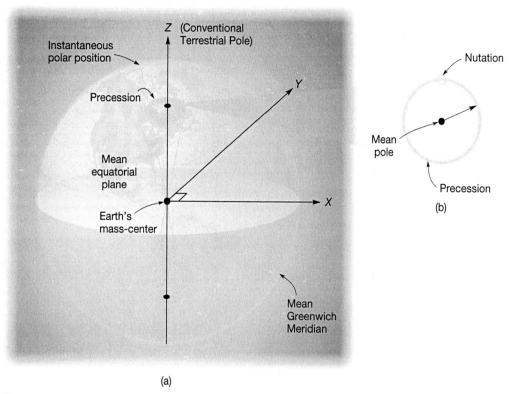

Figure 19.2 Motions of the Earth's polar axis: (a) three-dimensional and (b) plan view.

period of time. The pole makes a complete revolution approximately once every 26,000 years. Additionally, the pole wanders in much smaller radial arcs that are superimposed upon precession. These smaller circles are known as a nutation, and are completed about once every 18.6 years. By international convention, the mean rotational axis of the Earth was defined as the "mean" position of the pole between the years of 1900.0 and 1905.0. This position is known as the *Conventional Terrestrial Pole* (CTP). This pole was determined by astro-geodetic techniques. Likewise, the Earth's equatorial plane wobbles along with the polar axis. By defining the CTP, our coordinate systems also use a mean equatorial plane. In 1984 the Bureau International de l'Heure (BIH) redefined the pole using geodetic coordinates. This position is known as BIH 1984.0. This redefinition was adopted for use by the Department of Defense for the global positioning system (see Chapters 13, 14, and 15) as well as by the *International Earth Rotation and Reference Systems Service* (IERS), which defines global terrestrial reference systems.

Due to deflection of the vertical (see Section 19.5) at Greenwich Royal Observatory and because of the difference between astronomic positioning and geodetic positioning, the position of zero longitude shifted from the Airy Transit Circle, its astronomic position, east by approximately 102 m to its geodetic position giving the Airy Transit Circle a geodetic position of approximately 0°00′05.3″ W (Maly et al., 2015). These differences are strictly due to deflection of the vertical at the Royal Observatory.

BIH 1984.0 defines the Z-axis of a geodetic three-dimensional global Cartesian coordinate system with the northern portion being positive. The positive X-axis lies in the mean equatorial plane, begins at the mass center of the Earth, and passes through the mean Greenwich meridian. Finally, the Y-axis also lies in the mean equatorial plane, and creates a right-handed Cartesian coordinate system. This coordinate system, which is Earth-centered and Earth-fixed (ECEF),[2] is known as the *Conventional Terrestrial System* (CTS). The CTS is shown in Figure 19.2.

Since 1988, IERS[3] has monitored the instantaneous position of the pole with respect to the BIH 1984.0 using observations made by participating organizations employing advanced space methods including *Very Long Baseline Interferometry* (VLBI), and lunar and satellite laser ranging. Consequently, the CTS is now defined by a global set of stations through their instantaneous spatial coordinate positions known as the *International Terrestrial Reference Frame* (ITRF). This system is used in the computation of precise satellite orbits (see Chapters 13, 14, and 15) and is referenced through time-dependent models (see Section 19.7.1) to other coordinate systems.

The instantaneous position of the pole is given in (x, y) coordinates with respect to the BIH 1984.0. An application of these positions can be seen in the reduction of astronomical azimuths (see Appendix C) where the observed astronomical azimuth with respect to the instantaneous position of the pole can be related to the BIH 1084.0 by

$$Az_A = Az_{obs} - (x \sin \lambda + y \cos \lambda) \sec \varphi \qquad (19.3)$$

where Az_A is the astronomic azimuth related to the CTP, Az_{obs} the observed astronomical azimuth made with relation to the instantaneous pole, (x, y) the coordinates in units of arcseconds of the instantaneous pole, and (φ, λ) the geodetic latitude and longitude, respectively, of the observing station. Future references to the polar axis of the Earth will implicitly refer to the BIH 1984.0.

■ 19.4 GEODETIC POSITION AND ELLIPSOIDAL RADII OF CURVATURE

Figure 19.3 shows a three-dimensional view of the ellipsoid and illustrates a point P on the surface of the Earth, which in this illustration is shown to exist at a distance of h_P above the ellipsoid. Point P' is on the ellipsoid along the *normal* through P. (The normal is defined in the following text.) The geodetic position of point P is given by its *geodetic latitude* φ_P, *geodetic longitude* λ_P, and *geodetic height* h_P. To define these three terms, it is necessary to first define *meridians* and *meridian planes*. Meridians are great circles on the circumference of the ellipsoid, which pass through the north and south poles. Any plane containing a meridian and the polar axis is a meridian plane. The angle in the plane of the Equator from the Greenwich meridian plane to the meridian plane passing through point P defines the geodetic longitude λ_P of the point. The plane defined by the vertical

---

[2]As an example, NAD 27 was not Earth centered and Earth fixed since its origin was defined by the control station Meades Ranch and its orientation by the azimuth to station Waldo.

[3]The instantaneous positions of the Earth's pole can be found on the IERS website at https://www.iers.org/.

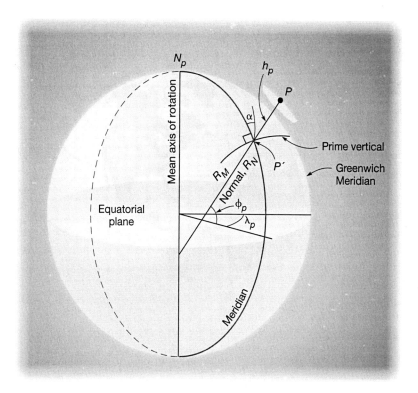

Figure 19.3
Different radii on the ellipsoid.

circle that passes through point P, *perpendicular to the meridian plane* on the ellipsoid, is called the plane of the *prime vertical* (also known as the *normal section*). The radius of the prime vertical at point P, R_N is also called the *normal* since it is perpendicular to a plane that is tangent to the ellipsoid at P. The geodetic latitude φ_P is the angle, in the meridian plane containing P, between the equatorial plane and the normal at P.

Both latitude and longitude are typically expressed with 4 or 5 decimals. The rationale behind this comes from the definition of the nautical mile where 1 nautical mile (1852 m) equals 1' of longitudinal arc on the equator. Thus, 60 sec is equal to approximately 6000 ft, or 1 sec about 100 ft. Finally, to express a coordinate to 0.01 ft, the latitude and longitude should be stated to 0.0001" and to the nearest millimeter as 0.00001". Thus, latitude and longitude are typically written with 4 or 5 decimal places.

To uniquely define the location of point P on the surface of the Earth, geodetic height h_P must be included. Geodetic height is the distance measured along the extension of the normal from P' on the ellipsoid to P on the Earth's surface. Geodetic height is not equivalent to elevation (orthometric height) determined by differential leveling. (These differences were described in Section 13.4.3 and will be discussed further in Section 19.5.)

Because the Earth is approximated by an ellipsoid and not a sphere, the great circle that defines the prime vertical at P has a radius R_N that is different than the radius in the meridian R_M at P.[4] The lengths of these two radii, which are

---

[4]Note that R_N is often referred to as N, and R_M is similarly designated as M.

collinear at any point, are used in many geodetic computations. Figure 19.3 shows R_N. Additionally, the radius R_α of a great circle at any azimuth α to the meridian is different from either R_N or R_M. These three radii are computed as

$$R_N = N = \frac{a}{\sqrt{1 - e^2 \sin^2 \varphi}} \tag{19.4}$$

$$R_M = M = \frac{a(1 - e^2)}{(1 - e^2 \sin^2 \varphi)^{3/2}} \tag{19.5}$$

$$R_\alpha = \frac{R_N R_M}{R_N \cos^2 \alpha + R_M \sin^2 \alpha} \tag{19.6}$$

where a and e are parameters for the ellipsoid as defined in Section 19.2, and φ is the geodetic latitude of the station for which the radii are computed. From an analysis of Equations (19.4) and (19.5), it can readily be shown that R_N equals R_M at the poles where φ is 90°. Furthermore, since the quantity $(1 - e^2)$ is less than one, the radius of the prime vertical R_N is greater than the radius of the meridian R_M at every location other than the pole where φ is equal to 90°. As will be discussed in Chapter 20, another the radius known as the *Gaussian mean radius* is sometimes used in the reduction of distances. It is computed as

$$R_G = \sqrt{MN} = \frac{a\sqrt{1 - e^2}}{1 - e^2 \sin^2 \varphi} \tag{19.7}$$

■ Example 19.2

Using the GRS80 ellipsoidal parameters, what are the radii for the meridian and prime vertical for a point of latitude 41°18′15.0132″ N. What is the radius of the great circle that is at an azimuth of 142°14′36″ at this point?

Solution

From Example 19.1, $e^2 = 0.081819191^2 = 0.00669438$

By Equation (19.4), the radius of the prime vertical is

$$R_N = \frac{6378137.0}{\sqrt{1 - e^2 \sin^2(41°18'15.0132'')}} = 6{,}387{,}458.536 \text{ m}$$

By Equation (19.5), the radius of the meridian is

$$R_M = \frac{6378137(1 - e^2)}{[1 - e^2 \sin^2(41°18'15.0132'')]^{\frac{3}{2}}} = 6{,}363{,}257.346 \text{ m}$$

By Equation (19.6), the radius of the great circle at an azimuth of 142°14′36″ is

$$R_\alpha = \frac{R_N R_M}{R_N \cos^2(41°18'15{,}0132'') + R_M \sin^2(41°18'15{,}0132'')} = 6{,}372{,}309.401 \text{ m}$$

19.5 GEOID UNDULATION AND DEFLECTION OF THE VERTICAL

As discussed earlier, the geoid is an equipotential surface defined by gravity. If the Earth was a perfect ellipsoid without internal density variations, the geoid would match the ellipsoid perfectly. However, this is not the case, and thus the geoid is a least-squares best fit of the ellipsoid. Current geoids depart from some ellipsoids by as much as 100 m or more in certain regions of the Earth.

Since traditional surveying instruments are oriented with respect to gravity, observations obtained with them are typically made with respect to the geoid, and are known as astronomic observations. As can be seen in Figure 19.4, and discussed in Section 13.4.3, the separation between the geoid and the ellipsoid creates a difference between the height of a point above the ellipsoid (*geodetic height*) and its orthometric height above the geoid. This difference, known as *geoid height*,[5] can often be observed when comparing the geodetic height of a point derived by GNSS surveys, with its elevation as determined by differential leveling. The relationship between the orthometric height H and geodetic height h at any point is

$$h = H + N \qquad (19.8)$$

where N is the geoid height at that point.

The GRS 80 and WGS 84 ellipsoids were both developed with the intent of providing a good fit to the geoid worldwide. However, they yield a relatively poor fit to the geoid within the continental United States, where the average geoidal height is approximately −30 m (the negative sign means the geoid is below the ellipsoid in the conterminous United States). The Clarke 1866 ellipsoid, on the

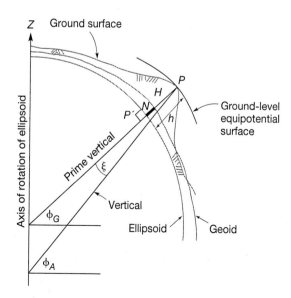

Figure 19.4
Relationships between the ellipsoid and geoid.

---

[5]The National Geodetic Survey regularly publishes geoid models for the United States. Its latest version is GEOID18. This model can be obtained from the NGS website at http://www.ngs.noaa.gov/GEOID/GEOID18/.

other hand, provided a very close fit to the geoid in the United States, that is, geoidal heights are typically only a few meters.

In general, equipotential gravitational surfaces are not parallel to either the geoid or ellipsoid. This is partly due to the rotation of the Earth, which causes the surfaces to separate as they approach the equator, and partly due to density anomalies near the surface of the Earth. As a result, certain inconsistencies can occur in different types of field observations, and hence corrections must be made. Some of the more significant of these corrections are discussed in Section 19.14.

As illustrated in Figure 19.5, the *deflection of the vertical* (also called *deviation of the vertical*) at any ground point P is the angle between the vertical (direction of gravity) and the normal to the ellipsoid. This angle is generally reported by giving two components: its orthogonal projections onto the meridian and normal planes. In the figure, the zenith of the ground-level equipotential surface is called the astronomical zenith Z_a since it corresponds to the direction of gravity (zenith) of a leveled instrument during astronomical observations. Also in Figure 19.5, Z_G is the normal at point P. The projected components of the total deviation of the vertical onto the meridian and normal planes are called ξ (xi) and η (eta), respectively.[6] The relationships between the astronomic latitude (φ_A), astronomic longitude (λ_A), and astronomic azimuth (Az_A); the geodetic latitude (φ_G), geodetic longitude (λ_G), and geodetic azimuth (Az_G); and ξ and η are as follows:

$$\xi = \varphi_A - \varphi_G \tag{19.9}$$

$$\eta = (\lambda_A - \lambda_G) \cos \varphi = (Az_A - Az_G) \cot \varphi \tag{19.10}$$

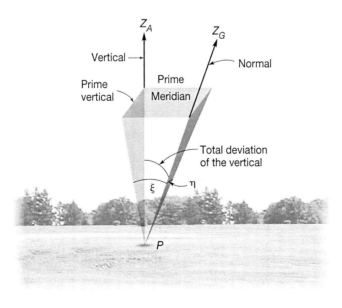

Figure 19.5
ξ and η components of deflection of the vertical.

---

[6]As with geoidal undulations, values for ξ and η can also be obtained through the National Geodetic Survey's deflection of vertical models. The latest version, DEFLEC18, can be obtained from the NGS website at http://www.ngs.noaa.gov/GEOID/DEFLEC18/.

In Equation (19.10), φ can be either the astronomic or geodetic latitude. From this equation, the so-called *Laplace equation* can be derived as

$$Az_G = Az_A - (\lambda_A - \lambda_G) \sin \varphi = Az_A - \eta \tan \varphi \quad (19.11)$$

Stations at which the necessary parameters are known, such that Equation (19.11) can be formed are called *Laplace stations*. At these stations it is possible using Equations (19.9) to (19.11) to compute ξ and η. Note that in Equation (19.10), for points near the equator, latitude φ approaches 0°, and the astronomic and geodetic azimuths become essentially the same. As will be discussed in Section 19.15.3, additional corrections are needed to properly reduce an observed azimuth to its geodetic equivalent on the ellipsoid.

19.6 U.S. REFERENCE FRAMES

Traditionally, horizontal and vertical datums consist of a network of control monuments and benchmarks whose horizontal positions and/or orthometric heights have been determined by precise geodetic control surveys. These monuments serve as reference points for originating subordinate surveys of all types and as such are known as *reference frames*. The horizontal and vertical reference systems used in the immediate past and at present in the United States are described in the following subsections.

19.6.1 North American Horizontal Datum of 1927 (NAD 27)

In 1927, a least-squares adjustment was performed, which incorporated all horizontal geodetic surveys that had been completed up to that date. This network of monumented points included in the adjustment, together with their adjusted geodetic latitudes and longitudes, was referred to as the *North American Datum of 1927* (NAD 27). The adjustment utilized the Clarke ellipsoid of 1866 and held fixed the latitude and longitude of an "initial point," station *Meades Ranch* in Kansas, along with the azimuth to nearby station *Waldo*. The project yielded adjusted latitudes and longitudes for some 25,000 monuments existing at that time. Until the advent of the NAD 83 datum, positions of stations established after 1927 were adjusted in processes that held NAD 27 monuments fixed.

19.6.2 North American Horizontal Datum of 1983 (NAD 83)

The National Geodetic Survey (NGS) began a program in 1974 to perform another general adjustment of the North American horizontal datum. The adjustment was deemed necessary because of the multitude of post-1927 geodetic observations that existed and because many inconsistencies had been discovered in the NAD 27 network. The project was originally scheduled for completion in 1983, hence its name *North American Datum of 1983* (NAD 83), but it was not actually finished until 1986, and is now referred to as NAD 83 (1986). The adjustment was a huge undertaking, incorporating approximately 270,000 stations and all geodetic surveying observations on record—nearly 2 million of them! About 350 person-years of effort were required to accomplish the task.

The initial point in the new adjustment is not a single station such as Meades Ranch in Kansas; rather, the Earth's mass center as known at that time and numerous other points whose latitudes and longitudes had been precisely established using radio astronomy and satellite observations were used. The GRS80 ellipsoid was employed since, as noted earlier, it fits the Earth, in a global sense, more accurately than the Clarke ellipsoid of 1866.

Within the United States, the adjusted latitudes and longitudes of all monuments in NAD 83 differ from their NAD 27 values. These differences result primarily because of the different ellipsoids and origins used, but part of the change is also due to the addition of the many post-NAD 27 observations in the NAD 83 adjustment. The approximate magnitudes of these changes in the conterminous United States, expressed in meters, are illustrated in Figure 19.6. Of course, these differences have a significant impact on all existing control points and map products that were based on NAD 27. Various mathematical models were developed to transform NAD 27 values to their NAD 83 positions (see Section 19.7).

Figure 19.6 Approximate changes in latitude and longitude (in meters) in the conterminous United States from NAD 27 to NAD 83. Upper figure: Latitude. Lower figure: Longitude. (Adapted from National Geodetic Survey maps.)

19.6.3 Later Realizations of NAD 83

The internal consistency of first-order points adjusted in NAD 83 was specified to be at least 1:100,000, but tests have verified that on average it is probably 1:200,000 or better. However, there are some areas where relative accuracies fall below 1:100,000. Since GNSS survey accuracies are often better than 1:100,000, there was concern among GNSS users about how to fit their observations to a network of reference points whose inherent accuracies were less.

State governments and the National Geodetic Survey cooperatively sought to solve this problem by adding *High-Accuracy Reference Networks* (HARNs) to the National Spatial Reference System (see Section 19.9). From 1987 to 1997, HARN networks were created in every state. When each state's HARN was completely observed, an adjustment of these new stations was performed. This created an interim reference frame that was available to surveyors using GPS. This second version of NAD 83 is known as NAD 83 (HARN). For these regional reference systems, NGS retained the location of the mass center of the Earth and orientation of the Cartesian coordinate axes,[7] but introduced a new scale that was consistent with the International Terrestrial Reference Frame of 1989 (ITRF89).

With the introduction of the CORS network in 1994 (see Section 14.3.5), the NGS was again faced with the problem of having stations of higher accuracy that were being adjusted using a different reference frame. Thus, a third realization of NAD 83 was obtained using the ITRF93 reference frame. This transformation created another datum involving only the CORS stations, and was known as NAD 83 (CORS93).

In the spring of 1996, the NGS computed the positional coordinates for all existing CORS stations using ITRF94. This created the fourth realization of NAD 83, and was known as NAD 83 (CORS94). In 1998, the NGS computed positional coordinates for all existing CORS sites using ITRF96 as their reference frame. This version of NAD 83 is known as NAD 83 (CORS96). Each datum differs slightly from the previous definition of NAD 83. For instance, the positions of sites in NAD 83 (CORS96) differ by a maximum of about 2 cm in horizontal and 4 cm in vertical from their equivalent NAD 83 (CORS94) values. Furthermore, the difference between any NAD 83 CORS adjustment and NAD 83 (HARN) is under 10 cm in horizontal and 20 cm in vertical.

With completion of the last statewide HARN in 1997, the NGS had three spatial reference systems with the original NAD 83 (1986) for points determined by conventional surveys, NAD 83 (HARN) for HARN stations, and NAD 83 (CORS96) for CORS stations. Since GPS technologies and related accuracies had improved over the time of the HARN creation, the NGS decided in 1998 to reobserve all HARN stations. This process known as the *Federal Base Network* (FBN) survey was initiated in 1999. At this time, they also made the decision to not include the optically surveyed stations from the original NAD 83 adjustment in any subsequent adjustments. In 2007, a simultaneous readjustment of all HARN and CORS observations was completed. This created a new

---

[7]Using high-precision surveys, it was determined that the NAD 83 (1986) Cartesian coordinate axes were misaligned by 0.03″, and its scale differed by 0.0871 ppm from the true definition of the meter.

definition of NAD 83 known as the NAD 83 (2007). This system is connected to the *International Terrestrial Reference Frame* (ITRF) using the ITRF coordinates of the CORS sites. This datum removed problems of having two different reference frames available for use in GNSS surveys but still leaves surveyors with the NAD 83 (1986) triangulation adjustment and NAD 83 (2007). In 2011, the NGS readjusted the CORS and HARN stations using the *International GNSS Service reference system* of 2008 (IGS2008). It now reports station positions as both IGS2008 in epoch 2005.0 and NAD 83 (2011) epoch 2010.0.

The plan by the NGS is to have GNSS surveys replace the need for conventional static monumentation. As time passes, the ground monumentation will eventually be lost through attrition. The advantage to surveyors is that GNSS allows someone to create accurate positional coordinates, which are tied to the National Spatial Reference System, as required in any project.

19.6.4 National Geodetic Vertical Datum of 1929 (NGVD 29)

Vertical datums for referencing benchmark elevations are based on a single equipotential surface. Prior to the NAVD 88 readjustment (see Section 19.6.5), the vertical datum used in the United States was the *National Geodetic Vertical Datum of 1929* (NGVD 29). The NGVD 29 was obtained from a best fit of mean sea level observations taken at 26 tidal gage stations in the United States and Canada, and thus is often referred to as "mean sea level (MSL)." Unfortunately, the use of the term "mean sea level" is still used today when expressing elevations of benchmarks. As will be discovered in the next subsection, the use of "mean sea level" to define the elevation of a station is incorrect since the current datum was defined using a single benchmark. This important distinction should not be forgotten and future references to the elevation of a station should be called its *orthometric height* and never referred to as a mean sea level height.

19.6.5 North American Vertical Datum of 1988 (NAVD 88)

Between 1929 and 1988 more than 625,000 km of additional control leveling lines had been run. Furthermore, crustal movements and subsidence had changed the elevations of many benchmarks. To incorporate the additional leveling, and correct elevations of erroneous benchmarks, a general vertical adjustment was performed. This adjustment included the new observational data, as well as an additional 81,500 km of releveled lines, and leveling observations from both Canada and Mexico. It was originally scheduled for completion in 1988 and named the *North American Vertical Datum of 1988* (NAVD 88), but it was not actually released to the public until 1991. This adjustment shifted the position of the reference equipotential surface from the mean of the 26 tidal gage stations used in NGVD 29 to a single tidal gage benchmark known as *Father Point*, which is in Rimouski, Canada on the mouth of the Saint Lawrence Seaway in Quebec, Canada. As a result of these changes, published elevations of benchmarks in NAVD 88 have shifted from their NGVD 29 values. The magnitudes of these changes in the conterminous United States, expressed in millimeters, are shown

Figure 19.7 Approximate shift in vertical datum (in millimeters). Values shown are NAVD 88 minus NGVD 29. (Adapted from National Geodetic Survey map.)

in Figure 19.7. Note that the changes are largest in the western half of the country, with shifts of more than 1.5 m occurring in the Rocky Mountain region.[8]

It should be recognized that geoid models are directly related to an associated geometric reference frame by Equation (19.7). That is, as different realizations of NAD 83 were created, different geoid models followed, which were based on the geodetic heights of benchmarks in their new realization versus their orthometric heights (elevations). Thus, one should use different geoid models with the different realizations of NAD83. For example, with NAD83 (HARN) use GEIOD03, with NAD83 (2007) use GEOID09, and with NAD 83 (2011) uses the latest in a sequence of geoid models starting with GEOID12b and proceeding to GEOID18.

19.6.6 Future Reference Frames in the United States

In the late 1990s, the National Geodetic Survey decided to discontinue maintenance of the conventional surveying monumentation, which are known as *passive marks*, that make up the original NAD 83 and NAVD 88 reference frames. This was due to the facts that crustal plate motions cause horizontal positions and heights of stations to change with time, there was no congressional funding to maintain marks, marks were lost due to construction and vandalism, and also

---

[8]Those wishing to convert NGVD 29 benchmark elevations to NAVD 88 values can use the software VERTCON available from the NGS on their website at http://www.ngs.noaa.gov/TOOLS/Vertcon/vertcon.html.

that GNSS surveys can quickly and reliably establish control for any project in extremely short periods of time. Thus, traditional monumentation is no longer required to obtain precise coordinates in the national reference system. However, as was mentioned in Chapter 13, GNSS surveys provide geodetic heights, which are referenced to the ellipsoid. On the other hand, orthometric heights (elevations) are referenced to the geoid. In the United States, a hybrid geoid is used currently, which was created to match the existing benchmark monumentation at the time of the creation of NAVD 88. However, this hybrid geoid has a half meter bias and a 1-m slope going from east to west across the continent when compared to the gravimetric geoid, which means that discrepancies between geodetic heights derived by GNSS receivers varied from the orthometric heights at existing benchmarks. Thus, the NGS created a series of hybrid geoids to compensate for this difference. They included GEOID96, GEOID99, GEOID03, GEOID09, GEOID12B, and GEOID18.

By 2022 the NGS plans to have orthometric heights established using GNSS observations and a time-dependent gravimetric geoid model, which provides a least-squares best fit of the mean global sea surface of the Earth. Canada and Mexico are also involved in the redefinition of the vertical datum. It has been agreed that the datum will be defined by the 62,636,856.00 m^2/s^2 equipotential surface (see Section 19.15.4). The goal is to have a geoid model that has a differential height accuracy of 1 cm over any distance throughout the United States. To support this effort, the NGS obtained gravity values using long-wavelength gravity observations from the *Gravity Recovery and Climate Experiment* (GRACE) mission and medium-wavelength gravity observations from the GRAV-D airborne mission.

Figure 19.8(a) depicts the velocity vectors of the NOAA CORS network (NCN) in relation to ITRF2014. Note that with the exception of the west coast in the United States, which is on the Pacific plate, most of the velocity vectors show a similar rotation with respect to the International Terrestrial Reference Frame of 2018 (ITRF2018). This rotation can be described as a constant rotation about what is called an *Euler pole*. The rotation is mathematically removed by defining a new set of XYZ coordinate axes, which deviate slightly over time from ITRF2018, with the resulting residuals shown in Figure 19.8(b). Such a new set of coordinate axes is often called a "plate fixed frame," since it has the effect of making the plate appear to be not rotating, leaving only small residual motions (except on the west coast, in this case). These residual motions will then be handled with an Intra-Frame Velocity Model (IFVM). The IFVM is time variant, and thus it will be important to not only know the coordinates of a station but also the epoch date when the data used to compute the coordinates were collected. However, with this information, it will be possible to compare the coordinates of a station where the data was collected at different epochs in time. For example, if a GNSS survey was used to determine the coordinates of a station in 2022, these coordinates can be compared to coordinates determined by a GNSS survey of the same station in 2032. Current plans are to estimate new coordinates of stations every five years using the Euler Pole Parameters 2022 model (EPP2022) and the IFVM2022 model. Any passive marks, which are not resurveyed and submitted to the NGS during this time period will also have their coordinates updated but

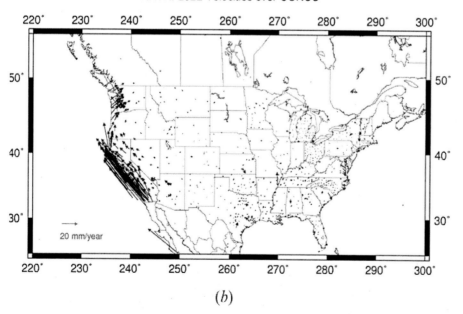

Figure 19.8
(a) Depicts the velocity vectors of the continental United States with respect to their ITRF2014 coordinates.
(b) Depicts the resulting residuals after the EPP2022 model is applied to the coordinates. (Courtesy National Geodetic Survey.)

with an increasing level of uncertainty in their coordinates. Thus, it will be important for future surveyors to occupy such passive marks with GNSS according to NGS requirements and submit this data to the NGS if they wish to maintain the accuracy of any passive marks.

Since parts of the United States are on four different plates, the 2022 National Spatial Reference System will consist of four different reference frames called the North American Terrestrial Reference Frame of 2022 (NATRF2022), Pacific Terrestrial Reference Frame of 2022 (PATRF2022), Mariana Terrestrial Reference Frame of 2022 (MATRF2022), and the Caribbean Terrestrial Reference Frame of 2022 (CATRF2022), which are defined with four different Euler poles. Thus, the horizontal reference frame will be aligned to an International Terrestrial Reference Frame system, and GNSS-derived geodetic positions and heights that make up the National Spatial Reference System (NSRS) will be more consistent with coordinates obtained from GNSS receivers. This new system will be a true four-dimensional coordinate system with the fourth dimension being time.

19.7 TRANSFORMING COORDINATES BETWEEN REFERENCE FRAMES

Historically, a goal of geodesy has been to obtain one common reference frame for coordinates. However realistically, each country or region has often developed its reference frame independently. Additionally, many surveyors today start surveys with arbitrary coordinates. Thus, we often need to transform station coordinates between those derived using GNSS surveys and those developed in some regional or arbitrary reference system. In Section 15.9, this process was introduced as *localization*. Some approaches require stations with known geodetic coordinates in both reference frames. Others allow GNSS-derived coordinates to be transformed into any local coordinate system including arbitrary coordinates. This section will discuss the mathematics and surveying procedures behind these transformations.

19.7.1 Helmert Transformation and Its Variant

If sufficient common stations are known, a three-dimensional coordinate transformation (see Section 19.18) can be used to convert the coordinates of stations from one reference frame into another. Since most reference frames have nearly aligned coordinate axes with rotation angles in the milli-arcseconds, the three-dimensional coordinate transformation can be simplified to the so-called *Helmert transformation*. The Helmert transformation is mathematically expressed as

$$\begin{bmatrix} X \\ Y \\ Z \end{bmatrix}_2 = (1 + \Delta S) \begin{bmatrix} X \\ Y \\ Z \end{bmatrix}_1 + \begin{bmatrix} 0 & R_Z & -R_Y \\ -R_Z & 0 & R_X \\ R_Y & -R_X & 0 \end{bmatrix} \begin{bmatrix} X \\ Y \\ Z \end{bmatrix}_1 + \begin{bmatrix} T_X \\ T_Y \\ T_Z \end{bmatrix} \quad (19.12)$$

where $[X \ Y \ Z]_1^T$ and $[X \ Y \ Z]_2^T$ are the geocentric coordinates of the common stations in the two reference frames derived using Equation (13.1); ΔS is the scale factor change between reference frames; R_X, R_Y, and R_Z the rotations in radian units for the X, Y, and Z axes, respectively; T_X, T_Y, and T_Z the translations between the two reference frames. Table 19.2 shows the parameters to transform PZ-90.11 coordinates into WGS 84 (G1762) coordinates. To perform this transformation using stations with known geodetic coordinates, a minimum of two stations known in horizontal position and three stations in elevation are required. However as shown in Figure 15.5, it is recommended that common points be obtained in all four quadrants of the survey near the boundaries of the survey. Additionally any

TABLE 19.2 HELMERT TRANSFORMATION PARAMETERS TO ITRF 2008

Reference Frame	T_x (mm) S	T_y (mm) S	T_z (mm) S	ΔS (ppb) S	R_X (msec) S	R_Y (msec) S	R_Z (msec) S
WGS 84 (G1674)	0	0	0	0	0	0	0
PZ-90.11	3	−1	0	0	19	−42	2
	±2	±2	±2	±0.0003	±72	±73	±90

crucial engineering points, such as benchmarks on bridge abutments are included in the transformation. When sufficient common stations are not available, it is possible to perform this transformation using only the translations. However, this produces lower-quality results since it only accounts for the differences in the origins of the reference systems and not their scaling or misalignment differences.

Besides different reference frames, the crustal plates of the Earth are constantly in motion. For example, some parts of California are moving at a rate of 4 cm per year on average while stations on the east coast are moving an average of 1 to 2 cm per year. Additionally because of post-glacial rebound, the vertical surface of the Earth is rising in areas that were covered by glaciers during the last ice age. Thus, the coordinates of points in any reference frame must be tied to a specific moment in time, or epoch. The National Geodetic Survey has combined the Helmert transformation with the velocity vectors of the crustal plates to produce a 14-parameter coordinate transformation software package known as *Horizontal Time-Dependent Positioning* (HTDP) software.[9] Thus, in this software each of the transformation parameters is modified by adding a velocity component. For example, $R_X = R'_X + v_{Rx}(t - t_0)$ where R'_X was the computed rotational parameter of the YZ plane around the X axis at epoch t_0, v_{Rx} is the rotational velocity of the YZ plane, and t is the current or selected epoch in time for the computed coordinates. This operation is similarly performed for each of the seven parameters involved in the Helmert transformation. This software allows users to transform coordinates across epochs in time and between reference frames. Computations for the Helmert transformation as used in the HTDP software are demonstrated in a Mathcad worksheet on the companion website.

19.7.2 The Two Plus One Approach

It is also possible to perform the transformation in two separate transformations (horizontal and vertical), which is typical. This is especially useful when the design coordinates are arbitrarily assigned. In this case, the geodetic coordinates derived from GNSS survey are converted to coordinates in a map projection system (see Chapter 20). Since the horizontal and vertical transformations are separated, a two-dimensional coordinate transformation in Equation (11.41) is used to transform the map projection coordinates of the GNSS-derived positions into their equivalent regional or arbitrary coordinate system. This process is demonstrated in the following example.

---

[9]The HTDP software can be found at http://www.ngs.noaa.gov/TOOLS/Htdp/Htdp.html.

Example 19.3

A surveyor establishes a control network of points using an arbitrary coordinate system. In preparation for staking out the project using kinematic survey (see Chapter 15), the surveyor reoccupies each station with a receiver. The resulting GNSS-derived coordinates are transformed into a two-dimensional map projection coordinate system with the common station coordinate values in both systems listed in Table 19.3. Determine the rotation and scale between the arbitrary and worldwide reference frames.

TABLE 19.3 COORDINATE VALUES OF COMMON STATIONS

	Arbitrary Reference Frame		GPS Reference Frame	
Station	X (ft)	Y (ft)	E (m)	N (m)
A	5000.00	5000.00	635797.076	464685.605
B	1978.54	6075.88	625530.377	462379.464
C	6328.46	5983.64	637760.165	469740.901
D	6058.04	5000.00	638732.517	466538.417

Solution

An oblique stereographic map projection (see Section 20.12.1) was used to transform the observed geodetic coordinates derived by the application software to a coordinate system with the centroid of the project as its origin.

These values are then used to translate the arbitrary coordinates to a common origin resulting in the following set of coordinates. The local orthometric height and GNSS-derived orthometric height are also shown.

	Arbitrary			GNSS		
Station	X (ft)	Y (ft)	H (m)	$e' = E - E_0$ (m)	$n' = N - N_0$ (m)	H (m)
A	5000.00	5000.00	282.486	45.212	−157.888	282.476
B	1978.54	6075.88	296.577	−869.005	−188.611	296.571
C	6328.46	5983.64	313.819	456.132	133.694	313.814
D	6058.04	5000.00	304.191	367.660	−164.417	304.205

Since the two coordinate systems share a common origin, Equation (11.41) is modified as

$$\begin{bmatrix} a & -b \\ b & a \end{bmatrix} \begin{bmatrix} e' \\ n' \end{bmatrix} = \begin{bmatrix} x' \\ y' \end{bmatrix} + \begin{bmatrix} v_x \\ v_y \end{bmatrix} \qquad (19.13)$$

Substituting the above coordinates into this equation results in

$$A = \begin{bmatrix} 45.212 & 157.888 & 1 & 0 \\ -157.888 & 45.212 & 0 & 1 \\ -869.005 & -188.611 & 1 & 0 \\ 188.611 & -869.005 & 0 & 1 \\ 456.132 & -133.694 & 1 & 0 \\ 133.694 & 456.132 & 0 & 1 \\ 367.660 & 164.417 & 1 & 0 \\ -164.417 & 367.660 & 0 & 1 \end{bmatrix} \quad X = \begin{bmatrix} a \\ b \\ c \\ d \end{bmatrix} \quad L = \begin{bmatrix} 5000.00 \\ 5000.00 \\ 1978.54 \\ 6075.88 \\ 6328.46 \\ 5983.64 \\ 6058.04 \\ 5000.00 \end{bmatrix}$$

Using Equation (16.6), the solution of this system $(AX = L + V)$ results in $a = 3.27987, b = 0.066308, c = 4841.261,$ and $d = 5514.880$. Recognizing that $\tan \theta = \dfrac{b}{a}$, the scale and rotation between the two systems of coordinates are $s = 3.28054$ and $\theta = 1°09'29.4''$, respectively. These values are now used in conjunction with Equation (19.12) to transform the GNSS-derived map projection coordinates into the arbitrary local coordinate system for any additional points.

Note that the scale factor is approximately equal to the conversion factor going from meters to feet of 3.2808, and that the computed residuals for the observations (in units of feet) are 0.020, 0.027, −0.011, −0.003, −0.009, −0.016, −0.001, and −0.007, respectively, which are within the observational precisions of kinematic survey for horizontal work.

The GNSS-derived heights can also be transformed onto a local level plane. This process must account for the translation between the two systems and the obliquity of the two planes. The obliquity of the two level surfaces is corrected applying two rotations in the cardinal north and east directions to bring the GNSS-derived level surface parallel to the local horizontal plane. This is performed as

$$T_0 + r_e N_{GNSS} + r_n E_{GNSS} = H_L - H_G + v \qquad (19.14)$$

where T_0 is the translation between the two level planes; N_{GNSS} and E_{GNSS} the northing and easting of map projection coordinates derived from the GNSS application software (see Example 19.3), respectively; r_e and r_n the rotations about the x and y axes, respectively; H_L the local height of the control points used in the project design; and H_G is either the geodetic height of the point as derived by GNSS or the orthometric height of the point as derived by the combination of GNSS-derived geodetic heights and a geoid model. It should be noted here that the geoid model should always be applied since this would represent a systematic error in the adjustment if geodetic heights were used. Additionally, since there is no scale factor in Equation (19.14), the units of the local and GNSS heights must match.

As can be seen, Equation (19.14) involves three unknown parameters, $T_0, r_e,$ and r_n. Thus, a minimum of three benchmarks with local heights must be known. Since scale is not part of this transformation, it is important that the units of the stations in both the GNSS system and local system are the same. However as discussed in Section 15.9, it is wise to always have a fourth benchmark for the purposes of redundancy and a check. Again, when less than three benchmarks are known, the translation can be computed from a single station. However, the accuracy of this transformation will decrease significantly.

■ Example 19.4

Using the data given in Example 19.3, determine the three transformation parameters of Equation (19.13).

Solution

Using the northing (N), easting (E), and heights from Example 19.3 in concert with Equation (19.14) yields the following observation equations:

$$A = \begin{bmatrix} 1 & -157.888 & 45.212 \\ 1 & 188.611 & -869.005 \\ 1 & 133.694 & 456.132 \\ 1 & -164.417 & 367.660 \end{bmatrix} \quad X = \begin{bmatrix} T_0 \\ r_e \\ r_n \end{bmatrix} \quad L = \begin{bmatrix} 0.010 \\ 0.006 \\ 0.005 \\ 0.014 \end{bmatrix}$$

Solving the system of equations, $AX = L + V$, using Equation (16.6) yields $T_0 = 0.002$, $r_e = 3.5''$, and $r_n = -0.9''$. With these transformation parameters and the GNSS-derived map projection coordinates, a geodetic height can be transformed into a local height. The resulting residuals for the observations (in SI units) are $-0.011, 0.003, -0.003$, and 0.011, respectively. Again, these values are well within the vertical accuracy of GNSS-observed heights.

To have the distances obtained from GNSS surveys match their equivalent ground distances, the map projection plane is brought to the surface using an appropriate scaling factor. As discussed in Section 20.12.1, the oblique stereographic map projection has a defining scale factor of k_0 at its origin, which is the center of the projection. This makes the oblique stereographic map projection the preferred projection for this process. If this value is set to an appropriate scale, the plane surface of the map projection will be coincident with the average elevation of the project. As an example, a scale factor of

$$k_0 = 1 + \frac{H_{\text{average}}}{R_e} \tag{19.15}$$

This scale factor is often used as one of the defining parameters for the oblique stereographic map projection system (see Section 20.12.1). The lengths of the distances are further refined with scaling factor as derived from the two-dimensional conformal coordinate transformation as defined in Example 19.3. The combination of these scales applied to the observed and transformed geodetic coordinates will result in distances that closely match their equivalent ground values. Computations for this problem are demonstrated in the Mathcad worksheets *Helmert.xmcd* and *C19-6.xmcd*, which are on the companion website.

19.8 ACCURACY STANDARDS AND SPECIFICATIONS FOR CONTROL SURVEYS

The required accuracy for a control survey depends primarily on its purpose. Some major factors that affect accuracy are type and condition of equipment used, field procedures adopted, and the experience and capabilities of personnel employed. In 1984, the Federal Geodetic Control Subcommittee (FGCS) published the Standards and Specifications for Geodetic Networks. In 1998, the Federal Geographic Data Committee (FGDC) published a different sets of detailed standards of accuracy and specifications for geodetic surveys.[10] The rationale for both sets of standards is twofold: (1) to provide a uniform set of standards specifying minimum acceptable accuracies of control surveys for various purposes and (2) to establish specifications for instrumentation, field procedures, and misclosure checks to ensure that the intended level of accuracy is achieved.

Table 19.4 lists the 1998 FGDC accuracy standards for control points. These standards are independent of the method of survey and based on a 95% confidence level (see Section 16.12). In order to meet these standards, control points in the survey must be consistent with all other points in the geodetic control network and not merely those within that particular survey. In Table 19.4, for horizontal surveys the accuracy standard specifies the radius of a circle within which the true or theoretical location of the survey point falls 95% of the time. To compute the 95% error circle for a point a constant C is determined for each station using its semimajor (S_u) and semiminor (S_v) axes of the standard error ellipse from a least-squares adjustment (see Section 16.10) as

$$C = \frac{S_v}{S_u} \quad (19.16)$$

Then using a polynomial from Leenhouts (1985), the 95% radial error of a point is approximated as

$$r_{95\%} = S_u(1.960790 + 0.004071C + 0.114276C^2 + 0.371625C^3) \quad (19.17)$$

The vertical accuracy standard specifies a linear value (plus or minus) within which the true or theoretical location of the point falls 95% of the time using the normal distribution multiplier of 1.9599 (see Section 3.16). Procedures leading to classification according to these standards involve four steps:

1. The survey observations, field records, sketches, and other documentation are examined to insure their compliance with specifications for the intended accuracy of the survey.
2. A minimally constrained least-squares adjustment of the survey observations is analyzed to guarantee that the observations are free from blunders and have been correctly weighted.

---

[10]The 1998 FGDC standards are titled *Geospatial Positioning Accuracy Standards*. They can be downloaded at the following website: http://www.fgdc.gov. The 1984 FGCS standards, published in a booklet entitled *Standards and Specifications for Geodetic Control Networks*, are available from the National Geodetic Information Center, NOAA, National Geodetic Survey, N/CG17, SSMC3 Station 09535, 1315 East-West Highway, Silver Spring, MD 20910.

3. The accuracy of control points in the local existing network to which the survey is tied is computed by random error propagation and weighted accordingly in the least-squares adjustment of the survey network.
4. The survey accuracy is checked at the 95% confidence level by comparing minimally constrained adjustment results against established control. The comparison takes into account the network accuracy of the existing control as well as systematic effects such as crustal motion or datum distortion.

Because many existing products, including control datasheets in the NAD 83 datum, refer to the 1984 FGCS standards, these will also be described. This earlier set of standards established three distinct *orders of accuracy* to govern traditional control surveys, given in descending order: *first-order*, *second-order*, and *third-order*. For horizontal control surveys, second-order and third-order each have two separate accuracy categories, *class I* and *class II*. For vertical surveys, first-order and second-order each have class I and class II accuracy divisions. In 1985, three new orders of accuracy were defined for GNSS surveys (see Section 14.5.1). These were orders AA, A, and B. Another lower order of accuracy for GNSS surveys, identified as Order C, was also specified in these standards. It overlaps the three orders of accuracy applied to traditional horizontal surveys (see Tables 19.4 and 14.5).

Triangulation, trilateration, and traverse surveys are included in the 1984 horizontal control standards and specifications, and differential leveling is covered in the vertical control section.

Tables 19.5 and 19.6 give the 1984 and 1985 FGCS accuracy standards required for the various orders and classes of horizontal and vertical control surveys, respectively. Values in Table 19.5 are ratios of allowable relative positional errors of a pair of horizontal control points, to the horizontal distance separating

TABLE 19.4 1998 FGDC ACCURACY STANDARDS: HORIZONTAL, ELLIPSOID HEIGHT, AND ORTHOMETRIC HEIGHT

Accuracy Classifications	95% Confidence Less Than or Equal to
1 millimeter	0.001 meters
2 millimeters	0.002 meters
5 millimeters	0.005 meters
1 centimeter	0.010 meters
2 centimeters	0.020 meters
5 centimeters	0.050 meters
1 decimeter	0.100 meters
2 decimeters	0.200 meters
5 decimeters	0.500 meters
1 meter	1.000 meter
2 meters	2.000 meters
5 meters	5.000 meters
10 meters	10.000 meters

TABLE 19.5 1984 AND 1985 FGCS HORIZONTAL CONTROL SURVEY ACCURACY STANDARDS

GPS Order*	Traditional Surveys Order and Class**	Relative Accuracy Required Between Points
Order AA		1 part in 100,000,000
Order A		1 part in 10,000,000
Order B		1 part in 1,000,000
Order C-1	First Order	1 part in 100,000
	Second Order	
Order C-2-I	Class I	1 part in 50,000
Order C-2-II	Class II	1 part in 20,000
	Third Order	
Order C-3	Class I	1 part in 10,000
	Class II	1 part in 5000

*Published in 1985.
**Published in 1984.

TABLE 19.6 1984 FGCS VERTICAL CONTROL SURVEY ACCURACY STANDARDS

Order and Class	Relative Accuracy Required Between Benchmarks*
First Order	
Class I	0.5 mm $\times \sqrt{K}$
Class II	0.7 mm $\times \sqrt{K}$
Second Order	
Class I	1.0 mm $\times \sqrt{K}$
Class II	1.3 mm $\times \sqrt{K}$
Third Order	2.0 mm $\times \sqrt{K}$

*K is distance between benchmarks, in kilometers.

them. Thus, two first-order stations located 100 km (60 mi) apart are expected to be correctly located with respect to each other to within ± 1 m.

Table 19.6 gives maximum relative elevation errors allowable between two benchmarks, as determined by a weighted least-squares adjustment (see Chapter 16). Thus, elevations for two benchmarks 25 km apart, established by second-order class I standards, should be correct to within $\pm 1.0\sqrt{25} = \pm 5$ mm. These standards are not the same as the maximum allowable loop misclosures for the five classes of leveling given in Section 5.5. Table 19.6 values specify relative accuracies of benchmarks after adjustment, whereas loop misclosures enable assessment of results in differential leveling prior to adjustment.

The ultimate success of any engineering or mapping project depends on appropriate survey control. The higher the order of accuracy demanded, the more time

564 CONTROL SURVEYS AND GEODETIC REDUCTIONS

and expense are required. It is therefore important to select the proper order of accuracy for a given project and carefully follow the specifications. Note that no matter how accurately a control survey is conducted, errors will exist in the computed positions of its stations, but a higher order of accuracy presumes smaller errors.

■ 19.9 THE NATIONAL SPATIAL REFERENCE SYSTEM

To meet the various local needs of surveyors, engineers, and scientists, the federal government has established a National Spatial Reference System (NSRS) consisting of more than 270,000 horizontal control monuments and approximately 600,000 benchmarks throughout the United States. The National Geodetic Survey (NGS), which began control surveying operations as the *Survey of the Coast* in 1807, changed to *Coast Survey* in 1836, to *Coast and Geodetic Survey* in 1878, and to a division of the *National Ocean Survey* (NOS) in 1970, has primary responsibility for the NSRS. Today the NGS is a division of the National Ocean and Atmospheric Administration (NOAA). It continues to assist with, and to coordinate, geodetic control surveying activities with other agencies and with all states to establish new NSRS control stations and upgrade and maintain existing ones. It also disseminates a variety of publications and software related to geodetic surveying.

The NSRS is split into horizontal and vertical divisions. All control within each part is classified in a ranking scheme based on purpose and order of accuracy. These are described in the following two sections.

■ 19.10 HIERARCHY OF THE NATIONAL HORIZONTAL CONTROL NETWORK

The hierarchy of control stations within the NSRS Horizontal Control Network, from highest to lowest order, and their primary uses are as follows:

Global-regional geodynamics consist of GNSS-surveyed points that meet the Order AA accuracy requirements. These are primarily used for international deformation studies.

Primary control consists of GNSS-surveyed points that meet the Order A accuracy requirements. These points are used for regional-local geodynamic and deformation studies.

Secondary control densifies the network within areas surrounded by primary control, especially in high-value land areas and for high-precision engineering surveys. Secondary control surveys are executed to GNSS Order B standards.

Terrestrial-based control is used for dependent control surveys to meet mapping, land information systems, property survey, and engineering needs. This network consists principally of stations set by traverse and triangulation to first- and second-order standards, and stations set using GNSS surveys to Order C standards.

Local control provides reference points for local construction projects and small-scale topographic mapping. These surveys are referenced to higher-order control monuments and, depending on accuracy requirements, may be third-order class I, or third-order class II.

■ 19.11 HIERARCHY OF THE NATIONAL VERTICAL CONTROL NETWORK

The scheme of benchmarks within the National Vertical Control Network is classified as follows:

Basic framework is a uniformly distributed nationwide network of benchmarks whose elevations are determined to the highest order of accuracy. It consists of nets *A* and *B*. In net *A*, adjacent level lines are ideally spaced an average of about 100 to 300 km apart using first-order class I standards; in net *B* the average separation is ideally about 50 to 100 km, and first-order class II standards are specified. Benchmarks are placed intermittently along the level lines at convenient locations.

Secondary network densifies the basic framework, especially in metropolitan areas and for large engineering projects. It is established to second-order class I standards.

General area control consists of vertical control for local engineering, surveying, and mapping projects. It is established to second-order class II standards.

Local control provides vertical references for minor engineering projects and small-scale topographic mapping. Benchmarks in this category satisfy third-order standards.

■ 19.12 CONTROL POINT DESCRIPTIONS

To obtain maximum benefit from control surveys, horizontal stations and benchmarks are placed in locations favorable to their subsequent use. The points should be permanently monumented and adequately described to ensure recovery by future potential users. Reference monuments placed by the NGS are marked by bronze disks about 3.5 in. in diameter set in concrete or bedrock. Figure 19.9 shows two types of these disks.

Procedures for establishing permanent monuments vary with the type of soil or rock, climatic conditions, and intended use for the monument. In cases where

Figure 19.9 Bronze disks used by the National Geodetic Survey to mark horizontal and vertical control stations.

soil can be excavated, monuments are commonly set in concrete that goes a foot or more below the local maximum frost depth. The bottom of the excavation is generally wider than the top to maximize monument stability during periods of freeze and thaw. Another option commonly used today is to drive a stainless steel rod to refusal using powered tools. Driving depths of 10 or more feet are common when using this technique. In bedrock, holes are often drilled into the rock and the monument is simply cemented into the hole. Other variations for monumenting can be used, if the resulting objects will remain stable in their positions.

The NGS ranks their monuments in their database by their stability. Quality code A monuments are the most reliable and are expected to hold a precise elevation. These monuments are typically rock outcrops, bedrock, and similar features plus massive structures with deep foundations; large structures with foundations on bedrock; or sleeved deep settings (10 ft or more) with galvanized steel pipe or galvanized steel, stainless steel, or aluminum rods. Quality code B monuments are those that will probably hold a precise elevation. Examples include unsleeved deep settings (10 ft or more) with galvanized steel pipe or galvanized steel, stainless steel, or aluminum rods; massive structures other than those listed under Quality code A, massive retaining walls, abutments and piers of large bridges or tunnels, unspecified rods or pipe in a sleeve less than 10 ft, or sleeved copper-clad steel rods. Quality code C monuments are those that may hold a precise elevation but are subject to ground movement. Examples of these monuments include metal rods with base plates less than 10 ft deep, concrete posts (3 ft or more deep), unspecified rods or pipe more than 10 ft deep, large boulders, retaining walls for culverts or small bridges, footings or foundation walls of small to medium-size structures, or foundations such as landings, platforms, or steps. Quality code D monuments are those of questionable stability. Examples include objects of unknown character, shallow set rods or pipe (less than 10 ft), light structures, pavements such as street, curbs, or aprons, piles and poles such as spikes in utility poles, masses of concrete, or concrete posts less than 3 ft deep.

The NGS makes complete descriptions of all its control stations available to surveyors. As an example, a partial listing from an NGS horizontal control station description is given in Figure 19.10. These descriptions give each station's general placement in relation to nearby towns, instructions on how to reach the station following named or numbered roads in the area, and the monument's precise location by means of distances and directions to several nearby objects. The station's specific description, such as "a triangulation disk set in drill hole in rock outcrop," is given, along with a record of recovery history. Data supplied with horizontal control-point descriptions include the datum(s) used and the station's geodetic latitude and longitude. Also given are the state plane coordinates, convergence angle and scale factor, UTM coordinates (see Chapter 20), and approximate elevation and geoidal height (in meters).

Some station descriptions give geodetic and grid azimuths (see Chapter 20) to a nearby station or stations. Geodetic and grid azimuths differ by an amount equal to the convergence angle, and therefore the appropriate azimuth must be selected for the particular surveying methods used.

As shown in Figure 19.11, published benchmark data include approximate station locations and adjusted elevations in both meters and feet. Again the relevant datum is identified.

```
LZ1878    National Geodetic Survey, Retrieval Date = MARCH 9, 2015
LZ1878    ******************************************************************
LZ1878    DESIGNATION -   HAYFIELD NE 1974
LZ1878    PID -           LZ1878
LZ1878    STATE/COUNTY-   PA/LUZERNE
LZ1878    COUNTRY -       US
LZ1878    USGS QUAD -     HARVEYS LAKE (1983)
LZ1878
LZ1878                              *CURRENT SURVEY CONTROL
LZ1878    _____
LZ1878*   NAD 83(1992) POSITION-     41 18 20.25432(N)   076 00 57.00156(W)   NO
                                                                              CHECK
LZ1878*   NAVD 88      ORTHO HEIGHT-       398.7 (meters)       1308. (feet)  VERTCON
LZ1878    _____
LZ1878    GEOID HEIGHT -              -31.73  (meters)        GEOID12A
LZ1878    LAPLACE CORR -                0.27  (seconds)       DEFLEC12A
LZ1878    HORZ ORDER   -       SECOND
LZ1878
LZ1878.   The horizontal coordinates were established by classical geodetic methods and ad-
LZ1878.   justed by the National Geodetic Survey in June 2002.
LZ1878.
LZ1878    .No horizontal observational check was made to the station.
LZ1878.
LZ1878.   The NAVD 88 height was computed by applying the VERTCON shift value to the NGVD
LZ1878.   29 height (displayed under SUPERSEDED SURVEY CONTROL.)
LZ1878
LZ1878.   The Laplace correction was computed from DEFLEC12A derived deflections.
LZ1878
LZ1878.   The following values were computed from the NAD 83(1992) position.
LZ1878
LZ1878;                       North          East      Units  Scale Factor   Converg.
LZ1878;   SPC  PA  N   -  127,939.407    745,212.656    MT    0.99995873    +1 08 50.0
LZ1878;   SPC  PA  N   -  419,747.87   2,444,918.52    sFT    0.99995873    +1 08 50.0
LZ1878;   UTM      18  - 4,573,182.995   414,962.095    MT    0.99968900    -0 40 14.0
LZ1878
LZ1878                    -  Elev Factor  x  Scale Factor   =  Combined Factor
LZ1878    SPC  PA  N      -  0.99994244   x  0.99995873     =  0.99990118
LZ1878    UTM      18     -  0.99994244   x  0.99968900     =  0.99963146
LZ1878
LZ1878                              SUPERSEDED SURVEY CONTROL
LZ1878    LZ1878 NAD 83(1986)- 41 18 20.25410(N)   076 00 57.00239(W) AD( ) 2
LZ1878    NAD 27 - 41 18 19.96597(N)               076 00 58.27835(W) AD( ) 2
LZ1878    NGVD 29 (07/19/86) 398.9 (m)             1309. (f) VERT ANG
LZ1878
LZ1878.   Superseded values are not recommended for survey control.
(continues ...)
```

Figure 19.10 Partial listing of station data sheet in the National Spatial Reference System for horizontal control station Hayfield NE. (Courtesy National Geodetic Survey.)

568 CONTROL SURVEYS AND GEODETIC REDUCTIONS

```
NB0293     ***********************************************************
NB0293     DESIGNATION -  F 137
NB0293     PID -          NB0293
NB0293     STATE/COUNTY-  NY/TIOGA
NB0293     COUNTRY -      US
NB0293     USGS QUAD -    ENDICOTT (1976)

NB0293
NB0293                          *CURRENT SURVEY CONTROL
NB0293     _____
NB0293*    NAD 83(1986)  42 04 09.40   (N)        076 07 03.43  (W)     HD_HELD1
           POSITION-
NB0293*    NAVD 88 OR-         252.471  (meters)          828.32 (feet)
           THO HEIGHT-
ADJUSTED
NB0293     _____
NB0293     GEOID HEIGHT-       -32.77   (meters)                         GEOID12A
NB0293     DYNAMIC HEIGHT -    252.373  (meters)  827.99 (feet)          COMP
NB0293     MODELED GRAVITY -   980,231.5 (mgal)                          NAVD 88
NB0293
NB0293     VERT ORDER -        FIRST CLASS II

NB0293
NB0293.The horizontal coordinates were determined by differentially corrected
NB0293.hand held GPS observations or other comparable positioning techniques
NB0293.and have an estimated accuracy of +/- 3 meters.
NB0293.
NB0293.The orthometric height was determined by differential leveling and
NB0293.adjusted by the NATIONAL GEODETIC SURVEY
NB0293.in June 1991.
NB0293
NB0293.The dynamic height is computed by dividing the NAVD 88
NB0293.geopotential number by the normal gravity value computed on the
NB0293.Geodetic Reference System of 1980 (GRS 80) ellipsoid at 45
NB0293.degrees latitude (g = 980.6199 gals.).
NB0293
NB0293.The modeled gravity was interpolated from observed gravity values.
NB0293
NB0293;          North       East      Units  Estimated Accuracy
NB0293; SPC NY C - 229,893.4 288,540.5  MT    (+/- 3 meters HH1 GPS)
(continues ...)
```

Figure 19.11 An excerpt from a NGS data sheet for benchmark F 137. (Courtesy National Geodetic Survey.)

Besides control within the national network set by the NGS, additional marks have been placed in various parts of the United States by other federal agencies such as the USGS, Corps of Engineers, and Tennessee Valley Authority. State, county, and municipal governments may have also added control. When this work is coordinated through the NGS, the descriptions of the stations involved are distributed by the NGS.

As noted earlier, complete descriptions for all points in the National Spatial Reference System can be obtained from the NGS.[11] This includes horizontal, vertical, and GNSS control points. The descriptions can be obtained in hard copy or digital form.

19.13 FIELD PROCEDURES FOR CONVENTIONAL HORIZONTAL CONTROL SURVEYS[12]

As noted earlier, despite the increasing prominence of GNSS surveys, horizontal-control surveys over limited areas are still being accomplished by the traditional methods of triangulation, trilateration, precise traverse, or a combination of these techniques. These methods are described briefly in the subsections that follow.

Traditional methods in horizontal control surveys require observations of horizontal distances, angles, and observations of astronomic azimuths. Basic theory, equipment, and procedures for making these observations have been covered in earlier chapters. The following sections concentrate on procedures specific to control surveys, and on matters related to obtaining the higher orders of accuracy generally required for these types of surveys. Readers interested in performing traditional geodetic surveys should refer to the FGCS manuals.

19.13.1 Triangulation

Prior to the emergence of electronic distance-measuring equipment, triangulation was the preferred and principal method for horizontal-control surveys, especially if extensive areas were to be covered. Angles could be more easily observed compared with distances, particularly where long lines over rugged and forested terrain were involved, by erecting towers to elevate the operators and their instruments. Triangulation possesses many inherent checks and closure conditions that help detect blunders and errors in field data, and increase the possibility of meeting a high standard of accuracy.

As implied by its name, triangulation utilizes geometric figures composed of triangles. Horizontal angles and a limited number of sides called *baselines* are observed for length. By using the angles and baseline lengths, triangles are solved trigonometrically and positions of stations (vertices) calculated.

Different geometric figures are employed for control extension by triangulation but chains of quadrilaterals called arcs (Figure 19.12) have been most common. They are the simplest geometric figures that permit rigorous closure

---

[11]It is possible to obtain *data sheets* directly from the NGS Internet site at http://www.ngs.noaa.gov/datasheet.html. This website allows the user to search for data sheets of control points based on the station's name, *permanent identifier* (PID), and perform radial and rectangular searches from a location, or from a clickable image map.

[12]Conventional as used here implies non-GNSS ground-surveying methods.

570 CONTROL SURVEYS AND GEODETIC REDUCTIONS

Figure 19.12
Chain of quadrilaterals.

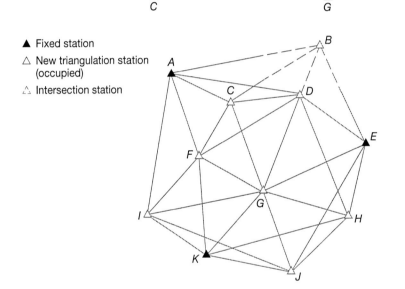

Figure 19.13
Triangulation network for a metropolitan control survey.

checks and adjustments of field observational errors, and they enable point positions to be calculated by two independent routes for computational checks. More complicated figures like that illustrated in Figure 19.13 have frequently been used to establish horizontal control by triangulation in metropolitan areas.

In executing triangulation surveys, *intersection stations* can be located as part of the project. In this process, angles are observed from as many occupied points as possible to tall prominent objects in the area such as church spires, smokestacks, or water towers. The intersection stations are not occupied, but their positions are calculated; thus, they become available as local reference points. An example is station *B* in Figure 19.13.

To compensate for the errors that occur in the observations, triangulation networks must be adjusted. The most rigorous method utilizes least squares (see Chapter 16). In that procedure all angle, distance, and azimuth observations are simultaneously included in the adjustment and given appropriate relative weights based on their precisions. The least-squares method not only yields the most probable adjusted station coordinates for a given set of data and weights, but also gives their precisions.

19.13.2 Precise Traverse

Precise traversing is common among local surveyors for horizontal-control extension, especially for small projects. Fieldwork consists of two basic parts: observing horizontal angles at the traverse hubs and observing distances between stations. With total station instruments, these observations can be observed simultaneously. Precise traverses always begin and end on stations established by equal- or higher-order surveys.

Unlike triangulation, in which stations are normally widely separated and placed on the highest ridges and peaks in an area, traverse routes generally follow the cleared rights-of-way of highways and railroads, with stations located relatively close together. Besides easing fieldwork, this provides a secondary benefit in accessibility to the stations. Traverses lack the automatic checks inherent in triangulation, and extreme observational caution must be exercised therefore to avoid blunders. Also, since traverses usually run along single lines, they are generally not as good as triangulation for establishing control over large areas.

Control traverses can be strengthened to provide additional checks in the data by establishing "offset stations" such as A', C', and E' of Figure 19.14. An offset station is set near every-other primary traverse station. In performing the field observations, instrument setups are made only at the primary traverse stations. All possible angles are observed with horizon closures at each station; thus, four angles are determined at interior single primary stations and two angles are observed at primary stations with nearby offset stations. This observation scheme is shown in Figure 19.13. Additionally, all distances are observed, that is, at station 1 distances $1A$ and $1A'$ are observed, at station A, lengths $A1$, AA', and AB are observed, and so on. When the field observations have been completed, the network can be adjusted using all observations in a least-squares adjustment, thereby providing geometric checks for all angle and distance observations in the traverse. Additional geometric strength in the figure could be obtained by also observing all of the angles at the offset stations.

In traversing to gain overall project efficiency and improve angle accuracy, it is always preferable to have long sight distances. Also to avoid mistakes it is advisable to avoid having nearly "flat" angles (values near 180°) whenever

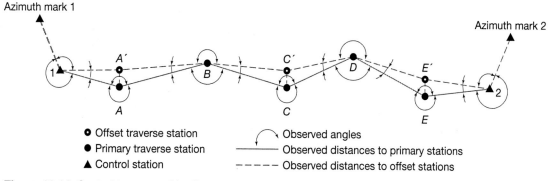

Figure 19.14 Control traverse with offset stations.

572 CONTROL SURVEYS AND GEODETIC REDUCTIONS

possible. To accomplish this, presurvey reconnaissance is recommended. An oft-made mistake is to construct the traverse while collecting the observations. This technique works in low-order surveys but frequently results in poorly designed control traverses.

For long traverses, checks on the observed horizontal angles can be obtained by making periodic astronomical azimuth observations (see Appendix C). These should agree with the values computed from the direction of the starting line and the observed horizontal angles. However, if a traverse extends an appreciable east–west direction, as illustrated in Figure 19.15, meridian convergence will cause the two azimuths to disagree. For example, in Figure 19.15, azimuth *FG* obtained from direction *AB* and the observed horizontal angles should equal astronomic azimuth *FG* + θ, where θ is the meridian convergence. A good approximation for meridian convergence between two points on a traverse is

$$\theta'' = \frac{\rho d \tan \varphi}{R_e} \qquad (19.18)$$

where θ'' is meridian convergence in seconds; d the east–west distance (departure) between the two points in meters; R_e the mean radius of the Earth (6,371,000 m); ϕ the mean latitude of the two points; and ρ the number of seconds per radian (206,265). Because of meridian convergence, forward and back azimuths of long east–west lines do not differ by exactly 180°, but rather by 180° ± θ. (If the

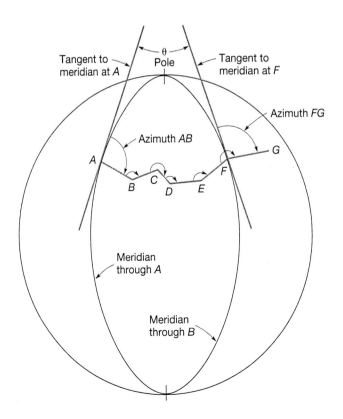

Figure 19.15
Meridian convergence on long east–west traverses.

traverse proceeds in an easterly direction the sign of θ is positive, if it goes westerly θ is negative. A sketch will clarify the situation.) From Equation (19.18) an east–west traverse of 1 mi length at latitude 30° produces a convergence angle of approximately 30″. At latitude 45°, convergence is approximately 51″/mi east–west. These calculations illustrate that the magnitude of convergence can be appreciable and must be considered when astronomic or GNSS observations are made in connection with plane surveys, which assume the y-axis parallel throughout the project area.

Procedures for precise traverse computation vary depending on whether a geodetic or a plane reference coordinate system is used. In either case, it is necessary first to eliminate mistakes and compensate for systematic errors. In the adjustment, closure conditions are enforced for (1) azimuths or angles, (2) departures, and (3) latitudes. The most rigorous process, the least-squares method (see Chapter 16), should be used because it simultaneously satisfies all three conditions and gives residuals having the highest probability.

19.13.3 Trilateration

Trilateration, a method for horizontal control surveys based exclusively on observed horizontal distances, has gained acceptance because of electronic distance measuring capability. Both triangulation and traversing require time-consuming horizontal angle measurement. Hence, trilateration surveys often can be executed faster and produce equally acceptable accuracies.

The geometric figures used in trilateration, although not as standardized, are similar to those employed in triangulation. Stations should be intervisible and are therefore placed in elevated locations, sometimes using towers to elevate instruments and observers when necessary.

Because of intervisibility requirements and the desirability of having essentially square networks, trilateration is ideally suited to densify control in metropolitan areas and on large engineering projects. In special situations where topography or other conditions require elongated narrow figures, reading some horizontal angles can strengthen the network. Also, for long trilateration arcs, astronomic azimuth observations can help prevent the network from deforming in direction.

As in triangulation, surveys by trilateration can be extended from one or more monuments of known position. If only a single station is fixed, at least one azimuth must be known or observed.

Trilateration computations consist of reducing observed slope distances to horizontal lengths, then to the ellipsoid and finally to grid lengths if the calculations are being done in state plane coordinate systems (see Chapter 20). Observational errors in trilateration networks must be adjusted, preferably by the least-squares method.

19.13.4 Combined Networks

With the ability to easily observe both distances and angles in the field, networks similar to that shown in Figure 19.13 are becoming increasingly popular. In a combined network, many or all angles and distances are observed. These surveys provide the greatest geometric strength and the highest coordinate accuracies for traditional survey techniques. As described in Section 19.15, all observations must be corrected to the ellipsoid or a mapping grid (see Chapter 20). The least-squares method as described in Chapter 16 is used to adjust the observations.

19.14 FIELD PROCEDURES FOR VERTICAL-CONTROL SURVEYS

Vertical-control surveys are generally run by either differential leveling or trigonometric leveling. The method selected will depend primarily on the accuracy required, although the type of terrain over which the leveling will be done is also a factor. Differential leveling, described in Section 5.4 produces the highest order of accuracy typically. The GNSS surveys can be used for lower-order vertical control surveys, but to get accurate elevations using this method, *geoidal heights* in the area must be known and applied (see Sections 13.4.3 and 19.5).

Although trigonometric leveling produces a somewhat lower order of accuracy than differential leveling, the method is still suitable for many projects such as establishing vertical control for topographic mapping or for lower-order construction stakeout. It is particularly convenient in hilly or mountainous terrain where large differences in elevation are encountered. Field procedures for trigonometric leveling and methods for reducing the data are discussed in Section 4.5.4.

Differential leveling can produce varying levels of accuracy, depending on the precautions taken. In this section only *precise differential leveling*, which produces the highest quality results, is considered. The procedures for precise leveling are discussed in Section 5.8.

As noted in Section 19.8 and Table 19.6, the FGCS established accuracy standards and specifications for various orders of differential leveling. To achieve the higher orders, special care must be exercised to minimize errors, but the same basic principles apply.

Special level rods are needed for precise work. They have scales graduated on Invar strips, which are only slightly affected by temperature variations. Precise level rods are equipped with rod level bubbles to facilitate plumbing and special braces aid in holding the rod steady. They usually have two separate graduated scales. One type of rod is divided in centimeters on an Invar strip on the rod's front side, with a scale in feet painted on the back for checking readings and minimizing blunders. A second type of rod, shown in Figure 19.16, has two sets of centimeter graduations on the Invar strip with the right one precisely offset from the left by a constant, thereby giving checks on readings.

Cloudy weather is preferable for precise leveling, but an umbrella can be used on sunny days to shade the instrument and prevent uneven heating, which causes the bubble to run. (One design encases the vial in a Styrofoam shield.) Automatic levels are not as susceptible to errors caused by differential heating. Precise work should not be attempted on windy days. For best results, short and approximately equal backsight and foresight distances are recommended.

Figure 19.16
Reticle of a precise level shown with dual metric-scale precise leveling rod.

Table 19.7 lists the maximum sight distances and allowable differences between backsight and foresight lengths for first-, second-, and third-order leveling. Rodpersons can pace or count rail lengths or highway slab joints to set sight distances, which are then checked for accuracy by three-wire stadia methods (see Section 5.8). Precise leveling demands good-quality turning points. Lines of sight should not pass closer than about 0.5 m from any surface, for example, the ground, to minimize refraction. Readings at any setup must be completed in rapid succession; otherwise changes in atmospheric conditions might significantly alter refraction characteristics between them.

Three-wire leveling has been employed for much of the precise work in the United States. In this procedure, as described in Section 5.8, rod readings at the upper, middle, and lower cross hairs are taken and recorded for each backsight and foresight. The difference between the upper and middle readings is compared with that between the middle and lower values for a check, and the average of the three readings is used. A sample set of field notes for three-wire leveling is illustrated in Figure 5.10. When using a digital level, the elevation difference along with the backsight or foresight length can be digitally recorded for each sight. A second technique in precise leveling employs the parallel-plate micrometer attached to a precise leveling instrument and a pair of precise rods like those described earlier.

TABLE 19.7 RECOMMENDED FIELD CONDITIONS FOR PRECISE LEVELING

Order Class	First I	First II	Second I	Second II	Third
Maximum sight length (m)	50	60	60	70	90
Difference between foresight and backsight lengths never to exceed					
per setup (m)	2	5	5	10	10
per section (m)	4	10	10	10	10

576 CONTROL SURVEYS AND GEODETIC REDUCTIONS

It is generally advisable to design large level networks so that several smaller circuits are interconnected. This enables making checks that isolate blunders or large errors. For example in Figure 19.17, it is required to determine the elevations of points X, Y, and Z by commencing from BM A and closing on BM B. As a minimum, running level lines 1 through 4 could do this, but if an unacceptable misclosure were obtained at BM B, it would be impossible to discover which line the blunder occurred. If additional lines 5, 6, and 7 are run, calculating differences in elevation by other routes through the network should isolate the blunder. Furthermore, by including the supplemental observations, precisions of the resulting elevations at X, Y, and Z are increased.

For long lines, one procedure used to help isolate mistakes and minimize field time is to run small loops with approximately five setups between temporary benchmarks. In this procedure as each loop is completed, it is checked for acceptable closure before proceeding forward to the next loop. This procedure increases the number of observations and helps minimize the amount of time that is required to uncover mistakes. Each smaller loop is connected to subsequent loops until the entire network is observed. Figure 19.18 depicts this procedure.

In precise differential leveling, frequent calibration of the leveling instrument is necessary to determine its collimation error. A collimation error exists if, after leveling the instrument, its line of sight is inclined or declined from horizontal. This causes errors in determining elevations when backsight and foresight

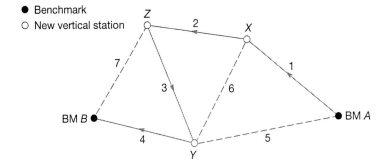

Figure 19.17
Interconnecting level network.

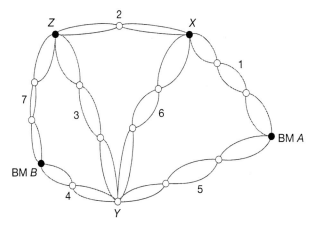

Figure 19.18
Leveling of network performed in small closing loops.

distances are not equal. But they can be eliminated if the magnitude of the collimation error is known.

A method that originated at the National Geodetic Survey can be used to determine the collimation error. It requires a baseline approximately 300 ft (90 m) long. Stakes are set at each end of the line and at two intermediate stations located approximately 20 ft (6 m) and 40 ft (12 m) from the two ends. Figure 19.19 shows an example set of field notes for determining the collimation error and includes a sketch illustrating the baseline layout. With the instrument at station 1, middle wire r_1 and R_1 are observed on stations A and B, respectively. If there were no collimation error, the true elevation difference ΔH from these observations would be $r_1 - R_1$. However, if a collimation error is present, each observation must be corrected by adding an amount proportional to the horizontal distance from the level to the rod. The horizontal distance is measured by the stadia interval (see Section 5.4). Introducing collimation corrections, the true elevation difference ΔH is

$$\Delta H = [r_1 + C(i_1)] - [R_1 + C(I_1)] \qquad (a)$$

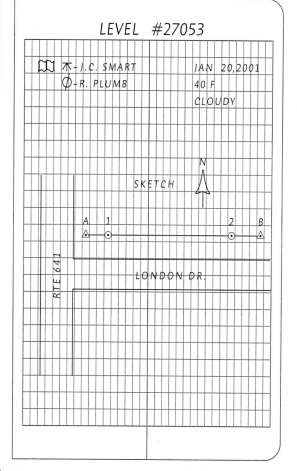

STA	A	INT	B	INT
	4.516		5.567	
1	4.414	0.102	4.644	0.923
	4.312	0.102	3.721	0.923
	r_1	i_1	R_1	I_1
	4.414	0.204	4.644	1.846
	4.699		4.182	
2	3.825	0.874	4.042	0.140
	2.950	0.875	3.901	0.141
	R_2	I_2	r_2	i_2
	3.825	1.749	4.042	0.281

LEVEL CALIBRATION

LEVEL #27053

$$C = \frac{(4.414 + 4.042) - (4.644 + 3.825)}{(1.846 + 1.749) - (0.204 + 0.281)}$$

$$C = \frac{-0.013}{3.110} = -0.0042 \ m/m$$

Figure 19.19 Field notes for determining collimation factor.

In Equation (a), i_1 and I_1 are the stadia intervals (differences between top and bottom wire values) for the rod readings on stations A and B, respectively, and C is the collimation factor (in feet per foot, or meters per meter, of stadia interval). A similar equation for the true elevation difference can be written for rod readings R_2 and r_2 taken on stations A and B, respectively, from station 2, or

$$\Delta H = [R_2 + C(I_2)] - [r_2 + C(i_2)] \qquad \text{(b)}$$

Note that in Equations (a) and (b), uppercase R and I apply to the longer sights, and lowercase r and i are for the shorter sights. Equating the right sides of Equations (a) and (b), and reducing yields

$$C = \frac{(r_1 + r_2) - (R_1 + R_2)}{(I_1 + I_2) - (i_1 + i_2)} \qquad (19.19)$$

As previously noted, the units of the collimation factor calculated by Equation (19.19) are either in feet per feet, or meters per meter, of stadia interval. Computation of the factor is illustrated in Figure 19.19 for the data of the field notes (given in feet).

Because the collimation correction increases linearly with distance, it is unnecessary to apply it to each plus and minus sight. Rather the corrected elevation difference $\Delta H'$ for any loop or section leveled is computed as

$$\Delta H' = \Sigma BS - \Sigma FS + C(\Sigma I_{BS} - \Sigma I_{FS}) \qquad (19.20)$$

where ΣBS is the sum of the middle cross-wire readings of plus sights in the loop or section; ΣFS is the middle cross-wire total of minus sights, and ΣI_{BS} and ΣI_{FS} are the sums of the stadia intervals for the plus sights and minus sights, respectively.

■ Example 19.5

The section from BM A to BM 3 is leveled using the instrument whose collimation factor of -0.00012 m/m of interval was determined in the field notes in Figure 19.18. The sum of the plus sights is 125.590 m, and the sum of the minus sights is 88.330 m. Backsight stadia intervals total 351.52 m, while the sum of foresight intervals is 548.40 m. Find the corrected elevation difference.

Solution

By Equation (19.20)

$$\Delta H' = (125.590 - 88.330) + (-0.00012)(351.52 - 548.40)$$
$$= 37.260 + 0.024 = 37.284 \text{ m}$$

Regardless of precautions taken in field observations, errors accumulate in leveling and must be adjusted to provide perfect mathematical closure. For simple level loops, adjustment procedures presented in Section 5.6 can be followed; for interconnected level networks such as those of Figures 19.17 and 19.18, the method of least squares is preferable. An example least-squares adjustment of an interconnected network is given in Section 16.6.

19.15 REDUCTION OF FIELD OBSERVATIONS TO THEIR GEODETIC VALUES

Traditional surveying instruments, such as levels and total stations, are oriented with respect to the local gravity surface. In geodetic work, since horizontal surveys are referenced to an ellipsoid and vertical surveys to the geoid, corrections must be made to field observations to obtain their equivalent geodetic values. The following subsections discuss some of these corrections. Many of these computations are demonstrated in the Mathcad worksheet *geodobs.xmcd*, which is available on the companion website. Additionally, WOLFPACK, which is available from the same website, has options to perform these computations.

19.15.1 Reduction of Distance Observations Using Elevations

In geodetic control survey computations, observed slope distances (sometimes called *slant* distances) must first be reduced to the surface of the ellipsoid. Observed distances in geodetic surveys are often long, and thus the short-line reduction techniques given in Section 6.13 do not provide satisfactory accuracy. This is especially true for long lines that are steeply inclined.

A procedure for reducing long slope distances to their ellipsoidal (geodetic) lengths is discussed here. The method is based on elevation differences between the end points of the sloping line. In Figure 19.20, an EDM instrument is at A, a reflector is at B, and S is the observed slope distance from A to B. (Assume that S has been corrected for meteorological conditions.) Length D_1 is the "mark-to-mark" distance between stations A and B. Mark-to-mark distances apply for EDM calibration lines, as well as for GNSS baselines. Length D_2 is the arc distance on the ellipsoid, which is known as the *geodetic distance*. It is the length required for most geodetic computations. Distance D_3 is the ellipsoidal chord length between stations A and B.

In Figure 19.20, let $h'_1 = h_1 + hi$ and $h'_2 = h_2 + hr$, where hi and hr are instrument and reflector heights, respectively, above stations A and B, and h_1 and h_2 are the geodetic heights at A and B, respectively. Expressing the relationship of the three sides of triangle ABO using the law of cosines [see Equation (11.2)] gives

$$S^2 = (R_\alpha + h'_1)^2 + (R_\alpha + h'_2)^2 - 2(R_\alpha + h'_1)(R_\alpha + h'_2) \cos \theta \quad (19.21)$$

where R_α is the radius of the Earth in the azimuth of the distance from point A as defined by Equation (19.6), and θ the angle subtended by the verticals from points A and B. Substituting the trigonometric identity of $\cos \theta = 1 - 2\sin^2(\theta/2)$ into Equation (19.21), and expanding yields

580 CONTROL SURVEYS AND GEODETIC REDUCTIONS

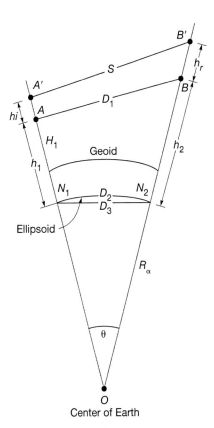

Figure 19.20
Reduction of long lengths to the ellipsoid based on elevations.

$$S^2 = (h_2' - h_1')^2 + 4R_\alpha^2\left(1 + \frac{h_1'}{R_\alpha}\right)\left(1 + \frac{h_2'}{R_\alpha}\right)\sin^2\left(\frac{\theta}{2}\right) \quad (19.22)$$

Substituting $\Delta h' = h_2' - h_1'$ and $D_3 = 2R_\alpha \sin(\theta/2)$ into Equation (19.22), the expression reduces to

$$S^2 = \Delta h'^2 + \left(1 + \frac{h_1'}{R_\alpha}\right)\left(1 + \frac{h_2'}{R_\alpha}\right)D_3^2 \quad (19.23)$$

Solving Equation (19.23) for D_3 yields the following expression for the ellipsoidal chord length:

$$D_3 = \sqrt{\frac{S^2 - \Delta h'^2}{\left(1 + \frac{h_1'}{R_\alpha}\right)\left(1 + \frac{h_2'}{R_\alpha}\right)}} \quad (19.24)$$

The arc length on the ellipsoid (*geodetic distance* or *geodetic length*) can be computed from this chord distance as

$$D_2 = 2R_\alpha \sin^{-1}\left(\frac{D_3}{2R_\alpha}\right) \quad (19.25)$$

Equations (19.24) and (19.25) can be used to compute the distance on any level surface by simply modifying the heights of the endpoints as appropriate. It is important to realize that the unit of the arcsine in Equation (19.25) is radians. To compute the chord distance between two points at different elevations, for example D_1 in Figure 19.19, the following equation is used:

$$D_1 = \sqrt{D_3^2\left(1 + \frac{h_1}{R_\alpha}\right)\left(1 + \frac{h_2}{R_\alpha}\right) + (h_2 - h_1)^2} \quad (19.26)$$

■ Example 19.6

A slope distance of 5000.000 m is observed between two points A and B whose orthometric heights are 451.200 and 221.750 m, respectively. The geoidal undulation at point A is −29.7 m and is −29.5 m at point B. The height of the instrument at the time of the observation was 1.500 m and the height of the reflector was 1.250 m. What are the geodetic and mark-to-mark distances for this observation? (Use a value of 6,386,152.318 m for R_α in the direction AB.).

Solution

By Equation (19.8), the geodetic heights at points A and B are

$$h_A = 451.200 - 29.7 = 421.500 \,\text{m}$$
$$h_B = 221.750 - 29.5 = 192.250 \,\text{m}$$

Thus, $h'_A = 421.500 + 1.500 = 423.000$ m, $h'_B = 192.250 + 1.250 = 193.500$ m, $\Delta h' = 193.500 - 423.000 = -229.500$ m, and by Equation (19.22), the ellipsoidal chord distance D_3 is

$$D_3 = \sqrt{\frac{5000^2 - 229.5^2}{\left(1 + \frac{423.0}{R_\alpha}\right)\left(1 + \frac{193.5}{R_\alpha}\right)}} = 4994.489 \,\text{m}$$

By Equation (19.25), the reduced ellipsoidal arc, or geodetic length, for the line AB is

$$D_2 = 2R_\alpha \sin^{-1}\left(\frac{4994.489}{2R_\alpha}\right) = 4994.489 \,\text{m}$$

Finally by Equation (19.26), the mark-to-mark distance is

$$D_1 = \sqrt{4994.489^2\left(1 + \frac{421.500}{R_\alpha}\right)\left(1 + \frac{192.250}{R_\alpha}\right) + (192.25 - 421.5)^2}$$
$$= 4999.987 \,\text{m}$$

Note that the ellipsoid arc and chord lengths are the same to the nearest millimeter. However as lines become longer, this will not necessarily be the case. Nevertheless, for most geodetic observations, these arc and chord values will generally be nearly the same. Note also that the observed slope distance differs from the mark-to-mark distance by 13 mm. Finally, if the short-line reduction procedure of Section 6.13 had been used, an error of more than 0.2 m would have resulted.

19.15.2 Reduction of Distance Observations Using Vertical Angles

Figure 19.21 illustrates a slope distance S observed from A to B. Points A and B represent an EDM instrument and a reflector, respectively, O is the Earth's center, and R_α its radius in the direction of the azimuth as defined by Equation (19.6). Vertical angles α and β were observed at A and B, respectively. Arc AB_2, which is closely approximated by its chord, is the required horizontal distance. If the short-line reduction procedures given in Section 6.13 were used, horizontal distance AB_1 would result, which would be in error by $B_1 B_2$. Arc $A'B'$ is the required ellipsoid distance.

From Figure 19.20 the following equations can be written to compute required horizontal (chord) distance AB_2

$$\delta = \frac{\alpha - \beta}{2} \tag{19.27}$$

$$AB_1 = S \cos \delta \tag{19.28}$$

$$BB_1 = S \sin \delta \tag{19.29}$$

$$\psi = \frac{AB_1}{R_\alpha + h_A}\left(\frac{180°}{\pi}\right) \text{(approx.)} \tag{19.30}$$

$$B_1 B_2 = BB_1 \tan\left(\frac{\psi}{2}\right) \tag{19.31}$$

$$AB_2 = AB_1 - B_1 B_2 \tag{19.32}$$

Finally, ellipsoidal (chord) length $A'B'$ can be computed from

$$A'B' = AB_2 \left(\frac{R_\alpha}{R_\alpha + h_A}\right) \tag{19.33}$$

where h_A is the ellipsoidal height.

■ Example 19.7

In Figure 19.20, slope distance L and vertical angles α and β were observed as 14,250.590 m, 4°32'18", and −4°38'52", respectively. If the geodetic height at A is 438.4 m, what is distance $A'B'$ reduced to the ellipsoid? (Use the mean radius of 6,371,000 m for R_α.)

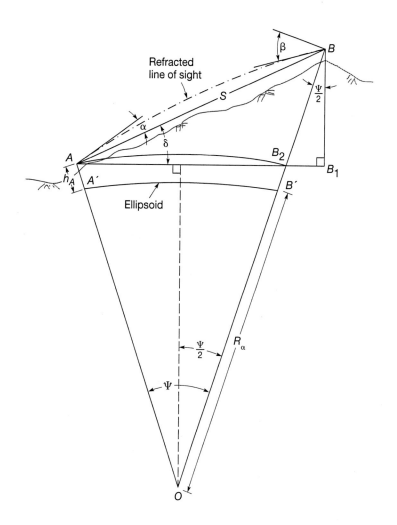

Figure 19.21
Reduction of long lengths to the ellipsoid based on vertical angles.

Solution

Solving Equations (19.27) through (19.33) in sequence

$$\delta = \frac{4°32'18'' - (-4°38'52'')}{2} = 4°35'35''$$

$$AB_1 = 14{,}250.590 \cos 4°35'35'' = 14{,}204.826 \text{ m}$$

$$BB_1 = 14{,}250.590 \sin 4°35'35'' = 1141.160 \text{ m}$$

$$\psi = \frac{14{,}204.826}{6{,}371{,}000 + 438.4}\left(\frac{180°}{\pi}\right) = 0.127738° = 0°07'40''$$

$$B_1B_2 = 1141.160 \tan\left(\frac{0°07'40''}{2}\right) = 1.272 \text{ m}$$

$$AB_2 = 14{,}204.826 - 1.272 = 14{,}203.554 \text{ m}$$

$$A'B' = 14{,}203.554 \frac{6{,}371{,}000}{6{,}371{,}000 + 438.4} = 14{,}202.577 \text{ m}$$

(Note that if the short-line reduction procedure of Section 6.13 had been used, an error equal to B_1B_2 or 1.272 m, would have resulted.)

In the foregoing computations, the effects of refraction were eliminated by averaging reciprocal vertical angles α and β in Equation (19.26). This procedure yields the best results. If the vertical angle were observed at only one end of the line, as angle α at A, then because refraction is approximately 1/7 curvature, or $\psi/7$, a correction for its effect can be applied. In that case instead of using Equation (19.26), angle δ is computed as

$$\delta = \alpha + \frac{\psi}{2} - \frac{\psi}{7} = \alpha + \frac{5\psi}{14} \tag{19.34}$$

where

$$\psi = \frac{S \cos \alpha}{R_e + h_A}\left(\frac{180°}{\pi}\right)$$

Then Equations (19.28), (19.29), and (19.31) through (19.33) are solved as before.

■ Example 19.8

Compute the geodetic (ellipsoidal) length of the line in Example 19.7 using only the observed vertical angle at A.

Solution

$$\psi = \frac{14{,}250.590 \cos 4°32'18''}{6{,}371{,}000 + 438.4}\left(\frac{180°}{\pi}\right) = 0.127748° = 0°07'40''$$

By Equation (19.34)

$$\delta = 4°32'18'' + \frac{5}{14}(0°07'40'') = 4°35'02''$$

Then Equations (19.28), (19.29), and (19.31) through (19.32) are solved in order

$$AB_1 = 14{,}250.590 \cos 4°35'02'' = 14{,}205.008 \text{ m}$$

$$BB_1 = 14{,}250.590 \sin 4°35'02'' = 1138.888 \text{ m}$$

$$B_1B_2 = 1138.888 \tan\frac{0°07'40''}{2} = 1.270 \text{ m}$$

$$AB_2 = 14{,}205.008 - 1.270 = 14{,}203.738 \text{ m}$$

$$A'B' = 14{,}203.738 \frac{6{,}371{,}000}{6{,}371{,}000 + 438.4} = 14{,}202.761 \text{ m}$$

Note that this answer differs by 0.184 m from the one obtained in Example 19.7. This can be expected, because refraction varies with atmospheric conditions, and the correction $\psi/7$ only approximates its effects. Thus, as stated earlier, it is best to measure the vertical angles at both ends of the line when possible.

19.15.3 Reduction of Directions and Angles

As was discussed in Section 19.5, deflection of the vertical varies at different locations on the surface of the Earth. Because of this, during the angle observing process both the horizontal and vertical circles of a total station instrument are, in general, misaligned with the horizontal and vertical of the ellipsoid. Thus, for geodetic calculations, the observed directions must be corrected according to Equation (19.12). Additionally, because of the sphericity of the Earth, the normals at the observing and target stations are skewed with respect to each other, and hence two additional corrections may be necessary. First, if the height of the target above the ellipsoid is substantial enough, this may necessitate a correction. Second, if the latitudes of the occupied and sighted stations differ significantly, this may also require a correction. However, for the target heights and relatively small latitude differences that apply in most surveys, corrections to azimuths for these conditions are often smaller than the observational errors. Thus, except for very precise geodetic work they are not made. Finally, a correction that stems from deflection of the vertical is often significant and the procedures for making it are given next.

Correction for deflection of the vertical Since the instrument is set up with respect to the local vertical, angular measurements in the vertical and horizontal will both be affected by deflection of the vertical. The correction C''_{defl} in an observed horizontal direction in units of arc seconds for deflection of the vertical is

$$C''_{defl} = Az - \alpha = -\eta \tan \varphi - (\xi \sin Az - \eta \cos Az) \cot z \quad \textbf{(19.35)}$$

where Az is the astronomic azimuth of the observed direction, α is the geodetic azimuth, and z is the zenith angle.

Adding the correction determined in Equation (19.35) yields the corrected geodetic azimuth of an observed direction as

$$\alpha = Az_A + C''_{defl} \quad \textbf{(19.36)}$$

For angles, the backsight and foresight directions that compose the angle are each corrected according to Equation (19.36) and subtracted according to Equation (11.11). This is demonstrated in Example 19.9.

Example 19.9

As shown in Figure 19.22, a horizontal angle BIF of $57°44'06''$ is observed at instrument station I, whose latitude is $41°13'24.67''$ as scaled from a topographic map. Orthometric heights were also scaled from a topographic map, and estimated as 613.8 m at backsight station B and 853.9 m at foresight station F. The geoid height is estimated using the NGS software GEOID18 as −29.45 m at station B and −29.84 m at station F. The reflector height hr at both target stations is set to 1.650 m. The azimuth from the observing station to station B is $23°16'24''$. The components of the deflection of the vertical at the observing station are estimated using the NGS software DEFLEC18 as $\eta = 4.82''$ and $\xi = 0.29''$. The geodetic distance IB, as defined by Equation (19.21) is 975.548 m, and for IF it is 883.49 m. The zenith angle to station B is $71°30'56''$ and to station F is $57°21'46''$. What is the corrected geodetic angle BIF?

Solution

In this solution, the individual correction components are independently computed for each sight of the angle and subtracted to determine the corrected angle.

Geodetic heights: Using Equation (19.8), the geodetic heights of h_B at the backsight station, and h_F

$$h_B = 613.8 - 29.45 + 1.65 = 586.00 \text{ m}$$
$$h_f = 853.9 - 29.84 + 1.65 = 825.71 \text{ m}$$

Correction for deflection of the vertical: By Equation (19.35)

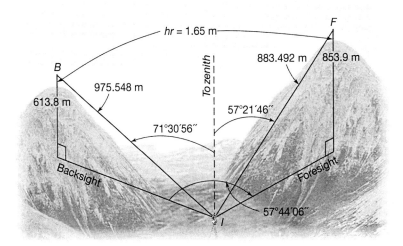

Figure 19.22
Figure for Example 19.9.

Backsight:

$$C''_{defl} = -4.82'' \tan(\varphi) - [0.29'' \sin(Az_B) - 4.82'' \cos(Az_B)] \cot(71°30'56'')$$
$$= -2.78''$$

where ϕ and Az_B are $41°13'24.67''$ and $23°16'24''$, respectively.

Foresight:

$$Az_F = 23°16'24'' + 57°44'06'' = 81°00'30''$$

$$C''_{defl} = -4.82'' \tan(\varphi) - [0.29'' \sin(Az_F) - 4.82'' \cos(Az_F)] \cot(57°21'46'')$$

where Az_F is $81°00'30''$.

Corrected azimuths:

$$\text{Backsight}: 23°16'24'' - 2.78'' = 23°16'21.22''$$

$$\text{Foresight}: 81°00'30'' - 3.92'' = 81°00'26.08''$$

By Equation (11.11), the corrected angle is: $81°00'26.08'' - 23°16'21.22'' = 57°44'04.9''$.

Note that the correction to the angle over these short distances is $-1.1''$. Also note in Equation (19.35) that $\eta \tan \varphi$ is the same for both the backsight and foresight directions of an angle. Thus, it did not need to be included in correction of directions for angles. The correction for deflection of vertical in angles could be rewritten as

$$C''_z = (\xi \sin Az_{FS} - \eta \cos Az_{FS}) \cot z_{FS} - (\xi \sin Az_{BS} - \eta \cos Az_{BS}) \cot z_{BS}.$$

where *FS* represents the foresight values and *BS* represents the backsight values.

For precise geodetic control surveys, a correction due to deflection of the vertical must also be made to observed vertical angles in order to obtain their geodetic equivalents. For zenith angles, the following equation applies:

$$z_C = z_{obs} + \xi \cos Az + \eta \sin Az \qquad (19.37)$$

where z_C is the corrected zenith angle, z_{obs} the observed zenith angle, and Az the azimuth of the line of sight when the zenith angle is observed.

19.15.4 Leveling and Orthometric Heights

Distances observed along plumb lines (elevation differences) between equipotential gravitational surfaces provide the basis for specifying orthometric heights. One of these equipotential surfaces, the geoid, is defined as the datum for observing these heights. For example, in Figure 19.23, the orthometric heights of points A and B are H_A and H_B, respectively.

The Earth spins about its rotational axis in approximately a 24-h period. From physics it is known that this spin results in a centrifugal acceleration that acts on all bodies which take part in the spin. The *gravitational force*[13] is a combination of the attractive force of the Earth's mass, and centrifugal force that acts on a body on the surface of the Earth. Since these two forces are in different directions, their combined effect results in a gravitational force that is less than the attractive force, and is directed radially toward the mass center of the Earth. Thus, points on the equator, which experience the greatest rotational velocity, have the weakest gravitational force. Conversely, points at the poles that are not subject to rotational velocity, do not experience any centrifugal acceleration component, and thus experience the maximum gravitational force. It should be pointed out here that the difference between gravitational force at the pole and at the equator is only about 5 gals.[14]

As stated earlier, an equipotential surface has the same gravitational potential throughout its extent. From physics, potential can be defined as

$$W = mar \tag{19.38}$$

where m is the mass of the body, a the acceleration of the body, and r the radial distance from the mass center of the Earth. In the case of orthometric elevation differences, the acceleration a in Equation (19.38) can be replaced by the gravitational acceleration $-g$ acting on the point,[15] or

$$W = -mgr \tag{19.39}$$

From Equation (19.39) and the previous discussion concerning changes in the values of gravitational force on the Earth, it can be stated that as gravitational force increases, the radial distance must decrease in order that the potential of points on a given equipotential surface remain equal. Thus, equipotential surfaces are not concentric, and instead they converge as they approach the poles, as shown in Figure 19.22.

The difference in potential is defined as

$$\delta W = -g\,\delta l \tag{19.40}$$

where δW is the change in potential and δl the separation between the two equipotential surfaces. A proper expression for the potential of a point on an equipotential surface is the *geopotential number C* of a point, where C is defined as a negative difference in the potential of a point P and the potential of the geoid, and is mathematically defined as

$$C = -(W - W_0) = \int_{P_0}^{P} g\,dl \tag{19.41}$$

---

[13]Force is defined from physics as a product of mass and acceleration.
[14]A gal is a unit of acceleration where 1 gal = 1 cm/sec^2; 1 kgal = 1000 gal; 1 mgal = 0.001 gal.
[15]A negative sign is introduced in Equation (19.37) to account for the fact that gravitational attraction points radially inward while increases in elevation occur radially outward.

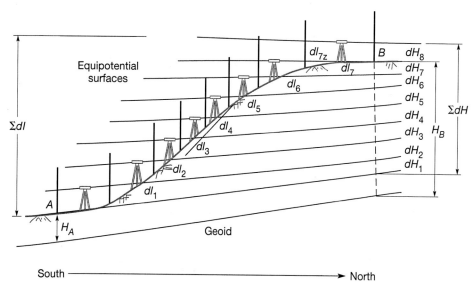

Figure 19.23 Leveled height versus orthometric height.

where W_0 is the potential of a point P_0 on the geoid. The units of the geopotential number are kgal-meters (kgals-m) where 1 kgal-m is called a *geopotential unit* (GPU). In regards to the new vertical datum to be defined by the NGS (see Section 19.6.6), the zero surface will be set coincident with the equipotential surface having 6,263,685.600 GPU. Since neither l nor g is known as a continuous function, in practice the difference in geopotential number is approximated as

$$\Delta C_{ij} = \sum_{k=i}^{j} \bar{g}_k dl_k \qquad (19.42)$$

where $\bar{g}_k$ is the average gravitational attraction between two adjacent benchmarks i and j and δl_k the observed level difference between the two adjacent benchmarks. Because it is impractical to observe g at every benchmark, and since the units of geopotential numbers are unfamiliar to surveyors, *dynamic heights* were introduced. Dynamic heights are the geopotential number divided by a *reference gravity*, or

$$H_i^D = \frac{C_i}{g_R} \qquad (19.43)$$

where H_i^D is the dynamic height of a point, C_i the geopotential number of the point, and g_R the appropriate gravitational constant. In the United States, the NGS has adopted a reference gravity value of 980.6199 gal. The reader should note that the units of dynamic height are traditional distance units. In Figure 19.10, the dynamic height of point F 137 is 252.373 m where the reference gravity is 0.9806199 kgal. Thus, by rearranging Equation (19.43), the geopotential number of point F 137 is 0.9806199(252.373) = 247.482 kgal-m = 247.482 GPU.

Without the inclusion of gravity observations, height differences determined by leveling do not produce true orthometric height differences, and thus a correction must be applied. As can be seen in Figure 19.22, as the leveling process proceeds from station A to B, the equipotential surfaces converge. Letting dl_i represent the leveled height differences and dH_i represent the orthometric height differences between incremental equipotential surfaces, it is apparent that the sum of $dl_1, dl_2, dl_3, dl_4, dl_5, dl_6, dl_7$, and dl_8 (Σdl_i) will not equal the sum of $dH_1, dH_2, dH_3, dH_4, dH_5, dH_6, dH_7$, and dH_8 (ΣdH_i) because of convergence of the equipotential surfaces. The difference between the leveled height difference and the orthometric height difference is called the *orthometric correction*. The orthometric correction O_c is added to the leveled height to obtain the orthometric height, or

$$dH = dl + O_c \tag{19.44}$$

where dH represents the orthometric height difference between two points and dl represents the leveled height difference between the two points.

Because equipotential surfaces converge near the Earth's poles, and orthometric corrections are a function of gravity values observed along the leveling lines, the largest corrections occur along lines that are run in the north–south direction. When running a line of levels between two NSRS benchmarks, it is possible to estimate the orthometric correction from data that is published on the data sheets. Refer again to Figure 19.10, which shows an excerpt from the data sheet for benchmark F 137. Note that the NAVD 88 (orthometric) height is 252.471 m, the modeled gravity value is 980,231.5 mgals, and the modeled geoid height is −32.70 m. To compute the leveled height difference, the potential heights for two control benchmark stations must be computed as

$$H_C(A) = H_A\left(g_A + \frac{0.0424}{1,000,000}H_A\right) \tag{19.45}$$

where $HC(A)$ is the potential height of station A in units of kgal-meters, g_A the modeled gravity value at station A in units of kgals, and H_A the orthometric height of the station. Following this, the difference in the potential heights is computed and divided by the average gravity value for the two benchmarks, or

$$dl = \frac{2[H_C(B) - H_C(A)]}{g_A + g_B} \tag{19.46}$$

■ Example 19.10

Given the following information from the control data sheets for Stations F 137 and J 231, what should be the leveled height difference between stations?

Station	Orthometric Height (m)	Gravity (mgal)
F 137	252.471	980,231.5
J 231	294.548	980,143.5

Solution

By Equation (19.43)

$$H_C(F137) = 252.471\left[0.9802315 + \frac{0.0424}{1{,}000{,}000}252.471\right] = 247.4827\,\text{GPU}$$

$$H_C(J237) = 294.548\left[0.9801435 + \frac{0.0424}{1{,}000{,}000}294.548\right] = 288.7030\,\text{GPU}$$

By Equation (19.46)

$$dl(F,J) = \frac{2(288.7030 - 247.4827)}{0.9802315 + 0.9801435} = 42.053\,\text{m}$$

Note that in Example 19.10, the difference in orthometric heights is 294.548 − 252.471 = 42.077 m, but the leveled height difference is 42.053 m yielding a difference of 2.4 cm. This difference represents the orthometric correction for the leveled line and would be part of the misclosure of the line if this computation is not considered. In this example, Stations F 137 and J 231 are approximately 120 km (74.5 mi) apart in the north–south direction. As can be seen by example, the convergence of the equipotential surfaces is very modest over long distances. Thus, it is only considered in high precision surveys involving long north–south extents.

After applying the orthometric correction, the resultant misclosure in the leveling circuit can be adjusted using least squares. However, for the most precise surveys, the gravity values at the intermediate benchmarks must also be considered. Readers, who wish to learn more on this topic, should consult the references at the end of this chapter.

■ 19.16 GEODETIC POSITION COMPUTATIONS

Geodetic position computations involve two basic types of calculations, the *direct* and the *inverse* problems. In the direct problem, given the latitude and longitude of station A and the geodetic length and azimuth of line AB, the latitude and longitude of station B are computed. In the inverse problem, given the latitudes and longitudes of stations A and B, the geodetic length of AB and its forward and back azimuths are calculated.

For long lines it is necessary to account for the ellipsoidal shape of the Earth in these calculations to maintain suitable accuracy. Many formulas are available for making direct and inverse calculations, some of which are simplified approximations that only apply for shorter lines. This book will present those developed by Vincenty (1975). The procedures presented in the following subsections have a stated accuracy of a few centimeters for lines up to 20,000 km in length. These computations are demonstrated in the Excel spreadsheet *vincenty.xls*, which is available on the companion website for this book.

19.16.1 Direct Geodetic Problem

In the direct problem, φ_1 and λ_1 represent the latitude and longitude, respectively, s the geodetic length from station 1 to station 2, and α_1 the forward azimuth from station 1 to station 2. The variables a, b, and f are the defining parameters of the ellipsoid as presented in Section 19.2. The unknowns in the problem are φ_2 and λ_2, the geodetic latitude and longitude of the sighted station, and α_2 the azimuth of the line from station 2 to station 1. Note that the observations used in this computation must be corrected to the ellipsoid using procedures outlined in Section 19.15.

The computational steps are as follows:[16]

1. $\tan U_1 = (1 - f)\tan \varphi_1$

2. $\tan \sigma_1 = \tan U_1 / \cos \alpha_1$

3. $u = e' \cos \alpha$

4. $\sin \alpha = \cos U_1 \sin \alpha_1$

5. $A = 1 + \dfrac{u^2}{16{,}384}\{4096 + u^2[-768 + u^2(320 - 175u^2)]\}$

6. $B = \dfrac{u^2}{1024}\{256 + u^2[-128 + u^2(74 - 47u^2)]\}$

7. $2\sigma_m = 2\sigma_1 + \sigma$; where the first iteration uses $\sigma = \dfrac{s}{bA}$

8. $\Delta\sigma = B \sin\sigma \left\{\cos(2\sigma_m) + \dfrac{B}{4}\left[\cos\sigma(-1 + 2\cos^2\sigma_m)\right.\right.$
   $\left.\left. - \dfrac{B}{6}\cos(2\sigma_m)(-3 + 4\sin^2\sigma)(-3 + 4\cos^2\sigma_m)\right]\right\}$

9. $\sigma = \dfrac{s}{bA} + \Delta\sigma$

10. Repeat steps 7, 8, and 9 until the $\Delta\sigma$ becomes negligible.

11. $\tan \varphi_2 = \dfrac{\sin U_1 \cos\sigma + \cos U_1 \sin\sigma \cos\alpha_1}{(1-f)\sqrt{\sin^2\alpha + (\sin U_1 \sin\sigma - \cos U_1 \cos\sigma \cos\alpha_1)^2}}$

12. $\tan \lambda = \dfrac{\sin\sigma \sin\alpha_1}{\cos U_1 \cos\sigma - \sin U_1 \sin\sigma \cos\alpha_1}$

13. $C = \dfrac{f}{16}\cos^2\alpha[4 + f(4 - 3\cos^2\alpha)]$

14. $L = \lambda - (1 - C)f\sin\alpha\{\sigma + C\sin\sigma[\cos 2\sigma_m + C\cos\sigma(-1 + 2\cos^2 2\sigma_m)]\}$

15. $\lambda_2 = \lambda_1 + L$

16. $\tan \alpha_2 = \dfrac{\sin\alpha}{-\sin U_1 \sin\sigma + \cos U_1 \cos\sigma \cos\alpha_1}$

---

[16] For the derivation of this formulation, and that of the inverse problem which follows, consult the publication by T. Vincenty cited in this chapter's bibliography.

19.16.2 Inverse Geodetic Problem

In the inverse geodetic problem, φ_1, λ_1 and φ_2, λ_2 represent the latitude and longitude of the first and second stations, respectively. In the solution, the geodetic length, s, between the two points, and the forward and back azimuths of the line, α_1 and α_2, respectively, are determined. The variables a, b, and f again are the defining parameters of the ellipsoid as presented in Section 19.2.

Steps:

1. $L = \lambda = \lambda_2 - \lambda_1$
2. $\tan U_1 = (1 - f) \tan \varphi_1$
3. $\tan U_2 = (1 - f) \tan \varphi_2$
4. Starting with λ equal to L repeat steps 7 and 16 until changes in λ become negligible.
5. $\sin^2 \sigma = (\cos U_2 \sin \lambda)^2 + (\cos U_1 \sin U_2 - \sin U_1 \cos U_2 \cos \lambda)^2$
6. $\cos \sigma = \sin U_1 \sin U_2 + \cos U_1 \cos U_2 \cos \lambda$
7. $\sin \alpha = \cos U_1 \cos U_2 \sin \lambda / \sin \sigma$
8. $\cos 2\sigma_m = \cos \sigma - 2 \sin U_1 \sin U_2 / \cos^2 \alpha$
9. $C = \dfrac{f}{16} \cos^2 \alpha [4 + f(4 - 3 \cos^2 \alpha)]$
10. $s = bA(\sigma - \Delta\sigma)$ where $\Delta\sigma$ comes from the steps 12 to 15 below.
11. $u = e' \cos \alpha$
12. $A = 1 + \dfrac{u^2}{16{,}384}\{4096 + u^2[-768 + u^2(320 - 175u^2)]\}$
13. $B = \dfrac{u^2}{1024}\{256 + u^2[-128 + u^2(74 - 47u^2)]\}$
14. $\Delta\sigma = B \sin \sigma \left\{ \cos(2\sigma_m) + \dfrac{1}{4} B \left[\cos \sigma(-1 + 2 \cos^2 2\sigma_m) \right. \right.$
    $\left. \left. - \dfrac{1}{6} B \cos(2\sigma_m)(-3 + 4 \sin^2\sigma)(-3 + 4 \cos^2 2\sigma_m) \right] \right\}$
15. $\lambda = L - (1 - C)f \sin \alpha \{\sigma + C \sin \sigma[\cos 2\sigma_m$
    $+ C \cos \sigma(-1 + 2 \cos^2 2\sigma_m)]\}$
16. $\tan \alpha_1 = \dfrac{\cos U_2 \sin \lambda}{\cos U_1 \sin U_2 - \sin U_1 \cos U_2 \cos \lambda}$
17. $\tan \alpha_2 = \dfrac{\cos U_1 \sin \lambda}{-\sin U_1 \cos U_2 + \cos U_1 \sin U_2 \cos \lambda}$

The software WOLFPACK on the companion can be used to do both computations. Figure 19.24 shows the data entry screen for the direct geodetic

Figure 19.24
Data entry screen for forward computation from WOLFPACK.

problem. A similar data screen is available in WOLFPACK to compute the inverse geodetic problem. An Excel spreadsheet and Mathcad worksheet that demonstrate these computations are also provided on the companion website.

To simplify position calculations for long lines, and yet maintain geodetic accuracy, state plane coordinate systems have been developed. These are described in Chapter 20.

■ 19.17 THE LOCAL GEODETIC COORDINATE SYSTEM

GNSS surveys (see Chapters 13, 14, and 15) yield three-dimensional baseline components (ΔX, ΔY, ΔZ). These vector components are in the geocentric coordinate system (see Section 13.4.3). It is a common practice to transform these geocentric coordinate vector components into a *local geodetic system of easting* (Δe), *northing* (Δn), and *local up* (Δu). The two coordinate systems are illustrated in Figure 19.25, where XYZ represents the geocentric system and e, n, u is the local geodetic system. The local geodetic system is user oriented in that the e and n axes are in a local horizontal plane (u is coincident with the normal at the origin of the local coordinate system) and n is in the direction of local north.

To perform a transformation from the geocentric coordinate system to local geodetic, a set of three-dimensional rotation matrices must be employed. These rotation matrices are similar to their two-dimensional counterparts as shown in Equation (11.40). In the transformation process, rotations occur about each of the three coordinate axes. Letting the rotation angle around the X-axis be θ_1, the rotation angle around the Y-axis be θ_2, and the rotation angle around the Z-axis be θ_3, the three-dimensional rotation matrices are

$$R_X(\theta_1) = \begin{bmatrix} 1 & 0 & 0 \\ 0 & \cos\theta_1 & \sin\theta_1 \\ 0 & -\sin\theta_1 & \cos\theta_1 \end{bmatrix} \quad (19.47)$$

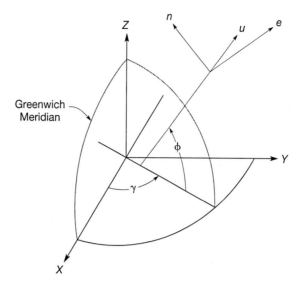

Figure 19.25
The relationship between the geocentric coordinate system and the local geodetic coordinate system.

$$R_Y(\theta_2) = \begin{bmatrix} \cos\theta_2 & 0 & -\sin\theta_2 \\ 0 & 1 & 0 \\ \sin\theta_2 & 0 & \cos\theta_2 \end{bmatrix} \qquad (19.48)$$

$$R_Z(\theta_3) = \begin{bmatrix} \cos\theta_3 & \sin\theta_3 & 0 \\ -\sin\theta_3 & \cos\theta_3 & 0 \\ 0 & 0 & 1 \end{bmatrix} \qquad (19.49)$$

As shown in Figure 19.25, a rotation about the Z-axis by an amount of $\lambda - 270°$ must occur to align the X-axis with the local e-axis. Following this rotation, the once-rotated Z-axis is brought into coincidence with the u-axis by a rotation of $90° - \varphi$ about the once rotated X-axis.[17] The resultant expression is

$$\begin{bmatrix} \Delta e \\ \Delta n \\ \Delta u \end{bmatrix} = R_x(90° - \varphi) R_Z(\lambda - 270°) \begin{bmatrix} \Delta X \\ \Delta Y \\ \Delta Z \end{bmatrix} \qquad (19.50)$$

Performing the proper trigonometric substitutions, and rearranging the equations to place them in the standard order of (n, e, u), the final transformation equations are

---

[17] A three-dimensional animation viewable in a web browser having the appropriate plug-in demonstrates the rotation of the local geodetic coordinate system axes into a geocentric coordinate system. The animation *LOC2GEO.wrl* is available on the companion website.

$$\begin{bmatrix} \Delta n \\ \Delta e \\ \Delta u \end{bmatrix} = \begin{bmatrix} -\sin\varphi & 0 & \cos\varphi \\ 0 & 1 & 0 \\ \cos\varphi & 0 & \sin\varphi \end{bmatrix} \begin{bmatrix} \cos\lambda & \sin\lambda & 0 \\ -\sin\lambda & \cos\lambda & 0 \\ 0 & 0 & 1 \end{bmatrix} \begin{bmatrix} \Delta X \\ \Delta Y \\ \Delta Z \end{bmatrix}$$

$$= \begin{bmatrix} -\sin\varphi\cos\lambda & -\sin\varphi\sin\lambda & \cos\varphi \\ -\sin\lambda & \cos\lambda & 0 \\ \cos\varphi\cos\lambda & \cos\varphi\sin\lambda & \sin\varphi \end{bmatrix} \begin{bmatrix} \Delta X \\ \Delta Y \\ \Delta Z \end{bmatrix} = R(\varphi,\lambda) \begin{bmatrix} \Delta X \\ \Delta Y \\ \Delta Z \end{bmatrix}$$

(19.51)

19.18 THREE-DIMENSIONAL COORDINATE COMPUTATIONS

Sometimes it is advantageous to compute three-dimensional changes in a local geodetic coordinate system from reduced field observations. In Figure 19.26, the geodetic values for the observations of azimuth Az, ellipsoid distance s, and altitude angle α can be used to derive the changes in the local geodetic coordinate system as

$$\begin{bmatrix} \Delta n \\ \Delta e \\ \Delta u \end{bmatrix} = \begin{bmatrix} -\sin\varphi & 0 & \cos\varphi \\ 0 & 1 & 0 \\ \cos\varphi & 0 & \sin\varphi \end{bmatrix} \begin{bmatrix} \cos\lambda & \sin\lambda & 0 \\ -\sin\lambda & \cos\lambda & 0 \\ 0 & 0 & 1 \end{bmatrix} \begin{bmatrix} \Delta X \\ \Delta Y \\ \Delta Z \end{bmatrix}$$

$$= \begin{bmatrix} -\sin\varphi\cos\lambda & -\sin\varphi\sin\lambda & \cos\varphi \\ -\sin\lambda & \cos\lambda & 0 \\ \cos\phi\cos\lambda & \cos\varphi\sin\lambda & \sin\varphi \end{bmatrix} \begin{bmatrix} \Delta X \\ \Delta Y \\ \Delta Z \end{bmatrix}$$

$$= R(\varphi,\lambda) \begin{bmatrix} \Delta X \\ \Delta Y \\ \Delta Z \end{bmatrix}$$

(19.52)

Figure 19.26 Reduction of observations in a local geodetic coordinate system. (Courtesy National Geodetic Survey.)

Equations (19.52) can be modified by Equation (8.2) to incorporate a zenith angle z as

$$\Delta n = s \sin z \cos Az$$
$$\Delta e = s \sin z \sin Az \quad (19.53)$$
$$\Delta u = s \cos z$$

From Figure 19.25, the following inverse relationships can be developed:

$$s = \sqrt{\Delta n^2 + \Delta e^2 + \Delta u^2}$$
$$Az = \tan^{-1}\frac{\Delta e}{\Delta n} \quad (19.54)$$
$$\alpha = \sin^{-1}\frac{\Delta u}{\Delta s} \text{ or } z = \cos^{-1}\frac{\Delta u}{\Delta s}$$

By combining Equation (19.51) with Equation (19.54), the geodetic values for the observations can be obtained directly from changes in geocentric coordinates as

$$s = \sqrt{\Delta X^2 + \Delta Y^2 + \Delta Z^2}$$

$$Az_1 = \tan^{-1}\left(\frac{-\Delta X + \Delta Y^2 + \Delta Z^2}{-\Delta X \sin\varphi_1 \cos\lambda_1 - \Delta Y \sin\varphi_1 \sin\lambda_1 + \Delta Z \cos\varphi_1}\right)$$

$$z_1 = \cos^{-1}\frac{\Delta X \cos\varphi_1 \cos\lambda_1 + \Delta Y \cos\varphi_1 \sin\lambda_1 + \Delta Z \sin\varphi_1}{s} \quad (19.55)$$

It is important to note that Equation (19.55) uses the latitude and longitude of the observation station P_1 in Figure 19.25. These values can be computed from the geocentric coordinate values of P_1 based on Equations (13.3) through (13.7).

■ Example 19.11

The mark-to-mark distance from station Bill to station Red is 568.138 m, and the zenith angle and azimuth of this course are 92°14′25″ and 40°36′23″, respectively. If the geodetic coordinates of station Bill are 61°10′42.1058″ N latitude and 149°11′12.1033″ W longitude, what are the changes in the geocentric coordinate system?

Solution

Using Equation (19.53), the changes in the local coordinate system are

$$\Delta n = 568.138 \sin(92°14'25'') \cos(40°36'23'') = 431.000 \text{ m}$$
$$\Delta e = 568.138 \sin(92°14'25'') \sin(40°36'23'') = 369.495 \text{ m}$$
$$\Delta u = 568.138 \cos(92°14'25'') \qquad\qquad\quad = -22.209 \text{ m}$$

To solve for the changes in geocentric coordinates in Equation (19.51), the inverse of $R(\varphi, \lambda)$ must be determined. Since $R(\varphi, \lambda)$ is an orthogonal rotation matrix, its inverse is $R^T(\varphi, \lambda)$. Thus, the changes in geocentric coordinates for these observations are

$$\begin{bmatrix} \Delta X \\ \Delta Y \\ \Delta Z \end{bmatrix} = \begin{bmatrix} -\sin \varphi \cos \lambda & -\sin \lambda & \cos \varphi \cos \lambda \\ -\sin \varphi \sin \lambda & \cos \lambda & \cos \varphi \sin \lambda \\ \cos \varphi & 0 & \sin \varphi \end{bmatrix} \begin{bmatrix} 431.0000 \\ 369.4950 \\ -22.2087 \end{bmatrix}$$

$$= \begin{bmatrix} 522.773 \\ -118.425 \\ 188.321 \end{bmatrix}$$

Note that the changes in geodetic coordinates computed in Example 19.11 should closely agree with GNSS baseline vector observations that would be obtained if the baseline between Bill and Red was observed. Equations (19.53) can be used to determine the mark-to-mark distance, geodetic azimuth, and zenith angle of the observations that would be obtained if a survey were conducted at station Bill. An advantage of local coordinate systems is that unlike plane coordinate systems as presented in Chapter 20, all three dimensions are simultaneously computed using standard observations collected with a total station. Additionally, these local systems can be integrated with GNSS observations for simultaneous GNSS and terrestrial least-squares adjustments. However, it should always be remembered that the computed values would differ from the field-observed values if the appropriate corrections discussed in Section 19.14 were not applied.

19.19 SOFTWARE

In this chapter, an introductory discussion of geodetic computations was presented, horizontal and vertical reference frames used in the United States were briefly discussed, and transformations between surveyed, geocentric, and geodetic observations presented. WOLFPACK, which is available on the companion, contains options to perform many of the computations presented in this chapter. Figure 19.23 shows the entry screen in WOLFPACK for the forward geodetic problem. Readers wishing to see these reductions in a higher-level programming language should explore the Mathcad worksheets, which are on the companion website. Readers who wish to explore these topics in more detail should refer to the bibliography at the end of this chapter.

PROBLEMS

Asterisks (*) indicate problems that have answers given in Appendix G.

19.1 What do horizontal and vertical control surveys establish?

19.2 Define the geoid?

19.3 Draw a sketch showing geodetic latitude and longitude.

19.4* What is the difference between the equatorial circumference of the Clarke 1866 ellipsoid and that of the WGS 84 ellipsoid?

19.5 Determine the first and second eccentricities for the WGS 84 ellipsoid.

19.6 Why is it necessary to define the terrestrial pole?

19.7 The observed azimuth of a line from a station with geodetic coordinates of (N 41°23′14.09274″, W 78°53′42.21864″, 215.234 m) was 12°06′48.6″. At the time of the observation, the instantaneous position of the pole was (0.014067″, 0.290203″). What is the astronomical azimuth of the line?

19.8 What are the radii in the meridian and prime vertical for a station with latitude 23°48′36.84098″ using the GRS80 ellipsoid?

19.9 For the station listed in Problem 19.8, what is the radius of the great circle at the station that is at an azimuth of 316°27′01″ using the GRS80 ellipsoid?

19.10* What are the radii in the meridian and prime vertical for a station with latitude 42°37′26.34584″ using the GRS80 ellipsoid?

19.11 For the station listed in Problem 19.9, what is the radius of the great circle at the station that is at an azimuth of 4°38′26″ using the GRS80 ellipsoid?

19.12* The orthometric height at Station Y 927 is 304.517 m, and the modeled geoid height at that station −31.893 m. What is its geodetic height?

19.13 The geodetic height at Station Apple is 56.947 m, and the modeled geoid height is 34.598 m. What is its orthometric height?

19.14 The orthometric height of a benchmark is 845.37 ft, and the modeled geoid height at the station is −30.873 m. What is the geodetic height of the benchmark? Draw a sketch depicting the geoid, ellipsoid, and benchmark.

19.15* The deflection of the vertical components ξ and η are −2.85″ and −5.94″, respectively, at a station with geodetic coordinates of (29°37′23.0823″ N, 108°56′01.0089″ W). The observed zenith angle is 42°36′58.8″ along a line with azimuth of 88°52′36.7″. What is the geodetic zenith angle?

19.16 Repeat Problem 19.15 with ξ and η are −3.06″ and 5.24″, respectively, and an observed zenith angle of 49°26′33″ along a line with an azimuth of 115°36′09″. What is the geodetic zenith angle?

19.17 What is a Laplace station?

19.18 Using the data from Problem 19.7, what is the geodetic azimuth of the line where the zenith angle is 48°56′43″ and the deflection of the vertical components, ξ and η are −3.06″ and 5.24″, respectively?

19.19* Given the following information for stations JG00050 and KG0089, what should be the leveled height difference between them?

Station	Height (m)	Gravity (mgal)
JG0050	474.442	979,911.9
KG0089	440.552	979,936.2

19.20 Similar to Problem 19.19 except that the station data for R 164 and C 65 near Fairbanks, AK.

Station	Height (m)	Gravity (mgal)
R 164	453.699	982,192.3
C 65	185.018	982,197.2

19.21 Similar to Problem 19.19 except that the station data for W 368 and C 593 near Gainesville, FL.

Station	Height (m)	Gravity (mgal)
W 368	47.853	979,329.6
C 593	32.110	979,245.5

19.22* A slope distance of 2458.663 m is observed between stations *Gregg* and *Brian*, whose orthometric heights are 458.966 m and 566.302 m, respectively. The geoid heights are −25.66 and −25.06 m at *Gregg* and *Brian*, respectively. The height of the instrument at station *Gregg* at the time of the observation was 1.525 m and the height of the reflector at station *Brian* was 1.603 m. What are the geodetic and mark-to-mark distances for this observation? (Use an average radius for the Earth of 6,371,000 m for R_α.)

19.23 If the latitude of station *Gregg* in Problem 19.22 was 28°15′32.46158″ and the azimuth of the line was 25°36′29″, what are the geodetic, and mark-to-mark distances for this observation? (Use the GRS80 ellipsoid.)

19.24 A slope distance of 3854.698 m is observed between two stations *A* and *B* whose orthometric heights are 523.487 m and 226.097 m, and geoid heights are −30.087 m and −30.125 m, respectively. The height of the instrument at the time of the observation was 1.544 m, and the height of the reflector was 2.000 m. The latitude of Station *A* is 41°26′35.10876″ and the azimuth of *AB* is 283°26′02″. What are the geodetic, and mark-to-mark distances for this observation?

19.25 Define what is meant by dynamic height.

19.26* Compute the back azimuth of a line 5863 m long at a mean latitude of 45°01′32.0654″ whose forward azimuth is 88°16′33.2″ from north. (Use an average radius for the Earth of 6,371,000 m.)

19.27 Compute the back azimuth of a line 4876.313 m long at a mean latitude of 39°26′48.07891″ whose forward azimuth is 105°32′27″ from north. (Use an average radius for the Earth of 6,371,000 m.)

19.28 In Figure 19.14, azimuth of *AB* is 102°36′20″ and the angles to the right observed at *B*, *C*, *D*, *E*, and *F* are 132°01′05″, 241°45′12″, 141°15′01″, 162°09′24″, and 202°33′19″, respectively. An astronomic observation yielded an azimuth of 82°24′03″ for line *FG*. The mean latitude of the traverse is 39°58′20″, and the total departure between points *A* and *F* was 24,986.26 ft. Compute the angular misclosure and the adjusted angles. (Assume the angles and distances have already been corrected to the ellipsoid.)

19.29 In Figure 19.19 slope distance *S* is observed as 5208.643 m. The orthometric elevations of points *A* and *B* are 268.794 m and 286.349 m, respectively, and the geoid heights at both stations is −32.846 m. The instrument and reflector heights were both set at 1.500 m. Calculate geodetic distance *A′B′* (Use an average radius for the Earth of 6,371,000 m.)

19.30 In Figure 19.20, slope distance S and zenith angle α at station A were observed as 2072.33 m and 80°56′07″, respectively, to station B. If the elevation of station A is 435.967 m and the geoid height at stations A and B are both −28.04 m, what is ellipsoid length $A'B'$? (Use an average radius for the Earth of 6,371,000 m.)

19.31* Components of deflection of the vertical at an observing station of latitude 43°15′47.5864″ are $\xi = -6.87''$ and $\eta = -3.24''$ If the observed zenith angle on a course with an astronomic azimuth of 204°32′44″ is 80°56′07″, what are the azimuth and zenith angles corrected for deviation of the vertical?

19.32 At the same observation station as for Problem 19.31, the observed zenith angle on a course with an azimuth of 16°33′42″ is 82°01′15″, what are the azimuth and zenith angles corrected for deviation of the vertical?

19.33 What is a Euler pole?

19.34 What is the difference between the Helmert transformation and the Horizontal Time-Dependent Positioning (HTDP) software?

19.35 What advantages does the two-plus-one approach have over the Helmert transformation?

19.36 Compute the collimation correction factor C for the following field data, taken in accordance with the example and sketch in the field notes of Figure 19.18. With the instrument at station 1, high, middle, and low crosshair readings were 6.334, 5.501, and 4.668 ft on station A and 4.978, 3.728, and 2.476 ft on station B. With the instrument at station 2, high, middle, and low readings were 7.304, 6.053, and 4.803 ft on A and 5.111, 4.279, and 3.446 ft on B.

19.37 A leveling instrument having a collimation factor of 0.00005 m/m of interval was used to run a section of three-wire differential levels from BM A to BM B. Sums of backsights and foresights for the section were 2506.837 m and 2538.464 m, respectively. Backsight stadia intervals totaled 3408.9, while the sum of foresight intervals was 2486.7. What is the corrected elevation difference from BM A to BM B?

19.38 Discuss the advantages of breaking a long leveling line into shorter loops.

19.39 The baseline components of a GNSS baseline vector observed at a station A in meters are (808.364, 264.879, 286.497). The geodetic coordinates of the first base station are 36°17′53.08349″ N latitude and 89°23′04.64831″ W longitude. What are the changes in the local geodetic coordinate system of $(\Delta n, \Delta e, \Delta u)$ in meters?

19.40 In Problem 19.39, what are the slant distance, zenith angle, and azimuth for the baseline vector?

19.41 If the slant distance between two stations is 456.982 m, the zenith angle between them is 82°15′22″ and the azimuth of the line is 304°14′54″, what are the changes in the local geodetic coordinates?

BIBLIOGRAPHY

Doyle, D., et al. 2012. "Definition and Desification of the Puerto Rico Vertical Datum of 2002." *Surveying and Land Information Science* 72 (No. 3): 109.

Evan-Tzur, G. 2011. "Updating the Semi-Dynamic Datum of Israel." *Surveying and Land Information Science* 71 (No. 2): 41.

Ghilani, C. 2014. "Geodetic Observations, Part 1." *Professional Surveyor* 34 (No. 5): 36.

_____. 2014. "Geodetic Observations, Part 2." *Professional Surveyor* 34 (No. 3): 36.

_____. 2014. "Low-Distortion Map Projections." *XYHT* 1 (No. 6): 37.

_____. 2014. "Transformations and Reference Frames, Part 2." *Professional Surveyor* 34 (No. 1): 34.

_____. 2014. "Type of Distances for Geodetic Observations." *XYHT* 1 (No. 2): 42.

_____. 2014. "Using a Project Factor." *XYHT* 2 (No. 3): 41.

Heiskanen, W. A. and H. Moritz. 1967. *Physical Geodesy*. San Francisco, CA: W. H. Freeman and Co.

Leenhouts, P. (1985). On the Computation of Bi-Normal Radial Error. Navigation 32 (No. 1): 16.

Maly, S., J. Seago, N. Pavlis, P. Seidelmann, and G. Kaplan. 2015. "Why the Greenwich Meridian Moved." *Journal of Geodesy* 89: 1263. https://doi.org/10.1007/s00190-015-0844-y.

Meyer, T., et al. 2010. "Assessing the Accuracy of Geoid03 Geoid09 in Connecticut." *Surveying and Land Information Science* 70 (No. 2): 89.

Naberezny, B. 2014. "National Surveyors Week 2014." *Professional Surveyor* 34 (No. 5): 8.

Nicholson, H. 2015. "Feeling the Effects of Glacial Isostatic Adjustment." *Surveying and Land Information Science* 74 (No. 1): 37.

Olaleye, J., et al. 2010. "Establishment of an Empirical Geoid Model for Small Geographic Area: A Case Study of Port Harcourt, Nigeria." *Surveying and Land Information Science* 70 (No. 1): 39.

Silver, M. 2015. "Geodetic Preppers: Surveying the Next OPUS Disaster." *The American Surveyor* 12 (No. 3): 47.

Smith, D., et al. 2015. "Recent Activities at the National Geodetic Survey: GRAV-D." *The American Surveyor* 12 (No. 2): 22.

_____. 2015. "Recent Activities at the National Geodetic Survey: Updating NGS 58 and NGS 59." *The American Surveyor* 12 (No. 3): 22.

Smith, D. A. and D. R. Doyle. 2006. "The Future Role of Geodetic Datums in Control Surveying in the United States." *Surveying and Land Information Science* 66 (No. 2): 101.

Sutkowski, S. 2010. "Defining the Geoid Using the Stokes Formula." *Surveying and Land Information Science* 70 (No. 1): 49.

Véronneau, et al. 2014. "Canada's New Vertical Datum: CGVD2013, Part 1." *XYHT* 1 (No. 4): 43.

_____. 2014. "Canada's New Vertical Datum: CGVD2013, Part 2." *XYHT* 1 (No. 5): 41.

Vincenty, T. 1975. "Direct and Inverse Solutions of the Geodesics on the Ellipsoid with Application of Nested Equations." *Survey Review* 22 (No. 176): 88.

20
State Plane Coordinates and Other Map Projections

■ 20.1 INTRODUCTION

Most surveys of small areas assume that the Earth's surface is a plane. However as explained in Chapter 19 for large-area surveys, it is necessary to consider Earth curvature. Unfortunately, the calculations necessary to determine geodetic positions from survey observations and get distances and azimuths from them are lengthy, and practicing surveyors often are not familiar with these procedures. Clearly, a system for specifying positions of geodetic stations using plane rectangular coordinates is desirable, since it allows computations to be made using simple coordinate geometry formulas, such as those presented in Chapter 11. The National Geodetic Survey (NGS) met this need by developing a state plane coordinate system for each state.

The state plane coordinate system is a map projection that provides a common datum of reference for horizontal control of all surveys in a large area, just as the geoid furnishes a single datum for vertical control. It eliminates having individual surveys based on different assumed coordinates, unrelated to those used in other adjacent work. State plane coordinates are published for all control points in the National Spatial Reference System (NSRS) and for many other control points as well. They are widely used as the reference points for initiating surveys of all types, including those for highway construction projects, property boundary delineation, and photogrammetric mapping.

There are many examples illustrating the value of state plane coordinates. They make it possible for extensive surveys on highway projects to begin on one control station and close on another that is tied to the same coordinate system. On boundary surveys, if a parcel's corners are referenced to the state plane coordinate system, their locations are indestructible. The iron pipes, posts, or other monuments marking their positions may disappear, but their original locations can be restored from surveys initiated at other nearby monuments referenced to the state plane coordinate system or by GNSS surveys. For this reason, some states require that state plane coordinates be included on all new subdivisions. State plane coordinates are highly recommended as the reference frame for entering maps and other data into Land and Geographic Information Systems. This allows all data to be referenced to a common system, and thus can be accurately registered and overlaid for analysis purposes.

20.2 PROJECTIONS USED IN STATE PLANE COORDINATE SYSTEMS

To convert geodetic positions of a portion of the Earth's surface to plane rectangular coordinates, points are projected mathematically from the ellipsoid to some imaginary *developable surface*—a surface that can be developed or "unrolled and laid out flat." A rectangular grid can be superimposed on the *developed* plane surface, and the positions of points in the plane specified with respect to X and Y grid axes. A plane grid developed using this mathematical process is called a *map projection*.

There are several types of map projections, with the oldest known projections dating back to the times of the ancient Greeks. Today, two of the most commonly used map projections are the *Lambert conformal conic* and the *Transverse Mercator* projections. Johann Heinrich Lambert initially developed both of these projections. The Transverse Mercator projection was further developed and redefined for the ellipsoid by Carl Friedrich Gauss in 1825 and Johann Heinrich Louis Krüger in 1912, and thus is also known as the *Gauss–Krüger* projection. These two projections are used in state plane coordinate systems. The Lambert conformal conic projection utilizes an imaginary cone as its developable surface, and the Transverse Mercator employs an imaginary cylinder. These are shown in Figure 20.1(a) and (b). The cone and cylinder are usually secant to the ellipsoid in the state plane coordinate systems; that is, they intersect the ellipsoid along two arcs AB and CD as shown. With this placement, the conical and cylindrical surfaces conform better to the ellipsoid over larger areas than they would if placed tangent. However, as warned by Snyder (1987, p. 5), doing this means that secant conic projections may not be geometrically correct. This will be discussed in more detail in Section 20.13.1.

Figure 20.1(c) and (d) illustrate plane surfaces "developed" from the cone and cylinder. Here, points are projected mathematically from the ellipsoid to the surface of the imaginary cone or cylinder based on their geodetic latitudes and longitudes. For discussion purposes only, this may be considered a radial projection from the Earth's center. Figure 20.2 illustrates this process diagrammatically and displays the relationship between the length of a line on the ellipsoid and its extent when projected onto the surface of either a cone or a cylinder. Note that distance $a'b'$ on the projection surface is greater than ab on the ellipsoid, and similarly $g'h'$ is longer than gh. From this observation it is clear that map projection scale is larger than true ellipsoid scale where the cone or cylinder is outside the ellipsoid. Conversely distance $d'e'$ on the projection is shorter than de on the ellipsoid, and thus map scale is smaller than true ellipsoid scale when the projection surface is inside the ellipsoid. Points c and f occur at the intersection of the projection and ellipsoid surfaces, and therefore map scale equals true ellipsoid scale

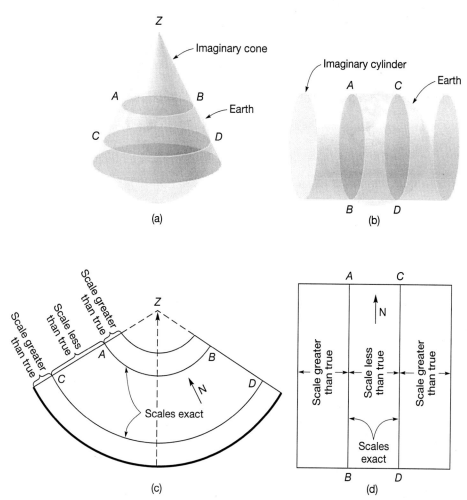

Figure 20.1 Surfaces used in state plane coordinate systems.

along the lines of intersection. These relationships of map scale to true ellipsoid scale for various positions on the two projections are indicated in Figure 20.1(c) and (d). As will be discussed later, these length differences are accounted for by means of a *scale factor*.

From the foregoing discussion it should be clear that points could not be projected from the ellipsoid to developable surfaces without introducing distortions in the lengths of lines or the shapes of areas. However, these distortions are held to a minimum by selected placement of the cone or cylinder secant to the ellipsoid, by choosing a *conformal projection* (one that preserves true angular relationships around points in a limited region), and also by limiting the zone size or extent of coverage on the Earth's surface for any particular projection. If the width of zones is held to a maximum of 158 mi (254 km), and if approximately two thirds of this zone width is between the secant lines, distortions (differences in line lengths on the two surfaces, the reference ellipsoid and the grid) are kept to 1 part in 10,000 or less. The NGS intended this accuracy in its development of the state plane coordinate systems for most state zones. For small states such as Connecticut and Delaware, one state plane coordinate zone is sufficient to

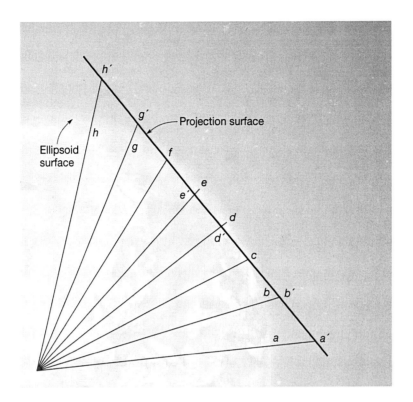

Figure 20.2
Method of projection.

Figure 20.3
Coverages and overlap of zones in Pennsylvania's Lambert conformal conic state plane coordinate system.

cover the entire state. Larger states require several zones to encompass them; for example, Alaska has 10, California 6, and Texas 5. Where multiple zones are needed to cover a state, the 1:10,000 design for adjacent zones overlap each other. As explained in Section 20.10, this is important when lengthy surveys extend from one zone to another. Figures 20.3 and 20.4 show the coverage of zones in Pennsylvania and Indiana, respectively. Both states have two zones, with Pennsylvania using the Lambert conformal conic projection and Indiana the Transverse Mercator projection.

When working with state plane coordinates, it is important to properly apply the appropriate conversion factor when converting coordinates values from

20.2 Projections Used in State Plane Coordinate Systems 607

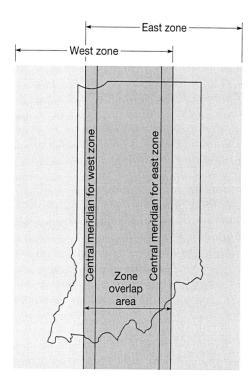

Figure 20.4
Coverages and overlap of zones in Indiana's transverse Mercator state plane coordinate system.

metric to English units or vice versa. Depending on the state, different definitions of the foot were legislated for SPCS83. Six states, Arizona, Michigan, Montana, North Dakota, Oregon, and South Carolina use the international foot (ift) definition. Four states, Alabama, Alaska, Hawaii, and Missouri did not include a definition of the foot in their legislation; Alabama and Hawaii have no SPCS83 legislation at all. The remaining states all adopted the survey foot definition (sft). Two SPCS83 zones are also defined for territories, one for Puerto Rico and the U.S. Virgin Islands, and one for Guam. Legislation for both of these zones does not specify the linear unit.

It should be noted that in 2022, the NGS is deprecating the survey foot definition, which means it will no longer be used on their datasheets. Thus, all English units for coordinates will be given using the international foot definition and simply be listed as feet (ft). This change does not affect typical boundary surveys where instrumentation and software is already using the international foot definition. However, when converting large numbers, such as state plane coordinates, the difference in the definition of the foot can result in large discrepancies. Due to the size of state plane coordinates, it is important to always use exact conversions for the foot. For example, when converting a coordinate value of 650,000.000 m to survey feet one should always use the value of 39.37/12 in states that adopted the survey foot definition. A common mistake by beginners is to always use the truncated value of 3.2808 in the conversion. Doing this results an error in the coordinate in English units of almost 22 sft where the correctly converted coordinate is 2,132,541.67 sft compared to the incorrect value of 2,132,520.00 sft. This can be explained by significant figures (see Chapter 3) where the value to convert has nine significant figures but the conversion factor has only five. The exact conversion for states using the international foot definition is 0.3048 ift/m. Thus, the proper conversion of 650,000.000 m to international feet is 650,000.000/0.3048 = 2,132,545.932 ift. It is important to note that this book always uses the survey foot definition when converting units from the metric system (SI) to English units or vice versa. However, the reader should adopt the definition that is appropriate for their location and time of the survey.

20.3 LAMBERT CONFORMAL CONIC PROJECTION

The Lambert conformal conic projection, as its name implies, is a projection onto the surface of an imaginary cone. The term *conformal*, as noted earlier, means that true angular relationships are retained around all points in a limited region. More specifically, the angles at the intersection of two curves on a reference ellipsoid will equal those of their images on the grid produced by a conformal map projection, and vice versa. Scale on a Lambert projection varies from north to south but not from east to west, as shown on Figure 20.1(c). Zone widths in the projection are therefore limited north–south, but not east–west. Thus, the Lambert projection is ideal for mapping states that are narrow north–south but which extend long distances in an east–west direction—for example, Kentucky, Montana, Pennsylvania, and Tennessee.

Figure 20.5 shows the portion of the developed cone of a Lambert projection covering an area of interest. In the Lambert projection, the cone intersects the ellipsoid along two parallels of latitude, called *standard parallels*, at about one sixth of the zone width from the north and south zone limits. All meridians

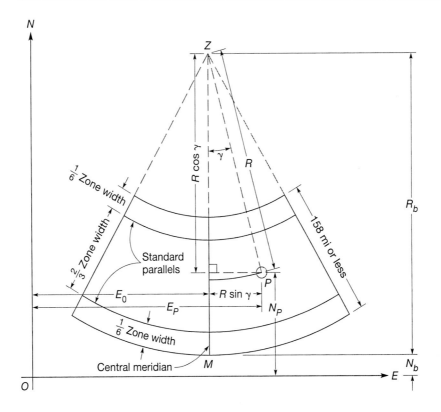

Figure 20.5
The Lambert conformal conic projection (SPCS 83 symbology).

are straight lines converging at Z, the apex of the cone. An example is ZM, which is the *central meridian*. All parallels of latitude are the arcs of concentric circles having centers at the apex. The projection is located in a zone in an east–west direction by assigning the central meridian a longitudinal value that is near the middle of the area to be covered. The direction of the central meridian on the projection establishes grid north. All lines parallel with the central meridian point in the direction of grid north. Therefore, except at the central meridian, directions of "geodetic" and "grid" north do not coincide because geodetic meridians converge. As shown in Figure 20.5, the *easting* (E) and *northing* (N) coordinates of points are measured perpendicular and parallel to the central meridian, respectively, from a reference E–N axis system that is offset to the west and south.

20.4 TRANSVERSE MERCATOR PROJECTION

The transverse Mercator projection is also a conformal projection but is based on an imaginary cylinder as its developable surface. As illustrated in Figure 20.1(d), scale in the transverse Mercator projection varies from east to west, but not from north to south. Thus, this projection is used for states like Illinois, Indiana, and New Jersey, which are narrow east–west and longer north–south.

In developing the transverse Mercator projection, the axis of the imaginary cylinder is placed in the plane of the Earth's equator. The cylinder cuts the spheroid along two elliptical curves approximately equidistant from the central meridian. On the developed plane surface (Figure 20.6) all parallels of

Figure 20.6
The transverse Mercator projection.

latitude, and all meridians except for the central meridian are curves (shown in light broken lines). As with the Lambert conformal conic projection, the central meridian establishes the direction of grid north, and the zone is centered over an area of interest by assigning the central meridian a longitude value that applies approximately at the center of the region to be mapped. Also, the E and N coordinates of points are measured perpendicular to and parallel with the central meridian, respectively, from an *E–N* axis system offset to the west and south.

■ 20.5 STATE PLANE COORDINATES IN NAD 27 AND NAD 83

The first state plane coordinate system was developed by the NGS in 1933 for the State of North Carolina. Systems for all other states followed within one year. As noted earlier, state plane coordinates of points are computed from geodetic latitudes and longitudes. Section 19.4 and Figure 19.3 define and illustrate these two terms and explain that geodetic latitudes and longitudes are defined with respect to a reference ellipsoid and associated datum. From 1927 until the inception of NAD 83, the reference datum used in the conterminous United States and mainland Alaska was NAD 27, based on the Clarke 1866 ellipsoid. All original state plane coordinates were therefore developed by the NGS in accordance with that ellipsoid and datum. This system is referred to as the *State Plane Coordinate System of 1927* (SPCS 27).

As noted in Chapter 19, NAD 83 employs a different set of defining parameters than NAD 27, and it uses a reference surface of different dimensions, the GRS 80 ellipsoid. Thus, the latitudes and longitudes of points in NAD 83 are somewhat different from their values in NAD 27. Because of these changes, the constants and variables that define the state plane coordinate systems also changed. Thus, following completion of NAD 83, it was necessary to develop a new state plane coordinate system. This system is called the *State Plane Coordinate System of 1983* (SPCS 83).

In the SPCS 83 most states retained the same projections they had used in SPCS 27, with the same central meridian and standard parallel positioning. There were changes, however; some major ones being (1) Montana reduced its number of zones from three to one, (2) Nebraska and South Carolina reduced their number of zones from two to one, and (3) Kentucky adopted a single statewide zone in addition to its two previous ones (for a total of three zones). In SPCS 83, 29 states (plus Puerto Rico) employ the Lambert conformal conic projection, 18 (plus Guam) use the transverse Mercator, and Alaska, Florida, and New York utilize both. In addition to using eight transverse Mercator zones for its mainland and a Lambert conformal conic for the Aleutian Islands, Alaska also employs an *oblique Mercator* projection for the southeast portion of the state (see Section 20.11), which is known as the panhandle of Alaska.

Although the same fundamental approach was used to develop both the SPCS 27 and the SPCS 83, as noted above, the parameters that define the two systems are different. Accordingly, different symbols are used, and the equations for computing coordinates in SPCS 83 have been changed. Thus, slightly different procedures are used in making computations in the two systems. For those wishing to review the procedures used in SPCS 27, please refer to an earlier edition of this book.

20.6 COMPUTING SPCS 83 COORDINATES IN THE LAMBERT CONFORMAL CONIC SYSTEM

A procedure for computing E and N coordinates of points from their geodetic latitudes and longitudes in the Lambert conformal conic system are illustrated in Figure 20.5. This process of converting from geodetic to plane coordinates is called the *direct problem*. The fundamentals described herein apply to both SPCS 27 and SPCS 83, although the symbology used in Figure 20.5 and this section is applicable only to SPCS 83. Computation in the reverse manner can also be performed; that is, calculating geodetic latitude and longitude from a given set of E, N state plane coordinates. This reverse form of calculation is called the *inverse problem*.

Most state plane coordinate computations are performed using software. However, the use of tables is often simpler when handheld calculators are used. For these reasons both methods of computations will be presented.

20.6.1 Zone Constants

A zone is defined in the Lambert conformal conic map projection by the selection of four sets of parameters. They are the defining ellipsoidal parameters a and f, grid origin (φ_0, λ_0), latitudes of the northern and southern standard parallels φ_N and φ_S, and false easting and northing of the grid's origin (E_0, N_b). From these defining parameters a set of zone constants are mathematically defined that are used in both the direct and inverse problems. Common functions used in the definition of the map projection are

$$W(\varphi) = \sqrt{1 - e^2 \sin^2 \varphi} \tag{20.1}$$

$$M(\varphi) = \frac{\cos \varphi}{W(\varphi)} \tag{20.2}$$

$$T(\varphi) = \sqrt{\left(\frac{1 - \sin \varphi}{1 + \sin \varphi}\right)\left(\frac{1 + e \sin \varphi}{1 - e \sin \varphi}\right)^e} \tag{20.3}$$

where e is the eccentricity of the ellipse as defined by Equation (19.2a). The equations for the defining zone constants for a Lambert conformal conic map projection are:

$$w_1 = W(\varphi_S) \tag{20.4}$$

$$w_2 = W(\varphi_N) \tag{20.5}$$

$$m_1 = M(\varphi_S) \tag{20.6}$$

$$m_2 = M(\varphi_N) \tag{20.7}$$

$$t_0 = T(\varphi_0) \tag{20.8}$$

$$t_1 = T(\varphi_S) \tag{20.9}$$

$$t_2 = T(\varphi_N) \tag{20.10}$$

$$n = \sin \varphi_0 = \frac{\ln(m_1) - \ln(m_2)}{\ln(t_1) - \ln(t_2)} \tag{20.11}$$

$$F = \frac{m_1}{n \cdot t_1^n} \tag{20.12}$$

$$\rho(t) = aFt^n \tag{20.13}$$

$$R_b = \rho(t_0) \tag{20.14}$$

20.6.2 The Direct Problem

In Figure 20.5, line ZM is the central meridian of the projection, the N axis (line ON) is parallel to ZM, and point O is the origin of the rectangular coordinate system. A constant E_0 is adopted to offset the N grid axis from the central meridian and make E coordinates of all points positive. Similarly, a constant N_b can be adopted to offset the E grid axis from the southern edge of the projection. The N coordinate of the cone's apex is the constant $(R_b + N_b)$, the numerical values of these terms being such that all N coordinates are positive.

The coordinates E and N of point P, whose geodetic latitude φ_p and geodetic longitude λ_p are known, are to be determined. Line ZP represents a portion of the meridian through point P with its length designated as R. Angle λ between the central meridian and meridian ZP represents the amount of convergence between these two meridians. In SPCS 83 it is termed the *convergence angle*. In SPCS 27 it was called the *mapping angle*.

From Figure 20.5, the following equations of the direct problem can be solved for the *easting* (E) and *northing* (N) coordinates of point P:

$$\begin{aligned} E_p &= R \sin \gamma + E_0 \\ N_P &= R_b - R \cos \gamma + N_b \end{aligned} \tag{20.15}$$

To solve Equations (20.15), values for E_0, N_b, R_b, R, and λ must be known. The quantities E_0 and N_b are constants for any zone. In many states, the SPCS 83 value for E_0 has been assigned a value of 600,000 m, and N_b assigned a value of 0.000 m; R_b is a computed zone constant defined by Equation (20.14). Values for R and λ also depend on the ellipsoid used and vary with changing locations of points in the zone; R changes with latitude, λ with longitude. The convergence angle λ can be computed as

$$\gamma = (\lambda_{CM} - \lambda) \tag{20.16}$$

$$t = T(\varphi_P) \tag{20.17}$$

$$R = \rho(t) \tag{20.18}$$

TABLE 20.1 EXCERPT FROM THE PENNSYLVANIA NORTH ZONE TABLES

Zone constants: $N_b = 0.000$ m $E_0 = 600{,}000.000$ m; $\lambda_{CM} = 77°45'$;
$R_b = 7{,}379{,}348.3668$ m $\sin\varphi = 0.661539733812$

Latitude (φ)	R (m)	Tab. Diff.	k
41°10'	7268294.836	30.84819	0.99996637
41°11'	7266443.945	30.84824	0.99996514
41°12'	7264593.050	30.84830	0.99996400
41°13'	7262742.152	30.84836	0.99996295
41°14'	7260891.251	30.84842	0.99996198
41°15'	7259040.346	30.84848	0.99996109
41°16'	7257189.437	30.84855	0.99996029
41°17'	7255338.524	30.84862	0.99995957
41°18'	**7253487.607**	**30.84869**	**0.99995893**
41°19'	7251636.685	30.84876	0.99995838

where Equation (20.15) is adjusted for western longitudes. On the companion website for the book is an Excel® spreadsheet *map_projections.xls* and Mathcad® worksheet *Lambert.xmcd* that demonstrate these computations.

To aid in hand solutions of Equations (20.15), the NGS has computed and published individual SPCS 83 booklets of projection tables for each state. These give the constants for each zone and tabulate R and scale factor k values versus latitude. Thus, given the geodetic latitude of any point, R for that point can be interpolated from the tables for use in Equations (20.15). Table 20.1 shows an excerpt from the north zone of the Pennsylvania tables. The column labeled *Tab. Diff.* provides the change in radius R per second of latitude. The use of this table is demonstrated in Example 20.1. The equations are solved using a calculator and the computational procedure is called the *tabular method*.

■ Example 20.1

Using values in Table 20.1, what are the coordinates E (easting) and N (northing) for station "Hayfield NE" that lies in Pennsylvania's north Lambert conformal conic zone? The station's geodetic latitude is 41°18'20.25410"N and its geodetic longitude is 76°00'57.00239"W.

Solution

Step 1: Determine the radius to Hayfield NE.
    The tabular difference (Tab. Diff. column) listed in Table 20.1 for

latitude of 41°18′ is 30.84869. Thus, the change in the radius ΔR from 41°18′ to 41°18′20.25410″ is

$$\Delta R = 20.25410''(30.84869) = 624.8125 \text{ m}$$

As latitude increases, the radius decreases, and thus ΔR must be subtracted from the tabulated R of 7,253,487.607 m. Thus, the radius to Hayfield NE is

$$R = 7{,}253{,}487.607 - 624.8125 = 7{,}252{,}862.794 \text{ m}$$

The equivalent linear interpolation formula for the radius is

$$R = 7{,}253{,}487.607 + (7{,}251{,}636.685 - 7{,}253{,}487.607)20.25410''/60''$$
$$= 7{,}252{,}862.794 \text{ m}$$

Step 2: Compute the convergence angle, γ, using Equation (20.16) as

$$\gamma = (77°45' - 76°00'57.00239'')\,0.661539733812 = 1°08'49.991''$$

Step 3: Solve Equations (20.15) as

$$E = 7{,}252{,}862.795 \sin 1°08'49.991'' + 600{,}000.000$$
$$= 745{,}212.637 \text{ m}$$
$$N = 7{,}379{,}348.3668 - 7{,}252{,}862.795 \cos 1°08'49.991'' + 0.000$$
$$= 127{,}939.400 \text{ m}$$

20.6.3 The Inverse Problem

The inverse problem in state plane coordinate computations is the determination of the geodetic latitude and geodetic longitude of a station based on its state plane coordinates and zone. The inverse equations also utilize the zone constants computed in Section 20.6.1 or given in Table 20.1. The remaining equations for accomplishing this can be derived from Equations (20.15) as

$$\begin{aligned} N' &= R_b - (N - N_b) \\ E' &= E - E_0 \\ \gamma &= \tan^{-1}\frac{E'}{N'} \\ R &= \frac{N'}{\cos \gamma} = \sqrt{E'^2 + N'^2} \\ \lambda &= \lambda_{CM} + \frac{\gamma}{n} \end{aligned} \quad (20.19)$$

The solution for the latitude of the station is iterative. This process is

$$t = \left(\frac{R}{aF}\right)^{\frac{1}{n}} \quad (20.20)$$

$$\chi = \frac{\pi}{2} - 2\tan^{-1} t \quad (20.21)$$

$$\varphi = \frac{\pi}{2} - 2\tan^{-1}\left[t\left(\frac{1 - e\sin\varphi_p}{1 + e\sin\varphi_p}\right)^{e/2}\right] \quad (20.22)$$

Iterate Equation (20.22) starting with φ_p equal to χ on the first iteration, using the previously computed φ for φ_p in subsequent iterations, and continuing until changes in φ are negligible. This procedure is demonstrated in the previously mentioned Excel spreadsheet and Mathcad worksheet on the companion website for the book.

When using the tables, the latitude φ_p of the station can be interpolated, versus R, from the tables. The procedure is demonstrated in Example 20.2.

■ Example 20.2

What is the geodetic latitude and geodetic longitude for station Hayfield NE given the following SPCS 83 coordinates? (The station lies in the north zone of Pennsylvania's Lambert conformal conic projection.)

$$\text{Easting} = 745{,}212.637 \text{ m}$$
$$\text{Northing} = 127{,}939.400 \text{ m}$$

Solution

Using Table 20.1 and Equations (20.18)

$$E' = 745{,}212.637 - 600{,}000.000 = 145{,}212.637 \text{ m}$$
$$N' = 7{,}279{,}348.3668 - (127{,}939.400 - 0.000) = 7{,}251{,}408.9668 \text{ m}$$
$$\gamma = \tan^{-1}\left(\frac{145{,}212.637}{7{,}251{,}408.9668}\right) = 1°08'49.991''$$
$$R = \sqrt{145{,}212.637^2 + 7{,}251{,}408.9668^2} = 7{,}252{,}862.7943 \text{ m}$$
$$\lambda = 77°45' - \frac{1°08'49.99''}{0.661539733812} = 76°00'57.00239''$$

Now the latitude of the station can be interpolated from the values in Table 20.1. As can be seen in the table, the computed radius R is between 41°18' and 41°19'. To determine the number of arc seconds to be added to 41°18', the difference between the tabulated radius for 41°18' of 7,253,487.607 and the computed radius R for the station is evaluated and divided by the tabulated difference of 30.84869. That is

$$\Delta\varphi'' = \frac{7{,}253{,}487.607 - 7{,}252{,}862.7943}{30.84869} = 20.25411''$$

Thus, the latitude and longitude of the station are computed as N41°18'20.25411" and W76°00'57.00239", respectively. Rounding errors in both the forward and inverse problems caused the slight difference of 0.00001" in the computed and given latitude from Example 20.1.

616 STATE PLANE COORDINATES AND OTHER MAP PROJECTIONS

Figure 20.7
WOLFPACK data entry screen for direct SPCS 83 computations of Example 20.1.

To aid in the solution of state plane coordinate computations, the NGS has published a booklet[1] entitled "State Plane Coordinate System of 1983" (Stem, 1989), which gives the zone constants and formulas for every zone in the United States and its territories. These constants are repeated in the Excel spreadsheet on the companion website for the book. The computer program WOLFPACK, also on the companion website for this book, contains a state plane coordinate option under its *map projections* menu. The direct data entry screen from WOLFPACK for Example 20.1 is shown in Figure 20.7. Of course, the state and zone within that state, must also be specified.

It should be emphasized that computation of coordinates in the SPCS 83 should only be performed using points whose geodetic positions are given in NAD 83.

20.7 COMPUTING SPCS 83 COORDINATES IN THE TRANSVERSE MERCATOR SYSTEM

The NGS has also published SPCS 83 Transverse Mercator state plane coordinate zones for those states that use this projection. All necessary constants and variables are computed and tabulated, and instructions together with sample problems are given to illustrate the computational procedures.

20.7.1 Zone Constants

A zone is defined in the Transverse Mercator map projection by the selection of four sets of parameters. They are the defining ellipsoidal parameters a and f, grid origin (φ_0, λ_0), scale factor at the central meridian λ_0, and false easting and

---

[1]Individual booklets for each state for making SPCS 83 tabular solutions, and NOAA Manual NOS NGS 5 can be obtained from the National Geodetic Information Center, NOAA, National Geodetic Survey, N/CG174, SSMC3 Station 09535, 1315 East West Highway, Silver Spring, MD 20910, telephone (301) 713-3242 or at https://geodesy.noaa.gov/library/pdfs/NOAA_Manual_NOS_NGS_0005.pdf. Similar products are also available for SPCS 27.

northing (E_0, N_b). From these defining parameters a set of zone constants are defined mathematically, which are used in both the direct and inverse problems. Common functions used in the definition of the map projection are

$$C(\varphi) = e'^2 \cos^2 \varphi \quad T(\varphi) = \tan^2 \varphi$$

$$M(\varphi) = a \left[\begin{array}{c} \left(1 - \dfrac{e^2}{4} - \dfrac{3e^4}{64} - \dfrac{5e^6}{256}\right)\varphi - \left(\dfrac{3e^2}{8} + \dfrac{3e^4}{32} + \dfrac{45e^6}{1024}\right) \sin 2\varphi \\ + \left(\dfrac{15e^4}{256} + \dfrac{45e^6}{1024}\right) \sin 4\varphi - \left(\dfrac{35e^6}{3072}\right) \sin 6\varphi \end{array} \right] \quad (20.23)$$

where e and e' are the first and second eccentricities of the ellipse as defined by Equations (19.2a) and (19.2b), respectively.

20.7.2 The Direct Problem

The computations for the direct problem using the map projection formulas are

$$m_0 = M(\varphi_0) \quad (20.24)$$
$$m = M(\varphi) \quad (20.25)$$
$$t = T(\varphi) \quad (20.26)$$
$$c = C(\varphi) \quad (20.27)$$
$$A = (\lambda_0 - \lambda) \cos \varphi \quad (20.28)$$

where Equation (20.27) has been adjusted for western longitudes.

$$E = k_0 R_N \left[A + (1 - t + c)\dfrac{A^3}{6} + (5 - 18t + t^2 + 72c - 58e'^2)\dfrac{A^5}{120} \right] + E_0 \quad (20.29)$$

$$N = k_0 \left\{ m - M_0 + R_N \tan \varphi \left[\dfrac{A^2}{2} + (5 - t + 9c + 4c^2)\dfrac{A^4}{24} \right. \right.$$
$$\left. \left. + (61 - 58t + t^2 + 600c - 330e'^2)\dfrac{A^6}{720} \right] \right\} + N_b \quad (20.30)$$

where R_N is the radius in the prime vertical for the latitude φ as defined by Equation (19.4). The convergence angle λ is computed as

$$c_2 = \dfrac{1 + 3c + 2c^2}{3} \quad c_3 = \dfrac{2 - \tan^2 \varphi}{15}$$
$$\gamma = A \tan \varphi [1 + A^2(c_2 + c_3 A^2)] \quad (20.31)$$

Computations utilizing the Equations (20.24) to (20.31) are demonstrated in an Excel spreadsheet *map_projections.xls* and Mathcad worksheet *TM.xmcd*, which are on the companion website for the book.

TABLE 20.2 EXCERPT FROM THE TRANSVERSE MERCATOR PROJECTION TABLES FOR NEW JERSEY FOR THE DIRECT PROBLEM

Zone constants: $E_0 = 150{,}000$ m $\varphi_b = 38°50'$ $\lambda_0 = 74°30'$ $N_b = 0.000$ m $k_0 = 0.9999$

Latitude	(I) / (IV)	Difference 1″	(II) / (V)	Difference 1″	(III) / (VI)
39°00′	18500.4650	30.834594	3670.4645	0.007613	1.902
	240604.8369	−0.910942	19.8284	−0.000975	−0.026
39°01′	20350.5407	30.834682	3670.9213	0.007592	1.901
	240548.3803	−0.941283	19.7699	−0.000975	−0.026
39°02′	**22200.6216**	**30.834771**	**3671.3768**	**0.007572**	**1.900**
	240491.9034	**−0.941623**	**19.7114**	**−0.000974**	**−0.026**
39°03′	24050.7079	30.834859	3671.8311	0.007551	1.899
	240435.4060	−0.941964	19.6529	−0.000974	−0.026
39°04′	25900.7994	30.834947	3672.2842	0.007530	1.898
	240378.8882	−0.942304	19.5945	−0.000974	−0.026

Second-difference corrections

	00″	10″	20″	30″		00″	10″	20″	30″
	60″	50″	40″	30″		60″	50″	40″	30″
(I)	0.0000	−0.0004	−0.0006	−0.0007	(IV)	0.0000	0.0014	0.0023	0.0025

With reference to Figure 20.6 and appropriate Transverse Mercator projection tables such as those shown in Table 20.2, the following SPCS 83 equations for hand computations yield the solution of the *direct problem*, that is, obtaining the E_P and N_P coordinates of any point P from its geodetic coordinates:

$$\Delta\lambda = (\lambda_0 - \lambda)3600''/°$$
$$p = 10^{-4}\Delta\lambda''$$
$$N_P = (\text{I}) + (\text{II})p^2 + (\text{III})p^4 + N_b$$
$$E_P = (\text{IV})p + (\text{V})p^3 + (\text{VI})p^5 + E_0$$

(20.32)

In Equations (20.32), $\Delta\lambda''$ is the difference in longitude between the central meridian and the point in seconds, and N and E are the state plane coordinates of the point in meters. The values for λ_0 and E_0 are zone constants, and are supplied with the table as shown in Table 20.2. The values for Roman numerals (I), (II), (III), (IV), (V), and (VI) are interpolated from the Transverse Mercator tables using the appropriate value given in the *Difference* column. For the first (I) and fourth (IV) column values, a small second-difference correction must also be interpolated using the numbers given at the bottom of the Table. Example 20.3 demonstrates the use of Equations (20.32) and Table 20.2.

■ Example 20.3

The geodetic latitude and geodetic longitude of station Stone Harbor in the state of New Jersey, which uses the Transverse Mercator projection, are 39°02′21.63632″ and 74°46′08.80133″, respectively. What are the station's SPCS 83 coordinates?

Solution

The determination of the second differences and (III) and (VI) column values involve a linear interpolation. For example, the second difference for column (I) is determined as

$$-0.0006 + [-0.0007 - (-0.0006)]1.63632''/10'' = -0.00062$$

where 1.63632″ comes from the second's portion of the station's latitude. The three values (III), (IV), and (VI) are interpolated in similar fashion and shown in the computations that follow. Using Table 20.2, the appropriate column values are

$$\text{I} = 22{,}200.6216 + 30.834771(21.63632) + (-0.00062) = 22{,}867.77195$$
$$\text{II} = 3671.3768 + 0.007572\,(21.63632) = 3671.54063$$
$$\text{III} = 1.900 + (1.899 - 1.900)(21.63632/60) = 1.89964$$
$$\text{IV} = 240{,}491.9034 + (-0.941623)(21.63632) + 0.00233 = 240{,}471.53247$$
$$\text{V} = 19.7114 + (-0.000974)(21.63632) = 19.690326$$
$$\text{VI} = -0.026 + (-0.026 + 0.026)(21.63632/60) = -0.026$$

Substituting these values into Equations (20.32) yields

$$\Delta\lambda'' = (74°30' - 74°46'08.80133'')3600''/° = -968.80133''$$
$$p = -968.80133 \times 10^{-4} = -0.096880133$$
$$N = 22{,}867.77197 + 3671.54063p^2 + 1.89964p^4 + 0 = 22{,}902.2323 \text{ m}$$
$$E = 240{,}471.5324p + 19.690326p^3 - 0.026p^5 + E_0 = 126{,}703.0680 \text{ m}$$

20.7.3 The Inverse Problem

The Transverse Mercator inverse problem is solved as

$$E' = E - E_0 \quad (20.33)$$
$$N' = N - N_b \quad (20.34)$$
$$e_1 = \frac{1 - \sqrt{1 - e^2}}{1 + \sqrt{1 - e^2}} \quad (20.35)$$

$$m = m_0 + \frac{N'}{k_0} \tag{20.36}$$

$$\chi = \frac{m}{a\left(1 - \frac{e^2}{4} - \frac{3e^4}{64} - \frac{5e^6}{256}\right)} \tag{20.37}$$

The footpoint latitude ϕ_f is

$$\varphi_f = \chi + \left(\frac{3e_1}{2} - \frac{27e_1^3}{32}\right)\sin 2\chi + \left(\frac{21e_1^2}{16} - \frac{55e_1^4}{32}\right)\sin 4\chi$$
$$+ \left(\frac{151e_1^3}{96}\right)\sin 6\chi + \left(\frac{1097e_1^4}{512}\right)\sin 8\chi \tag{20.37}$$

$$c_1 = C(\varphi_f) \tag{20.38}$$

$$t_1 = T(\varphi_f) \tag{20.39}$$

$$N_1 = R_N \text{ evaluated using Equation (19.4) with } \varphi_f \tag{20.40}$$

$$M_1 = R_M \text{ evaluated using Equation (19.5) with } \varphi_f \tag{20.41}$$

$$D = \frac{E'}{N_1 k_0} \tag{20.42}$$

$$B = \frac{D^2}{2} - (5 + 3t_1 + 10c_1 - 4c_1^2 - 9e')\frac{D^4}{24}$$
$$+ (61 + 90t_1 + 298c_1 + 45t_1^2 - 252e' - 3c_1^2)\frac{D^6}{720} \tag{20.43}$$

$$\varphi = \varphi_f - \left(\frac{N_1 \tan \varphi_f}{R_1}\right)B \tag{20.44}$$

Computations using Equations (20.37) to (20.44) are demonstrated in and Excel® spreadsheet *map_productions.xls* and Mathcad® worksheet *NJ_Table.xmcd* on the companion website. Similar to Table 20.2, Table 20.3 contains the necessary parameters and second differences to compute the inverse problem. The formulas necessary to perform this calculation are

$$\begin{aligned} E' &= E - E_0 \\ N' &= N - N_b \\ q &= E' \times 10^{-6} \\ \varphi_f &= \varphi_{\text{table}} + \Delta\varphi \text{ (interpolated from table)} \\ \Delta\varphi''_f &= -(\text{VII})q^2 + (\text{VIII})q^4 \\ \Delta\lambda'' &= -(\text{IX})q + (\text{X})q^3 + (\text{XI})q^5 \\ \varphi &= \varphi_f + \Delta\varphi''_f \\ \lambda &= \lambda_b + \Delta\lambda'' \end{aligned} \tag{20.45}$$

20.7 Computing SPCS 83 Coordinates in the Transverse Mercator System

TABLE 20.3 EXCERPT FROM THE TRANSVERSE MERCATOR PROJECTION TABLES FOR NEW JERSEY FOR THE INVERSE PROBLEM

Zone constants: $E_0 = 150{,}000$ m $\varphi_b = 38°50'$ $\lambda_0 = 74°30'$ $N_b = 0.000$ m $k_0 = 0.9999$

Latitude	(I) (IX)	Difference 1"	(VII) (X)	Difference 1"	(VIII) (XI)
39°00'	18500.4650	30.834594	2056.2443	0.020257	29.191
	41561.9242	0.162576	393.3224	0.005941	7.024
39°01'	20350.5407	30.834682	2057.4597	0.020267	29.218
	41571.6788	0.162711	393.6788	0.005949	7.035
39°02'	**22200.6216**	**30.834771**	**2058.6757**	**0.020276**	**29.245**
	41581.4415	**0.163846**	**394.0357**	**0.005957**	**7.047**
39°03'	24050.7079	30.834859	2059.8923	0.020286	29.272
	41591.2122	0.162982	394.3931	0.005965	7.058
39°04'	25900.7994	30.834947	2061.1094	0.020295	29.299
	41600.9911	0.163117	394.7510	0.005973	7.069

Second-difference corrections

	00"	10"	20"	30"		00"	10"	20"	30"
	60"	50"	40"	30"		60"	50"	40"	30"
(I)	0.0000	−0.0004	−0.0006	−0.0007	(IX)	−0.0000	−0.0006	−0.0009	−0.0010

In Equations (20.46) E' and N' are in meters, $\Delta\varphi''$ and $\Delta\lambda''$ are in seconds, and φ_f, φ, and λ are in degrees, minutes, and seconds. The value of φ_f is interpolated from column (I) in Table 20.3 and is known as the *foot point latitude*. Notice that in this table, the first (I) and the ninth (IX) values have second-difference corrections. Example 20.4 demonstrates the use of these tables in the inverse problem.

■ Example 20.4

What are the geodetic coordinates of station Stone Harbor if its SPCS 83 easting and northing coordinates are 126,703.0681 m and 22,902.2323 m, respectively? (The station lies in New Jersey's Transverse Mercator zone.)

Solution

Step 1: Calculate E', N', and q as

$$E' = 126{,}703.0680 - 150{,}000.000 = -23{,}296.9320 \text{ m}$$
$$N' = 22{,}902.2323 - 0.000 = 22{,}902.2323 \text{ m}$$
$$q = -23{,}296.9320 \times 10^{-6} = -0.023296932$$

Step 2: Looking at Table 20.3, it can be seen that the northing coordinate lies between the (I)-values of 39°02′ and 39°03′. Thus $\Delta\varphi_f''$ can be interpolated from column (I) as

$$\Delta\varphi_f'' = (22{,}902.2323 - 22{,}200.6216)/30.834771 = 22.75388068''$$

Hence the footpoint latitude is

$$\varphi_f = 39°02' + 22.75388068'' = 39°02'22.75388068''$$

Step 3: Using $\Delta\varphi_f''$, evaluate the tabular values for VII, VIII, IX, X, and XI. Note in this procedure, the values for VIII and XI must be linearly interpolated using the tabular values for 39°02′ and 39°03′.

$$\text{VII} = 2058.6757 + 0.020276(22.75388068) = 2059.13706$$
$$\text{VIII} = 29.245 + (29.272 - 29.245)(22.75388068/60) = 29.25524$$
$$\text{IX} = 41{,}581.4415 + 0.162846(22.75388068) - 0.0009 = 41{,}585.14688$$
$$\text{X} = 394.0357 + 0.005957(22.75388068) = 394.17124$$
$$\text{XI} = 7.047 + (7.058 - 7.047)(22.75388068/60) = 7.05117$$

Step 4: Applying the computed tabular values of step 3, compute the latitude and longitude of station Stone Harbor using Equations (20.45) as

$$\Delta\varphi'' = -2059.13706q^2 + 29.25524q^4 = -1.11758''$$
$$\Delta\lambda'' = -41{,}585.14688q + 394.17124q^3 + 7.05117q^5 = 968.80135''$$
$$\varphi = 39°02'22.75388068'' - 1.11758'' = 39°02'21.6363''$$
$$\lambda = 74°30' + 968.80135'' = 74°46'08.8013''$$

Again, except for small rounding errors from both the forward and inverse computations, the solution produces the geodetic latitude and longitude of Stone Harbor that were given for Example 20.3.

As with the Lambert conformal conic map conversions, the solutions of the direct and inverse problems are typically performed with computers. Several programs have been written that allow for the easy conversion from geodetic to state plane coordinates and vice versa. WOLFPACK, which is available on the companion website for the book, contains this option under its coordinate computation menu.

20.8 REDUCTION OF DISTANCES AND ANGLES TO STATE PLANE COORDINATE GRIDS

Ground-surveyed distances and angles must undergo corrections prior to using them in making computations in state plane coordinate systems. As shown in Figure 20.8, distances must first be reduced from their ground-surveyed lengths to their ellipsoidal equivalents. Furthermore, as shown in Figure 20.2, these ellipsoidal

20.8 Reduction of Distances and Angles to State Plane Coordinate Grids

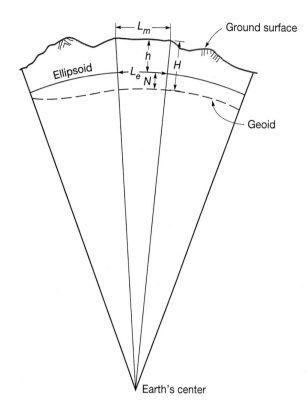

Figure 20.8
Reduction of lengths from surface observations to the ellipsoid.

distances must then be reduced to the developable surface of the state plane system being used. Also, whenever angles or azimuths are used in these computations, they can be reduced to their grid equivalents. Once these reductions have been completed, traverse computations (see Chapter 10), and adjustment procedures given in both Chapters 10 and 16 may be performed. This section describes the processes of reducing these observations to state plane grids.

20.8.1 Grid Reduction of Distances

The reduction of distances is normally done in two steps: (1) reduce the observations from their ground lengths to ellipsoid lengths (geodetic distance) and (2) reduce the ellipsoid lengths to their grid equivalents. The most precise formulas for reduction of slope distances to the ellipsoid were given in Section 19.15. However, in most local surveys, the lengths of the distances are short and simpler methods of reduction can be used since the horizontal distance L_m closely matches the arc distance at the surface of the Earth. With this simplification, the relationship between the ground-surveyed length and the ellipsoid length L_e is

$$L_e = L_m\left(\frac{R_\alpha}{R_\alpha + H + N}\right) \quad (20.47a)$$

alternatively,

$$L_e = L_m\left(\frac{R_\alpha}{R_\alpha + h}\right) \quad (20.47b)$$

where R_a is the radius of the ellipsoid in the azimuth of the line as given by Equation (19.6), H the average orthometric height of the observed line above the geoid, N the average geoid height, and h is the average geodetic height of the line. The ratios $R_a/(R_a + H + N)$ and $R_a/(R_a + h)$ are commonly called the *elevation factor* of the line. For all but the most rigorous surveys, acceptable results can be obtained from Equation (20.47) by substituting the average radius of the Earth (20,902,200 ft or 6,372,000 m) for R_a. However, for higher accuracy surveys geodetic reduction of the observed distance should be performed as described in Sections 19.15.1 or 19.16.

After a distance has been reduced to its ellipsoidal equivalent, it must then be scaled to its grid equivalent. This is accomplished by multiplying the ellipsoidal length of the line by an appropriate scale factor. For Lambert conformal conic mapping projections, the scale factor for latitude ϕ can be computed as

$$m = M(\varphi) \tag{20.48}$$

$$k = \frac{Rn}{am} \tag{20.49}$$

where M, R, and n are previously defined in this chapter for the Lambert conformal conic map projection, and a is the length of the ellipsoid's semi-major axis. The scale factor at a point can be also interpolated from tables.

■ Example 20.5

What is the scale factor for Hayfield NE of Example 20.1? (This station is in the north zone of Pennsylvania's Lambert conformal conic projection.)

Solution

From Example 20.1, the geodetic latitude of station Hayfield NE is 41°18′20.25410″. Using the tabulated scale factors for latitudes of 41°18′ and 41°19′ from Table 20.1, the interpolated scale factor is

$$k = 0.99995893 + (0.99995838 - 0.99995893)\left(\frac{20.25410''}{60''}\right) = 0.999958744$$

For a Transverse Mercator map projection, the scale factor k for any point is computed as

$$k = k_0 \left[1 + (1 + c)\frac{A^2}{2} + (5 - 4t + 42c + 13c^2 - 28e'^2)\frac{A^4}{24} \right.$$
$$\left. + (61 - 148t + 16t^2)\frac{A^6}{720} \right] \tag{20.50}$$

20.8 Reduction of Distances and Angles to State Plane Coordinate Grids

TABLE 20.4 EXCERPT FROM THE TRANSVERSE MERCATOR PROJECTION TABLES FOR NEW JERSEY FOR COMPUTATION OF THE SCALE FACTOR. (NOTE: COLUMNS ARE REARRANGED FOR PUBLICATION PURPOSES ONLY.)

Zone constants: $E_0 = 150{,}000$ m $\varphi_b = 38°50'$ $\lambda_b = 74°30'$ $N_b = 0.000$ m $k_0 = 0.9999$

Latitude	(I) (XIV)	Difference 1"	(XII) (XV)	Difference 1"	(XIII) (XVI)
39°00'	18500.4650	30.834594	6293.2039	0.037673	3.014
	26155.7664	0.258924	353.110		0.012311
39°01'	20350.5407	30.834682	6295.4643	0.037664	3.014
	26171.3018	0.259046	353.486		0.012311
39°02'	**22200.6216**	**30.834771**	**6297.7241**	**0.037655**	**3.014**
	26186.8446	**0.259167**	**353.863**		**0.012311**
39°03'	24050.7079	30.834859	6299.9834	0.037646	3.013
	26202.3946	0.259289	354.240		0.012310
39°04'	25900.7994	30.834947	6302.2422	0.037637	3.013
	26217.9520	0.259412	354.618		0.012310

Second-difference corrections

	00"	10"	20"	30"		00"	10"	20"	30"
	60"	50"	40"	30"		60"	50"	40"	30"
(I)	0.0000	−0.0004	−0.0006	−0.0007	(XIV)	−0.0000	−0.0005	−0.0008	−0.0009

where k_0 is a zone constant and c, A, and t are previously defined for the transverse Mercator map projection. To compute the scale factor k using tables, the following formula is used:

$$k = k_0[1 + (\text{XVI})q^2 + 0.00003q^4] \qquad (20.51)$$

where q is defined in Equations (20.45), and (XVI) comes from Table 20.4.

■ Example 20.6

Compute the scale factor for station Stone Harbor of Example 20.3. (This station lies in New Jersey's Transverse Mercator projection.)

Solution

From Example 20.3, the geodetic latitude of station Stone Harbor is 39°02'21.63632". Using Equation (20.51), the scale factor for this station is

$$k = 0.9999[1 + 0.0123106q^2 + 0.00003q^4] = 0.99990668$$

where q is −0.023296932 as determined in Example 20.4, and the value for (XVI) is interpolated from the values at the bounding latitudes of 39°02' and 39°03'.

A line consists of an infinite number of points, and thus there are several approaches to computing the scale factor k of a line. The easiest and least precise involves determining an average scale factor for an entire survey project and applying this single value to all reduced distances. This approach is adequate for low-accuracy surveys and for surveys that cover small areas. To achieve a higher level of accuracy, an "average scale factor" can be applied to each individual line. In this method, the values are obtained by averaging the scale factors of the end points of the lines. This method works well for moderately long distances. However, for the most precise surveys, an additional scale factor at the midpoint of the line should be computed. Then an improved scale factor k_{12} is determined as

$$k_{12} = \frac{k_1 + 4k_m + k_2}{6} \qquad (20.52)$$

where k_1 and k_2 are scale factors for the end points of the line, and k_m is the scale factor for the midpoint of the line. Obviously, this method requires that coordinates and scale factor k_m for the midpoint of the line be computed.

The product of the elevation factor and the scale factor is the so-called *combined factor* and mathematically expressed as

$$\text{combined factor} = \text{elevation factor} \times \text{scale factor} \qquad (20.53)$$

In NAD 27, the combined factor was known as the *grid factor*.

It is common in lower-order surveys, and surveys that cover small areas, to use a single combined factor for the entire survey. Often data collectors allow for the entry of this value, so that coordinates are computed directly on the grid while performing the survey. With surveys that cover larger areas, or require more rigorous procedures, the scale factor and elevation factors for each line should be computed. The reduced grid distance is

$$\text{grid distance} = \text{ground distance} \times \text{combined factor} \qquad (20.54)$$

■ Example 20.7

What are the grid lengths for the observed distances in Figure 20.9? The scale factor of station Hayfield NE in the figure was determined in Example 20.5.

Solution

For this example, elevation factors were computed using Equation (20.47a) and employing the given average geoidal separation of −29.8 m and the mean radius of the Earth. Using computational procedures as given in Chapter 12, approximate coordinate values for each station were determined, and then used to compute the scale factor at each station (see Example 20.5). For this survey, it was appropriate to use scale factors obtained by averaging the end-point values. Elevation factors were obtained by using average elevations for each line in Equation (20.47a). Grid factors for each line were then determined by multiplying the elevation factor by the average scale factor. Finally grid distances were computed by multiplying each observed distance by its corresponding combined factor. The results of these calculations are listed in Table 20.5.

20.8 Reduction of Distances and Angles to State Plane Coordinate Grids

Station	Elevation (m)
Hayfield NE	380
B	301
C	292
D	354
21002	390

Average geoid height = –29.8 m

CONTROL STATION COORDINATES

Station	Northing	Easting
Hayfield NE	127,939.400	745,212.637
21002	123,131.289	760,208.805

Figure 20.9 Field observed traverse.

TABLE 20.5 REDUCED GRID DISTANCES FOR FIGURE 20.9

Station	Elev. (m)	Distance (m)	Elev. Factor	k	k_{avg}	Combined Factor	Grid Dist. (m)
Hayfield NE	380			0.99995874			
		3732.564	0.99995123		0.99995854	0.99990978	3732.2272
B	301			0.99995834			
		5175.575	0.99995814		0.99995895	0.99991709	5175.1459
C	292			0.99995956			
		4116.475	0.99995398		0.99996014	0.99991412	4116.1215
D	354			0.99996072			
		4994.073	0.99994629		0.99996068	0.99990698	4993.6084
21002	390			0.99996065			

20.8.2 Grid Reduction of Azimuths and Angles

As shown in Figure 20.10, all grid meridians are parallel while all geodetic meridians converge to a single point. The primary difference between these directions is the convergence angle γ. Computation of the convergence angle for the Lambert conformal conic was demonstrated in both Examples 20.1 and 20.2. The convergence angle (in seconds) for the Transverse Mercator projection is computed as

$$c_2 = \frac{1 + 3c + 2c^2}{3}$$

$$c_3 = \frac{2 - \tan^2\varphi}{15}$$

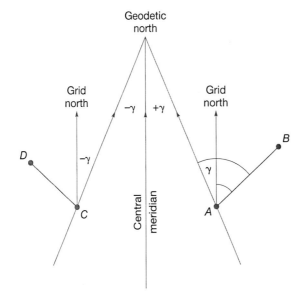

Figure 20.10
Relationship between geodetic azimuth, grid azimuth, and convergence angle λ

$$\gamma = A \tan \phi [1 + A^2(c_2 + c_3 A^2)] \quad (20.55)$$

where A and c are previously defined in this chapter for the transverse Mercator map projection. It can be computed using tables as

$$\gamma'' = (\text{XII})p + (\text{XIII})p^3 = (\text{XIV})q - (\text{XV})q^3 \quad (20.56)$$

Using the value for q determined in Example 20.4 and Equation (20.55), the convergence angle at station Stone Harbor in New Jersey is

$$\gamma'' = (26{,}192.4512)q - (353.9989482)q^3 = -610.2'' = -0°10'10.2''.$$

where (XIV) and (XV) were interpolated from Table 20.4.

Another factor that affects the reduction of azimuths is the projection of the geodetic azimuth onto a developable mapping surface. As can be seen in Figure 20.11 the projection of geodetic azimuths onto a flat surface results in an arc between the occupied and sighted stations. Letting the grid azimuth be t and the geodetic azimuth be T, the difference in these values is known as the *second-term correction*, also called the *arc-to-chord* correction, and is designated as δ.

The sign of the arc-to-chord correction is given by the location of the line, and the type of map projection. For the Lambert conformal conic, the projected geodetic arc is always concave toward the central parallel of the zone. For the Transverse Mercator, the projected geodetic arc is always concave toward the central meridian for the zone. In Lambert conformal conic projections, the value for N_0 (the north–south center of the zone) can be determined from the derived zone constant $\sin(\phi_0)$ as

$$N_0 = R_b + N_b - R_0 \quad (20.57)$$

where R_b and N_b are zone constants, and R_0 can be interpolated from the table using the latitude derived from the $\sin(\phi_0)$, which is the gives the latitude of the

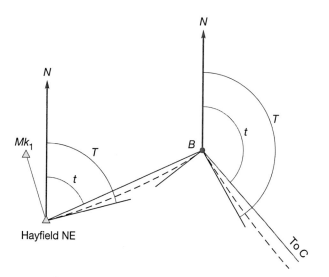

Figure 20.11
Part of Figure 20.9 showing arc-to-chord $(t - T)$ correction at stations *Hayfield NE* and *B*.

central parallel. Generally, the precise numerical value for N_0 is not necessary for computations because the line's location in relation to the zone's north–south center, and the latitude φ_0 as derived from $\sin(\varphi_0)$ is all that is needed to determine the concavity of the projected geodetic arc. For example in the north zone of Pennsylvania, the $\sin\varphi_0$ is given as 0.661539733812 (see Table 20.1), which yields a central parallel of approximately 41°25′03″. This value is sufficient to determine the concavity of the projected geodetic azimuth. Table 20.6 shows the sign of δ'' based on these criteria.

The arc-to-chord correction is a function of the positions of the end points of the line and the type of map projection. For a Lambert conformal conic map projection δ is given as

$$\delta_{12} = 0.5(\sin \varphi_3 - \sin \varphi_0)(\lambda_1 - \lambda_2) \tag{20.58}$$

where (φ_1, λ_1) and (φ_2, λ_2) are the geodetic positions of the end points of the line using positive values for western longitudes, $\sin\varphi_0$ is a zone constant, and $\varphi_3 = (2\varphi_1 + \varphi_2)/3$. For the Transverse Mercator projection, δ can be computed from the northing and easting coordinates for the end points of the line as

$$E' = E - E_0 \tag{20.59a}$$
$$\Delta N = N_2 - N_1 \tag{20.59b}$$
$$N_m = 0.5(N_1 + N_2) \tag{20.59c}$$
$$n = f/(2 - f) \tag{20.59d}$$
$$r = a(1 - n)(1 - n^2)(1 + 9n^2/4 + 225n^4/64)$$
$$= 6{,}367{,}449.14577 \text{ m (GRS 80)} \tag{20.59e}$$
$$\omega = (N_m - N_b + S_0)/(k_0 r) \tag{20.59f}$$
$$\varphi_f = \omega + V_0 \sin \omega \cos \omega \tag{20.59g}$$
$$\eta_f = e'^2 \cos^2 \varphi_f \tag{20.59h}$$

TABLE 20.6 THE SIGN OF THE δ CORRECTION

Map Projection		Azimuth of the Line from North	
Lambert	Sign of $N - N_0$	0° to 180°	180° to 360°
	Positive	+	−
	Negative	−	+
Transverse Mercator	Sign of $E - E_0$ or E_3	270° to 90°	90° to 270°
	Positive	−	+
	Negative	+	−

$$F = (1 - e^2 \sin^2 \varphi_f)(1 + \eta_f^2)/(k_0 a)^2 \tag{20.59i}$$

$$E_3 = 2E_1' + E_2' \tag{20.59j}$$

$$\delta_{12} = -\frac{1}{6} \Delta N E_3 F \left(1 - \frac{1}{27} E_3^2 F\right) \tag{20.59k}$$

where S_1 is tabulated in Appendix F for each Transverse Mercator zone, r the radius of the rectifying sphere, V_1 a constant of 0.005022893948 for GRS 80, (N_1, E_1) and (N_2, E_2) are the northing and easting state plane coordinates of the occupied and sighted stations, respectfully, and other variables are as previously defined. An approximate value for δ_{12} can be computed as $-25.4 \, \Delta N(E_3/3)10^{-10}$ seconds where the coordinate values are in meters.

Combining the two aforementioned corrections to the grid azimuth Az, the geodetic azimuth α can be computed as

$$\alpha_{12} = Az_{12} + \gamma_1 - \delta_{12} \tag{20.60}$$

In Example 20.8, the arc-to-chord correction is computed for the lines in Figure 20.9 to demonstrate the procedure. The arc-to-chord correction is generally ignored for short lines since it is typically below the error of the angle or azimuth observation. The National Geodetic Survey recommends that this correction need only be applied for lines longer than 8 km. However, each surveyor should determine the maximum arc-to-chord correction acceptable for their survey, which can be based on the ISO 17123-3 standard of the instrument or the intended accuracy of the final product from the survey.

■ Example 20.8

Evaluate the arc-to-chord correction for the lines in Figure 20.9.

Solution

Using the observed angles and computational procedures as demonstrated in Chapter 10, approximate coordinates were computed for each station. From these approximate coordinate values, the magnitudes of the second-term corrections were calculated using Equation (20.59). Note as can be shown from Table 20.6, the sign of all the corrections is negative since $N - N_0$ is negative for all lines in this example. The results of these computations are shown in Table 20.7.

The values in Table 20.7 are used to correct geodetic directions between the stations. Since Figure 20.9 has two grid azimuths, these azimuths do not need corrections. However, when the lines are over 8 km or 5 mi in length, or a rigorous reduction is desired, the observed angles should be corrected. The arc-to-chord correction to observed angles can be found by taking the difference in the forward and back azimuths. The corrections for the backsight azimuths have the same magnitude, but opposite signs of the computed foresight corrections in Table 20.7. For example, the correction for the azimuth from station C to B is 1.42″. Since the sight distances to both the azimuth marks was short, the corrections were assumed to be zero. The reduction of the observed angles is shown in Table 20.8. Note that if the instrument used in the survey had an ISO 17123-3 pointing and reading uncertainty of $\pm 1''$, the surveyor may choose to compute the arc-to-chord correction for each angle. However, this decision should also consider the intended accuracy of final product. If this is a high-accuracy control survey, this correction might be necessary to achieve the highest results. However, if this is a control survey for a mapping project or a boundary survey, this correction is likely unnecessary.

As can be seen in Table 20.8, the corrections are small. Often the arc-to-chord correction is ignored for traverses involving lines under 8 km, and for lower-order surveys. However, the reduction of observed distances to the mapping grid is generally significant for most traverse surveys. Failure to account for these corrections will result in incorrect misclosures and subsequent incorrect adjustments and coordinate values. When these corrections are properly performed, the resulting adjustments will yield results like those achieved with geodetic computations. If adjusted ground distances are needed after an adjustment, a rearranged form of Equation (20.54) may be used to determine their values.

TABLE 20.7 COMPUTATION OF THE ARC-TO-CHORD CORRECTION FOR FIGURE 20.9

Station	Approx. Northing	Approx. Easting	Approx. φ	Approx. λ	φ_3	δ
Hayfield NE	127,939.400	745,212.637	41°18′20.2541″	76°00′57.0024″		
					41°18′35.2908″	−0.10″
B	129,400.865	748,646.890	41°19′05.3642″	75°58′28.1096″		
					41°18′24.1269″	−0.11″
C	125,657.841	752,220.832	41°17′01.6524″	75°55′57.8495″		
					41°16′30.5490″	−0.12″
D	122,842.331	755,223.511	41°15′28.3421″	75°53′51.4269″		
					41°15′30.2932″	−0.22″
21002	123,131.289	760,208.805	41°15′34.1953″	75°50′17.0477″		

TABLE 20.8 ARC-TO-CHORD CORRECTION FOR ANGLES IN FIGURE 20.9

Station	Obs. Angle	Backsight δ	Foresight δ	Total δ	Corr. Angle
Hayfield NE	84°58′58″	0.00″	−0.10″	−0.10″	84°58′57.9″
B	249°22′17″	0.10″	−0.11″	−0.21″	249°22′16.8″
C	176°50′04″	0.11″	−0.12″	−0.23″	176°50′03.8″
D	133°31′46″	0.12″	−0.22″	−0.34″	133°31′45.7″
21002	101°53′19″	0.22″	0.00″	−0.22″	101°53′18.8″

■ 20.9 COMPUTING STATE PLANE COORDINATES OF TRAVERSE STATIONS

Determining state plane coordinates of new traverse stations is a problem routinely solved by surveyors. Normally it requires only that traverses (or triangulation or trilateration surveys) start and end on existing stations having known state plane coordinates, and from which known grid azimuths have been established. Generally, these data are available for immediate use, but if not, they can be calculated as indicated in Sections 20.6 and 20.7 when geodetic latitude and geodetic longitude are known. State plane coordinates and the grid azimuths to a nearby azimuth mark are published by the NGS for most stations in the national horizontal network. In most areas, many other stations set by local surveyors exist that also have state plane coordinates and reference grid azimuths.

It is important to note that if a survey begins with a given grid azimuth and ties into another, directions of all intermediate lines are automatically grid azimuths. Thus, corrections for convergence of meridians are not necessary when the state plane coordinate system is used throughout the survey. However, as demonstrated in Section 20.8.2, the arc-to-chord correction should be considered and applied to observed angles when appropriate. Assuming that starting stations meeting the above described conditions are available, then there isn't any difference between making traverse computations in state plane coordinates and the procedures given for plane surveys in Chapter 10.

To illustrate the procedure of computing a traverse in the SPCS 83, the following example is solved step by step.

■ Example 20.9

The traverse illustrated in Figure 20.9 originates from station Hayfield NE and closes on station 21002, both in the North zone for Pennsylvania. The reduction of both the distances and angles to the SPCS 83 grid were demonstrated in Section 20.8. Using these values, compute and adjust the traverse, and determine the state plane coordinates of all traverse stations.

Solution

1. The computed azimuth of the line station 21002 to $AZ\ Mk_2$ is compared with its fixed control value. The difference (+9.0″) represents the traverse angular misclosure. This misclosure is divided by the number of angles (five) to get the correction per angle (−1.8″). [This calculation is shown at the bottom of Table 20.9.] It should be interesting to note that if the observed angles had been used in the computations, the angular misclosure would have been +10″ or 1″ greater than is appropriate. Also note that for lines this short the arc-to-chord corrections are minimal and could easily have been avoided without appreciably affecting the final solution.
2. Traverse computations are performed using the same steps as described in Chapter 10. The procedure, shown in Figure 20.12, includes (1) calculating departures and latitudes [columns (1) and (2)], (2) adjusting the departures and latitudes [columns (3) and (4)], and (3) determining the station coordinates [columns (5) and (6)]. Adjustment of departures and latitudes in this example has been done by compass rule, but any method could be used including least squares. In the adjustment, the differences in eastings (X) and northings (Y) between control points were computed and checked against their fixed values to obtain the misclosures in departure (+0.164 m) and latitude (+0.284 m). An adjustment was then made to correct these computed differences to the required totals. The relative precision of the traverse was

TABLE 20.9 REDUCED HORIZONTAL DISTANCES, ANGLES TO THE RIGHT, ANGLE MISCLOSURE, AND ADJUSTED AZIMUTHS FOR EXAMPLE 20.9

Station	Reduced Horizontal Distance (m)	Corrected Angle to the Right	Preliminary Azimuth	Adjusted Azimuth
$AZ\ Mk_1$			161°58′03.0″ (fixed)	161°58′03.0″ (fixed)
HAYFIELD NE		84°58′57.9″		
	3732.227		66°57′00.9″	66°56′59.1″
B		249°22′16.8″		
	5175.146		136°19′17.7″	136°19′14.1″
C		176°50′03.8″		
	4116.122		133°09′21.5″	133°09′16.1″
D		133°31′45.7″		
	4993.608		86°41′07.2″	86°41′00.0″
21002		101°53′18.8″		
			8°34′26.0″	8°34′17.0″ (fixed)
$AZ\ Mk_2$				

Angular misclosure = 8°34′26.0″ = +9″
Correction per angle = − 9.0″/5 = −1.8″

634 STATE PLANE COORDINATES AND OTHER MAP PROJECTIONS

```
                                          Unbalanced
   Course    Length      Azimuth         Dep         Lat

    A-B     3,732.227   66°56'59.1"   3434.2528    1461.3098
    B-C     5,175.146  136°19'14.1"   3574.0730   -3742.7447
    C-D     4,116.122  133°09'16.1"   3002.7618   -2815.2943
    D-E     4,993.608   86°41'00.0"   4985.2439     288.9023
            ---------                 ---------    ---------
    Sum = 18,017.103                 14996.3515  -4807.8269


  Misclosure in Departure =  14,996.3315 -  14,996.1680 = 0.1635
  Misclosure in Latitude  =  -4,807.8269 -  -4,808.1110 = 0.2841


              Balanced                       Coordinates
        Dep          Lat        Point       X             Y

      3434.2189   1461.2509      A      745,212.637   127,939.400
      3574.0261  -3742.8264      B      748,646.856   129,400.651
      3002.7245  -2815.3592      C      752,220.882   125,657.825
      4985.1986    288.8236      D      755,223.606   122,842.465
                                 E      760,208.805   123,131.289


      Linear misclosure   = 0.3278
      Relative Precision  = 1 in 55,000
```

Figure 20.12
Modified excerpt of compass rule adjustment done using WOLFPACK.

1:55,000. Had the original distance and angle observations been used in the computations instead of their reduced equivalents, the relative precision of the traverse would have been only 1:10,000. This demonstrates the importance of making proper observational reductions before attempting an adjustment.

In summary, the following steps are necessary for performing traverse computations in state plane coordinates:

1. Obtain a starting and closing azimuth, and, if necessary, reduce them to grid azimuths.
2. Analyze the scale factor for the project. A mean of the published scale factors may be adequate for the project. This can be done by analyzing the number of significant figures in the longest measured length versus the number of common digits in the scale factors. To avoid rounding errors there should be one more common digit in the scale factors than there is in the longest observed distance.
3. Analyze the elevation factor for the project. A mean factor may be adequate in terrain with small relief. Again the elevation factor for the highest and lowest station elevations in the project should have one more common digit than the number of significant figures in the longest observed length to avoid rounding errors.
4. If a project scale factor and elevation factor can be used, compute a combined factor for the project, otherwise reduce all distances to grid using one of the previously described methods.
5. Reduce all horizontal distances to their grid equivalents.

For lines under 8 km and lower-order surveys, steps 6 through 8 are typically not followed, however, the decision to follow or not follow steps 6 through 8 should be based on the intended accuracy of the survey.

6. Using preliminary azimuths derived from unreduced angles and grid distances, compute approximate coordinates.
7. Analyze the magnitude of the arc-to-chord correction for each line using the approximate coordinates.
8. Apply the arc-to-chord corrections to the observed angles.
9. Compute and adjust the traverse.
10. Compute the final adjusted SPCS 83 coordinates for the new stations. If adjusted ground distances are required, apply the inverse of the combined factor to each line.

20.10 SURVEYS EXTENDING FROM ONE ZONE TO ANOTHER

Surveys in border areas often cross into different zones or even abutting states. However, this does not represent a problem because adjacent zones overlap by appreciable distances, as shown in Figures 20.3 and 20.4.

The general procedure for extending surveys from one zone to another requires that the survey proceed from the first zone into the overlap region with the second. Then the geodetic latitudes and longitudes are computed for two intervisible stations using their grid coordinates in the first zone. (Recall that this conversion is called the *inverse problem.*) Using the geodetic positions of the two points, their state plane coordinates in the new zone are then computed. (This is the *direct problem.*) Finally, the grid azimuth for the line in the new zone can be obtained from the new coordinates of the two points.

Suppose, for example, that a survey being computed in SPCS 83 originates in southern Wisconsin, which uses the Lambert conformal conic projection, and extends into the western zone in northern Illinois, which uses the Transverse Mercator grid. With Wisconsin south zone SPCS 83 coordinates of two intervisible points in the overlap region of the two zones known, Equations (20.3) are solved for the geodetic latitudes and longitudes of the points.

With the geodetic latitudes and longitudes of the two points known, Equations (20.4) are solved using constants for the appropriate Illinois zone entered to obtain their E and N coordinates in that zone. From these coordinates, the grid azimuth of the line joining the intervisible points can be calculated by using Equation (11.5), and the survey can continue into Illinois.

If immediate coordinate values in the new zone are not required, the entire survey can be computed in one zone. The inverse problem can then be used to compute geodetic coordinate values of the points followed by the direct problem to compute grid coordinates in the second zone. For example, the survey in the previous paragraph could be computed entirely in the Wisconsin South Zone if the control stations in the Illinois Western Zone are inversed to obtain their geodetic values and then converted into Wisconsin South Zone coordinate values. This can be done using the inverse problem to obtain the geodetic coordinates for the control stations in northern Illinois. These geodetic coordinates are then converted into their equivalent Wisconsin South Zone values. Additionally, any control azimuths in the Illinois Western Zone must be converted to their geodetic equivalent values and then to their Wisconsin South Zone values. That is, all control used in the survey must be converted into their common, single zone values. Following the adjustment of the survey, the stations in Illinois can be converted from their adjusted Wisconsin South Zone values to their equivalent values in the Illinois Western Zone following the same procedure.

It should be remembered that the zone limits do not mark the end of the map projection, but simply the extents of the zone where 1:10,000 precisions are maintained between grid and geodetic lengths. If the proper reductions as presented in Section 20.8 are performed to the observed distances, the use of a single zone for computational purposes can be extended well into neighboring zones without loss of accuracy to the survey. For example, the zones in Pennsylvania can be used to perform traverse computations in neighboring New Jersey, Ohio, Maryland, New York, and so on. Once the Pennsylvania grid coordinates for the points are determined, their geodetic equivalents can be determined using inverse computations and converted to the appropriate state zone with direct computations.

Solving the direct and inverse problems that are necessary in this procedure is most conveniently handled using the computer programs described previously.

20.11 THE UNIVERSAL TRANSVERSE MERCATOR PROJECTION

The Universal Transverse Mercator (UTM) system is another important map projection that has worldwide use. Originally developed by the Department of Defense primarily for artillery use, it provides worldwide coverage from 80°S latitude to 80°N latitude. Each zone has a 6° longitudinal width; thus 60 zones are required to encircle the globe. The UTM system is a Transverse Mercator map projection, and thus uses the equations presented earlier in this chapter. It has recently taken on added importance for surveyors, since UTM coordinates in metric units are now being included along with state plane and geodetic coordinates for all published NAD 83 station descriptions. UTM grids are also being included on all maps in the national mapping program, and UTM coordinates are being used more frequently for referencing positions of data entered into Land and Geographic Information Systems.

UTM zones are numbered easterly from 1 through 60, beginning at longitude 180°W. The conterminous United States (CONUS) is covered from zone 10 (west coast) through zone 20 (east coast). The central meridian for each zone is assigned a false easting E_0 of 500,000 m. A false northing N_b of zero is applied for the northern hemisphere of each zone, and 10,000,000 m is assigned for the southern hemisphere to avoid negative Y coordinates. To specify the position of any point in the UTM system, the zone number must be given as well as its northing and easting.

In the UTM system, each zone overlaps adjacent ones by 0°30′. Since zone widths of 6° are considerably larger than those used in state plane systems, lower accuracies result. The scale distortion is deliberately 1 part in 2500 ($k_0 = 0.9996$) too small along the central meridian. At 1800 m from the central meridian there is no distortion. Beyond this point the scale is too large, and on the Equator at the zone edges, the distortion is 1:1000. Equations for calculating X and Y coordinates in the UTM system are the same as those for the Transverse Mercator projection. As with the state plane systems, tables giving formulas and constants for the system are available. Also, like the state plane coordinate system, the datum of reference must be specified. Because UTM coordinates are available for all points within the NAD 83, calculations between very widely spaced points can readily be made. This is convenient and entirely consistent with current capabilities for conducting surveys of global extent with new devices such as GNSS receivers. The Universal Transverse Mercator map projection spreadsheet, which contains a graphic showing zone boundaries, is included in the Excel spreadsheet *map_projections.xls*, which is on the companion website.

■ 20.12 OTHER MAP PROJECTIONS

The Lambert conformal conic and Transverse Mercator map projections are designed to cover areas extensive in east–west and north–south directions, respectively. However, these systems do not conveniently cover circular areas or long strips of the Earth that are skewed to the meridians. Two other systems, the *oblique stereographic* and the *Oblique Mercator* projections, satisfy these problems.[2]

20.12.1 Oblique Stereographic Map Projection

The oblique stereographic projection can be divided into two classes: *tangent plane* and *secant plane*. In either case, as illustrated in Figure 20.13, the projection point P (the origin) is on the ellipsoid where a line perpendicular to the map plane and passing through center point O intersects the ellipsoid. In the tangent plane system, ellipsoid points a and b are projected outward to a' and b', respectively, on the map plane. For the secant plane system, ellipsoid points c and d are projected inward to c' and d' on the map plane. (If they were outside of the secant points, projection would be outward.) The oblique stereographic map projection is conformal, and thus preserves the shapes of objects.

Oblique stereographic projections are not employed in the United States typically but are used in Canada and other parts of the world. As discussed in

---

[2] The oblique Mercator projection is also called the Hotine skew orthomorphic projection, named after the English geodesist Martin Hotine.

638 STATE PLANE COORDINATES AND OTHER MAP PROJECTIONS

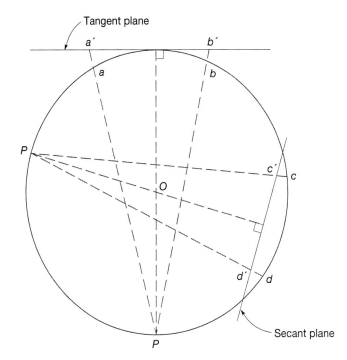

Figure 20.13
Tangent plane and secant plane horizon stereographic map projections.

Section 19.7.2, these projections are also used to convert geodetic coordinates determined by GNSS surveys into map projection coordinates for use in the localization process. If point P is the North or South Pole, the projection is called *polar stereographic*; if it is on the equator, *equatorial stereographic*.

The defining parameters of this projection are the latitude and longitude of the grid origin (φ_0, λ_0) and the scale factor at the grid origin k_0. It uses common functions

$$\chi(\varphi) = 2\tan^{-1}\left[\tan\left(\frac{\pi}{4} + \frac{\varphi}{2}\right)\left(\frac{1 - e\sin\varphi}{1 + e\sin\varphi}\right)^{e/2}\right] - \frac{\pi}{2} \tag{20.61}$$

$$M(\varphi) = \frac{\cos\varphi}{\sqrt{1 - e^2\sin^2\varphi}} \tag{20.62}$$

where e is the eccentricity of the ellipsoid as defined by Equation (19.2a). Using Equation (20.61), the zone constants for the projection are

$$\chi_0 = X(\varphi_0) \tag{20.63}$$

$$m_0 = M(\chi_0) \tag{20.64}$$

Using the latitude φ and longitude λ of the point, and the semimajor axis a for the ellipsoid, the equations for the direct problem are

$$\chi = X(\varphi) \tag{20.65}$$

$$m = M(\varphi) \tag{20.66}$$

$$A = \frac{2ak_0m_0}{\cos\chi_0[1 + \sin\chi_0\sin\chi + \cos\chi_0\cos\chi\cos(\lambda - \lambda_0)]} \tag{20.67}$$

$$E = A\cos\chi\sin(\lambda - \lambda_0) \tag{20.67}$$

$$N = A[\cos\chi_0\sin\chi - \sin\chi_0\cos\chi\cos(\lambda - \lambda_0)] \tag{20.68}$$

$$k = \frac{A\cos\chi}{am} \tag{20.69}$$

■ Example 20.10

The following geodetic coordinates are observed using GNSS methods. What are the oblique stereographic map projection coordinates for Station A using a grid origin of (41°18′15″N, 76°00′00″W) and $k_0 = 1$? (Use WGS84 ellipsoidal parameters.)

Station	Latitude	Longitude	Height (m)
A	41°18′09.88223″N	75°59′58.05637″W	282.476
B	41°18′21.11176″N	76°00′37.35445″W	296.571
C	41°18′19.33293″N	75°59′40.39279″W	313.814
D	41°18′09.67030″N	75°59′44.19645″W	304.205

Solution

The zone constants are
By Equation (20.63):

$$\chi_0 = 2\tan^{-1}\left[\tan\left(\frac{\pi}{4} + \frac{41°18'15''}{2}\right)\left(\frac{1 - e\sin 41°18'15''}{1 + e\sin 41°18'15''}\right)^{e/2}\right] - \frac{\pi}{2}$$
$$= 41°06'48.66298''$$

By Equation (20.64):

$$m_0 = \frac{\cos 41°18'15''}{\sqrt{1 - e^2\sin^2 41°18'15''}} = 0.752314$$

By Equation (20.65):

$$\chi = 2\tan^{-1}\left[\tan\left(\frac{\pi}{4} + \frac{41°18'09.88223''}{2}\right)\left(\frac{1 - e\sin 41°18'09.88223''}{1 + e\sin 41°18'09.88223''}\right)^{e/2}\right]$$
$$= 41°06'43.54972''$$

By Equation (20.66):

$$m = \frac{\cos 41°18'09.88223''}{\sqrt{1 - e^2 \sin^2 41°18'09.88223''}} = 0.752330$$

By Equation (20.67):

$$A = \frac{2(6{,}378{,}137)(1)0.752330}{\cos\chi_0[1 + \sin\chi_0 \sin\chi + \cos\chi_0 \cos\chi \cos(-75°59'58.05637'' + 76°00'00'')]}$$
$$= 6{,}368{,}873.344\,\text{m}$$

By Equation (20.68):

$$E = A \cos 41°06'43.54972'' \sin(-75°59'58.05637'' + 76°00'00'') = 45.218\,\text{m}$$

By Equation (20.69):

$$N = A[\cos\chi_0 \sin\chi - \sin\chi_0 \cos\chi \cos(-75°59'58.05637'' + 76°00'00'')]$$
$$= -157.869\,\text{m}$$

By Equation (20.70):

$$k = \frac{6{,}368{,}873.344 \cos 41°06'43.54972''}{6{,}378{,}137(0.752330)} = 1.00000$$

Note in the example that the direct problem was not defined with false easting or northings, and thus one of the resultant coordinates has a negative value. For localization, as discussed in Section 19.7.2, this is not a problem since the user never views these coordinates. However, a false easting and northing could be added to Equations (20.68) and (20.69) to provide positive coordinates values in a limited region.

The inverse problem converts the map projection coordinates of (N, E) to geodetic latitude and longitude. The equations used in the inverse problem for the oblique stereographic map projection are

$$\rho = \sqrt{E^2 + N^2} \tag{20.71}$$

$$c = 2 \tan^{-1}\left(\frac{\rho \cos\chi_0}{2ak_0 m_0}\right) \tag{20.72}$$

$$\chi = \sin^{-1}\left(\cos c \sin\chi_0 + \frac{N \sin c \cos\chi_0}{\rho}\right) \tag{20.73}$$

$$\lambda = \lambda_0 + \tan^{-1}\left(\frac{E \sin c}{\rho \cos\chi_0 \cos c - N \sin\chi_0 \sin c}\right) \tag{20.74}$$

$$\varphi = 2 \tan^{-1}\left[\tan\left(\frac{\pi}{4} + \frac{\chi}{2}\right)\left(\frac{1 + e \sin\varphi}{1 - e \sin\varphi}\right)^{e/2}\right] - \frac{\pi}{2} \tag{20.75}$$

Using $\varphi = \chi$ in the first iteration, Equation (20.75) is iterated until the change in φ becomes negligible. The scale factor is computed using Equation (20.70). Computations for the oblique stereographic map projection are demonstrated in a Mathcad worksheet *oblique.xmcd* on the companion website.

20.12.2 Oblique Mercator Map Projection

The oblique Mercator projection is designed for areas whose major extent runs obliquely to a meridian, such as northwest to southeast. The projection is conformal and is developed by projecting points from the ellipsoid to an imaginary cylinder that is oriented with its axis skewed to the equator. It is used as the state plane coordinate projection for the southeast portion of Alaska and is being considered for states such as Florida. Computations for the aforementioned map projections are demonstrated in the Excel spreadsheet and Mathcad worksheet on the companion website for this book.

Also, on the companion website are an Excel spreadsheet, *map_projections.xls*, which contains all the map projections presented in this chapter.

20.13 GROUND VERSUS GRID PROBLEM

As discussed in Section 10.10 surveyors often determine the length of a line by inversing coordinates. However, when using map projection coordinates such as state plane coordinates, the inversing of coordinates results in what is known as a *grid length*, which is a distance on the map projection surface between the two points. In SPCS 27 and SPCS 83, the map projection surfaces are designed to be secant to the ellipsoid, and thus the grid lengths often do not match ground lengths very well due to the height of the lines. Since the typical distance needed by surveyors is a horizontal ground length, Equation (20.54) must be rearranged to convert the grid distance to a ground distance as

$$\text{ground distance} = \frac{\text{grid distance}}{CF} \quad (20.76)$$

where *CF* is the combined factor for the line. However, when this operation is performed in the field, the appropriate combined factor for the line may not be known. This is known as the *ground versus grid* problem.

Similarly, GNSS software typically publishes a mark-to-mark distance (see Section 19.15.1), geodetic distance, or grid distance. Again, since surveyors typically need a horizontal ground distance, the combined factor must be known to apply Equation (20.76). One solution to this problem is to design a map projection that is near the topographic surface, and thus provide grid distances that very closely match ground distances and not its geodetic distance. These are known as *low-distortion projections* (LDP). Another solution is to provide the survey controller with an average combined factor for the project, which is known as a *project factor*. This section discusses both methods and looks at the advantages and disadvantages of each method.

20.13.1 Low-Distortion Projections

To understand what a low-distortion projection does consider the Transverse Mercator map projection. Its defining parameters are the defining parameters for the ellipsoid, such as a and f, grid origin (φ_0, λ_0), false northing and easting (N_b, E_0), and the scale factor k_0 at the central meridian. If k_0 is set equal to one, the map projection surface, which is a cylinder, will be tangent to the ellipsoid. In the state plane coordinate system, (SPCS) the scale factor is set to 0.9999 typically to maintain a 1:10,000 precision between the grid distance and the geodetic distance. This scale factor makes the cylinder secant to the ellipsoid. However, if the scale factor is greater than one, the map projection surface will be above the ellipsoid. Thus, the precision between a grid and geodetic distance will decrease but the precision between the grid distance and observed horizontal ground distance will increase. By using a scale factor that is based by the average ellipsoid (geodetic) height of the project site, the map projection surface can be brought to this height from its geodetic equivalent. This can be seen by rearranging Equation (20.46) as

$$L_m = L_e\left(1 + \frac{h_{\text{avg}}}{R_e}\right) \qquad (20.77)$$

where L_m and L_e are the ground and geodetic distances, respectively, h_{avg} is the average geodetic height of the project site, and R_e is the average radius of the Earth at the project site. The parenthetical () portion of Equation (20.77) is the scale factor, k_{LDP}, needed to bring the map projection surface tangent to the Earth's surface at the geodetic height of h_{avg} where $k_{LDP} = 1 + h_{\text{avg}}/R_e$. This entire process can conceptually be thought of as raising the developable surface to a height near the average ground height. However, as was warned by Snyder (see Section 20.2) and will be mentioned later, this process can introduce some problems mathematically with angles.

It is known that precision can be increased throughout the region by making the map projection surface secant to the projected surface. For the Lambert conformal conic map projection, this is achieved by having the two standard parallels separated so that they contain about two-thirds of the defined zone limits for the projection as was discussed in Section 20.3. Thus, when using a Lambert Conformal Conic map projection to create an LDP, the scale factor k_{LDP} determined from Equation (20.77) is used to scale all coordinates as defined by Equation (20.14). This means that the LDP coordinates for the Lambert conformal conic map projection are

$$\begin{aligned} E_{LDP} &= (R\sin\gamma + E_0)k_{LDP} \\ N_{LDP} &= (R\cos\gamma + N_b)k_{LDP} \end{aligned} \qquad (20.78)$$

where E_{LDP} and N_{LDP} are the coordinates of the point on the low-distortion projection.

A better approach to defining a low-distortion projection using the Lambert conformal conic projection is to define the projection with a single central parallel and a scale factor as discussed in Section 20.14. If this approach is used, the map designer only needs to define the central parallel for the projection and the scale factor at the central parallel, which can be determined as presented. When the scale factor is equal to one, the projection is tangential to the ellipsoid. If it is less than one, the projection is secant to the ellipsoid, and when it is greater than one, the projection is above the ellipsoid. This approach removes the need for defining standard parallels at approximately two-thirds the zone width. Additionally, it scales the radii in the projection rather than the coordinates.

As discussed earlier, to make the Transverse Mercator projection secant to the ellipsoid, a scale factor k_0 of 0.9999 was typically used for states that use the Transverse Mercator map projection for their state plane coordinate system (SPCS). But this scale factor would only provide a 1:10,000 distance precision with observed distances at the average geodetic height of the surface. Actually, the precision is even less than 1:10,000 in most cases, since the average geodetic height of topography in the conterminous U.S. is about 750 m, which increases the scale error to about 1:4600 (and even more in higher areas). Since today's instruments can provide higher precisions than 1:10,000 for the typical survey, it is recommended to use a scale factor that is higher than that obtained by a typical survey. For example, a scale factor of 0.99999 would provide a distance precision of 1:100,000 when combined with the parenthetical expression $(1 + h_{avg}/R_e)$. This would create a map projection surface secant to the region at the average geodetic height of the project site. Thus, the defining k_{LDP} for the Transverse Mercator map projection in this example could be

$$k_{LDP} = 0.99999\left(1 + \frac{h_{avg}}{R_e}\right) \qquad (20.79)$$

This will provide a cylinder that is secant to the average geodetic height near the surface of the Earth and may provide distance precisions near or better than 1:100,000 between and a distance obtained by inversing its LDP coordinates and the horizontal distance observed with a total station. The reason for saying that it may provide this precision is due to the elevation variation in the project site. Similar to state plane coordinates, the actual precision will vary with the discrepancies between the average geodetic height and the actual geodetic height of the line as defined by Equation (20.47). As the discrepancies in elevation between the average geodetic height and height of the line increases, the precision between the map projection length and observed horizontal distance between two points will decrease. This is true whether the line is above or below the average geodetic height.

Besides being a method to solve the ground versus grid problem, a low-distortion projection provides a means through the use of the inverse problem as discussed in Sections 20.6.3 and 20.7.3 to convert their coordinates back into geodetic latitudes and longitudes, which can then be converted to state plane coordinates and connected to the National Spatial Reference System as discussed in Section 19.6 or any other globally recognized coordinate systems such as UTM discussed in Section 20.11.

Advantages of low-distortion projections over scaling state plane is that they can be designed to minimize distortion over reasonably sized regions, and that they facilitate data transferability since they are defined in a way that other software can understand (most notably GIS).

However, the ability to convert to other coordinate systems is only true if the defining parameters for the low-distortion projection are saved in perpetuity so that geodetic coordinates can be determined from LDP coordinates. Assume that a company returns to a job site where they had monumented project control on the site a few years earlier. They could tie into this control using a GNSS receiver if the defining parameters for the low-distortion projection have been saved. However, if these parameters are unavailable for the coordinates of the control, it would be impossible to perform the inverse problem, and thus the coordinates would become an arbitrary coordinate system. In this case, the GNSS operator would need a localization on the project control using the methods discussed in Section 19.7.2.

Another disadvantage is that all controllers used by field crews on the project must have the low-distortion projection parameters entered in the software and loaded in the project. If this does not occur, large discrepancies will probably occur between the LDP coordinates and what the field crew establishes or observes without these parameters. Additionally, to achieve higher precisions than state plane coordinates, the size of the region covered by the low-distortion projection must be limited. Thus, it is more likely that procedures to transfer surveys from one projection zone to another will occur more frequently, and the procedures discussed in Section 20.10 will be needed to resolve the differences in previously established low-distortion projections. As implied by Snyder (1987) on page 5, secant constructions of this kind are not geometrically correct. Finally, Rollins and Meyer (2019) show in detail that the above procedures do not produce conformal projections of surfaces of constant height-above-the-ellipsoid. Such surfaces provide rigor to the concept of horizontal-distance-at-elevation, whereas the secant developable surfaces do not. Thus, the above LDP method is nonconformal, meaning that angles and shapes of objects will be distorted. While this distortion is slight, Rollins and Meyer show that it can be as high as 2 ppm, which is similar to the difference between the survey and international foot definitions.

■ Example 20.11

A surveying firm has several contracts in a city that is bounded by the geodetic coordinates given in the following table. They decide to use a single low-distortion projection in the city to resolve the grid-versus-ground problem. Design a suitable low-distortion projection using the GRS80 ellipsoid. Determine the map precision of a line that is 1000 m long having endpoint geodetic coordinates of (41°16′28.45321″N, 76°54′29.34085″W, 289.041 m) and (41°16′28.45098″N, 76°55′12.30619″W, 289.121 m).

Corner	Latitude	Longitude	height (ft)
SW	41°12′46″N	77°05′33″W	700
NE	41°16′34″N	76°55′12″W	868

Solution

The region to be covered is long in the east–west direction. Thus, the Lambert conformal conic map projection is the appropriate choice for use in this city. The difference in latitude between the northeast (NE) and southwest (SW) corner is 228″. One-sixth of the zone limits should be outside the standard parallels to ensure a uniform precision. One-sixth of 228″ is 38″. Thus, the south standard parallel should be 38″ north of 41°12′46″N, or it should be at 41°13′24″N. Similarly, the north standard parallel should be 38″ south of 41°16′34″N, or it should be at 41°15′56″N. The grid origin is typically placed on the southern zone limit and at the average of the two longitudes. Thus, the grid origin should be at (41°12′46″N, 77°00′22″W). Note that fractional seconds have been avoided in defining the parameters for this projection. This should always be the case. To avoid negative coordinates, the false northing and easting are set to 10,000 m and 20,000 m. The average geodetic height for the city is 784 ft or 239 m. Using Equation (19.7), a rounded Gaussian mean radius at the latitude of 41°15′56″ is about 6,375,000 m. Using these values, the appropriate scale factor for the low-distortion projection is determined as

$$k_{LDP} = 1 + \frac{239}{6{,}375{,}000} = 1.000037$$

Using these defining parameters the low-distortion projection *ne* coordinates for the endpoints of the line in meters are (18,208.349, 26,868.243) and (17,208.352, 26,867.116). Inversing these coordinates, we find that the grid distance is 999.9975 m, which results in a precision with the observed ground distance of 1000.000 m of over 1:402,000. Since survey precisions are less than this, the grid distance can be considered to be the same as the horizontal ground distance at the precision of the survey.

When developing a low-distortion projection, it is still possible that there are areas in the project region that have grid distance precisions that are less than what is necessary for the project during the layout due to height variations. In these cases, the computations as defined by Equations (20.54) and (20.76) are still applicable for typical surveying distances.

20.13.2 Using a Single Project Factor

Another approach to the ground versus grid problem is available in today's survey controllers. When establishing a project, part of the project definition is a project factor. This is a scale factor used by the survey controller to reduce observed horizontal distances to grid distances using Equation (20.54) or determine ground lengths using Equation (20.76) when coordinates are inversed to obtain a grid distance. Like using LDPs, when using state plane coordinates the project factor should be set to an average SPCS combined factor for the project site. This would involve an average scale factor and average elevation factor. By providing the survey controller with an appropriate average combined factor, the coordinates computed by the controller will be on the SPCS map projection surface and any grid distance to be staked out will be converted to a horizontal ground length.

Again, since it is working with an average combined factor instead of one unique combined factor for each line, precisions will decrease as the discrepancy between the average geodetic height and the actual geodetic height of the line increase. The advantage of this approach is that state plane coordinates are generated for the project. The parameters for these map projections are built into most software today, and thus there is no need save these parameters. In fact, it is possible to recreate the average combined factor later if the one used to create the SPCS coordinates is lost. Thus, if the resulting precision is higher than that obtainable by the survey, the coordinates generated, or the distance staked out, will be performed at the precision of the survey. However, this approach does have one disadvantage, which is common to state plane coordinates as previously discussed; if the coordinates are inversed without using Equation (20.76), the precision between the grid distance and horizontal ground distance will decrease as the discrepancy between the height used to compute the project factor and the actual line height increases. However, this is also true using LDPs.

Using the data given in Example 20.11, the average scale factor for the project area is 0.99996153. Using the average geodetic height of 239 m, the average elevation factor is 0.99996249. Thus, the average combined factor for the project area is 0.99992402, which would be entered as the project factor in the survey controller. The region in Example 20.11 lies in the Pennsylvania North zone, and the SPCS *ne* coordinates for the endpoints of the line are (123,379.452, 670,530.504) and (123,369.733, 669,530.637). Thus, the SPCS grid distance is 999.914 m. Applying Equation (20.76) to this distance using the project factor, the resulting distance is 999.990 m. This results in a distance precision over 1:100,000, which is better than the typical survey precision, and thus is an acceptable length for stakeout. Again, if the average combined factor is entered into a survey controller for the project factor, these computations will occur without interaction from the user. If the project factor had not been applied to the grid distance, the precision of the line would only be 1:11,800, which is almost ten times worse than the converted distance. Again, the difference between the grid length and ground length of a line will increase as discrepancies between the design height of 239 m and the actual height of the line increases.

It should be pointed out that Example 20.11 was developed to produce a worst-case scenario of precision in both instances for this region. Higher precisions will typically be seen in the project area as elevations approach the average elevation or the scale factor of a line approached the average scale factor for the project area. The use either method shown in 20.13.1 and 20.13.2 is crucial to surveying in areas of higher elevations such as in Denver, CO or Flagstaff, AZ where heights are 1 mi. and greater above the ellipsoid. Without these approaches and by ignoring proper distance reductions, the elevations may result in distance precisions that are 1:4000 or less.

20.14 PROPOSED CHANGES TO SPCS IN 2022

As discussed in Section 20.13, linear distortions can increase dramatically as the height of the observed distance above the ellipsoid increases. To help reduce this distortion, zones in the State Plane Coordinate System of 2022 (SPCS 2022) will be designed to reduce linear distortion at the topographic surface, rather than at the ellipsoid surface (as done for SPCS 27 and 83). In addition, states will be allowed to select low-distortion projections for their SPCS 2022 zones. However, it is unlikely that NGS will have the resources to design low-distortion projections, so states must design their own (upon approval from NGS).

SPCS 2022 zones not designed by states will be designed by NGS. The default NGS designs in most cases will use the same zone extents and projection type as used for SPCS 83 but will differ in how the projection axis is handled. The *projection axis* is the line or curve along which the linear distortion with respect the ellipsoid is minimum and constant. For SPCS projections, it is the central meridian of the transverse Mercator, the central parallel of the Lambert conformal conic, and the skew axis of the oblique Mercator (distortion is nearly but not quite constant along the skew axis).

There are two main ways that the projection axis will be treated differently for SPCS 2022. One concerns how its location and scale factor are determined, and the other is that Lambert conformal conic projections will be based on a one-parallel rather than two-parallel definition. The location of the projection axis will be selected to minimize the *range* in linear distortion within a zone at the topographic surface (the range is only affected by axis location, not by its scale). Its scale factor will be determined through a combination of maximizing the area and population within the zone distortion design criterion, while also attempting to balance positive and negative distortion within the zone.

For states using the transverse Mercator, the design process is as described above. First the central meridian location is determined such that the range in linear distortion is minimized; in many zones the central meridian will shift by one to several arc-minutes from its SPCS 83 location. Then a scale factor is assigned to achieve a distortion design criterion at the topographic surface, such as 1:10,000 or ± 100 parts per million (ppm). In many cases, the resulting scale factor will be greater than that used for SPCS 83 since the topography is usually above the ellipsoid and increasing the scale in essence raises the developable surface. Whether the scale is greater than, less than, or equal to 1 depends on both the topographic height and the width of the zone.

As an example, consider the New Mexico Central Zone show in Figure 20.14. For the preliminary SPCS 2022 default design, its central meridian shifts eastward from E 253°45′ to E 253°54′ because elevations increase significantly from east to west in that zone, and the central meridian scale increases from 0.9999 to 1.00022 because of its overall high elevation (average ellipsoid height of about 1800 m), so in this case the developable surface is above the ellipsoid. For this design, 72% of the zone area and 92% of the population are within ± 100 ppm ppm (1:10,000) linear distortion. Compare to the SPCS 83 design, where 0% of the area and population are within ± 100 ppm distortion.

648 STATE PLANE COORDINATES AND OTHER MAP PROJECTIONS

Figure 20.14
Map showing linear distortions in (a) SPCS 83 New Mexico Central zone and (b) a preliminary SPCS 2022 New Mexico Central zone design.
(Courtesy National Geodetic Survey.)

It should be realized that while this process will reduce the grid versus ground problem, it will not remove it since over large areas, many states have too much height variation to achieve such results.

In contrast to New Mexico, for the preliminary SPCS 2022 default design of the Mississippi East Zone, the central meridian scale increases only slightly, from 0.99995 to 0.99997, so that the developable surface remains secant. For the Georgia West Zone, the central meridian scale increases from 0.9999 to exactly 1, so the developable surface becomes tangent to the ellipsoid.

The situation and design process are the same for Lambert conformal conic zones in SPCS 2022. But, as mentioned above, this projection will be defined using one central parallel, rather than a north and south standard parallel as done in SPCS 27 and 83. Note that one- and two-parallel Lamberts are mathematically identical. In fact, the central parallel is the projection axis for either type, so in essence the one-parallel definition is more consistent with the mathematical definition (and somewhat simpler computationally). The only difference is that the one-parallel form requires that the projection axis scale be specified explicitly, whereas for the two-parallel form, the scale is defined implicitly by the separation between the north and south standard parallels. The main reasons for requiring the one-parallel form are (1) the defining scale is much more readily apparent since it is stated rather than inferred; (2) it is easier to design zones with respect to distortion at the topographic surface with one parallel than it is with two parallels; (3) it is consistent with the transverse and oblique Mercator projections, which both use explicit projection axis scales; (4) when used for most low-distortion projections they must be one-parallel, and so using one-parallel for all Lambert zones also improves system consistency; and (5) the number of defining parameters is reduced to five (the same as for a transverse Mercator), if the central parallel is set equal to the latitude of grid origin, which is required for SPCS 2022.

Although mathematically identical, it is worthwhile to discuss how the implementations differ. In SPCS 83, the Lambert conformal conic projection is defined by the ellipsoid (see Section 19.2), its grid origin (φ_0, λ_0), north and south standard parallels (φ_N, φ_S), and false easting and northing of E_0 and N_b, respectively. In SPCS 2022, the north and south standard parallels will no longer be defined. Instead these two parameters will be replaced by the latitude of the central parallel and the scale factor at the central parallel, which allows for a more simplistic definition. It should be realized that if the scale at the central parallel is equal to one, the map projection surface is tangent to the ellipsoid, when it is less than one, the surface is secant to the ellipsoid, and when it is greater than one, the surface is above the ellipsoid.

For example, in the Pennsylvania north zone, SPCS 83 currently defines the south and north standard parallels as N 40°53′ and N 41°57′, which results in a 0.9999568402 scale factor at a central parallel of N 41°25′02.66745″ where the central parallel is not exactly the average of the standard parallels at N 41°25′ to enforce the conformality condition. Thus, the use of two standard parallels can be replaced by the latitude of the central parallel and a scale factor, which would result in the same zone definition. When the Lambert conformal conic map projection is defined, as it currently is, with a north and south standard parallel, the latitude of the central parallel is computed as

$$\varphi_{CP} = \sin^{-1}\left\{\frac{2\ln\left(\dfrac{\cos\varphi_S}{\cos\varphi_N}\right)\sqrt{\dfrac{1 - e^2\sin^2\varphi_N}{1 - e^2\sin^2\varphi_S}}}{\ln\left(\dfrac{1 + \sin\varphi_N}{1 - \sin\varphi_N}\right) - \ln\left(\dfrac{1 + \sin\varphi_S}{1 - \sin\varphi_S}\right) + e\left[\ln\left(\dfrac{1 + e\sin\varphi_S}{1 - e\sin\varphi_S}\right) - \ln\left(\dfrac{1 + e\sin\varphi_N}{1 - e\sin\varphi_N}\right)\right]}\right\}$$

(20.80)

where e is the eccentricity of the ellipse and φ_N and φ_S are the north and south standard parallels of the two-parallel zone.

Although Equation (20.80) is useful for computing the central parallel of a two-parallel Lambert, it is not needed for a one-parallel version. Instead, the latitude of the central parallel (φ_{CP}) and the scale factor at the central parallel (k_0) are specified as defining parameters, with modifications to Equations 20.4 through 20.11, as follows. Since the north and south standard parallels will no longer be defined, the computation of w_1 and w_2, for example, in Equations (20.4) and (20.5), will no longer be necessary nor possible. Instead these two equations will be replaced by one equation defined as

$$w_1 = W(\varphi_{CP}) \qquad (20.81)$$

where φ_{CP} is the latitude of the central parallel. Similarly, Equations (20.6) and (20.7) and Equations (20.9) and (20.10) are replaced by the single equations

$$m_1 = M(\varphi_{CP}) \qquad (20.82)$$

$$t_1 = T(\varphi_{CP}) \qquad (20.83)$$

Additionally, Equation (20.11) is computed as

$$n = \sin\varphi_{CP} \qquad (20.84)$$

and Equation (20.13) is modified as

$$\rho(t) = k_0 a F t^n \qquad (20.85)$$

where k_0 is the scale factor at the central parallel.

The remaining equations are computed as previously defined in Section 20.6. Note that less computations are required for the one-parallel Lambert, since three equations (for w_2, m_2, and t_2) are not evaluated, and a simpler equation is used for n. Also, t_0 is not evaluated if the central parallel equals the latitude of grid origin, as will be required for SPCS 2022.

Conceptually the modification to Equation 20.85 can be thought of as enlarging or reducing the developing cone. These equations are the same as for a tangential developable surface, which has scale factor of one, where the desired scale factor can either produce a state plane zone that is comparable to those in SPCS 83 ($k_0 < 1$) or a low-distortion projection (usually $k_0 > 1$). This equivalence with the SPCS 83 PA North zone computations as demonstrated in Example 20.1 is shown in the following example and repeated in the MathCad file *Lambert-Single Parallel.xmcd*.

■ Example 20.12

Using the central parallel for the PA SPCS 83 North zone of N 41°25′02.66745″, the computed scale factor, k_0, for this latitude of 0.9999568402, and the station given in Example 20.1, compute the coordinates of the station using a scaled tangent Lambert conformal conic projection.

Solution

Using Equations 20.81–20.85, the compute zone constants are:

$$w_1 = W(\varphi_{CP}) = 0.9985340786$$

$$m_1 = M(\varphi_{CP}) = 0.7510110381$$

$$t_0 = T(\varphi_0) = 0.4665502273$$

$$t_1 = T(\varphi_{CP}) = 0.4533396213$$

$$n = \sin\varphi_{CP} = 0.6615397338$$

$$F = \frac{m_1}{nt_1^n} = 1.9159306051$$

$$R_b = \rho(t_0) = 7{,}379{,}348.367 \text{ m}$$

The direct problem is then solved as:

$$t = T(41°18′20.25410″) = 0.4545150743$$

$$m = M(41°18′20.25410″) = 0.7522972496$$

$$R = \rho(t) = 7{,}252{,}862.7946$$

$$\gamma = n(77°45′ - 76°00′57.00239″) = 1°08′49.991″$$

$$E = R\sin\gamma + E_0 = 745{,}212.637 \text{ m}$$

$$N = R_b - R\cos\gamma + N_b = 127{,}939.400 \text{ m}$$

Note that these results match those in Example 20.1 exactly. When performing the inverse problem, the only modification is to Equation (20.18) where R is divided by the defining scale factor and is computed as

$$R = \frac{N'}{k_0 \cos \gamma} = \frac{\sqrt{E'^2 + N'^2}}{k_0} \quad (20.86)$$

One of the requirements for SPCS 2022 is that all defining latitude and longitude parameters be specified to the nearest arcminute and that the scale factor be limited to six places in the decimals. As with the transverse Mercator, the central parallel scale can be less than, greater than, or equal to 1, and in some cases the central parallel can be shifted a substantial amount, especially if the topography overall slopes north or south.

Another notable characteristic of SPCS 2022 is that states with multiple zones (whether low-distortion projections or "traditional" zones) will have an additional statewide zone that co-exists as another "layer" with the multiple "subzones." As mentioned in Section 20.5, such a layered system exists in SPCS 83 for Kentucky, which has a north and south subzone, along with a statewide zone (all three are Lambert conformal conic projections). A preliminary SPCS2022 statewide zone design for Pennsylvania has $\varphi_{CP} = $ N 40°57′ and $k_0 = 0.99993$, and it has 65% of the area and 88% of the population within ±100 ppm (1:10,000) linear distortion.

The main purpose of statewide zones is to provide a projected coordinate system for statewide GIS applications. Many analyses and operations performed in GIS require a single, consistent geometry, and statewide zones provide this in an optimal way. This is preferable to using existing large-area systems, such as UTM, which are not located or oriented appropriately for most states.

20.15 MAP PROJECTION SOFTWARE

The computations presented in this chapter are often difficult to perform correctly using a calculator due to their length and the magnitude of the values involved. Thus, programs have been developed that allow users to easily and conveniently perform these computations. On the companion website for the book are three such software packages. As discussed earlier, WOLFPACK can perform direct and inverse computations for any 1983 state plane coordinate system zone, universal transverse Mercator zone, or a user-specified oblique stereographic map projection. The software accepts manually entered coordinate values or can read a file of coordinates. Figure 20.15 shows the format for a file of geodetic coordinates as presented in Table 20.7 in the upper-half of the window. After performing direct problem in the Pennsylvania North zone, a portion of the resulting output file is shown in the lowerhalf of the screen. A complete explanation of the input file is discussed in the help file that accompanies the software. WOLFPACK can also perform reductions on the distance and azimuth observations. That is, it can scale an observed horizontal distance to its *grid* equivalent and reduce a geodetic azimuth to

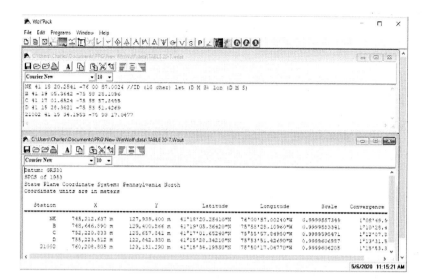

Figure 20.15 WOLFPACK with file for direct problem computations in Pennsylvania north zone shown in upper-half of screen and the resulting computed file shown in the lower-half.

its equivalent grid value. Additionally, the software can reduce a file of observations. This option allows for 12 different file entry formats that are discussed in the help file, which accompanies this software. Note that this option uses Equation (20.51) to reduce lengths of all horizontal distances. These options, shown in Figure 20.16, are under the *map projections* submenu in the *programs* menu item.

Additionally, on the companion website is the Excel spreadsheet, *map_projections.xls*. This spreadsheet demonstrates the computations discussed in this chapter. For those interested in programming these computations using a higher-level language, there are several Mathcad worksheets on the companion website that can be explored. The file *Lambert.xmcd* demonstrates the direct and inverse computations using both Lambert conformal conic map projection. The file *TM.xmcd* contains the direct and inverse computations for the transverse Mercator map projection and file *oblique.xmcd* contains the direct and inverse computations for the oblique Mercator map projection used in Alaska. Finally, the file *ostereo.xmcd* contains the programming for the direct and inverse computations using an oblique stereographic map projection. For those wishing to compare their intermediate computations using tables with software-derived values, the files *PA_Table.xmcd* and *NJ_Table.xmcd* provide this feature.

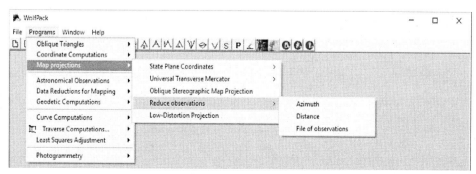

Figure 20.16 WOLFPACK menu items for map projection computations.

Finally, the file *GridObs.xmcd* demonstrates the reduction of observations as presented in this chapter. These additional features are provided as learning aids so that the user can check their values when exploring the example problems presented in this chapter.

PROBLEMS

Asterisks (*) indicate problems that have answers given in Appendix G.

20.1 Discuss the advantages of placing surveys on state plane coordinate systems.
20.2 What are developable surfaces?
20.3 What observation types are distorted when employing a Lambert conformal conic projection?
20.4 Why are only conformal projections used in surveying?
20.5 Why is it important to use exact conversions when converting state plane coordinates to from metric to English units?
20.6 What states use the international definition of the foot?
20.7 How will the defining parameters change for the Lambert conformal conic in SPCS 2022?
20.8 Define the direct problem in state plane coordinates.
20.9 Define the inverse problem in state plane coordinates.
20.10 What is true about all grid azimuths?
20.11* The Pennsylvania North Zone SPCS 83 state plane coordinates of points A and B are as follows:

Point	E (m)	N (m)
A	541,983.399	115,702.804
B	541,457.526	115,430.257

Calculate the grid length and grid azimuth of line AB.

20.12 Similar to Problem 20.11, except points A and B have the following New Jersey SPCS 83 state plane coordinates:

Point	E (m)	N (m)
A	376,652.268	37,592.043
B	376,667.000	37,837.239

20.13 What are the SPCS83 coordinates (in sft) and convergence angle for a station in the North zone of Pennsylvania with geodetic coordinates of 41°59′27.85406″ N and 80°00′17.08943″ W?
20.14* Similar to Problem 20.13 except that the station's geodetic coordinates are 40°47′41.11083″ N and 77°51′45.57113″ W.
20.15 What is the scale factor for the station in Problem 20.13?
20.16 What is the scale factor for the station in Problem 20.14?
20.17* What are the SPCS 83 coordinates for a station in New Jersey with geodetic coordinates of 40°44′32.73687″ N and 74°10′45.47356″ W?
20.18 Similar to Problem 20.17 except that the geodetic coordinates of the station are 40°44′32.74244″ N and 74°10′36.34765″ W.

20.19 What are the convergence angle and scale factor at the station in Problem 20.17?

20.20 What are the convergence angle and scale factor at the station in Problem 20.18?

20.21 What is the grid azimuth and grid distance in meters between the points in Problems 20.17 and 20.18?

20.22 If the average geodetic height of the line between the points in Problems 20.17 and 20.18 is 38 sft, what is the combined factor for the line, geodetic azimuth, and ground distance in meters? (Use an average radius for the Earth of 6,371,000 m)

20.23* What are the geodetic coordinates for a point A in Problem 20.11?

20.24 Similar to Problem 20.23 except for point B in Problem 20.11.

20.25* What are the geodetic coordinates for a point A in Problem 20.12?

20.26 Similar to Problem 20.25 except for point B in Problem 20.12.

20.27 In computing state plane coordinates for a project area who's mean orthometric height is 856.47 sft, an average scale factor of 0.99998347 was used. The average geoid height for the area is −31.284 m. The given distances between points in this project area were computed from SPCS83 state plane coordinates. What horizontal length would have to be observed to lay off these lines on the ground? (Use 6,371,000 m for an average radius for the Earth.)
 (a)* 567.79 sft
 (b) 265.054 m
 (c) 2438.25 sft

20.28 Similar to Problem 20.27, except that the mean project area elevation was 4506.89 sft, the geoidal separation −28.334 m, the scale factor 0.99996594, and the computed lengths of lines from SPCS83 were:
 (a) 176.349 m
 (b) 2648.97 ift
 (c) 493.578 m

20.29 The horizontal ground lengths of a three-sided closed polygon traverse were measured in feet as follows: $AB = 201.32$, $BC = 176.38$, and $CA = 181.58$ ft. If the average scale factor is 0.99998467, orthometric height of the area is 2241.56 sft, and the average geoid height is −30.513 m, calculate grid lengths of the lines suitable for use in computing SPCS83 coordinates. (Use 6,371,000 m for an average radius for the Earth.)

20.30 For the traverse of Problem 20.29, the grid azimuth of a line from A to a nearby azimuth mark was 9°22′06″ and the clockwise angle measured at A from the azimuth mark to B, 14°26′18″. The measured interior angles were $A = 54°34′03″$, $B = 57°00′47″$, and $C = 68°25′13″$. Balance the angles and compute grid azimuths, latitudes and departures balanced latitude and departures, linear misclosure, and relative precision for the traverse.

20.31* What is the combined factor for the traverse of Problem 20.30, and what distance precision does this yield when grid distances are compared to ground distances?

20.32 In what region is an oblique transverse Mercator map projection used?

20.33 What is the ground versus grid problem and what two methods can be used to solve this problem?

20.34 What scale factor should be used with the Lambert conformal conic map projection when creating an LDP that is secant to the Earth at the height of h_{avg}? . . . for the Transverse Mercator map projection?

20.35 The average geodetic height in a project area is 6892.76 sft. Using an average radius of the Earth of 6,371,000 m, what is the appropriate scale factor for an LDP using the Transverse Mercator map projection?

20.36 The traverse in Problem 10.21 was performed in the Pennsylvania North Zone of SPCS83. The average elevation for the area was 241.894 m and the average geoidal height was −31.244 m. Using the data in Table 20.1 and a mean radius for the Earth, compute a project factor based on the control coordinates, reduce the observations to grid, and adjust the traverse. Compare this solution with that obtained in Chapter 10. (Use 6,371,000 m for an average radius of the Earth.)

20.37 What changes are being proposed for the computation of state plane coordinates in 2022?

20.38* If the geodetic azimuth of a line is 205°06′36.2″ the convergence angle is −0°42′26.1″ and the arc-to-chord correction is +0.8″ what is the equivalent grid azimuth for the line?

20.39 If the geodetic azimuth of a line is 243°06′34.5″ the convergence angle is −2°06′34.2″ and the arc-to-chord correction is +0.7″, what is the equivalent grid azimuth for the line?

20.40 Using the values given in Problems 20.38 and 20.39, what is the acute grid angle between the two azimuths?

20.41 The grid azimuth of a line is 158°13′26″. If the convergence angle at the endpoint of the azimuth is −1°58′02.9″ and the arc-to-chord correction is +1.5″, what is the geodetic azimuth of the line?

20.42 Similar to Problem 20.41, except the convergence angle is +2°16′32.7″ and the arc-to-chord correction is +1.6″.

20.43 A project is bounded by the geodetic coordinates in the southwest corner of (32°15′37″ N, 106°48′49″W, 1180 m) and in the northeast corner of (32°21′43″ N, 106°43′56″W, 1290 m). Should the transverse Mercator or the Lambert conformal conic map projection be used in the design of an LDP for the region?

20.44 In Problem 20.43, what scale factor (k_{LDP}) be used in the design of the LDP? (Use an average radius for the Earth.)

20.45 The region in problem 20.44 is surrounds Las Cruces, NM, which is in the New Mexico Central Zone of the SPCS 1983. What is the appropriate project factor for this region?

20.46 Compute the central parallel and its scale factor for the SPCS 83 zone you reside in.

BIBLIOGRAPHY

Dennis, M. L. 2018. "The State Plane Coordinate System History, Policy, and Future Directions." *NOAA Special Publication NOS NGS 13*. Silver Spring, MD: National Geodetic Survey.

Ghilani, C. 2015. "Transformation of Observation—Part 3: Using a Project Factor." *XYHT* 2 (No. 3): 41.

GIA. 2006. "How Things Work: Scale, Elevation, Grid, and Combined Factors Used in Instrumentation." *Professional Surveyor* 26 (No. 2): 47.

Hartzell, P., L. Strunk, and C. Ghilani. 2002. "Pennsylvania State Plane Coordinate System: Converting to a Single Zone." *Surveying and Land Information Science* 62 (No. 2): 95.

Rollins, Craig M. and Thomas H. Meyer. 2019. "Four Methods for Low-Distortion Projections." *Journal of Surveying Engineering* 145 (No. 4) (accessed at https://ascelibrary.org/doi/pdf/10.1061/%28ASCE%29SU.1943-5428.0000295 on 9/21/2019).

Snay, R. A. 1999. "Using the HTDP Software to Transform Spatial Coordinates Across Time and Reference Frames." *Surveying and Land Information Systems* 59 (No. 1): 15.

Snyder, J. P. 1987. *Map Projections—A Working Manual*. Washington, DC: U.S. Government Printing Office.

Stachurski, R. 2002. "History of American Projections: The American Projection, Part 1." *Professional Surveyor* 22 (No. 4): 16.

———. 2002. "History of American Projections: The American Projection, Part 2." *Professional Surveyor* 22 (No. 5): 32.

Stem, J. E. 1990. "State Plane Coordinate System of 1983." *NOAA Manual NOS NGS 5*. Rockville, MD: National Geodetic Information Center.

21
Boundary Surveys

■ 21.1 INTRODUCTION

The oldest types of surveys in recorded history are boundary surveys, which date back to about 1400 B.C., when plots of ground were subdivided in Egypt for taxation purposes. Boundary surveys still are one of the main areas of surveying practice. From Biblical times[1] when the death penalty was assessed for destroying corners, to the colonial days of George Washington[2] who was licensed as a land surveyor by William and Mary College of Virginia, and through the years to the present, natural objects (i.e., trees, rivers, rock outcrops, etc.), and manmade objects (i.e., fences, wooden posts, iron, steel, or concrete markers, etc.), have been used to identify land parcel boundaries.

As property increased in value and owners disputed rights to land, the importance of more accurate surveys, monumentation of the boundaries, and written records became obvious. When Texas became a state in 1845, its public domain amounted to about 172,700,000 acres, which the U.S. government could have acquired by payment of the approximately $13,000,000 in debts accumulated by the Republic of Texas. However, Congress allowed the Texans to retain their land and pay their own debts—a good bargain at roughly 7.6 cents/acre!

The term *land tenure system* applies to the manner in which rights to land are held in any given country. Such a system, as a minimum, must provide (1) a means for transferring or changing the title and rights to the land, (2) permanently monumented or marked boundaries, which enable parcels to be found on the ground, (3) officially retained records defining who possesses what rights to

---

[1]"Cursed be he that removeth his neighbor's landmark. And all the people shall say Amen." Deut. 27:17.
[2]"Mark well the land, it is our most valuable asset." George Washington.

the land, and (4) an official legal description of each parcel. In the United States a two-tier land tenure system exists. At the federal level, records of surveys and rights to federal land are maintained by the U.S. *Bureau of Land Management* (*BLM*). At the state and local levels, official records concerning land tenure are held in county court houses.

Land titles in the United States are now transferred by written documents called *deeds* (*warranty*, *quit claim*, or *agreement*), which contain a description of the property. Property descriptions are prepared as the result of a land survey. The various methods of description include (1) metes and bounds, (2) block-and-lot number, (3) coordinate values for each corner, and (4) township, section, and smaller subdivisions of the United States *Public Land Survey System* (*PLSS*) commonly referred to as the *aliquot part*. Often a property description will combine two or more of these methods. The first three methods are discussed briefly in this chapter; the PLSS is covered in Chapter 22.

21.2 CATEGORIES OF LAND SURVEYS

Activities involved in the practice of land surveying can be classified into three categories: (1) *original surveys* to subdivide the remaining unsurveyed U.S. public lands, most of which are in Alaska, (2) *retracement surveys* to recover and monument or mark boundary lines that were previously surveyed, and (3) *subdivision surveys* to establish new smaller parcels of land within lands already surveyed. The last two categories are described in this chapter; the first is discussed in Chapter 22.

In establishing new property lines, and especially in retracing old ones, surveyors must exercise acute judgment based on education, practical experience, and knowledge of land laws. They must also be accurate and articulate in making observations. This background must be bolstered by tenacity in searching the records of all adjacent property as well as studying descriptions of the land in question. In fieldwork, surveyors must be untiring in their efforts to find points called for by the deed. Often it is necessary to obtain *parol evidence*; that is, testimony from people who have knowledge of accepted land lines and the location of corners, reference points, fences, and other information about the correct lines.

Modern-day land surveyors are confronted with a multitude of problems, created over the past two centuries under different technology and legal systems that now require professional solutions. These include defective compass and chain surveys; incompatible descriptions and plats of common lines for adjacent tracts; lost or obliterated corners and reference marks; discordant testimony by local residents; questions of riparian rights; and a tremendous number of legal decisions on cases involving property boundaries.

The responsibility of a professional surveyor is to weigh all evidence and try to establish the originally intended boundary between the parties involved in any property-line dispute, although without legal authority to force a compromise or settlement. Fixing title boundaries must be done by agreement of adjacent owners or court action. Surveyors are often called upon to serve as expert witnesses in proceedings to establish boundaries, but to do so they should be registered.

Because of the complicated technical judgment decisions that must be made, the increasing cost of land surveyors' professional liability insurance for "errors and omissions" has become a major part of operating expenses. Some states demand that a surveyor have it for the protection of all parties.

21.3 HISTORICAL PERSPECTIVES

In the eastern part of the United States, individuals acquired the first land titles by gifts or purchase from the English Crown. Surveys and maps were completely lacking or inadequate, and descriptions could be given in only general terms. The remaining land in the 13 colonies was transferred to the states at the close of the Revolutionary War. Later this land was parceled out to individuals, generally in irregular tracts. Boundary lines were described by *metes and bounds* (see Section 21.4).

Many original transfers and subsequent ownerships and subdivisions were not recorded. Those that were usually had scanty or defective descriptions, since land was cheap and abundant. Trees, rocks, and natural landmarks defining the corners, as in the first example metes-and-bounds description (see Section 21.4), were soon disturbed. The intersection of two property lines might be described only as "the place where John killed a bear" or "the bend in a footpath from Jones's cabin to the river."

Numerous problems in land surveying stem from the confusion engendered by early property titles, descriptions, and compass surveys. The locations of thousands of corners have been established by compromise after resurveys, or by court interpretation of all available evidence pertinent to their original or intended positions. *Squatters' rights*, *adverse possession*, and *riparian changes* have fixed other corners. Many boundaries are still in doubt, particularly in areas having marginal land where the cost of a good retracement survey exceeds the property's value.

The fact that four corners of a field can be found and the distances between them agree with the "calls" in a description does not necessarily mean that they are in the proper place. Title or ownership is complete only when the land covered by a deed is positively identified and located on the ground.

Land law from the time of the Constitution has been held as a state's right, subject to interpretation by state court systems. Many millions of land parcels have been created in the United States over the past four centuries under different technology and legal systems. Some of the countless problems passed on to today's professional surveyors equipped with immensely improved equipment are discussed in this chapter and in Chapter 22.

Land surveying measurements and analysis follow basic plane surveying principles. But a land surveyor needs years of experience in a given state to become familiar with local conditions, basic reference points, and legal interpretations of complicated boundary problems. Methods used in one state for prorating differences between recorded and measured distances may not be acceptable in another. Rules on when and how fences determine property lines are not the same in all states or even in adjacent ones.

The term *practical location* is used by the legal profession to describe an agreement, either explicit or implied, in which two adjoining property owners mark out an ambiguous boundary or settle a boundary dispute. Fixed principles enter the process, and the boundary established, called an *agreed-upon boundary*, can become permanent.

Different interpretations are given locally to (1) the superiority or definiteness of one distance over another associated with it, (2) the position of boundaries shown by occupancy, (3) the value of corners in place in a tract and its subdivisions, and (4) many other factors. Registration of land surveyors is therefore required in all states to protect the public interest.

21.4 PROPERTY DESCRIPTION BY METES AND BOUNDS

As noted earlier, metes and bounds is one of the methods commonly used in preparing legal descriptions of property. Descriptions by metes (to measure or assign by measure) and bounds (boundary lines or property limits) have a *point of commencement* (POC) such as a nearby existing corner of the PLSS (see Chapter 22). Commencing at this point, successive lengths and directions of lines are given that lead to the *point of beginning* (POB). The POB is usually a fence post, iron or steel rod, or some natural feature, which marks one corner of the property. Lengths and directions (bearings or azimuths) of successive lines from the point of beginning that enclose or bound the property are then given. Early distance units of chains, poles, and rods are now replaced by feet and decimals and sometimes by metric units. Bearings or azimuths may be geodetic, astronomic, magnetic, or grid. Care must be exercised to indicate clearly which of these is the basis of directions so that no confusion arises. In the past, *assumed* bearings or azimuths have sometimes been used, but many states no longer allow them because they are not readily reproducible. In some states, survey regulations call for exterior lines of new subdivisions to be based on the astronomic, grid, or geodetic meridian.

Surveyors write metes-and-bounds property descriptions, and they are included in the legal documents that accompany the transfer of title to property. In preparing the descriptions, extreme care must be exercised. A single mistake in transcribing a numerical value, or one incorrect or misplaced word or punctuation mark, may result in litigation for more than a generation, since the intentions of the *grantor* (person selling property) and the *grantee* (person buying property) may then be unclear. If numbers are both spelled out and given as figures, words control in the case of conflicts unless other proof is available. There is a greater likelihood of transposing than misspelling—and lawyers prefer words!

The importance of permanent monuments to mark property is evident. In fact, some states require pipes, iron pins, and/or concrete markers set deep enough to reach below the frost line at all property corners before surveys will be accepted for recording. Almost any suitable marker could be called for as a monument. A map attached to the description will contain a legend, which identifies all monuments. By scaling from the map, a rough check on the distances and directions in the description can be obtained.

To increase precision in property surveys, and to simplify the process of preparing property descriptions and relocating the corners of established parcels, large cities and some states have established a network of control monuments to supplement stations of the National Spatial Reference System (NSRS).

These points are then available as control for initiating topographic and construction surveys and can also serve as POCs for commencing property descriptions. Some states have embarked on statewide programs in which a relatively dense grid of control monuments is being set on a county-by-county basis. For example, the State of Florida established a 6-mi grid. State plane coordinates are determined for the monumented positions, and they are tied to corners of the PLSS.

Description of land by metes and bounds in a deed should always contain the following information in addition to the recital:

1. *Point of commencement (POC)*. This is an established reference point such as a corner of the PLSS or NSRS monument to which the property description is tied or referenced. It serves as the starting point for the description.
2. *Point of beginning (POB)*. This point must be identifiable, permanent, well referenced, and one of the property corners. Coordinates, preferably state plane, should be given if known or computable. Note that a POB is no more important than others and a called-for monument in place at the next corner establishes its position, even though bearing and distance calls to it may not agree.
3. *Definite corners*. Such corners are clearly defined points with coordinates if possible.
4. *Lengths and directions of the property sides*. All lengths in feet and decimals (or metric units), and directions by angles, true bearings, or azimuths must be stated to permit computation of any misclosure error. Omitting the length or direction of a closing line to the POB and substituting a phrase "and thence to the point of beginning" is not acceptable. The survey date is required and particularly important if directions are referred to magnetic north.
5. *Names of adjoining property owners*. These are helpful to show the intent of a deed in case an error in the description leaves a gap or creates an overlap. However, called-for monuments in place will control title over calls for adjoiners unless precluded by *senior rights* (see Section 21.7).
6. *Areas*. The included area is normally given as an aid in the valuation and identification of a piece of property. Areas of rural land are given in acres or hectares and those of city lots in square feet or square meters. Because of differences in measurements and depending on the adjustment method used for a traverse (compass rule, least squares, etc.), one surveyor's calculated area, directions, and distances may differ slightly from another's. The expression "more or less," which may follow a computed area, allows for minor errors, and avoids nuisance suits for insignificant variations. Additionally, as discussed in Section 12.5, random errors propagate from the surveying observations to the computed area. As a rule of thumb, the uncertainty in the area of a parcel can be computed using Equation (12.9). The number of significant figures in the area of a parcel should not be expressed with more digits than the precision of the measurements warrants (see Section 12.5).

Note that items 4, 5, and 6 above establish the *shape*, *intent*, and *size*, respectively, of the parcel. Without item 5, the description would be strictly "metes" and would lack intent. Also if items 4 and 6 were excluded, the description would be purely "bounds," and would lack the important measurements needed to establish

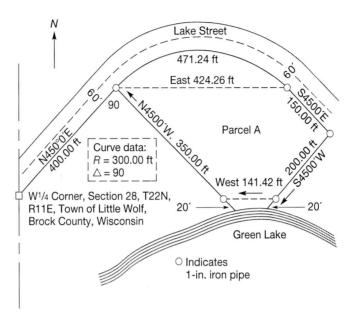

Figure 21.1
Metes-and-bounds tract.

the shape and size of the parcel. A partial metes-and-bounds description for the tract shown in Figure 21.1 is given as an example.

> That part of the SW 1/4 of the NW 1/4 of Section 28, T 22 N, R 11 E, 4′th P.M., Town of Little Wolf, Brock County, Wisconsin, described as follows: Starting at the point of commencement, which is a stone monument at the W 1/4 corner of said Section 28; thence N 45 degrees 00 minutes E, four hundred (400.00) feet along the Southeasterly R/W line of Lake Street to a 1-inch iron pipe at the point of beginning of this description, said point also being the point of curvature of a tangent curve to the right having a central angle of 90 degrees 00 minutes and radius of three hundred (300.00) feet; thence Easterly, four hundred seventy one and twenty four hundredths (471.24) feet along the arc of the curve, the long chord of which bears East, four hundred twenty four and twenty six hundredths (424.26) feet, to a 1-inch iron pipe at the point of tangency thereof, said arc also being the aforesaid Southwesterly R/W line of Lake Street; thence continuing along the Southerly R/W line of Lake Street, S 45 degrees 00 minutes E, one hundred fifty (150.00) feet to a 1-inch iron pipe; thence S 45 degrees 00 minutes W, two hundred (200.00) feet to a 1-inch iron pipe located N 45 degrees 00 minutes E, twenty (20) feet, more or less, from the water's edge of Green Lake, and is the beginning of the meander line along the lake; thence West one hundred forty one and forty two hundredths (141.42) feet along the said meander line to a 1-inch iron pipe at the end of the meander line; said pipe being located N 45 degrees 00 minutes W, twenty (20) feet, more or less from the said water's edge; thence N 45 degrees 00 minutes W, three hundred fifty (350.00) feet to a 1-inch iron pipe at the point of beginning... including all lands lying between the meander line herein described and the Northerly shore of Green Lake, which lie between true extensions of the Southeasterly and Southwesterly boundary lines of the parcel herein described, said parcel containing 2.54 acres, more or less. Bearings are based on astronomic north.

Many metes-and-bounds descriptions have been prepared, which because of a variety of shortcomings, have created later problems. To illustrate, portions of two *old* metes-and-bounds descriptions from the eastern United States follow. The first, part of an early deed registered in Maine, is:

> Beginning at an apple tree at about 5 minute walk from Trefethen's Landing, thence easterly to an apple tree, thence southerly to a rock, thence westerly to an apple tree, thence northerly to the point of beginning.

With numerous apple trees and an abundance of rocks in the area, the dilemma of a surveyor trying to retrace the boundaries many years later is obvious.

The second, a more typical old description of a city lot showing lack of comparable precision in angles and distances, follows:

> Beginning at a point on the west side of Beech Street marked by a brass plug set in a concrete monument located one hundred twelve and five tenths (112.5) feet southerly from a city monument No. 27 at the intersection of Beech Street and West Avenue; thence along the west line of Beech Street S 15 degrees 14 minutes 30 seconds E fifty (50) feet to a brass plug in a concrete monument; thence at right angles to Beech Street S 74 degrees 45 minutes 30 seconds W one hundred fifty (150) feet to an iron pin; thence at right angles N 15 degrees 14 minutes 30 seconds W parallel to Beech Street fifty (50) feet to an iron pin; thence at right angles N 74 degrees 45 minutes 30 seconds E one hundred fifty (150) feet to place of beginning; bounded on the north by Norton, on the east by Beech Street, on the south by Stearns, and on the west by Weston.

■ 21.5 PROPERTY DESCRIPTION BY BLOCK-AND-LOT SYSTEM

As urban areas grow, adjoining parcels of land are subdivided to create streets, blocks and lots according to an orderly and specific plan. Each new subdivided parcel, called a *subdivision*, is assigned a name and annexed by the city. Block-and-lot number, tract and lot number, or subdivision name and lot number identify the individual lots within the subdivided areas. Examples are:

> Lot 34 of Tract 12314 as per map recorded in book 232, pages 23 and 24 of maps, in the office of the county recorder of Los Angeles County.
>
> Lot 9 except the North twelve (12) feet thereof, and the East twenty-six (26) feet of Lot 10, Broderick's Addition to Minneapolis. [Parts of two lots are included in the parcel described.]
>
> That portion of Lot 306 of Tract 4178 in the City of Los Angeles, as per Map recorded in Book 75, pages 30 to 32 inclusive of maps in the office of the County Recorder of said County, lying Southeasterly of a line extending Southwesterly at right angles from the Northeasterly line of said lot, from a point in said Northeasterly line distant Southeasterly twenty-three and seventy-five hundredths (23.75) feet from the most Northerly corner of said Lot.

The block-and-lot system is a short and unique method of describing property for transfer of title. Standard practice calls for a map or plat of each subdivision to be filed with the proper office. The plat must show the types and

664 BOUNDARY SURVEYS

locations of monuments, dimensions of all blocks and lots, and other pertinent information such as the locations and dimensions of streets and easements, if any. These subdivision plats are usually kept in map books in the city or county recorder's office.

In most cases, the lot-and-block descriptions are created simultaneously and thus are not subject to *junior* or *senior rights* (described in Section 21.7). In performing resurveys to find or reestablish lot corners, therefore, any excess or deficiency found in the measurements is prorated equally among the lots.

Figure 21.2 is an example of a small hand-drafted block-and-lot subdivision. Figure 18.13 shows a portion of a subdivision map of blocks and lots produced using a CADD system.

NO	DELTA (I)	TAN	ARC	RADIUS	CHORD	CHORD BEAR
1	12°37'	67.00'	133.46'	606.07'	133.19'	N65°00'30"E
2	5°49'40"	7.67'	15.33'	150.72'	15.32'	N28°23'10"W
3	90°00'	25.00'	39.27'	25.00'	35.36'	N13°42'E
4	12°37'	105.57'	210.28'	954.93'	209.85'	S65°00'30"W
5	5°50'15"	33.59'	67.11'	658.72'	67.08'	N68°23'30"E
6	97°40'27"	28.60'	42.62'	25.00'	37.64'	S65°41'05"E
7	75°32'50"	19.37'	32.96'	25.00'	30.63'	N20°55'35"E
8	22°20'50"	19.75'	39.00'	100.00'	38.76'	S5°40'25"E
9	67°39'10"	16.75'	29.52'	25.00'	27.83'	N50°40'25"W
10	29°12'20"	36.38'	71.16'	139.62'	70.41'	S80°53'50"W

Figure 21.2 Small subdivision plat.

■ 21.6 PROPERTY DESCRIPTION BY COORDINATES

State plane coordinate systems provide a common reference system for surveys in large regions, even entire states (see Chapter 20). Several advantages result from using them on property surveys. One of the most significant is that they greatly facilitate the relocation of lost and obliterated corners. Every monument, which has known state plane coordinates, becomes a "witness" to other corner markers whose positions are given in the same system. State plane coordinates also enable the evaluation of adjoiners with less fieldwork. As cities and counties develop *land information systems* (LISs) and *geographic information systems* (GISs) (see Chapter 28), descriptions by coordinates are becoming the norm. Many local and state governments currently require state plane coordinates on the corners of recorded subdivision plat boundaries, and on the exteriors of large boundary surveys. It is only a matter of time before state plane coordinates are required on all descriptions of land.

A coordinate description of corners may be used alone, but is usually prepared in conjunction with an alternative method. An example of a description by coordinates of a parcel in California follows.

> A parcel of tide and submerged land, in the state-owned bed of Seven Mile Slough, Sacramento County, California, in projected Section 10, T 3 N, R 3 E, Mt. Diablo Meridian, more particularly described as follows:
>   BEGINNING at a point on the southerly bank of said Seven Mile Slough, which bears S62°37′ E, 860 feet from a California State Lands Commission brass cap set in concrete stamped "JACK 1969," said point having coordinates of X = 2,106,973.68 and Y = 164,301.93 as shown on Record of Survey of Owl Island, filed October 6, 1969, in Book 27 of Surveys, Page 9, Sacramento County Records, thence to a point having coordinates of X = 2,107,196.04 and Y = 164,285.08; thence to a point having coordinates of X = 2,107,205.56, Y = 164,410.72; thence to a point having coordinates of X = 2,106,983.20, Y = 164,427.57; thence to the point of beginning. Coordinates, bearings, and distances in the above description are based on the California Coordinate System, Zone II.

When the preceding description was prepared, the writer could not have anticipated that an ambiguity would later arise concerning the datum of reference. From the dates given in the description, of course it can be concluded that the coordinates are referred to NAD 27. In preparing coordinate descriptions nowadays, the date of the survey datum upon which the coordinates are based should be identified as being either in NAD 27, NAD 83 (####), or as appropriate, along with the epoch date of the coordinates to avoid any later confusion.

Earthquakes in Alaska, California, and Hawaii and the subsidence caused by the withdrawal of oil and groundwater in many states, have caused ground shifts that move corner monuments and thereby change their coordinates. If undisturbed relative to their surroundings, the monuments, rather than the coordinates, then have greater weight in ownership rights.

■ 21.7 RETRACEMENT SURVEYS

Retracement surveys are run for the purpose of relocating or reestablishing previously surveyed boundary lines. *They are perhaps the most challenging of all types of surveys.* The rules used in retracement surveys are guided by case law and as such can vary from state to state and with time. However, the fundamental precept governing retracement surveys is that the monuments as originally placed and agreed to by the grantee and grantor constitutes the correct boundary location. It is always important to remember that surveyors can set marks on the ground but their acceptance cannot be forced by a surveyor—any disagreements must be resolved in the courts. *Thus, the objective of resurveys is to restore boundary markers to their original locations, not to correct them*, and this should guide all of the surveyor's actions.

In making a retracement survey, written evidence of title for the parcel involved should first be obtained. This will normally be in the form of a deed but could also be obtained from an abstract or title policy. Even if the deed is available, it is a good practice to trace it back to its creation in order to ensure that no transcription errors have been made, and to check for possible modifiers (such as the term "surface measure"). Deeds of all adjoining properties should also always be obtained and matched to (1) determine if any gaps or overlaps exist, and (2) understand any junior-senior rights (defined below) that might possibly apply. The possibility that the written title could be supplanted by an *unwritten conveyance* must also be investigated. In the absence of any alterations of the written title by unwritten means, an evaluation of all evidence related to the written conveyance should be made in order to properly establish the property boundaries.

Various types of evidence are considered and used when retracing boundary surveys. When conflicts exist between the different types, the order of importance, or weight, generally assigned in evaluating that evidence, is as follows:

1. *Senior rights.* When parcels of land are conveyed in sequence, the one created first (senior) receives all that was specified in the written documents, and in case of any overlap of descriptions, the second (junior) receives the remainder. In case of overlaps in other subsequent conveyances, the elder receives the benefits.
2. *Intent of the parties.* The intent of the grantee and grantor at the time of conveyancing must be considered in resurveys. Usually the best evidence of intent is contained in the written documents themselves.
3. *Call for a survey.* If the written documents describe a survey, an attempt should be made to locate the stakes or monuments placed as a result of that survey.
4. *Monuments.* If the written documents describe original monuments that were set to mark the boundaries, these must be searched out. When there are conflicts in monuments, natural ones such as trees or streams receive more weight than artificial ones such as stakes and iron pipes.
5. *Measurements.* Courts have consistently ruled that measurements called for in a description merely describe the positions of the corners. Consequently, they generally receive the least weight in interpreting a conveyance. When measurements are evaluated as evidence, the order of importance that generally applies to them is (1) distance, (2) direction, (3) area, and (4) coordinates. However in some states, the order of distances and directions is reversed.

It should be noted that the foregoing are "general rules" for evaluating conflicting evidence, but it is possible, in certain circumstances, for a supposedly inferior element to control a superior one.

Good judgment is especially important when old lines are being restored and the original corners are lost. Then all possible evidence that applies to the original location must be found, and a decision made as to what evidence is good and what can or should be discarded. All found evidence should be recorded in the resurvey field notes, and reasons for using or rejecting any of it noted. Then if the survey is questioned in the courts, all actions taken can be justified. Familiarity with state and local laws, and past court decisions affecting surveys in an area, is valuable when making evaluations of evidence.

A corner that has been preserved is the best evidence of the original location of a line in question. The original notes are used as guides in locating these monuments, but a marker's actual position is the governing factor. Thus, if a monument is found that is and has been accepted for years as the location of a corner, the surveyor should also accept it.

A monument should not be assumed lost unless every possible source of information has been exhausted and no trace of it can be found. Even then the surveyor should hesitate before disturbing settled possessions. It may be possible, for example, that where one or more corners are lost, all concerned parties have acquiesced in lines or corners based upon some other corner or landmark. These acquiesced corners may not be the original ones, but it would be unreasonable to discredit them when the people concerned do not question them. In a legal controversy, the law as well as common sense will normally declare that a boundary line, long acquiesced in, is better evidence of where the real line should be than one established by a survey done long after the original monuments have disappeared.

In retracement surveys, a determination must be made as to whether directions cited in the documents are astronomic or magnetic. If they are magnetic, the declination at the time of the original survey must be determined so that astronomic bearings or azimuths can be determined, and lines retraced following them. A good assumption to adopt in retracing lines is that the boundaries are where the description says they are. If they are not, some indications of their actual locations will probably be found when the distances and directions in the description are followed.

Testimony of persons who remember boundary locations is always valuable, but not always reliable. Therefore, when such testimony is taken, a careful search must be made to find some corroborative evidence. Old fences or decayed wood at the location where a stake was purportedly originally set are examples of extremely valuable corroborative evidence. When two possibilities exist for establishing a boundary, evidence for rejecting one is often as important as evidence for accepting the other.

In retracement surveys, measurements made with a total station instrument between found monuments may not agree with distances on record. This situation provides a real test for surveyors. Perhaps the original chain or tape had an uncorrected systematic error, or a mistake was made. Alternatively, the markers may have been disturbed, or are the wrong ones. If differences exist between distances of record and measured values for found monuments, it may be helpful to determine a "scale factor,"[3] which relate original recorded distances to actual

---

[3]A "scale factor" can be obtained by carefully measuring the distance between two found monuments whose positions appear to be undisturbed and reliable, and dividing the distance between these monuments as recorded on the original survey by the measured distance. As an example, if the recorded distance between two found monuments was 750.00 ft, and the measured distance was 748.62 ft, then the scale factor would be 748.62/750.00 = 0.99816.

measurements. In searching for monuments and evidence, this scale factor can then be applied to lay off the actual distances noted by the original surveyor.

Measurements may indicate that bearings and angles between adjacent sides also do not fit those called for in the writings. These discrepancies could be due to a faulty original compass survey, incorrect corner marks, or other causes. If, while conducting a resurvey, discrepancies are found between adjacent descriptions, which cannot be resolved, they should be called to the attention of the owners so that steps can be taken to harmonize the land actually owned with that contained in the titles. Once the location of the original boundary is decided upon, it should be carefully marked and witnessed so that it can be easily found in the future.

21.8 SUBDIVISION SURVEYS

Subdivision surveys consist in establishing new smaller parcels of land within larger previously surveyed tracts. In these types of surveys, one or only a few new parcels may be created, in which case they may be described using the metes-and-bounds system. Conversely, in areas where new housing is planned, a block-and-lot subdivision survey can be conducted, thus creating many small lots simultaneously. Laws governing subdivision surveys vary from state to state and surveyors must follow them carefully in performing these types of surveys.

Whether a single new lot is created or a block-and-lot survey performed, generally a closed traverse is first run around the larger tract, with all corners being occupied if possible. Fences, trees, shrubbery, hedges, common (party) walls, and other obstacles may necessitate running the traverse either inside or outside the property. If corners cannot be occupied, stub (side shot) measurements can be made to them and their coordinates computed, from which side lengths and bearings can be computed (see Section 10.12). All measurements should be made with a precision suited to the specifications and land values.

Before traversing around the larger tract, it is necessary to first establish a reference direction for one of its lines. Also, if state plane coordinates are to be used in the new parcel's description, starting coordinates must be determined for one of the parcel's corners. A reference direction can be established by making an astronomic observation. Alternatively, as illustrated in Figure 21.3, both a reference direction and coordinates can be transferred to the parcel by including one of its lines in a separate traverse initiated at existing nearby control monuments.

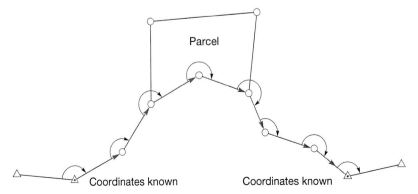

Figure 21.3
Transfer of direction and coordinates to a parcel by traverse.

A check is secured on the traverse by closing back on the starting station, or on a second nearby monument, as indicated in the figure.

After the traverse around the larger exterior parcel has been completed and adjusted, the new parcels can be surveyed. Where a single new parcel is being created, its corners are established according to the owner's specifications within requirements of the statutes. From the survey, the parcel description is prepared, certified, and recorded.

Block-and-lot subdivisions must not only conform to state statutes, but in addition, many municipalities have laws covering these types of surveys. Regulations may specify minimum lot size, allowable misclosures for surveys, types of corner marks to be used, minimum width of streets and the procedure for dedicating them, rules for registry of plats, procedures for review, and other matters. Often several jurisdictions and agencies, each with its own laws and regulations, may have authority over the subdivision of land. In these cases, if standards conflict, the most stringent usually applies. The mismatched street and highway layouts of today could have largely been eliminated by suitable subdivision regulations and by a thorough review of them in past years.

A portion of a small block-and-lot subdivision is shown in Figure 21.2. (Some lot areas have been deleted so that their calculations can become end-of-chapter problems.) Computers with appropriate software such as coordinate geometry (see Chapter 11) and CADD greatly reduce the labor of computing subdivisions. They are especially valuable for large plats and designs with curved streets. Automatic plotters using stored computer files make drafting the final plat map accurate, simple, and fast.

Critical subdivision design and layout considerations include creating good building sites, an efficient street and utility layout, and assured drainage. Furthermore, subdivision rules and regulations must be followed and the developer's desires met as nearly as possible. A subdivision project involves a survey of the *exterior* boundaries of the tract to be divided, followed by a topographic survey, design of the subdivision, and layout of the interior of the tract. Following is a brief outline of the steps to be accomplished in these procedures.

1. Exterior survey
   (a) Obtain recorded deed descriptions of the parent tract of land to be subdivided, and of all adjoiners, from the Register of Deeds office. Note any discrepancies between the parent tract and its adjoiners.
   (b) Search for monuments marking corners of the parent tract, those of its adjoiners where necessary, and, where appropriate, for U.S. Public Land Survey monuments to which the survey may be referred or tied. Resolve any discrepancies with adjoiners.
   (c) Make a closed survey of the parent tract and adequately reference it to existing monuments.
   (d) Compute departures, latitudes, and misclosures to see whether the survey meets requirements. Balance the survey if the misclosure is within allowable limits.
   (e) Resolve, if possible, any encroachments on the property, or differences between occupation lines and title lines, so there will be no problems later with the final subdivision boundary.

2. Interior survey, design, and layout
   (a) Perform a topographic survey of the area within the tract. Include existing utility infrastructure, which may extend into the subdivision. From the survey data, prepare a topographic map for evaluating drainage, and for use in preliminary street layout and lot design.
   (b) Develop a preliminary plat showing the streets and the blocks and lots into which the tract is to be subdivided. Compute departures and latitudes on every block and lot to ensure their mathematical closures.
   (c) Obtain the necessary approvals of the preliminary plat.
   (d) Prepare a final plat map that will conform to state and city platting regulations; and include whatever certificates may be required.
   (e) Set block-and-lot corners of the tract according to the approved preliminary plat. Block corners should be located first, and lot corners set by measuring along lines between block ends. (In some cases, final stakeout may be delayed until utilities are placed so corners are not destroyed in this process.)
   (f) Have the certificates executed and witnessed, and the final plat approved and recorded.

21.9 PARTITIONING LAND

A common problem in property surveys is partitioning land into two or more parcels for sale or distribution to family members, heirs, etc. Prior to partitioning, a boundary survey of the parent tract is run, departures and latitudes computed, the traverse balanced, and the total enclosed area calculated. Computational procedures involved in partitioning land vary depending on conditions. Some parent tract shapes and partitioned parcel requirements permit formula solutions, often by using analytical geometry. Others require trial-and-error methods. These procedures were discussed in Chapter 11 and Section 12.8 where examples were computed to illustrate the procedures.

Typical partitioning problems consist of cutting a certain number of acres from a parent parcel, or dividing the parent parcel into halves, thirds, and so on. Required *cutoff lines* to separate a certain area from the parcel may have (1) a specific starting point (distance from one corner of the tract polygon and run to the midpoint, or any other location on the opposite side); or (2) a required direction (parallel with, perpendicular to, or on a designated bearing angle from a selected line). These cases can often be handled by trial-and-error solutions involving an initial assumption such as the cutoff line direction or the starting point. Certain problems are amenable to solution using coordinate formulas for the intersection of two lines (see Chapter 11).

Other partitioning problems may call for dividing a parcel into an easterly and westerly half, or a northerly and southerly half. But such descriptions could be ambiguous as illustrated in Figure 21.4. This figure shows that a statement "the southerly half" of tract *ABCD* can have a number of meanings—the most southerly half, half the frontage, or half the actual acreage. An important consideration in land partitioning is the final shape of each lot. Connecting midpoints *G* and *H* leaves the southerly "half" smaller than the northerly "half," but provides equal frontage for both parts on the two streets. Course *EF* parallel with *AD*, produces

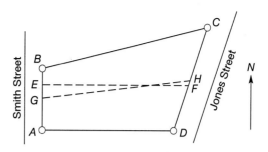

Figure 21.4 "The southerly half."

one trapezoidal lot but a poorly shaped northerly parcel with meager frontage on Smith Street. In either case, the intent of the deed must be clearly stated. This is best accomplished by describing the dividing line by metes-and-bounds, which clearly depicts the intent.

21.10 REGISTRATION OF TITLE

To remedy difficulties arising from inaccurate descriptions and disputed boundary claims, some states provide for registration of property titles under rigid rules. The usual requirements include marking each corner with standardized monuments referenced to established points, recording a plat drawn to scale, and containing specified items. The court under certain conditions then guarantees titles.

A number of states have followed Massachusetts' example and maintain separate land courts dealing exclusively with *land titles*. As the practice spreads, the accuracy of property surveys will be increased and transfer of property simplified.

Title insurance companies search, assemble, and interpret official records, laws, and court decisions affecting ownership of land, and then insure purchasers against loss regarding title defects and recorded liens, encumbrances, restrictions, assessments, and easements. Defense in lawsuits is provided by the company against threats to a clear title from claims shown in public records and not exempted in the policy. The locations of corners and lines are not guaranteed; hence it is necessary to establish, on the ground, the exact boundaries called for by the deed and title policy. Close cooperation between surveyors and title insurance companies is necessary to prevent later problems for their clients.

Many technical and legal problems are considered before title insurance is granted. In some states, title companies refuse to issue a policy covering a lot if fences in place are not on the property line and exclude from the contract "all items that would be disclosed by a property survey."

To guide surveyors in their conduct of land title survey the American Land Title Association (ALTA) and the National Society of Professional Surveyors (NSPS) have established a set of standards. Named the "ALTA/NSPS Land Title Surveys,"[4] they set forth concise guidelines on what must be included in property surveys for title insurance purposes and also state the following regarding point positional tolerance: *Relative Positional Accuracy* may be tested by (1) comparing the relative location of points in a survey as measured by an independent survey of

---

[4]A copy of the current ALTA/NSPS Land Title Surveys is available at http://www.nsps.us.com/?page=ALTAACSMStandards

higher accuracy, or (2) the results of a minimally constrained, correctly weighted least-squares adjustment of the survey. *Allowable Relative Positional Accuracy for Measurements Controlling Land Boundaries on ALTA/NSPS Land Title Surveys* is $\pm[0.07$ ft (20 mm) $+ 50$ ppm$]$ at the 95% confidence level. Other accuracy criteria specified relate to the required precisions of instruments used and acceptable field procedures. Benefits derived from these guidelines are clarification of the exact requirements of land title surveys so that uniformly high-quality results are obtained.

21.11 ADVERSE POSSESSION AND EASEMENTS

Adverse rights can generally be applied to gain ownership of property by occupying a parcel of land for a period of years specified by state law, and performing certain acts.[5] To claim land or rights to it by *adverse possession*, its occupation or use must be (1) actual, (2) exclusive, (3) open and notorious, (4) hostile, and (5) continuous. It may also be necessary for the property to be held under *color of title* (a claim to a parcel of real property based on some written instrument, though a defective one). In some states all taxes must be paid. The time required to establish a claim of adverse possession varies from a minimum of 5 years in California to a maximum of 60 years for urban property in New York. The customary period is 20 years.

The occupation and use of land belonging to a neighbor but outside his or her apparent boundary line as defined by a fence may lead to a claim of adverse possession. Continuous use of a street, driveway, or footpath by an individual or the general public for a specified number of years results in the establishment of a right-of-way privilege, which cannot be withheld by the original owner.

An *easement* is a right, by grant or agreement, which allows a person or persons to use the land of another for a specific purpose. It always implies an interest in the land on which it is imposed. *Black's Law Dictionary* lists and defines 18 types of easements; hence the exact purpose of an easement should be clearly stated. The discussion of property surveys has necessarily been condensed in this text, but it provides helpful information to readers while deterring inexperienced people from attempting to run boundary lines. For more extensive coverage, references are listed in the Bibliography.

21.12 CONDOMINIUM SURVEYS

The word condominium is derived from the prefix *con* meaning "together," and from classical Roman law, *dominium* meaning "ownership." In the United States, the term *condominium* refers to a type of property ownership, where individual units within a multiple-unit building are owned separately. Every unit owner receives a deed describing their property, and is able to buy, mortgage, or sell their unit independent of the other owners. Thus, legal descriptions based on surveys are required. The condominium concept of ownership is relatively new in the United States, as compared with other countries. They have been present in Europe since the Middle Ages, and appeared in this country in the later part of the 19th century. The number of condominiums in the United States has been growing rapidly as

---

[5]Except in a few special circumstances, adverse rights cannot be claimed against public lands.

more families discover the many benefits this type of living offers. Condominium ownership has tax advantages, investment benefits, and most of all, eliminates rent increases. This form of ownership can be an economical solution to rising land values, building costs, and maintenance expenses. It can also provide shared recreational facilities and other amenities that might otherwise be unaffordable.

A *condominium association* is the entity responsible for the operation of a condominium and the unit owners are members of the association. The document creating the association is called the *Articles of Incorporation*, which describes the purpose, powers, and duties of the association. The *By-laws* provide for the administration of the association, including meetings, quorums, voting, and other rules. The document that establishes a condominium is known as the *Condominium Declaration*, and once it is filed in the public records, the condominium is legally created. The Declaration contains important information including a legal description of the property, descriptions of the units, designation of *common elements* (those jointly owned and used by all units such as sidewalks, stairways, swimming pool, tennis courts, etc.), and identification of *limited common elements* (those reserved for the exclusive use of a particular unit such as a designated parking space). It also describes any *covenants* or restrictions on the use of the units, common elements, and limited common elements.

Although condominium ownership often applies to multistory residential buildings, it is also used in commercial and industrial situations. The condominium concept has been applied to mobile home lots, travel trailer and camper sites, boat slips and docks, horse stables, shopping centers, and other types of properties. Special types of condominiums include: *timeshare*, in which the owner purchases an interest in a unit for a specified time period each year; *mixed-use*, which includes both residential and commercial units; and *multiple condominium community*, which is a development containing several separate condominiums that share a single common recreation area.

Condominium surveys differ from ordinary land surveys in several ways. They also bear many similarities with some property surveys, especially those for creating subdivision plats. Each state adopts statutory laws and promulgates rules that govern the procedures and requirements for creating condominiums. In many states, the statute is known as the *Condominium Act*. Preparation of the required materials for a condominium project is a joint effort, which typically includes an architect, engineer, attorney, and surveyor. The architect prepares the building plans and specifications; the engineer designs the construction plans; and the attorney creates the legal documentation for the condominium and the association. The surveyor assembles necessary information; prepares the required condominium plat, graphic plans, and descriptions; and performs the surveys needed for describing the parcel boundary and for locating the "as-built" improvements. It is important for the information shown on the graphic plans to agree with the provisions described in the Declaration.

Figures 21.5 through 21.8 illustrate a four-sheet sample set of plans for a proposed multistory residential condominium. The main purpose of these graphic plans is to clearly and accurately represent the locations of the units, common areas, and limited common areas of the condominium parcel. Figure 21.5 is the *Boundary Survey and Plot Plan*. It shows the exterior boundary survey of the condominium parcel, gives a description for the parcel, and locates the proposed improvements with dimensions given from parcel boundary lines. Also included on this diagram are some general notes and the Surveyor's *Certificate* for the boundary survey. The Boundary Survey and Plot Plan is typically a small-scale drawing and is therefore generally unsuitable for showing sufficient detail and dimensions for all improvements. Thus, it is necessary to attach additional sheets at larger scales.

Figure 21.6 is the *Building and Carport Plot Plan*. It is a larger-scale graphic that not only depicts the residential building and carport areas, but also delineates the common and limited common elements, and labels the individual units with identification numbers. Again, some general notes are included for clarification purposes. Although the scale of this drawing is larger than the Boundary Survey and Plot Plan, it is still too small to effectively show necessary details and dimensions for the individual unit areas. Thus, *Typical Unit Plans* are prepared at a still larger scale, as shown in Figure 21.7. These show the interior floor plans of the units, and their *perimetrical boundaries* (the perimeter or horizontal dimensions encompassing the vertical planes of the interior surfaces of the walls bounding the unit). In addition, this drawing includes a *Typical Wall Section* showing the elevation of the ground floor and heights of the units and building. Typically, the elevation of the ground floor of the building is referenced to a well-established vertical datum. Also included on the plan is a description of the boundaries for the units, and some general notes.

Figure 21.8 details a *Typical Carport and Storage Plan*, with dimensions showing the sizes of the storage and parking areas assigned to individual units. A tabular list of the *Undivided Share of Common Elements* for each unit is also shown on this sample plan, but alternatively it could be included only in the declaration. Computation of the undivided shares is sometimes prorated on the basis of a unit's area to the total area of all units. Another method uses the number of bedrooms in the unit relative to the total number of bedrooms for all units. Figure 21.8 also shows an *As-Built Dimension Table* used to record the actual field measurements of selected portions of each unit or limited common element. An ideal time to measure as-built dimensions is during construction just after the exterior unit walls are completed, but before the interior room partitions are added. Not all as-built dimensions are tabularized. Some measurements, such as the building ties from the boundary lines, may be revised directly on the appropriate plan sheets to reflect the as-built location of the building. If the difference between a measured distance and its corresponding proposed distance is within the construction tolerance, then the dimension need not be revised. For example, if the measured width of a driveway pavement was within ± 0.1 ft of the proposed width, then the proposed dimension generally would not be revised. Should the size or location of an improvement differ substantially from that proposed, then it would be changed on the graphic to reflect the as-built condition.

The last item on Figure 21.8 is the sample *Surveyor's Certificate*. This is executed only upon "substantial completion" of construction of the proposed improvements. Definitions for "substantial completion" vary. As a general rule, if a local building department issues a certificate of occupancy or other similar permit, then this signifies the construction of improvements as being substantially complete. However, if the certificate of occupancy is issued only for the building and a proposed improvement, for example, the pool and deck, is not substantially complete, then the surveyor's certificate should exclude those improvements from the certificate and label those amenities as "under construction" or "proposed."

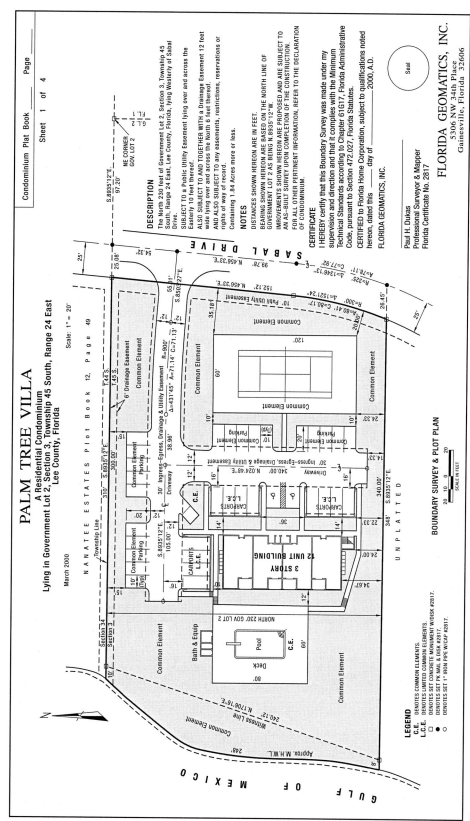

Figure 21.5 Condominium boundary survey and proposed plot plan. (Reprinted with permission from Paul Dukas, Professional Surveyor and Mapper, Geomatics Consultant.)

Figure 21.6 Condominium building and carport plan. (Reprinted with permission from Paul Dukas, Professional Surveyor and Mapper, Geomatics Consultant.)

Figure 21.7 Typical unit plans and wall section. (Reprinted with permission from Paul Dukas, Professional Surveyor and Mapper, Geomatics Consultant.)

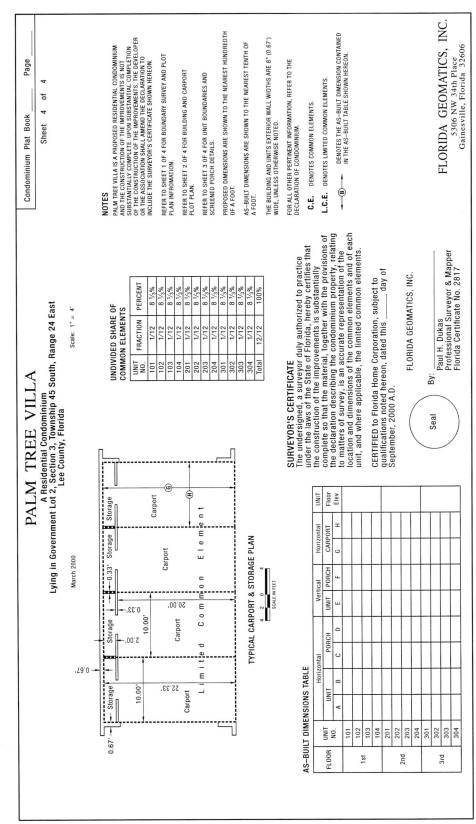

Figure 21.8 Typical carport plan, shares of common element, as-built data and certificate. (Reprinted with permission from Paul Dukas, Professional Surveyor and Mapper, Geomatics Consultant.)

Because this sample condominium is in the proposed stage, a statement of that fact is necessary. Note that this statement appears in the "Notes" of both Figures 21.5 and 21.8. For convenience many declarations include a reduction of the full-sized condominium plat and graphic plans. In this case, careful consideration of the graphic scales, text sizes, and line weights should be exercised when preparing the original full-sized drawings.

21.13 GEOGRAPHIC AND LAND INFORMATION SYSTEMS

As noted in Chapter 28, surveyors are playing a major role in the development and implementation of modern *Geographic Information Systems* (GISs) and *Land Information Systems* (LISs), and this activity will continue in the future. These systems include computerized data banks that contain descriptive information about the land such as its shape, size, location, topography, ownership, easements, zoning, flood plain extent if any, land use, soil types, existence of mineral and water resources, and much more. The information is available for rapid retrieval and is invaluable to surveyors, government officials, lawyers, developers, planners, environmentalists, and others.

Boundary surveys are fundamental to GISs and LISs since knowledge about the land is meaningless unless its position on Earth is specified. In most modern systems being developed, positional data are established by associating attributes about the land with individual lots or tracts of ownership. Perhaps the most fundamental information associated with each individual parcel is its legal description, which gives the unique location on the earth of the parcel and thus provides the positional information needed to support the system. As described in previous sections, legal descriptions are written instruments, based on measurements and prepared to exacting standards and specifications. Thus, the land surveyor's role in modern LISs and GISs is an important one.

21.14 SOURCES OF ERROR IN BOUNDARY SURVEYS

Some sources of error in boundary surveys follow:

1. Errors in measured distances and directions.
2. Corners not defined by unique monuments.
3. Judgment errors in evaluating evidence.

21.15 MISTAKES

Some typical mistakes in connection with boundary surveys are:

1. Failure to perform closed traverse surveys around parcels, or not closing on a control station.
2. Not properly adjusting errors of closure.
3. Use of the wrong corner marks.
4. Failure to check deeds of adjacent property as well as the description of the parcel in question.

5. Failure to describe "intent" in deed descriptions, or preparing ambiguous deed descriptions.
6. Omission of the length or direction of the closing line in parcel descriptions.
7. Magnetic bearings not properly corrected to the date of the new survey.

PROBLEMS

Asterisks (*) indicate problems that have answers given in Appendix G.

21.1 Define the following terms:
  (a) Aliquot part
  (b) Retracement survey
  (c) Original survey
  (d) Adverse possession

21.2 What are the primary components of a land tenure system?

21.3 Visit your county courthouse and obtain a copy of a metes-and-bounds property description. Write a critique of the description, with suggestions on how the description could have been improved.

21.4 How are land titles in the United States transferred?

21.5 Write a metes-and-bounds description for the exterior boundary of lot 15 in Figure 21.2.

21.6 Write a metes-and-bounds description for the house and lot where you live. Draw a map of the property.

21.7 What are the essential elements required when writing a deed description by coordinates?

21.8 From the metes-and-bounds description of the lot in the Town of Little Wolf, described in Section 21.4, compute the lot's misclosure.

21.9 What is the point of commencement in a property description?

21.10 What is the point of beginning in a property description?

21.11 What is the primary objective in performing a retracement surveys?

21.12* List in their order of importance the following types of evidence when conducting retracement surveys: (a) measurements, (b) call for a survey, (c) intent of the parties, (d) monuments, and (e) senior rights.

21.13 In performing retracement surveys, list in their order of importance, the four different types of measurements called for in a description for your state.

21.14 List in order the steps that must be performed in making subdivision surveys?

21.15 Identify all types of pertinent information or data that should appear on the plat of a completed property survey.

21.16 Why are lot-and-block descriptions not subject to junior and senior rights typically?

21.17 Two disputing neighbors employ a surveyor to check their boundary line. Discuss the surveyor's authority if (1) the line established is agreeable to both clients and (2) the line is not accepted by one or both of them.

21.18 What is the purpose of retracement surveys?

21.19 Compute the misclosure of lot 18 in Figure 21.2. On the basis of your findings, would this plat be acceptable for recording? Explain.

21.20* Compute the area of lot 18 of Figure 21.2.

21.21 Determine the misclosure of lot 16 of Figure 21.2, and compute its area.

21.22 For the accompanying figure; using a line perpendicular to AB through x, divide the parcel into two equal parts, and determine lengths xy and By.

21.23 For the figure of Problem 21.22, calculate the length of line de, parallel to BA, which will divide the tract into two equal parts. Give lengths Bd, de, and eA.

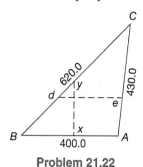

Problem 21.22

21.24 Prepare a metes-and-bound description for the parcel shown in the accompanying figure. Assume all corners are marked with 1-in.-diameter steel pins, and a 20-ft meander line setback from Indian Lake.

21.25 Draw a plat map of the parcel in Problem 21.24 at a convenient scale. Label all monuments and the lengths and directions of each boundary line on the drawing. Include a title, scale, north arrow, and legend.

Problem 21.24

21.26 Prepare a metes-and-bounds description for the property shown in the accompanying figure. Assume all corners are marked with 2-in.-diameter iron pipes.

21.27 Create a 1.25-acre tract on the westerly side of the parcel in Problem 21.26 with a line parallel to the westerly property line. Give the lengths and bearings of all lines for both new parcels.

Problem 21.26

21.28 Discuss the ownership limits of a condominium unit.
21.29 Define *common elements* and *limited common elements* in relation to condominiums. Give examples of each.
21.30 What types of measurements are typically made by surveyors in performing work for condominium developments?
21.31 What is the objective of a resurvey of an existing property?

BIBLIOGRAPHY

Blake, L. 2014. "Best Practices in Land Surveying: Creating Superior Land Description Plats, Part 1." *Professional Surveyor* 34 (No. 3): 32.
_____. 2014. "Best Practices in Land Surveying: Creating Superior Land Description Plats, Part 2." *Professional Surveyor* 34 (No. 6): 28.
Erickson, C. and L. Erickson. 2014. "Claim Jumping Surveyors, Meet the Drop Kickers, Part 3." *The American Surveyor* 11 (No. 6): 34.
Ghilani, C. 2011. "A Simulation Study in the Precision Criterion for the 2011 ALTA-ACSM Land Title Survey Standards." *Surveying and Land Information Science* 71 (No. 1): 49.
Marsico, S. A. 2009. "Deeds: Types, Formalities, and Warranties." *Surveying and Land Information Science* 69 (No. 3): 121.
U.S. Department of the Interior, 2009. *Manual of Surveying Instructions: For the Survey of the Public Lands of the United States.* Bureau of Land Management. Denver, CO: U.S. Government Printing Office.

22
Surveys of the Public Lands

22.1 INTRODUCTION

The term *public lands* is applied broadly to the areas that have been subject to administration, survey, and transfer of title to private owners under the public lands laws of the United States since 1785. These lands include those turned over to the federal government by the colonial states, and the larger areas acquired by purchase from (or treaty with) the Native Americans or foreign powers that had previously exercised sovereignty.

Thirty states, including Alaska, constitute the public land survey states that have been, or are being, subdivided into rectangular tracts (see Figure 22.1). The area of these states represents approximately 72% of the United States. Title to the vacant lands, and therefore direction over the surveys within their own boundaries, was retained by the colonial states, the other New England and Atlantic coast states (except Florida), and later by the states of West Virginia, Kentucky, Tennessee, Texas, and Hawaii. In these areas, the U.S. public land laws have not been applicable.

The beds of navigable bodies of water are not public domain, and are not subject to survey and disposal by the United States. Sovereignty is in the individual states.

The survey and disposition of the public lands were governed originally by two factors:

1. A recognition of the value of grid-system subdivision based on experience in the colonies and another large-scale systematic boundary survey—the 1656 Down Survey in Ireland.
2. The need of the colonies for revenue from the sale of public lands. Monetary returns from their disposal were disappointing, but the planners' farsighted vision of a grid system of subdivision deserves commendation.

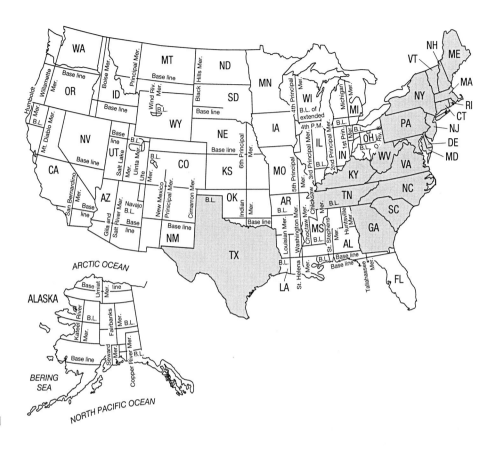

Figure 22.1
Areas covered by the public land surveys, with principal meridians shown. Areas excluded are shaded. (Hawaii, although not shown on this map, would also be shaded. Texas has a rectangular system similar to the U.S. public land system.)

Although nearly a billion acres of public lands has either been sold or granted since 1785, approximately one third of the area of the country is still federally owned. The U.S. *Bureau of Land Management* (BLM), within the Department of the Interior, was created in 1946 as a merger of the U.S. *Grazing Service* and the U.S. *General Land Office*, and is responsible for surveying and managing a significant portion of these federal lands.

22.2 INSTRUCTIONS FOR SURVEYS OF THE PUBLIC LANDS

The U.S. *Public Land Survey System* (PLSS) was inaugurated in 1784, and the territory immediately northwest of the Ohio River in what is now eastern Ohio served as a test area. Sets of instructions for the surveys were issued in 1785 and 1796. Manuals of instructions were later issued in 1855, 1881, 1890, 1894, 1902, 1930, 1947, 1973, and 2009.

In 1796 General Rufus Putnam was appointed as the first U.S. *Surveyor General*, and at that time the numbering of sections changed to the system now in use (see the excerpt from the 2009 *Manual of Surveying Instructions* below, and Figure 22.8). Supplementary rules were promulgated by each succeeding surveyor general "according to the dictates of his own judgment" until 1836, when the General Land Office (GLO) was reorganized. Copies of changes and instructions for local use were not always preserved and sent to the GLO in Washington. As a result, no office in the United States has a complete set of instructions under which the original surveys were supposed to have been made.

Many of the later public land surveys have been run by the procedures to be described in this chapter, or variations of them. The task of present-day surveyors consists primarily of retracing the original lines set by earlier PLSS surveyors and/or further subdividing sections. To accomplish these tasks, they must be thoroughly familiar with the rules, laws, equipment, and conditions that governed their predecessors in a given area.

The rules of survey stated in the 2009 *Manual of Surveying Instructions* are as follows:

> "The law provides that (1) the public lands of the United States shall be divided by lines intersecting true north and south at right angles so as to form townships 6 miles squares; (2) the townships shall be marked with progressive numbers from the beginning; (3) the townships shall be subdivided into 36 sections, each 1 mile square containing 640 acres as nearly as may be; (4) the sections shall be numbered, respectively, beginning with the number 1 in the northeast section, and proceeding west and east alternately through the township with progressive numbers to and including 36; and (5) a fair plat describing the subdivisions and marks shall be made and recorded at a designated office (Rev. Stat. 2395; 43 U.S.C. 751)."

Additional rules of survey covering field books, subdivision of sections, adjustment for excess and deficiency, and other matters are given in the manuals and special instructions. Private surveyors, on a contract basis, were paid $2/mi of line run until 1796 and $3/mi thereafter. Sometimes the amount was adjusted in accordance with the importance of a line, the terrain, location, and other factors. From this, meager fee surveyors had to pay and feed a party of at least four while on the job, and in transit to and from distant points. They had to brush out and *blaze* (mark trees by scarring the bark) the line, set corners and other marks, and provide satisfactory notes and one or more copies of completed plats. The contract system was completely discarded in 1910. Public land surveyors are now appointed.

As discussed in Chapters 19 and 20, meridians converge at the poles. Thus, it is evident that the requirements that lines shall conform to the true meridians and townships shall be 6 mi square are mathematically impossible. An elaborate system of subdivision was therefore worked out as a practical solution.

Two principles furnished the legal background for stabilizing lines on the land:

1. Boundaries of public lands established by duly appointed surveyors are unchangeable.
2. Original township and section corners established by surveyors must stand as the true corners they were intended to represent, whether or not in the place shown by the field notes.

Expressed differently, the original surveyors had an official plan with detailed instructions for its layout, and presumably set corners to the best of their ability. After title passed from the United States, their established corners (monuments), regardless of errors, became the lawful ones. Therefore, if monuments have disappeared, *the purpose of resurveys is to determine where they were, not where they should have been.* Correcting mistakes or errors now would disrupt too many accepted property lines and result in an unmanageable number of lawsuits.

In general, the procedure in surveying the public lands provides for the following subdivisions:

1. Division into quadrangles (tracts) approximately 24 mi on a side (after about 1840)
2. Division of quadrangles into townships (16), approximately 6 mi on a side
3. Division of townships into sections (36), approximately 1 mi square
4. Subdivision of sections (usually by local surveyors)

It will be helpful to keep in mind that the purpose of the grid system was to obtain sections 1 mi on a side. To this end, surveys proceeded from south to north and east to west, and all discrepancies were thrown into the sections bordering the north and west township boundaries to get as many regular sections as possible.

Although the general method of subdivision outlined above was normally followed, detailed procedures were altered in surveys made at different times in various regions of the country. As examples, instructions for New Mexico said only township lines were to be run where the land was deemed unfit for cultivation, and in Wisconsin, the first four correction lines north of the baseline were 60 mi apart rather than 24.

Another current example relates to the surveys in Alaska, where the area's sheer vastness requires changes. When Alaska gained statehood in 1958, only 2% of its 375 million acres had been surveyed. Priorities were set for conducting the remaining surveys and plans developed that called for subdividing some 155 million acres to be transferred to that state and native Alaskans. To accelerate the project, the 18,651 townships in Alaska were first established on protraction diagrams, and latitude and longitude determined for each corner. In executing the surveys, markers are being set at 2-mi intervals in most areas, and GNSS surveys are being utilized extensively. But even with this modern technology and relaxed procedures, with so much area involved, it will take many years to complete the job.

Distances given in the instructions were in *chains* and miles. The particular chain referred to is the *Gunter's chain*, which was 66-ft long and was introduced as a unit of length in Section 2.2. It was selected for two reasons: (1) it was the best measuring device available to surveyors in the United States at the inception of the PLSS and (2) it had a convenient relationship to the rod, mile, and acre, that is, 1 chain (ch) = 4 rods, 80 ch = 1 mile, and 10 square chains (ch^2) = 1 acre.

Figure 22.2 illustrates a Gunter's chain. It had 100 links, each link equal to 0.66 ft or 7.92 in. The links were made of heavy wire, had a loop at each end, and were joined by three rings. The outside ends of the handles fastened to the end links were the 0 and 66 ft marks. Tags with one, two, three, and four teeth were fastened to the 10th, 20th, 30th, and 40th links, respectively, from both ends of the chain. The 50th link was marked with a round tag. These tags saved time when measuring partial chain lengths. With its many connecting link and ring surfaces subject to frictional wear, hard use elongated the chain, and its length had to be adjusted by means of bolts in the handles. Distances measured with Gunter's chains were recorded either in chains and *links* or in chains and

Figure 22.2
Gunter's chain.

decimals of chains—for example, 7 ch 94.5 lk or 7.945 ch. Decimal parts of links were estimated.

Gunter's chains are no longer manufactured and are seldom, if ever, used today. Nevertheless, the many chain surveys on record oblige the modern practitioner to understand the limits of accuracy possible with this equipment, and the conversion of distances recorded in chains and links to feet or meters. Descriptions of field procedures for conducting PLSS surveys given in the following sections of this chapter are taken from the *2009 Manual of Surveying Instructions*. Again, because lengths are frequently given in chains, familiarity with this unit of measure is essential to understanding the material presented. Of course, surveyors involved in PLSS work today would likely use either total station instruments and measure distances electronically or employ GNSS equipment, but the cited distances and the same basic principles described still apply.

22.3 INITIAL POINT

Thomas Jefferson recognized the importance of property surveys and served as chairman of a committee to develop a plan for locating and selling the western lands. His report to the Continental Congress in 1784, adopted as an ordinance on May 20, 1785, called for survey lines to be run and marked before land sales. Many of today's property disputes would have been eliminated if all property lines had been resurveyed and monuments checked and/or set before sales became final!

Subdivision of the public lands became necessary in many areas as settlers moved in and mining or other land claims were filed. The early hope that surveys would precede settlement was not fulfilled.

As settlers pressed westward, in each area where a substantial amount of surveying was needed, an *initial point* was established within the region to be surveyed.

Figure 22.3
Survey of quadrangles. (Only a few of the standard corners and closing corners are identified.)

It was located by astronomic observations. The manual of 1902 was the first to specify an indestructible monument, preferably a copper bolt, firmly set in a rock ledge, if possible, and witnessed by rock bearings. In all, thirty-seven initial points have been set, five of them in Alaska. An initial point is illustrated near the center of Figure 22.3.

22.4 PRINCIPAL MERIDIAN

From each initial point, a true north–south line called a *principal meridian* (Prin. Mer. or PM) was run north and/or south to the limits of the area to be covered. Generally, a solar attachment—a device for solving mechanically the mathematics of the astronomical triangle—was used. Monuments were set for section and quarter-section corners every 40 ch, (1/2 mi) and at the intersections with all meanderable bodies of water (streams 3 ch or more in width, and lakes covering 25 acres or more).

The line was supposed to be within 3′ of the cardinal direction. Two independent sets of linear measurements were required to check within 20 lk (13.2 ft)/80 ch, which corresponds to a precision ratio of only 1/400. The allowable difference between sets of measurements is now limited to 7 lk/80 ch (precision ratio of 1/1140).

Areas within a principal meridian system vary greatly as depicted in Figure 22.1.

22.5 BASELINE

From the initial point, a baseline was extended east and/or west as a true parallel of latitude to the limits of the area to be covered. As required on the principal meridian, monuments were set for section and quarter-section corners every 40 ch, and at the intersections with all meanderable bodies of water. Permissible closures were the same as those for the principal meridian.

Baselines are circular curves on the Earth's surface, and that were run with chords of 40 ch by the (1) solar method, (2) tangent method, or (3) secant method. These are briefly described as follows:

1. **Solar method.** An observation is made on the Sun to determine the direction of astronomic north. A right angle is then turned off and a line extended 40 ch, where the process is repeated. The series of lines so established, with a slight change in direction every half mile, closely approaches a true parallel. Obviously, if the Sun is obscured, the method cannot be used.
2. **Tangent method.** This method of laying out a true parallel is illustrated in Figure 22.4. A 90° angle is turned to the east or to the west, as may be required from an astronomic meridian, and corners are set every 40 ch. At the same time, proper offsets, which increase with increasing latitudes, are taken from *Standard Field Tables* issued by the BLM and measured north from the tangent to the parallel. In the example shown, the offsets in links are 1, 2, 4, 6 1/2, 9, 12 1/2, 16 1/2, 20 1/2, 25 1/2, 31, and 37. The error resulting from taking right-angle offsets instead of offsets along the converging lines is negligible. The main objection to the tangent method is that the parallel departs considerably from the tangent, so both the tangent and the parallel must sometimes be brushed out to clear sight lines.
3. **Secant method.** This method of laying out an astronomic parallel is shown in Figure 22.5. It is a modification of the tangent method in which a line parallel to the tangent at the 3 mi (center) point is passed through the 1 and 5 mi points to produce minimum offsets.

Fieldwork includes establishing a point on the true meridian, south of the beginning corner, at a distance taken from the *Standard Field Tables* for the

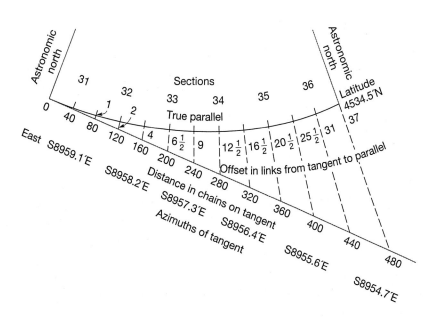

Figure 22.4
Layout of parallel by tangent method. (Adapted from 1973 *Manual of Surveying Instructions*.)

Figure 22.5
Layout of parallel by secant method. (Adapted from 1973 *Manual of Surveying Instructions.*)

latitude of a desired parallel. The proper bearing angle from the same table is turned to the east or west from the astronomic meridian to define the secant, which is then projected 6 mi. Offsets, which also increase with increasing latitudes, are measured north or south from the secant to the parallel. Advantages of the secant method are that the offsets are small and can be measured perpendicular to the secant without appreciable error. Thus, the amount of clearing is reduced.

22.6 STANDARD PARALLELS (CORRECTION LINES)

After the principal meridian and the baseline have been run, *standard parallels* (Stan. Par. or SP), also called *correction lines*, are run as true parallels of latitude 24 mi apart in the same manner as was the baseline. All 40 ch corners are marked. Standard parallels are shown in Figure 22.3. In some early surveys, standard parallels were placed at intervals of 30, 36, or 60 mi.

Standard parallels are numbered consecutively north and south of the baseline; examples are first standard parallel north and third standard parallel south.

22.7 GUIDE MERIDIANS

Guide meridians (GM) are run due north (astronomic) from the baseline and the standard parallels at intervals of 24-mi east and west of the principal meridian, in the same manner as was the principal meridian, and with the same limits of error. Before work is started, the chain or tape must be checked by measuring 1 mi on the baseline or standard parallel. All 40-ch corners are marked.

Because meridians converge, a *closing corner* (CC) is set at the intersection of each guide meridian and standard parallel or baseline (see Figure 22.3). The distance from the closing corner to the *standard corner* (SC which was set when the parallel was run, is measured and recorded in the notes as a check. Any error in the 24-mi-long guide meridian is put in the northernmost half mile.

Guide meridians are numbered consecutively east and west of the principal meridian; examples are first guide meridian west and fourth guide meridian east. Correction lines and guide meridians, established according to instructions, created *quadrangles* (or tracts) whose nominal dimensions are 24 mi on a side. These are shown in Figure 22.3.

22.8 TOWNSHIP EXTERIORS, MERIDIONAL (RANGE) LINES, AND LATITUDINAL (TOWNSHIP) LINES

Division of a quadrangle, or tract, into townships is accomplished by running *range* (R) and *township* (T or Tp) lines

Figure 22.6
Order of running lines for the subdivision of a quadrangle into townships.

Range lines are astronomic meridians through the standard township corners previously established at intervals of 6 mi on the baseline and standard parallels. They are extended north to intersect the next standard parallel or baseline and closing corners set (see Figures 22.3 and 22.6). Township lines are east–west lines that connect township corners previously established at intervals of 6 mi on the principal meridian, guide meridians, and range lines.

The angular amount by which two meridians converge is a function of latitude and the distance between the meridians. The linear amount of convergence is a function of the same two variables, plus the length that the meridians are extended. Formulas for angular and linear convergence of meridians (derived in various texts on geodesy), are as follows:

$$\theta = 52.13 \, d \tan \varphi \qquad (22.1)$$

and

$$c = \frac{4}{3} Ld \tan \varphi \quad \text{(slight approximation)} \qquad (22.2)$$

where θ is the angle of convergence (in seconds); d the distance between meridians (in miles), on a parallel; φ the mean latitude; c the linear convergence (in feet); and L the length of meridians (in miles).

■ Example 22.1

Compute the angular convergence at 40°25′ North latitude between two adjacent guide meridians.

Solution

By Equation (22.1) (guide meridians are 24 mi apart)

$$\theta = 52.13(24)\tan 40°25' = 1065'' = 17'45''$$

■ Example 22.2

Determine the distance that should exist between the standard corner and its closing corner (if there were no surveying errors) for a range line 12 mi east of the principal meridian, extended 24 mi north at a mean latitude of 43°10′.

Solution

By Equation (22.2)

$$c = \frac{4}{3}(24)(12)\tan 43°10' = 360.18\,\text{ft}$$

■ 22.9 DESIGNATION OF TOWNSHIPS

A township is identified by a unique description based on the principal meridian governing it.

North and south rows of townships are called *ranges* and numbered in consecutive order east and west of the principal meridian as indicated in Figure 22.3.

East and west rows of townships are named *tiers* and numbered in order north and south of the baseline. By common practice, the term tier is usually replaced by the township in designating the rows.

An individual township is identified by its number north or south of the baseline, followed by the number east or west of the principal meridian. An example is Township 7 South, Range 19 East, of the Sixth Principal Meridian. Abbreviated, this becomes T 7 S, R 19 E, 6th PM.

■ 22.10 SUBDIVISION OF A QUADRANGLE INTO TOWNSHIPS

The method to be used in subdividing a quadrangle into townships is fixed by regulations in the *Manual of Surveying Instructions*. Under the old regulations, township boundaries were required to be within 21 min of the cardinal direction. Later this was reduced to 14 min to keep interior lines within 21 min of the cardinal direction.

The detailed procedure for subdividing a quadrangle into townships can best be described as a series of steps designed to ultimately produce the maximum number of regular sections with minimum unproductive travel by the field party. The order of running the lines is shown by consecutive numbers in Figure 22.6. Some details are described in the following steps:

1. Begin at the southeast corner of the southwest township, point A, after checking the chain or tape against a 1 mi measurement on the standard parallel.

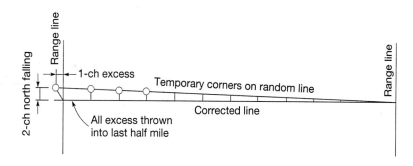

Figure 22.7
Correction of random line for excess and falling.

2. Run north on the astronomic meridian for 6 mi (line 1 of Figure 22.6), setting alternate section and quarter-section corners every 40 ch. Set township corner *B*.
3. From *B*, run a random line (line 2 of Figure 22.6) due west to intersect the principal meridian. Set temporary corners every 40 ch.
4. If the random line has an excess or deficiency of 3 ch or less (allowing for convergence) and a falling north or south of 3 ch or less, the line is accepted. It is then corrected back (line 3), and all corners are set in their proper positions. Any excess or deficiency is thrown into the most westerly half mile. The method of correcting a random line having an excess of 1 ch and a north falling of 2 ch is shown in Figure 22.7.
5. If the random line misses the corner by more than the permissible 3 ch, all four sides of the township must be retraced.
6. The same procedure is followed until the southeast corner *D* of the most northerly township is reached. From *D*, range line 10 is continued as an astronomic meridian to intersect the standard parallel or baseline, where a closing corner is set. All excess or deficiency in the 24 mi is thrown into the most northerly half mile.
7. The second and third ranges of townships are run the same way, beginning at the south line of the quadrangle.
8. While the third range is being run, random lines are also projected to the east and corrected back and any excess or deficiency is thrown into the most westerly half mile. (On this line, all points may have to be moved diagonally to the corrected line, instead of just the last point, as in Figure 22.7.)

22.11 SUBDIVISION OF A TOWNSHIP INTO SECTIONS

Sections are now numbered from 1 to 36, beginning in the northeast corner of a township and ending in the southeast corner, as shown in Figure 22.8. The method used to subdivide a township can be described most readily as a series of steps to produce the maximum number of regular sections 1 mi on a side. Lines were run in the following order:

1. Set up at the southeast corner of the township, point *A*, and observe the astronomic meridian. Retrace the range line northward and the township line westward for 1 mi to compare the meridian, needle readings, and taped distances with those recorded.

2. From the southwest corner of section 36, run north *parallel with the east boundary of the township*. Set quarter-section and section corners on line 1 (Figure 22.8).
3. From the section corner just set, run a random line *parallel with the south boundary of the township* eastward to the range line. Set a temporary quarter corner at 40 ch.
4. If the 80-ch distance on the random line is within 50 lk, falling or distance, the line is accepted. The correct line is calculated and a quarter corner located at the *midpoint* of line BC connecting the previously established corner C and the new section corner B.
5. If the random line misses the corner by more than the permissible 50 lk, the township lines must be rechecked and the source of the error determined.
6. The east range of sections is run in a similar manner until the southwest corner of section 1 is reached. From this point a continuation of the northward line is run to connect with the north township line section corner. A quarter corner is set 40 ch from the south section corner (on line 17, corrected back by later manuals). All discrepancies in the 6 mi are thrown into the last half mile.
7. Successive ranges of sections across the township are run until the first four have been completed. All north-south lines are parallel with the township east side. All east-west lines are run randomly parallel with the south boundary line and then corrected back.
8. When the fifth range is being run, random lines are projected to the west as well as to the east. Quarter corners in the west range are set 40 ch from

Figure 22.8 Order of running lines for the subdivision of a township into sections.

the east side of the section, with all excess or deficiency resulting from the errors and convergence being thrown into the most westerly half mile.
9. If the north side of the township is a standard parallel, the northward lines, which are run parallel with the east township boundary, are projected to the correction line and closing corners set. The distance to the nearest corner is measured and recorded.
10. Bearings of interior north–south section lines for any latitude can be obtained by applying corrections from tables for the convergence at a given distance from the east boundary.

By placing the effect of meridian convergence into the westernmost half mile of the township and all errors to the north and west, 25 regular sections nominally $1\,\text{mi}^2$ are obtained. Also, the south half of sections 1, 2, 3, 4, and 5; the east half of sections 7, 18, 19, 30, and 31; and the southeast quarter of section 6 are all regular size.

■ 22.12 SUBDIVISION OF SECTIONS

A section was the basic unit of the General Land Office system but land was often patented in parcels smaller than a section. Local surveyors and others performed subdivision of sections as the owners took up the land. This subdivision of regular sections using the midpoint protractions of the sections is known as dividing the section into its aliquot parts. Thus, the NE 1/4 of Section 8 is called an aliquot part of Section 8. The BLM provides guidelines on the proper and intended way a section should be subdivided. To divide a section into quarter sections (nominally 160 acres), straight lines are run between opposite quarter-section corners previously established or reestablished. This rule holds whether or not the quarter-section corners are equidistant from the adjacent-section corners. Due principally to underground ore deposits, which caused large local attraction errors in compass directions, one-quarter section in a Wisconsin township contains 640 acres!

To divide a quarter section into quarter–quarter sections (nominally 40 acres), straight lines are run between opposite quarter–quarter-section corners established at the midpoints of the four sides of the quarter section. The same procedure is followed to obtain smaller subdivisions.

If the quarter sections are on the north or west side of a township, the quarter–quarter-section corners are placed 20 ch from the east or south quarter-section corners—or by single proportional measurement (see Section 22.19) on line if the total length on the ground is not equal to that on record.

■ 22.13 FRACTIONAL SECTIONS

In sections made fractional by rivers, lakes, or other bodies of water, lots are formed bordering on the body of water and numbered consecutively through the section (see Section 8 in Figure 22.9). Boundaries of lots usually follow the quarter section and quarter–quarter-section lines, but extreme lengths or narrow widths are avoided, as are areas of fewer than 5 acres or more than 45 acres.

Quarter sections along the north and west boundaries of a township, made irregular by discrepancies of measurements and convergence of the range lines, are usually numbered as lots (see Figure 22.9). Lot lines are not actually run in the

696 SURVEYS OF THE PUBLIC LANDS

Figure 22.9 Subdivision of regular and fractional sections.

field. Like quarter–section lines, they are merely indicated on the plats by *protraction* (subdivisions of parcels on paper only). Areas needed for selling the lots are computed from the plats.

22.14 NOTES

Specimen field notes for each of the several kinds of lines to be run are shown in various instruction manuals. Actual recording had to closely follow the model sets. The original notes, or copies of them, are maintained in a land office in each state for the benefit of all interested persons.

22.15 OUTLINE OF SUBDIVISION STEPS

Pertinent points in the subdivision of quadrangles into townships, and townships into sections, are summarized in Table 22.1.

TABLE 22.1 Subdivision Steps

Item	Subdivision of a Quadrangle	Subdivision of a Township
Starting point	SE corner of SW township	SW corner of SE section (36)
Meridional lines		
Name	Range line	Section line
Direction	Astronomic North	North, parallel with east range line
Length	6 mi = 480 ch	1 mi = 80 ch
Corners set	Quarter-section and section corners at 40 and 80 ch alternately	Quarter-section corner at 40 ch; section corner at 80 ch

TABLE 22.1 (CONTINUED)

Item	Subdivision of a Quadrangle	Subdivision of a Township
Latitudinal lines		
Name	Township line	Section line
Direction of random	True east–west parallel	East, parallel with south side of section
Length	6 mi less convergence	1 mi
Permissible error	3 ch, length or falling	50 lk, length of falling
Distribution of error		
Falling	Corners moved proportionately from random to true line	Corner moved proportional from random to true line
Distance	All error thrown into west quarter section	Error divided equally between quarter sections

(Work repeated until north side of area is reached. Subdivision of last area on the north of the range of townships and sections follows.)

Case I. When Line on the North Is a Standard Parallel

Item	Subdivision of Quadrangle	Subdivision of Township
Direction of line	Astronomic North	North, parallel with east range line
Distribution of error in length	Placed in north quarter section	Placed in north quarter section
Corner placed at end	Closing corner	Closing error
Permissible errors	Specified in *Manual of Surveying Instructions*	Specified in *Manual of Surveying Instructions*

Case II. When Line on the North Is Not a Standard Parallel

Item	Subdivision of a Quadrangle	Subdivision of a Township
Direction of line	No case	Random north and correct back to section corner already established
Distribution of error in length		Same as Case I

[Other Ranges of Townships and Sections Continued Until All but Two Are Laid Out.]

	Subdivision of a Quadrangle	Subdivision of a Township
Location of last two ranges	On east side of tract	On west side of township
Next-to-last range subdivided	As before	As before
Last range		
Direction of random	True east	Westerly, parallel with south side of section
Nominal length	6 mi less convergence	1 mi less convergence
Correction of temporary corners	Corners moved proportionately from random to true line	Corners moved proportionately from random to true line
Distribution of closure error	Corners moved westerly (or easterly) to place error in west quarter section	Corners are placed on the true line so total error falls in west quarter section

22.16 MARKING CORNERS

Various materials were approved and used for monuments in the original surveys. These included pits and mounds, stones, wooden posts, charcoal, and bottles. A zinc-coated, alloyed iron pipe with 2 1/2 in. outside diameter, 30 in. long, with brass cap is now standard. The bottom end of the pipe is split for several inches to form flanges that help hold it in place in the ground. Substitutions for this standard monument are permitted when authorized. Specially manufactured markers are now becoming commonplace. One type has a breakaway top so that if it is accidentally hit, for example by a plow or bulldozer, the upper part of the stake will break off but the lower part will remain in place. Another type uses a small, attached magnetic device to aid in recovery with a metal detector. In rock outcrops, a 3-1/4 in. diameter brass tablet with 3-1/2 in. stem is specified.

Stones and posts were marked with one to six notches on one or two faces. The arrangements identify a monument as a particular section or township corner. Each notch represents 1 mi of distance to a township line or corner. Quarter-section monuments were marked with the fraction "1/4" on a single face. In prairie country, where large stones and trees were scarce, a system of pits and mounds was used to mark corners. Different groupings of pits and mounds, 12 in. deep and 18 in. square, designated corners of several classes. However, unless some other type of mark perpetuated these markers, these corners were unfortunately lost in the first plowing.

22.17 WITNESS CORNERS

Whenever possible, monuments were witnessed by two or three adjacent objects such as trees and rock outcrops. *Bearing trees* were blazed on the side facing the corner and marked with scribing tools.

When a regular corner fell in a creek, pond, swamp, or other place where it was impractical to place a mark, *witness corners* (WC) were set on a line(s) leading to the corner. Letters WC were added to all witness corner markers, and the witness corners were also witnessed.

22.18 MEANDER CORNERS

A *meander corner* (MC) was established on survey lines intersecting the bank of a stream having a width greater than 3 ch, or a lake, bayou, or other body of water of 25 acres or more. The distance to the nearest section corner or quarter-section corner was measured and recorded in the notes. A monument was set and marked MC on the side facing the water, and the usual witness noted. If practical, the line was carried across the stream or other body of water by triangulation to another corner set in line on the farther bank.

A traverse joined successive meander corners along the banks of streams or lakes and followed as closely as practical the sinuosities of the bank. Calculating the position of the new meander corner and comparing it with its known position on a surveyed line checked the traverse.

Meander lines follow the mean high-water mark and are used only for plotting and protraction of the area. They are not boundaries defining the limits of property adjacent to the water.

22.19 LOST AND OBLITERATED CORNERS

A common problem in resurveys of the public lands is the replacement of lost or obliterated corners. This difficult task requires a combination of experience, hard work, and ample time to reestablish the location of a wooden stake or post, set perhaps 150 years ago, and with all witness trees long since cut or burned by apathetic owners.

An *obliterated corner* is one for which there are no remaining traces of the monument or its accessories, but whose location has been perpetuated or can be recovered beyond reasonable doubt. The corner may be restored from the acts or testimony of interested landowners, surveyors, qualified local authorities, witnesses, or from written evidence. Satisfactory evidence has value in the following order:

1. Evidence of the corner itself.
2. Bearing trees or other witness marks.
3. Fences, walls, or other evidence showing occupation of the property to the lines or corners.
4. Testimony of living persons.

A *lost corner* is one whose position cannot be determined beyond reasonable doubt, either from traces of the original marks or from acceptable evidence or testimony that bears on the original position. It can be restored only by rerunning lines from one or more independent corners (existing corners that were established at the same time and with the same care as the lost corner). *Proportionate measurements* distribute the excess or deficiency between a recently measured distance d separating the nearest found monuments that straddle the lost point and the record distance D given in the original survey notes between these monuments. Then the distance x from one of the found monuments required to set the lost point is calculated by proportion as $x = X(d/D)$, where X is the record distance from that monument.

Single-proportionate measurement follows the procedure just described and is used to relocate lost corners that have a specific alignment in one direction only. These include standard corners on baselines and standard parallels, intermediate section corners on township boundaries, all quarter-section corners, and meander corners established originally on lines carried across a meanderable body of water. Corners on true lines of latitude such as baselines and standard parallels must be offset (south) from the proportion line to maintain the curvature of the latitudinal line.

■ Example 22.3

Figure 22.10 illustrates a lost quarter-section corner a on the line between sections 2 and 3. Section corners b and c have been found. The record distances

700 SURVEYS OF THE PUBLIC LANDS

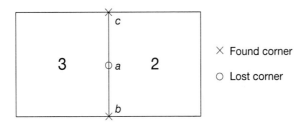

Figure 22.10
Example of single-proportionate measurement.

for lines *ba* and *ac* are 40.00 and 39.57 ch, respectively. The observed distance between found corners *b* and *c* was 5246.25 ft. Describe the process for restoring lost corner *a*.

Solution

1. Since point *a* is a quarter-section corner, it is replaced on line *bc* by single-proportionate measurement.
2. Distance *ba* that must be laid off from section corner *b* to restore lost corner *a* is

$$(ba) = \frac{40}{79.57} \times 5246.25 = 2637.30 \, \text{ft}$$

Double-proportionate measurements are used to establish lost corners located originally by specific alignment in two directions, such as interior section corners and corners common to four townships. The general procedure for single-proportionate measurement is used, but in two directions. It will establish two temporary points: one on the north–south line and another on the east–west line. The lost corner is then located where lines run from the two points, in the cardinal directions of north–south and east–west, intersect.

■ Example 22.4

Figure 22.11 illustrates lost corner *a* that is common to sections 22, 23, 26, and 27. Corners *b, c, d*, and *e* have been found. Record and measured distances are as follows:

Record		Measured	
Line	Distance (ch)	Line	Distance (ft)
ba	40.00	bd	7925.49
ca	40.00	ec	5293.24
da	79.20		
Ea	39.72		

Describe the process of restoring lost corner *a*.

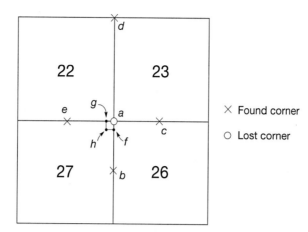

Figure 22.11
Example of double-proportionate measurement.

Solution

1. Corner *a* is an interior section corner, which is constrained in alignment in two directions. Thus, it must be restored by double-proportionate measurement.
2. First establish a temporary point *f* by laying off distance *bf* along line *bd*, where *bf* is computed as

$$bf = \frac{40.00}{119.20} 7925.49 = 2659.56 \text{ ft}$$

3. Then locate temporary point *g* by laying of distance *eg* along line *ec*, where *eg* is computed as

$$eg = \frac{39.72}{79.72} 5293.24 = 2637.32 \text{ ft}$$

4. Establish point *h*, the restored lost corner, where an east–west line through *f* intersects a north–south line through *g*.

When the original surveys were run, *topographic calls* (distances along each line from the starting corner to natural features such as streams, swamps, and ridges) were recorded. Using the recorded distances to any of these features found today, and applying single- or double-proportionate measurements to them may help locate an obliterated corner or produce a more reliable reestablished lost corner.

22.20 ACCURACY OF PUBLIC LAND SURVEYS

The accuracy required in the early surveys was a very low order. Frequently it fell below what the notes indicated. A small percentage of the surveys were made by unscrupulous surveyors drawing on their imaginations in the comparative comfort of a tent; no monuments were set, and the notes serve only to confuse the situation for present-day surveyors and land owners. A few surveyors threw in an extra chain length at intervals to assure a full measure.

Many surveys in one California county were fraudulent. In another state, a meridian 108 mi in length was run without including the chain handles in its 66 ft length. When discovered, it was rerun without additional payment.

The poor results obtained in various areas were primarily caused by the following:

1. Lack of trained personnel; some contracts were given to persons with no technical training.
2. Very poor equipment by today's standards.
3. Work done by contract at low prices.
4. Surveys made in piecemeal fashion as the Indian titles and other claims were extinguished.
5. Conflicts with native people.
6. Swarms of insects, dangerous animals, and reptiles.
7. Lack of appreciation for the need to do accurate work.
8. Erratic or missing field inspection.
9. Magnitude of the problem.

In general, considering the handicaps listed, the work was reasonably well done in most cases.

22.21 DESCRIPTIONS BY TOWNSHIP SECTION, AND SMALLER SUBDIVISION

Description by the U.S. Public Land Survey System offers a means of defining boundaries uniquely, clearly, and concisely. Several examples of acceptable descriptions are listed.

Sec. 6, T 7 S, R 19 E, 6th PM.
Frac. Sec. 34, T 2 N, R 5 W, Ute Prin. Mer.
The $SE\frac{1}{4}$, $NE\frac{1}{4}$, Sec. 14, Tp. 3 S, Range 22 W, SBM [San Bernardino Meridian].
$E\frac{1}{2}$ of $N\frac{1}{4}$ of Sec. 20, T 15 N, R 10 E, Indian Prin. Mer.
E 80 acres of $NE\frac{1}{4}$ of Sec. 20, T 15 N, R 10 E, Indian Prin. Mer.

Note that the last two descriptions do not necessarily describe the same land. A California case in point occurred when the owner of a southwest quarter section, nominally 160 acres but actually 162.3 acres, deeded the westerly portion as "the West 80 acres" and the easterly portion as "the East 1/2."

Sectional land that is privately owned may be partitioned in any legal manner at the option of the owner. The metes-and-bounds form is preferable for irregular parcels. In fact, metes and bounds are required to establish the boundaries of mineral claims and various grants and reservations.

Differences between the physical and legal (or record) ground locations and areas may result because of departures from accepted procedures in description writing, loose and ambiguous statements, or dependence on the accuracy of early surveys.

22.22 BLM LAND INFORMATION SYSTEM

As noted in Section 22.1, approximately one third of the United States land area remains in the public domain. The Bureau of Land Management (BLM) is responsible for managing a significant portion of this vast acreage. Rapid access to accurate information related to this public land is in demand now more than ever before. As an example of its importance, consider that millions of dollars worth of oil and mineral royalties can he gained or lost by relatively small changes in property boundaries.

To assist in the monumental task of managing this enormous quantity of diverse land, the BLM and Forest Service provide business solutions for the management of cadastral records and land parcel information in a Land Information System (LIS—see Chapter 28) called the *National Integrated Land System* (NILS).[1] The LIS can be used to aid in making resource management decisions such as processing applications for mineral leases, designating utility corridors, issuing land use permits, locating wildlife habitat improvements, preparing timber sales, evaluating alternatives in environmental assessments and land use plans, and generating reports and maps.

The NILS concept provides users with tools to manage land records and cadastral data. The NILS project has four major components: (1) *survey management*, (2) *measurement management*, (3) *parcel management*, and (4) the *GeoCommunicator*. The survey management component consists of an integrated set of automation objects that will be embedded into compatible data collection software packages. This software will support the capture and input of feature measurements, and metadata (see Section 28.8) directly into a GIS database format. The measurement management system will allow users to further enhance the data set by performing a weighted least-squares adjustment (see Chapter 16) of new features. This will enable creation of a higher-quality network database in both PLSS and metes-and-bounds environments. The parcel management system will provide a process for managing land records and cadastral data stored in the database model. The GeoCommunicator is a website for sharing information about data and activities of interest to land managers. This system allows users to discover information that meets their needs with the goal of GeoCommunicator to facilitate data sharing and collaborative efforts among land managers.

As part of the NILS project, the BLM created a *Geographic Coordinate Data Base* (GCDB).[2] This database contains a digital layer of information on the

---

[1]Information on this project can be obtained at http://www.blm.gov/pgdata/etc/medialib/blm/wo/MINERALS__REALTY__AND_RESOURCE_PROTECTION_/energy/renewable_references.Par.29372.File.dat/NILS_Reference_links.pdf.

[2]More information on the GCDB can be found on the BLM web page http://www.blm.gov/wo/st/en/prog/more/gcdb.html.

U.S. Public Land Survey System (PLSS) and provides the positional components necessary for correlating all other information in the LIS. Included in the GCDB are geographic coordinates of all PLSS corners and estimates of their reliability, identifications of the surveyors who set the corners, the types of corners set and dates placed, any records of resurveys, a full record of ownership of each parcel, ownership of abutting parcels, and much more information. This effort was started in 1989 and continues today. The GCDB files for many townships are available for download via BLM websites. The national GCDB website is http://www.blm.gov/wo/st/en/prog/more/gcdb.html.

22.23 SOURCES OF ERROR

Some of the many sources of error in retracing the public land surveys are:

1. Discrepancy between length measured with an early surveyor's chain and one obtained with present-day equipment.
2. Change in magnetic declination, local attraction, or both.
3. Lack of agreement between field notes and actual measurements.
4. Change in watercourses.
5. Nonpermanent objects used for corner marks.
6. Loss of witness corners.

22.24 MISTAKES

Some typical mistakes in the retracement of boundaries in public land surveys are:

1. Failing to follow the general rules and special instructions of procedure governing the original survey.
2. Neglecting to calibrate measured lengths against record distances for marks in place.
3. Treating corners as lost when they are actually obliterated.
4. Resetting corners without exhausting every means of relocating the original corners.
5. Failing to recognize that restored corners must be placed in their original locations regardless of deviations by the original surveyor from the general rules and special instructions.

PROBLEMS

Asterisks (*) indicate problems that have answers given in Appendix G.

22.1* Convert 41.3 chains to feet.
22.2 What steps in the subdivision of public lands is left to local surveyors?
22.3 In what states are public land surveys not applicable?
22.4 Define aliquot part.
22.5 Why are the boundaries of the public lands established by duly appointed surveyors unchangeable, even though incorrectly set in the original surveys?

22.6 What is the convergence in feet of meridians for the following conditions:
(a)* 24 mi apart, extended 24 mi, at mean latitude 45°20′N.
(b) 6 mi apart, extended 24 mi, at mean latitude 29°45′N

22.7 What is the angular convergence, in seconds, for the two meridians defining a township exterior at mean latitude of:
(a) 43°33′N
(b) 32°41′N

22.8 What is the nominal distance in miles between the following?
(a)* First Guide Meridian East, and the west Range Line of R 8 E.
(b) SE corner of Sec. 14, T 6 S, R 5 E, Salt River PM, and the NW corner of Sec. 23, T 6 S, R 3 E, Salt River PM.

22.9 Discuss when meander corners are set in a public land survey.

Sketch and label pertinent lines and legal distances, and compute nominal areas of the parcels described in Problems 22.10 through 22.12.

22.10 E 1/2, SE 1/4, Sec. 6, T 2 S, R 3 E, Boise PM
22.11 SE 1/4, NW 1/4, Sec. 15, T 1 N, R 2 E, Fairbanks PM.
22.12 NE 1/4, SE 1/4, SE 1/4, Sec. 30, T 1 S, R 4 E, 6th PM.
22.13 What are the nominal dimensions and acreages of the following parcels:
(a) NE 1/4, NE 1/4, Sec. 22.
(b) S 1/2, SE 1/4, Sec. 9.
(c) SW 1/4, NW 1/4, SW 1/4, Sec. 12.

22.14 How many rods of fence are required to enclose the following:
(a)* A parcel including the NE 1/4, NE 1/4, Sec. 32, and the NW 1/4, NW 1/4, Sec. 33, T 2 N, R 3 E?
(b) A parcel consisting of Secs. 8, 9, and 16 of T 2 N, R 1 W?

22.15 What lines of the U.S. public land system were run as random lines?

22.16 In subdividing a township, which section line is run first? Which last?

22.17 Corners of the SE 1/4 of the NW 1/4 of Sec. 22 are to be monumented. If all section and quarter-section corners originally set are in place, explain the procedure to follow, and sketch all lines to be run and corners set.

22.18 The quarter-section corner between Secs. 15 and 16 is found to be 40.28 ch from the corner common to Secs. 9, 10, 15, and 16. Where should the quarter-quarter-section corner be set along this line in subdividing Sec. 15?

22.19 As shown in the figure, in a normal township the exterior dimensions of Sec. 6 on the west, north, east, and south sides are 81, 78, 82, and 79 ch, respectively. Explain with a sketch how to divide the section into quarter sections.

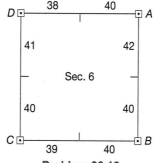

Problem 22.19

22.20 The problem figure shows original record distances. Corners $A, B, C,$ and D are found, but corner E is lost. Measured distances are $AB = 7035.6$ ft and $CD = 10{,}718.4$ ft. Explain how to establish corner E. (See the following figure.)

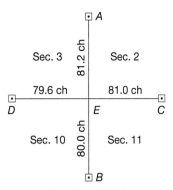

Problem 22.20

To restore the corners in Problems 22.21 through 22.24, which method is used, single proportion or double proportion?

22.21* Township corners on guide meridians; section corners on range lines.
22.22 Section corners on section lines; township corners on township lines.
22.23 Quarter-section corners on range lines.
22.24 Quarter–quarter-section corners on section lines.
22.25 What is a closing corner?
22.26 Explain the difference between an "obliterated corner" and a "lost corner."
22.27 Why are meander lines not accepted as the boundaries defining ownership of lands adjacent to a stream or lake?
22.28 The southern boundary of a township lies on a standard parallel at latitude 38°30′N. By how much is the northern border shorter?
22.29 Why are the areas of many public land sections smaller than the nominal size?

BIBLIOGRAPHY

Dalager, B. 2010. "The Largest Non-BLM Cadastral Retracement in History." *The American Surveyor* 7 (No. 5): 28.

Erickson, C. and L. Erickson. 2014. "Land Locators & Claim Jumpers, Part 1." *The American Surveyor* 11 (No. 4): 22.

Foose, J. 2014. "Bona Fide 'Wrongs' & 'Fixings' the Footsteps." *The American Surveyor* 11 (No. 3): 23.

Griffin, C. D. 2014. "The Importance of Thomas Freeman's Surveys in the Alabama Territory." *Surveying and Land Information Science* 73 (No. 2): 49.

_____ 2014. "The Importance of Thomas Freeman's Surveys in the Mississippi Territory." *Surveying and Land Information Science* 73 (No. 1): 31.

Nash, S. 2014. "The Initial Point of Montana." *The American Surveyor* 11 (No. 2): 34.

Penry, J. 2014. "The McShane Corner." *The American Surveyor* 11 (No. 7): 44.

Stahl, J. B. 2009. "Marking a Point." *Point of Beginning* 35 (No. 3): 28.

U.S. Department of the Interior, 2009. *Manual of Surveying Instructions: For the Survey of the Public Lands of the United States.* Bureau of Land Management. Denver, CO: U.S. Government Printing Office.

Zimmer, R. and S. Kirkpatrick. 2009. "GIS Data Integration with the GCDB." *The American Surveyor* 6 (No. 5): 48.

23
Construction Surveys

■ 23.1 INTRODUCTION

Construction is one of the largest industries in the United States, and thus surveying, as the basis for it, is extremely important. It is estimated that 60% of all hours spent in surveying are on location-type work, giving line and grade. Nevertheless, insufficient attention is frequently given to this type of survey.

An accurate control, topographic survey, and site map are the first requirements in designing streets, sewer and water lines, and structures. Surveyors then lay out and position these facilities according to the design plan. A final "as-built" map, incorporating any modifications made to the design plans, is prepared during and after construction and filed. Such maps are extremely important, especially where underground utilities are involved, to assure that they can be located quickly if trouble develops, and that they will not be disturbed by later improvements.

Traditionally, construction surveying involves establishing both *line* and *grade* by means of stakes and reference lines that are placed on the construction site. These guide the contractor so that proposed facilities are constructed according to a plan. Placement of the stakes is most often done by making the fundamental observations of horizontal distances, horizontal and vertical angles, and differences in elevation using the basic equipment and methods described in earlier chapters of this text. However, the global navigation satellite system (GNSS) is also being used with increasing frequency for construction surveys (see Section 23.11). Other specialized equipment, including laser alignment devices and reflectorless electronic distance measuring equipment (see Section 23.2), are also used, which greatly facilitates construction surveying. These newer technologies have led to machine guidance and control eliminating the need for extensive staking.

All surveyors, engineers, and architects who may be involved with planning, designing, and building constructed facilities should be familiar with the fundamental procedures involved in construction surveying. In smaller projects, traditional methods of staking out the construction details are still performed. However, in larger projects, machine guidance and control using robotic total stations and/or GNSS surveying equipment has replaced many of the staking requirements. This chapter describes procedures applicable for some of the more common types of construction projects using both traditional methods and machine control methods. Chapters 24, 25, and 26 cover the subjects of horizontal curves, vertical curves, and volume computations, respectively. These topics are all pertinent to construction surveys, particularly those for transportation routes.

Common steps for surveying engineers in any construction project consist of (1) placement of horizontal and vertical control, (2) a topographic survey used in the creation of an existing surface map, (3) staking of the design, which may include physically staking the design on the ground to guide the equipment operators or calibration of the equipment and uploading and maintenance of the multiple design surfaces into a machine control system, (4) periodic checks on the layout of the design determination of quantities moved or placed during the construction process, and (5) a final as-built survey of the project. Construction surveying is perhaps best learned on the job and consists in adapting fundamental principles to the undertaking at hand. Since each project may involve unique conditions and present individual problems, coverage in this text is limited to a discussion of the fundamentals.

23.2 SPECIALIZED EQUIPMENT FOR CONSTRUCTION SURVEYS

As noted above, the placement of stakes for line and grade to guide construction operations is often accomplished using the surveyor's standard equipment—levels, tapes, total station instruments, and GNSS receivers. However, there are some additional instruments that improve, simplify, and greatly increased the speed with which certain types of construction surveying are accomplished. *Visible laser-beam* alignment instruments, *reflectorless total stations*, and laser scanners are among these. These are described briefly in the subsections that follow.

23.2.1 Visible Laser-Beam Instruments

The fundamental purpose of laser instruments is to create a visible line of known orientation, or a plane of known elevation, from which measurements for line and grade can be made. Two general types of lasers are described here:

Single-beam lasers, as shown in Figures 23.1 and 23.2, project visible reference lines ("string lines" or "plumb lines") that are utilized in linear and vertical alignment applications such as tunneling, sewer pipe placement, and building construction. The instrument shown in Figure 23.1 is a single-beam-type laser, which can be placed on the inside of a pipe to ensure correct grade of the pipe. The instrument is self-leveling in a wide range, and the design-grade of the pipe can be entered into its user-interface. It can sit in a pipe as small as 4-in. (100 mm) or be mounted on a tripod for above ground use. Thus, it provides a string line on a construction site without the string.

Rotating-beam lasers are merely single-beam lasers with spinning optics that rotate the beam in azimuth, thereby creating planes of reference. They expedite

Figure 23.1 Single-beam pipeline laser level. (Courtesy Leica Geosystems.)

Figure 23.2 Rotating laser providing finished grades to machine control system in grader. (Courtesy Topcon Positioning Systems.)

the placement or eliminate the need for grade stakes over large and small areas. The instrument shown in Figure 23.2 has a horizontal working range of 2400 m and a vertical working range of 40 m. When combined with GNSS and machine guidance and control technologies, it can provide millimeter accuracies in elevation during the construction of highway alignments and large areas such as airports, parking lots, and subdivisions.

Figure 23.3 shows a rotating-beam-type laser that is typically seen on a building construction site. It projects a beam up to 350 m while rotating at 600 rpm. One or more receivers attached to grade rods or staffs can pick up the laser signal. The instrument is self-leveling and is quickly set up. If somehow bumped out of level, the laser beam shuts off and does not come back on until it is releveled. It can be operated with the laser plane oriented horizontally for setting footings, floors, etc., or the beam can be turned 90° and used vertically for plumbing walls or columns.

Because laser beams are not readily visible to the naked eye in bright sunlight, special detectors attached to a hand-held rod are often used. To lay out horizontal planes with either of these devices, the height of the instrument above datum, *HI*, must be established. Then the height on a graduated rod that a reference mark or detector must be set is the difference between the *HI* and the plane's required elevation.

Figure 23.3
Sokkia LP30 laser plane level. (Courtesy Sokkia Corporation.)

23.2.2 Reflectorless EDMs

As described in Chapter 8, total station instruments simultaneously observe horizontal and vertical angles as well as slope distances. Their built-in microcomputers reduce the observed slope distances to their horizontal and vertical components and display the results in real time. These features make total stations very convenient for construction stakeout.

Reflectorless total stations include an electronic distance measurement (EDM) that does not require a reflector. Rather, these instruments employ a laser light, which is capable of measuring distances of up to 100 m or more to objects without the use of a reflector. Alternatively, they can be used with reflectors, or reflective sheets that can be applied to surfaces, a procedure that enables them to measure longer distances. Figure 23.4(a) shows a total station instrument equipped with this technology, and Figure 23.4(b) depicts a similar hand-held pulsed laser EDM instrument. Both devices are useful in observing distances to inaccessible locations, a feature that is particularly useful in assembling and checking the placement of structural members in bridges, buildings, and other large fabricated objects as well as establishing control to support laser scanning (see Sections 17.9.5 and 23.13).

23.2.3 Laser Scanners

As shown in Figure 23.5 and discussed further in Section 23.13, laser scanners are often used to capture data at sites where detailing with traditional surveying instrumentation would be hazardous or time consuming. In the figure, the details of a bridge are being captured safely from the side of a divided highway. The scanner, described in Section 17.9.5, can create point clouds with at varying densities where each point has a three-dimensional coordinate. After data capture is complete, the point clouds can be used to obtain various measurements on the structure. For example, the point cloud of the refinery in Figure 17.11 allowed engineers to find a route for a new pipe to be added to the site. Additionally, scanners can be combined with GNSS receivers and other positioning technology such as total stations to create scenes like that shown in Figure 17.12, which provide details

(a)   (b)

Figure 23.4 Pulsed laser instruments: (a) total station and (b) hand-held EDM instrument. (Courtesy Leica Geosystems.)

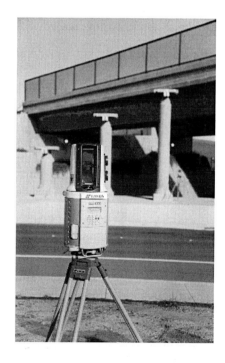

Figure 23.5 Laser scanner being used to capture detail on a divided highway bridge. (Courtesy Topcon Positioning Systems.)

captured as the vehicle (see Figure 1.5) moves through the area. This technology is expected to play a larger role in construction surveys in the future. There are also scanners that are mounted on a vehicle that scan the surface below a vehicle. They can capture the quality of the road surface while driving at highway speeds.

23.3 HORIZONTAL AND VERTICAL CONTROL

The importance of a good framework of horizontal and vertical control in a project area cannot be overemphasized. It provides the basis for positioning structures, utilities, roads, etc., in each of the stages of planning, design, and construction. Too often surveyors and engineers have skimped on establishing a suitable network of control points, and they have also failed to preserve them through proper monumentation, references, and ties.

The surveyor in charge should receive copies of the plans well in advance of construction to become familiar with the job, and have time to "tie out" or "transfer" any established control points that might be destroyed during building operations. The methods shown in Figures 9.4 and 9.5 are especially applicable, and should be used with intersection angles as close to 90° as possible.

On most projects, additional horizontal and vertical control is required to supplement any control already available in the job area. The control points must be:

1. Convenient for use, that is, located sufficiently close to the item being built so that work is minimized and accuracy enhanced in transferring alignment and grade.
2. Far enough from the actual construction to ensure working room for the contractor and to avoid possible destruction of stakes.
3. Clearly marked and understood by the contractor in the absence of a surveyor.
4. Supplemented by guard stakes to deter removal and referenced to facilitate restoring them. Contracts usually require the owner to pay the cost of setting initial control points and the contractor to replace damaged or removed ones.
5. Suitable for securing the accuracy agreed on for construction layout (which may be to only the nearest foot for a manhole, 0.01 ft for an anchor bolt, or 0.001 ft for a critical feature).

Construction stakes can be set in their required horizontal positions by making observations of horizontal angles or horizontal distances from established control points. Radial stakeout by angle and distance from one control point is often most expedient, but the choice will depend on the project's nature and extent. Frequent checks should be made on points set. This can be done with observations from other control stations or by checking distances from nearby points to verify their correctness in position.

Grade stakes and reference elevations are most often set using a leveling instrument whose *HI* has been established by differential leveling. For convenience, enough benchmarks are generally placed on construction sites so that at least one is readily accessible at any location in the area. Then the *HI* of the level can be established with just a single backsight to the benchmark. After grade stakes are set, a closing foresight is taken back to the benchmark for a check. However, this practice can be dangerous since the instrument operator will tend to expect the closing foresight to agree with the initial backsight, and therefore could read it carelessly. As a result, a serious mistake in the initial backsight could go undetected, resulting in a faulty setting of grade stakes. Therefore, even though it requires more time, it is recommended that level circuits for setting grade stakes always begin on one benchmark and close on another. As a rule of thumb when using any project control for stakeout, the control should be tied to another control point in the project before any grade stakes are set. This will ensure that the control is consistent and suitable for use later in the project.

■ 23.4 STAKING OUT A PIPELINE

Pipelines are used to carry water for human consumption, storm water, sewage, oil, natural gas, and other fluids. Pipes, which carry storm runoff, are called *storm sewers*; those that transport sewage, *sanitary sewers*. Flow in these two types of sewers is usually by gravity, and therefore their alignments and grades must be set carefully. Flow in pipes carrying city water, oil, and natural gas is generally under pressure, so usually they need not be aligned to as high an order of accuracy.

In pipeline construction, trenches are usually opened along the required alignment to the prescribed depth (slightly undercut if pipe bedding is required), the pipe is installed according to plan, and the trench backfilled. Pipeline grades are fixed by a variety of existing conditions, topography being a critical one. A profile like that of Figure 5.13 is usually used to analyze the topography and assist in designing the grade line for each pipe segment. To minimize construction difficulties and costs, excavation depths are minimized but at the same time a certain minimum cover over the pipeline must be maintained to protect it from damage by heavy loading from above and to prevent freezing in cold climates. Minimum grades also become an important design factor for pipes under gravity flow. Accordingly, a grade of at least 0.5% is recommended for storm sewers, but slightly higher grades are needed for sanitary sewers. In designing pipe grade lines, other existing underground elements often must be avoided, and due regard must also be given to the grades of connecting lines and the vertical clearances needed to construct manholes, catch basins, and outfalls.

Prior to staking a pipeline, the surveyor and contractor should discuss details of the project. An understanding must be reached concerning the planned trench width, where the installation equipment will be placed, and how and where the excavated material will be stockpiled. Then a reference offset line can be appropriately established that will (1) meet the contractor's needs, (2) be safe from destruction, and (3) not interfere with operations.

The alignment and grade for the pipeline are taken from the plans. An offset reference line parallel to the required centerline is established, usually at 25- or 50-ft stations when the ground is reasonably uniform. Marks should be closer together on horizontal and vertical curves than on straight segments. For pipes of large diameter, stakes may be placed for each pipe length—say, 6 or 8 ft. On hard surfaces where stakes cannot be driven, points are marked by paint, spikes, or scratch marks.

Either *batter boards* or laser beams guide precise alignment and grade for pipe placement. Figure 23.6 shows one arrangement of a batter board for a sewer line. It is constructed using 1 × 4 in., 1 × 6 in., 2 × 4 in., or boards nailed to 2 × 4 in. posts, which have been pointed and driven into the ground on either side of the trench. Depending upon conditions, these may be placed at 50 ft, 25 ft, or any other convenient distance along the sewer line. The top of the batter board

714 CONSTRUCTION SURVEYS

Figure 23.6
Batter board for sewer line.

is generally placed a full number of feet above the *invert* (flow line or lower inside surface) of the pipe. Nails are driven into the board tops so a string stretched tightly between them will define the pipe centerline. A graduated pole or special rod, often called a *story pole*, is used to measure the required distance from the string to the pipe invert. Thus, the string gives both line and grade. It can be kept taut by hanging a weight on each end after wrapping it around the nails.

In Figure 23.6, instead of a fixed batter board, a two-by-four carrying a level vial can be placed on top of the offset-line stake, which has been set at some even number of feet above the pipe's invert elevation. Measurement is then made from the underside of the leveled two-by-four with a tape or graduated pole to establish the flow line.

If laser devices are used for laying pipes, the beam is oriented along the pipe's planned horizontal alignment and grade, and the trench opened. Then with the beam set at some even number of feet above the pipe's planned invert, measurements can be made using a story pole to set the pipe segments. Thus, the laser beam is equivalent to a batter-board string line. On some jobs that have a deep wide cut, the laser instrument is set up in the trench to give line and grade for laying pipes. And, if the pipe is large enough, the laser beam can be oriented inside it.

■ 23.5 STAKING PIPELINE GRADES

Staking pipeline grades is essentially the reverse of running profiles, although in both operations the centerline must first be marked and stationed in horizontal location. The actual profiling and staking are on an offset line.

Information conveyed to the contractor on stakes for laying pipelines usually consists of two parts: (1) giving the depth of cut (or fill), normally only to the nearest 0.1 ft, to enable a rough trench to be excavated; and (2) providing precise grade information, generally to the nearest 0.01 ft, to guide in the actual placement of the pipe invert at its planned elevation. Cut (or fill) values for the first part are vertical distances from ground elevation at the offset stakes to the pipe invert. After the pipe's grade line has been computed and the offset line run, cuts (or fills) can be determined by a leveling process, illustrated in Figure 23.7 and the corresponding field notes given in Plate B.6 in Appendix B. The process is summarized as follows:

1. List the stations staked on the pipeline (column 1 of the field notes).
2. Compute the flow line or invert elevation at each station (column 6).
3. Set up the level and get an *HI* by reading a plus sight on a BM; for example, $HI = 2.11 + 100.65 = 102.76$ (see Plate B.6 in Appendix B and Figure 23.7).
4. Obtain the elevation at each station from a rod reading on the ground at every stake (column 4)—for example, 4.07 at station $1 + 00$ (see Plate B.6 and Figure 23.7)—and subtract it from the *HI* (column 5); for example, $102.76 - 4.07 = 98.69$ at station $1 + 00$.
5. Subtract the pipe elevation from the ground elevation to get cut (+) or fill (−) (column 7); for example, $98.69 - 95.34 = C3.35$ (see Plate B.6 in Appendix B and Figure 23.7).
6. Mark the cut or fill (using a permanent marking felt pen or keel) on an offset stake facing the centerline; the station number is written on the other side.

In another variation, which produces the same results, *grade rod* (difference between *HI* and pipe invert) is computed, and *ground rod* (reading with rod held at stake) is subtracted from it to get cut or fill. For station $1 + 00$, grade rod $= 102.76 - 95.34 = 7.42$, and $7.42 - 4.07 = C3.35$.

After the trench has been excavated based on cuts and fills marked on the stakes, batter boards are set. Marks needed to place them can be made with a

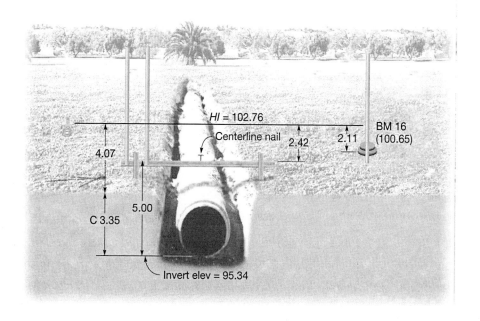

Figure 23.7
Leveling process to determine cut or fill and set batter boards for laying pipelines. (Courtesy Topcon Positioning Systems.)

pencil or felt pen on the offset stakes during the same leveling operation used to obtain cut and fill information. Figure 23.7 also illustrates the process. Suppose that at station 1 + 00, the batter board will be set so its top is exactly 5.00 ft above the pipe invert. The rod reading necessary to set the batter board is obtained by subtracting the pipe invert elevation plus 5.00 ft from the *HI*; thus 102.76 − (95.34 + 5.00) = 2.42 ft (see Figure 23.7). The rod is held at the stake and adjusted in vertical position by commands from the level operator until a rod reading of 2.42 ft is obtained; then a mark is made at the rod's base on the stake. (To facilitate this process, a rod target or a colored rubber band can be placed on the rod at the required reading.) The board is then fastened to the stake with its top at the mark using nails or C clamps, and a carpenter's level is used to align it horizontally across the trench. A nail marking the pipe centerline is set by measuring the stake's offset distance along the board.

If a laser is to be employed, this same leveling procedure can be used to establish the elevation of the laser beam at some desired vertical offset distance above the pipe's invert. The procedure is used to establish the height of the laser instrument and also to set another identical offset elevation at a station forward on line. Then the laser beam is aimed at that target to establish the required grade line.

■ 23.6 COMPUTING THE BEND ANGLES IN PIPELINES

In water and sewer lines, angle points are usually made using angle connectors with water lines and manholes in sewer lines. However, when pipelines are used in the transportation of other fluids such as gas and oil, a change in direction is often made by bending the pipe. When the change in direction is only horizontal or vertical, the angular change in direction is simply used in bending of the pipe. Conversely when the particular location involves both a change in horizontal direction and grade, the combination of these two angles yields a bend angle that is different from the design horizontal angle. While this latter case is not common, it does occur in practice.

The bend angle in this situation can be determined using vector algebra and the dot product where *xyz* coordinates are replaced by the standard vector nomenclature of *ijk*. The bend angle from the dot product of two vectors $V_1 = a_1 i + b_1 j + c_1 k$ and $V_2 = a_2 i + b_2 j + c_2 k$ where (i, j, k) are the vector components of each segment of the pipe with lengths of l_1 and l_2, respectively, is computed as

$$\theta = \cos^{-1}\frac{a_1(a_2) + b_1(b_2) + c_1(c_2)}{l_1(l_2)} \tag{23.1}$$

where the origin is at the bending point and the vector components for the first and second sections of the pipe are computed as follows:

$$a_1 = 0 \text{ and } b_1 = l_2 \sin\theta$$
$$b_1 = -l_1 \text{ and } b_2 = l_2 \cos\theta$$
$$c_1 = -g_1 l_1 \text{ and } c_2 = g_2 l_2$$

and g_1 and g_2 the respective grades of the first and second sections of the point in percentage points.

Example 23.1

Assume there is a pipe where a change in direction of 30° occurs 40 ft from the beginning of a 60-ft pipe. At this same location, the grade changes from +1% to +15%. What is the appropriate bend angle in the pipe to the nearest decimal degree to accommodate the change in horizontal direction and change in grade?

Solution

As shown in Figure 23.8, using only the horizontal change in direction, the pipe would be bent to 30°. However with the additional change in grade, the proper bend angle is given by Equation (23.1). To use Equation (23.1), the *ijk* components for each segment of the pipe must first determined. To do this assume that the first 40-ft segment of the pipe defines the *J* axis. The (a_1, b_1, c_1) vector components for this segment are simply $(0, -40, -0.4)$ where the *k* component is computed using the 1% rise in grade over the 40 ft, or $-0.01(40)$, which equals -0.4 ft. The *ijk* vector components for the 20-ft segment can be computed for the 20-ft segment horizontal change in the direction of 30° for θ. From this we see that the vector components for the 20-ft segment are

$$a_2 = 20 \sin 30° = 10 \quad b_2 = 20 \cos 30° = 17.32 \quad c_2 = 0.15(20) = 3$$

Using Equation (23.1), the angle between the two vectors is

$$\theta = \cos^{-1}\frac{0(10) - 40(17.32) - 0.4(3)}{40(20)} = 150.2°$$

Thus the pipe should only be bent by 29.8° (180°−150.2°) from straight to accommodate both the change in horizontal and vertical direction.

23.7 STAKING OUT A BUILDING

The first task in staking out a building is to locate it properly on the correct lot by making measurements from the property lines. Most cities have an ordinance establishing setback lines from the street and between houses to improve appearance and provide fire protection.

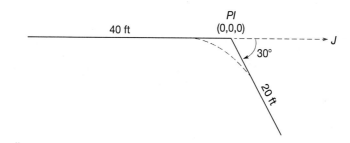

Figure 23.8
Bend angle of a straight pipe.

Stakes may be set initially at the exact building corners as a visual check on the positioning of the structure, but obviously such points are lost immediately when excavation is begun on the footings. A set of batter boards and reference stakes, placed as shown in Figure 23.9, is therefore erected near each corner, but out of the way of construction. The boards are nailed a full number of feet above the footing base, or at first-floor elevation. (The procedure of setting boards at a desired elevation was described in the preceding section.) Nails are driven into the batter board tops so that strings stretched tightly between them define the outside wall or form line of the building. The layout is checked by measuring diagonals and comparing them with each other (for symmetric layouts) or to their computed values. Figure 23.10 illustrates the placement on a lot and staking of a slightly more complicated building. The following are recommended steps in the procedure:

1. Set hubs A and B 5.00 ft inside the east lot line, with hub A 20.00 ft from the south lot line and hub B 70.00 ft from A. Mark the points precisely with nails.
2. Set a total station instrument over hub A, backsight on hub B, and turn a clockwise angle of 270° to set batter-board nails 1 and 2 and stakes C and D.
3. Set the instrument over hub B, backsight on hub A, and turn a 90° angle. Set batter-board nails 3 and 4 and stakes E and F.
4. Measure diagonals CF and DE and adjust if the error is small or restake if large.
5. Set the instrument over C backsight on E and set batter-board nail 5. Plunge the instrument and set nail 6.
6. Set the instrument over D, backsight on F, and set nail 7. Plunge and set nail eight.

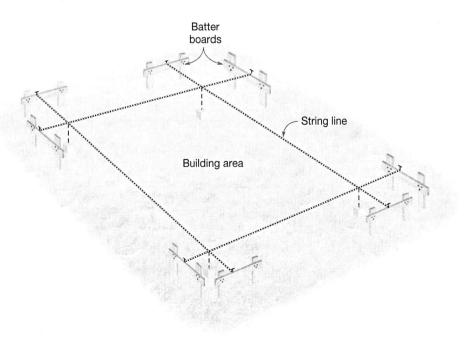

Figure 23.9
Batter boards for building layout.

Figure 23.10
Location of building on lot and batter-board placement.

7. Set batter-board nails 9, 10, 11, 12, 13, and 14 by measurements from established points.
8. Stretch the string lines to create the building's outline and check all diagonals.

(*Note:* When each batter board is set, it must be placed with its top at the proper elevation.)

As an alternative to this building stakeout procedure, radial methods (described in Section 9.9) can be used. This can substantially reduce the number of instrument setups and stakeout time required. In the radial method, coordinates of all building corners are computed in the same coordinate system as the lot corners. Then the total station instrument is set on any convenient control point and oriented in azimuth by sighting another intervisible control point. Angles and distances, computed from coordinates, are then laid off to mark each building corner. Measuring the distances between adjacent points, and also the diagonals, checks the layout. (An example illustrating radial stakeout of a circular curve is given in Section 24.13.) After constructing the batter boards and setting the cross pieces at the desired elevations, the alignment nails on the batter boards can be set by pulling taut string lines across established corners. For example in Figure 23.10, with corners *D* and *F* marked, a line stretched across these two points enables placing nails 7 and 8 on the boards. With the strings in place after setting batter-board nails, diagonals between corners and wall lengths should be checked.

Another method of laying out buildings, is to stake two points on the building, occupy one of them with the total station instrument, take a backsight on the other, and stake all (or many) of the remaining points from that setup using precalculated angles and distances. In some cases, advantage can be taken of symmetrical layouts to save considerable time. Figure 23.11 shows an unusual symmetrical building shape, which was laid out rapidly using only two setups (at points *A* and *O*). With this choice of stations, half the corners could be set from each setup, and the same calculated angles and distances could be used (see the field notes of Figure 23.12). Again, if this method is used, it is essential that enough dimensions be checked between marked corners to ensure that no large errors or mistakes were made.

To control elevations on a building construction site, a benchmark (two or more for large projects) should be set outside the construction limits but within easy sight distance. Rotating lasers (see Section 23.2.2) can be used to control elevations for the tops of footings, floors, and so on.

Permanent foresights are helpful in establishing the principal lines of the structure. Targets or marks on nearby existing buildings can be used if movement due to thermal effects or settlement is considered negligible. On formed concrete structures, such as retaining walls, offset lines are necessary because the outside wall face is obstructed. Two-by-two in. hubs with tacks can first mark the positions of such things as interior footings, anchor bolts for columns, and special piping or equipment. Survey disks, scratches on bolts or concrete surfaces, and steel pins can also be used. Batter boards set inside the building dimensions for column footings have to be removed as later construction develops.

On multistory buildings, care is required to ensure vertical alignment in the construction of walls, columns, elevator shafts, structural steel, and so on. One method of checking plumbness of constructed members is to carefully aim a total station's line of sight on a reference mark at the base of the member. The line of

Figure 23.11
Building layout.

BUILDING LAYOUT

Sta.	Angle	To the	Distance
⊼ @ Point A			
O	0°00′	Rt	220.00 ft
F	90°00′	Rt	98.00 ft
G	90°00′	Rt	135.00 ft
E	120°00′	Rt	169.74 ft
D	150°00′	Rt	196.00 ft
C	180°00′	Rt	169.74 ft
B	210°00′	Rt	98.00 ft
⊼ @ Point O			
A	0°00′	Lt	220.00 ft
J	90°00′	Lt	98.00 ft
H	90°00′	Lt	135.00 ft
K	120°00′	Lt	169.74 ft
L	150°00′	Lt	196.00 ft
M	180°00′	Lt	169.74 ft
N	210°00′	Lt	98.00 ft

Figure 23.12 Precalculated angles and distances for building layout.

sight is then raised to its top. For an instrument that has been carefully leveled and that is in proper adjustment, the line of sight will define a vertical plane as it is raised. It should not be assumed that the instrument is in good adjustment; therefore, the line should be raised in both the direct and reversed positions. It is necessary to check plumbness in two perpendicular directions when using this procedure. To guide construction of vertical members in real time, two instruments can be set up with their lines of sight oriented perpendicular to each other, and verticality monitored as construction progresses. Alternatively, lasers can be used to guide and monitor vertical construction.

If the surveyor does not give sufficient forethought to the basic control points required, the best method to establish them, and the most efficient approach to staking out a building, the job can be a time-consuming and difficult process. The number of instrument setups should be minimized to conserve time and calculations made in the office if possible, rather than in the field while a survey party waits.

■ 23.8 STAKING OUT HIGHWAYS

Alignments for highways, railroads, and other transportation routes are designed after careful study of existing maps, aerial photography, and preliminary survey data of the area. From alternative routes, the one that best meets the overall objectives while minimizing costs, environmental impacts, and other factors is selected. Before construction can begin, the surveyor must transfer that alignment (either the centerline or an offset reference line) to the ground.

Normally staking will commence at the initial point where the first straight segment (*tangent*) is run, placing stakes at *full stations* (100-ft intervals) if the

English system of units is used, or at perhaps 30- or 40-m spacing if the metric system is employed. Stationing (this subject is described in Section 5.9.1) continues until the planned alignment changes direction at the first *point of intersection* (PI). The deflection angle is measured there and the second tangent stationed forward to the next PI, where the deflection angle there is measured. The process continues to the terminal point. Staking continuously from the initial point to the terminus may result in large amounts of accumulated error on long projects. Therefore, work should be checked by making frequent ties to intermediate horizontal control points, and adjustments should be made as necessary. Alternatively, on smaller projects the alignments can be run from both ends to a point near the middle.

After tangents are established, horizontal curves (usually circular arcs) are inserted at all PIs according to plan. The subject of horizontal alignments, including methods for computing and laying out horizontal curves, is discussed in detail in Chapter 24. Vertical alignments are described in Chapter 25.

After the centerline or reference line (including curves) has been established, the PIs, intermediate *points on tangent* (POTs) on long tangents, and points where horizontal curves begin (PCs), and end (PTs), are referenced using procedures described in Section 9.5. Points used in referencing must be located safely outside the construction limits. Referencing is important because the centerline points will be destroyed during various phases of construction and will need to be replaced several times. Benchmarks are also established at regular spacing (usually not more than about 1000 ft apart) along the route. These are placed on the right of way, far enough from the centerline to be safe from destruction, but convenient for access.

After the centerline or reference line has been established, stakes marking the right-of-way should be set. This is normally done by carefully measuring perpendicular offsets from the established reference line. The right-of-way is staked at every change in its width, at all changes in alignment, including each PC and PT, and at sufficient other intermediate points along the tangents so that it is clearly delineated.

When the reference line and right-of-way have been staked, the limits of actual construction are marked so that the contractor can clear the area of obstructions. Following this, some contractors want points set on the right-of-way with subgrade elevations, showing cut or fill to a given elevation, for use in performing rough grading and preliminary excavation of excess material.

To guide a contractor in making final excavations and embankments, *slope stakes* are driven at the *slope intercepts* (intersections of the original ground and each side slope), or offset a short distance, perhaps 4 ft (see Figure 23.13). The cut or fill at each location is marked on the slope stake. Note that there is no cut or fill at a slope stake—the value given is the vertical distance from the ground elevation at the slope stake to grade.

Grade stakes are set at points that have the same ground and grade elevation. This happens when a grade line changes from cut to fill, or vice versa. As shown in Figure 23.14, three *transition sections* normally occur in passing from cut to fill (or vice versa), and a grade stake is set at each one. A line connecting grade stake, perhaps scratched out on the ground, defines the change from cut to fill, as we will see in line *ABC* in Figure 26.1.

Slope stakes can be set at slope intercept locations predetermined in the office from cross-sectional data. (Methods for determining slope intercepts from cross

Figure 23.13
Slope stakes (shoulders and ditches not shown).

Figure 23.14
Grade points at transition sections.

sections are described in Chapter 26.) If predetermined slope intercepts are used, the ground elevation at each stake must still be checked in the field to verify its agreement with the cross section. If a significant discrepancy in elevation exists, the stake's position must be adjusted by a trial-and-error method, as shown in Example 23.2. The amount of cut or fill marked on the stake is computed from the actual difference in elevation between the ground at the slope stake and grade elevation.

If slope intercepts have not been precalculated from cross-section data, slope stakes are located by a trial-and-error method based on mental calculations involving the *HI*, grade rod, ground rod, half roadway width, and side slopes. One or two trials are generally sufficient to fix the stake position within an allowable error of 0.3–0.5 ft for rough grading. The infinite number of ground variations prohibits use of a standard formula in slope staking. An experienced surveyor employs only mental arithmetic, without scratch paper or hand calculator. Whether using the method described in Example 23.2 or any other, systematic procedures must be followed to avoid confusion and mistakes.

Example 23.2 lists the sequential steps to be taken in slope staking, *assuming for simplicity, academic conditions of a level roadway*. In practice, travel lanes and shoulders of modern highways have lateral slopes for drainage, then a steeper slope to a ditch in cut, and another slope up the hillside to the slope intercept. These *design templates* of the roadbed can often be loaded into a data collector for field stakeout. Transition sections may have half-roadway widths in cuts different from those in fills to accommodate ditches, and flatter side slopes for fills that tend to be less stable than cuts. But the same basic steps still apply and can be extended by students after learning the fundamental approach.

■ Example 23.2

List the field procedures, including calculations, necessary to set slope stakes for a 40-ft wide level roadbed with side slopes of 1:1 in cut and 1-1/2: 1 in fill (see Figures 23.12 and 23.13).

Solution

1. Compute the cut at the centerline stake from profile and grade elevations ($603.0 - 600.0 = C3.0$ in Figure 23.13). Check in field by grade rod minus ground rod = $7.8 - 4.8 = C3.0$ ft. Mark the stake C 3.0/0.0. (On some jobs the center stake is omitted, and stakes are set only at the slope intercepts.)
2. Estimate the difference in elevation between the left-side slope-stake point (20 + ft out) and the center stake. Apply the difference—say, +0.5 ft—to the center cut and get an estimated cut of 3.5 ft.
3. Mentally calculate the distance out to the slope stake, $20 + 1(3.5) = 23.5$ ft, where 1 is the side slope.
4. Hold the zero end of a cloth tape at the center stake while the rodperson goes out at right angles with the other end and holds the rod at 23.5 ft. [The right angle can be established by prism (see Figure 17.8) or by using a total station instrument.]
5. *Forget all previous calculations to avoid confusion of too many numbers and remember only the grade-rod value.*
6. Read the rod with the level and get the cut from grade rod minus ground rod, perhaps $7.8 - 4.0 = C3.8$ ft.
7. Compute the required distance out for this cut, $20 + 1(3.8) = 23.8$ ft.
8. Check the tape to see what is being held and find it is 23.5 ft.

9. The distance is within a few tenths of a foot and close enough. Move out to 23.8 ft if the ground is level and drive the stake. Move farther out if the ground slopes up, since a greater cut would result, and thus the slope stake must be beyond the computed distance, or not so far if the ground has begun to slope down, which gives a smaller cut.
10. If the distance has been missed badly, make a better estimate of the cut, compute a new distance out, and take a reading to repeat the procedure.
11. In going out on the other side, the rodperson lines up the center and left-hand slope stake to get the right-angle direction.
12. To locate grade stakes at the road edge, one person carries the zero end of the tape along the centerline while the rodperson walks parallel, holding the 20-ft mark until the required ground-rod reading is found by trial. Note that the grade rod changes during the movement but can be computed at 5- or 10-ft intervals. The notekeeper should have the grade rod listed in the field book for quick reference at full stations and other points where slope stakes are to be set.
13. Grade points on the centerline are located using a starting estimate determined by comparing cut and fill at back and forward stations.

Practice varies for different organizations, but often the slope stake is set 4 ft beyond the slope intercept. It is marked with the required cut or fill, distanced out from the centerline to the slope-stake point, side-slope ratio, and base half-width noted on the side facing the centerline. Stationing is given on the backside. A reference stake having the same information on it may also be placed 6 ft or farther out of the way of clearing and grading. On transition sections, grade-stake points are marked.

Total station instruments, with their ability to automatically reduce measured slope distances to horizontal and vertical components, speed slope staking significantly, especially in rugged terrain where slope intercept elevations differ greatly from centerline grade. When the data collector allows the user to input the design template (see Section 26.3), it can rapidly determine the positions of the slope stakes using field observed data. GNSS receivers operating in the real-time kinematic mode (see Chapter 15) can also be used advantageously in these types of terrain if satellite visibility exists. Combined with a machine control system (see Section 23.12), GNSS receivers allow heavy equipment operators to shape the design without the need for stakes.

Slope staking should be done with utmost care, for once cut and fill embankments are started, it is difficult and expensive to reshape them if a mistake is discovered.

After rough grading has shaped cuts and embankments to near final elevation, finished grade is constructed more accurately from *blue tops* (stakes whose tops are driven to grade elevation and then marked with a blue keel or spray paint). These are not normally offset, but rather driven directly on centerline or shoulder points. The procedure for setting blue tops at required grade elevation is described in Section 25.8.

Highway and railroad grades can often be rounded off to multiples of 0.05 or 0.10% without appreciably increasing earthwork costs or sacrificing good drainage. Streets need a minimum 0.50% grade for drainage from intersection to intersection, or from midblock both ways to the corners. They are also crowned to provide for lateral flow to gutters. Drainage profiles, prepared to verify or construct drainage cross sections, can be used to locate drainage structures and easements accurately. An experienced engineer when asked a question regarding the three most important items in highway work, thoughtfully replied "drainage, drainage, and drainage." Good surveying and design must satisfy this requirement.

To ensure unobstructed drainage after construction, culverts must be placed in most fill sections so that water can continue to flow in its normal pattern from one side of the embankment to the other. In staking culverts, their locations, skew angles, if any, lengths, and invert elevations are taken from the plans. Required pipe alignments and grades are marked using stakes, offset from each end of the pipe's extended centerline. The invert elevation (or an even number of feet above or below it) is noted on the stake. This field procedure, like setting slope stakes, requires marking a point on the stake where a rod reading equals the difference between the required grade and the current *HI* of a leveling instrument.

After the subgrade has been completed, if the highway is being surfaced with rigid concrete pavement, paving pins will be necessary to guide this operation. They are usually about 1/2-in. diameter steel rods, driven to mark an offset line parallel to one edge of the required pavement. This line is usually staked at 50-ft increments, but closer spacing may be used on sharp curves. The finished grade (or one parallel to it but offset vertically above) is marked on the pins using tape, or affixing a special string line holding device. Again in this operation, elevations are set by marking the stake where a rod reading equals the difference between the finished grade (or a vertically offset one) and a current *HI*. (The need for frequent project benchmarks at convenient locations is obvious.)

Utility relocation surveys may be necessary in connection with highway construction; for example, manhole or valve-box covers have to be set at correct grade before earthwork begins so they will conform to finished grade. Here differential elevation resulting from the transverse surface slope must be considered. Utilities are located by centerline station and offset distance.

Location staking for railroads, rapid-transit systems, and canals follows the same general methods outlined above for highways.

23.9 OTHER CONSTRUCTION SURVEYS

For planning and constructing causeways, bridges, and offshore platforms, it is often necessary to perform hydrographic surveys (see Section 17.14). These types of projects require special procedures to solve the problem of establishing horizontal positions and depths where it is impossible to hold a rod or reflector. Modern surveying equipment and procedures, and sonar mapping devices, are used to plot dredging cross sections for underwater trenching and pipe layout. Today, more pipelines are crossing wider rivers, lakes, and bays than ever before. Mammoth pipeline projects now in progress to transport crude oil, natural gas, and water have introduced numerous new problems and solutions. Permafrost, extremely low temperatures, and the need to provide animal crossings are examples of special problems associated with the Alaskan pipeline construction projects.

Large earthwork projects such as dams and levees require widespread permanent control for quick setups and frequent replacement of slope slakes, all of which may disappear under fill in one day. Fixed signals for elevation and alignment painted or mounted on canyon walls or hillsides can mark important reference lines. Failures of some large structures demonstrate the need for monitoring them periodically so that any necessary remedial work can be done.

Underground surveys in tunnels and mines necessitate transferring lines and elevations from the ground above, often down shafts. Directions of lines in mine tunnels can be most conveniently established using north-seeking gyros. In another, and still practiced method, two heavy plumb bobs hung on wires (and damped in oil or water) from opposite sides of the surface opening can be aligned by total station there and in the tunnel. (A vertical collimator will also provide two points on line below ground.) A total station or laser is "wiggled-in" (see Section 8.16) on the short line defined by the two plumb-bob wires, a station mark set in the tunnel ceiling above the instrument, and the line extended. Later setups are made beneath spads (surveying nails with hooks) anchored in the ceiling. Elevations are brought down by taping or other means. Benchmarks and instrument stations are set on the ceiling, out of the way of equipment.

Surveys are run at intervals on all large jobs to check progress for periodic payments to the contractor. And finally, an *as-built* survey is made to determine compliance with plans, note changes, make terminal contract payment, and document the project for future reference.

Airplane and ship construction requires special equipment and methods as part of a unique branch of surveying called *optical tooling*. The precise location and erection of offshore oil drilling platforms many miles from a coast utilizes new surveying technology, principally the global navigation satellite systems.

■ 23.10 CONSTRUCTION SURVEYS USING TOTAL STATION INSTRUMENTS

The procedures described here apply to most total station instruments, although some may require interfaced data collectors to perform the operations described.

Before using a total station for stakeout, it is necessary to orient the instrument. Depending on the type of project, *horizontal* or both *horizontal and vertical* orientation may be needed. For example, if just the lot corners of a subdivision are being staked then only horizontal orientation (establishing the instrument's position and direction of pointing) is needed. If grade stakes are to be set then the instrument must also be oriented vertically (its *HI* determined).

With total station instruments, three methods are commonly used for horizontal orientation: (1) *azimuth*, (2) *coordinates*, and (3) *resection*. The first two apply where an existing control point is occupied, and the latter is used when the instrument is set up at a noncontrol point. In azimuth orientation, the coordinates of the occupied control station and the known azimuth to a backsight station are entered into the instrument. If the occupied station's coordinates have been downloaded into the instrument prior to going into the field, it is only necessary to input its point number. The backsight station is then sighted, and when completed, the azimuth of the line is transferred to the total station by a keyboard stroke, whereupon it appears in the display.

The coordinate method of orientation uses the same approach, except that the coordinates of both the occupied and the backsight station are entered. Again, these data could have been downloaded previously so that it would only be necessary to key in the numbers identifying the two stations. The instrument computes the backsight line's azimuth from the coordinates, displays it, and prompts the operator to sight the backsight station. Upon completion of the backsight, the azimuth is transferred to the instrument with a keystroke, and it appears on the display.

In the resection procedure, a station whose position is unknown is occupied and the instrument's position determined by sighting two or more control stations. This is very convenient on projects where a certain point of high elevation in an open area gives good visibility to all (or most) points to be staked. As noted, two or more control points must be sighted. Observations of angles, or of angles and distances, are made to the control stations. Since a least-squares resection solution can provide checks on the consistency of the control and the resection solution, three or more control stations is always recommended. The microprocessor then computes the instrument's position by the methods discussed in Sections 11.7 or 11.10.

Project conditions will normally dictate which orientation procedure to use. Regardless of the procedure selected, after orientation is completed, a check should be made by sighting another control point and comparing the observed azimuth and distance against their known values. If there is a discrepancy, the orientation procedure should be repeated. It is also a good idea to recheck orientation at regular intervals after stakeout has commenced, especially on large projects. In fact, if possible, a reflector should be left on a control point just for that purpose.

Vertical orientation of a total station (i.e., determining its *HI*) can be achieved using one of two procedures. The simplest case occurs if the elevation of the occupied station is known, as then it is only necessary to *carefully* measure and add the *hi* (height of instrument above the point) to the elevation of the point. If the occupied station's elevation is unknown, then another station of known elevation must be sighted. The situation is illustrated in Figure 23.15, where the instrument is located at station *A* of unknown elevation, and station *B* whose elevation is known is sighted. From slope distance *S* and zenith angle *z* the instrument computes *V*. Then its *HI* is

$$HI = elev_B + h_r - V \qquad (23.2)$$

where h_r is the reflector height above station *B*. As with horizontal orientation, it is a good practice to check the instrument's vertical orientation by sighting a second vertical control point.

Once orientation is completed, project stakeout can begin. In general, staking is either a two- or three-dimensional problem. Staking lots of a subdivision or layout of horizontal construction alignments is generally two dimensional. Slope staking, blue-top setting, pipeline layout, and batter-board placement require both horizontal position and elevation and are therefore three dimensional.

For two-dimensional stakeout, after the file of coordinates for all control stations and points to be staked is downloaded and the instrument is oriented horizontally, the identifying number of a point to be staked is entered into the instrument through the keyboard. The microprocessor immediately calculates the horizontal

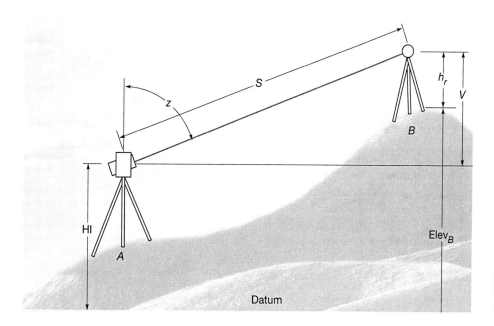

Figure 23.15
Vertical orientation of total station.

distance and azimuth required to stake the point. The difference between the instrument's current direction of pointing and that required is displayed. The operator turns the telescope until the difference becomes zero to achieve the required direction. With total stations having robotic capabilities, the instrument will swing in direction to the proper azimuth without any further operator intervention.

Following azimuth alignment, the distance to the point must be laid out. To do this, the reflector is directed onto the azimuth alignment and a horizontal distance reading taken, whereupon the difference between it and that required is displayed. The reflector is then directed inward or outward, as necessary, until the distance difference is zero and the stake placed there. A two-way radio is invaluable for communicating with the reflector person in this operation. Additionally, a small tape measure can often be used to speed the process of locating the rod in its correct position. This procedure for stakeout is discussed further in Section 24.13, and an example problem presented.

Special tracking systems have been developed to aid the reflector person in getting on line with the station. For example, some total stations utilize "constraint" and "flashing" lights to indicate whether the reflector is left or right of the line of sight while others use lights of different colors. The prism person, upon seeing these lights, immediately knows what direction to move to get on line.

For three-dimensional staking, the total station must be oriented vertically as well as horizontally. The initial part of three-dimensional stakeout is exactly like that described for the two-dimensional procedure; that is, the horizontal position of the stake is set first. Then simultaneously with the observation of the stake's horizontal position, its vertical component, and thus its elevation is determined. The difference ΔZ between the required elevation and the stake's elevation is displayed with a plus or minus sign, the former indicating fill, the latter cut.

This information is communicated to the reflector person for marking the stake, or incrementally driving it further down until the required grade, or if desired, some even number of feet above or below grade is reached. From these descriptions it is easy to see why a robotic total station with remote keypad (see Figure 8.7) will greatly simplify these operations.

With the high order of accuracy possible using total station instruments, stakes quite distant from the instrument can be laid out, and thus many points set from a single setup. Often, in fact, an entire project can be staked from one location. This is made possible in many cases because of the flexibility that resection orientation provides in instrument placement. It should be remembered that if long distances are involved in three-dimensional staking, Earth curvature and refraction should be considered (see Section 4.4). Also with total stations, each point is set independently of the others, and thus no inherent checks are available. Checks should therefore be made by either repeating the observations, checking placements from different control stations, or measuring between staked stations to ascertain their relative accuracies.

23.11 CONSTRUCTION SURVEYS USING GNSS EQUIPMENT

Any of the GNSS surveying methods discussed in Chapters 14 and 15 could be used on construction projects. Specifically, static surveys can be used to establish project control and kinematic surveys can be used to produce maps for planning and design. Finally, real-time kinematic (RTK) or Network (RTN) surveys (see Chapter 15) can be used to locate construction stakes.

The base receiver does not have to occupy a station with known GNSS coordinates. Instead the application software can be instructed to determine its position in *autonomous* mode. This process determines the base station receiver coordinates using point-positioning methods yielding low-accuracy coordinates. However, all points determined from this station will have GNSS-type accuracies relative to the base station coordinates. The localization process discussed in Sections 15.9 and 19.7.2 transforms this set of low-accuracy GNSS coordinates into the project control reference frame eliminating the inaccuracies of the autonomous base station coordinates.

As discussed in Sections 15.9 and 19.7, care must be taken to ensure that points located using GNSS are placed in the same reference frame as the project coordinates. As discussed in Section 15.9, sufficient project control known in the local reference frame must be established at the perimeter of the construction project. Then prior to staking any points, the GNSS receiver must occupy this control. Using the project coordinates and the GNSS coordinates, transformation parameters (see Section 19.7.2) are computed so that the GNSS-derived coordinates can be transformed into the local project reference frame. It is important that this transformation occur only once in a project and include important control in the transformation. That is, if a benchmark on a bridge abutment was used to design a replacement structure, then this benchmark should be included in the localization process regardless of its location in the project. The localization process should only occur once during a construction project to avoid the introduction of varying orientation parameters caused by random errors. Once the localization is accepted, the transformation parameters should be distributed amongst the various GNSS receivers involved in the project.

Additional control points must also be established at critical locations in the construction project to provide convenient location of base station receivers, total stations (in canopy conditions), and laser levels for finish work. While the type of radio and antenna will determine the range of the base station radio, base station radios typically have a maximum range of about 10 km. Thus, sufficient horizontal control must be established to support the radio's range. However, vertical control is often limited by the range of the laser level and thus benchmarks are often required every 500 to 1500 ft.

In construction staking using RTK surveying methods, a minimum of two receivers are needed. Each is equipped with a radio. One receiver occupies a nearby control station and the other, called the "rover," is moved from one point (to be set) to another. The points set must have their project coordinates known before setting any stakes. The base radio broadcasts the raw satellite data from the base receiver to the rover. At the rover, the application software processes the data from both receivers in real time using relative positioning techniques (see Section 13.9). This determines the location of the rover relative the base station. If the observed coordinates do not agree with the required values for the point being staked, the GNSS controller will indicate the direction and distance that the rover must be moved. The rover's position is adjusted until agreement is reached, and the stake is set at this location.

Although excellent horizontal accuracies can be achieved using GNSS, elevations are less reliable. GNSS-determined ellipsoid heights are typically accurate to within a few centimeters. But to get an orthometric height (elevation related to datum), the geoid height must be applied as discussed in Sections 13.4.3 and 19.5. Geoid heights are not precisely known, but models are available, which yield values that are generally accurate to within a few centimeters in most regions of the United States. However, they can be off by several decimeters in mountainous regions. For this reason, if very precise elevations are required in construction staking, GNSS surveys are unsatisfactory. However, it does provide sufficient accuracy for lower-accuracy jobs, such as slope staking, assuming corrections are made for geoid heights by applying the geoid model. The GRAV-D program (see Section 19.6.6) in the National Geodetic Survey is currently working on developing a geoid model accurate to within 1 cm for the entire United States around 2022.

As shown in Figure 23.16, string lines have traditionally been used to guide finishing work. However by combining a rotating laser level, as shown in Figure 23.2, with a GNSS machine control package it is possible to obtain millimeter accuracy in both horizontal and vertical positioning. Another manufacturer solution is to mount a GNSS receiver on a robotic total station, which allows the robotic total station to be used in areas where previous control points were not established.

GNSS surveys are particularly useful in staking widely spaced points, especially in areas where terrain or vegetation make it difficult to conduct traditional ground surveys. Staking subdivisions containing large parcels in rugged terrain and setting slope stakes in rugged areas where deep cuts and fills exist, are examples of situations where GNSS surveys can be very convenient for construction surveying. Of course, GNSS surveys require an obstruction-free view of the satellites.

732 CONSTRUCTION SURVEYS

Figure 23.16
A string line guiding a paver. (Courtesy Topcon Positioning Systems.)

23.12 MACHINE GUIDANCE AND CONTROL

In recent years, research has led to *stakeless* construction where GNSS receivers, robotic total stations, and laser levels are used to guide earth-moving equipment in real time. The major difference between machine control and machine guidance is that the machine control system actually controls the heavy equipment on the job site through the use of hydraulics whereas machine guidance informs the operator to take action to either change the direction of the equipment or the level of the cutting edge to meet the desired design. Data necessary for this machine guidance and control include a *digital elevation model* (DEM) (see Section 17.8) and digital design plans with their alignments, grades, and design templates developed in the same three-dimensional coordinate system such as shown in Figure 18.10. With GNSS receivers, robotic total stations, sonic receptors, and lasers to guide the equipment operators, and an on-board computer that continually updates cut and fill information, the grading can be accomplished without the need for construction stakes and with limited assistance of grade foremen. Machine guidance and control has been implemented on dozers, hoes, pans, graders, and trucks.

Using machine guidance or control, the surveyor's role in construction surveying shifts to tasks such as establishing the project reference coordinate system and control, creating a DEM (see Section 18.14) of the existing surface for the design and grading work, managing the electronic design on the job site, calibrating the surveying equipment with respect to the construction site, providing for the calibration of the cutting surfaces of the heavy equipment with respect to the surveying control, and developing the necessary digital data for the system operation.

As discussed in the previous section, the project design is normally performed in a project reference coordinate system. Thus, the GNSS receiver must be localized (see Sections 15.9 and 19.7.2) before any excavation is performed. It is important in this process to have control surrounding the job as well as including any control crucial to the design of any structures. Once localization is performed and accepted, the application software will convert the GNSS coordinates of latitude, longitude, and height to the project's horizontal and vertical reference frames. Additionally, the GNSS receivers must be referenced to the cutting edges of the construction equipment. Since the GNSS receiver is often placed on the blade of the construction vehicle, this often means measuring the height of the antenna reference point above the cutting edge of the blade on a regular basis to account for wear.

GNSS surveys can provide heights to a few centimeters. Thus, it is sufficiently accurate for rough grading. However in finished grading, a robotic total station or laser level is required. One manufacturer has combined a laser level with the GNSS receiver to provide millimeter accuracies in both horizontal and vertical locations. The range of the vehicle from the laser level is often limited, and thus additional vertical control will be required to control the final grades.

Three-dimensional machine control is also possible with a robotic total station. In this system, a multifaceted, 360° prism replaces the GNSS receiver on the construction vehicle. Similar to using GNSS, a DEM and grading plan of the site are loaded into the system. The robotic total station then tracks the prism mounted on the construction vehicle and provides the system with the prism's position and elevation. This in turn guides the operator during the excavation and finishing process. Again, the offset from the prism to the cutting edge of the equipment must be measured and entered into the system on a regular basis to account for wear.

Since robotic total stations typically work in a local coordinate system, there is no need for localization using a robotic total station. However, the range of a robotic total station is generally limited to only 1000 ft (300 m). Additionally there must be continuous line of sight between the robotic total station and construction vehicle. Thus, many more control stations must be added to the site when using a robotic total station.[1] Another drawback is that a robotic total station must be dedicated to each piece of construction equipment. When using GNSS equipment for machine control, a single base station can serve as the control for many pieces of construction equipment and is only limited by the range of its radio. For example, to cover the typical maximum range (10 km) of a GNSS base station radio using a robotic total station would require 33 control stations.[2] Another drawback of using robotic total stations in machine control is that each total station must be located and oriented for each construction vehicle. While a GNSS system requires

---

[1]To reduce the need for physical monuments, one manufacturer has integrated a GNSS receiver with their robotic total station so that the total station can be located anywhere within range of the GNSS base station without the need of a physical monument.

[2]The range of a GNSS base station can be extended with data modems where cell coverage is available.

an additional machine control system for each construction vehicle in the base station's range, it does require continuous line of sight from the base station or any additional base station receivers and radios unless the distance from the base becomes too far for a successful radio link with the base. In this instance, repeater radios maybe used, or a second base can be established. Thus, necessary control monumentation is reduced. Additionally, since the surveyor is no longer required to place stakes in the path of the construction vehicles, safety is also increased.

Similar to machine control, site communication applications on trucks allow managers to monitor quantities of excavation, hauling distances, and any surplus or fill earthwork. The system provides managers with daily reports on earthwork quantities, machine maintenance, and timetable for gains or loses. Because of these features, many companies are finding that they can complete projects in a timely fashion often receiving bonuses for completing the project on or ahead of schedule. Additionally, as stated in Chapter 15, RTN GNSS should not be used to guide machines since data latency can cause problems in following engineering designs.

23.13 AS-BUILT SURVEYS WITH LASER SCANNING

As mentioned in Section 23.9, as-built surveys are performed at the completion of a construction project to ensure that project specifications are met and note any changes to plans. In many cases these surveys are performed with traditional surveying equipment. However, in projects that involve extensive detail, danger to instrument operator, or interruption of daily commerce, laser scanning can provide superior results in a fraction of the time. Figure 23.17 shows a rendered image of a bridge that was surveyed for renovations. In the bridge survey, enormous quantities of data were collected from an on-shore location. The digital image of the bridge is shown in the lower-right inset. The rendered, three-dimensional image of the bridge allows designers to obtain accurate measurements between points in the image. Figure 17.11 depicts the point-cloud image of a refinery with the path of a new pipe shown in white. This three-dimensional image allowed engineers to design the new pipe alignment so that existing obstructions were cleared. A traditional survey would have either lacked the detail provided by the three-dimensional laser-scanned image or cost considerably more to locate all the existing elements. Using laser-scanning technology in these projects saved thousands of dollars and provided safe conditions for the field crews.

Figure 23.17 Rendered and rotated image of bridge shown in lower-right corner. (Courtesy Christopher Gibbons, Leica Geosystems, Inc.)

■ 23.14 SOURCES OF ERROR IN CONSTRUCTION SURVEYS

Important sources of error in construction surveys are:

1. Inadequate number and/or location of control points on the job site.
2. Errors in establishing control.
3. Observational errors in layout.
4. Failure to double-center in laying out angles or extending lines, and failure to check vertical members by plunging the instrument.
5. Careless referencing of key points.
6. Movement of stakes and marks.

■ 23.15 MISTAKES

Typical mistakes often made in construction surveys are:

1. Lack of foresight as to where construction will destroy points.
2. Notation for cut (or fill) and stationing on stake not checked.
3. Wrong datum for cuts, whether cut is to finished grade or subgrade.
4. Arithmetic mistakes, generally due to lack of checking.
5. Use of incorrect elevations, grades, and stations.
6. Failure to check the diagonals of a building.
7. Carrying out computed values to too many decimal places (one good hundredth is better than all the bad thousandths).
8. Reading the rod on top of stakes instead of on the ground beside them in profiling and in slope staking.

PROBLEMS

Asterisks (*) indicate problems that have partial answers given in Appendix G.

23.1 What do construction surveys typically establish?
23.2 List the common steps for surveying engineers in a construction project.
23.3 In what types of surveys are single-beam laser used?
23.4 What is a point cloud?
23.5 What properties must control points have when being established in a construction project?
23.6 What is a story pole, and how is it used in pipeline layout?
23.7 Where should stakes be set closer on the stake out of a pipeline and why?
23.8 What information is typically conveyed to the contractor on stakes for laying a pipeline?
23.9 A sewer pipe is to be laid from station 10+00 to station 12 + 50 on a +0.25% grade, starting with invert elevation 845.22 ft at 10 + 00 Calculate invert elevations at each 50-ft station along the line.
23.10* A sewer pipe must be laid from a starting invert elevation of 784.58 ft at station 9 + 25 to an ending invert elevation 782.48 ft at station 12 + 75. Determine the uniform grade needed, and calculate invert elevations at each 50-ft station.
23.11 Grade stakes for a pipeline running between stations 0 + 00 and 5 + 64 are to be set at each full station. Elevations of the pipe invert must be 160.85 ft at station 0 + 00

and 159.72 ft at 5 + 64, with a uniform grade between. After staking an offset centerline, an instrument is set up nearby, and a plus sight of 4.06 taken on BM A (elevation 163.42 ft). The following minus sights are taken with the rod held on ground at each stake: (0 + 00, 5.31); (1 + 00, 5.32); (2 + 00, 5.35); (3 + 00, 5.85); (4 + 00, 5.84); (5 + 00, 6.73); (5 + 64, 6.73) and (A, 4.06). Prepare a set of suitable field notes for this project (see Plate B.6 in Appendix B) and compute the cut required at each stake. Close the level circuit on the benchmark.

23.12 If batter boards are to be set exactly 3.00 ft above the pipe invert at each station on the project of Problem 23.11, calculate the necessary rod readings for placing the batter boards. Assume the instrument has the same HI as in Problem 23.11.

23.13 Assume there is a pipe where a change in direction of 20° occurs 30 ft from the beginning of an 80-ft pipe. At this same location, the grade changes from +0.5% to +15%. What is the appropriate bend angle in the pipe to the nearest decimal degree to accommodate the change in horizontal direction and change in grade?

23.14 What details should a surveyor and contractor discuss?

23.15 By means of a sketch, show how and where batter boards should be located: (**a**) for an I-shaped building (**b**) for an L-shaped structure.

23.16 A building in the shape of an L must be staked. Corners $ABCDEF$ all have right angles. Proceeding clockwise around the building, the required outside dimensions are AB = 80.00 ft, BC = 30.00 ft, CD = 40.00 ft, DE = 40.00 ft, EF = 40.00 ft, and FA = 70.00 ft. After staking the batter boards for this building and stretching string lines taut, check measurements of the diagonals should be made. What should be the values of AC, AD, AE, FB, FC, FD, and BD?

23.17* Compute the floor area of the building in Problem 23.16.

23.18* The design floor elevation for a building to be constructed is 1069.31 ft. An instrument is set up nearby, leveled, and a plus sight of 5.34 ft taken on BM A whose elevation is 1070.58 ft. If batter boards are placed exactly 1.00 ft above floor elevation, what rod readings are necessary on the batter board tops to set them properly?

23.19 Compute the diagonals necessary to check the stakeout of the building in Figure 23.8.

23.20 What information is typically conveyed on stakes when laying a pipeline?

23.21 Where is the invert of a pipe measured?

23.22 Describe a method of checking verticality on a tall structure.

23.23 How can horizontal orientation be achieved with a total station on a construction site?

23.24 How can vertical orientation be achieved with a total station on a construction site?

23.25 Discuss the advantages of combining digital elevation models with design templates in staking out highway alignments with a data collector.

23.26 The base receiver in a GNSS survey is started in autonomous mode. What must occur before the roving receivers can be used to stakeout points?

23.27 What is a slope stake?

23.28 A highway centerline subgrade elevation is 406.38 ft at station 12 + 00 and 410.25 ft at 17 + 00 with a smooth grade in between. To set blue tops for this portion of the centerline, a level is setup in the area and a plus sight of 5.46 ft taken on a benchmark whose elevation is 405.68 ft. From that HI, what rod readings will be necessary to set the blue tops for the full stations from 12 + 00 through 17 + 00?

23.29* Similar to Problem 23.27, except the elevations at stations 12 + 00 and 17 + 00 are 405.65 and 410.63 ft, respectively, the BM elevation is 406.36 ft, and the backsight is 5.66 ft.

23.30 Discuss the checks that should be made when laying out a building using coordinates.
23.31 What are the advantages of an as-built survey using a laser scanner?
23.32 Describe how control can be brought quickly into a deep open-pit mine.
23.33 How should finished grades be established in machine control projects?
23.34 What is the minimum number of control points needed to establish finish grades using a robotic total station on a machine-controlled project that is 12 mi in length?
23.35 What is the minimum number of control points needed to guide machines using a GNSS receiver on a machine-guidance project that is 6 mi in length?
23.36 Review an article on an application of machine guidance or control.

BIBLIOGRAPHY

Bernardini, F. 2014. "Challenge Beneath San Giusto." *The American Surveyor* 11 (No. 2): 10.
Binder, N. 2015. "Moving London Safely Forward." *The American Surveyor* 12 (No. 2): 14.
Carter, N. F. 2009. "Establishing Vertical Control on the Hoover Dam Bypass–Colorado River Bridge." *Surveying and Land Information Science* 69 (No. 1): 53.
Forslin, K. 2014. "Mapping the Centerline." *Professional Surveyor* 34 (No. 3): 16.
Gakstatter, E. 2009. "Rebuilding the Greens at the Olympic Club Lake Course." *The American Surveyor* 6 (No. 9): 8.
Luccio, M. 2014. "Dredge. Survey. Repeat." *XYHT* 1 (No. 6): 22.
Stenmark, J. 2009. "Imaging Goes Underground." *Point of Beginning* 35 (No. 2): 18.
Stokes, J. 2014. "A New Approach for Sewer Mapping." *The American Surveyor* 11 (No. 7): 34.
Trojak. L. 2014. "Road Scholars." *The American Surveyor* 11 (No. 3): 42.
Wilson, G. 2014. "High Accuracy for Sanitary Lines." *XYHT* 1 (No. 2): 36.

24 Horizontal Curves

24.1 INTRODUCTION

Straight (tangent) sections of most types of transportation routes, such as highways, railroads, and pipelines, are connected by curves in both the horizontal and vertical planes. An exception is a transmission line, in which a series of straight lines is used with abrupt angular changes at tower locations when needed.

Curves used in horizontal planes to connect two straight tangent sections are called *horizontal curves*. Two types are used: *circular arcs* and *spirals*. Both are readily laid out in the field with surveying equipment. A *simple curve* [Figure 24.1(a)] is a circular arc connecting two tangents. It is the type most often used. A *compound curve* [Figure 24.1(b)] is composed of two or more circular arcs of different radii tangent to each other, with their centers on the same side of the alignment. The combination of a short length of tangent (less than 100 ft) connecting two circular arcs that have centers on the same side [Figure 24.1(c)] is called a *broken-back* curve. A *reverse curve* [Figure 24.1(d)] consists of two circular arcs tangent to each other, with their centers on opposite sides of the alignment. Compound, broken-back, and reverse curves are unsuitable for modern high-speed highway, rapid transit, and railroad traffic, and should be avoided if possible. However, they are sometimes necessary in mountainous terrain to avoid excessive grades or very deep cuts and fills. Compound curves are often used on exit and entrance ramps of interstate highways and expressways, although easement curves are generally a better choice for these situations.

Easement curves are desirable, especially for railroads and rapid transit systems, to lessen the sudden change in curvature at the junction of a tangent and a circular curve. A *spiral* makes an excellent easement curve because its radius decreases uniformly from infinity at the tangent to that of the curve it meets. Spirals are used to connect a tangent with a circular curve, a tangent with a tangent (double spiral), and a circular curve with a circular curve. Figure 24.2 illustrates these arrangements.

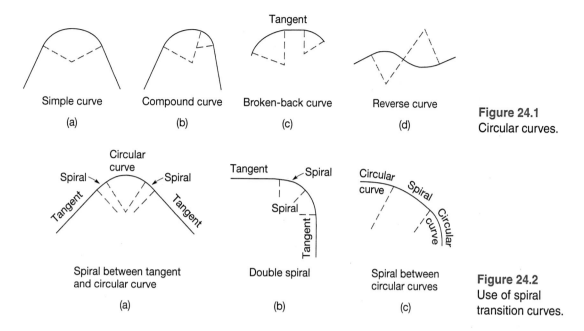

Figure 24.1 Circular curves.

Figure 24.2 Use of spiral transition curves.

The effect of centrifugal force on a vehicle passing around a curve can be balanced by *superelevation*, which raises the outer rail of a track or outer edge of a highway pavement. Correct transition into superelevation on a spiral increases uniformly with the distance from the beginning of the spiral, and is in inverse proportion to the radius at any point. Properly superelevated spirals ensure smooth and safe riding with less wear on equipment. As noted, spirals are used for railroads and rapid-transit systems. This is because trains are constrained to follow the tracks, and thus a smooth, safe, and comfortable ride can only be assured with properly constructed alignments that include easement curves. On highways, spirals are less frequently used because drivers are able to overcome abrupt directional changes at circular curves by steering a spiraled path as they enter and exit the curves. However with today's computational capabilities, easement curves should be used whenever possible for highways also.

Although this chapter concentrates on circular curves, methods of computing and laying out spirals are introduced in Section 24.19.

24.2 DEGREE OF CIRCULAR CURVE

The rate of curvature of circular curves can be designated either by their *radius* (e.g., a 1500-m curve or a 1000-ft curve), or by their *degree of curve*. There are two different designations for degree of curve, the *arc definition* and the *chord definition*, both of which are defined using the English system of units. By the arc definition, degree of curve is the central angle subtended by a circular *arc* of 100 ft [see Figure 24.3(a)]. This definition is preferred for highway work. By the chord definition, degree of curve is the angle at the center of a circular arc subtended by a chord of 100 ft [see Figure 24.3(b)]. This definition is convenient for very gentle curves and hence is preferred for railroads. The formulas relating radius R and degree D of curves for both definitions are shown next to the illustrations.

740 HORIZONTAL CURVES

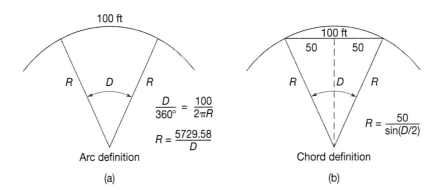

Figure 24.3
Degree of circular curve.

Using the equations given in Figure 24.3, radii of arc and chord definition curves for values of D from 1° to 10° have been computed and are given in columns (2) and (5) of Table 24.1. Although radius differences between the two definitions appear to be small in this range, they are significant.

When metric units are used, degree of curve can still be specified. As an example, a curve having a radius of exactly 700 m would have a degree of curve (arc definition) of

$$\frac{5729.58}{700(3.28083)} = 2°29'41''$$

Arc-definition curves have the advantage that computations are somewhat simplified as compared to the chord definition and, as will be shown later, the formula for curve length is exact, which simplifies preparing right-of-way descriptions. A disadvantage with the arc definition is that most measurements between full stations are shorter than a full 100-ft tape length, but this is of little significance

TABLE 24.1 FUNCTIONS OF CIRCULAR CURVES (LENGTHS IN FEET)

	Arc Definition			Chord Definition		
Degree of Curve D (1)	Radius R (2)	True Chord Full Sta (3)	True Chord Half-Sta (4)	Radius R (5)	Arc Length Full Sta (6)	True Chord Half-Sta (7)
1	5729.58	100.00	50.00	5729.65	100.00	50.00
2	2864.79	99.99	50.00	2864.93	100.01	50.00
3	1909.86	99.99	50.00	1910.08	100.01	50.00
4	1432.39	99.98	50.00	1432.69	100.02	50.01
5	1145.92	99.97	50.00	1146.28	100.03	50.01
6	954.93	99.95	50.00	955.37	100.05	50.02
7	818.51	99.94	50.00	819.02	100.06	50.02
8	716.20	99.92	49.99	716.78	100.08	50.03
9	636.62	99.90	49.99	637.27	100.10	50.04
10	572.96	99.88	49.98	573.69	100.13	50.05

if a total station instrument is used for stakeout. With the chord definition, full stations are separated by chords of exactly 100 ft regardless of the value of D.

For a given value of D, arc and chord definitions give practically the same result when applied to the flat curves common on modern highways, railroads, and rapid-transit systems. However, as degree of curve increases, the differences become greater.

24.3 DEFINITIONS AND DERIVATION OF CIRCULAR CURVE FORMULAS

Circular curve elements are shown in Figure 24.4. The *point of intersection* PI, of the two tangents is also called the *vertex V*. In stationing, the *back tangent* precedes the PI, the *forward tangent* follows it. The beginning of the curve, *point of curvature* PC, and the end of the curve, or *point of tangency* PT, are also sometimes called BC and EC, respectively. Other expressions for these points are tangent to curve, TC, and curve to tangent, CT. The curve radius is R. Note that the radii at the PC and PT are perpendicular to the back tangent and forward tangent, respectively.

The distance from PC to PI and from PI to PT is called the *tangent distance*, T. The line connecting the PC and PT is the *long chord* LC. The *length of the curve*, L, is the distance from PC to PT, *measured along the curve for the arc definition, or by 100-ft chords for the chord definition*.

The external distance E is the length from the PI to the curve midpoint on a radial line. The middle ordinate M is the (radial) distance from the midpoint of the long chord to the curve's midpoint. Any *point on curve* is POC; any *point on tangent*, POT. The degree of any curve is D_a (arc definition) or D_c (chord definition). The change in direction of two tangents is the intersection angle I, which is also equal to the central angle subtended by the curve.

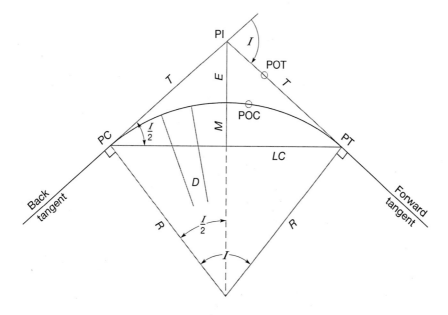

Figure 24.4
Circular curve elements.

By definition, and from inspection of Figure 24.4, relations for the arc definition follow:

$$L = RI \quad (I \text{ in radians}) \tag{24.1}$$

$$L = 100\frac{I°}{D°} \text{ (ft)} \tag{24.2a}$$

$$L = \frac{I°}{D°} \text{ (sta)} \tag{24.2b}$$

$$R = \frac{5729.58}{D} \text{ (ft)} \tag{24.3}$$

$$T = R \tan \frac{I}{2} \tag{24.4}$$

$$LC = 2R \sin \frac{I}{2} \tag{24.5}$$

$$\frac{R}{R+E} = \cos \frac{I}{2} \text{ and } E = R\left[\frac{1}{\cos(I/2)} - 1\right] \tag{24.6}$$

$$\frac{R-M}{R} = \cos \frac{I}{2} \text{ and } M = R\left(1 - \cos \frac{I}{2}\right) \tag{24.7}$$

Other convenient formulas that can be derived are:

$$E = T \tan \frac{I}{4} \tag{24.8}$$

$$M = E \cos \frac{I}{2} \tag{24.9}$$

Although curves are normally calculated by computers, if a handheld calculator is used, it is expedient to compute R, T, E, and M in the sequence of Equations (24.3), (24.4), (24.8), and (24.9), because the previously computed value is in the calculator and available for each succeeding calculation.

The formulas for T, L, LC, E, and M also apply to a chord-definition curve. However, L calculated by Equation (24.2a) implies the total length as if observed along the 100-ft chords of an inscribed polygon. The following formula is used for relating R and D for a chord definition curve:

$$R = \frac{50}{\sin(D/2)} \text{ (ft)} \tag{24.10}$$

Note that for the equations given previously, Equations (24.2a), (24.2b), (24.3), and (24.10) involve degree of curve, and thus assume distances in feet, while either metric or English units can be used in all others.

24.4 CIRCULAR CURVE STATIONING

Normally, an initial route survey consists of establishing the PIs according to plan, laying out the tangents, and establishing continuous *stationing* along them from the start of the project, through each PI, to the end of the job. (Stationing was described in Section 5.9.1.) The beginning point of any project is assigned a station value, and all other points along the reference line are then related to it. If the beginning point is also the end point of a previous adjacent project, its station value may be retained and the new job referenced to that stationing. Otherwise, an arbitrary value such as 100 + 00 for English unit stationing, or 10 + 000 for metric stationing is assigned. Assigning a starting stationing of 0 + 00 is generally not done to avoid the possibility that future revisions to the project could extend it back beyond the starting point and hence result in negative stationing. In the English system, staking is usually in *full stations* (100 ft apart), although *half stations* (50 ft apart), or even *quarter stations* (25 ft apart), can be set depending on conditions. In metric stationing, full stations are generally 1 km apart, but stakes may be set at 40, 30, 20, or even 10 m apart, depending on conditions. Staking at the closer spacing is usually done in urban situations, on sharp curves, or in rugged terrain, while the stakes are placed farther apart in relatively flat or gently rolling rural areas.

After the tangents have been staked and stationed, the intersection angle (I) is observed at each PI, and curves computed and staked. The station locations of points on any curve are based upon the stationing of the curve's PI. To compute the PC station, tangent distance T is subtracted from the PI station, and to calculate the PT station, curve length L is added to the PC station.

■ Example 24.1

Assume that $I = 8°24'$, the station of the PI is 64 + 27.46, and terrain conditions require the minimum radius permitted by the specifications of, say, 2864.79 ft (arc definition). Calculate the PC and PT stationing, and the external and middle ordinate distances for this curve.

Solution

By Equation (24.1) $\quad L = 2864.79(8°24')\dfrac{\pi}{180} = 420.00$ ft

By Equation (24.2a) $\quad D° = 100\dfrac{8°24'}{420} = 2°00'$

By Equation (24.4) $\quad T = 2864.79 \tan \dfrac{8°24'}{2} = 210.38$ ft

Calculate stationing

$$
\begin{aligned}
\text{PI station} &= 64 + 27.46 \\
-T &= -2 + 10.38 \\
\text{PC station} &= 62 + 17.08 \\
+L &= 4 + 20.00 \\
\text{PT station} &= 66 + 37.08
\end{aligned}
$$

Also by Equation (24.5) $LC = 2(2864.79)\sin\dfrac{8°24'}{2} = 419.62$ ft

And by Equation (24.8) $E = 210.38 \tan\dfrac{8°24'}{4} = 7.71$ ft

Finally by Equation (24.9) $M = 7.71 \cos\dfrac{8°24'}{2} = 7.69$ ft

---

Calculation for the stations of the PC and PT should be arranged as shown. Note that the stationing of the PT cannot be obtained by adding the tangent distance to the station of the PI, although the location of the PT on the ground is determined by measuring the tangent distance from the PI. Points representing the PC and PT must be carefully marked and placed exactly on the tangent lines at the correct distance from the PI so other computed values will fit their fixed positions on the ground.

If route surveys are originally staked as a series of tangents having continuous stationing, as described above, then an adjustment has to be made at each PT after curves are inserted. This is necessary because the length around the curve from PC to PT is shorter than the distance along the tangents from the PC to the PI to the PT. Thus for final stationing at the PT, there is a "station equation," which relates the stationing back along the curve to that forward along the tangent. For Example 24.1, it would be $66 + 37.08$ back $= 66 + 37.84$ ahead, where $66 + 37.08 = PI - T + L$, and $66 + 37.84 = PI + T$. The difference between the ahead and back stations represents the amount the route was shortened by inserting the curve. If the curves are run in and stationed at the time of staking the original alignment, continuous stationing along the route results and station equations at PTs are avoided.

The curve used in a particular situation is selected to fit ground conditions and specification limitations of minimum R or maximum D. Normally the value of the intersection angle, I, and the station of the PI are available from field observations on the preliminary line. Then a value of R or D, suitable for the highway or railroad, is chosen. Most of today's highways are designed using a minimum value for the radius R. Sometimes the distance E or M required to miss a stream or steep slope outside or inside the PI is observed, and D or R computed holding that distance fixed. Tangent distance governs infrequently. (One exception is to make a railroad, bus, or subway station fall on a tangent rather than a superelevated curve.) The length of curve practically never governs.

■ 24.5 GENERAL PROCEDURE OF CIRCULAR CURVE LAYOUT BY DEFLECTION ANGLES

Except for unusual cases, the radii of curves on route surveys are too large to permit swinging an arc from the curve center. Circular curves are therefore laid out by more practical methods, including (1) deflection angles, (2) coordinates, (3) tangent offsets, (4) chord offsets, (5) middle ordinates, and (6) ordinates from the long chord. Layout by deflection angles has been the standard approach, although with the advent of total station instruments, the coordinate method has replaced the deflection angle method in most instances.

Layout of a curve by deflection angles can be done by either the *incremental chord method* or the *total chord method*. In the past, the incremental chord method was almost always used as it could be readily accomplished with a theodolite and tape. The method can still be used when a total station instrument is employed, although then the distances are observed by taping rather than electronically. (Taping is still efficient in staking the stations along alignments because of the relatively short distance increments involved.) The total chord method was not practical until the advent of total stations, but with these instruments it is now conveniently employed even though longer distance measurements are involved.

The incremental chord method is illustrated in Figure 24.5. Assume that the instrument is set up over the PC (station 62 + 17.08 in Example 24.1). For this illustration, assume that each full station is to be marked along the curve, since cross sections are normally taken, construction stakes set, and computations of earthwork made at these points. (Half-stations or any other critical points can also be established, of course.) The first station to be set in this example is 63 + 00. To mark that point from the PC, a backsight is taken on the PI with zero set on the instrument's horizontal circle. Deflection angle δ_a to station 63 + 00 is then turned and two tapepersons measure chord c_a from the PC and set 63 + 00 at the end of the chord on the instrument's line of sight. With station 63 + 00 set, the tapepersons next measure the chord length c from it and stake station 64 + 00, where the line of sight of the instrument, now set to δ_{64}, intersects the end of that chord. This process is repeated until the entire curve is laid out. In this procedure, it is seen that the accuracy in the placement of each succeeding station depends on the accuracies of all those stations previously set.

The total chord method can also be described with reference to Figure 24.5. In this procedure, a total station instrument is set up at the PC, a backsight taken on the PI, and zero indexed on the horizontal circle. To set station 63 + 00, deflection angle δ_a is turned with the instrument, the reflector placed on line and adjusted until its distance from the instrument is c_a, and the stake set. To set

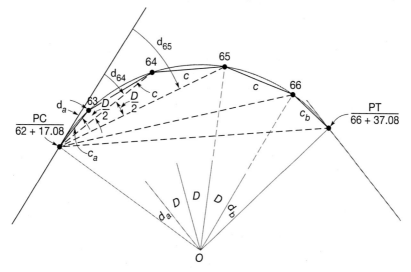

Figure 24.5
Circular curve layout by deflection angles.

station 64 + 00, deflection angle δ_{64} is turned, the reflector placed on this line of sight and adjusted in position until the total chord from the PC to station 64 + 00 is obtained, and the stake set.

This procedure is repeated, with each station being set independently of the others, until the entire curve is staked. This method of staking a curve has some drawbacks. One is that in some areas vegetation or other obstructions can block sight lines along the chords. Another is that each station is set independently, and thus there is no check at the end of the curve. However, incremental chords could also be measured with a tape to provide a check at each station.

24.6 COMPUTING DEFLECTION ANGLES AND CHORDS

From the preceding discussion it is clear that deflection angles and chords are important values that must be calculated if a curve is to be run by the deflection-angle method. To stake the first station, which is normally an odd distance from the PC (shorter than a full-station increment), *subdeflection angle* δ_a and *subchord* c_a are needed. These are shown in Figure 24.6. In this figure, central angle d_a subtended by arc s_a from the PC to 63 + 00 is calculated by proportion according to the definition of D as

$$\frac{d_a}{s_a} = \frac{D}{100} \text{ from which } d_a = \frac{s_a D}{100} \text{(degrees)} \qquad (24.11a)$$

where s_a is the difference in stationing between the two points. Equation (24.11a) is based on curves defined in English units using the arc definition for degree of curvature. For curves computed in English or metric units, the equivalent expression is

$$\frac{d_a}{s_a} = \frac{I}{L} \text{ from which } d_a = \frac{s_a I}{L} \text{(degrees)} \qquad (24.11b)$$

A fundamental theorem of geometry helpful in circular curve computation and stakeout is that *the angle at a point between a tangent and any chord is equal to*

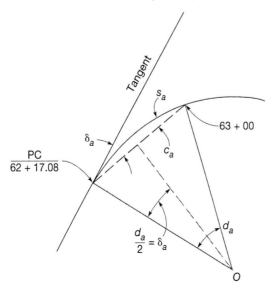

Figure 24.6
Subchords and subdeflections.

half the central angle subtended by the chord. Thus, subdeflection angle δ_a needed to stake station 63 + 00 is $d_a/2$, or

$$\delta_a = \frac{s_a D}{200} \text{ (degrees)} \tag{24.12a}$$

Also recognizing that I and L are constants for any particular curve, Equation (24.11b) can be rewritten as

$$\delta_a = s_a k \text{(degrees)} \tag{24.12b}$$

where
$$k = I/(2L)$$

The length of the subchord c_a can be represented in terms of δ_a and the curve radius as

$$\sin \delta_a = \frac{c_a}{2R} \text{ from which } c_a = 2R \sin \delta_a \tag{24.13}$$

Since the arc between full stations subtends the central angle D, from the earlier stated geometric theorem, deflection angles to each full station beyond 63 + 00 are found by adding $D/2$ to the previous deflection angle. Full chord c, which corresponds to 100 ft of curve length, is calculated using Equation (24.13), except that $D/2$ is substituted for δ_a. Equations (24.12) and (24.13) are also used to compute the last subdeflection angle δ_b and subchord c_b, but the difference in stationing s_b between the last full station and the PT replaces arc length s_a.

Equations (24.12) and (24.13) are also used for computing deflection angles and total chords for the total chord method of staking. Here, s is simply the difference in stationing between the station being set and the PC.

For curves up to about 2°00' (arc definition), the lengths of arcs and their corresponding chords are nearly the same. On sharper curves, chords are shorter than corresponding arc lengths. This is verified by the data in columns (3) and (4) of Table 24.1, which give true chord lengths for full- and half-station increments for varying values of D (arc definition).

Computations for deflection angles and chords on chord-definition curves use the same formulas, but R is calculated by Equation (24.10). Note that for a given degree of curve, R is longer for chord definition than for arc definition. Also arc lengths for full stations are longer than their nominal 100.00 ft value, and true subchords are longer than their *nominal values* (differences in stationing). A check of columns (6) and (7) in Table 24.1 verifies these facts.

■ Example 24.2

Compute subdeflection angles and subchords δ_a, c_a, δ_b, and c_b, and calculate chord c of Example 24.1.

Solution

By Equation (24.12a)

$$\delta_a = 82.92(0.0100°) = 0.8292° = 0°49'45''$$

$$\delta_b = 37.08(0.0100°) = 0.3708° = 0°22'15''$$
(Note that $D/200 = 2°/200 = 0.01°$)

By Equation (24.13)

$$c_a = 2(2864.79) \sin 0°49'45'' = 82.92 \text{ ft}$$
$$c_b = 2(2864.79) \sin 0°22'15'' = 37.08 \text{ ft}$$
$$c = 2(2864.79) \sin 1°00'00'' = 99.99 \text{ ft}$$

24.7 NOTES FOR CIRCULAR CURVE LAYOUT BY DEFLECTION ANGLES AND INCREMENTAL CHORDS

Based on principles discussed, the deflection angle and incremental chord data for stakeout of the complete curve of Examples 24.1 and 24.2 have been computed and listed in Table 24.2. Normally, as has been done in this case, the data are prepared for stakeout from the PC, although field conditions may not allow the curve to be completely run from there. This problem is discussed in Section 24.9.

Values of deflection angles are normally carried out to several decimal places for checking purposes, and to avoid accumulating small errors when D is a noninteger number, such as, perhaps, $3°17'24''$. Note in Table 24.2 that the deflection angle to the PT is $4°12'$, exactly half the I angle of $8°24'$. This comparison affords an important check on the calculations of all deflection angles.

Field notes for the curve of this example are recorded in Figure 24.7 as they would appear in a field book. Notes run up the page to simplify sketching while looking in a forward direction. Computers can conveniently perform all necessary calculations and list the notes for curve stakeout by deflection angles.

In many cases, it is desirable to "back in" a curve by setting up over the PT instead of the PC. One setup is thereby eliminated and the long sights are taken on

TABLE 24.2 DEFLECTION ANGLE AND INCREMENTAL CHORD DATA FOR EXAMPLE CURVE

Station	Incremental Chord	Deflection Increment	Deflection Angle
66 + 37.08 (PT)	37.08	0°22'15"	4°12'00"✓
66 + 00	99.99	1°00'00"	3°49'45"
65 + 00	99.99	1°00'00"	2°49'45"
64 + 00	99.99	1°00'00"	1°49'45"
63 + 00	82.92	0°49'45"	0°49'45"
62 + 17.08 (PC)			

24.8 Detailed Procedures for Circular Curve Layout by Deflection Angles and Incremental Chords 749

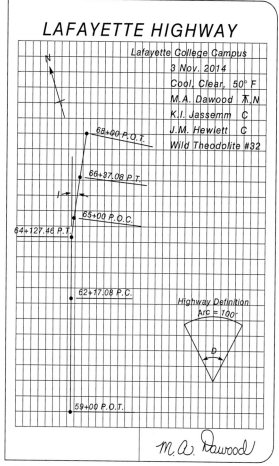

Figure 24.7 Field notes for horizontal curve in Examples 24.1 and 24.2.

the first measurements. In precise work it is better to run in the curve from both ends to the center, where small errors can be adjusted more readily. On long or very sharp circular curves, or if obstacles block sights from the PC or PT, setups on the curve are necessary (see Section 24.9).

24.8 DETAILED PROCEDURES FOR CIRCULAR CURVE LAYOUT BY DEFLECTION ANGLES AND INCREMENTAL CHORDS

Regardless of the method used to stake intermediate curve points, the first steps in curve layout are: (1) establishing the PC and PT, normally by measuring tangent distance T from the PI along both the back and forward tangents and (2) measuring the total deflection angle at the PC from PI to PT. This latter step should be performed whenever possible, since the observed angle must check $I/2$; if it does not, an error exists in either observation or computation, and time should not be wasted running an impossible curve.

It is also good practice to stake the midpoint of a curve before beginning to set intermediate points, especially on long curves. The midpoint can be set by bisecting the angle $180° - I$ at the PI and laying off the external distance from there. A check of the deflection angle from the PC to the midpoint should yield $I/4$. When staking intermediate points along the curve has reached the midpoint of the curve, a chord check measurement should be made to it.

The remaining steps in staking intermediate curve points by the deflection angle incremental chord method are presented with reference to the curve of Examples 24.1 and 24.2. With the instrument set up and leveled over the PC, it is oriented by backsighting on the PI, or on a point along the back tangent, with $0°00'$ on the circle. The subdeflection angle of $0°49'45''$ is then turned. Meanwhile, the 17-ft mark of the tape is held on the PC. The zero end of the (add) tape is swung until the line of sight hits a point 0.08 ft ahead from the zero mark. This is station $63 + 00$. To stake station $63 + 00$. the rear tapeperson next holds the 99-ft mark on station 63 and the forward tapeperson sets station 64 at distance 99.99 ft by direction from the instrument operator, who has placed an angle of $1°49'45''$ on the circle. An experienced forward tapeperson will walk along the extended first full chord, know or estimate the chord offset, and from an outside-the-chord position will be holding the tape end and a stake within a foot or so of the correct location when the instrument operator has the deflection angle ready.

After placing the final full station ($66 + 00$ in this example), to determine any misclosure in staking a curve, the closing PT should be staked using the final deflection angle and subchord. This will rarely agree with the PT established by measuring distance T along the forward tangent from the PI because of accumulated errors. Any misclosure ("falling") should be observed; then the field precision can be expressed as a numerical ratio like that used in traverse checks. The observed falling distance is the numerator, and $L + 2T$ the denominator. If the misclosure of this example was 0.25 ft, the precision would be $0.25/[420.00 + 2(210.38)] = 1/3300$.

24.9 SETUPS ON CURVE

Obstacles that prevent visibility along curve chords and extremely long sight distances sometimes make it necessary to set up on the curve after it has been partially staked. The simplest procedure to follow is one that permits use of the same notes computed for running the curve from the PC. In this method, the following rule applies: *The instrument is moved forward to the last staked point, backsighted on a station with the telescope inverted and the circle set to the deflection angle from the PC for the station sighted. The telescope is then plunged to the normal position, and deflection angles previously computed from the PC for the various stations are used.*

The example of the preceding sections is used to illustrate this rule. If a setup is required at station 65, place $0°00'$ on the instrument and sight the PC with the telescope inverted, plunge, set the circle to read the deflection angle $3°49'45''$ and stake station 66.

To prove this geometric rule, assume the instrument were set on station $65 + 00$ and the PC backsighted with $0°00'00''$ on the circle. If the telescope were plunged and turned $2°49'45''$ in azimuth, the line of sight would then be tangent to the curve. To stake the next station ($66 + 00$), an additional angle of $D/2$ or $1°00'00''$ would need to be turned. The sum of $2°49'45''$ and $1°00'00''$ is of course the deflection angle to station 66 + 00 from the PC. The rule also applies for backsights to any previously set stations, not just the PC. Thus, with the instrument at 65 + 00, station 63 + 00 could be sighted with 0°49′45″ on the circle and then turned to 3°49′45″ to set station 66 + 00. Further study of the geometry illustrated in Figure 24.5 should clarify this procedure.

24.10 METRIC CIRCULAR CURVES BY DEFLECTION ANGLES AND INCREMENTAL CHORDS

Most countries and many highway departments in the United States use metric units for observations and stationing on their projects. As noted earlier, in the metric system, circular curves are designated by their radius values rather than degree of curve. Otherwise, as illustrated by the following example, computations for laying out a curve using metric units by the deflection angle incremental chord method follow the same procedures as for the English system of units and stationing.

Example 24.3

Assume that a metric curve will be used at a PI where $I = 8°24′$. Assume also that the station of the PI is 6 + 427.464, and that terrain conditions require a minimum radius of 900 m. Calculate the PC and PT stationing, and other defining elements of the curve. Also compute notes for staking the curve using 20-m increments.

Solution

By Equation (24.1) $L = 900\left(8°24′\dfrac{\pi}{180°}\right) = 131.947$ m

By Equation (24.4) $T = 900 \tan\left(\dfrac{8°24′}{2}\right) = 66.092$ m

Calculate stationing

$$
\begin{aligned}
\text{PI station} &= 6 + 427.464 \\
-T &= 066.092 \\
\text{PC station} &= 6 + 361.372 \\
+L &= 131.947 \\
\text{PT station} &= 6 + 493.319
\end{aligned}
$$

Also by Equation (24.5) $LC = 2(900) \sin\left(\dfrac{8°24′}{2}\right) = 131.829$ m

And by Equation (24.8) $E = 66.092 \tan\left(\dfrac{8°24′}{4}\right) = 2.423$ m

Finally by Equation (24.9) $M = 2.423 \cos\left(\dfrac{8°24′}{2}\right) = 2.416$ m

The arc distance from the PC to the station 6 + 380 is (6380 − 6361.372) = 18.628 m. The arc distance for the final stationing is 6493.319 − 6480 = 13.319 m.

TABLE 24.3 DEFLECTION ANGLE AND INCREMENTAL CHORD DATA FOR EXAMPLE CURVE

Station	Incremental Chord	Deflection Increment	Deflection Angle
6 + 493.319 (PT)	18.628	0°25′26.2″	4°12′00″ ✓
6 + 480	19.999	0°38′11.8″	3°46′34″
6 + 460	19.999	0°38′11.8″	3°08′22″
6 + 440	19.999	0°38′11.8″	2°30′10″
6 + 420	19.999	0°38′11.8″	1°51′58″
6 + 400	19.999	0°38′11.8″	1°13′46″
6 + 380	13.319	0°35′34.6″	0°35′35″
6 + 361.372 (PC)			

All other stations have 20-m stationing intervals. Tables 24.3 and Figure 24.8 depict the curve data and field notes necessary to stake the curve in this example.

By Equation (24.12b)

$$k = 8°24′/[2(131.947)] = 0.03183096°$$
$$\delta_a = 0.03183096(18.628) = 0.59295° = 0°35′34.6″$$
$$\delta = 0.03183096(20) = 0.63662° = 0°38′11.8″$$
$$\delta_b = 0.03183096(13.319) = 0.42396° = 0°25′26.2″$$

ALIGNMENT OF LAFAYETTE HIGHWAY

Station	Chord	Total Def.	Curve	Data
(PT)		° ′ ″		
6+493.319		4 12 00	I = 8°24′	
	13.318		R = 900 m	
6+480		3 46 34	k = 0.03183096	
	19.999		L = 131.947 m	
6+460		3 08 22	T = 66.092 m	
	19.999		LC = 131.829 m	
6+440		2 30 10	E = 2.423 m	
	19.999		M = 2.416 m	
6+420		1 51 58		
	19.999			
6+400		1 13 46		
	19.999			
6+380		0 35 35		
	18.628			
6+361.372				
(PC)				

Figure 24.8 Field notes for horizontal curve in Example 24.3.

By Equation (24.13)

$$c_a = 2(900) \sin 0°35'34.6'' = 18.628 \text{ m}$$
$$c = 2(900) \sin 0°38'11.8'' = 19.999 \text{ m}$$
$$c_b = 2(900) \sin 0°25'26.2'' = 13.318 \text{ m}$$

■ 24.11 CIRCULAR CURVE LAYOUT BY DEFLECTION ANGLES AND TOTAL CHORDS

If field conditions permit, and a total station instrument is available, curves may be conveniently laid out by deflection angles and total chords. By using this method, the field party size is reduced from three to two, or possibly even a single person if a robotic total station instrument is available. Deflection angles are calculated and laid off as in the preceding example, but the chords are all measured electronically as radial distances (total chords) from the PC or other station where the instrument is placed. If stakeout is planned from the PC, total chords from there are the dashed lines of Figure 24.5. They are calculated by Equation (24.13), except that the deflection angle for each station is substituted for δ_a to obtain the corresponding chord. The total chords necessary to stake the curve of Example 24.2 using a total station instrument set up at PC are 82.92 ft for 63 + 00, 182.89 ft for 64 + 00, 282.80 ft for 65 + 00, 382.63 ft for 66 + 00, and 419.62 ft for the PT, which is the long chord LC given by Equation (24.5). The same deflection angles given in Table 24.2 apply.

To stake curves using a total station, the instrument is placed in its tracking mode. The deflection angle to each station is turned and the required chord to that station entered in the instrument. The instrument operator directs the person with the reflector to the proper alignment. The reflector is then moved forward or back as necessary, until the proper total chord distance is achieved, where the stake is set. It is often convenient to carry a short tape when staking out stations to quickly move to the final position from a nearby trial position. If intermediate setups are required on the curve using this method, the instrument is oriented as described in Section 24.9. New radial chords to be measured from the intermediate station would then have to be calculated.

Although curves can be staked rapidly with total stations using this method, as noted earlier an associated danger is that each stake is set individually, and therefore does not depend on previous stations. Thus, a check at the end of the curve is not achieved as in the incremental chord method and mistakes in angles or distances at intermediate stations could go undetected. Large blunders can usually be discovered by visual inspection of the curve stakes, but quickly taping the incremental chords between adjacent stations provides a better check.

■ 24.12 COMPUTATION OF COORDINATES ON A CIRCULAR CURVE

Today, because of the availability of total station instruments with data collectors, circular curves are often staked using the coordinate method. For this procedure, coordinates of the points on the curve to be staked must first be determined in some

754 HORIZONTAL CURVES

reference coordinate system. Although they are most often based upon an established map projection[1] such as the State Plane Coordinate System or the Universal Transverse Mercator projection (see Chapter 20), often an arbitrary project coordinate system will suffice. It should be remembered that the map projections typically used in surveying and elevation of the ground distort distances. The inverse of two map projection coordinates yields a grid distance. However, a ground distance must be laid out. Thus, grid lengths must be converted to ground lengths as discussed in Section 20.13. Typically, a project factor as discussed in Section 20.13.2 is determined and entered into the survey controller before layout of the curve is started. However, a low-distortion projection as discussed in Section 20.13.1 can also be used. This section describes the process of determining coordinates for stations on circular curves.

In Figure 24.9, assume that the azimuth of the back tangent going from A to V is known, the coordinates of the PI (point V) are known, and that the defining parts of the curve have been computed using Equations (24.1) through (24.10). Using the tangent distance and azimuth of the back tangent, the departure and latitude are computed by Equations (10.1) and (10.2), where Az_{VA} is the back azimuth of line AV. The coordinates of A (the PC) are then

$$X_A = X_V + T \sin Az_{VA}$$
$$Y_A = Y_V + T \cos Az_{VA} \qquad (24.14)$$

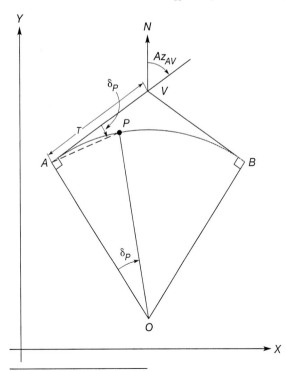

Figure 24.9
Geometry for computing coordinates of curve points.

---

[1]From Equation (24.27), the angle at the TS in triangle TS-SPI-SC of Figure 24.18 is $\Delta_S/3$. Also from Figure 24.18, the angle at the SPI in this triangle is $180° - \Delta_S$. It follows therefore that with the instrument at the SC, after backsighting the TS, the angle that must be turned to get the line-of-sight tangent to the spiral and circular curve is $2\Delta_S/3$.

With the coordinates of the PC known, coordinates of points on the curve can be computed using the same deflection angles and subchords used to stake out the curve by the total chord method. Deflection angles are added to the azimuth of AV to get azimuths of the chords to all stations to be set. Using the total chord length and chord azimuth for each station, departures and latitudes are calculated, and added to the coordinates of A (the PC) to get the station coordinates. With coordinates known for all curve points, they can be staked with the total station occupying any convenient point whose coordinates are also known in the same reference system. The PC, PT, PI, or curve midpoints are points that are often used.

It is sometimes convenient to stake a circular curve by placing the instrument at the center of the curve, that is, point O of Figure 24.9. In this case, the coordinates of the curve's center point are computed, and then the coordinates of the stations to be set can be conveniently computed using radial lines from that point. From Figure 24.9, the azimuth of the radius going from A to the center of the curve is

$$Az_{AO} = Az_{AV} + 90° \qquad (24.15a)$$

Equation (24.15a) is valid for a curve that lies right of the back tangent. For a curve that bends to the left, the proper expression is

$$Az_{AO} = Az_{AV} - 90° \qquad (24.15b)$$

Using the appropriate azimuth from Equations (24.15), and the radius of the curve, R, the coordinates of center point O of Figure 24.9 are

$$X_O = X_A + R \sin Az_{AO}$$
$$Y_O = Y_A + R \cos Az_{AO} \qquad (24.16)$$

The azimuth of the radius line from O to any station P on the curve is

$$Az_{OP} = Az_{OA} + d_P \qquad (24.17)$$

where d_P is determined in Equation (24.11). Then the coordinates of P are

$$X_P = X_O + R \sin Az_{OP}$$
$$Y_P = Y_O + R \cos Az_{OP} \qquad (24.18)$$

Example 24.4 in the following section demonstrates the method of computing coordinates of curve points using deflection angles and total chords.

24.13 CIRCULAR CURVE LAYOUT BY COORDINATES

The coordinate method can be used to advantage in staking circular curves, especially if a total station instrument or global navigation satellite system (GNSS) receivers are employed. In this procedure, the coordinates of each curve station to be staked are calculated as described in the preceding section. The total station instrument can then be placed at the PC, PT, curve midpoint, curve center point, or any other nearby control station, which gives a good vantage point of the entire area where the curve will be laid out.

756 HORIZONTAL CURVES

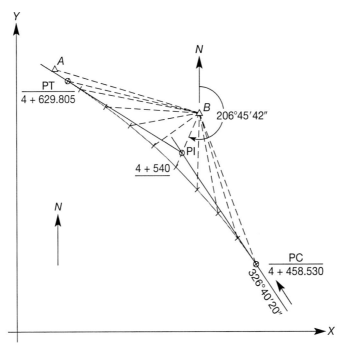

Figure 24.10
Layout of circular curve by coordinates with a total station instrument.

Azimuths and distances to each station are computed by inversing, using the coordinates of the occupied station and those of each curve station. The instrument is oriented by backsighting another visible control station. Then each curve point is staked by laying out its computed distance along its calculated azimuth. Again care must be exercised to ensure that ground lengths are being used and not grid lengths.

Figure 24.10 illustrates a situation where a curve is being staked by the coordinate method. The total station instrument is placed at control station B because all curve points are visible from there. After a backsight on control station A, distances and directions are used to stake all curve points. The computations necessary for staking this curve by the coordinate method are illustrated with the following example.

■ Example 24.4

Two tangents intersect at a PI station of 4 + 545.500 whose coordinates are $X = 5723.183$ and $Y = 3728.947$ m. The intersection angle is 24°32′ left, and the azimuth of the back tangent is 326°40′20″. A curve with radius, R of 400 m will be used to join the tangents. Compute the data needed to stake the curve at 20 m increments by coordinates using a total station instrument. For staking, the instrument will be set at station B, whose coordinates are $X = 5735.270$ m and $Y = 3750.402$ m, and a backsight taken on station A, whose coordinates are $X = 5641.212$ m and $Y = 3778.748$ m.

Solution

By Equation (24.1) $\quad L = 400 \times 24°32' \left(\dfrac{\pi}{180°}\right) = 171.275$ m

By Equation (24.4) $\quad T = 400.000 \tan(12°16') = 86.970$ m

Curve stationing

$$PI = 4 + 545.500$$
$$-T = 86.970$$
$$PC = 4 + 458.530$$
$$+L = 171.275$$
$$PT = 4 + 629.805$$

A tabular solution for curve point coordinates is given in Table 24.4. The differences in stationing from one curve point to the next are listed in column (2). Total deflection angles are calculated using Equation (24.12b) and tabulated in column (3). Total chords, computed from Equation (24.13) using these total deflection angles, are tabulated in column (4). From the azimuth of the back tangent and the deflection angles, the azimuth of each total chord is calculated and tabulated in column (5). The coordinates of the PC are computed using Equations (24.14) as

$$X_{PC} = 5723.183 + 86.970 \sin(326°40'20'' - 180°) = 5770.967 \text{ m}$$

$$Y_{PC} = 3728.947 + 86.970 \cos(326°40'20'' - 180°) = 3656.280 \text{ m}$$

Using their lengths and azimuths, the departure ΔX and latitude ΔY of each total chord are calculated. These are added to the coordinates of the PC to obtain the coordinates of the curve points. Values of ΔX and ΔY are tabulated in columns (6) and (7), and the X and Y coordinates are listed in columns (8) and (9) of Table 24.4.

A check on the PT coordinates of Table 24.4 can be obtained by computing them independently using the azimuth and tangent length of the forward tangent.

TABLE 24.4 COMPUTATIONS FOR STAKING CURVE OF EXAMPLE 24.4 BY COORDINATES

Station (1)	Station Difference (2)	Total Deflection (3)	Total Chord (4)	Chord Azimuth (5)	ΔX (6)	ΔY (7)	X (8)	Y (9)
4 + 458.530							5770.967	3656.280
4 + 460	1.470	0°06'19"	1.470	326°34'01"	−0.810	1.227	5770.157	3657.507
4 + 480	21.470	1°32'16"	21.468	325°08'04"	−12.272	17.614	5758.695	3673.894
4 + 500	41.470	2°58'12"	41.452	323°42'08"	−24.539	33.408	5746.428	3689.688
4 + 520	61.470	4°24'09"	61.410	322°16'11"	−37.580	48.569	5733.387	3704.849
4 + 540	81.470	5°50'06"	81.330	320°50'14"	−51.362	63.060	5719.605	3719.340
4 + 560	101.470	7°16'02"	101.199	319°24'18"	−65.851	76.843	5705.116	3733.123
4 + 580	121.470	8°41'59"	121.004	317°58'21"	−81.011	89.885	5689.956	3746.165
4 + 600	141.470	10°07'55"	140.734	316°32'25"	−96.803	102.153	5674.164	3758.433
4 + 620	161.470	X	160.376	315°06'28"	−113.189	113.616	5657.777	3769.896
4 + 629.805	171.275	12°16'00"	169.970	314°24'20"	−121.427	118.933	5649.540	3775.213

Note: All lengths and coordinate values are in meters.

The azimuth of the forward tangent is calculated by subtracting the I angle from the back-tangent's azimuth

$$Az = 326°40'20'' - 24°32' = 302°08'20''$$

Then the X and Y coordinates of the PT are

$$X_{PT} = 5723.183 + 86.970 \sin(302°08'20'') = 5649.540 \text{ m} \checkmark$$

$$Y_{PT} = 3728.947 + 86.970 \cos(302°08'20'') = 3775.213 \text{ m} \checkmark$$

Computations for the lengths and azimuths of the radial lines needed to stake curve points from station B are listed in Table 24.5. Column (1) gives the curve stations, and columns (2) and (3) list the differences ΔX and ΔY between each curve point's X and Y coordinates and those of station B. Radial lengths L computed from Equation (11.4) and azimuths computed by Equation (11.5) are tabulated in columns (4) and (5). While the calculations needed to stake a curve by coordinates may seem rather extensive, they are routinely handled by computers.

For orienting the instrument it is necessary to calculate the azimuth of line BA. This, also by Equation (11.5), is

$$Az_{BA} = \tan^{-1}\left(\frac{5641.212 - 5735.270}{3778.748 - 3750.402}\right) + 360° = 286°46'16''$$

After backsighting station A, $286°46'16''$ is indexed on the total station's horizontal circle. Then each curve point is staked by measuring its radial distance and azimuth taken from Table 24.5. The radial lines are shown dashed in Figure 24.11. Note that to stake station $4 + 540$, for example, a distance of 34.789 m is observed on an azimuth of $206°45'42''$, as shown in the figure.

TABLE 24.5 COMPUTATIONS FOR RADIAL LENGTHS AND AZIMUTHS FOR STAKING CURVE OF EXAMPLE 24.4 BY COORDINATES

Station (1)	ΔX (2)	ΔY (3)	L (4)	Az (5)
4 + 458.530 (PC)	−85.730	24.811	89.248	286°08'27"
4 + 460	−77.493	19.494	79.907	284°07'13"
4 + 480	−61.106	8.031	61.631	277°29'14"
4 + 500	−45.314	−4.238	45.512	264°39'25"
4 + 520	−30.154	−17.280	34.754	240°11'05"
4 + 540	−15.665	−31.063	34.789	206°45'42"
4 + 560	−1.883	−45.553	45.592	182°22'01"
4 + 580	11.158	−60.714	61.731	169°35'11"
4 + 600	23.425	76.508	80.014	162°58'36"
4 + 620	34.887	92.895	99.230	159°24'58"
4 + 629.805 (PT)	35.697	94.122	100.664	159°13'48"

Note: The units of ΔX, ΔY, and L are meters.

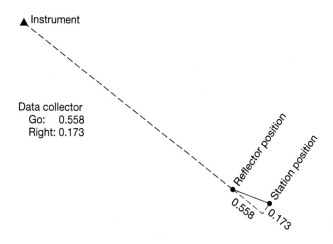

Figure 24.11
Setting a station with a tape from a "stake-out" position.

The process of staking a curve can be greatly simplified by using a data collector equipped with a "stakeout" option. When operating in this mode, before going into the field a file of point identifiers (PIDs) and their corresponding coordinates for the job are downloaded into the data collector. In the field, the operator must input (1) the occupied station PID, (2) the backsight station PID (or the azimuth of the backsight line), and (3) the PID of the point to be staked. The backsight is then taken, which automatically orients the instrument. Next, the prism person walks to the estimated location of the station to be set, and the instrument operator sights the reflector. The stakeout software determines the coordinates of the reflector, and informs the operator of the distances and directions that the reflector must be moved, from the prism-person's perspective, to establish the station. For example, the software may say, "GO 0.558 RIGHT 0.173." This indicates that the rodperson should move the reflector 0.558 m from the instrument and 0.173 m to their right as they face the instrument. Or the data collector may say "COME 0.558 LEFT 0.173." This means that the rod should be moved 0.558 m toward the instrument and 0.173 m to the left. A short tape is useful in quickly determining the location of the station to be set. Figure 24.11 shows the final measurements needed to set a station with a stakeout command of "GO 0.558 RIGHT 0.173." If a robotic total station and RPU (see Section 8.6) are available to set a curve, the instrument will automatically swing in azimuth to the desired line of sight. This equipment enables one person to lay out a curve.

As noted earlier, in staking circular curves by the coordinate method, any point can be selected for the instrument station as long as its coordinates are known. It can be a point on curve, another established control point, or a new point can be set by traversing (see Chapter 9). Alternatively, the instrument can be set at any point of unknown location that provides good vantage and its position can quickly be determined by the resection method (see Sections 11.7 and 11.10). In any case, after the instrument has been set up and oriented, a check should be made to assure its accuracy by measuring to another station of known coordinates. Any significant discrepancy should be reconciled before the curve is laid out.

It is important to note once again that each curve point is staked independently of the others, and thus there is no misclosure at the end to verify the accuracy of the work. Thus the work must be done very carefully, and the layout checked.

Figure 24.12
Data entry screen for horizontal curve computations in WOLFPACK.

To check, chord distances between successive stations can be quickly measured with a tape or all stations could be checked from an instrument setup at a second station of known coordinates.

The WOLFPACK software, which can be found on the companion, can be used to establish curve-staking notes. The data entry screen for the horizontal curve computations option is shown in Figure 24.12. With this software the user can compute coordinates for the curve and have these coordinates saved to a coordinate file for uploading to a data collector. The figure shows a completed data entry screen for computing the curve in Example 24.4. Two *Computation Options* are selected: "Compute coordinates" and "Create coordinate file." As additional options are selected, additional data entry boxes will be displayed. The help file that comes with the software describes each option.

24.14 CURVE STAKEOUT USING GNSS RECEIVERS AND ROBOTIC TOTAL STATIONS

As discussed in Sections 23.10 and 23.11, horizontal curves can also be staked out using kinematic (RTK or RTN) GNSS surveying methods. However, when doing this, it is important to perform a localization procedure first as discussed in Sections 15.9 and 23.11 to place the satellite-derived coordinates into the project coordinate system. This procedure requires control points known in the project coordinate system that encompass the project area. As discussed in Section 23.11, the base station coordinates can be established initially using the autonomous mode or by a prior GNSS survey (see Chapters 14 and 15). Since only relative positions of stations are required, the transformation process will remove any inaccuracies in the base station coordinates. However, failure to apply or perform this transformation when using GNSS in stakeout, or extrapolation of stakeout points beyond the area encompassed by the project control, can result in serious errors. Additionally, this procedure should only be performed once for any project. Repeated localizations will result in varying solutions due to random errors, which, in turn, will propagate into positional errors. Using kinematic surveying methods, the surveyor is guided by the GNSS survey controller to each stake location, where it is witnessed with a hub.

Horizontal layout is also possible using a GNSS machine guidance and control system. As discussed in Section 23.11, a GNSS receiver is located on the construction vehicle and used in conjunction with a DEM and finished grading plan to guide the vehicle through the construction process and control earthwork. A base station must be located within radio range of the rover, typically less than 10 km, to supply the rover with the base receiver observations.

Machine control systems are also available for robotic total stations. In this process, the total station is set up on a point with known coordinates and referenced to another. A DEM and grading plan are loaded into the machine control system and the robotic total station guides the construction equipment by tracking a multifaceted, 360° prism (see Figure 8.7) mounted on the construction vehicle.

The robotic total station requires sufficient control stations in the construction site to provide both location and orientation. Since the typical range of this system is about 1000 ft (300 m), and since there must be continuous line of sight between the total station and prism, this system requires more control stations in the project area than does a GNSS machine control system. Additionally, a robotic total station must be dedicated to each construction vehicle. Thus, total stations are typically only used where canopy conditions prohibit the use of GNSS. The robotic total station provides both horizontal and vertical positioning to the machine control system. Additionally, it offers the benefit of having sufficient accuracy to control finish grading without the need of a laser level. Machine guidance and control for this system is so accurate that it has provided guidance to curbing machines without the use of a stringline.

24.15 CIRCULAR CURVE LAYOUT BY OFFSETS

For short curves, when a total station instrument is not available, and for checking purposes, one of four offset-type methods can be used for laying out circular curves: *tangent offsets* (TO), *chord offsets* (CO), *middle ordinates* (MO), and *ordinates from the long chord*. Figure 24.13 shows the relationship of CO, TO, and MO. Visually and by formula comparison, the chord offset for full station is

$$CO = 2c \sin\frac{D}{2} = \frac{c^2}{R} = TO \approx 8\ MO \qquad (24.19)$$

Since $\sin 1° = 0.0175$ (approx.), $CO = c(0.0175)D$, where D is in degrees and decimals.

The middle ordinate m for any subchord is $R(1 - \cos \delta)$, with δ being the deflection angle for the chord. A useful equation in the layout or checking of curves in place is

$$D(\text{degrees}) = m(\text{inches}) \text{ for a 62-ft chord(approx.)} \qquad (24.20)$$

762 HORIZONTAL CURVES

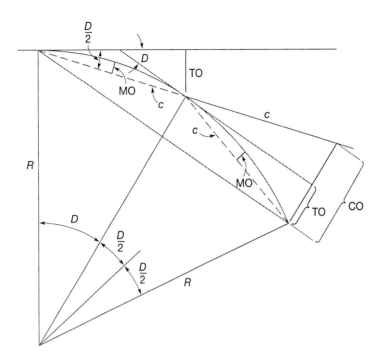

Figure 24.13
Circular curve offsets.

The geometry of the tangent-offset method is shown in Figure 24.14. The figure illustrates that the curve is most conveniently laid out in both directions from PC and PT to a common point near the middle of the curve. This procedure avoids long observations and provides a check where small adjustments can be made more easily if necessary. To lay out a curve using this method, tangent distances are measured to established temporary points A, B, and C in Figure 24.14. From these points, right-angle observations (tangent offsets) are made to set the curve stakes. Tangent distances (TD) and tangent offsets (TO) are calculated using chords and angles in the following formulas:

$$\text{TD} = c \cos \delta \qquad (24.21)$$

$$\text{TO} = c \sin \delta \qquad (24.22)$$

where δ angles are calculated using either Equation (24.12a) or (24.12b), and chords c are determined from Equation (24.13). These procedures are seldom used by today's surveyors (geomaticians).

■ Example 24.5

Compute and tabulate the data necessary to stake, by tangent offsets, the half stations of a circular curve having $I = 11°00'$, $D_c = 5°00'$ (chord definition), and PC = 77 + 80.00.

Solution

By Equation (24.2a), the curve length is $L = 100(11/5) = 220$ ft.

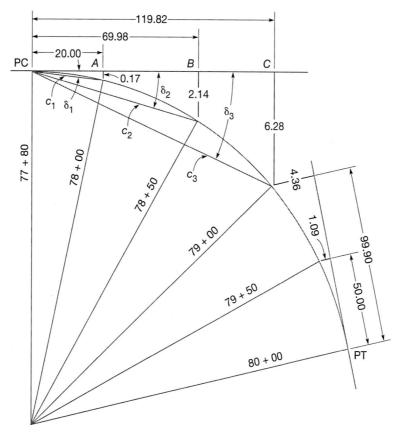

Figure 24.14
Circular curve layout by tangent offsets.

Therefore, the PT station is $(77 + 80) + (2 + 20) = 80 + 00$. Intermediate stations to be staked are $78 + 00$, $78 + 50$, $79 + 00$, and $79 + 50$, as shown in Figure 24.14.

By Equation [(24.12(a))], δ angles from the PC are

$$\delta_1 = 0.025(20) = 0.50° = 0°30'$$
$$\delta_2 = 0.025(70) = 1.75° = 1°45'$$
$$\delta_3 = 0.025(120) = 3.00° = 3°00'$$

where $D/200 = 0.025$.

By Equation (24.10), the radius is

$$R = \frac{50}{\sin 2°30'} = 1146.28 \text{ ft}$$

By Equation (24.13), chords from the PC are

$$c_1 = 2(1146.28) \sin 0°30' = 20.00 \text{ ft}$$
$$c_2 = 2(1146.28) \sin 1°45' = 70.01 \text{ ft}$$
$$c_3 = 2(1146.28) \sin 3°00' = 119.98 \text{ ft}$$

TABLE 24.6 TANGENT OFFSET DATA FOR EXAMPLE 24.5

Station	Deflection Angle δ	Chord c	Tangent Offset $c \cos \delta$	Tangent Distance $c \sin \delta$
80 + 00 (PT)				
79 + 50	1°15'	50.01	50.00	1.09
79 + 00	2°30'	100.00	99.90	4.36
79 + 00	3°00	119.98	119.82	6.28
78 + 50	1°45'	70.01	69.98	2.14
78 + 00	0°30'	20.00	20.00	0.17
77 + 80 (PC)				

Now, using Equations (24.21) and (24.22), tangent distances and tangent offsets are calculated. Chords, angles, tangent distances, and tangent offsets to stake points from the PT are computed in the same manner. All data for the problem are listed in Table 24.6. The tangent distances tabulated are lengths from the PC or PT that must be measured to establish points A, B, C, etc., and the tangent offsets are distances from these points needed to locate the curve stakes. Accurate curve layout by tangent offsets generally requires a total station to turn the right angles from the tangent. This involves more instrument setups and a greater amount of time than stakeout by deflection angles or coordinates. However, rough layouts can be done using a tape and right-angle prism.

24.16 SPECIAL CIRCULAR CURVE PROBLEMS

Many special problems arise in the design and computation of circular curves. Three of the more common ones are discussed here, and each can be solved using the coordinate geometry formulas given in Chapter 11.

24.16.1 Passing a Circular Curve Through a Fixed Point

One problem that often occurs in practice is to determine the radius of a curve connecting two established tangents and going through a fixed point such as an underpass, overpass, or existing bridge. The problem can be solved by establishing an XY coordinate system, as shown in Figure 24.15 where the origin occurs at V (the PI) and X coincides with the back tangent. Coordinates of the radius point in this system are $X_0 = -R \tan(I/2)$ and $Y_0 = R$. From observations of distance PV and angle θ the coordinates X_P and Y_P of point P can be determined. Then the following equation for a circle, obtained by substitution into Equation (11.9), can be written as:

$$R^2 = \left(X_P + R \tan \frac{I}{2}\right)^2 + (Y_P + R)^2 \qquad (24.23)$$

With X_P, Y_P, and $I/2$ known, a solution for R can be found. The equation is quadratic and can be solved using Equation (11.3).

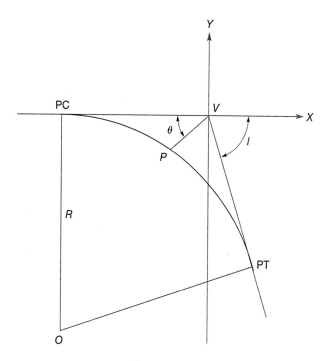

Figure 24.15
Passing a circular curve through a point.

24.16.2 Intersection of a Circular Curve and a Straight Line

Another frequently encountered problem involving curve computation is the determination of the intersection point of a circular curve and a straight line. An example is illustrated in Figure 11.6. In typical cases, coordinates X_A, Y_A, X_B, and Y_B are known, as well as R. Procedures for solving the problem are outlined in Section 11.5, and a worked example is presented.

24.16.3 Intersection of Two Circular Curves

Figure 11.7 illustrates another common problem: computing the intersection point of two circular curves. This can be handled by coordinate geometry, as discussed in Section 11.6. Coordinates X_A, Y_A, X_B, and Y_B are typically determined through survey, and R_1 and R_2 are selected based on design or topographic constraints. A worked example is given in Section 11.6.

The problems described in this section and the preceding one arise most often in the design of subdivisions and interchanges, and in calculating right-of-way points along highways and railroads.

24.17 COMPOUND AND REVERSE CURVES

Compound and reverse curves are combinations of two or more circular curves. They should be used only for low-speed traffic routes, and in terrain where simple curves cannot be fitted to the ground without excessive construction costs since the rapid change in curvature causes unsafe driving conditions. Special formulas have been derived to facilitate computations for such curves and are demonstrated in texts on route surveying. A compound curve can be staked with instrument

24.18 SIGHT DISTANCE ON HORIZONTAL CURVES

Highway safety requires certain minimum sight distances in zones where passing is permitted, and in nonpassing areas to assure a reasonable stopping distance if there is an object on the roadway. Specifications and tables list suitable values based on vehicular speeds, the perception and reaction times of an average individual, the braking distance for a given coefficient of friction during deceleration, and type and condition of the pavement.

A minimum stopping sight distance of 450 to 550 ft is desirable for a speed of 55 mi/h. An approximate formula for determining the available horizontal sight distance on a circular curve can be derived from Figure 24.16, in which the clear sight distance past an obstruction is the length of the long chord AS, denoted by C; and the required clearance is the middle ordinate PM, denoted by m. Then in similar triangles SPG and SOH

$$\frac{m}{SP} = \frac{SP/2}{R} \text{ and } m = \frac{(SP)^2}{2R}$$

Usually m is small compared with R, and SP may be assumed equal to $C/2$. Then

$$C = \sqrt{8mR} \qquad (24.24)$$

If distance m from the centerline of a highway to the obstruction is known or can be measured, the available sight distance C is calculated from the formula. Actually cars travel on either the inside or the outside lane, so sight distance AS is not exactly the true stopping distance, but the computed length is on the safe side and satisfactory for practical use. If the calculation reveals a sight distance restriction, the obstruction might possibly be removed, or a safe speed limit posted in the area.

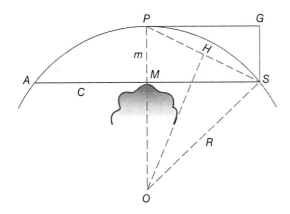

Figure 24.16
Horizontal sight distance on a circular curve.

24.19 SPIRALS

As noted in Section 24.1, spirals are used to provide gradual transitions in horizontal curvature. Their most common use is to connect straight sections of alignment with circular curves, thereby lessening the sudden change in direction that would otherwise occur at the point of tangency. Since spirals introduce curvature gradually, they afford the logical location for introducing superelevation to offset the centrifugal force experienced by vehicles entering curves.

24.19.1 Spiral Geometry

Figure 24.17 illustrates the geometry of spirals connecting tangents with a circular curve of radius R and degree of curvature D. The entrance spiral at the left begins on the back tangent at the *TS* (*tangent to spiral*) and ends at the *SC* (*spiral to curve*). The circular curve runs from the *SC* to the beginning of the exit spiral at the *CS* (*curve to spiral*), and the exit spiral terminates on the forward tangent at the *ST* (*spiral to tangent*).

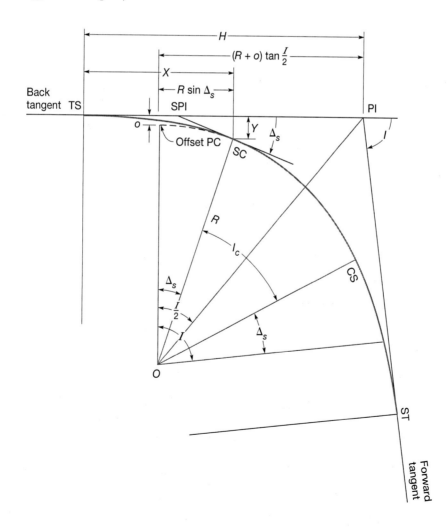

Figure 24.17
Spiral geometry.

768 HORIZONTAL CURVES

The entrance and exit spirals are geometrically identical. Their length L_S, is the arc distance from the *TS* to the *SC*, or *CS* to *ST*. The designer selects this length to provide sufficient distance for introducing the curve's superelevation.

If a tangent to the entrance spiral (and curve) at the SC is projected to the back tangent, it locates the *spiral point of intersection* (SPI). The angle at the SPI between the two tangents is the spiral angle Δ_S. From the basic property of a spiral, that is, *its radius changes uniformly from infinity at the TS to the radius of the circular curve at the SC*, it follows that the spiral's degree of curve changes uniformly from $0°$ at the *TS* to *D* at the *SC*. Since the change is uniform, the average degree of curve over the spiral's length is $D/2$. Thus, from the definition of degree of curve, spiral angle Δ_S is

$$\Delta_S = L_S \frac{D}{2} \tag{24.25}$$

where L_S is in stations and Δ_S and D are in degrees.

Assume in Figure 24.18 that *M* is at the midpoint of the spiral, so its distance from the *TS* is $L_S/2$. If this reasoning is continued, the degree of curvature at *M* is $D/2$, the average degree of curvature from the *TS* to *M* is $(D/2)/2$ and the spiral angle Δ_M is

$$\Delta_M = \frac{L_S}{2} \frac{D}{4} = \frac{L_S D}{8} \tag{a}$$

Solving for *D* in Equation (24.25) and substituting into Equation (a) gives

$$\Delta_M = \frac{L_S}{8} \frac{2\Delta_S}{L_S} = \frac{\Delta_S}{4} \tag{b}$$

According to Equation (b), at $L_S/2$, the spiral angle is $\Delta_S/4$. This demonstrates another basic property of a spiral: *spiral angles at any point are proportional to the square of the distance from TS to the point*, or

$$\Delta_P = \left(\frac{L_P}{L_S}\right)^2 \Delta_S \tag{24.26}$$

where Δ_P is the spiral angle at any point *P* whose distance from the TS is L_P.

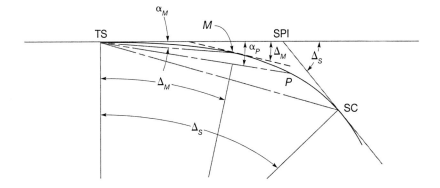

Figure 24.18
Spiral Deflection Angles.

24.19.2 Spiral Calculation and Layout

Deflection angles and chords in a manner similar to that used for circular curves can be used to lay out spirals. In Figure 24.18, α_P is the deflection angle from the TS to any point P. From calculus it can be shown that for the gradual spirals used on transportation routes, deflection angles are very nearly one-third their corresponding spiral angles. Thus

$$\alpha_P = \left(\frac{L_P}{L_S}\right)^2 \frac{\Delta_S}{3} \tag{24.27}$$

In Equation (24.27), L_P is the distance from the TS to P, which is simply the difference in stationing from the TS to point P.

In Figure 24.17, coordinates X and Y give the position of the SC. In this coordinate system, the origin is at the TS, and the X-axis coincides with the back tangent. Approximate formulas for computing X and Y are

$$X = L_S(100 - 0.0030462\ \Delta_S^2)\ (\text{ft}) \tag{24.28}$$

$$Y = L_S(0.58178\Delta_S - 0.000012659\ \Delta_S^3)\ (\text{ft}) \tag{24.29}$$

where L_S is in stations and Δ_S is in degrees. More accurate formulas for computing X and Y coordinates of any station, P, which is a distance L_P along the spiral, are

$$X_P = L_P\left(1 - \frac{\Delta_P^2}{10} + \frac{\Delta_P^4}{216} - \frac{\Delta_P^6}{9360} + \frac{\Delta_P^8}{685,440}\right)$$

$$Y_P = L_P\left(\frac{\Delta_P}{3} - \frac{\Delta_P^3}{42} + \frac{\Delta_P^5}{1320} - \frac{\Delta_P^7}{75,600} + \frac{\Delta_P^9}{6,894,720}\right) \tag{24.30}$$

where Δ_P, defined in Equation (24.26), is expressed in radian units and L_P is the stationing distance, in either feet or meters, from the TS to point P. If L_S and Δ_S are substituted for L_P and Δ_P, respectively, in Equations (24.30), then coordinates X and Y of the SC will result. Equations (24.30) can be used to compute spiral coordinates in either metric or English units. The spiral could be staked by offsets using these coordinates, or they could be used with Equations (11.4) and (11.5) to calculate the deflection angles and chord distances necessary to stake out the spiral.

When a spiral is inserted ahead of a circular curve, as illustrated in Figure 24.17, the circular curve is shifted inward by an amount o, which is known as the *throw*. This can be visualized by constructing the circular curve back from the SC to the offset PC (the point where a tangent to the curve is parallel to the back tangent). The perpendicular distance from the offset PC to the back tangent is the throw, which from Figure 24.17 is

$$o = Y - R(1 - \cos \Delta_S) \tag{24.31}$$

To lay out a spiral in the field, distance H of Figure 24.17 from PI to TS is measured back along the tangent to locate the TS. Then the SC can be established by laying out X and Y. From the geometry of Figure 24.17 distance H is

770 HORIZONTAL CURVES

$$H = X - R \sin \Delta_S + (R + o) \tan \frac{I}{2} \qquad (24.32)$$

The following example will illustrate the computations required in laying out a spiral by the deflection-angle method.

■ Example 24.6

A spiral of 300-ft length is used for transition into a 3°00′ circular curve (arc definition). Angle I at the PI station of 20 + 00 is 60°00′. Compute and tabulate the deflection angles and chords necessary to stake out this spiral at half stations.

Solution

By Equation (24.3) $R = 5729.58/3.00 = 1909.86$ ft
By Equation (24.25) $\Delta_S = 3(3.00)/2 = 4.5° = 4°30′$
By Equations (24.28) and (24.29)

$$X = 3[100 - 0.0030462(4.5)^2] = 299.81 \text{ ft}$$

$$Y = 3[0.58178(4.5) - 0.000012659(4.5)^3] = 7.86 \text{ ft}$$

[*Note:* These same results can be obtained using Equations (24.30), with $L_P = 300.00$ ft, and $\Delta_P = (4.5° \times \pi/180°)$.]

By Equation (24.31)

$$o = 7.86 - 1909.86 \, (1 - \cos 4°30′) = 1.97 \text{ ft}$$

By Equation (24.32)

$$H = 299.81 - 1909.86 \sin 4°30′ + (1909.86 + 1.97) \tan 30° = 1253.75 \text{ ft}$$

Calculate stationing

$$
\begin{aligned}
\text{PI station} &= 20 + 00.00 \\
-H &= -12 + 53.75 \\
\text{TS station} &= 7 + 46.25 \\
+L_S &= 3 + 00.00 \\
\text{SC station} &= 10 + 46.25
\end{aligned}
$$

The deflection angles calculated using Equation (24.27) are listed in column (3) of Table 24.7. The values for L_P used in the equation are given in column (2). If a total station is used for stakeout with a setup at the *TS*, the total chords of column (2) are used with the deflection angles of column (3). If a theodolite and tape were used, the deflection angles, and incremental chords listed in column (4) would be used. The chords in columns (2) and (4) are simply station differences between points on the spiral, and are nearly exact for the relatively flat curvature of highway and railroad spirals. After progressing through the spiral, the *SC* is finally staked. The falling between its position and the *SC* set by coordinates should be measured, the precision calculated, and a decision made to accept or repeat the work.

TABLE 24.7 DATA FOR STAKING SPIRAL OF EXAMPLE 24.6

Station (1)	Total Chord Distance from TS (ft) (2)	Deflection Angle (3)	Incremental Chord (ft) (4)
10 + 46.25 (SC)	300.00	1°30.0'	46.25
10 + 00	253.75	1°04.4'	50.00
9 + 50	203.75	0°41.5'	50.00
9 + 00	153.75	0°23.6'	50.00
8 + 50	103.75	0°10.8'	50.00
8 + 00	53.75	0°02.9'	50.00
7 + 50	3.75	0°00.0'	3.75
7 + 46.25 (TS)			

To continue staking the alignment, the instrument is moved forward to the SC. With $2\Delta_S/3$ set on the horizontal circle, a backsight is taken on the TS and the line of sight plunged.[1] Turning the instrument to 0°00'00" orients the line of sight tangent to the circular curve, ready for laying off deflection angles. The circular curve is computed and laid out by methods given in earlier sections of this chapter, except that its intersection angle I_c, as illustrated in Figure 24.17, is given by

$$I_c = I - 2\Delta_S \qquad (24.33)$$

When staking reaches the circular curve's end, the CS is set. The exit spiral is calculated by the same methods described for the entrance spiral and laid out by staking back from the ST.

Spirals can be computed and staked by several different methods. In this brief treatment, only one commonly applied procedure has been discussed. Students interested in further study of spirals can consult a text on route surveying listed in the bibliography at the end of this chapter.

■ 24.20 COMPUTATION OF "AS-BUILT" CIRCULAR ALIGNMENTS

Most highways that exist today were carefully designed, and then constructed according to plan. Therefore, their centerlines "as-built" are precisely known and coordinates for critical points on their alignments are on file for future use. However, some roads have their origins from "cartways" that through the years were periodically upgraded and improved in place. Thus, it is possible that no formal plans or records of their alignments exist. Yet the boundary lines of adjoining properties may be referenced to the centerline, and thus it becomes

---

[1]From Equation (24.27), the angle at the TS in triangle TS-SPI-SC of Figure 24.18 is $\Delta_S/3$. Also from Figure 24.18, the angle at the SPI in this triangle is $180° - \Delta_S$. It follows therefore that with the instrument at the SC, after backsighting the TS, the angle that must be turned to get the line-of-sight tangent to the spiral and circular curve is $2\Delta_S/3$.

important for it to be precisely established. Also, it is sometimes desirable to determine the parameters of an as-built roadway to check adherence to contract specifications. In these cases, the coordinates of critical points on the approximate centerline of the facility, both on the curves and on the tangents, must be determined. One further important application of this type of problem relates to railroad abandonment programs. Here the rails have served for years as monuments for delineating right of way lines. Therefore, before their removal it is important to obtain the coordinates of important points along their alignments for use in future work related to establishing property lines adjoining the railroad right of way.

The procedure of establishing the coordinates of an existing or as-built alignment is illustrated in Figure 24.19. In the figure, assume that a traverse was used to establish coordinates for points A through F along the existing alignment, as shown. From points B thru E, a least-squares fit of points to Equation (11.10) can be performed, which will establish the coordinates of the center point O and the radius R of the circle. In this example, the matrices of coefficients, unknowns, and observations are

$$A = \begin{bmatrix} 2X_B & 2Y_B & -1 \\ 2X_C & 2Y_C & -1 \\ 2X_D & 2Y_D & -1 \\ 2X_E & 2Y_E & -1 \end{bmatrix} \quad X = \begin{bmatrix} X_O \\ Y_O \\ f \end{bmatrix} \quad L = \begin{bmatrix} X_B^2 + Y_B^2 \\ X_C^2 + Y_C^2 \\ X_D^2 + Y_D^2 \\ X_E^2 + Y_E^2 \end{bmatrix} \quad (24.34)$$

Using Equation (16.6) or (16.7), the least-squares solution for X_O, Y_O, and f are determined. The radius R is computed using

$$R = \sqrt{X_O^2 + Y_O^2 - f} \quad (24.35)$$

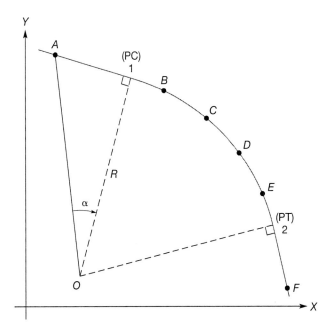

Figure 24.19
Geometry of an "as-built" survey.

Example 24.7

In reference to Figure 24.19, the following coordinates were determined during an "as-built" survey. What are the defining parameters for the curve, and the coordinates of the PC and PT?

Point	X (ft)	Y (ft)	Point	X (ft)	Y (ft)
A	5354.86	7444.14	D	8084.03	6071.29
B	6975.82	6947.93	E	8431.38	5542.00
C	7577.11	6572.60	F	8877.96	4268.90

Solution

Substituting the coordinate values for points B thru E into Equation (24.34) yield the matrices A and L as

$$A = \begin{bmatrix} 2(6975.82) & 2(6947.93) & -1 \\ 2(7577.11) & 2(6572.60) & -1 \\ 2(8084.03) & 2(6071.29) & -1 \\ 2(8431.38) & 2(5542.00) & -1 \end{bmatrix} = \begin{bmatrix} 13{,}951.64 & 13{,}895.86 & -1 \\ 15{,}154.22 & 13{,}145.20 & -1 \\ 16{,}168.06 & 12{,}142.58 & -1 \\ 16{,}862.76 & 11{,}084.00 & -1 \end{bmatrix}$$

$$L = \begin{bmatrix} 6975.82^2 + 6947.93^2 \\ 7577.11^2 + 6572.60^2 \\ 8084.03^2 + 6071.29^2 \\ 8431.38^2 + 5542.00^2 \end{bmatrix} = \begin{bmatrix} 96{,}935{,}795.96 \\ 100{,}611{,}666.71 \\ 102{,}212{,}103.30 \\ 101{,}801{,}932.70 \end{bmatrix}$$

Substituting the A and L matrices into Equation (16.6) yields the solution

$$X_O = 5587.375$$
$$Y_O = 4053.992$$
$$f = 37{,}351{,}007.416$$

And finally by Equation (24.35), the radius of the circle is

$$R = \sqrt{(5587.375)^2 + (4053.992)^2 - 37{,}351{,}007.416} = 3209.766$$

Figure 24.20 shows the data file and the results of the least-squares fit of points to a circle from the software WOLFPACK, which is available on the companion website.

From this solution, the coordinates of the PC can be determined by computing the length and azimuth of AO of Figure 24.19 using Equations (11.4) and (11.5), and then solving right triangle $AO1$ for angle α using the cosine function. For this example, the values are

By Equations (11.4) and (11.5) and

$$OA = \sqrt{(5354.86 - 5587.375)^2 + (7444.14 - 4053.992)^2}$$

$$Az_{OA} = \tan^{-1}\left(\frac{5354.86 - 5587.375}{7444.14 - 4053.992}\right) + 360° = 356°04'35''$$

774 HORIZONTAL CURVES

Figure 24.20
Data file and adjustment results for Example 24.7 using WOLFPACK.

```
Data file for Example 24.7 using WOLFPACK
B 6975.82 6947.93 **Point id, X, Y
C 7577.11 6572.60
D 8084.03 6071.29
E 8431.38 5542.00

Results of Least Squares Adjustment of Points
*****   Least Squares Fit Of Circle   *****
Center of Circle at: X = 5,587.375
                     Y = 4,053.992
With A Radius of:        3,209.766

Misfit of Points to Best Fit Curve
Station     Distance From Curve
====================
B                0.009
C               -0.026
D                0.028
E               -0.010
```

From the cosine function

$$\alpha = \cos^{-1}\left(\frac{3209.766}{3398.11}\right) = 19°09'56''$$

From Figure 24.19, the azimuth of line $O1$ is

$$Az_{O1} = 356°04'35'' + 19°09'56'' - 360° = 15°14'31''$$

Now the coordinates of the PC can be determined using the following equations

$$X_{PC} = X_O + R \sin Az_{O1} = 5587.375 + 843.838 = 6431.21$$

$$Y_{PC} = Y_O + R \cos Az_{O1} = 4053.992 + 3096.859 = 7150.85$$

In similar fashion, the coordinates for the PT can be computed. From the azimuths of lines $O1$ and $O2$, the intersection angle can be determined using Equation (11.11). Additionally, the long chord can be determined using Equation (11.4). From these parameters, the tangent, external, and middle ordinate distances can be computed using Equations (24.4), (24.8), and (24.9), respectively. These steps are left as an exercise for the student.

■ 24.21 SOURCES OF ERROR IN LAYING OUT CIRCULAR CURVES

Some sources of error in curve layout are:

1. Errors in setting up, leveling, and reading the instrument.
2. Total station (or theodolite) out of adjustment.
3. Bull's-eye bubble out of adjustment on prism pole used for stakeout with a total station.
4. Measurement errors in laying out angles and distances.

24.22 MISTAKES

Typical mistakes that occur in laying out a curve in the field are:

1. Failure to take equal numbers of direct and reversed measurements of the deflection angle at the PI before computing or laying out the curve.
2. Using 100.00-ft chords to lay out arc-definition curves having D greater than 2°.
3. Taping subchords of nominal length for chord-definition curves having D greater than 5° (a nominal 50-ft subchord for a 6° curve requires a measurement of 50.02 ft).
4. Failure to check curve points after staking them using either the total chord or the coordinate method.
5. A mistake in the backsight when staking by the coordinate method.
6. Failure to stake the PT independently by measuring the tangent distance forward from the PI.
7. Incorrect orientation of the instrument's horizontal circle.
8. Failure to properly establish the setup information in an automatic data collector.
9. Misidentification of a station, or stations.

PROBLEMS

Asterisks (*) indicate problems that have answers given in Appendix G.

24.1 What is the purpose of a horizontal curve?
24.2 What is a compound curve?
24.3 What is an easement curve?
24.4 How can the effect of centrifugal force on a vehicle be balanced?
24.5 For the following circular curves having a radius R, what is their degree of curvature by (1) arc definition and (2) chord definition?
*(a) 500.00 ft
(b) 900.00 ft
(c) 2500.00 ft

Compute L, T, E, M, LC, R, and stations of the PC and PT for the circular curves in Problems 24.6 through 24.8. Use the chord definition for the railroad curve and the arc definition for the highway curves.

24.6* Railroad curve with $D_c = 4°00'$, $I = 24°00'$, and PI station = 36 + 45.00 ft.
24.7 Highway curve with $D_a = 4°00'$, $I = 15°30'$, and PI station = 15 + 22.00 ft.
24.8 Highway curve with $R = 500.000$ m, $I = 28°00'$, and PI station = 1 + 984.000 m.

Tabulate R or D, T, L, E, M, PC, PT, deflection angles, and incremental chords to lay out the circular curves at full stations (100 ft or 30 m) in Problems 24.9 through 24.14.

24.9 Highway curve with $D_a = 5°00'$, $I = 20°30'$, and PI station = 63 + 25.53 ft.
24.10 Railroad curve with $D_c = 2°30'$, $I = 12°00'$, and PI station = 154 + 26.80 ft.
24.11 Highway curve with $R = 800$ m, $I = 20°00'$, and PI station = 3 + 290.600 m.
24.12 Highway curve with $R = 1000$ m, $I = 16°30'$, and PI station = 3 + 264.759 m.
24.13 Highway curve with $R = 1000$ ft, $I = 25°00'$, and PI station = 26 + 40.00 ft.

24.14 Highway curve with $R = 270$ m, $I = 15°00'$, and PI station $= 6 + 500.000$ m.

In Problems 24.15 through 24.18 tabulate the curve data, deflection angles, and incremental chords needed to lay out the following circular curves at full-station increments using a total station instrument set up at the PC.

24.15 The curve of Problem 24.9.
24.16 The curve of Problem 24.10.
24.17 The curve of Problem 24.11.
24.18 The curve of Problem 24.12.
24.19 A rail line on the center of an 80-ft street makes a 75°24′ turn into another street of equal width. The corner curb line has $R = 12$ ft. What is the largest R that can be given a circular curve for the track centerline if the law requires it to be at least 10 ft from the curb?

Tabulate all data required to lay out by deflection angles and incremental chords, at the indicated stationing, for the circular curves of Problems 24.20 and 24.21.

24.20 A highway survey PI falls in a pond, so a cut off line $AB = 275.21$ ft is run between the tangents. In the triangle formed by points A, B, and PI, the angle at $A = 16°28'$ and at $B = 22°16'$. The station of A is $54 + 92.30$. Calculate and tabulate curve notes to run, by deflection angles and incremental chords, a 4°30′ (arc definition) circular curve at full-station increments to connect the tangents.

24.21 In the figure, a single circular highway curve (arc definition) will join tangents XV and VY and also be tangent to BC. Calculate R, L, and the stations of the PC and PT.

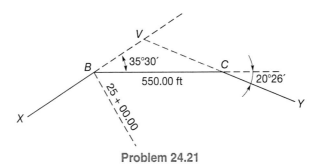

Problem 24.21

24.22* Compute R_x to fit requirements of the figure and make the tangent distances of the two curves equal.

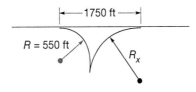

Problem 24.22

24.23 After a backsight on the PC with 0°00′ set on the instrument, what is the deflection angle to the following circular curve points?
   ***(a)** Setup at curve midpoint, deflection to the PT.
   (b) Instrument at curve midpoint, deflection to 3/4 point.
   (c) Setup at 1/4 point of curve, deflection to 3/4 point.
24.24 In surveying a construction alignment, why should the I angle be measured using both faces of the instrument?

24.25 A highway curve (arc definition) to the right, having $R = 500$ m and $I = 15°30'$, will be laid out by coordinates with a total station instrument setup at the PI. The PI station is $3 + 855.200$ m, and its coordinates are $X = 75,428.863$ m and $Y = 36,007.434$ m. The azimuth (from north) of the back tangent proceeding toward the PI is $45°00'00''$. To orient the total station, a backsight will be made on a POT on the back tangent. Compute lengths and azimuths necessary to stake the curve at 30-m stations.

24.26 In Problem 24.25, compute the XY coordinates at 30-m stations.

24.27 An exercise track must consist of two semicircles and two tangents and be exactly 500 m along its centerline. The two tangents are 150.000 m each. Calculate the radius for the curves.

What sight distance is available if there is an obstruction on a radial line through the PI inside the curves in Problems 24.28 and 24.29?

24.28* For Problem 24.9, obstacle 15 ft from curve.

24.29 For Problem 24.11, obstacle 10 m from curve.

24.30 If the misclosure for the curve of Problem 24.9, computed as described in Section 24.8, is 0.10 ft, what is the field layout precision?

24.31 Assume that a 50-ft entry spiral will be used with the curve of Problem 24.9. Compute and tabulate curve notes to stake out the alignment from the TS to ST at full stations using a total station and the deflection-angle, total chord method.

24.32 Same as Problem 24.31, except use a 150-ft spiral for the curve of Problem 24.10.

24.33 Same as Problem 24.31, except for the curve of Problem 24.11, with a 50-m entry spiral using stationing of 30 m and a total station instrument.

24.34 Compute the area bounded by the two arcs and tangent in Problem 24.22.

24.35 In an as-built survey, the XY coordinates in meters of three points on the centerline of a highway curve are determined to be A: (3770.52, 4913.84); B: (3580.80, 4876.37); C: (3399.27, 4809.35). What are the radius, and coordinates for the center of the curve in meters?

24.36 In Problem 24.35, if the (x, y) coordinates in meters of two points on the centerline of the tangents are (3042.28, 4616.77) and (4435.66, 4911.19), what are the coordinates of the PC, PT, and the curve parameters L, T, and I?

BIBLIOGRAPHY

American Association of State Highway and Transportation Officials. 2001. *Guidelines for Geometric Design of Very-Low Volume Local Roads (ADT≤400)*. Washington, DC: AASHTO.

American Association of State Highway and Transportation Officials. 2004. *A Policy on Geometric Design of Highways and Streets*. Washington, DC: AASHTO.

Battjes, N., et al. 2007. "Railroad Surveys: History and Curve Computations." *Surveying and Land Information Science* 67 (No. 3): 137.

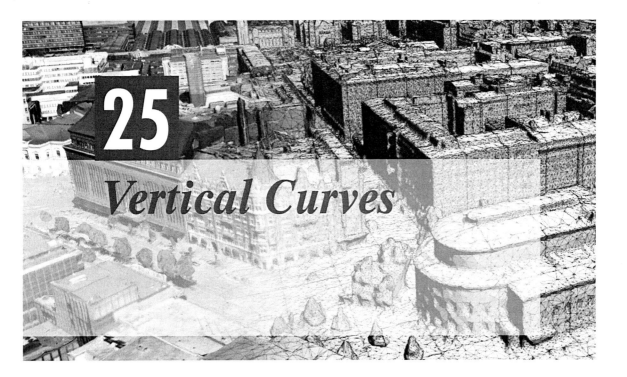

25 Vertical Curves

■ 25.1 INTRODUCTION

Curves are needed to provide smooth transitions between straight segments (tangents) of grade lines for highways and railroads. Because these curves exist in vertical planes, they are called vertical curves. An example is illustrated in Figure 25.1, which shows the profile view of a proposed section of highway to be constructed from A to B. A grade line consisting of three tangent sections has been designed to fit the ground profile. Two vertical curves are needed: curve a to join tangents 1 and 2, and curve b to connect tangents 2 and 3. The function of each curve is to provide a gradual change in grade from the initial (back) tangent to the grade of the second (forward) tangent. Because parabolas provide a constant rate of change of grade, they are ideal and almost always applied for vertical alignments used by vehicular traffic.

Two basic types of vertical curves exist, *crest* and *sag*. These are illustrated in Figure 25.1. Curve a is a crest type, which by definition undergoes a negative change in grade; that is, the curve turns downward. Curve b is a sag type, in which the change in grade is positive and the curve turns upward.

There are several factors that must be taken into account when designing a grade line of tangents and curves on any highway or railroad project. They include (1) providing a good fit with the existing ground profile, thereby minimizing the depths of cuts and fills, (2) balancing the volume of cut material against fill, (3) maintaining adequate drainage, (4) not exceeding maximum specified grades, and (5) meeting fixed elevations such as intersections with other roads. In addition, the curves must be designed to (a) fit the grade lines they connect, (b) have lengths sufficient to meet specifications covering a maximum rate of change of grade (which affects the comfort of vehicle occupants), and (c) provide sufficient sight distance for safe vehicle operation (see Section 25.11).

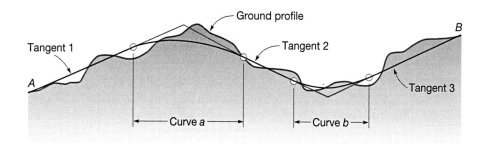

Figure 25.1
Grade line and ground profile of a proposed highway section.

Elevations at selected points (e.g., full or half stations in the English system of stationing, or 20, 30, or 40 m in the metric system) along vertical parabolic curves are usually computed by the *tangent-offset* method. It is simple, straightforward, conveniently performed with calculators and computers, and self-checking. After the elevations of curve points have been computed, they are staked in the field to guide construction operations so the route can be built according to plan.

25.2 GENERAL EQUATION OF A VERTICAL PARABOLIC CURVE

The general mathematical expression of a parabola, with respect to an XY rectangular coordinate system, is given by

$$Y_P = a + bX_P + cX_P^2 \tag{25.1}$$

where Y_P is the ordinate at any point p on the parabola located at distance X_P from the origin of the curve, and a, b, and c are constants. Figure 25.2 shows a parabola in a XY rectangular coordinate system, and illustrates the physical significance

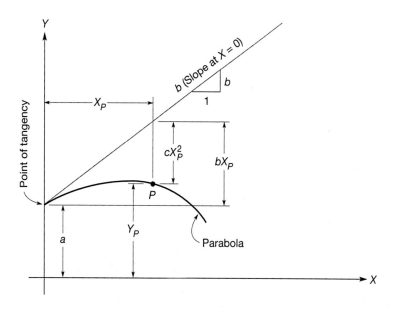

Figure 25.2
Terms for a parabola.

of the terms in Equation (25.1). Note from the figure that a is the ordinate at the beginning of the curve where $X = 0$, b is the slope of a tangent to the curve at $X = 0$, bX_P the change in the ordinate along the tangent over distance X_P, and cX_P^2 the parabola's departure from the tangent (*tangent offset*) in distance X_P. When the terms a, bX_P, and cX_P^2 are combined as in Equation (25.1) and shown in Figure 25.2, they produce Y_P, the elevation on curve at X_P. For the crest curve of Figure 25.2, b has positive algebraic sign and c is negative.

■ 25.3 EQUATION OF AN EQUAL TANGENT VERTICAL PARABOLIC CURVE

Figure 25.3 shows a parabola that joins two intersecting tangents of a grade line. The parabola is essentially identical to that in Figure 25.2, except that the terms used are those commonly employed by surveyors and engineers. In the figure, BVC denotes the beginning of vertical curve, sometimes called the *VPC* (vertical point of curvature); V is the vertex, often called the VPI (vertical point of intersection); and EVC denotes the end of vertical curve, interchangeably called the *VPT* (vertical point of tangency). The percent grade of the back tangent (straight segment preceding V) is g_1, that of the forward tangent (straight segment following V) is g_2. The curve length L is the horizontal distance (in stations) from the BVC to the EVC. The curve of Figure 25.3 is called equal tangent because the horizontal distances from the BVC to V and from V to the EVC are equal, each being $L/2$. Proof of this is given in Section 25.5.

On the XY axis system in Figure 25.3, X values are horizontal distances measured from the BVC, and Y values are elevations measured from the vertical datum of reference. By substituting this surveying terminology into Equation (25.1), the parabola can be expressed as

$$Y = Y_{BVC} + g_1 X + cX^2 \qquad (25.2)$$

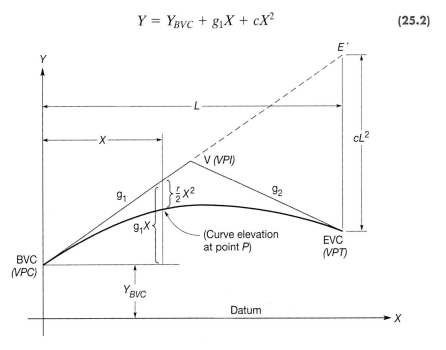

Figure 25.3
Vertical parabolic curve relationships.

In Equation (25.2), if the English system of units is used, Y is in ft, g_1 is percent grade, and for consistency of units, X must therefore be in 100-ft stations. If the metric system is used, Y is in meters, g_1 is again in percent grade, and thus X must be in units of 100 m, or 1/10th stations, where full stations are 1 km apart. The correspondence of terms in Equation (25.2) to those of Equation (25.1) is $a = Y_{BVC}$ (elevation of BVC) and $bX_P = g_1X$ (change in elevation along the back tangent with increasing X). To express the constant c of Equation (25.2) in surveying terminology, consider the tangent offset from E' on the extended back tangent to the EVC (dashed line in Figure 25.3). Its value, which is negative for the crest curve shown, is cL^2 where L, the horizontal distance from BVC to E', is substituted for X. From the figure, cL^2 can be expressed in terms of horizontal lengths (in stations) and percent grades as follows:

$$cL^2 = g_1\left(\frac{L}{2}\right) + g_2\left(\frac{L}{2}\right) - g_1 L \quad \text{(a)}$$

Solving Equation (a) for constant c gives

$$c = \frac{g_2 - g_1}{2L} \quad \text{(b)}$$

Substituting Equation (b) into Equation (25.2) results in the following equation for an equal-tangent vertical curve in surveying terminology:

$$Y = Y_{BVC} + g_1 X + \left(\frac{g_2 - g_1}{2L}\right)X^2 \quad \textbf{(25.3)}$$

The *rate of change of grade*, r, for an equal-tangent parabolic curve equals the total grade change from BVC to EVC divided by length L (in stations for the English system, or 1/10th stations for metric units), over which the change occurs, or

$$r = \frac{g_2 - g_1}{L} \quad \textbf{(25.4)}$$

As mentioned earlier, the value r, which is negative for a crest curve and positive for a sag type, is an important design parameter because it controls the rate of curvature and hence rider comfort. To incorporate it in the equation for parabolic curves, Equation (25.4) is substituted into Equation (25.3)

$$Y = Y_{BVC} + g_1 X + \left(\frac{r}{2}\right)X^2 \quad \textbf{(25.5)}$$

Figure 25.3 illustrates how the terms of Equation (25.5) combine to give the curve elevation at point P. Because the last term of the equation is the curve's offset from the back tangent, this formula is commonly called the *tangent-offset equation*.

25.4 HIGH OR LOW POINT ON A VERTICAL CURVE

To investigate drainage conditions, clearance beneath overhead structures, cover over pipes, or sight distance, it may be necessary to determine the elevation and location of the low (or high) point on a vertical curve. At the low or high point, a tangent to the curve will be horizontal, and its slope equal to zero. Based on this fact, by taking the derivative of Equation (25.3), and setting it equal to zero, the following formula is readily derived:

$$X = \frac{g_1 L}{g_1 - g_2} \quad (25.6)$$

where X is the distance from the BVC to the high or low point of the curve (in full stations in the English system of units, and in 1/10th stations in the metric system), g_1 the tangent grade through the BVC, g_2 the tangent grade through the EVC, and L the curve length (in stations or 1/10th stations).

If g_2 is substituted for g_1 in the numerator of Equation (25.6), distance X is measured back from the EVC. By substituting Equation (25.4) into Equation (25.6), the following alternate formula for locating the high or low point results:

$$X = \frac{-g_1}{r} \quad (25.7)$$

25.5 VERTICAL CURVE COMPUTATIONS USING THE TANGENT-OFFSET EQUATION

In designing grade lines, the locations and individual grades of the tangents are normally selected first. This produces a series of intersection points V, each defined by its station and elevation. A curve is then chosen to join each pair of intersecting tangents. The parameter selected in vertical-curve design is length L. Having chosen it, the station of the BVC is obtained by subtracting $L/2$ from the vertex station. Adding L to the BVC station then determines the EVC station.

Stationing for the points on curve that are computed and staked are those that are *evenly divisible by the selected staking increment*. Thus, if a full station is the staking increment selected for a curve in the English system of units, each full station, that is, $10 + 00, 11 + 00, 12 + 00$, etc., would be staked, but $10 + 50, 11 + 50$, etc., would not be staked. For example, if the staking increment is half stations in the English system, then $15 + 00, 15 + 50, 16 + 00, 16 + 50$, and so on, would be staked, and not $15 + 25, 15 + 75, 16 + 25, 16 + 75$, etc. In the metric system, if 40 m is the staking increment, then $2 + 400, 2 + 440, 2 + 520$, and so on would be staked, but not $2 + 420, 2 + 460, 2 + 500$, etc.

Computations for vertical parabolic curves are normally done in tabular form.

25.5.1 Example Computations Using the English System of Units

Following are example computations for an equal-tangent vertical curve in the English system of units. The curve is a crest type.

Example 25.1

A grade g_1 of +3.00% intersects grade g_2 of −2.40% at a vertex whose station and elevation are 46 + 70 and 853.48 ft, respectively. An equal-tangent parabolic curve 600 ft long has been selected to join the two tangents. Compute and tabulate the curve for stakeout at full stations. (Figure 25.4 shows the curve.)

Solution

By Equation (25.4)

$$r = \frac{-2.40 - 3.00}{6} = -0.90\% \text{ station}$$

Stationing

$$V = 46 + 70$$
$$-L/2 = \underline{3 + 00}$$
$$BVC = 43 + 70$$
$$+L = \underline{6 + 00}$$
$$EVC = 49 + 70$$

$$\text{Elev}_{BVC} = 853.48 - 3.00(3) = 844.48 \text{ ft}$$

The remaining calculations utilize Equation (25.5) and are listed in Table 25.1.

A check on curve elevations is obtained by computing the first and second differences between the elevations of full stations, as shown in the right-hand columns of the table. Unless disturbed by rounding errors, all second differences (rate of change) should be equal. For curves in the English system of units at full-station increments, the second differences should equal r; for half-station increments they should be $r/4$.

It is sometimes desirable to calculate the elevation of the curve's center point. This can be done using $X = L/2$ in Equation (25.5). For Example 25.1, it is

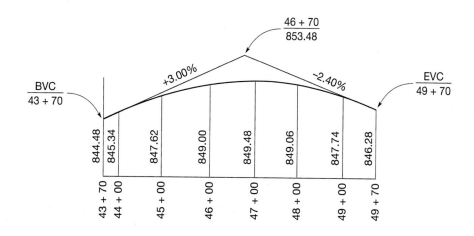

Figure 25.4
Crest curve of Example 25.1.

TABLE 25.1 NOTES FOR CURVE OF EXAMPLE 25.1

Station	X (sta)	$g_1 X$	$\dfrac{rX^2}{2}$	Curve Elevation	First Difference	Second Difference
49 + 70 (EVC)	6.0	18.00	−16.20	846.28		
49 + 00	5.3	15.90	−12.64	847.74	−1.32	−0.90
48 + 00	4.3	12.90	−8.32	849.06	−0.42	−0.90
47 + 00	3.3	9.90	−4.90	849.48	0.48	−0.90
46 + 00	2.3	6.90	−2.38	849.00	1.38	−0.90
45 + 00	1.3	3.90	−0.76	847.62	2.28	
44 + 00	0.3	0.90	−0.04	845.34		
43 + 70 (BVC)	0.0	0.00	−0.00	844.48		

Check: $EVC = ELEV_V - g_2\left(\dfrac{L}{2}\right) = 853.48 - 2.40(3) = 846.28$ ✓

$$Y_{center} = 844.48 + 3.00(3) - \left(\dfrac{0.90}{2}\right)(3)^2 + 849.43 \text{ ft}$$

This can be checked by employing the property of a parabolic curve, which is *the curve center falls halfway between the vertex and the midpoint of the long chord* (line from BVC to EVC). The elevation of the midpoint of the long chord (LC) is simply the average of the elevations of the BVC and EVC. For Example 25.1, it is

$$Y_{midpoint\ LC} = \dfrac{844.48 + 846.28}{2} = 845.38 \text{ ft}$$

By the property just stated, the elevation of the curve center for Example 25.1 is the average of the vertex elevation and that of the midpoint of the long chord, or

$$Y_{center} = \dfrac{845.38 + 853.48}{2} = 849.43 \text{ ft} ✓$$

25.5.2 Example Computations Using the Metric System

The following example illustrates the computations for an equal-tangent vertical curve when metric units are used. The curve is a sag type.

■ Example 25.2

A grade g_1 of −3.629% intersects grade g_2 of 0.151% at a vertex whose station and elevation are 5 + 265.000 and 350.520 m, respectively. An equal-tangent parabolic curve of 240-m length will be used to join the tangents. Compute and tabulate the curve for staking at 40 m increments.

25.5 Vertical Curve Computations Using the Tangent-Offset Equation

TABLE 25.2 NOTES FOR CURVE OF EXAMPLE 25.2

Station	$X\left(\dfrac{m}{100}\right)$	$g_1 X$	$\dfrac{rX^2}{2}$	Curve Elevation (m)	First Difference	Second Difference
5 + 385.000 (EVC)	2.400	−8.710	4.536	350.701		
5 + 360.000	2.150	−7.802	3.640	350.713	−0.223	0.252
5 + 320.000	1.750	−6.351	2.412	350.936	−0.475	0.252
5 + 280.000	1.350	−4.899	1.435	351.411	−0.727	0.252
5 + 240.000	0.950	−3.448	0.711	352.138	−0.979	0.252
5 + 200.000	0.550	−1.996	0.238	353.117	−1.232	
5 + 160.000	0.150	−0.544	0.018	354.348		
5 + 145.000 (BVC)	0.000	−0.000	0.000	354.875		

Check: EVC = ELEV$_{VPI}$ − $g_2\left(\dfrac{L}{2}\right)$ = 350.520 + (0.151 × 120/100) = 350.701 ✓

Solution

By Equation (25.4)

$$r = \dfrac{0.151 + 3.629}{2.4} = 1.575$$

[Note that L used in Equation (25.4) is in units of m/100, or 1/10th stations.]

Stationing

$$\text{VPI Station} = 5 + 265$$
$$-L/2 = 120$$
$$\text{BVC Station} = 5 + 145$$
$$+L = 240$$
$$\text{EVC Station} = 5 + 385$$
$$\text{Elev}_{BVC} = 350.520 + 3.629(120/100) = 354.875 \checkmark$$

The remaining calculations employ Equation (25.5) and are listed in Table 25.2.
Note that the second differences are all equal, which checks the computations. [The value of 0.252 is $r/6.25$, where 6.25 is $(100 \text{ m}/40 \text{ m})^2$.]

■ Example 25.3

Compute the station and elevation of the curve's high point in Example 25.1.

Solution

By Equation (25.7), $X = -3.00/-0.90 = 3.3333$ stations

Then the station of the high point is

$$\text{station}_{\text{high}} = (43 + 70) + (3 + 33.33) = 47 + 03.33$$

By Equation (25.3), the elevation at this point is

$$844.48 + 3.00(3.333) + \frac{-2.40 - 3.00}{2(6)}(3.3333)^2 = 849.48 \text{ ft}$$

Note that in using Equation (25.7) and all other equations of this chapter, correct algebraic signs must be applied to grades g_1 and g_2.

By applying the same equations and procedures, the station and elevation of the low point of the curve of Example 25.2 are $5 + 375.413$ and 350.694 m, respectively. The calculations are left as an exercise.

25.6 EQUAL TANGENT PROPERTY OF A PARABOLA

The curve defined by Equations (25.3) and (25.5) has been called an equal-tangent parabolic curve, which means the vertex occurs a distance $X = L/2$ from the BVC. Proof of this property is readily made with reference to Figure 25.5, which illustrates a sag curve. In the figure, assume the horizontal distance from BVC to V is an unknown value X; thus, the remaining distance from V to EVC is $L - X$. Two expressions can be written for the elevation of the EVC. The first, using Equation (25.3) with $X = L$, yields

$$Y_{\text{EVC}} = Y_{\text{BVC}} + g_1 L + \left(\frac{g_2 - g_1}{2L}\right) L^2 \tag{c}$$

The second, using changes in elevation that occur along the tangents, gives

$$Y_{\text{EVC}} = Y_{\text{BVC}} + g_1 X + g_2(L - X) \tag{d}$$

Equating Equations (c) and (d) and solving, $X = L/2$. Thus, distances BVC to V and V to EVC are equal—hence the term *equal-tangent parabolic curve*.

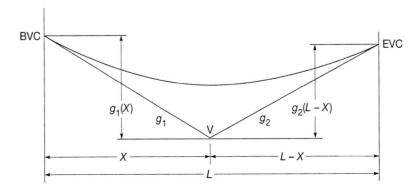

Figure 25.5
Proof of equal-tangent property of a parabola.

25.7 CURVE COMPUTATIONS BY PROPORTION

The following basic property of a parabola can be used to simplify vertical curve calculations: offsets from a tangent to a parabola are proportional to the squares of the distances from the point of tangency. Calculating the offset E at the midpoint of the curve, and then computing offsets at any other distance X from the BVC by proportion according to the following formula conveniently utilize this property:

$$\text{offset}_X = E\left(\frac{X}{\frac{L}{2}}\right)^2 \qquad (25.8)$$

The value of E in Equation (25.8) is simply the difference in elevation from the curve's midpoint to the VPI. Computation of the curve's midpoint elevation was discussed in Section 25.4, and for Example 25.1 it was 849.43 ft. To illustrate the use of Equation (25.8), the offset from the tangent to the curve for station 47 + 00, where $X = 3.3$ stations, will be computed

$$\text{offset}_{47+00} = (849.43 - 853.48)\left(\frac{3.3}{3}\right)^2 = -4.90 \text{ ft}$$

In the previous calculation, 3.3 in the numerator is X in stations, and 3 in the denominator is $L/2$ also in stations. The value -4.90 checks the tangent offset listed in Table 25.1 for station 47 + 00. This simplified procedure is convenient for computing vertical curves in the field with a handheld calculator.

25.8 STAKING A VERTICAL PARABOLIC CURVE

Prior to initiating construction on a route project, the planned centerline, or an offset one, will normally be staked at full or half stations, as well as other critical horizontal alignment points such as PCs and PTs. Then, as described in Section 23.7, slope stakes will be set out perpendicular to the centerline at or near the slope intercepts to guide rough grading. Excavation and embankment construction then proceed and continue until the grade is near plan elevation. The centerline stations are then staked again, this time using sharpened 2-in. square wooden hubs, usually about 10 in. long. These are known as "blue tops," so called because when their tops are driven to grade elevation, they are colored blue. Contractors request blue tops when excavated areas are still slightly high and embankments somewhat low. After blue tops are set to mark the precise grade, final grading is completed.

To set blue tops at grade, a circuit of differential levels is run from a nearby benchmark to establish the *HI* of a leveling instrument in the project area. The difference between the *HI* and any station's grade is the required rod reading on that stake. Assume, for example, a *HI* of 856.20 ft exists, and station 45 + 00 of Example 25.1 is to be set. The required rod reading is 856.20 − 847.62 = 8.58 ft. With the stake initially driven firmly into the ground, and the rod held on its top, suppose a reading of 8.25 is obtained. The stake then must be driven down an additional 8.58 − 8.25 = 0.33 ft further and the notetaker so indicates. After the stake is driven, a distance estimated to be somewhat less than 0.33 ft, the rod reading is checked. This is repeated until the required reading of 8.58 is achieved, whereupon the stake is colored blue using keel or spray paint.

This process is continued until all stakes are set. The required rod reading at station 46 + 00, for an *HI* of 856.20, is 7.20 ft. If a rock is encountered and the stake cannot be driven to grade, a vertical offset of, say, 1.00 ft above grade can be marked and noted on the stake.

When the level is too far away from the station being set, a turning point is established, and the instrument is brought forward to establish a new *HI*. Whenever possible, level circuit checks should be made by closing on nearby benchmarks as blue top work on the project progresses. Also, when quitting for the day, or when the job is finished, the level circuit must always be closed to verify that no mistakes were made.

25.9 MACHINE CONTROL IN GRADING OPERATIONS

As stated in Section 23.12, GNSS methods provide sufficient vertical accuracy for rough grading operations. However, it is not sufficient to provide final grades on most projects. Thus, a GNSS machine control project must be augmented with laser levels to provide vertical accuracies under 3 mm. The laser level requires a sensor on the construction vehicle and a rotating laser (see Figure 23.2). The level must be positioned over a known calibrated point within 1 foot horizontally. The laser covers a radius of about 1500 ft. Continuous line of sight must be maintained between the rotating laser level and the sensor on the construction vehicle.

As stated in Section 23.12, a robotic total station with multifaceted 360° prism (see Figure 8.7) can be used also to establish horizontal and vertical positioning. This system is limited to a range of about 1000 ft from the total station to the construction vehicle. However, it can provide both horizontal and vertical accuracies at a sufficient level for most construction applications. One manufacturer has included combined special laser level with their GNSS unit to provide what is called "mm-GPS." Both of these units provide sufficient vertical accuracy to guide finish grades.

25.10 COMPUTATIONS FOR AN UNEQUAL TANGENT VERTICAL CURVE

An unequal-tangent vertical curve is used to enable the vertical curve to fit closely the ground conditions, which, in turn, minimizes excessive cut or fill quantities. This vertical curve is simply a pair of equal-tangent curves, where the EVC of the first curve is the BVC of the second. This point is called CVC, which represents the point of compound vertical curvature. In Figure 25.6, a −2.00% grade intersects a +1.60% grade at station 87 + 00 and elevation 743.24 ft. A vertical curve of length $L_1 = 400$ ft is to be extended back from the vertex, and a curve of length $L_2 = 600$ ft run in forward, which closely fits the ground conditions.

To perform calculations for this type of curve, connect the midpoints of the tangents for the two curves, stations 85 + 00 and 90 + 00, to obtain line *AB*. Point *A* is the vertex of the first curve and is located a distance of $L_1/2$ back from V. Point *B* is the vertex of the second curve which is $L_2/2$ forward from V. Compute elevations for *A* and *B* and, using them, calculate the grade of *AB* by dividing the difference in elevation between *B* and *A* by the distance in stations separating these two points. From grade *AB*, determine the CVC elevation.

Now compute two equal-tangent vertical curves, one from BVC to CVC and another from CVC to EVC, by the methods of Section 25.4. Since both curves are tangent to the same line AB at point CVC, they will be tangent to each other and form a smooth curve.

■ Example 25.4

For the configuration of Figure 25.6, compute and tabulate the notes necessary to stake the unequal-tangent vertical curve at full stations.

Solution

1. Calculate elevations of BVC, EVC, A, B, and CVC, and grade AB

$$\text{Elev}_{BVC} = 743.24 + 4(2.00) = 751.24 \text{ ft}$$

$$\text{Elev}_A = 743.24 + 2(2.00) = 747.24 \text{ ft}$$

$$\text{Elev}_{EVC} = 743.24 + 6(1.60) = 752.84 \text{ ft}$$

$$\text{Elev}_B = 743.24 + 3(1.60) = 748.04 \text{ ft}$$

$$\text{Grade}_{AB} = \left(\frac{748.04 - 747.24}{5}\right) = +0.16\%$$

$$\text{Elev}_{CVC} = 747.24 + 2(0.16) = 747.56 \text{ ft}$$

These elevations are shown in Figure 25.6.

2. In computing the first curve, the grade of AB will be g_2 in the formulas, and for the second curve it will be g_1. The rates of change of grade for the two curves are, by Equation (25.4)

$$r_1 = \frac{0.16 - (-2.00)}{4} = +0.54\%/\text{station}$$

$$r_2 = \frac{1.60 - 0.16}{6} = +0.24\%/\text{station}$$

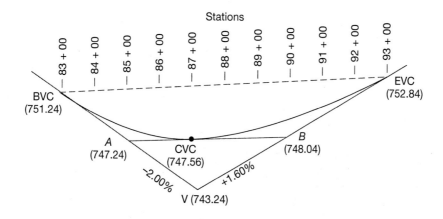

Figure 25.6
Unequal-tangent vertical curve.

TABLE 25.3 Notes for Curve of Example 25.3

Station	X (sta)	$g_1 X$	$rX^2/2$	Curve Elevation	First Difference	Second Difference
93 + 00 (EVC)	6	0.96	4.32	752.84✓	1.48	
92 + 00	5	0.80	3.00	751.36	1.24	0.24
91 + 00	4	0.64	1.92	750.12	1.00	0.24
90 + 00	3	0.48	1.08	749.12	0.76	0.24
89 + 00	2	0.32	0.48	748.36	0.52	0.24
88 + 00	1	0.16	0.12	747.84	0.28	0.24
87 + 00 (CVC)	4	−8.00	4.32	747.56✓	−0.11	
86 + 00	3	−6.00	2.43	747.67	−0.65	0.54
85 + 00	2	−4.00	1.08	748.32	−1.19	0.54
84 + 00	1	−2.00	0.27	749.51	−1.73	0.54
83 + 00 (BVC)	0	0.00	0.00	751.24		

3. Equation (25.5) is now solved in tabular form, and the results are listed in Table 25.3.

Vertical curve computations by themselves are quite simple, hardly a challenge to a modern computer. But when vertical curves are combined with horizontal curves, spirals, and superelevation in complex highway designs, programming saves time. Figure 25.7 shows a completed data entry box for computing the first vertical curve of Example 25.3 using the WOLFPACK software, which is on the companion website. The software employs the equations discussed in this chapter to compute staking notes for a vertical curve. The computed results are written to a file in table form like those given in Tables 25.1, 25.2, and 25.3, so they can be printed and taken into the field for staking purposes.

■ 25.11 DESIGNING A CURVE TO PASS THROUGH A FIXED POINT

The problem of designing a parabolic curve to pass through a point of fixed station and elevation is frequently encountered in practice. For example, it occurs where a new grade line must meet an existing railroad or highway crossings or when a minimum vertical distance must be maintained between the grade line and underground utilities or drainage structures.

Given the station and elevation of the VPI, and grades g_1 and g_2 of the back and forward tangents, respectively, the problem consists of calculating the curve length required to meet the fixed condition. It is solved by substituting known quantities into Equation (25.3) and reducing the equation to its quadratic form containing only L as an unknown. Two values will satisfy the quadratic equation, but the correct one will be obvious.

25.11 Designing a Curve to Pass Through a Fixed Point

Figure 25.7
Data entry screen in WOLFPACK for computation of first vertical curve staking notes in Example 25.3.

■ Example 25.5

In Figure 25.8, grades $g_1 = -4.00\%$ and $g_2 = +3.80\%$ meet at VPI station $52 + 00$ and elevation 1261.50. Design a parabolic curve to meet a railroad crossing, which exists at station $53 + 50$ and elevation 1271.20.

Solution

In referring to Figure 25.8 and substituting known quantities into Equation (25.3), the following equation is obtained:

$$1271.20 = \left[1261.50 + 4.00\left(\frac{L}{2}\right)\right] + \left[-4.00\left(\frac{L}{2} + 1.5\right)\right]$$
$$+ \left[\frac{3.80 + 4.00}{2L}\left(\frac{L}{2} + 1.5\right)^2\right]$$

In this expression, the value of X for the railroad crossing is $L/2 + 1.5$ and the terms within successive brackets are Y_{BVC}, $g_1 X$, and $(r/2)X^2$, respectively. Reducing the equation to quadratic form gives

$$0.975L^2 - 9.85L + 8.775 = 0$$

Figure 25.8
Designing a parabolic curve to pass through a fixed point.

Solving by use of Equation (11.3) for L gives 9.1152 stations. To check the solution, $L = 9.1152$ stations and $X = [9.1152/2 + 1.5]$ stations are used in Equation (25.3) to calculate the elevation at station $53 + 50$. A value of 1271.20 checks the computations.

25.12 SIGHT DISTANCE

The vertical alignments of highways should provide ample sight distance for safe vehicular operation. Two types of sight distances are involved: (1) stopping sight distance (the distance required, for a given "design speed,"[1] to safely stop a vehicle thus avoiding a collision with an unexpected stationary object in the roadway ahead) and (2) passing sight distance (the distance required for a given design speed, on two-lane two-way highways to safely overtake a slower moving vehicle, pass it, and return to the proper lane of travel leaving suitable clearance for an oncoming vehicle in the opposing lane). For either condition, as speed increases, required sight distance also increases. All highways should provide safe stopping sight distances for their entire extent at their given design speed, and if this cannot be achieved in certain sections, signs must be posted to reduce travel speeds to levels consistent with the available sight distances. Passing sight distance should be provided at frequent intervals along any section of highway to allow faster moving vehicles to pass slower moving ones. In sections of highway that do not provide ample passing sight distances, appropriate centerline markings and signs are used to inform drivers of this condition. The American Association of State Highway and Transportation Officials (AASHTO) has recommended minimum sight distances for both stopping and passing for various design speeds. Table 25.4 lists these values for some commonly used design speeds.

Sight distances must be carefully considered in the design of vertical alignments of highway projects. Given the grades of two intersecting tangent sections, the length of vertical curve used to provide a transition from one to the other fixes the sight distance. A longer curve provides a greater sight distance.

TABLE 25.4 MINIMUM SIGHT DISTANCES FOR VARYING DESIGN SPEEDS ON LEVEL SECTIONS

Design Speed (mph)	Stopping Sight Distance (ft)	Passing Sight Distance (ft)
30	200	1090
40	305	1470
50	425	1835
60	570	2135
70	730	2480

---

[1]Design speed is defined as the maximum safe speed that can be maintained over a specified section of highway when conditions are so favorable that the design features of the highway govern. Once selected, all of the pertinent features of the highway, especially those involving safety, should be related to that speed.

The formula for length of curve L necessary to provide sight distance S on a *crest vertical curve, where S is less than L* is

$$L = \frac{S^2(g_1 - g_2)}{2(\sqrt{h_1} + \sqrt{h_2})^2} \qquad (25.9)$$

In Equation (25.9), the units of S and L are stations if the English system is used, and one-tenth stations in the metric system. Also the units of h_1 (the height of the driver's eye) and h_2 (the height of an object sighted on the roadway ahead) are in feet for the English system and meters for the metric system. For design, AASHTO recommends 3.5 ft (1080 mm) for h_1. Recommended values for h_2 are 2.0 ft (600 mm) for stopping and 4.25 ft (1.300 m) for passing. The lower value for h_2 represents the size of an object that would damage a vehicle and the higher value represents the height of an oncoming car.

Then for a crest curve having grades of $g_1 = +1.40\%$ and $g_2 = -1.00\%$, by Equation (25.9) the length of curve needed to provide a 570-ft stopping sight distance is

$$L = \frac{(5.70)^2(1.40 + 1.00)}{2\left(\sqrt{3.50} + \sqrt{2.00}\right)^2} = 3.61 \text{ stations}$$

Since S is greater than L, and thus not in agreement with the assumption used in deriving the formula, a different expression must be employed. If the vehicle is off the curve but on the tangent leading to it and *S is greater than L*, the applicable sight distance formula is

$$L = 2S - \frac{2\left(\sqrt{h_1} + \sqrt{h_2}\right)^2}{g_1 - g_2} \qquad (25.10)$$

Then in the preceding example, the length of curve necessary to provide 570 ft of stopping sight distance is

$$L = 2(5.70) - \frac{2(\sqrt{3.50} + \sqrt{2.0})^2}{1.40 + 1.00} = 2.41 \text{ stations}$$

In this solution, the sight distance of 570 feet is greater than the computed curve length of 2.41 stations (241 ft), and thus the conditions are met.

Sag vertical curves also limit sight distances because they reduce lengths ahead that can be illuminated by headlights during night driving. Equations that apply in computing lengths of sag vertical curves based upon headlight criteria are:

(a) *S* less than *L*

$$L = \frac{S^2(g_2 - g_1)}{4 + 3.5S} \qquad (25.11)$$

(d) *S* greater than *L*

$$L = 2S - \frac{4 + 3.5S}{|g_2 - g_1|} \tag{25.12}$$

As discussed in Section 24.17, horizontal curves may also limit visibility and sight distances for them can be computed as well. For a combined horizontal and vertical curve, the sight distance that governs is the smaller of the two values computed independently for each curve.

For a complete discussion on sight distances for the design of highways and streets, the reader should refer to the AASHTO publication *A Policy on Geometric Design of Highways and Streets*, which is cited in the bibliography at the end of this chapter.

25.13 SOURCES OF ERROR IN LAYING OUT VERTICAL CURVES

Some sources of error in staking out parabolic curves are:

1. Making errors in measuring distances and angles when staking the centerline.
2. Not holding the level rod plumb when setting blue tops.
3. Using a leveling instrument that is out of adjustment.

25.14 MISTAKES

Some typical mistakes made in computations for vertical curves include the following:

1. Arithmetic mistakes.
2. Failure to properly account for the algebraic signs of g_1 and g_2.
3. Subtracting offsets from tangents for a sag curve, or adding them for a crest curve.
4. Failure to make the second-difference check.
5. Not completing the level circuit back to a benchmark after setting blue tops.

PROBLEMS

Asterisks (*) indicate problems that have partial answers given in Appendix G.

25.1 What is the definition of a crest vertical curve?

25.2 Identify the type of vertical curve (sag or crest) by the following rates of change in grade:
*(a) −2.3%
(b) +4.3%
(c) −0.5%

25.3 What common geometric figure is used in vertical design?

Tabulate station elevations for an equal-tangent parabolic curve for the data given in Problems 25.4 through 25.9, and check the table by second differences.

25.4* A -2.50% grade meets a $+1.75\%$ grade at station $44 + 25$ and elevation 3386.96 ft, 400-ft curve, stakeout at half stations.

25.5 A -3.5% grade meets a $+2.50\%$ grade at station $4 + 200$ and elevation 154.657 m, 180-m curve, stakeout at 30-m increments.

25.6 A 500-ft curve, grades of $g_1 = +2.00\%$ and $g_2 = -2.50\%$, VPI at station $96 + 80$, and elevation 845.26 ft, stakeout at full stations.

25.7 A 600-ft curve, grades of $g_1 = -1.50\%$ and $g_2 = -2.25\%$, VPI at station $38 + 00$, and elevation 160.00 ft, stakeout at full stations.

25.8 A 500-ft curve, $g_1 = -2.50\%$, $g_2 = +3.50\%$, VPI station $= 98 + 10$, VPI elevation $= 5378.26$ ft, stakeout at full stations.

25.9 A 80-m curve, $g_1 = 2.25\%$, $g_2 = 1.50\%$, VPI station $= 12 + 280.15$, VPI elevation $= 4855.600$ m, stakeout at 10-m increments.

Field conditions require a highway curve to pass through a fixed point. Compute a suitable equal-tangent vertical curve and full-station elevations for Problems 25.10 through 25.12.

25.10* Grades of $g_1 = -3.50\%$ and $g_2 = +0.50\%$, VPI elevation 750.00 ft at station $30 + 00$, and a fixed elevation 753.00 ft at station $30 + 00$.

25.11 Grades of $g_1 = -2.50\%$ and $g_2 = +1.50\%$, VPI elevation 860.00 ft at station $22 + 00$, and a fixed elevation 864.00 ft at station $20 + 50$.

25.12 Grades of $g_1 = +1.500\%$ and $g_2 = +2.500\%$ VPI station $4 + 600$ and elevation 155.460 m and a fixed elevation 157.968 m at station $4 + 700$. (Use 100-m stationing.)

25.13 A -1.10% grade meets a $+0.60\%$ grade at station $36 + 00$ and elevation 800.00 ft. The $+0.60\%$ grade then joins a $+2.40\%$ grade at station $39 + 00$. Compute and tabulate the notes for an equal-tangent vertical curve, at half-stations, that passes through the midpoint of the 0.60% grade.

25.14 When is it advantageous to use an unequal-tangent vertical curve instead of an equal-tangent one?

Compute and tabulate full-station elevations for an unequal-tangent vertical curve to fit the requirements in Problems 25.15 through 25.17.

25.15 A $+3.50\%$ grade meets a -3.25% grade at station $60 + 00$ and elevation 1310.00 ft. Length of first curve 600 ft, second curve 400 ft.

25.16 Grade $g_1 = +2.25\%$, $g_1 = +3.75\%$, VPI at station $62 + 00$ and elevation 850.00 ft, $L_1 = 700$ ft and $L_2 = 500$ ft.

25.17 Grades g_1 of $+5.00\%$ and g_2 of -1.50% meet at the VPI at station $4 + 300$ and elevation 154.960 m. Lengths of curves are 200 m and 350 m. (Use 30-m stationing.)

25.18* A manhole is 12 ft from the centerline of a 30-ft wide street that has a 6-in. parabolic crown. The street center at the station of the manhole is at elevation 612.58 ft. What is the elevation of the manhole cover?

25.19 A 50-ft wide street has an average parabolic crown from the center to each edge of 1/4 in./ft. How much does the surface drop from the street center to a point 4 ft from the edge?

25.20 Determine the station and elevation at the low point of the curve in Problem 25.4.

25.21* Calculate the station and elevation at the high point of the curve in Problem 25.6.

25.22 Compute the station and elevation at the low point of the curve of Problem 25.8.

25.23 What are the station and elevation of the low point of the curve of Problem 25.13?

25.24 What are the requirements for sight distances on a vertical curve?

25.25* Compute the passing sight distance available in Problem 25.6. (Assume $h_1 = 3.50$ ft and $h_2 = 4.25$ ft.)

25.26 Similar to Problem 25.25, except for a stopping sight distance with an $h_2 = 2.00$ ft.

25.27 In determining sight distances on vertical curves, how does the designer determine whether the cars or objects are on the curve or tangent?

What is the minimum length of a vertical curve to provide a required sight distance for the conditions given in Problems 25.28 and 25.29?

25.28* Grades of +3.00% and −2.00%, sight distance 600 ft, $h_1 = 3.50$ ft and $h_2 = 2.00$ ft.

25.29 A crest curve with grades of +2.50% and −1.00% sight distance 500 ft, $h_1 = 3.50$ ft and $h_2 = 4.25$ ft.

25.30* A backsight of 4.86 ft is taken on a benchmark whose elevation is 33.86 ft. What rod reading is needed at that *HI* to set a blue top at grade elevation of 34.80 ft?

25.31 A backsight of 4.52 ft is taken on a benchmark whose elevation is 658.28 ft. A foresight of 5.04 ft and a backsight of 6.04 ft are then taken in turn on TP_1 to establish a *HI*. What rod reading will be necessary to set a blue top at a grade elevation of 657.96 ft?

BIBLIOGRAPHY

American Association of State Highway and Transportation Officials. 2001. *Guidelines for Geometric Design of Very-Low Volume Local Roads (ADT ≤ 400)*. Washington, DC: AASHTO.

American Association of State Highway and Transportation Officials. 2004. *A Policy on Geometric Design of Highways and Streets*. Washington, DC: AASHTO.

26
Volumes

26.1 INTRODUCTION

Persons engaged in surveying (geomatics) are often called on to determine volumes of various types of material. For example, quantities of earthwork and concrete are needed on many types of construction projects. Volume computations are also required to determine the capacities of bins, tanks, reservoirs, and buildings, and to check stockpiles of coal, gravel, and other materials. The determination of quantities of water discharged by streams and rivers, per unit of time, is also important.

The most common unit of volume is a cube having edges of unit length. Cubic feet, cubic yards, and cubic meters are used in surveying calculations, with cubic yards and cubic meters being most common for earthwork. (Note: 1 yd^3 = 27 ft^3; 1 m^3 = 35.31445 ft^3.) The *acre-foot* (the volume equivalent to an acre of area, 1-ft deep) is commonly used for large quantities of water, while cubic feet per second (ft^3/sec) and cubic meters per second (m^3/sec) are the usual units for water-flow measurement.

26.2 METHODS OF VOLUME MEASUREMENT

Direct measurement of volumes is rarely made in surveying, since it is difficult to actually apply a unit of measure to the material involved. Instead, indirect measurements are obtained by measuring lines and areas that have a relationship to the volume desired.

Three principal systems are used: (1) the cross-section method, (2) the unit-area (or borrow-pit) method, and (3) the contour-area method.

■ 26.3 THE CROSS-SECTION METHOD

The cross-section method is employed almost exclusively for computing volumes on linear construction projects such as highways, railroads, and canals. In this procedure, after the centerline has been staked, ground profiles called cross sections are taken (at right angles to the centerline), usually at intervals of full or half stations if the English system of units is being used, or at perhaps 10, 20, 30, or 40 m if the metric system is being employed. Cross-sectioning consists of observing ground elevations and their corresponding distances left and right perpendicular to the centerline. Readings must be taken at the centerline, at high and low points, and at locations where slope changes occur to determine the ground profile accurately. This can be done in the field using a level, level rod, and tape. Plate B.5 in Appendix B illustrates a set of field notes for cross-sectioning.

Much of the fieldwork formerly involved in running preliminary centerline, getting cross-section data, and making slope-stake and other measurements on long-route surveys is now being done more efficiently by photogrammetry. Research has shown that earthwork quantities computed from photogrammetric cross-sectioning agree to within about 1 or 2% of those obtained from good-quality field cross sections. It is not intended to discuss photogrammetric methods in this chapter; rather, basic field and office procedures for determining and calculating volumes will be presented briefly. Chapter 27 discusses the subject of photogrammetry.

In Section 17.8, the subject of terrain representation by means of digital elevation models (DEMs) was introduced, and concepts for deriving triangulated irregular networks (TINs) from DEMs were presented. It was noted that once a TIN model is created for a region, profiles and cross sections anywhere within the area could be readily derived using a computer. This can be of significant advantage where the general location of a proposed road or railroad has been decided upon, but the final alignment is not yet fixed. In those situations, controlling points and breaklines can be surveyed in the region where the facility is expected to be located, and a DEM for the area generated. Either ground or photogrammetric methods can be employed for deriving the terrain data. From the DEM information, a TIN model can be created, and then the computer can provide cross sections for the analysis of any number of alternate alignments automatically.

After cross sections have been taken and plotted, *design templates* (outlines of base widths and side slopes of the planned excavation or embankment) are superimposed on each plot to define the excavation or embankment to be constructed at each cross-section location. Areas of these sections, called *end areas*, are obtained by computation or by planimeter (see Section 12.9.4). Nowadays, using computers, end areas are calculated directly from field cross-section data and design information. From the end areas, volumes are determined by the *average-end-area*, or *prismoidal* formula, discussed later in this chapter.

Figure 26.1 portrays a section of planned highway construction, and illustrates some of the points just discussed. Centerline stakes are shown in place, with their stationing given in the English system of units. They mark locations where cross sections are taken, in this instance at full stations. End areas, based on the planned grade line, size of roadway, and selected embankment and excavation slopes, are superimposed at each station and are shown shaded. Areas of

Figure 26.1
Section of roadway illustrating excavation (cut) and embankment (fill).

these shaded sections are determined, whereupon volumes are computed using formulas given in Section 26.5 or 26.8. Note in the figure that embankment, or *fill*, is planned from stations 10 + 00 through 11 + 21, a transition from fill to excavation, or *cut*, occurs from station 11 + 21 to 11 + 64, and cut is required from stations 11 + 64 through 13 + 00.

26.4 TYPES OF CROSS SECTIONS

The types of cross sections commonly used on route surveys are shown in Figure 26.2. In flat terrain the *level section* (a) is suitable. The *three-level section* (b) is generally used where ordinary ground conditions prevail. Rough topography may require a *five-level section* (c), or more practically an *irregular section* (d). A *transition section* (e) and a *side-hill section* (f) occur when passing from cut to fill and on side-hill locations. In Figure 26.1, transition sections occur at stations 11 + 21 and 11 + 64, while a *side-hill section* exists at 11 + 40.

The width of base b, the finished roadway, is fixed by project requirements. As shown in Figure 26.1, it is usually wider in cuts than on fills to provide for drainage ditches. The side slope s [the horizontal dimension required for a unit vertical rise and illustrated in Figure 26.2(a)] depends on the type of soil encountered.

800 VOLUMES

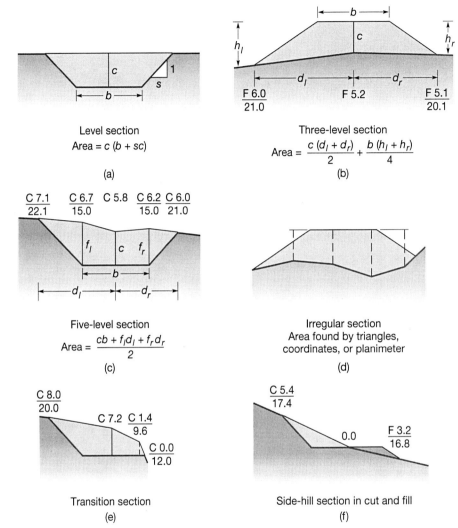

Figure 26.2
Earthwork sections.

Side slopes in fills usually are flatter than those in cuts where the soil remains in its natural state.

Cut slopes of 1:1 (1 horizontal to 1 vertical) and fill slopes of 1-1/2:1 might be satisfactory for ordinary loam soils, but 1-1/2:1 in excavation and 2:1 in embankment are common. Even flatter proportions may be required—one cut in the Panama Canal area was 13:1—depending on soil type, rainfall, and other factors. Formulas for areas of sections are readily derived and listed with some of the sketches in Figure 26.2.

■ 26.5 AVERAGE-END-AREA FORMULA

Figure 26.3 illustrates the concept of computing volumes by the average-end-area method. In the figure, A_1 and A_2 are end areas at two stations separated by a horizontal distance L. The volume between the two stations is equal to the average of the end areas multiplied by the horizontal distance L between them.

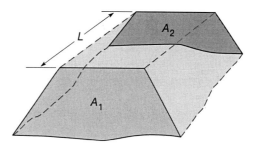

Figure 26.3
Volume by average-end-area-method.

Thus,

$$V_e = \frac{A_1 + A_2}{2} \times \frac{L}{27} \text{ (yd}^3\text{)} \quad \text{(26.1a)}$$

or

$$V_e = \frac{A_1 + A_2}{2} \times L \text{ (m}^3\text{)} \quad \text{(26.1b)}$$

In Equation (26.1a), V_e is the average end-area volume in cubic yards, A_1 and A_2 are in square feet, and L is in feet. In Equation (26.1b), A_1 and A_2 are in m^2, L is in m, and V_e is in m^3. Equation (26.1b) also applies to computing volumes in acre-ft, where A_1 and A_2 are in acres, and L is in ft.

If L is 100 ft, as for full stations in the English system of units, Equation (26.1a) becomes

$$V_e = 1.852(A_1 + A_2) \text{ yd}^3 \quad \text{(26.2)}$$

Equations (26.1) and (26.2) are approximate and give answers that generally are slightly larger than the true prismoidal volumes (see Section 26.8). They are used in practice because of their simplicity, and contractors are satisfied because pay quantities are generally slightly greater than true values. Increased accuracy is obtained by decreasing the distance L between sections. When the ground is irregular, cross sections must be taken closer together.

■ Example 26.1

Compute the volume of excavation between station 24 + 00, with an end area of 711 ft², and station 25 + 00, with an end area of 515 ft².

Solution

By Equation (26.2), $V = 1.852(A_1 + A_2) = 1.852(711 + 515) = 2270$ yd³.

■ 26.6 DETERMINING END AREAS

End areas can be determined either graphically or by computation. In graphic methods, the cross section and template are plotted to scale on grid paper; then the number of small squares within the section can be counted and converted to

802 VOLUMES

area, or the area within the section can be measured using a planimeter (see Section 12.9.4). Computational procedures consist of either dividing the section into simple figures such as triangles and trapezoids, and computing and summing these areas, or using the coordinate formula (see Section 12.5). These computational methods are discussed in the sections that follow. Most such calculations are now done by computer—usually by the coordinate method, which is general and readily programmed.

26.6.1 End Areas by Simple Figures

To illustrate the procedures of calculating end areas by simple figures such as triangles or trapezoids, assume the following excerpt of field notes (in the English system of units) applies to the cross section and end area, shown in Figure 26.4. In the notes, Lt indicates that the readings were started on the left side of the reference line as viewed facing in the direction of increasing stationing.

$$HI = 879.29 \text{ ft}$$

$$24 + 00 \text{ Lt} \quad \begin{array}{cccccc} 867.3 & 870.9 & 874.7 & 876.9 & 869.0 & 872.8 \\ \underline{12.0} & \underline{8.4} & \underline{4.6} & \underline{2.4} & \underline{10.3} & \underline{6.5} \\ 50 & 36 & 20 & \text{CL} & 12 & 50 \end{array}$$

In this excerpt of field notes, the top numbers are elevations (in ft) obtained by subtracting rod readings (middle numbers) from the leveling instrument's HI. Bottom numbers are distances from centerline (in ft), beginning from the left. Assume the design calls for a level roadbed of 30-ft width, cut slopes of 1-1/2:1, and a subgrade elevation at station 24 + 00 of 858.9 ft. A corresponding *design template* is superimposed over the plotted cross section in Figure 26.4. Subtracting the subgrade elevation from cross-section elevations at C, D, and E yields the ordinates of cut required at those locations. Elevations and distances out from centerline to the slope intercepts at L and R must be either scaled from the plot or computed. Assuming they have been scaled (methods for computing them are

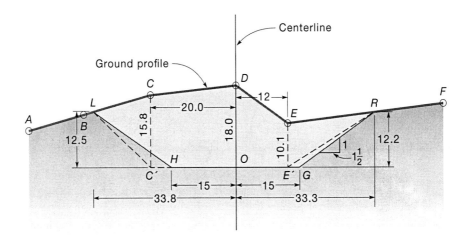

Figure 26.4
End-area computation.

given in Section 26.7), the following tabulation of distances from centerline and required cut ordinates at each point to subgrade elevation was made:

Station	H	L	C	D	E	R	G
24 + 00 (Lt)	0/15	C12.5/33.8	C15.8/20	C18.0/0	C10.1/12	C12.2/33.3	0/15

Numbers above the lines in the fractions (preceded by the letter C) are cut ordinates in feet; those below the lines are corresponding distances out from the centerline. Fills are denoted by the letter F. Using C instead of plus for cut, and F instead of minus for fill, eliminates confusion. From the cut ordinates and distances from centerline shown, the area of the cross section in Figure 26.4 is computed by summing the individual areas of triangles and trapezoids. A list of the calculations is given in Table 26.1. (Refer to Figure 26.4 for triangle and trapezoid designations.)

26.6.2 End Areas by Coordinates

The coordinate method for computing end areas can be used for any type of section, and has many engineering applications. The procedure was described in Section 12.5 as a way to determine the area contained within a closed polygon traverse.

To demonstrate the method in end-area calculation, the example of Figure 26.4 will be solved. Coordinates of each point of the section are calculated in an axis system having point O as its origin, using the earlier listed data on cuts and distances from centerline. In computing coordinates, distances to the right of centerline and cut values are considered plus values; distances left and fill values are considered minus values. Beginning with point O and proceeding clockwise around the figure, the coordinates of each point are listed in sequence. *Point O is repeated at the end* (see Table 26.2). Then Equation (12.7) is applied, with products of diagonals downward to the right ($\searrow$) considered minus, and diagonal products down to the left ($\swarrow$) plus. Algebraic signs of the coordinates must be considered. Thus, a plus ($\swarrow$) having a negative coordinate will be minus. The total

TABLE 26.1 END AREA BY SIMPLE FIGURES

Figure	Computation	Area
ODCC'	[(18.0 + 15.8)20]/2	338.0
C'CL	[(15.8)13.8]/2	109.0
HLC'	[−(5)12.5]/2	−31.2
ODEE'	[(18.0 + 10.1)12]/2	168.6
EE'R	[(10.1)21.3]/2	107.6
E'RG	[(3)12.2]/2	18.3
	Area =	710 ft²

804 VOLUMES

TABLE 26.2	END AREAS BY COORDINATES			
Point	X	Y	Plus+	Minus−
O	0	0	↙	↘
H	−15	0	0	0
L	−33.8	12.5	0	+188
C	−20	15.8	−250	+534
D	0	18.0	0	+360
E	12	10.1	216	0
R	33.3	12.2	336	−146
G	15	0	183	0
O	0	0	0	0
			+485	+936
			+936	
			Σ = 1421	
		Area = 1421 ÷ 2 = 710 ft²		(nearest ft²)

area is obtained by dividing the absolute value of the algebraic summation of all products by 2. The calculations are illustrated in Table 26.2.

It is necessary to make separate computations for cut and fill end areas when they occur in the same section (as at station 11 + 40 of Figure 26.1), since they must always be tabulated independently for pay purposes. Payment is normally made only for excavation (its unit price includes making and shaping the fills) except on projects consisting primarily of embankment such as levees, earth dams, some military fortifications, and highways built up by continuous fills in flat areas.

■ 26.7 COMPUTING SLOPE INTERCEPTS

The elevations and distances out from the centerline to the slope intercepts can be calculated using cross-section data and the cut or fill slope values. In Figure 26.4, for example, intercept R, occurs between ground profile point E (distance 12 ft right and elevation 869.0) and point F (distance 50 ft right and elevation 872.8). The cut slope is 1-1/2:1, or 0.67 ft/ft. A more detailed diagram, illustrating the geometry for calculating slope intercept R, is given in Figure 26.5.

The slope along ground line EF is $(872.8 − 869.0)/38 = 0.10$ ft/ft, where 38 ft is the horizontal distance between the points. The elevation of G' (point vertically above G) is $869.0 + 0.10(3) = 869.3$; thus, ordinate GG' is $(869.3 − 858.9) = 10.4$ ft. Lines EF and GR converge at a rate equal to the difference in their slopes (because they are both sloping upward), or $0.67 − 0.10 = 0.57$ ft/ft. Dividing ordinate GG' by this convergence yields horizontal distance GR, or $10.4/0.57 = 18.3$ ft. Adding 18.3 to distance OG yields $18.3 + 15 = 33.3$ ft, which is the distance from centerline to slope intercept R. Finally, to obtain the elevation of R, the increase in elevation from E to R is added to the elevation of E, or $0.10(21.3) + 869.0 = 871.1$. Thus, the cut ordinate

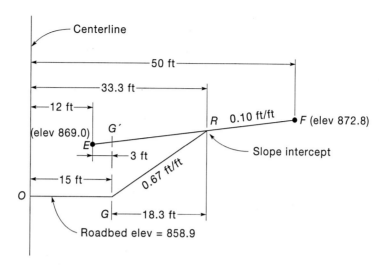

Figure 26.5 Computation of slope intercept R of Figure 26.4.

at R equals $871.1 - 858.9 = 12.2$ ft. Recall that 33.3 and 12.2 were the X and Y coordinates, respectively, used in the end-area calculations of Section 26.6.

The elevation and distance from centerline of the slope intercept L of Figure 26.4 are calculated in a similar manner, except the rate of convergence of lines CB and HL is the sum of their slopes because CB slopes downward and HL upward. Writing equations for the intersecting lines is another method used to compute slope intercepts. The equations are then set equal to each other and solved for x. This is demonstrated using the data in Figure 26.5 and Equation (11.6) in Example 26.2.

■ Example 26.2

Determine the coordinates of point R in Figure 26.5 using Equation (11.6).

Solution

The coordinates for the pertinent endpoints are:

Point	x	y
E	12	869.0
F	50	872.8
G	15	858.9

The equation for line EF is

$$y = \left(\frac{872.8 - 869.0}{50 - 12}\right)x + b$$
$$= 0.1x + b$$

Substituting in the coordinates for either E or F, we find that b is 867.8. Thus, the equation for line EF is

$$y = 0.1x + 867.8 \tag{a}$$

The slope intercept for the side slope equation can be determined using the slope of the side slope, which is 2/3, and the coordinates of point G as

$$b = 858.9 - (2/3)15 = 848.9$$

Thus, the line equation for the side slope is

$$y = (2/3)x + 848.9 \qquad \text{(b)}$$

Setting Equation (a) equal to (b) and solving for x yields

$$0.1x + 867.8 = \frac{2}{3}x + 848.9$$

$$18.9 = \left(\frac{2}{3} - 0.1\right)x$$

$$x = 33.3$$

Using either Equation (a) or (b), the elevation at 33.3 is

$$y = 0.1(33.3) + 867.8 = \left(\frac{2}{3}\right)33.3 + 848.9$$

$$= 871.1$$

Note that this procedure results in the same solution as that previously determined.

Calculations of slope intercepts are somewhat laborious, but routine when programmed for solution by computer. If a computer is not used for computing end areas and volumes, an alternate procedure is to plot the cross sections and templates, determine the end area by planimeter, and scale the slope intercepts from the plot. Slope intercepts are essential since the placement of slope stakes that guide construction operations is based on them.

26.8 PRISMOIDAL FORMULA

The prismoidal formula applies to volumes of all geometric solids that can be considered prismoids. A prismoid, illustrated in Figure 26.6, is a solid having ends that are parallel but not congruent, and trapezoidal sides that are also

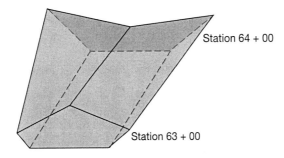

Figure 26.6
Sections for which the prismoidal correction is added to the end-area volume.

not congruent. Most earthwork solids obtained from cross-section data fit this classification. However, from a practical standpoint, the differences in volumes computed by the average-end-area method and the prismoidal formula are usually so small as to be negligible. Where extreme accuracy is needed, such as in expensive rock cuts or concrete placements, the prismoidal method can be used.

One arrangement of the prismoidal formula is

$$V_P = \frac{L(A_1 + 4A_m + A_2)}{6 \times 27} \text{ (yd}^3\text{)} \qquad (26.3)$$

where V_P is the prismoidal volume in cubic yards, A_1 and A_2 are areas of successive cross sections taken in the field, A_m is the area of a "computed" section midway between A_1 and A_2, and L is the horizontal distance between A_1 and A_2. Prismoidal volumes in m³ can be obtained by using a slight modification of Equation (26.3), that is, the conversion factor 27 in the denominator is dropped, and L is in meters, A_1, A_m, and A_2 are in m².

To use the prismoidal formula, it is necessary to know area A_m of the section halfway between the stations of A_1 and A_2. This is found by the usual computation *after averaging the heights and widths of the two end sections*. Obviously, the middle area is not the average of the end areas, since there would then be no difference between the results of the end-area formula and the prismoidal formula.

The prismoidal formula generally gives a volume smaller than that found by the average-end-area formula. For example, the volume of a pyramid by the prismoidal formula is $Ah/3$ (the exact value), whereas by the average-end-area method it is $Ah/2$. An exception occurs when the center height is great but the width narrow at one station, and the center height small but the width large at the adjacent station. Figure 26.6 illustrates this condition.

The difference between the volumes obtained by the average-end-area formula and the prismoidal formula is called the *prismoidal correction* C_p. Various books on route surveying give formulas and tables for computing prismoidal corrections, which can be applied to average-end-area volumes to get prismoidal volumes. A prismoidal correction formula, which provides accurate results for three-level sections is

$$C_P = \frac{L}{12 \times 27}(c_1 - c_2)(w_1 - w_2) \text{ (yd}^3\text{)} \qquad (26.4)$$

where C_P is the volume of the prismoidal correction in cubic yards, c_1 and c_2 are center heights in cut (or in fill), and w_1 and w_2 are widths of sections (from slope intercept to slope intercept) at adjacent sections. If the product of $(c_1 - c_2)(w_1 - w_2)$ is minus, as in Figure 26.6, the prismoidal correction is added rather than subtracted from the end-area volume. For sections other than three-level, Equation (26.4) may not be accurate enough, and therefore Equation (26.3) is recommended.

■ 26.9 VOLUME COMPUTATIONS

Volume calculations for route construction projects are usually done by computer and arranged in tabular form. To illustrate this procedure, assume that end areas listed in columns (2) and (3) of Table 26.3 apply to the section of roadway illustrated in Figure 26.1. By using Equation (26.1a), cut and fill volumes are computed and tabulated in columns (4) and (5).

The volume computations illustrated in Table 26.3 include the transition sections of Figure 26.1. This is normally not done when preliminary earthwork volumes are being estimated (during design and prior to construction) because the exact locations of the transition sections and their configurations are usually unknown until slope staking occurs. Thus, for calculating preliminary earthwork quantities, an end area of zero would be used at the station of the centerline grade point (station 11 + 40 of Figure 26.1), and transition sections (stations 11 + 21 and 11 + 64 of Figure 26.1) would not appear in the computations. After slope staking (procedures for slope staking are described in Section 23.8) the locations and end areas of transition sections are known, and they should be included in final volume computations, especially if they significantly affect the quantities for which payment is made.

In highway and railroad construction, excavation or cut material is used to build embankments or fill sections. Unless there are other controlling factors, a well-designed grade line should nearly balance total cut volume against total fill

TABLE 26.3 TABULAR FORM OF VOLUME COMPUTATION

Station (1)	End Area (ft^2) Cut (2)	End Area (ft^2) Fill (3)	Volume (yd^3) Cut (4)	Volume (yd^3) Fill (5)	Fill Volume +25% (yd^3) (6)	Cumulative Volume (yd^3) (7)
10 + 00		992				0
				2614	3268	
11 + 00		421				−3268
				190	238	
11 + 21	0	68				−3506
			12	29	37	
11 + 40	34	31				−3531
			79	14	17	
11 + 64	144	0				−3469
			553			
12 + 00	686					−3916
			2967			
13 + 00	918					+51

volume. To accomplish a balance, either fill volumes must be expanded or cut volumes shrunk.[1] This is necessary because, except for rock cuts, embankments are compacted to a density greater than that of material excavated from its natural state, and to balance earthwork this must be considered. (Rock cut expands to occupy a greater fill volume; thus, either the cut must be expanded or the fill shrunk to obtain a balance.) The rate of expansion depends on the type of material and can never be estimated exactly. However, samples and records of past projects in the immediate area are helpful in assigning reasonable factors. Column (6) of Table 26.3 lists expanded fills for the example of Figure 26.1, where a 25% factor was applied.

To investigate whether or not an earthwork balance is achieved, *cumulative volumes* are computed. This involves adding cut and expanded fill volumes algebraically from project beginning to end, with cuts considered positive and fills negative. Cumulative volumes are listed in column (7) of Table 26.3. In this example, there is a cut volume excess of 51 yd^3 between stations 10 + 00 and 13 + 00 or, in other words, there is a surplus of that much excavation.

To analyze the movement of earthwork quantities on large projects, *mass diagrams* are constructed. These are plots of cumulative volumes for each station as the ordinate, versus the stations on the abscissa. Horizontal (balance) lines on the mass diagram then determine the limit of economic haul and the direction of movement of material. Mass diagrams are described more thoroughly in books on route surveying. If there is insufficient material from cuts to make the required fills, the difference must be *borrowed* [obtained from borrow pits or other sources such as by "day-lighting" curves (flattening cut slopes to improve visibility)]. If there is excess cut, it is wasted or perhaps used to extend and flatten the fills.

For projects with more than a few cross sections, computer programs are available and are generally used for earthwork computations, but surveyors and engineers must still understand the basic methods.

26.10 UNIT-AREA, OR BORROW-PIT, METHOD

On many projects, except long linear-route constructions, the quantity of earth, gravel, rock, or other material excavated or filled can often best be determined by the borrow-pit method. The quantities computed form the basis for payment to the contractor or supplier of materials. The volume of coal or other loose materials in stockpiles can be found in the same way.

As an example, assume the area shown in Figure 26.7 is to be graded to an elevation of 358.0 ft for a building site. Notes for the fieldwork are shown in Plate B.2 of Appendix B. The area to be covered in this example is staked in squares of 20 ft, although 10, 50, 100, or more feet could be used, with the choice depending on project size and accuracy desired. A total station instrument and tape, or only

---

[1]Expansion of fill volumes is generally preferred, since payment is usually based on actual volumes of material excavated.

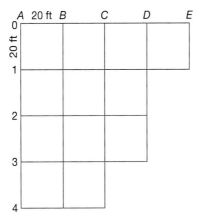

Figure 26.7
Borrow-pit leveling.

a tape may be used for the layout. A benchmark of known or assumed elevation is established outside the area in a place not likely to be disturbed.

After the area is laid out in squares, elevations are determined at all grid intersection points. For this, a level is set up at any convenient location, a plus sight taken on the benchmark, and minus sights read on each grid intersection. If the terrain is not too rough, it may be possible to select a point near the area center and take sights on all grid intersections from the same setup, as in the example of Plate B.2. For rough terrain, it may be most convenient to determine the elevations by radial surveying from one well-chosen setup using a total station instrument (see Section 17.9.1).

Letters and numbers designate grid intersection points, such as *A*-1, *C*-4, and *D*-2. For site grading to a specified elevation, say 358.0 ft, the amount of cut or fill at each grid square corner is obtained by subtracting 358.0 from its ground elevation. For each square, then, the average height of the four corners of each prism of cut or fill is determined and multiplied by the base area, 20×20 ft $= 400$ ft^2, to get the volume. The total volume is found by adding the individual values for each block and dividing by 27 to obtain the result in cubic yards.

To simplify calculations, the cut at each corner multiplied by the number of times it enters the volume computation can be shown in a separate column. The column sum is divided by 4 and multiplied by the base area of one block to get the volume. In equation form, this procedure is given as

$$V_P = \sum (h_{i,j} n) \left(\frac{A}{4 \times 27} \right) \text{yd}^3 \qquad (26.5)$$

where $h_{i,j}$ is the corner height in row i and column j, and n the number of squares to which that height is common. The corner at *C*-4, for example, is common to only one square, *D*-2 is common to two, *D*-1 is common to three, and *C*-1 is common to four. $\sum(h_{i,j} n)$ is the sum of the products of the height and the number of common squares, and A is the area of one square. An example illustrating the use of Equation (26.5) is given in the field notes of Plate B.2.

■ 26.11 CONTOUR-AREA METHOD

Volumes based on contours can be obtained from contour maps by using a planimeter to determine the area enclosed by each contour. Alternatively, *computer-aided drafting* (CAD) software can be used to determine these areas. Then the average area of the adjacent contours is obtained using Equation (26.1b), and the volume obtained by multiplying by the contour spacing (i.e., contour interval). Use of the prismoidal formula is seldom, if ever, justified in this type of computation. This procedure is the basis for volume computations in CAD software.

Instead of determining areas enclosed within contours by planimeter, they can be obtained using the coordinate formula [Equation (12.7) or (12.8)]. In this procedure, a tablet digitizer like the one shown in Figure 28.8 is first used to measure the coordinates along each contour at enough points to define its configuration satisfactorily.

The contour-area method is suitable for determining volumes over large areas, for example, computing the amounts and locations of cut and fill in the grading for a proposed airport runway to be constructed at a given elevation. Another useful application of the contour-area method is in determining the volume of water that will be impounded in the reservoir created by a proposed dam.

Figure 26.8
Determining the volume of water impounded in a reservoir by the contour-area method.

TABLE 26.4 VOLUME COMPUTATION BY CONTOUR-AREA METHOD

	Area		
Contour (1)	(in.²) (2)	(acres) (3)	Volume (acre-ft) (4)
910	1.683	9.659	—
920	5.208	29.889	197.7
930	11.256	64.598	472.4
940	19.210	110.246	874.2
			Σ = 1544.3

■ Example 26.3

Compute the volume of water impounded by the proposed dam illustrated in Figure 26.8. Map scale is 500 ft/in. and the proposed spillway elevation 940 ft.

Solution

The crosshatched portion of Figure 26.8 represents the area that will be inundated with water when the reservoir is full. The solution is presented in Table 26.4. Column (2) gives the area enclosed within each contour (determined by using a tablet digitizer and the coordinate method) in square inches, and in column (3) these areas have been converted to acres based on map scale, that is, 1 in.² = $[(500)^2]/43{,}560$ = 5.739 acres. Column (4) gives the volumes between adjacent contours, computed by Equation [26.1(b)]. The sum of column (4), 1544.3 acre-ft, is the volume of the reservoir.

■ 26.12 MEASURING VOLUMES OF WATER DISCHARGE

Volumes of water discharge in streams and rivers are a matter of vital concern, and must be monitored regularly. In the usual procedure, the stream's cross section is broken into a series of uniformly spaced vertical sections, as illustrated in Figure 26.9. The U.S. Geological Survey recommends using from 25 to 30 sections, with not more than 5% of the total flow occurring in any particular section. Depths and current velocities are measured at each ordinate using a *current meter*. (There are various types available.) The discharge volume for

Figure 26.9
Vertical sections for making stream discharge measurements.

each section is the product of its area and average current velocity. The sum of all section discharges is the total volume of water passing through the stream at the cross-section location. Units of section areas and current velocities can be either ft^2 and ft/sec, respectively, with the discharge in ft^3/sec; or in m^2 and m/sec, respectively, giving the volume in m^3/sec.

Current velocities can be measured at every 0.1 of the depth at each ordinate and the average taken. Alternatively, a good average results from the mean of the 0.2 and 0.8 depth velocities, or a single measurement at the 0.6 depth point. For depths up to 2-1/2 ft, the U.S. Geological Survey uses the 0.6 method; for deeper sections the 0.2 and 0.8 procedure is employed.

The cross section should be taken at right angles to the stream, and in a straight reach with solid bottom and uniform flow. In shallow streams, measurements can be made by wading, in which case, the current meter is held upstream free from eddies caused by the wader's legs. In deeper streams and rivers, measurements are taken from boats, bridges, or overhead cable cars. In these situations, the current meter, with a heavy weight attached to its bottom, is suspended by a cable and thus doubles as a lead line for measuring depths.

■ 26.13 SOFTWARE

As discussed in Section 26.7, the intersection of two lines can be easily programmed to determine the slope intercepts of an end area. As shown in Figure 26.10, WOLFPACK, which can be found on the companion, has been programmed to perform this operation. As shown in this figure, the user must supply the width of the bed, cut slope ratio, and fill slope ratio. In Section 26.7, the roadbed width was 30 ft and had a cut slope of 1-1/2:1. These values are entered in the dialog box as shown. Also note that only one set of design parameters may be used per file. The data file used by this option is shown in Figure 26.11. Each file begins with a title line, which can contain any information pertinent to the file. The title line is followed by the stationing of the first station in the alignment and its elevation. This line is followed by the cross-sectioning information. Distances are entered from left to right as viewed looking forward on the alignment. Cross-sectioning distances to the left of the centerline should be entered as negative values. The distances are followed by a line containing the height of the instrument (HI), and the minus sights that should match their respective distances from the centerline given in the previous line. If more than one set of cross-sectioning notes is available for the alignment, the data for the

Figure 26.10
End-area computations in WOLFPACK.

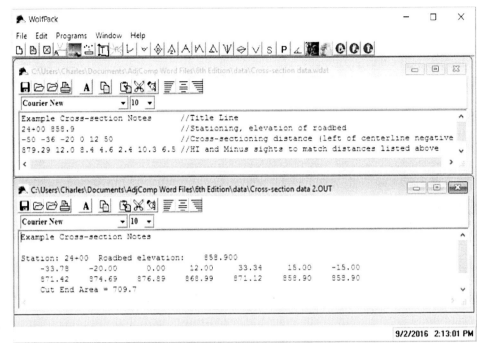

Figure 26.11
Data file and resultant end area computations file.

cross sections can follow the first station's set of notes in succession. At the bottom of Figure 26.11 is the resultant output file. As shown, the left intercept occurs at −33.78 with an elevation of 871.42 ft. The right intercept occurs at 33.34 ft with an elevation of 871.12 ft. The end area is a cut section with an area of 709.7 ft^2.

■ 26.14 SOURCES OF ERROR IN DETERMINING VOLUMES

Some common errors in determining areas of sections and volumes of earthwork are:

1. Making errors in measuring field cross sections, for example, not being perpendicular to the centerline.
2. Making errors in measuring end areas.
3. Failing to use the prismoidal formula where it is justified.
4. Carrying out areas of cross sections beyond the limit justified by the field data.

■ 26.15 MISTAKES

Some typical mistakes made in earthwork calculations are:

1. Confusing algebraic signs in end-area computations using the coordinate method.
2. Using Equation (26.2) for full-station volume computation when partial stations are involved.
3. Using average end-area volumes for pyramidal or wedge-shaped solids.
4. Mixing cut and fill quantities.

PROBLEMS

Asterisks (*) indicate problems that have partial answers given in Appendix G.

26.1 What are the principle methods used to determine volumes in surveying?

26.2 Where must ground elevations be read when using the cross-section method to determine volumes?

26.3 What are design templates?

26.4 Prepare a table of end areas versus depths of fill from 0 to 20 ft by increments of 2 ft for level sections, a 30-ft-wide level roadbed with side slopes of 1-1/2:1.

Draw the cross sections and compute V_e for the data given in Problems 26.5 through 26.8.

26.5* Two level sections 75 ft apart with center heights 4.8 and 7.2 ft in fill, base width 30 ft, side slopes 2:1.

26.6 Two level sections of 30-m stations with center heights of 3.44 and 2.58 m. in cut, base width 15 m, side slopes 1-1/2:1.

26.7 The end area at station 36 + 00 is 305 ft². Notes giving distance from centerline and cut ordinates for station 36 + 60 are C 4.8/17.2; C 5.3/0; C 6.2/10.2. Base is 30 ft.

26.8 An irrigation ditch with b = 12 ft. Notes giving distances from centerline and cut ordinates for stations 52 + 00 and 53 + 00 are C 2.4/10.8; C 3.0; C 3.7/13.4; and C 3.1/14.2; C 3.8; C 4.1/14.2.

26.9 Why must cut and fill volumes be totaled separately?

26.10* For the data tabulated, calculate the volume of excavation in cubic yards between stations 10 + 00 and 15 + 00.

Station	Cut End Area (ft²)
10 + 00	263
11 + 00	358
12 + 00	446
13 + 00	402
14 + 00	274
15 + 00	108

26.11 For the data listed, tabulate cut, fill, and cumulative volumes in cubic yards between stations 10 + 00 and 20 + 00. Use an expansion factor of 1.25 for fills.

Station	End Area (ft²)	
	Cut	Fill
10 + 00	0	
11 + 00	168	
12 + 00	348	
13 + 00	371	
14 + 00	146	
14 + 60	0	0
15 + 00		142
16 + 00		238
17 + 00		305
18 + 00		247
19 + 00		138
20 + 00		106

26.12 Calculate the section areas in Problem 26.5 by the coordinate method.

26.13 Determine the section areas for station 36+60 in Problem 26.7 by the coordinate method.

26.14 From the following excerpt of field notes, plot the cross section on graph paper and superimpose on it a design template for a 30-ft-wide level roadbed with fill slopes of 1-1/2:1 and a subgrade elevation at centerline of 3250.26 ft. Determine the end area graphically by counting squares.

HI = 3246.99 ft						
20 + 00 Lt	$\frac{4.8}{50}$	$\frac{3.6}{28}$	$\frac{5.4}{0}$	$\frac{5.3}{15}$	$\frac{5.8}{32}$	$\frac{6.3}{50}$

26.15 For the data of Problem 26.14, determine the end area by plotting the points in a CAD program, and listing the area.

26.16 For the data of Problem 26.14, calculate slope intercepts, and determine the end area by the coordinate method.

26.17 From the following excerpt of field notes, plot the cross section on graph paper or in a CAD program and superimpose on it a design template for a 30-ft-wide level roadbed with cut slopes of 1-1/2:1 and a subgrade elevation of 240.88 ft. Determine the end area graphically by counting squares.

HI = 252.66 ft						
46 + 00 Lt	$\frac{6.0}{50}$	$\frac{7.9}{27}$	$\frac{5.5}{10}$	$\frac{4.9}{0}$	$\frac{6.6}{24}$	$\frac{6.5}{50}$

26.18 For the data of Problem 26.17, calculate slope intercepts and determine the end area by the coordinate method.

26.19* Complete the following notes and compute V_e and V_P. The roadbed is level, the base is 30 ft.

Station 89 + 00	$\frac{C3.1}{24.3}$	$\frac{C4.9}{0}$	$\frac{C4.3}{35.2}$
Station 88 + 00	$\frac{C6.4}{34.2}$	$\frac{C3.6}{0}$	$\frac{C5.7}{32.1}$

26.20 Similar to Problem 26.19, except the base is 24 ft.

26.21 Calculate V_e and V_P for the following notes. Base is 24 ft.

12 + 90	$\frac{C6.4}{43.6}$	$\frac{C3.6}{0}$	$\frac{C5.7}{40.8}$
12 + 30	$\frac{C3.1}{30.4}$	$\frac{C4.9}{0}$	$\frac{C4.3}{35.2}$

26.22 Calculate V_e, C_P, and V_P for the following notes. The base in fill is 20 ft and base in cut is 30 ft.

46 + 00	$\frac{C3.4}{20.1}$	$\frac{C2.0}{0}$	$\frac{C0.0}{6.0}$	$\frac{F2.0}{13.0}$
45 + 00	$\frac{C2.2}{18.3}$	$\frac{0.0}{0}$	$\frac{F3.0}{14.5}$	

For Problems 26.23 and 26.24, compute the reservoir capacity (in acre-ft) between highest and lowest contours for areas on a topographic map.

26.23*

Elevation (ft)	860	870	880	890	900	910
Area (ft^2)	1370	1660	2293	2950	3550	4850

26.24

Elevation (ft)	1015	1020	1025	1030	1035	1040
Area (ft^2)	1850	1957	2088	2155	2236	2672

26.25 State two situations where prismoidal corrections are most significant.

26.26* Distances (ft) from the left bank, corresponding depths (ft), and velocities (ft/sec), respectively, are given for a river discharge measurement. What is the volume in ft^3/sec? 0, 1.0, 0; 10, 2.3, 1.30; 20, 3.0, 1.54; 30, 2.7, 1.90; 40, 2.4, 1.95; 50, 3.0, 1.60; 60, 3.1, 1.70; 74, 3.0, 1.70; 80, 2.8, 1.54; 90, 3.3, 1.24; 100, 2.0, 0.58; 108, 2.2, 0.28; 116, 1.5, 0.

BIBLIOGRAPHY

Chen, C. and H. Lin. 1990. "Estimating Pit Excavation Volume Using Cubic Spline Volume Formula." *ASCE, Journal of Surveying Engineering* 117 (No. 2): 51.

Chen, C. and H. Lin. 1992. "Estimating Excavation Volumes Using New Formulas." *Surveying and Land Information Systems* 52 (No. 2): 104.

Vijay, R., et al. 2005. "Computation of Reservoir Storage Capacity and Submergence Using GIS." *Surveying and Land Information Science* 65 (No. 4): 255.

27 Photogrammetry

27.1 INTRODUCTION

Photogrammetry may be defined as the science, art, and technology of obtaining reliable information from photographs. It encompasses two major areas of specialization: *metrical* and *interpretative*. The first area is of principal interest to those involved in surveying (geomatics), since it is applied in determining spatial information including distances, elevations, areas, volumes, cross sections, and data for compiling topographic maps from measurements made on photographs. *Aerial* photographs (exposed from aircraft) are normally used, although in certain special applications, *terrestrial* photos (taken from Earth-based cameras) are employed.

Interpretative photogrammetry involves recognizing objects from their photographic images and judging their significance. Critical factors considered in identifying objects are the shapes, sizes, patterns, shadows, tones, and textures of their images. This area of photogrammetry was traditionally called *photographic interpretation* because initially it relied on aerial photos. Now, other sensing and imaging devices such as multispectral scanners, thermal scanners, radiometers, and side-looking airborne radar are used, which aid greatly in interpretation. These instruments sense energy in wavelengths beyond those that the human eye can see, or standard photographic films can record. They are often carried in aircraft as remote as satellites; hence the term, *remote sensing*, is now generally applied to the interpretative area of photogrammetry.

In this chapter, metrical photogrammetry using aerial photographs will be emphasized because it is the area of specialization most frequently applied in surveying work. However, remote sensing has also become very important in small-scale mapping, and in monitoring our environment and managing our natural resources. This subject is discussed further in Section 27.19.

Metrical photogrammetry is accomplished in different ways depending upon project requirements and the type of equipment available. Simple analyses and computations can be made by making measurements on paper prints of aerial photos using engineer's scales, and assuming that the photos are "truly vertical," that is, the camera axis coincided with a plumb line at the time of photography. These methods produce results of lower order, but they are suitable for a variety of applications. Other more advanced techniques, including *analog*, *analytical*, and *softcopy* methods, do not assume vertical photos and provide more accurate determinations of the spatial locations of objects. The analog procedure relies on precise optical and mechanical devices to create models of the terrain that can be measured and mapped. The analytical method is based upon precise measurements of the photographic positions of the images of objects of interest, followed by a mathematical solution for their locations. Softcopy instruments utilize digital images in computerized procedures that are highly automated. While analog and analytical instruments may still exist in academic environments, softcopy instruments are most likely used exclusively in industry. For this reason, readers interested in the analog and analytical instruments should refer to earlier editions of this book or references listed in the bibliography.

27.2 USES OF PHOTOGRAMMETRY

Photography dates back to 1839, and the first attempt to use photogrammetry in preparing a topographic map occurred a year later. Photogrammetry is now the principal method employed in topographic mapping and compiling other forms of spatial data. For example, the U.S. Geological Survey uses the procedure almost exclusively in compiling its maps. Cameras and other photogrammetric instruments and techniques have improved continually, so that spatial data collected by photogrammetry today meets very high accuracy standards. Other advantages of this method are the (1) speed of collecting spatial data in an area, (2) relatively low cost, (3) ease of obtaining topographic details, especially in inaccessible areas, and (4) reduced likelihood of omitting details in spatial data collection.

Photogrammetry presently has many applications in surveying and engineering. For example, it is used in land surveying to compute coordinates of section corners, boundary corners, or points of evidence that help locate these corners. Large-scale maps are made by photogrammetric procedures for many uses, one being subdivision design. Photogrammetry is used to map shorelines in hydrographic surveying, to determine precise ground coordinates of points in control surveying and to develop maps and cross sections for route and engineering surveys. Photogrammetry is playing an increasingly important role in developing the necessary data for modern Land and Geographic Information systems.

Photogrammetry is also being successfully applied in many nonengineering fields, for example, geology, forestry, agriculture, conservation, planning, archeology, military intelligence, traffic management, and accident investigation. It is beyond the scope of this chapter to describe all the varied applications of photogrammetry.

27.3 AERIAL CAMERAS

Aerial mapping cameras are perhaps the most important photogrammetric instruments, since they expose the photographs on which the science depends. To understand photogrammetry, especially the geometrical foundation of its equations, it is essential to have a fundamental understanding of cameras and how they operate. Aerial cameras must be capable of exposing a large number of photographs in rapid succession while moving in an aircraft at high speed; so, a short cycling time, fast lens, efficient shutter, and large-capacity magazine or digital storage are required.

Single-lens frame cameras are the type most often used in metrical photogrammetry. These cameras expose the entire frame or format simultaneously through a lens held at a fixed distance from the focal plane. Generally, they have a format size of 9 × 9 in. (23 × 23 cm), and lenses with focal lengths of 6 in. (152.4 mm), although 3-1/2, 8-1/4, and 12 in. (90, 210, and 305 mm) focal lengths are also used. A single-lens frame camera, together with its viewfinder and electronic controls, is shown in Figure 27.1.

The principal components of a single-lens frame camera are shown in the diagram of Figure 27.2. These include the *lens* (the most important part), which gathers incoming light rays and brings them to focus on the focal plane: the *shutter* to control the interval of time that light passes through the lens; a diaphragm to regulate the size of lens opening; a *filter* to reduce the effect of haze and distribute light uniformly over the format; a *camera cone* to support the lens–shutter–diaphragm assembly with respect to the focal plane and prevent stray light from striking the film; a *focal plane*, the surface on which the film lies when exposed; *fiducial marks* (not shown in Figure 27.2 but illustrated later), four or eight in number, which are essential to define the geometry of the photographs; a *camera body* to house the drive mechanism that cocks and

Figure 27.1
Aerial camera with viewfinder and electronic controls. (From Elements of Photogrammetry: With Applications in GIS, by Wolf & Dewitt, 2000; Courtesy Carl Zeiss, Inc. and McGraw-Hill Book Co., Inc.)

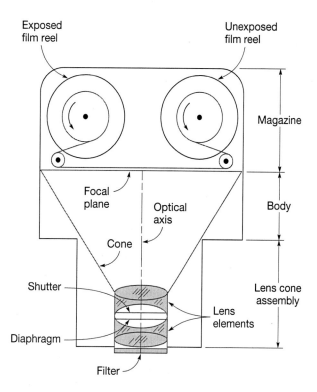

Figure 27.2 Principal components of a single lens frame aerial camera.

trips the shutter, flattens the film, and advances it between exposures; and *a magazine*, which holds the supply of exposed and unexposed film, or houses the digital storage device.

An aerial camera shutter can be operated manually by an operator, or automatically by the electronic control mechanism, so that photos are taken at specified intervals. A level vial attached to the camera enables an operator to keep the optical axis of the camera lens, which is perpendicular to the focal plane, nearly vertical in spite of any slight tip and tilt of the aircraft.

Images of the fiducial marks are printed on the photographs and lines joining opposite pairs intersect at or very near the *principal point*, defined as the point where a perpendicular from the emergent nodal point of the camera lens strikes the focal plane. Fiducial marks may be located in the corners, on the sides, or preferably in both places, as shown in Figures 27.4 and 27.5.

In digital cameras, an array of solid-state detectors, which are placed in the focal plane capture the image from the lens. The most common type of detector is the *charge-coupled device* (CCD). The array is composed of tiny detectors arranged in contiguous rows and columns, as shown in Figure 27.3. Each detector senses the energy received from its corresponding ground scene, and this constitutes one "picture element" *(pixel)* within the overall image. The principle of operation of CCDs is fundamentally quite simple. At any specific pixel location,

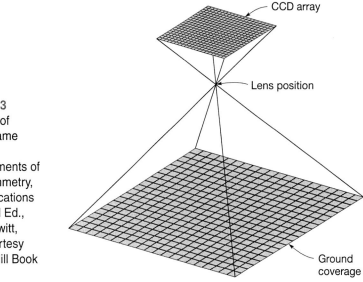

Figure 27.3
Geometry of a digital frame camera. (From Elements of Photogrammetry, With Applications in GIS, 3rd Ed., Wolf & Dewitt, 2000; Courtesy McGraw-Hill Book Co., Inc.)

the CCD element there is exposed to incident light energy, which builds up an electric charge proportional to the intensity of the incoming light. The electric charge is amplified, converted from analog to digital form and stored in a file together with its row and column location within the array. Currently, the sizes of the individual CCD elements being manufactured are in the range of from about 5 to 15 μm^2, and arrays may consist of from 500 rows and columns (250,000 pixels) for inexpensive cameras, to more than 4000 rows and columns. Obviously, significant storage and data handling capabilities are necessary in acquiring and processing digital images.

Aerial mapping cameras, whether the film- or digital-type, are calibrated to get precise values for the focal length and lens distortions. Flatness of the focal plane, relative position of the principal point with respect to the fiducial marks, and fiducial mark locations are also specified. These calibration data are necessary for precise photogrammetric work.

27.4 TYPES OF AERIAL PHOTOGRAPHS

Aerial photographs exposed with single-lens frame cameras are classified as *vertical* (taken with the camera axis aimed vertically downward or as nearly vertical as possible) and *oblique* (made with the camera axis intentionally inclined at an angle between the horizontal and vertical). Oblique photographs are further classified as *high* if the horizon shows on the picture and *low* if it does not. Figures 27.4 and 27.5 show examples of vertical and low oblique photographs, respectively. As illustrated by these examples, aerial photos clearly depict all natural and cultural features within the region covered such as roads, railroads, buildings, rivers, bridges, trees, and cultivated lands.

27.4 Types of Aerial Photographs   823

Figure 27.4
Vertical aerial photograph. (Courtesy Pennsylvania Department of Transportation.)

Figure 27.5
Low-oblique image of Harrisburg capitol building complex. (Courtesy Pennsylvania Department of Transportation.)

824 PHOTOGRAMMETRY

Vertical photographs are the principal mode of obtaining imagery for photogrammetric work. Oblique photographs are seldom used for metrical applications but are advantageous in interpretative work and for reconnaissance.

27.5 VERTICAL AERIAL PHOTOGRAPHS

A *truly vertical photograph* results if the axis of the camera is exactly vertical when exposure is made. Despite all precautions, small tilts, generally less than 1° and rarely greater than 3°, are invariably present, and the resulting photos are called *near-vertical* or *tilted* photographs. Although vertical photographs look like maps to laypersons, they are not true orthographic projections of the Earth's surface. Rather, they are perspective views, and the principles of perspective geometry must be applied to prepare maps from them. Figure 27.6 illustrates the geometry of a vertical photograph taken at exposure station L. The photograph, considered a *contact print positive*, is a 180° exact reversal of the negative. The positive shown in Figure 27.6 is used to develop photogrammetric equations in subsequent sections.

Distance oL (Figure 27.6) is the camera focal length. The x and y reference axes system for measuring photographic coordinates of images is defined by straight lines joining opposite-side fiducial marks shown on the positive of

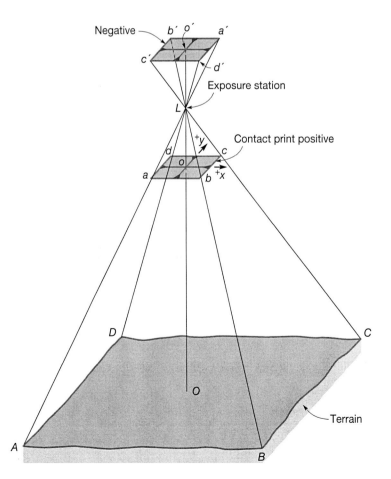

Figure 27.6
Geometry of a vertical aerial photograph.

Figure 27.6. The x axis, arbitrarily designated as the line most nearly parallel with the direction of flight, is positive in the direction of flight. Positive y is 90° counterclockwise from positive x.

Vertical photographs for topographic mapping are taken in strips, which normally run lengthwise over the area to be covered. The strips or *flight lines* generally have a *sidelap* (overlap of adjacent flight lines) of about 30%. *Endlap* (overlap of adjacent photographs in the same flight line) is usually about $60 \pm 5\%$. Figures 27.19(a) and (b) illustrate endlap and sidelap. An endlap of 50% or greater is necessary to assure that all ground points will appear in at least two photographs, and that some will show in three. Images common to three photographs permit *aerotriangulation* to extend or densify control through a strip or block of photographs using only minimal existing control.

27.6 SCALE OF A VERTICAL PHOTOGRAPH

Scale is ordinarily interpreted as the ratio of a distance on a map to that same length on the ground. On a map it is uniform throughout because a map is an orthographic projection. The scale of a vertical photograph is the ratio of a photo distance to the corresponding ground distance. Since a photograph is a perspective view, scale varies from point to point with variations in terrain elevation.

In Figure 27.7, L is the exposure station of a vertical photograph taken at an altitude H above datum. The camera focal length is f and o the photographic principal point. Points A, B, C, and D, which lie at elevations above datum of h_A, h_B, h_C, and h_D, respectively are imaged on the photograph at a, b, c, and d. The scale at any point can be expressed in terms of its elevation, the camera focal length, and the flying height above datum. From Figure 27.7, from similar triangles Lab and LAB, the following expression can be written:

$$\frac{ab}{AB} = \frac{La}{LA} \qquad (27.1a)$$

Also from similar triangles Loa and LOA, a similar expression results:

$$\frac{La}{LA} = \frac{f}{H - h_A} \qquad (27.1b)$$

Equating (a) and (b), recognizing that ab/AB equals photo scale at A and B, and considering AB to be infinitesimally short, the equation for the scale at A is

$$S_A = \frac{f}{H - h_A} \qquad (27.1c)$$

Scales at B, C, and D may be expressed similarly as $S_B = f/(H - h_B)$, $S_C = f/(H - h_C)$, and $S_D = f/(H - h_D)$.

It is apparent from these relationships that photo scale increases at higher elevations and decreases at lower ones. This concept is seen graphically in Figure 27.7, where ground lengths AB and CD are equal, but photo distances ab and cd are not, cd being longer and at larger scale than ab because of the higher elevation

826 PHOTOGRAMMETRY

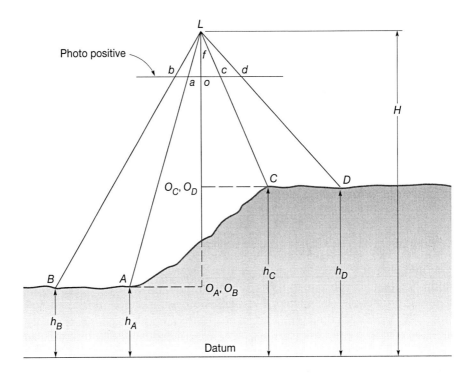

Figure 27.7
Scale of a vertical photograph.

of CD. In general, by dropping subscripts, the scale S at any point whose elevation above datum is h may be expressed as

$$S = \frac{f}{H - h} \tag{27.2}$$

where S is the scale at any point on a vertical photo, f is the camera focal length, H the flying height above datum, and h the elevation of the point.

Use of an average photographic scale is frequently desirable but must be accepted with caution as an approximation. For any vertical photographs taken of terrain whose average elevation above datum is h_{avg}, the average scale S_{avg} is

$$S_{avg} = \frac{f}{H - h_{avg}} \tag{27.3}$$

■ Example 27.1

The vertical photograph of Figure 27.7 was exposed with a 6-in. focal length camera at a flying height of 10,000 ft above datum. (1) What is the photo scale at point a if the elevation of point A on the ground is 2500 ft above datum?

(2) For this photo, if the average terrain is 4000 ft above datum, what is the average photo scale?

Solution

(a) From Equation (27.2),

$$S_A = \frac{f}{H - h_A} = \frac{6 \text{ in.}}{10{,}000 - 2500} = \frac{1 \text{ in.}}{1250 \text{ ft}} = 1{:}15{,}000$$

(b) From Equation (27.3),

$$S_{avg} = \frac{f}{H - h_{avg}} = \frac{6 \text{ in.}}{10{,}000 - 4000} = \frac{1 \text{ in}}{1000 \text{ ft}} = 1{:}12{,}000$$

The scale of a photograph can be determined if a map is available of the same area. This method does not require the focal length and flying height to be known. Rather, it is necessary only to measure the photographic distance between two well-defined points also identifiable on the map. Photo scale is then calculated from the equation

$$\text{photo scale} = \frac{\text{photo distance}}{\text{map distance}} \times \text{map scale} \tag{27.4}$$

In using Equation (27.4), the distances must be in the same units, and the answer is the scale at the average elevation of the two points used.

■ Example 27.2

On a vertical photograph, the length of an airport runway measures 4.24 in. On a map plotted to a scale of 1:9600, it extends 7.92 in. What is the photo scale at the runway elevation?

Solution

From Equation (27.4),

$$S = \frac{4.24}{7.92}\left(\frac{1}{9600}\right) = \frac{1}{17{,}930} \text{ or } 1 \text{ in.} = 1490 \text{ ft}$$

Note that the additional fourth digit in the scale is carried to limit errors due to rounding when these values are used in subsequent computations.

The scale of a photograph can also be computed readily if lines whose lengths are common knowledge appear in the photograph. Section lines, a football or baseball field, and so on, can be measured on the photograph and an approximate scale at that elevation ascertained as the ratio of measured photo distance to known ground length. With an approximate photographic scale known, rough determinations of the lengths of lines appearing in the photo can be made.

■ Example 27.3

On a certain vertical aerial photo, a section line (assumed to be 5280 ft long) is imaged. Its photographic length is 3.32 in. On this same photo, a rectangular parcel of land measures 1.74 by 0.83 in. Calculate the approximate ground dimensions of the parcel and its acreage.

Solution

1. Approximate photo scale

$$\frac{3.32}{5280} = \frac{1 \text{ in}}{1590 \text{ ft}} \text{ or } 1 \text{ in.} = 1590 \text{ ft}$$

2. Parcel dimensions and area:

$$\text{length} = 1590(1.74) = 2770 \text{ ft}$$
$$\text{width} = 1590(0.83) = 1320 \text{ ft}$$
$$\text{area} = \frac{2770(1320)}{43,560} = 84 \text{ acres}$$

■ 27.7 GROUND COORDINATES FROM A SINGLE VERTICAL PHOTOGRAPH

Ground coordinates of points whose images appear in a vertical photograph can be determined with respect to an arbitrary ground-axis system. The arbitrary X and Y ground axes are in the same vertical planes as photographic x and y, respectively, and the system's origin is in the datum plane vertically beneath the exposure station. Ground coordinates of points determined in this manner are used to calculate horizontal distances, horizontal angles, and areas.

Figure 27.8 illustrates a vertical photograph taken at flying height H above datum. Images a and b of ground points A and B appear on the photograph. The measured photographic coordinates are x_a, y_a, x_b, and y_b; the ground coordinates are X_A, Y_A, X_B, and Y_B. From similar triangles LO_AA' and Loa',

$$\frac{oa'}{O_AA'} = \frac{f}{H - h_A} = \frac{x_a}{X_A}$$

Then

$$X_A = \frac{(H - h_A)x_a}{f} \qquad (27.5)$$

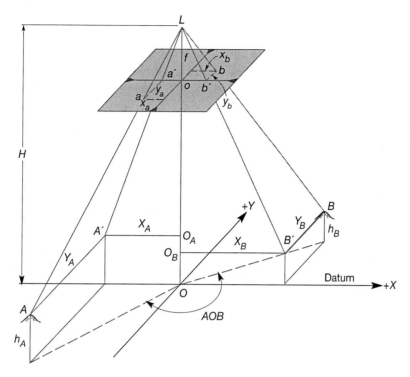

Figure 27.8
Ground coordinates from a vertical photograph.

Also, from similar triangles $LA'A$ and $La'a$,

$$\frac{a'a}{A'A} = \frac{f}{H - h_A} = \frac{y_a}{Y_A}$$

and

$$Y_A = \frac{(H - h_A)y_a}{f} \tag{27.6}$$

Similarly,

$$X_B = \frac{(H - h_B)x_b}{f} \tag{27.7a}$$

$$Y_B = \frac{(H - h_B)y_b}{f} \tag{27.7b}$$

Note that Equations (27.5) through (27.7) require point elevations h_A and h_B for their solution. These are normally either taken from existing contour maps, or they can be obtained by differential or trigonometric leveling. From the X and Y coordinates of points A and B, the horizontal length of line AB can be calculated using Equation (14.4).

If X and Y coordinates of all corners of a parcel are computed in this way, the parcel area can be determined from those coordinates by the method discussed in Chapter 12. The advantage of calculating lengths and areas by the coordinate formulas, rather than by average scale as in Example 27.3, is that better accuracy

830 PHOTOGRAMMETRY

results because differences in elevation, which cause the photo scale to vary, are more rigorously taken into account.

27.8 RELIEF DISPLACEMENT ON A VERTICAL PHOTOGRAPH

Relief displacement on a vertical photograph is the shift or movement of an image from its theoretical datum location caused by the object's relief—that is, its elevation above or below datum. Relief displacement on a vertical photograph occurs along radial lines from the principal point and increases in magnitude with greater distance from the principal point to the image.

The concept of relief displacement in a vertical photograph taken from a flying height H above datum is illustrated in Figure 27.9, where the camera focal length is f and o is the principal point. Points B and C are the base and top, respectively, of a pole with images at b and c on the photograph. A is an imaginary point on the datum plane vertically beneath B with corresponding imaginary position a on the photograph. Distance ab on the photograph is the image displacement due to h_B the elevation of B above datum, and bc is the image displacement because of the height of the pole.

From similar triangles of Figure 27.9, an expression for relief displacement is formulated. First, from triangles LO_AA and Loa,

$$\frac{r_a}{R} = \frac{f}{H}$$

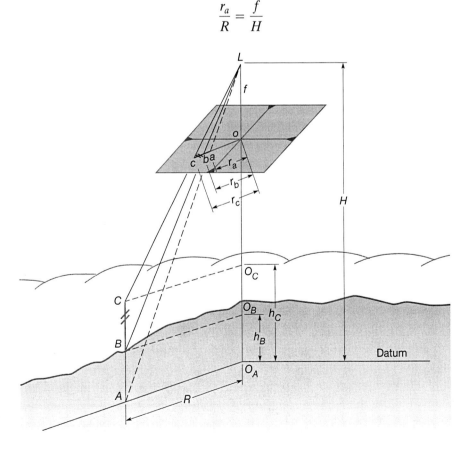

Figure 27.9
Relief displacement on a vertical photograph.

and rearranging,

$$r_a H = fR \tag{27.8a}$$

Also from similar triangles $LO_B B$ and Lob,

$$\frac{r_b}{R} = \frac{f}{H - h_B} \text{ or } r_b(H - h_B) = fR \tag{27.8b}$$

Equating (27.8a) and (27.8b),

$$r_a H = r_b(H - h_B)$$

and rearranging,

$$r_b - r_a = \frac{r_b h_B}{H}$$

If $d_b = r_b - r_a$ is the relief displacement of image b, then $d_b = r_b h_B / H$. Dropping subscripts, the equation can be written in general terms as

$$d = \frac{rh}{H} \tag{27.9}$$

where d is the relief displacement, r the photo radial distance from the principal point to the image of the displaced point, h the height above datum of the displaced point, and H the flying height above that same datum.

Equation (27.9) can be used to locate the datum photographic positions of images on a vertical photograph. True horizontal angles may then be scaled directly from the datum images, and if the photo scale at datum is known, true horizontal lengths of the lines can be measured directly. The datum position is located by scaling the calculated relief displacement d of a point along a radial line to the principal point (inward for a point whose elevation is above datum).

To determine heights using the equation, images of both the top and bottom of an object must be visible.

■ Example 27.4

In Figure 27.9, radial distance r_b to the image of the base of the pole is 75.23 mm, and radial distance r_c to the image of its top is 76.45 mm. The flying height H is 4000 ft above datum, and the elevation of B is 450 ft. What is the height of the pole?

Solution

The relief displacement is $r_c - r_b = 76.45 - 75.23 = 1.22$ mm. Selecting a datum at the pole's base and applying Equation (27.9),

$$d = \frac{rh}{H} \quad \text{so} \quad 1.22 = \frac{76.45h}{4000 - 450}$$

Then

$$h = \frac{3550(1.22)}{76.45} = 56.6 \text{ ft}$$

The relief displacement equation is particularly valuable to photo interpreters, who are usually interested in relative heights rather than absolute elevations. Figure 27.4 vividly illustrates relief displacements. This vertical photo shows the relief displacement of a water tower in the center-right-hand portion of the format. This displacement, as well as that of other buildings throughout the photograph, occurs radially outward from the principal point.

27.9 FLYING HEIGHT OF A VERTICAL PHOTOGRAPH

From previous sections it is apparent that the flying height above datum is an important parameter in solving basic photogrammetry equations. For rough computations, flying heights can be taken from altimeter readings if available. An approximate H can also be obtained by using Equation (27.2) if a line of known length appears on a photograph.

Example 27.5

The length of a section line (known to be 5280 ft) is measured on a vertical photograph as 4.15 in. Find the approximate flying height above the terrain if $f = 6$ in.

Solution

Assuming the datum at the section line elevation, Equation (27.2) reduces to

$$\text{scale} = \frac{f}{H} \quad \text{and} \quad \frac{4.15}{5280} = \frac{6}{H}$$

from which

$$H = \frac{5280(6)}{4.15} = 7630 \text{ ft above the terrain}$$

If the images of two ground control points a and b appear on a vertical photograph, the flying height can be determined more precisely from the Pythagorean theorem,

$$L^2 = (X_B - X_A)^2 + (Y_B - Y_A)^2$$

Substituting Equations (27.5) through (27.7) into this above expression,

$$L^2 = \left[\frac{(H - h_B)x_b - (H - h_A)x_a}{f}\right]^2 + \left[\frac{(H - h_B)y_b - (H - h_A)y_a}{f}\right]^2 \quad (27.10)$$

where L is the horizontal length of ground line AB, H the flying height above datum, h_A and h_B the elevations of the control points above datum, and x and y the measured photo coordinates of the control points.

In Equation (27.10) all variables except H are known. Hence a direct solution can be found for the unknown flying height. The equation is quadratic, so there are two solutions, but the incorrect one will be obvious and can be discarded.

27.10 STEREOSCOPIC PARALLAX

Parallax is defined as the apparent displacement of the position of an object with respect to a frame of reference due to a shift in the point of observation. For example, a person looking through the view finder of an aerial camera in an aircraft as it moves forward sees images of objects moving across the field of view. This apparent motion (parallax) is due to the changing location of the observer. By using the camera format as a frame of reference, it can be seen that parallax exists for all images appearing on successive photographs due to forward motion between exposures. Points closer to the camera (of higher elevation) will appear to move faster and have greater parallaxes than lower ones. For 60% endlap, the parallax of images on successive photographs should average approximately 40% of the focal plane width.

Parallax of a point is a function of its relief, and consequently measuring it provides a means of calculating elevations. It is also possible to compute X and Y ground coordinates from parallax.

Movement of an image across the focal plane between successive exposures takes place in a line parallel with the direction of flight. Thus to measure parallax, that direction must first be established. For a pair of overlapping photos, this is done by locating positions of the principal points and *corresponding principal points* (that is, principal points transferred to their places in the overlap area of the other photo). A line on each print ruled through these points defines the direction of flight. It also serves as the photographic x-axis for parallax measurement. The y-axis for making parallax measurements is drawn perpendicular to the flight line passing through each photo's principal point. The x coordinate of a point is scaled on each photograph with respect to the axes so constructed and the parallax of the point is then calculated from the expression

$$p = x - x_1 \quad (27.11)$$

Photographic coordinates x and x_1 are measured on the left-hand and right-hand prints, respectively, with due regard given for algebraic signs.

Figure 27.10 illustrates an overlapping pair of vertical photographs exposed at equal flight heights H above datum. The distance between exposure stations

Figure 27.10 Stereoscopic parallax relationships.

L and L_1 is called B, the *air base*. The inset figure shows the two exposure stations L and L_1 in superposition to make the similarity of triangles La'_1a' and $LA'L_1$ more easily recognized. When these two similar triangles are equated, there results

$$\frac{p}{f} = \frac{B}{H-h}$$

from which

$$H - h = \frac{Bf}{p} \tag{27.12a}$$

Also, from similar triangles LOA' and Loa',

$$X = \frac{x}{f}(H-h) \tag{27.12b}$$

Substituting Equation (27.12a) into (27.12b) gives

$$X = \frac{B}{p}x \tag{27.13a}$$

and from triangles LAA' and Laa', with substitution of Equation (27.12a) yields

$$Y = \frac{B}{p} y \qquad (27.13b)$$

In Equations (27.13a) and (27.13b), X and Y are ground coordinates of a point with respect to an origin vertically beneath the exposure station of the left photograph, with positive X coinciding with the direction of flight. Positive Y is 90° counterclockwise to positive X. The parallax of the point is p, x and y the photographic coordinates of a point on the left-hand print, H the flying height above datum, h the point's elevation above the same datum, and f the camera lens focal length.

Equations (27.12a) through (27.13b), commonly called the *parallax equations*, are useful in calculating horizontal lengths of lines and elevations of points. They also provide the fundamental basis for the design and operation of stereoscopic plotting instruments.

■ Example 27.6

The length of line AB and elevations of points A and B, whose images appear on two overlapping vertical photographs, are needed. The flying height above datum was 4050 ft, and the air base, 2410 ft. The camera had a 6-in. focal length. Measured photographic coordinates (in inches) on the left-hand image are $x_a = 2.10$, $x_b = 3.50$, $y_a = 2.00$, and $y_b = -1.05$; on the right-hand image, $x_{1a} = -2.25$ and $x_{1b} = -1.17$.

Solution

From Equation (27.11),

$$p_a = x_a - x_{1a} = 2.10 - (-2.25) = 4.35 \text{ in.}$$
$$p_b = x_b - x_{1b} = 3.50 - (-1.17) = 4.67 \text{ in.}$$

By Equations (27.13a) and (27.13b),

$$X_A = \frac{B}{p_a} x_a = \frac{2410(2.10)}{4.35} = 1160 \text{ ft}$$

$$X_B = \frac{2410(3.50)}{4.67} = 1810 \text{ ft}$$

$$Y_A = \frac{B}{p_a} y_a = \frac{2410(2.00)}{4.35} = 1110 \text{ ft}$$

$$Y_B = \frac{2410(-1.05)}{4.67} = -542 \text{ ft}$$

By Equation (11.4), length AB is

$$AB = \sqrt{(1810 - 1160)^2 + (-542 - 1110)^2} = 1780 \text{ ft}$$

By Equation (27.12a), the elevations of A and B are

$$h_A = H - \frac{Bf}{p_a} = 4050 - \frac{2410(6)}{4.35} = 726 \text{ ft}$$

$$h_B = H - \frac{Bf}{p_b} = 4050 - \frac{2410(6)}{4.67} = 954 \text{ ft}$$

27.11 STEREOSCOPIC VIEWING

The term *stereoscopic viewing* means seeing an object in three dimensions. This is a process that requires normal *binocular* (two-eyed) vision. In Figure 27.11, two eyes L and R are separated by a distance b called the *eye base*. When the eyes are focused on point A, their optical axes converge to form angle φ_1, and when sighting on B, φ_2 is produced. Angles φ_1 and φ_2 are called *parallactic angles*, and the brain associates distances d_A and d_B with them. The depth $d_B - d_A$ of the object is perceived from the brain's unconscious comparison of these parallactic angles.

If two photographs of the same subject are taken from two different perspectives or camera stations, the left print viewed with the left eye and simultaneously the right print seen with the right eye, a mental impression of a three-dimensional model results. In normal stereoscopic viewing (not using photos), the eye base gives a true impression of parallactic angles. While looking at aerial photographs stereoscopically, the exposure station spacing simulates an eye base so the viewer actually sees parallactic angles comparable with having one eye at each of the

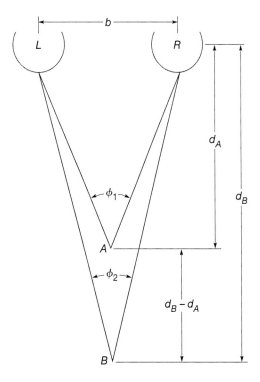

Figure 27.11
Parallactic angles in stereoscopic viewing.

Figure 27.12
Folding mirror stereoscope with parallax bar.

two exposure stations. This creates a condition called *vertical exaggeration*, which causes the vertical scale of the three-dimensional model to appear greater than its horizontal scale; that is, objects are perceived to be too tall. The condition is of concern to photo interpreters who often estimate heights of objects and slopes of surfaces when viewing air photos stereoscopically. The amount of vertical exaggeration varies with percent endlap and the camera's format dimensions and focal length. A factor of about 4 results if endlap is 60% and the camera has a 9-in. (23-cm) format with 6-in. (152-mm) focal length.

The *stereoscope* shown in Figure 27.12 permits viewing photographs stereoscopically by enabling the left and right eyes to focus comfortably on the left and right prints, respectively, assuming proper orientation has been made of the overlapping pair of photographs under the instrument. Correct orientation requires the two photographs to be laid out in the same order they were taken; with the stereoscope so set that the line joining its lens centers is parallel with the direction of flight. The print spacing is varied, carefully maintaining this parallelism, until a clear three-dimensional view (stereoscopic model) is obtained.

27.12 STEREOSCOPIC MEASUREMENT OF PARALLAX

The parallax of a point can be measured while viewing stereoscopically with the advantage of speed and, because binocular vision is used, greater accuracy. As the viewer looks through a stereoscope, imagine that two small identical marks etched on pieces of clear glass, called *half-marks*, are placed over each photograph. The viewer simultaneously sees one mark with the left eye and the other with the right eye. Then the positions of the marks are shifted until they seem to fuse together as one mark that appears to lie at a certain elevation. The height of the mark will vary or "float" as the spacing of the half-marks is varied; hence it is called the *floating mark*. Figure 27.13 demonstrates this principle and also illustrates that the floating mark can be set exactly on particular points such as A, B, and C by placing the half-marks (small black dots) at a and a', b and b', and c and c', respectively.

Based on the floating mark principle, the parallax of points is observed stereoscopically with a parallax bar, as shown beneath the stereoscope in

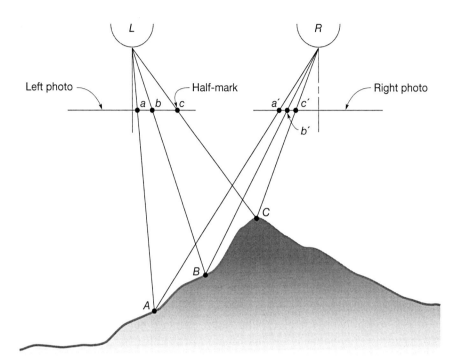

Figure 27.13
Principle of the floating mark.

Figure 27.12. It is simply a bar to which two half-marks are fastened. The right mark can be moved with respect to the left one by turning a micrometer screw to register the displacement on a dial. When the floating mark appears to rest on a point, a micrometer reading is taken and added to the parallax bar *setup constant* to obtain the parallax.

When a parallax bar is used, two overlapping photographs are oriented properly for viewing under a mirror stereoscope and fastened securely with respect to each other using drafting tape. The parallax bar constant for the setup is determined by measuring the photo coordinates for a discrete point and applying Equation (27.11) to obtain its parallax.

The floating mark is placed on the same point, the micrometer read, and the constant for the setup found by

$$C = p - r \qquad (27.14)$$

where C is the parallax bar setup constant, p the parallax of a point determined by Equation (27.11), and r the micrometer reading obtained with the floating mark set on that same point.

Once the constant has been determined, the parallax of any other point can be computed by adding its micrometer reading to the constant. Thus, a single measurement gives the parallax of a point. Each time another pair of photos is oriented for parallax measurements, a new parallax bar setup constant must be determined. A major advantage of the stereoscopic method is that parallaxes of nondiscrete points can be determined. Thus, elevations of hilltops, depressions, and so on, in fields can be calculated using Equations (27.12) and (27.13), even though their x coordinates cannot be measured for use in Equation (27.11).

27.13 ANALYTICAL PHOTOGRAMMETRY

Analytical photogrammetry involves the rigorous mathematical calculation of ground coordinates of points using computers. Input data consists of camera parameters (i.e., the lens focal length, its distortion characteristics, and the principal point location); observed photo coordinates of the images of all points whose ground coordinates are to be determined, as well as those of a limited number of well-distributed ground control points; and the ground coordinates of the control points. The photo coordinates are measured with respect to the coordinate system illustrated in Figure 27.6. Extremely precise instruments called *comparators* are used and values are recorded to the nearest micrometer. Unlike the elementary methods presented in earlier sections of this chapter that assume vertical photos and equal flying heights, analytical photogrammetry rigorously accounts for these variations.

Analytical photogrammetry generally involves the formation of large, rather complex, systems of redundant equations, which are then solved using the method of least squares. The concepts have existed for many years, but it was not until the advent of computers that the procedures became practical. The formation of the equations used in analytical photogrammetry is beyond the scope of this text, but interested students can find their derivations, and illustrations of their use, in books that specialize in photogrammetry.[1]

As noted above, accuracies attainable using analytical photogrammetry are very high and are frequently expressed as a ratio of the flying height of the photography used. Accuracies within about 1/10,000th to 1/15,000th of the flying height above ground are routinely obtained in computed X, Y, and Z coordinates. Thus, for photos taken from 6,000 ft above ground, coordinates accurate to within about ± 0.4 to 0.6 ft can be expected.

Analytical photogrammetry forms the basis for *softcopy stereoplotters*, which is discussed in the following section.

27.14 STEREOSCOPIC PLOTTING INSTRUMENTS

Stereoscopic plotting instruments, also simply called *stereoplotters*, are devices designed to provide accurate solutions for X, Y, and Z object space coordinates of points from their corresponding image locations on overlapping pairs of photos. The fact that the photos may contain varying amounts of tilt and have differing flying heights is of no consequence, because these instruments rigorously account for the position and orientation of the camera for each exposure. Stereoplotters are used to take cross sections, record digital elevation models, compile topographic maps, and generate other types of spatially related topographic information from overlapping aerial photographs.

Stereoplotters can be classified into four different categories: (1) *optical projection*, (2) *mechanical projection*, (3) *analytical*, and (4) *digital* or "softcopy" systems. Regardless of the type, all instruments contain optical and mechanical elements, and the newer versions have either built-in or interfaced computers. Only analytical and softcopy stereoplotters are now used in industry. For this reason,

---

[1] See Wolf, P. R., B. A. Dewitt, and B. E. Wilkinson. 2014. "Elements of Photogrammetry: With Applications in GIS," 4th ed., McGraw-Hill Book Co., Inc., New York.

840 PHOTOGRAMMETRY

only these two types of stereoplotters will be discussed herein. Readers can refer to previous editions of this book for information on the other type of plotters.

27.14.1 Basic Concepts in Stereoplotters

Figure 27.14 illustrates the basic design concepts of a stereoplotter. In Figure 27.14(a), an overlapping pair of aerial photos is exposed. *Diapositives* (positives developed on film or glass plates) are prepared to exacting standards from the negatives and placed in the projectors of the stereoplotter, as shown in Figure 27.14(b). With the diapositives in place, light rays are projected through them and the positions and orientations of the projectors are adjusted until the rays from corresponding images intersect below to form a model of the overlap area of the aerial photos. The model of the terrain is called a *stereomodel* and once formed, it can be viewed, measured, and mapped.

In order to accomplish the steps outlined in the preceding paragraph, it is necessary that stereoplotters incorporate the following components: (1) a projection system (to project the light rays that create a stereomodel), (2) a viewing system (which enables an operator to see the stereomodel in three dimensions), and (3) a measuring/tracing system (for measuring or mapping the stereomodel). Projectors used in optical projection stereoplotters function like ordinary slide projectors. However, they are much more precise and can be adjusted in angular orientation and position to recreate the spatial locations and attitudes that the aerial cameras had when the overlapping photos were exposed. This produces a "true" model of the terrain in the overlap area. The scale of the stereomodel is, of course, greatly reduced, and is the ratio of the *model base* (distance b between projector lenses) to the *air base* (actual distance B between the two exposure stations).

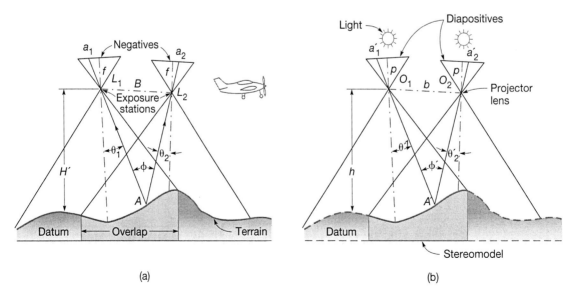

Figure 27.14 Fundamental concepts of stereoscopic plotting instrument design. (a) Aerial photography; (b) stereoscopic plotter.
(From *Elements of Photogrammetry: With Applications in GIS*, 3rd Ed., 2000, by Wolf & Dewitt, Courtesy McGraw-Hill Book Co., Inc.)

Plotter viewing systems must provide a stereoscopic view and hence be designed so that the left and right eyes see only projected images of the corresponding left and right diapositives. One method of accomplishing this is to project one image with a blue filter and the other with a red filter. The operator wears a pair of spectacles with corresponding blue and red lenses. Another method of separating the left and right images is to project them on half of the screen. The operator must then look through a viewing system similar to a stereoscope to restrict the left eye to view the left image only and the right eye the right. These systems of viewing stereoscopically are called the *anaglyphic* method. Other viewing systems are known as *passive* and *active*. In the passive viewing system, left and right images are projected in opposite polarity. The operator wears polarized spectacles with corresponding polarity in the lenses so that the left can only see the left image and the right eye can only see the right image. As shown in Figure 27.15, the active viewing system also projects the left and right images in opposite polarity. However, in the active system, the images are rapidly swapped in the projection system also. The operator's viewing spectacles are synchronized so that the left and right eyes can see the corresponding left and right images only.

A stereoplotter operator, preparing to measure or map a stereomodel, must go through a three-stage orientation process consisting of *interior orientation*, *relative orientation*, and *absolute orientation*. Interior orientation ensures that the light rays are geometrically correct, that is, angles θ'_1 and θ'_2 of Figure 27.14(b) (i.e., the angles between the light rays and the axis of the lens), must be identical to

Figure 27.15 Softcopy photogrammetric workstation.
(Courtesy Pennsylvania Department of Transportation.)

corresponding angles θ_1 and θ_2, respectively, in Figure 27.14(a) (i.e., the angles between the incoming light rays and the camera axis). Preparing the digital images to exacting specifications and precisely observing the images of the fiducial marks will accomplish this. The process called *relative orientation* is accomplished when parallactic angle φ' of Figure 27.14(b) for each corresponding pair of light rays will be identical to its corresponding parallactic angle φ of Figure 27.14(a) and a perfect three-dimensional model is formed. The model is brought to required scale by making the rays of at least two, but preferably three, ground control points intersect at their known positions at a desired scale. It is leveled by measuring image coordinates on a minimum of three, but preferably four, corner ground control points when the floating mark is set on them. *Absolute orientation* is a term applied to the processes of scaling and leveling the model.

When orientation is completed, a map can be made from the model, or cross sections and other spatial information compiled. In mapping, planimetric details are located first by bringing the floating mark into contact with objects in the model and tracing them. The positions of the details are digitally determined and recorded in a map file. Contours are traced by observing the positions and elevations of selected points throughout the model space. Then using procedures similar to those discussed in Section 18.14, a digital terrain model of the ground is created. Contours are determined from the triangulated model as discussed in Section 18.14. When a digital map is completed, it is examined for omissions and mistakes and field checked.

27.14.2 Softcopy Stereoplotters

Softcopy plotters are the latest development in the evolution of stereoscopic plotting instruments and have replaced the previous specialized systems. A softcopy photogrammetric system can be defined as a combined software and viewing system that allows for the precise measurement of photographic locations on a computer screen followed by a mathematical solution for their locations on the ground. Softcopy systems utilize digital images. These images can be acquired by using a digital camera of the type described in Section 27.3, or by scanning the negatives of images taken with film cameras. Scanners convert the contents of an image into an array of pixels arranged in rows and columns. Each pixel is identified by its row and column, and is assigned a digital number, which corresponds to its *gray level* or degree of darkness at that particular element. Figure 27.16 shows the PhotoScan, developed jointly by Carl Zeiss and Intergraph Corporation for digitizing photographs.

A softcopy photogrammetric system requires a computer with a high-resolution graphics display. The computer must be capable of efficiently manipulating large files of digital images. Also, the equipment must include a stereoscopic viewing mechanism, that is, one that enables the operator to see only the left image with the left eye and only the right image with the right eye. The operator moves the floating mark about the stereomodel and places it on any point whose position is desired. Once the mark is set, the row and column locations of the pixels at its location identify the photo coordinates of the point to the computer, which can then calculate, in real time, its ground X, Y, and Z coordinates by solving equations of analytical photogrammetry.

One type of softcopy photogrammetric workstation is shown in Figure 27.15. As seen in the figure, it displays the left and right images on one monitor

Figure 27.16 Zeiss/Intergraph PS1 PhotoScan. (Courtesy Carl Zeiss.)

and shows the generated map on another. With this device, the left and right photos are displayed simultaneously on the left and right portions, respectively, of the computer's monitor. An operator looking through an active pair of glasses can view the stereomodel in three dimensions. The floating marks, made of two reference pixels, one on each image, can also be moved about the screen using the cursor. While viewing stereoscopically, an operator shifts the imagery and moves the floating mark until it appears to rest exactly on the point of interest. This identifies to the computer the image pixels, which correspond to the point in both the left and right photos. The computer converts the pixel row and column locations to x and y photo coordinates and then calculates the point's ground coordinates by employing standard analytical photogrammetry equations.

Other softcopy photogrammetric systems are also available. Some rely on polarizing lenses for stereoviewing, while others, such as the Intergraph Image Station Z shown in Figure 1.6, use a system of electronically synchronized shutters. All are controlled by sophisticated software. Some advanced systems have become almost totally automated. They employ a process called *digital image correlation*. In this operation, the computer automatically matches points in the left photo of the stereopair to their corresponding or *conjugate* points in the right photo. This is done by comparing the patterns of image densities in the point's immediate area on both photos. Thus, an operator's process of making measurements while stereoviewing is eliminated.

Softcopy systems have replaced other forms of stereoplotters in the workplace. A simplistic softcopy photogrammetric system, which is implemented in WOLFPACK and discussed in Section 27.20 is available on the companion website at http://www.personhighered.com/ghilani.

27.15 ORTHOPHOTOS

As implied by their name, orthophotos, are orthographic representations of the terrain in picture form. They are derived from aerial photos in a process called *differential rectification*, which removes scale variations and image displacements due to relief and tilt. Thus, the imaged features are shown in their true planimetric positions.

Instruments used for differential rectification have varied considerably in their designs. The first-generation instruments were basically modified stereoscopic plotters with either optical or mechanical projection. Optical projection instruments derived an orthophoto by systematically *scanning* a stereomodel and photographing it in a series of adjacent narrow strips. *Rectification* (removal of tilt) was accomplished by leveling the model to ground control prior to scanning, and scale variations due to terrain relief were removed by varying the projection distance during scanning. As the instrument automatically traversed back and forth across the model, exposure was made through a narrow slit onto an orthonegative. An operator, viewing the model in three dimensions, continually monitored the scans and adjusted the projection distance to keep the exposure slit in contact with the stereomodel. Because the model itself had uniform scale throughout, the resulting *orthonegative* (from which the orthophoto was made) was also of uniform scale. Orthophoto systems based on modified mechanical projection stereoplotters functioned in a similar fashion. These mechanical instruments are seldom used today.

Contemporary orthophoto production is done using softcopy photogrammetric systems in a procedure called *digital image processing*. These systems employ digital images, which, as described in Section 27.14.4, may be obtained either by using digital cameras or by scanning negatives obtained with film cameras. As noted earlier, a digital image consists of a *raster* (grid) of tiny pixels, each of which is assigned a digital value corresponding to its gray level, and each having its photo location given in terms of its row and column within the raster. The digital image is input to the system's computer, which uses analytical photogrammetric equations to modify each pixel location according to the tilt in the photograph, and the scale at that point in the stereomodel. Through this process all pixels are modified to locations they would have on a truly vertical photo, and all are brought to a common scale. The modified pixels are then printed electronically to produce an orthophoto.

Orthophotos combine the advantages of both aerial photos and line maps. Like photos, they show features by their actual images rather than as lines and symbols, thus making them more easily interpreted and understood. Like maps, orthophotos show the features in their true planimetric positions. Therefore, true distances, angles, and areas can be scaled directly from them. *Orthophotomaps* (maps produced from orthophotos) are used for a variety of applications, including planning and engineering design. They have been particularly valuable in cadastral and tax mapping because the identification of property boundaries is greatly aided through visual interpretation of fence lines, roads, and other evidence. Because they are in digital form, they are also ideal for use as base maps and for analyses in geographic information systems.

Orthophotos can generally be prepared more rapidly and economically than line or symbol planimetric maps. With their many significant advantages, orthophotos have superseded conventional maps for many uses.

27.16 GROUND CONTROL FOR PHOTOGRAMMETRY

As pointed out in preceding sections, almost all phases of photogrammetry depend on ground control (points of known positions and elevations with identifiable images on the photograph). Whereas GNSS receivers registered to the camera have reduced the need for ground control, ground control is still used as a check on the accuracy of the system. Ground control can be basic control—traverse, triangulation, trilateration, or GNSS monuments already in existence and marked prior to photography to make them visible on the photos; or it can be *photo control*—natural points having images recognizable on the photographs, and positions that are subsequently determined by ground surveys originating from basic control. Instruments and procedures used in ground surveys were described in earlier chapters. Ordinarily, photo control points are selected after photography to ensure their satisfactory location and positive identification. Premarking points with artificial targets is sometimes necessary in areas that lack natural objects to provide definite images.

As discussed in Section 27.14, for mapping with stereoplotters scaling and leveling stereomodels require a practical minimum of three horizontal control points and four vertical points in each model. For large mapping jobs, therefore, the cost of establishing the required ground control is substantial. In these situations, *analytical aerotriangulation* (see Section 27.13) is used to establish many of the needed control points from only a sparse network of ground-surveyed points. This reduces costs significantly.

Currently GNSS survey methods are being used for real-time positioning of the camera at the instant each photograph is exposed. The kinematic GNSS surveying procedure is being employed (see Chapter 15), which requires two GNSS receivers. One unit is stationed at a ground control point, and the other is placed within the aircraft carrying the camera. The integer ambiguity problem is resolved using *on-the-fly* techniques (see Section 15.2). During the flight, camera positions are continuously determined at time intervals of a few seconds using the GNSS units, and precise timing of each photo exposure is also recorded. From this information, the precise location of each exposure station, in the ground coordinate system, can be calculated. Many projects have been completed using these methods, and they have produced highly accurate results, especially when supplemented with only a few ground control points. It is now possible to complete photogrammetric projects with only a few ground photo control points used for checking purposes.

27.17 FLIGHT PLANNING

Certain factors, depending generally on the purpose of the photography, must be specified to guide a flight crew in executing its mission of taking aerial photographs. Some of them are (1) boundaries of the area to be covered, (2) required scale of the photography, (3) camera focal length and format size, (4) endlap, and (5) sidelap. Once these elements have been fixed, it is possible to compute the entire flight plan and prepare a flight map on which the required flight lines have been delineated. The pilot then flies the specified flight lines by choosing and correlating headings on existing natural features shown on the flight map. In the most modern systems, the flight planning is done using a computer and the coordinates of flight lines are calculated. Then the aircraft is automatically guided by an on-board GNSS system along the planned flight lines.

The purpose of the photography is the paramount consideration in flight planning. For example, in taking aerial photos for topographic mapping using a stereoplotter, endlap should optimally be 60% and sidelap 30%. The required scale and contour interval of the final map must be evaluated to settle flying height. Enlargement capability from photo scale to map compilation scale is restricted for stereoplotting, and generally should not exceed about 5 if satisfactory accuracy is to be achieved. By these criteria, if required map scale is 200 ft/in., photo scale becomes fixed at 1000 ft/in. If the camera focal length is 6 in., flying height is established by Equation (27.2) at $6(1000) = 6000$ ft above average terrain. Some organizations may push this factor higher than 5, but it should be done with caution.

The *C factor* (the ratio of flight height above ground to contour interval that is practical for any specific stereoplotter) is a criterion often used to select the flying height in relation to the required contour interval. To ensure that their maps meet required accuracy standards, many organizations employ a *C* factor of about 1200–1500. Other organizations may push the value somewhat higher, but again this should be done with caution. By this criterion, if a plotter has a *C* factor of, say, 1200, and a map is to be compiled with a 5-ft contour interval, a flight height of not more than $1200(5) = 6000$ ft above the terrain should be sustained.

Information ordinarily calculated in flight planning includes (1) flying height above mean sea level, (2) distance between exposures, (3) number of photographs per flight line, (4) distance between flight lines, (5) number of flight lines, and (6) total number of photographs. A flight plan is prepared based on these items.

Example 27.7

A flight plan for an area 10-mi wide and 15-mi long is required. The average terrain in the area is 1500 ft above datum. The camera has a 6-in. focal length with 9×9 in. format. Endlap is to be 60%, sidelap 25%. The required scale of the photography is 1:12,000 (1000 ft/in.).

Solution

1. Flying height above datum from Equation (27.2):

$$\text{scale} = \frac{f}{H - h_{avg}} \quad \text{so} \quad \frac{1}{1000} = \frac{6}{H - 1500} \quad \text{and} \quad H = 7500 \text{ ft}$$

2. Distance between exposures, d_e: endlap is 60%, so the linear advance per photograph is 40% of the total coverage of 9 in. × 1000 ft/in. = 9000 ft. Thus, the distance between exposures is 0.40 × 9000 = 3600 ft.

3. Total number of photographs per flight line:

   length of each flight line = 15 mi (5280 ft/mi) = 79,200 ft

   $$\text{number of photos per flight line, } N_{\text{Photos/Line}} = \frac{79,200 \text{ ft}}{3600 \text{ ft/photo}} + 1 = 23$$

   Adding two photos on each end to ensure complete coverage, the total is 23 + 2 + 2 = 27 photos per flight line.

4. Distance between flight lines: sidelap is 25%, so the lateral advance per flight line is 75% of the total photographic coverage,

   $$\text{distance between flight lines, } d_s = 0.75(9000 \text{ ft}) = 6750 \text{ ft}$$

5. Number of flight lines:

   width of area = 10 mi(5280 ft/mi) = 52,800 ft

   $$\text{number of spaces between flight lines} = \frac{52,800 \text{ ft}}{6750 \text{ ft/line}} = 7.8 \text{ (say 8)}$$

   $$\text{total flight lines, } N_{\text{Lines}} = 8 + 1 = 9$$

   $$\text{planned spacing between flight lines} = \frac{52,800}{8} = 6600 \text{ ft}$$

   (*Note:* The first and last flight lines should either coincide with or be near the edges of the area, thus providing a safety factor to ensure complete coverage.)

6. Total number of photos required:

   total photos = 27 per flight line × 9 flight lines = 243 photos

Figures 27.17(a) and (b) illustrate endlap and sidelap, respectively, and (c) shows the flight map.

848 PHOTOGRAMMETRY

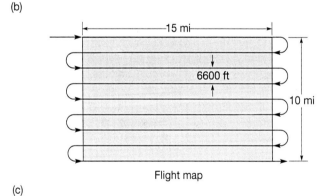

Figure 27.17
(a) Endlap,
(b) Sidelap, and
(c) Flight map.

27.18 AIRBORNE LASER-MAPPING SYSTEMS

Airborne laser-mapping systems called *LiDAR* (**Li**ght **D**etection **a**nd **R**anging) are commonly used today in traditional and hydrographic mapping. These systems, which are carried in airborne vehicles, consist of a laser scanning device, an inertial navigation system, a GNSS receiver, and computer. As the aircraft flies along its trajectory, laser pulses are transmitted toward the terrain below, reflected from the ground or other objects, and detected nearly instantaneously. From these pulses, distances and angles to reflective objects are determined. Concurrently, the inertial navigation device records the aircraft's attitude angles (pitch, yaw, and roll), and the GNSS receiver determines the X, Y, and Z positions of the detector. The computer processes all of this information to determine vector displacements (distances and directions) from known positions in the air, to unknown positions on the ground, and as a result it is able to compute the X, Y, and Z positions of the ground points.

The laser transmitter can generate pulses at an extremely rapid rate, that is, thousands per second, so that the coordinates of a dense pattern of ground points can be determined. Not only are positions of ground points determined, but an image of the ground is also generated. The data can be used to produce digital elevation models (DEMs) and from them contour maps and other topographic products can be produced. Accuracies possible with LiDAR devices are currently in the range of 10–15 cm, but with continued research and development, this is expected to improve. Using this technology, archeological sites have been mapped and, even sometimes, sites have been uncovered in locations that were previously unknown. This same technology has been mounted on ground-based vehicles such as that shown in Figure 1.4 in what is called a mobile mapping system. A similar system has been used by Google to create the street-level view on roads throughout the world.

27.19 SMALL UNMANNED AERIAL SYSTEMS

Small unmanned aerial systems (sUAS) combine small unmanned aerial vehicles (sUAV) such as those shown in Figure 1.1 with satellite positioning and digital cameras. These systems combined with a softcopy photogrammetric system allow surveyors to capture aerial imagery and produce maps from these images for small projects. Due to this capability, surveying firms that traditionally only provided ground control to photogrammetric firms are now able to capture and analyze aerial imagery that they captured on their own.

These aircraft have no pilot on board and can be flown either remotely, autonomously, or semiautonomously. In remote operation, a pilot controls the aircraft by means of a controller, which transmits radio signals to the aircraft. In autonomous mode, the aircraft has a series of waypoints (locations) uploaded to its flight controller, which it uses along with GNSS to fly a preprogrammed flight path. The semi-autonomous mode is a combination of the previously describe modes where the ground pilot takes control of the aircraft for a portion of the flight.

Although airspace regulations can vary from country to country, in the United States, the Federal Aviation Administration (FAA) requires that commercial operators of sUAV have a remote pilot's license. A sUAS must weigh less than 55 lbs. on takeoff. This would include all components including communication links, all components that control the aircraft and are required for safe and efficient operation of the aircraft, as well as its payload, which could be a digital camera, GNSS module, and an *inertial measurement unit* (IMU). It is also required that the remote pilot have a visual line of sight to the sUAV without any additional aids other than corrective lenses. Visibility must be over three statute miles, and the sUAV must be at least 500 ft below and 2000 ft horizontally from any cloud. However, a visual observer can be used in areas that the remote pilot cannot directly view but there must always be effective communication between the visual observer and the remote pilot at all times. Although not all sUAVs can reach this speed, the maximum speed of a sUAV is 100 miles per hour in the United States airspace. Additionally, sUAVs are restricted to Class G airspace[2], which typically means a maximum altitude is 400 ft unless the sUAV is flown within 400 ft of a structure where the maximum altitude is 400 ft above the structure. Additionally, for safety reasons, the sUAV cannot be flown in restricted airspace, near other

---

[2]Class G airspace is uncontrolled airspace in the United States.

aircraft (which might create a collision hazard), or over people who are not under a covered structure or in a vehicle unless the person is participating in the survey.

A sUAS can be adversely affected by weather conditions such as high winds or rain, and thus can only fly in fair weather conditions, which are defined by the manufacturer of the vehicle. These aircraft can either be fixed-wing or multi-rotor copters. The advantage of a fixed-wing aircraft is that it can carry heavier payloads, flies faster, and thus can perform a survey faster. Its disadvantage is that is must remain in motion during the entire flight, cannot hover over any location, and requires a large area to make turns on returning flight lines. Thus, it cannot be used in some areas due to airspace regulations and cannot due a vertical survey on objects such as tall buildings. On the other hand, the multi-rotor copter can hover over a location but typically carries a smaller payload, is slower, and has significantly shorter flight times. Obviously both systems have their advantages and disadvantages and both are the ideal platform depending on the type of survey.

The digital cameras, which can be used on small unmanned aerial systems, can vary in format size from the standard 35 mm, which is 36 × 24 mm, to as small as a 13.2 × 8.8 mm format. This variability in size must be taken into account during flight planning for an aerial mission. For example, a sUAV with a camera that has a 23.6 × 15.7 mm format will have a surface area of 23.6 × 15.7 mm^2 or 370.52 mm^2. If it is a 10.1 megapixel (mp) camera, the pixel size will be

$$\sqrt{\frac{370.52 \text{ mm}^2}{10.1(1024)^2}} = 0.00594 \text{ mm} = 5.91 \text{ } \mu\text{m}$$

If the focal length of the lens is 40 mm and the flight was at an altitude of 300 ft. or 91.44 m, the ground resolution of a pixel is determined using a variant of Equation (27.1) as

$$S = \frac{ab}{AB} = \frac{f}{H}, \text{ or}$$

$$AB = ab\left(\frac{H}{f}\right) = 0.00594\left(\frac{91.440}{0.040}\right) = 13.5 \text{ mm}$$

This means a ground resolution for this setup of 1.35 cm by 1.35 cm.

Similarly, the ground coverage of an image can be computed as

$$\text{Length}, G = \frac{15.7 \text{ mm} (91.440 \text{ m})}{40 \text{ mm}} = 35.890 \text{ m}$$

and

$$\text{Width}, W = \frac{23.6 \text{ mm} (91.440 \text{ m})}{40 \text{ mm}} = 53.950 \text{ m}$$

With a 70% end lap, the overlap between images would be

$$\text{overlap} = 0.7(G) = 0.7(35.890) = 25.123 \text{ m}$$

which results in an airbase, B, of

$$B = 35.890 - 25.123 = 10.767 \text{ m}$$

Thus, if assuming that the aircraft is flying at 11 meters per second (~40 km/h), the intervalometer setting (I) should be

$$I = \frac{10.767}{11} = 0.97 \text{ sec.}$$

Thus, the camera will be taking an image approximately every 0.97 seconds while in flight. For a color image, the camera must capture three color bands. Thus, for a 10.1 MP camera, each image involves 30.3 megabytes (MB) of data. This means that the storage card must be capable of writing more than 31.3 MB/s or 30.3 MB/0.97 s and be capable of storing all the images captured for the project.

■ 27.20 REMOTE SENSING

In general, remote sensing can be defined as any methodology employed to study the characteristics of objects using data collected from a remote observation point. More specifically, and in the context of surveying and photogrammetry, it is the extraction of information about the Earth and our environment from imagery obtained by various sensors carried in aircraft and satellites. Satellite imagery is unique because it affords a practical means of monitoring our entire planet on a regular basis.

Remote sensing imaging systems operate much the same as the human eye, but they can sense or "see" over a much broader range than humans. Cameras that expose various types of film are among the best types of remote sensing imaging systems. Nonphotographic systems such as *multispectral scanners* (MSS), *radiometers*, *side-looking airborne radar* (SLAR), and *passive microwave* are also employed. Their manner of operation, and some applications of the imagery, is briefly described below.

The sun and other sources emit a wide range of electromagnetic energy called the *electromagnetic spectrum*. X-rays, visible light rays, and radio waves are some familiar examples of energy variations within the electromagnetic spectrum. Energy is classified according to its wavelength (see Figure 27.18). Visible light (that energy to which our eyes are sensitive) has wavelengths from about 0.4 to 0.7 μm and thus, as illustrated in the figure, comprises only a very small portion of the spectrum.

Within the wavelengths of visible light, the human eye can distinguish different colors. The primary colors (blue, green, and red) consist of wavelengths in the ranges of 0.4–0.5, 0.5–0.6, and 0.6 − 0.7 μm, respectively. All other hues are combinations of the primary colors. To the human eye, an object appears a certain color because it reflects energy of wavelengths producing that color. If an object reflects all wavelengths of energy in the visible range, it will appear white, and if it absorbs

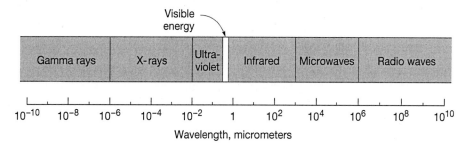

Figure 27.18 Classification of electromagnetic spectrum by wavelength.

all wavelengths, it will be black. If an object absorbs all green and red energy but reflects blue, that object will appear blue.

Just as the retina of the human eye can detect variations in wavelengths, photographic films or *emulsions* are also manufactured to have wavelength sensitivity variations. Normal color emulsions are sensitive to blue, green, and red energy; others respond to energy in the near-infrared range. These are called *infrared* (IR) emulsions. They make it possible to photograph energy that is invisible to the human eye. An early application of IR film was in camouflage detection, where it was found that dead foliation or green netting reflected infrared energy differently than normal vegetation, even though both appeared green to the human eye. Infrared film is now widely used for a variety of applications, such as detection of crop stress, and for identification and mapping of tree species.

Nonphotographic imaging systems used in remote sensing can detect energy variations over a broad range of the electromagnetic spectrum. *Multispectral scanner* systems (MSS), for example, are carried in satellites and can operate within wavelengths from about $0.3 - 14 \mu m$ In a manner similar to the way humans detect colors, MSS units isolate incoming energy into discrete spectral categories or *bands*, and then convert them into electric signals that can be represented by digits. These devices capture a digital image (see Section 27.3), that is, the scanned scene below the satellite's path is recorded as a series of contiguous rows and columns of pixels. The digits associated with each pixel represent intensities of the varying bands of energy within them. This digital format is ideal for computer processing and analysis and enables prints to be made by electronic processing. The bands of a scene can be analyzed separately, which is extremely useful in identifying and interpreting imaged objects. For certain applications, it is useful to combine two or more bands into a composite. The geometry of nonphotographic images differs from that of perspective photos, and thus methods for analyzing them also differ.[3]

Figure 27.19 shows an image obtained with the MSS system carried in an early Landsat satellite. It was taken at an altitude of 560 miles and shows a large portion of southeastern Wisconsin, including the city of Milwaukee. A portion of Lake Michigan is shown on the right side of the figure. Imagery of this type is useful for a variety of applications. As examples, geologic formations over large areas can be studied; the number of lakes in the area, their relative positions, shapes, and acreages can be determined readily; acreages of croplands and forests, with a breakdown of coniferous and deciduous tree types, can be obtained; and small-scale planimetric maps showing these different land-use classifications can be prepared. All of these tasks are amenable to computer processing.

The resolution of the earlier Landsat MSS imaging systems (like that used to acquire Figure 27.19) was 80 m, that is, each pixel represents a square of 80 m on the ground. This was improved to 30 m for the *Thematic Mapper* (TM) imaging systems carried in the second generation of Landsat satellites. Figure 27.20 shows an image taken by Landsat TM near Madison, Wisconsin. It clearly illustrates the improved resolution as compared to Figure 27.19. Note, for example, the clarity of the roads and the detail with which urban areas and agricultural crops are shown.

---

[3]Information on the geometry of nonphotographic imaging systems can be found in Lillesand and Kiefer, *Remote Sensing and Image Interpretation*. (See the bibliography at the end of this chapter.)

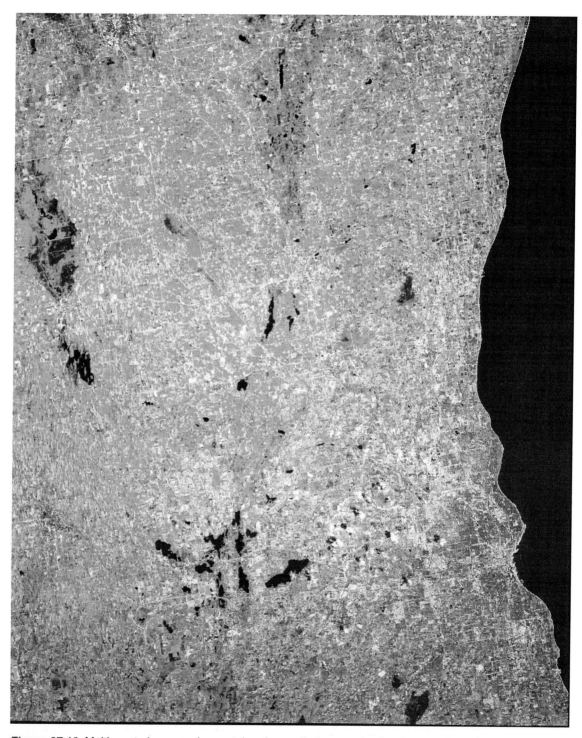

Figure 27.19 Multispectral scanner image taken from a first-generation Landsat satellite over Milwaukee, Wisconsin.

Figure 27.20 Landsat Thematic Mapper image taken near Madison, Wisconsin. (Note the improved resolution compared to the Landsat MSS image of Figure 27.19.) (Courtesy Environmental Remote Sensing Center, University of Wisconsin-Madison.)

Images of this type have been applied for land-use mapping; measuring and monitoring various agricultural crops; mapping soils; detecting diseased crops and trees; locating forest fires; studying wildlife; mapping the effects of natural disasters such as tornadoes, floods, and earthquakes; analyzing population growth and distribution; determining the locations and extent of oil spills; monitoring water quality and detecting the presence of pollutants; and accomplishing numerous other tasks over large areas for the benefit of humankind.

In recent years a significant amount of research and development has been directed toward the design of satellite imaging systems with improved resolution and geometric properties. The goals have been to enable smaller objects to be identified and analyzed, and to improve the mapping capabilities of the systems, thus making them useful for many additional types of applications. The *Enhanced Thematic Mapper Plus* (ETM+) imaging system aboard Landsat 7 satellite has a resolution capability of 15 m. In February of 2013, the most recent Landsat 8 satellite was launched. It added three new bands including a deep blue band for coastal/aerosol studies, a shortwave infrared band for cirrus cloud detection, and a quality assessment (QA) band. It has an *Operational Land Imager* (OLI), which is similar to Landsat's 7 ETM+ sensor but has two new spectral bands. Additionally, the *Thermal Infrared Sensor* (TIRS) collects two spectral bands in what was previously covered on the ETM+ sensor. Landsat 7 and 8 image the entire Earth every 16 days in an eight-day offset.[4] The French *Systeme Pour d'Observation de la Terre* (SPOT) satellite has an imaging system with 10 m resolution, and a nadir ground swath width of 37 miles. Its imaging system can be aimed at angles up to 27° off-nadir. With this feature, the same areas that were imaged on earlier passes can be covered again on subsequent passes from different orbits, thereby achieving stereoscopic coverage. This imagery is therefore suitable not only for small-scale planimetric mapping, but also for determining elevations.[5]

In September, 1999, Space Imaging launched the first commercial imaging satellite, IKONOS.[6] The most remarkable characteristic of its imaging system is its resolution of 1 m. This satellite is in orbit at an altitude of 423 miles above Earth, and the width of its image at the nadir is approximately 7 miles. The imaging system can be aimed off-nadir, either from side to side, or fore and aft. This not only enables pinpointing coverage on areas of interest, but it can also be used to obtain stereoscopic coverage. Thus, the imagery is suitable for both planimetric mapping and for determining elevations of ground points. Figure 27.21 is a 1-m image taken from IKONOS on October 11,1999 over the city of San Francisco. It features an Aquatic Park and a Fisherman's Wharf. Note that the level of detail that can be resolved from this image is far superior to those of Figures 27.19 or 27.20. Individual houses, trees, small boats, and even automobiles can be identified. This imagery is useful for many types of applications. To mention just a few possibilities of interest to surveyors and engineers, the imagery is suitable for (1) preparing site maps and preliminary plans for proposed construction projects; (2) planning for locations of GNSS stations in large-area control surveys, or planning aerial photography; (3) generating detailed information layers such as land cover, hydrography, transportation networks, etc., for use in geographic information systems; and (4) using the images as reference frames for performing GIS analyses. There are also a variety of additional applications in many other fields such as forestry, geology, agriculture, etc.

---

[4]Images from Landsat 8 and all previous Landsat satellites are available from the U.S. Geological Survey, Earth Science Information Center (ESIC), 12201 Sunrise Valley Drive, Reston, VA 20192. Contact can also be made by telephone at (888) ASK-USGS [(888) 275-8747], or at the following website: http://mapping.usgs.gov/esic/.

[5]Information on SPOT imagery can be obtained at the following website: http://www.spotimage.fr.

[6]Information about IKONOS satellite imagery can be obtained from Space Imaging by telephone at (800) 232-9037, or by visiting the following website: http://www.spaceimaging.com.

Figure 27.21
1-m resolution image obtained from the IKONOS satellite showing Aquatic Park and Fisherman's Wharf in San Francisco. (Courtesy NASA.)

In the future, those engaged in surveying (geomatics) will be called on to prepare maps and to extract a variety of other positional types of information from satellite images. Remote sensing will play a significant future role in providing data to assess the impacts of human activities on our air, water, and land resources. It can provide valuable information to assist in making sound decisions and formulating policies related to resource management, and land-use and land development activities.

27.21 SOFTWARE

As previously discussed, softcopy photogrammetric stereoplotters are efficient, as well as versatile. Not only are they capable of producing maps, cross sections, digital elevation models, and other digital topographic files, but they can also be employed for a variety of image interpretation problems, and they can support the production of mosaics and orthophotos (see Section 27.15). Also, digital maps produced by softcopy systems are created in a computer environment, and are therefore in formats compatible for CAD applications, and for use in the databases of Geographic Information Systems. Softcopy systems have the added advantage that their major item of hardware is a computer rather than an expensive singlepurpose stereoplotter, so it can be used for many other tasks in addition to stereoplotting. For these reasons, softcopy stereoplotters are taking over the industry.

As shown in Figure 27.22, a digital image viewing and measuring system has been incorporated in the WOLFPACK software, which is available on the

Figure 27.22
The observation of fiducial F3 in the WOLFPACK photogrammetric software.

companion website at http://www.pearsonhighered.com/ghilani. While this software is far from being a softcopy stereoplotter, it helps demonstrate some of the basic principles in analytical photogrammetry. The software utilizes digital images in the bitmap (bmp) format. The resulting image coordinates can be placed into the photo coordinate system using the *interior orientation* option. This option requires the calibrated fiducial coordinates of the camera and transforms the digitized image coordinates into the photo coordinate system. Furthermore, by observing a minimum of three imaged points whose ground coordinates are known, an *exterior orientation* can be performed. The exterior orientation determines the camera location and orientation parameters at the time of exposure. Finally, if this procedure is also performed on common points lying in the overlapping region of two photos, the ground coordinates of these points can be determined using the *space intersection* option. All the required points (fiducial points, ground control, and photo-identifiable points) for the analytical process should be observed in a single digitizing session so that repetition of the procedures is not necessary. This photogrammetric process is described in the help file that accompanies the WOLFPACK software.

While these functions only demonstrate the rudimentary operations of softcopy systems, it allows the reader to experience them. The companion website also contains suitable aerial imagery, a file of calibrated fiducial coordinates (camera.fid), and another of ground control (ground.crd). The fiducial coordinates on the image are labeled F1 through F8 and the ground control is also circled and labeled in the imagery to aid the user in identifying the points. The calibrated focal length of the aerial camera was 153.742 mm. Problems at the end of this chapter refer to this material. In Figure 27.22, fiducial F3, which is in the upper-left corner of the image, is being observed. Note that the user can monitor the image coordinates of the point and their standard deviations in the lower portion of the display. The thumbnail image in the lower portion of the screen allows the user to move rapidly around the image without using the scroll bars. The reset image button restores the image to its original position in the screen. In the lower-right corner of the display is an option box to modify the cursor to several different shapes. A circular cursor with a point at its center is being used to point on the fiducial mark.

27.22 SOURCES OF ERROR IN PHOTOGRAMMETRY

Some sources of error in photogrammetric work are:

1. Measuring instruments not standardized or calibrated.
2. Inaccurate location of principal and corresponding principal points.
3. Failure to use camera calibration data.
4. Assumption of vertical exposures when photographs are tilted.
5. Presumption of equal flying heights when they were unequal.
6. Disregard of differential shrinkage or expansion of photographic prints.
7. Incorrect orientation of photographs under a stereoscope or in a stereoscopic plotter.
8. Faulty setting of the floating mark on a point.

27.23 MISTAKES

Some mistakes that occur in photogrammetry are:

1. Incorrect reading of measuring scales.
2. Mistake of units—for example, inches instead of millimeters.
3. Confusion in identifying corresponding points on different photographs.
4. Failure to provide proper control or use of erroneous control coordinates.
5. Attachment of an incorrect sign (plus or minus) to a measured photographic coordinate.
6. Blunder in computations.
7. Misidentification of control-point images.

PROBLEMS

Asterisks (*) indicate problems that have partial answers given in Appendix G.

27.1 Define photogrammetry.
27.2 Define the term softcopy photogrammetry.
27.3 What are the advantages of photogrammetry?
27.4 Define the principal point of an image.
27.5 Define the term pixel.
27.6 What is the definition of a sUAS in the United States?
27.7 What is the difference between remote and autonomous operation of a sUAS?

27.8 The distance between two points on a vertical photograph is ab and the corresponding ground distance is AB. For the following data, compute the average photographic scale along the line ab.
  *(a) $ab = 2.41$ in., $AB = 4820$ ft
  (b) $ab = 5.47$ in.; $AB = 13{,}675$ ft
  (c) $ab = 86.00$ mm; $AB = 258$ m

27.9 On a vertical photograph of flat terrain, section corners, which are a nominal 1-mi apart, appear a distance d apart. If the camera focal length is f compute flying height above average ground in feet for the following data:
  (a) $d = 3.259$ in.; $f = 3 - 1/2$ in.
  (b) $d = 68.54$ mm; $f = 152.4$ mm

27.10 On a vertical photograph of flat terrain, the scaled distance between two points is ab. Find the average photographic scale along ab if the measured length between the same line is AB on a map plotted at a scale of S_{map} for the following data.
  (a) $ab = 1.86$ in.; $AB = 4.46$ in.; $S_{map} = 1{:}9600$
  (b) $ab = 116.04$ mm; $AB = 2.38$ mm; $S_{map} = 1{:}24{,}000$

27.11 What are the average scales of vertical photographs for the following data, given flying height above sea level, H, camera focal length, f, and average ground elevation h?
  *(a) $H = 7300$ ft; $f = 152.4$ mm; $h = 1210$ ft
  (b) $H = 5500$ ft; $f = 8\text{-}1/4$ in.; $h = 920$ ft
  (c) $H = 3525$ m; $f = 88.90$ mm; $h = 947$ m

27.12 The length of a football field from goal post to goal post scales 49.81 mm on a vertical photograph. Find the approximate dimensions (in meters) of a large rectangular building that also appears on this photo and whose sides measure 20.05 mm by 14.1 mm. (Hint: Football goal post are 120 yards apart.)

27.13* Compute the area in acres of a triangular parcel of land whose sides measure 48.78 mm, 84.05 mm, and 69.36 mm on a vertical photograph taken from 6050 ft above average ground with a 152.4 mm focal length camera.

27.14 Calculate the flight height above average terrain that is required to obtain vertical photographs at an average scale of S if the camera focal length is f for the following data:
  (a) $S = 1{:}10{,}000$; $f = 152.40$ mm
  (b) $S = 1{:}9600$; $f = 88.90$ mm

27.15 Determine the horizontal distance between two points A and B whose elevations above datum are $h_A = 1560$ ft. and $h_B = 1425$ ft and whose images a and b on a vertical photograph have photo coordinates $x_a = 2.95$ in., $y_a = 2.32$ in., $x_b = -1.64$ in., and $y_b = -2.66$ in. The camera focal length was 152.4 mm and the flying height above datum 7500 ft.

27.16* Similar to Problem 27.15, except that the camera focal length was 3-1/2 in., the flying height above datum 4075 ft, and elevations h_A and h_b 983 ft and 1079 ft, respectively. Photo coordinates of images a and b were $x_a = 108.81$ mm, $y_a = -73.73$ mm, $x_b = -87.05$ mm, and $y_b = 52.14$ mm.

27.17 On the photograph of Problem 27.15, the image c of a third point C appears. Its elevation $h_C = 1455$ ft and its photo coordinates are $x_c = 3.20$ in. and $y_c = -2.66$ in. Compute the horizontal angles in triangle ABC.

27.18 On the photograph of Problem 27.15, the image d of a third point D appears. Its elevation is $h_D = 1085$ ft and its photo coordinates are $x_d = 2.72$ in. and $y_d = 3.09$ in. Calculate the area, in acres, of triangle ABD.

27.19 Determine the height of a radio tower, which appears on a vertical photograph for the following conditions of flying height above the tower base H, distance on the photograph from principal point to tower base r_d and distance from principal point to tower top r_t.
  *(a) $H = 2425$ ft; $r_b = 3.18$ in.; $r_t = 3.34$ in.
  (b) $H = 6400$ ft; $r_b = 96.28$ mm; $r_t = 97.67$ mm

27.20 On a vertical photograph, images a and b of ground points A and B have photographic coordinates $x_a = 3.27$ in., $y_a = 2.28$ in., $x_b = -1.95$ in. and $y_b = -2.50$ in. The horizontal distance between A and B is 5350 ft, and the elevations of A and B above datum are 652 ft and 785 ft, respectively. Using Equation (27.9), calculate the flying height above datum for a camera having a focal length of 152.4 mm.

27.21 Similar to Problem 27.20, except $x_a = -52.53$ mm, $y_a = 69.67$ mm, $x_b = 26.30$ mm, $y_b = -59.29$ mm line length $AB = 4695$ ft. and elevations of points A and B are 925 and 875 ft, respectively.

27.22* An air base of 3205 ft exists for a pair of overlapping vertical photographs taken at a flying height of 5500 ft above MSL with a camera having a focal length of 152.4 mm. Photo coordinates of points A and B on the left photograph are $x_a = 40.50$ mm, $y_a = 42.80$ mm, $x_b = 23.59$ mm, and $y_b = -59.15$ mm. The x photo coordinates on the right photograph are $x_a = -60.68$ mm and $x_b = -70.29$ mm. Using the parallax equations, calculate horizontal length AB.

27.23 Similar to Problem 27.22, except the air base is 7450 ft, the flying height above mean sea level is 15,520 ft, the x and y photo coordinates on the left photo are $x_a = 37.98$ mm, $y_a = 50.45$ mm, $x_b = 24.60$ mm, and $y_b = -46.89$ mm, and the x photo coordinates on the right photo are $x_a = -52.17$ mm and $x_b = -63.88$ mm.

27.24* Calculate the elevations of points A and B in Problem 27.22.

27.25 Compute the elevations of points A and B in Problem 27.23.

27.26 List the different types of stereoscopic viewing systems.

27.27 Name the three stages in stereoplotter orientation, and briefly explain the objectives of each.

27.28 How can the operator's left and right eyes restricted to the left and right images, respectively?

27.29 What kind of images do softcopy stereoplotters require? Describe two different ways they can be obtained.

27.30 Compare an orthophoto with a conventional line and symbol map.

27.31 Discuss the advantages of orthophotos as compared to maps.

Aerial photography is to be taken of a tract of land that is X-mi square. Flying height will be H in feet above average terrain, and the camera has focal length f. If the focal plane opening is 9×9 in., and minimum sidelap is 30%, how many flight lines will be needed to cover the tract for the data given in Problems 27.32 and 27.33?

27.32* $X = 8$; $H = 4000$ ft; $f = 152.40$ mm
27.33 $X = 10$; $H = 7800$ ft; $f = 6.000$ in.

Aerial photography was taken at a flying height H in feet above average terrain. If the camera focal plane dimensions are 9×9 in, the focal length is f and the spacing between adjacent flight lines is X ft, what is the percent sidelap for the data given in Problems 27.34 and 27.35?

27.34* $H = 4500$; $f = 152.4$ mm; $X = 4700$.
27.35 $H = 6000$; $f = 88.90$ mm; $X = 10,000$

Photographs at a scale of S are required to cover an area X mi square. The camera has a focal length f and focal plane dimensions of 9×9 in. If endlap is 60% and sidelap 30%, how many photos will be required to cover the area for the data given in Problems 27.36 and 27.37?

27.36 $S = 1:6000$; $X = 6$; $f = 152.4$ mm
27.37 $S = 1:14,400$; $X = 40$; $f = 3.5$ in.
27.38 To what wavelengths of electromagnetic energy is the human eye sensitive? What wavelengths produce the colors blue, green, and red?

Problems 27.39 through 27.43 involve using WOLFPACK with images 5 and 6 on the companion website. The ground coordinates of the paneled points are listed in the file "ground.crd." The coordinates of the fiducials are listed in the file "camera.fid." To do these problems, digitize the eight fiducials and paneled points 21002, 4, 41, GYM, WIL1A, WIL1B, and RD on both images. After digitizing the points, perform an interior orientation to compute photo coordinates for the points on images 5 and 6. The focal length of the camera is 153.742 mm.

27.39 Using photo coordinates for points 4 and GYM on image 5, determine the scale of the photo.
27.40 Using photo coordinates for points 4 and GYM on image 5, determine the flying height of the camera at the time of exposure.
27.41 Using photo coordinates for points 4 and GYM on images 5 and 6, determine the ground coordinates of points WIL1A and WIL1B using Equation (27.12) and Equation (27.13).
27.42 Using the exterior orientation option in WOLFPACK, determine the exterior orientation elements for image 5.
27.43 Using the exterior orientation option in WOLFPACK, determine the exterior orientation elements for image 6.

BIBLIOGRAPHY

Ali, T. et al. 2015. "Semiautomatic Extraction of the Shoreline from High-Resolution Satellite Imagery and Coastal Terrain Model." *Surveying and Land Information Science* 74 (No. 1): 15.

American Society for Photogrammetry and Remote Sensing. 2004. *Manual of Photogrammetry*, 5th Ed. Bethesda, MD.

_____. 1997. *Manual of Remote Sensing, A Series, Earth Observing Platforms and Sensors*, 3rd Ed. (CD Rom), Bethesda, MD.

Barnes, G. and W. Volkmann. 2015. "High-Resolution Mapping with Unmanned Aerial Systems." *Surveying and Land Information Science* 74 (No 1): 5.

Battista T. and C. Taylor. 2015. "Finding Fishing Hot Spots." *XYHT* 2 (No. 3): 14.

Bernardini, F. 2015. "Power Scanning." *XYHT* 2 (No. 3): 33.

Luccio, M. 2014. "Extraterrestrial Photogrammetry." *XYHT* 1 (No. 1): 20.

_____. 2014. "From the Shore to the Floor." *XYHT* 1 (No. 4): 16.

_____. 2014. "From Traditional Photogrammetry to UAS." *XYHT* 1 (No. 2): 42.

Plaza, J. 2014. "Aerial Imaging in Antarctica." *XYHT* 1 (No. 3): 35.

Salmon, J. 2015. "The New Spacescape." *XYHT* 2 (No. 1): 14.

Waters, H. 2015. "UAVs for the Whole World." *XYHT* 2 (No. 1): 37.

Wolf, P. R., and B. A. Dewitt. 2000. *Elements of Photogrammetry: With Applications in GIS*, 3rd Ed. New York: McGraw-Hill.

28
Introduction to Geographic Information Systems

■ 28.1 INTRODUCTION

The term *Geographic Information System* (GIS) first appeared in published literature in the mid-1960s. However, many of its concepts have long been in existence. For example, the *map overlay* concept, which is one of the important tools used in GIS spatial analysis (see Section 28.9), was used by French cartographer Louis-Alexandre Berthier more than 200 years ago. He prepared and overlaid a series of maps to analyze troop movements during the American Revolution. In 1854, Dr. John Snow demonstrated another early example illustrating the use and value of the overlay concept. He overlaid a map of London showing where cholera deaths had occurred with another, giving locations of wells in that city to demonstrate the relationship between those two data sets. These early examples illustrate fundamentals that still comprise the basis of our modern GIS; that is, making decisions based on the simultaneous analysis of data of differing types, all located spatially in a common geographic reference system. However, the full capabilities and benefits of our modern GISs could not occur until the advent of the computer.

In general, a geographic information system can be defined as a system of hardware, software, data, and organizational structure for collecting, storing, manipulating, and spatially analyzing "geo-referenced" data, and displaying both graphical and descriptive information resulting from those processes. It allows someone to digitally create and manipulate spatial data for a multitude of applications so that users can make informed decision. To name a few, this includes applications in land management, political campaigns, monitoring of criminal activities, urban and regional planning, flooding, building management, wildlife and habitat management, and a host of other activities. A more detailed definition (Hanigan, 1988) describes a GIS as "any information management system that can:

1. Collect, store, and retrieve information based on its spatial location;
2. Identify locations within a targeted environment that meet specific criteria;

3. Explore relationships among data sets within that environment;
4. Analyze the related data spatially as an aid to making decisions about that environment;
5. Facilitate selecting and passing data to application-specific analytical models capable of assessing the impact of alternatives on the chosen environment; and
6. Display the selected environment both graphically and numerically either before or after analysis."

The thread that is common to both definitions just given is that in a GIS, decisions are made based on spatial analyses performed on data sets that are referenced in a common geographical system. The geographic referencing system used could be a state plane, geodetic (latitude and longitude), or UTM coordinate system, or other suitable local coordinate system. In any GIS, the accuracy of the spatial analyses and, hence, the validity of decisions reached as a result of those analyses, are directly dependent on the quality of the spatially related data used. It is therefore important to realize at the outset of this chapter that the surveyor's role in developing accurately positioned data sets is a critical one in GIS activity.

A generalized concept of how data of different types or "layers" are collected and overlaid in a GIS is illustrated in Figure 28.1. In that figure, maps A through G represent some of the different layers of spatially related information that can be digitally recorded and incorporated into a GIS database, and include parcels of different land ownership A, zoning B, floodplains C, wetlands D, land cover E, and soil types F. Map G is the geodetic reference framework, consisting of the network of survey control points in the area. Note that these control points

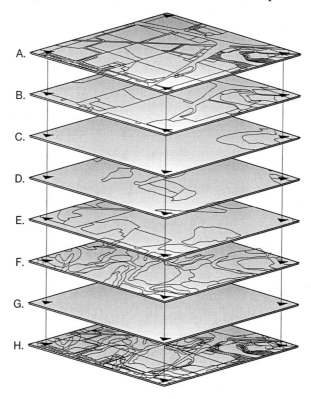

Figure 28.1
Concept of layers in a geographic information system. Map layers shown are (A) parcels; (B) zoning; (C) floodplains; (D) wetlands; (E) land cover; (F) soils; (G) reference framework; and (H) composite overlay. (Courtesy Land Information and Computer Graphics facility, College of Agricultural and Life Sciences, University of Wisconsin-Madison.)

are found in each of the other layers thereby providing the means for spatially locating all data in a common reference system. Thus, composite maps that merge two or more different data sets can be accurately created. For example in Figure 28.1, bottom map H is the composite of all layers.

A GIS merges conventional database management software with software for manipulating spatial data. This combination enables the simultaneous storage, retrieval, overlay, and display of many different spatially related data sets in the manner illustrated by Figure 28.1. These capabilities, coupled with sophisticated GIS software to analyze and query the data sets that result from these different overlay and display combinations, provide answers to questions that never before were possible to obtain. As a result, GISs have become extremely important in planning, design, impact assessment, predictive modeling, and many other applications.

GISs have been applied in virtually every imaginable field of activity, from engineering to agriculture, and from the medical science of epidemiology to wildlife management. Flood forecasting on a large regional basis, such as statewide, is one particular example that illustrates some of the benefits that can be derived from using GIS. Critical location-related data entered into the GIS to support statewide flood forecasting would include the state's topography; soil; land cover; number, sizes, and locations of drainage basins; existing stream network with stream gaging records; locations and sizes, of existing bridges, culverts, and other drainage structures; data on existing dams and the water impoundment capacities of their associated reservoirs; and records of past rainfall intensity and duration. Given these and other data sets, together with a model to estimate runoff, a computer can be used to perform an analysis and predict locations of potential floods and their severity. In addition, experiments can be conducted in which certain input can be varied. Examples may include (1) input of an extremely intense rainfall for a lengthy duration in a given area to assess the magnitude of the resulting flood, and (2) the addition of dams and other flood-detention structures of varying sizes at specific locations, to analyze their impact on mitigating the disaster. Many other similar examples could be given. Obviously, GISs are very powerful tools, and their use will increase significantly in the future.

The successful implementation of GISs relies on people with backgrounds and skills in many different disciplines, but none is more important than the contributions of those engaged in surveying (geomatics). Virtually every aspect of surveying, and thus all material presented in the preceding chapters of this book, bears upon GIS development, management, and use. However, of special importance are the global navigation satellite systems (Chapters 13 through 15), mapping surveys (Chapter 17), mapping (Chapter 18), control surveys and reference frames (Chapter 19), map coordinate systems (Chapter 20), boundary or cadastral surveys (Chapter 21), the U.S. Public Land Survey System (Chapter 22), and photogrammetry and remote sensing (Chapter 27). In addition to surveying (geomatics) specialists, personnel in the fields of computer science, geography, soil science, forestry, landscape architecture, and many others play important roles in GIS development.

28.2 LAND INFORMATION SYSTEMS

The terms *Geographic Information System* (GIS) and *Land Information System* (LIS) are sometimes used interchangeably. They do have many similarities, but the distinguishing characteristic between the two is that a LIS has its focus directed primarily toward land records data. Information stored within a LIS for a given locality would include a spatial database of land parcel information derived from property descriptions in the U.S. Public Land system; other types of legal descriptions such as metes and bounds or block and lot that apply to parcels in the area; and other cadastral data. It might include the actual deeds and other records linked to the spatial data. Information on improvements and parcel values would also be included.

Land information systems and geographic information systems can share data sources such as control networks, parcel ownership information, and municipal boundaries. However, a GIS will usually incorporate data over a broader range and might include layers such as topography, soil types, land cover, hydrography, depth to ground water, and so on. Because of its narrower focus, there is a tendency to consider a LIS as a subset of a GIS.

LISs are used to obtain answers to questions about who has ownership or interests in the land in a certain area, the particular nature of those interests, and the specific land affected by them. They can also provide information about what resources and improvements exist in a given area, and give their values. Answers to these questions are essential in making property assessments for taxation, transferring title to property, mortgaging, making investment decisions, resolving boundary disputes, and developing roads, utilities, and other services on the land that require land appraisal and property acquisitions. The data are also critical in policy development and land-use planning.

28.3 GIS DATA SOURCES AND CLASSIFICATIONS

As noted earlier, the capabilities and benefits of any GIS are directly related to the content and integrity of its database. Data that are entered into a GIS come from many sources and may be of varying quality. To support a specific GIS, a substantial amount of new information will generally need to be gathered expressly for its database. More than likely, however, some of the data will be obtained from existing sources such as maps, engineering plans, aerial photos, satellite images, and other documents and files that were developed for other purposes. Building the database is one of the most expensive and challenging aspects of developing a GIS. In fact, it has been estimated that this activity may represent about 60–80% of the total cost of implementing a GIS.

Two basic data classifications are used in GISs (1) *spatial* and (2) *nonspatial*. These are described in the sections that follow.

28.4 SPATIAL DATA

Spatial data, sometimes interchangeably called *graphic data* consists in general of natural and cultural features that can be shown with lines or symbols on maps, or seen as images on photographs. In a GIS these data must be represented and spatially located, in digital form, using a combination of fundamental elements called "simple spatial objects." The formats used in this representation are either *vector* or *raster*. The "relative spatial relationships" of the simple spatial objects are given by their *topology*.

866 INTRODUCTION TO GEOGRAPHIC INFORMATION SYSTEMS

These important GIS topics: (1) simple spatial objects, (2) data formats, and (3) topology are described in the following subsections.

28.4.1 Simple Spatial Objects

The simple spatial objects most commonly used in spatially locating data are illustrated in Figure 28.2 and described as follows:

1. *Points* define single geometric locations. They are used to locate features such as houses, wells, mines, or bridges [see Figure 28.2(a)]. Their coordinates give the spatial locations of points, commonly in state plane or UTM systems (see Chapter 20).
2. *Lines* and *strings* are obtained by connecting points. A line connects two points, and a string is a sequence of two or more connected lines. Lines and strings are used to represent and locate roads, streams, fences, property lines, etc. [see Figure 28.2(b)].
3. *Interior areas* consist of the continuous space within three or more connected lines or strings that form a closed loop [see Figure 28.2(c)]. For example, interior areas are used to represent and locate the limits of governmental jurisdictions, parcels of land ownership, different types of land cover, or large buildings.
4. *Pixels* are usually tiny squares that represent the smallest elements into which a digital image is divided [see Figure 28.2(d)]. Continuous arrays of pixels, arranged in rows and columns, are used to enter data from aerial photos, orthophotos, satellite images, etc. Assigning a numerical value to each pixel specifies the distributions of colors or tones throughout the image. Pixel size can be varied, and is usually specified by the number of *dots per inch* (dpi). As an example, 100 dpi would correspond to squares having dimensions of 1/100 in. on each side. Thus, 100 dpi yields 10,000 pixels per square inch.
5. *Grid cells* are single elements, usually square, within a continuous geographic variable. Similar to pixels, their sizes can be varied, with smaller cells yielding improved resolution. Grid cells may be used to represent slopes, soil types, land cover, water table depths, land values, population density, and so on. The distribution of a given data type within an area is indicated by assigning a numerical value to each cell; for example, showing soil types in an area using the number 2 to represent sand, 5 for loam, and 9 for clay, as illustrated in Figure 28.2(e).

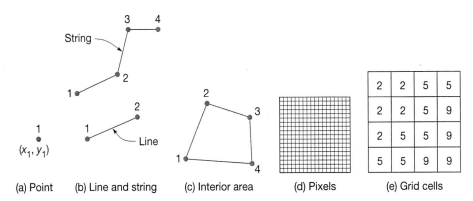

Figure 28.2
Simple spatial objects used to represent data digitally in a GIS.

(a) Point (b) Line and string (c) Interior area (d) Pixels (e) Grid cells

28.4.2 Vector and Raster Formats

The simple spatial objects described in Section 28.4.1 give rise to two different formats for storing and manipulating spatial data in a GIS—*vector* and *raster*. When data are depicted in the vector format, a combination of points, lines, strings, and interior areas is used. The raster format uses pixels and grid cells.

In the vector format, points are used to specify locations of objects such as survey control monuments, utility poles, or manholes; lines and strings depict linear features such as roads, transmission lines or boundaries; and interior areas show regions having common attributes; for example, governmental entities or areas of uniform land cover. An example illustrating the vector format is given with Figure 28.3 and Table 28.1. Figure 28.3 shows two adjacent land parcels, one designated parcel I, owned by Smith, and the other identified as parcel II, owned by Brown. As shown, the configuration consists of points, lines, and areas.

Vector representation of the data can be achieved by creating a set of tables, which list these points, lines, and areas (Table 28.1). Data within the tables are linked using *identifiers* and related spatially through the coordinates of points. As illustrated in Table 28.1(a), all points in the area are identified by a reference number. Similarly, each line is described by its endpoints, as shown in Table 28.1(b), and the endpoint coordinates locate the various lines spatially. As shown in Table 28.1(c), areas in Figure 28.3 are defined by the lines that enclose them. As before, coordinates of line endpoints locate the areas and enable determining their locations and computing their magnitudes.

Another data type can also be represented in vector format. For example, consider the simple case illustrated with the land cover map shown in Figure 28.4(a). In that figure, areas of different land cover (forest, marsh, etc.) are shown with standard topographic symbols (see Figure 18.5). A vector representation of this region is shown in Figure 28.4(b). Here lines and strings locate boundaries of regions having a common land cover. The stream consists of the string connecting points 1 through 10. By means of tables similar to Table 28.1 the data of this

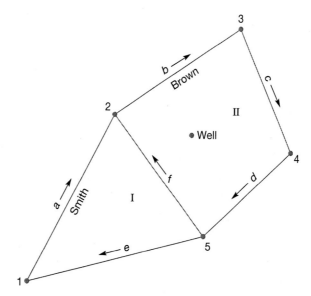

Figure 28.3
Vector representation of a simple graphic record.

TABLE 28.1 Vector Representation of Figure 28.3

(a)		(b)		(c)	
Point Identifier	Coordinates	Line Identifier	Points	Area Identifier	Lines
1	$(X, Y)_1$	a	1,2	I	a, f, e
2	$(X, Y)_2$	b	2,3	II	b, c, d, f
3	$(X, Y)_3$	c	3,4		
4	$(X, Y)_4$	d	4,5		
5	$(X, Y)_5$	e	5,1		
Well	$(X, Y)_{well}$	f	5,2		

figure can also be entered into a GIS using the vector format. Having considered these simple vector representations, imagine the magnitude and complexity of entering data in vector format to cover a much larger area such as that shown in the map of Figure 17.2.

As an alternative to the vector approach, data can be depicted in the raster format using grid cells (or pixels if the data are derived from images). Each equal-sized cell (or pixel) is uniquely located by its row and column numbers and is coded with a numerical value or *code* that corresponds to the properties of the specific area it covers. In the raster format, a point would be indicated with a single grid cell, a line would be depicted as a sequence (linear array) of adjacent grid cells having the same code, and an area having common properties would be shown as a group of identically coded contiguous cells. Therefore, it should be appreciated that in general the raster method yields a coarser level of accuracy or definition of points, lines, and areas than the vector method.

In the raster format, the size of the individual cells defines the *resolution*, or precision, with which data are represented. Smaller the area covered by each cell, the higher the resolution for any given image. Examples illustrating raster representation, and the degradation of resolution with increasing grid size, are given in Figure 28.4(c) and (d). In these figures the land cover data from Figure 28.4(a) have been entered as two different raster data sets. Each cell has been assigned a value representing one of the land cover classes, that is, *F* for forest, *G* for grassland, *M* for marsh, and *S* for stream. Figure 28.4(c) depicts the area with a relatively large-resolution grid and, as shown, it yields a coarse representation of the original points, lines, and areas. With a finer-resolution grid, such as in Figure 28.4(d), the points, lines, and areas are rendered with more precision. However, it is important to note, that as grid resolution increases, so does the volume of data (number of grid cells) required to enter the data.

Despite the coarser resolution present in a raster depiction of spatial features, this format is still often used in GISs. One reason is that many data are available in raster format. Examples include aerial photos, orthophotos, and satellite images. Another reason for the popularity of the raster format is the ease with which it enables collection, storage, and manipulation of data using computers. Furthermore, various refinements of raster images are readily made using

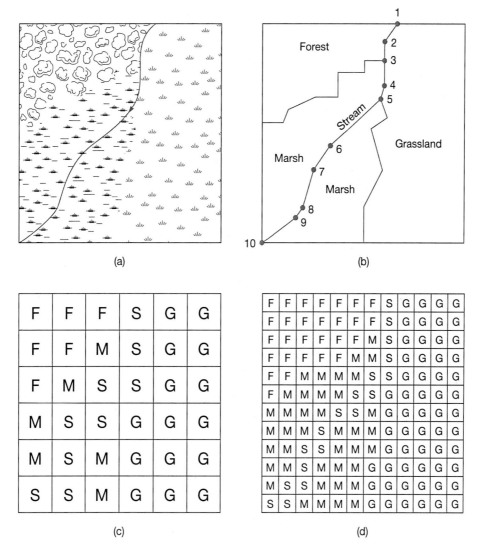

Figure 28.4 Land cover maps of a region. (1) The region using standard topographic symbols. (2) Vector representation of the same region. (3) Raster representation of the region using a coarse-resolution grid cell. (4) Raster representation using a finer-resolution grid cell.

available "image processing" software programs. Finally, for many data sets such as wetlands and soil types, boundary locations are rather vague, and the use of the raster format does not adversely affect the data's inherent accuracy.

28.4.3 Topology

Topology is a branch of mathematics that describes how spatial objects are related to each other. The unique sizes, dimensions, and shapes of the individual objects are not addressed by topology. Rather, it is only their *relative relationships* that are specified.

In discussing topology, it is necessary to first define *nodes*, *chains*, and *polygons*. These are some additional simple spatial objects that are commonly used for specifying the topological relationships of information entered into GIS databases. Nodes define the beginnings and endings of chains or identify the junctions of intersecting chains. Chains are similar to lines (or strings) and are used to define the limits of certain areas or delineate specific boundaries. Polygons are closed loops similar to areas, and are defined by a series of connected chains. Sometimes in topology, single nodes exist within polygons for labeling purposes.

In GISs, the most important topological relationships are:

1. *Connectivity*. Specifying which chains are connected at which nodes.
2. *Direction*. Defining a "from node" and a "to node" of a chain.
3. *Adjacency*. Indicating which polygons are adjacent on the left and which are adjacent on the right side of a chain.
4. *Nestedness*. Identifying what simple spatial objects are within a polygon. They could be nodes, chains, or other smaller polygons.

The topological relationships just described are illustrated and described by example with reference to Figure 28.3. For example, in the figure, through connectivity, it is established that nodes 2 and 3 are connected to form the chain labeled *b*. Connectivity would also indicate that at node 2, chains *a*, *b*, and *f* are connected. Topological relationships are normally listed in tables and stored within the database of a GIS. Table 28.2(a) summarizes all of the connectivity relationships of Figure 28.3.

Directions of chains are also indicated topologically in Figure 28.3. For example, chain *b* proceeds from node 2 to node 3. Directions can be very important in a GIS for establishing such things as the flow of a river or the direction traffic moves on one-way streets. In a GIS, often a consistent direction convention is followed, that is, proceeding clockwise around polygons. Table 28.2(b) summarizes the directions of all chains within Figure 28.3.

The topology of Figure 28.3 would also describe, through adjacency, that Smith and Brown share a common boundary, which is chain *f* from node 5 to node 2, and that Smith is in the left side of the chain and Brown is on the right. Obviously, the chain's direction must be stated before left or right positions can be declared. Table 28.2(c) lists the adjacency relationships of Figure 28.3. Note that a

TABLE 28.2 TOPOLOGICAL RELATIONSHIPS OF ELEMENTS IN THE GRAPHIC RECORD OF FIGURE 28.3

(a) Connectivity		(b) Direction			(c) Adjacency			(d) Nestedness	
Nodes	Chain	Chain	From Node	To Node	Chain	Left Polygon	Right Polygon	Polygon	Nested Node
1–2	a	a	1	2	a	0	I	II	Well
2–3	b	b	2	3	b	0	II		
3–4	c	c	3	4	c	0	II		
4–5	d	d	4	5	d	0	II		
5–1	e	e	5	1	e	0	I		
5–2	f	f	5	2	f	I	II		

zero has been used to designate regions outside of the polygons and beyond the area of interest. Nestedness establishes that the well is contained within Brown's polygon. Table 28.2(d) lists that topological information.

The relationships expressed through the identifiers for points, lines, and areas of Table 28.1, and the topology in Table 28.2, conceptually yield a "map." With these types of information available to the computer, the analysis and query processes of a GIS are made possible.

28.5 NONSPATIAL DATA

Nonspatial data, also often called *attribute* or *descriptive data*, describe geographic regions or define characteristics of spatial features within geographic regions. Nonspatial data are usually alphanumeric and provide information such as color, texture, quantity, quality, and value of features. Smith and Brown as the property owners of parcels I and II of Figure 28.3 and the land cover classifications of forest, marsh, grassland, and stream in Figure 28.4 are examples. Other examples could include the addresses of the owners of land parcels, their types of zoning, dates purchased, and assessed values; or data regarding a particular highway, including its route number, pavement type, number of lanes, lane widths, and year of last resurfacing. Nonspatial data are often derived from sources such as documents, files, and tables.

In general, spatial data will have related nonspatial attributes, and thus some form of linkage must be established between these two different types of information. Usually this is achieved with a *common identifier* that is stored with both the graphic and the nongraphic data. Identifiers such as a unique parcel identification number, a grid cell label, or the specific mile point along a particular highway may be used.

28.6 DATA FORMAT CONVERSIONS

In manipulating information within a GIS database, it is often necessary to either integrate vector and raster data, or convert from one form to the other. Integration of the two types of data, that is, using both types simultaneously, is usually accomplished by displaying vector data overlaid on a raster image background, as illustrated in Figure 28.5. In that figure, vector data representing the dwellings (points) that exist within the different subdivisions (areas) are overlaid on a satellite image of the same area. This graphic was developed as part of a population growth and distribution study being conducted for a municipality. The combination of vector and raster data is useful to provide a frame of reference and to assist GIS operators in interpreting displayed data. Sometimes it is necessary or desirable to convert raster data to vector format or vice versa. Procedures for accomplishing these conversions are described in subsections that follow.

28.6.1 Vector-to-Raster Conversion

Vector-to-raster conversion is also known as *coding* and can be accomplished in several ways, three of which are illustrated in Figure 28.6. Figure 28.6(a) is an overlay of the vector representation of Figure 28.4(b) with a coarse raster of grid cells. In one conversion method, called *predominant type coding*, each grid cell is assigned the value corresponding to the predominant characteristic of the area it

Figure 28.5 Vector data overlaid on raster image background. (Courtesy Pennsylvania Department of Transportation.)

covers. For example, the cell located in row 3, column 1 of Figure 28.6(a) overlaps two polygons, one of forest (type F) and one of marsh (type M). As shown in Figure 28.6(b), since the largest portion of this cell lies in forest, the cell is assigned the value F—the predominant type.

In another coding method, called *precedence coding*, each category in the vector data is ranked according to its importance or "precedence" with respect to the other categories. In other words, each cell is assigned the value of the highest ranked category present in the corresponding area of the vector data. A common example involves water. While a stream channel may cover only a small portion of a cell area, it is arguably the most important feature in that area. Also, it is important to avoid breaking up the stream. Thus for the cell in row 4, column 3 of Figure 28.6(a), which is illustrated in Figure 28.6(c), water would be given the highest precedence and the cell coded S even though most of the cell is covered by marsh.

Center-point coding is the third technique for converting from vector-to-raster data. Here a cell is simply assigned the category value at the vector location corresponding to its center point. An example is shown in Figure 28.6(d), which represents the cell in row 2, column 3 of Figure 28.6(a). Here, since marsh exists at the cell's center, the entire cell is designated as category M. Note that the grid cell of row 3, column 4 would be classified by predominant type as grassland, by precedence as stream, and by center point as marsh. This illustrates how different conversion processes can yield different classifications for the same data.

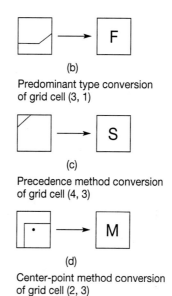

Figure 28.6 Methods for converting data from vector to raster format. (a) Vector representation of Figure 28.4. (b) Overlaid on a coarse raster format. (b), (c), and (d) Vector-to-raster conversion by predominant type, precedence, and center-point method, respectively.

The precisions of these vector-to-raster conversions depend on the size of the grid used. Obviously, using a raster of large cells would result in a relatively inaccurate representation of the original vector data. On the other hand, a fine-resolution grid can very closely represent the vector data, but would require a large amount of computer memory. Thus, the choice of grid resolution becomes a tradeoff between computing efficiency and spatial precision.

28.6.2 Raster-to-Vector Conversion

Raster-to-vector conversions are more vaguely defined than vector to raster. The procedure involves extracting lines from raster data, which represent linear features such as roads, streams, or boundaries of common data types. Whereas the approach is basically a simple one, and consists in identifying the pixels or cells through which vector lines pass, the resulting jagged- or "staircase"-type outlines are not indicative of true lines. One raster-to-vector conversion example is illustrated in Figure 28.7(a), which shows the cells identifying the stream line of Figure 28.4(b). Having selected these cells, a problem that must be resolved is how to fit a line to these jagged forms. One solution consists in simply connecting adjacent cell centers [see Figure 28.7(b)] with line segments. Note, however, that the resulting line [see Figure 28.7(c)] does not agree very well with the original stream of Figure 28.4(a). This example illustrates that some type of "line smoothing" is usually necessary to properly represent the gently curved boundaries that normally occur in nature. However, fitting smooth lines to the jagged cell boundaries is a complicated mathematical problem that does not necessarily have a unique solution. The decision ultimately becomes a choice between accuracy of representation and cost of computation.

No matter which conversion is performed, errors are introduced during the process, and some information from the original data is lost. Use of smaller grid cells improves the results. Nevertheless, as illustrated by the example of Figure 28.7, if a data set is converted from vector-to-raster and then back to vector

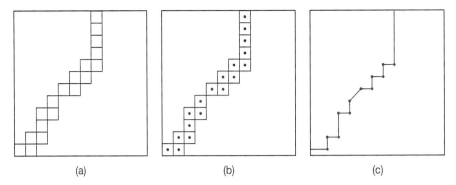

Figure 28.7 Conversion of data from raster-to-vector format. (a) Cells that identify the stream line of Figure 28.4(a). (b) Cell centers. (c) Vector representation of stream line recreated by connecting adjacent cell centers.

(or vice versa), the final data set will not likely match the original. Thus, it is important for GIS operators to be aware of how their data have been manipulated, and what can be expected if conversion is performed.

28.7 CREATING GIS DATABASES

Several important factors must be considered prior to developing the database for a GIS. These include the types of data that need to be obtained, optimum formats for these data, the reference coordinate system that will be used for spatially relating all data, and the necessary accuracy of each data type. Provisions for updating the database must also be considered. Having made these decisions, the next step is to locate data sources. Depending on the situation, it may be possible to utilize existing data, in which case a significant cost savings could result. However, in many cases, it is necessary to collect new data to meet the needs of the GIS. Different methods are available for generating the digital data needed to support a GIS. These are discussed in the subsections that follow. Regardless of the method used, *metadata* (see Section 28.8) should be included with each data file to document its source, the instruments and procedures used to collect it, its reference coordinate system and elevation datum, its accuracy, and any other information needed to qualify the data or describe its character.

28.7.1 Generating Digital Data from Field Surveys

Spatially related information needed to support a GIS is often generated by conducting new field surveys expressly for that purpose. Any of the equipment and procedures described in preceding chapters that are capable of locating objects in space can be used for this work. However, total station instruments interfaced with data collectors, GNSS equipment, LiDAR, and mobile mapping systems are particularly convenient because they can rapidly and efficiently provide coordinates of points directly in a reference coordinate system that is suitable for the GIS, and because necessary identification codes can also be entered at the time the data is collected. Both planimetric positions and elevation data (preferably in the form of digital elevation models) can be obtained with these instruments.

However, it must be remembered that elevations obtained with GNSS equipment are related to the ellipsoid, and thus must be converted to orthometric heights (see Section 13.4.3). The field data can be downloaded into a computer, processed using COGO (see Chapter 11) or other software if necessary, and entered directly into the GIS database. The cost of generating digital data in this manner is relatively high, but the data are generally very accurate.

Code-based GNSS receivers offer several advantages over other instruments in collecting of mapping data for a GIS. These instruments are relatively inexpensive, and when the data is reduced using differential techniques, the positional accuracies obtained are often sufficient for many but not all uses. Other advantages are speed, ease of use, and the ability to enter other ancillary data about a point. For instance when collecting the position of a utility pole for inclusion into a GIS, additional information such as type (electric, cable, telephone, etc.), pole number, diameter, height, and condition can be entered while its position is being determined. These code-based units can also be used to collect the alignment of a feature such as a utility line, road or sidewalk by simply entering the epoch rate and traversing the alignment either on foot or in a vehicle. Again, ancillary information about the alignment can be entered while the user traverses it. If additional accuracy is required, the surveyor can use carrier phase-shift GNSS receivers and kinematic reduction techniques to quickly obtain the data. However, as discussed in Chapter 15, in many situations kinematic surveys require project planning for successful completion due to obstructions and other factors.

Mobile mapping units that are discussed in Section 17.9.5 offer quick and efficient methods of collecting large amounts of georeferenced data, which is ready for a GIS entry. For example, the mobile mapping unit shown in Figure 1.4 can capture three-dimensional, georeferenced points at a rate of 1.3 million per second. It provides an integrated solution to quickly and safely collecting data for a GIS in urban areas where safety or speed are a major concern. These systems allow users to drive at highway speeds, and collect georeferenced point clouds of all objects within a specified range from the vehicle. Standalone terrestrial units can provide data capture at longer distances from the scanner. However, these systems must be georeferenced using either conventional or GNSS equipment. Additionally, obstructions can often result in incomplete data sets that must be later filled in by other surveying methods.

As discussed in Section 27.18, LiDAR can capture large amounts of data from mobile platforms. The systems can capture and provide inventories of objects such as vegetation, buildings, and other objects typically captured using aerial photography. The data can also be used create topographic maps of the region. One advantage of these systems is that additional features in the point clouds can be added later from the same data set if they are missed by the initial processing. Both ground-based and aerial scanners a means of capturing georeferenced point clouds from which inventories of objects can be assessed and exported for use in a GIS.

28.7.2 Digitizing from Aerial Photos with Stereoplotters

Information from aerial photographs can also be entered directly into a GIS database in digital form using a softcopy stereoplotter (see Section 27.14). In this procedure, both planimetric features and elevations can be recorded, and high accuracy can be achieved. The data are registered to the selected reference coordinate system and vertical datum by *orienting* the stereoplotter to ground control points prior to beginning the digitizing. Then to record planimetric features, an operator views the stereomodel, points to objects of interest, enters any necessary feature identifiers or codes, and depresses a key or foot pedal to transfer the information to a file in an interfaced computer. To digitize elevations, a DEM is read directly from the three-dimensional stereomodel and stored in a computer file.

Accuracies of data obtained using this procedure will depend mainly on the scale and quality of the aerial photography, the accuracy of the ground control points used to orient the stereoplotter, and the experience and capabilities of the stereoplotter operator. Other factors that may affect accuracy to a lesser degree include camera lens distortions, atmospheric refraction, differential shrinkage and expansion of the materials upon which the photographic products are printed, and optical and/or mechanical imperfections in the stereoplotting or digitizing equipment.

Data sets generated by digitizing stereomodels will usually need to be checked carefully to ensure that all desired features have been included. Also, the data must be corrected or "cleaned" before being used in a GIS. In this process, unwanted points and line portions must be removed and "unclosed" polygons, which result from imprecise pointing when returning to a polygon's starting node, must be closed (see Section 18.8.2). Finally, thin polygons or "slivers" created by lines being inadvertently digitized twice, but not precisely in the same location, must be eliminated. This editing process can be performed by the operator, or with a program that can find and remove certain features that fall within a set of user-defined tolerances.

Digitizing from stereomodels and editing the data can usually be accomplished in less time and with lower cost than obtaining the data by field surveys, especially where relatively large areas must be covered. Of course, some field surveying is needed to establish the ground control needed to orient the stereomodels. If experienced people do the work carefully, the resulting accuracy of data obtained by digitizing stereomodels is usually very good.

28.7.3 Digitizing Existing Graphic Materials

If sources such as maps, orthophotos, plans, diagrams, or other graphic documents already exist that will meet the needs of the GIS database, these can be conveniently and economically converted to digital files using a tablet digitizer. Many GIS software packages provide programs to support the procedure directly. A tablet digitizer, as shown in Figure 28.8, contains an electronic grid and an attached cursor. Movement of the cursor across the grid creates an electronic signal unique to the cursor's position. This signal is relayed to the computer, which records the digitizer's coordinates for the point. Data identifiers or attribute codes can be associated with each point through the computer's keyboard, or by pressing numerical buttons on the cursor.

The digitizing process begins by securing the source document to the tablet digitizer. If the document is a map, the next step is to register its reference coordinate system to that of the digitizer. This is accomplished by digitizing a series of

Figure 28.8
Tablet digitizer interfaced with personal computer. (Courtesy Tom Pantages.)

reference (tic) marks on the map for which geographic coordinates such as state plane, UTM, or latitude and longitude are precisely known. With both reference and digitizer coordinates known for these tic marks, a coordinate transformation (see Section 11.8) can be computed. This determines the parameters of scale change, rotation, and translation necessary to convert digitized coordinates into the reference geographic coordinate system. After this, any map features can be digitized, whereupon their coordinates are automatically converted to the selected system of reference, and they can then be entered directly into the database.

Both planimetric features and contours can be digitized from the map. Planimetric features are recorded by digitizing the individual points, lines, or areas that identify them. As described in Section 28.4.2, this process creates data in a vector format. Elevation data can be recorded as a digital elevation model (DEM) by digitizing critical points along contours. From these data, *triangulated irregular networks* (TIN models) can be derived using the computer (see Section 17.8). From the TIN models, point elevations, profiles, cross sections, slopes, aspects (slope directions), and contours having any specified contour interval can be derived automatically using the computer.

Data files generated in this manner can be obtained quickly and relatively inexpensively. Of course, the accuracy of the resultant data can be no better than the accuracy of the document being digitized, and its accuracy is further diminished by differential shrinkages or expansions of the paper or materials upon which the document is printed, and by inaccuracies in the digitizer and the digitizing process.

28.7.4 Keyboard Entry

Data can be entered into a GIS file directly using the keyboard on a computer. Often data input by this method are nonspatial, such as map annotations or numerical or tabular data. To facilitate keying in data, an intermediate file having a simple format is sometimes created. This file is then converted into a GIS-compatible format using special software. For example, metes-and-bounds descriptions (see Section 21.4) can also be computed using the coordinate geometry techniques discussed in Chapter 11. The resulting coordinates can be used to facilitate entry of the deed description into the GIS file.

28.7.5 Existing Digital Data Sets

Massive quantities of digital information are now being generated by a wide variety of offices and agencies involved in GIS activities. At the federal level, the U.S. Geological Survey, the National Oceanic and Atmospheric Administration, the Bureau of Land Management, the Environmental Protection Agency, and other organizations are developing digital information. The *digital line graphs* (DLGs) and *digital elevation models* (DEMs) produced by the U.S. Geological Survey (see Section 18.3) are examples of available digital files. In addition to federal agencies, offices of state governments, counties, and cities are involved in this work. As a result of this proliferation of information, an initiative known as the *National Spatial Data Infrastructure* (NSDI) has evolved at the federal level. The NSDI encompasses policies, standards, and procedures for organizations to cooperatively produce and share geographic data. The *National Geospatial Data Clearinghouse* is a component of the NSDI that provides a pathway to find information about available spatially referenced data.[1]

Of course, before using existing data, information about its content, source, date, accuracy, and other characteristics must be scrutinized to determine if it is suitable for the GIS at hand. This requirement underscores the need for maintaining good quality metadata (see Section 28.8) for all digital files. Also, existing digital data must often undergo conversion of file structures and formats to be usable with specific GIS software. Because of differences in the way data are represented by different software, it is possible that information can be lost, or that spurious data can creep in during this process.

28.7.6 Scanning

Scanners are instruments that automatically convert graphic documents into a digital format. As discussed in Section 27.14.4, they are used to digitize the contents of aerial images to support softcopy photogrammetry. In GIS work, scanners are used not only to digitize aerial photos, but also to convert larger documents such as maps, plans, and other graphics into digital form. The principal advantage of using scanners for this work is that the tedious work of manual digitizing is

---

[1]Information about available geospatial data can be obtained by writing to the National Spatial Data Infrastructure, U. S. Geological Survey, 508 National Center, Reston, VA 20192, or by contacting them through the internet at http://nsdi.usgs.gov.

eliminated, and the process of converting graphic documents to digital form is accelerated significantly.

Scanners accomplish their objective by measuring the amount of light reflected from a document and assigning this information to pixels. This is possible because different areas of a document will reflect light in proportion to their tones, from a maximum for white through the various shades of gray to a minimum for black. For example, the scanner of Figure 28.9 uses a linear array of light sensors to capture the varying intensities of reflected light, line by line, as the document is fed through the system. This creates a raster data set. Its pixel size can be varied and made as small as $1/500$ in.^{2} (500 dpi). Large complicated documents can be scanned in a matter of a few minutes. The data are stored directly on the hard drive of an interfaced computer and can be viewed on a screen, edited, and manipulated. Editing is an important and necessary step in the process, because the scanner will record everything, including blemishes, stains, and creases.

Documents such as subdivision plats, topographic maps, engineering drawings and plans, aerial photos, and orthophotos can be digitized using scanners. Then, if necessary, the raster data can be converted to vector form using techniques described in Section 28.6.2.

Accuracy of the raster file obtained from scanning depends somewhat on the instrument's precision, but pixel size or resolution is generally the major factor. A smaller pixel size will normally yield superior resolution. However, there are certain tradeoffs that must be considered. Whereas a large pixel size will result in a coarse representation of the original, it will require less scanning time and computer storage. Conversely, a fine resolution, which generates a precise depiction

Figure 28.9
Intergraph ANAtech Eagle 4050 scanner. (Courtesy Wisconsin Department of Transportation.)

of the original, requires more scanning time and computer storage. An additional problem is that at very fine resolution, the scanner will record too much "noise," that is, impurities such as specks of dirt. For these reasons and others, this is the least preferred method of capturing data in a GIS.

28.8 METADATA

Metadata, often simply defined as "data about data," describes the content, quality, condition, and other characteristics about geospatial data and provides a record of changes or modifications that have been made to that data. It normally includes information such as who originally created the data, when was it generated, what equipment and procedures were used in collecting the data, and what was its original scale and accuracy. As discussed in Section 18.17, it should include the reference datum for the coordinates, directions, and heights depicted on the product. Once created, data can travel almost instantaneously through a network and be transformed, modified, and used for many kinds of spatial analyses. It can then be re-transmitted to another user, and then to another, etc.[2] It is important that each change made to any data set be documented by updating its associated metadata.

Although generating the original metadata, and updating it as changes are made, may be burdensome and add cost, in the long run it is worth the effort because it preserves the value of the data and extends its useful life. If it is not done, prospective users may not trust the data, and as a result they may fail to take advantage of it and incur the cost of duplicate data collection.

The *Federal Geographic Data Committee* (FGDC) has developed metadata standards that provide a common set of terms and definitions for describing geospatial data, and outline a consistent and systematic approach to documenting data characteristics.[3] A portion of these standards are discussed in Chapters 18 and 19. The primary benefit to be realized by following these standards is that all users, regardless of their backgrounds or specialty areas, will have a common understanding of the source, nature, and quality of any data set.

28.9 GIS ANALYTICAL FUNCTIONS

Most GISs are equipped with a set of basic analytical functions that enable data to be manipulated, analyzed, and queried. These functions, coupled with appropriate databases, provide GISs with their powerful capabilities for supplying information that so significantly aids in planning, management, and decision-making.

The specific functions available within the software of any particular GIS system will vary. They enable data to be stored, retrieved, viewed, analyzed spatially and computationally, and displayed. Some of the more common and useful spatial analysis and computational functions are (1) proximity analysis, (2) boundary operations, (3) spatial joins, and (4) logical operations. These are briefly described in the subsections that follow.

---

[2]An example of a data exchange website, PASDA, with metadata on the Internet at http://www.pasda.psu.edu/.

[3]These metadata standards may be obtained from the FGDC Secretariat, U.S. Geological Survey, 590 National Center, Reston, VA 22092, or information can be obtained at the following website: http://www.fgdc.gov.

28.9.1 Proximity Analysis

This spatial analysis function creates new polygons that are geographically related to nodes, lines, or existing polygons, and usually involves processes called *buffering*. *Point buffering*, also known as *radius searching*, is illustrated in Figure 28.10(a). It involves the creation of a circular buffer zone of radius R around a specific node. Information about the new zone can then be gathered and analyses made of the new data. A simple example illustrates its value. Assume that well water that was polluted by an accidental spill has just been discovered. With appropriate databases available, all dwellings within a specified radius of the well can be located, the names, addresses, and telephone numbers of all individuals living within the point buffer zone tabulated, and the people quickly alerted to the possibility of their water also being polluted.

Line buffering, illustrated in Figure 28.10(b), creates new polygons along established lines such as streams and roads. To illustrate the use of line buffering, assume that to preserve the natural stream bank and prevent erosion, a zoning commission has set the construction setback distance from a certain stream at D. Line buffering can quickly identify the areas within this zone. *Polygon buffering*, illustrated in Figure 28.10(c), creates new polygons around an existing polygon. An example of its use could occur in identifying those landowners whose property lies within a certain distance D of the proposed site of a new industrial facility.

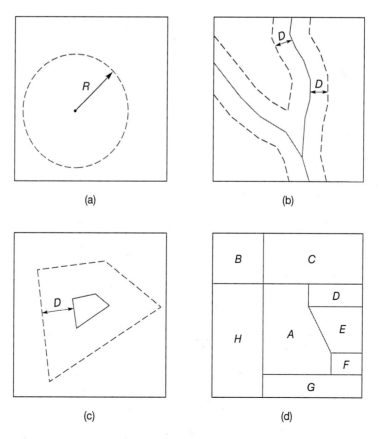

Figure 28.10 GIS spatial analysis functions. (a) Point buffering. (b) Line buffering. (c) Polygon buffering. (d) Adjacency analysis.

Many other examples could be given, which illustrate the value of buffering for rapidly extracting information to support management and decision-making.

28.9.2 Boundary Operations

If the topological relationships discussed in Section 28.4.3 have been entered into a database, certain analyses regarding relative positioning of features, usually called boundary operations, can be performed. *Adjacency* and *connectivity* are two important boundary operations that often assist significantly in management and decision-making. An example of adjacency is illustrated in Figure 28.10(d) and relates to a zoning change requested by the owner of parcel A. Before taking action on the request, the jurisdiction's zoning administrators are required to notify all owners of adjacent properties B through H. If the GIS database includes the parcel descriptions with topology and other appropriate attributes, an adjacency analysis will identify the abutting properties and provide the names and addresses of the owners.

Connectivity involves analyses of the intersections or connections of linear features. The need to repair a city's water main serves as an example to illustrate its value. Suppose that the decision has been made that these repairs will take place between the hours of 1:00 and 4:00 P.M. on a certain date. If infrastructure data are stored within the city's GIS database, all customers connected to this line whose water service will be interrupted by the repairs can be identified and their names and addresses tabulated. The GIS can even print a letter and address labels to facilitate a mailing announcing details of the planned interruption to all affected customers. Many similar examples could be given to illustrate benefits that can result from adjacency and connectivity.

28.9.3 Spatial Joins

Spatial joins, also called *overlaying*, are one of the most widely used spatial analysis functions of a GIS. As indicated in Figure 28.1, GIS graphic data are usually divided into layers, with each containing data in a single category of closely related features. Nonspatial data or "attributes" are often associated with each category. The individual layers are spatially registered to each other through a common reference network or coordinate system. Any number of layers can be entered into a GIS database, and could include parcels, municipal boundaries, public land survey system, zoning, soils, road networks, topography, land cover, hydrology, and many others.

Having these various sets of data available in spatially related layers makes the overlay function possible. Its employment in a GIS can be compared to using a collection of Mylar overlays in traditional mapping. However, much greater efficiency and flexibility are possible when operating in the computer environment of a GIS, and not only can graphic data be overlaid, but attribute information can be combined as well.

Many examples could be given to illustrate the applications and benefits of the GIS spatial join or overlay process. Consider one case where the land in a particular area suitable for development must be identified. To perform an in-depth analysis of this situation, the evaluation would normally have to consider numerous variables within the area, including the topography (slope and aspect of the terrain), soil type, land cover, land ownership, and others. Certain combinations of these variables could make land unsuitable for development. Figure 28.11 illustrates the simple case of land suitability analysis involving only two variables, slope and soil type. Figure 28.11(a) shows polygons within which the average slope is either 5 or 10%. Figure 28.11(b) classifies the soils in the area as E (erodible) or S (stable). The composite of the two data sets, which results from a *polygon-on polygon* overlay, is shown in Figure 28.11(c). It identifies polygon I, which combines 5% slopes and low-erodibility class S soils. Since this combination does not present potential erosion problems, considering those variables, the area within polygon I is suitable for development, while areas II, III, and IV are not.

Another GIS overlay function is that of point in polygon. Here the question involves which point features are in certain polygons where layers are combined. For example, to predict possible well contamination, a GIS operator may want to know which wells are located in an area of highly permeable soil. A similar overlay process, *line in polygon*, identifies specific linear features within polygons of interest. An example of its application would be the identification of all bituminous roads, paved more than 15 years ago, in townships whose roadway maintenance budgets are less than $250,000. Obviously such information would be valuable to support decisions concerning the allocation of state resources for local roadway maintenance.

The GIS functions just described can be used individually, as has been illustrated with the examples, or they can be employed in combination. The following example illustrates the simultaneous application of line buffering, adjacency, and overlay. The situation involves giving timely notice to affected persons of an impending flood that is predicted to crest at a specific stage above a particular river's normal elevation. Here it is first necessary to identify the lands that lie at or below the expected flood stage. This can be done using line buffering, where the thread of the river is the line of reference. However, the width of the buffer zone is variable and is determined by combining elevation data in the TIN model with the buffering process. The adjacency and overlay functions are then used to determine which landowners are next to or within the flood zone. Then the names, addresses, and telephone numbers of property owners and dwellers within and adjacent to the affected area can be tabulated. These people can then be notified of the impending situation, and emergency preparations

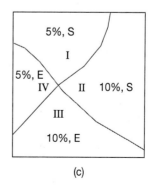

Figure 28.11 Example of GIS overlay function used to evaluate land suitability. (a) Polygons of differing slopes. (b) Varying soil types in the area. (c) Overlay of (a) and (b), identifying polygon I as an area combining lower slopes with stable soils that would be suitable for development.

such as construction of temporary levees can be performed, or, if necessary, the area can be evacuated. Should evacuation be required, the GIS may even be used to identify the best and safest escape routes. Data sets necessary for such analyses includes the topography in the area, including the river's location, normal stage, and floodplain cross sections; census data; property ownership; and the transportation network.

28.9.4 Logical Operations

Typically, attribute data that is related to the features present in the GIS are stored in a database. Thus, the database can be used to perform logical operations on the data. For instance, a city can construct a GIS database that contains the time streetlights are installed, and their rated life. The manager can then query the system to show all lights that have passed their rated life cycle, and schedule maintenance personnel to replace these lights. Today GISs are being used in large buildings to help managers keep records of maintenance and schedule routine maintenance jobs. The number of useful logical queries that can be performed in a GIS is limited only by the data contained in the GIS database, and the imagination of the user.

28.9.5 Other GIS Functions

In addition to the spatial analysis functions described in preceding subsections, many other functions are available with most GIS software. Some of these include the capability of computing (1) the number of times a particular type of point occurs in a certain polygon; (2) the distances between selected points, or from a selected line to a point; (3) areas within polygons; (4) locations of polygon centroids; and (5) volumes within polygons where depth or other conditions are specified. A variety of different mapping functions may be performed using GISs. These may include (a) performing map scale changes; (b) changing the reference coordinate system from, say, the state plane to the UTM system; (c) rotating the reference grid; and (d) changing the contour interval used to represent elevations.

Most GISs are also capable of performing several different digital terrain analysis functions. Some of these include (1) creating TIN models or other DEMs from randomly spaced XYZ terrain data; (2) calculating profiles along designated reference lines, and determining cross sections at specified points along the reference line; (3) generating perspective views where the viewpoint can be varied; (4) analyzing visibility to determine what can or cannot be seen from a given vantage point; (5) computing slopes and aspects; and (6) making sun intensity analyses.

Output from GISs can be provided in graphic form as charts, diagrams, and maps; in numerical form as statistical tabulations, or in other files that result from computations and manipulations of the geographic data. These materials can be supplied in either printed (hardcopy) or digital form.

■ 28.10 GIS APPLICATIONS

As stated earlier, and as indicated by the examples in preceding sections of this chapter, the areas of GIS applications are widespread. Further evidence of the diversity of GIS applications can be seen by reviewing the bibliography at the end of this chapter. The technology is being used worldwide, at all levels of government, in business and industry, by public utilities, and in private engineering and surveying offices. Some of the more common areas of application occur in (1) land-use planning; (2) natural resource mapping and management; (3) environmental impact assessment; (4) census, population distribution, and related demographic analyses; (5) route selection for highways, rapid-transit systems, pipelines, transmission lines, etc.; (6) displaying geographic distributions of events such as automobile accidents, fires, crimes, or facility failures; (7) routing buses or trucks in a fleet; (8) tax mapping and mapping for surveying and engineering purposes; (9) subdivision design; (10) infrastructure and utility mapping and management; (11) urban and regional planning; and many others.

As the use of GIS technology expands, there will be a growing need for trained individuals who understand the fundamentals of these systems. Users should be aware of the manner in which information is recorded, stored, managed, retrieved, analyzed, and displayed using a GIS. System users should also have a fundamental understanding of each of the GIS functions, including their bases for operation, their limits, and their capabilities. *Perhaps of most importance, users must realize that information obtained from a GIS can be no better than the quality of the data from which it was derived.*

From the perspective of those engaged in surveying (geomatics), it is important to underscore again the fact that the fundamental basis of GISs is a database of spatially related, digital data. Since accurate position determination and mapping are the surveyor's forte, in the future surveyors will continue to play key roles in designing, developing, implementing, and managing these systems. Their input will be particularly essential in establishing the necessary basic control frameworks, conducting ground and aerial surveys to locate geographic features and their attributes, compiling maps, and assembling the digital data files needed for these systems.

28.11 DATA SOURCES

The surveyor will find many data sets useful to aid in tasks that they perform.[4] These data sets include digital raster graphics (DRGs), digital elevation models (DEMs) (see Section 17.8), digital line graphs (DLGs) (see Section 18.3), digital ortho quarter quadrangles (DOQQs) (see Section 27.15), LiDAR data (see Section 27.19), land-use and land-cover (LULC) data, soil survey geographic (SSURGO) data, national wetlands inventory (NWI) data, and Federal Emergency Management Agency (FEMA) flood map data. These data sets are described as follows.

Digital raster graphics (DRGs) are scanned digital images of U.S. Geological Survey (USGS) 7-1/2 min topographic quadrangle sheets and are in Tagged Image File Format (TIFF). They contain all data that are located on the topographic maps only in digital format. Since they are images, they do not contain topology or elevation data. DRGs can be used as location maps for plans or to view topography, surrounding features, stream, and building locations before performing a field survey.

Digital elevation models (DEMs) are grid-based data where each grid cell has an average elevation of the area of the cell. The USGS has set the standard for DEM data with 10, 30, and 100 m cell size. DEMs are useful for spatial analyses

---

[4]For a more detailed description of data useful to the surveyor, and where to find it, refer Chapter 10 of Watersheds: Processes, Assessment, and Management (DeBarry, 2004).

and modeling. They can also be used to develop three-dimensional terrain models such as contours and triangulated irregular networks (see Section 17.8). From DEMs, flow direction, accumulation, and streams can be defined.

Digital line graphs are vector files containing the planimetric data such as U.S. public land survey system (PLSS) corners, survey control and markers, transportation, hydrography, vegetative surface cover, and so on. These data can be useful to begin a base map.

Digital ortho quarter quadrangles (DOQQs) are rectified digital images of color or black and white aerial photos, based upon one-quarter of the USGS 7-1/2 min quadrangles. As discussed in Section 27.15, rectification removes the edge distortion of typical aerial photographs. DOQQs are available in both GeoTIFF and native format, which consists of an ASCII keyword header followed by a series of 8-bit binary image lines for B/W and 24-bit band-interleaved-by-pixel (BIP) for color. DOQQs are in the Universal Transverse Mercator (UTM) map projection coordinate system (see Section 20.11) and are referenced to either the NAD 27 or NAD 83 (see Section 19.6).

LiDAR data (see Section 27.18) is used to produce various elevation data products including point-based digital terrain models (DTM), grid-based digital elevation models (DEM), and contours. In addition LiDAR processing generates raw point cloud, processed points, and breaklines. The accuracy of the data is much better than the data sets described above and can be utilized for accurate terrain and hydrologic modeling.

Land-use and land-cover data is polygon coverage of land cover based upon the Anderson (1976) method of land classification and describes water, vegetation, cultural and natural surface features. These data files can be used to help in hydrologic modeling, analyzing land use trends, etc.

Soil survey geographic data has been developed by the Natural Resources Conservation Service (NRCS), formerly the Soil Conservation Service (SCS), and contains the spatial and tabular data contained in the county soil surveys in digital format. Utilizing the attributes, it can be used to bring into a site plan and then determine the best location for on-lot septic systems, stormwater recharge areas, or used for erosion and sedimentation control plans.

The National Wetlands Inventory (NWI) data catalogued known wetlands on a USGS topographic quadrangle base map. The digital format can be brought in to site plans to get a preliminary indication if wetlands are present on a particular property. They were developed by the U.S. Fish and Wildlife Service at the national level, and therefore do not contain all wetlands, and the boundaries are not specific to a particular site. Field delineation and boundary survey must be conducted for site planning initiatives.

The Federal Emergency Management Agency (FEMA) is responsible for delineating regulatory floodplains as part of the National Flood Insurance Program (NFIP). The original flood insurance rate maps (FIRMs) were paper copies of floodplains on simple base maps showing just roads and streams. FEMA put these floodplains into digital format called Q3 data, which can be brought into the GIS and overlaid onto aerial photographs. FEMA is currently working on the map modernization program and placing digital floodplains on accurate digital aerial photographs (typically flown during LiDAR data collection). The floodplains are corrected where known errors occurred utilizing the LiDAR elevation data.

PROBLEMS

Asterisks (*) indicate problems that have partial answers given in Appendix G.

28.1 Define geographic information system.
28.2 List some areas where a GIS is used.
28.3 List the fundamental components of a GIS.
28.4 What is the difference between a GIS and a LIS?
28.5 Discuss the importance of metadata to a GIS.
28.6 Name and describe the different simple spatial objects used for representing graphic data in digital form. Which objects are used in raster format representations?
28.7 Discuss the differences between vector and raster formats for entering data into a GIS. Cite the advantages and disadvantages of each.
28.8 How many pixels are required to convert the following documents to raster form for the conditions given:
  *(a) A 384-square in. map scanned at 200 dpi.
  (b) A 9-in. square aerial photo scanned at 300 dpi.
  (c) An orthophoto of 11 × 17 in. dimensions scanned at 300 dpi.
28.9 Define the term topology.
28.10 For what types of data is the vector format best suited?
28.11 Discuss the compromising relationships between grid cell size and resolution in raster data representation.
28.12 Define the term connectivity in relation to a GIS.
28.13 Develop identifier and topology tables similar to those of Tables 28.1 and 28.2 in the text for the vector representation of (see the following figures):
  (a) Problem 28.13(a)
  (b) Problem 28.13(b)

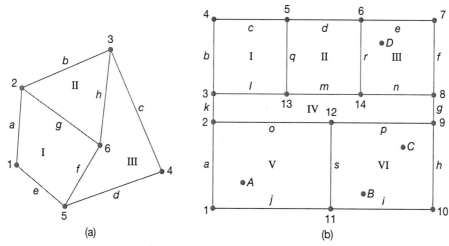

Problem 28.13

28.14 Compile a list of linear features for which the topological relationship of adjacency would be important.
28.15 Prepare a raster (grid cell) representation of the sample map of:
  (a) Problem 28.15(a), using a cell size of 0.10-in. square (see accompanying figure).
  (b) Problem 28.15(b), using a cell size of 0.20- in. square (see accompanying figure).

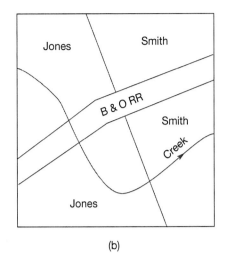

(a)   (b)

Problem 28.15

28.16 Discuss the advantages and disadvantages of using the following equipment for converting maps and other graphic data to digital form:
 *(a) tablet digitizers
  (b) scanners.
28.17 Explain the concepts of the following terms in GIS spatial analysis, and give an example illustrating the beneficial application of each: (1) adjacency and (2) connectivity.
28.18 If data were being represented in vector format, what simple spatial objects would be associated with each of the following topological properties?
  (a) Connectivity
  (b) Direction
  (c) Adjacency
  (d) Nestedness
28.19 Prepare a transparency having a 0.10-in grid, overlay it onto Figure 28.4(a), and indicate the grid cells that define the stream. Now convert this raster representation to vector using the method described in Section 28.6.2. Repeat the process using a 0.20-in grid. Compare the two resulting vector representations of the stream and explain any differences.
28.20 Discuss how spatial and nonspatial data are related in a GIS.
28.21 What are the actual ground dimensions of a pixel for the following conditions:
  (a) A 1:5000 scale orthophoto scanned at 300 dpi.
 *(b) A 1:24,000 map scanned at 200 dpi.
28.22 Describe the following GIS functions, and give two examples where each would be valuable in analysis:
  (a) line buffering, and
  (b) spatial joins.
28.23 Define metadata.
28.24 Compile a list of data layers and attributes that would likely be included in an LIS.

28.25 Compile a list of data layers and attributes that would likely be included in a GIS for:
  (a) Selecting the optimum corridor for constructing a new rapid-transit system to connect two major cities
  (b) Choosing the best location for a new airport in a large metropolitan area.
  (c) Routing a fleet of school buses.
  (d) Selecting the fastest routes for reaching locations of fires from various fire stations in a large city.
28.26 Discuss how LiDAR can be used to aid in the development of a GIS.
28.27 Consult the literature on GISs and, based on your research, describe an example that gives an application of a GIS in:
  (a) Natural resource management.
  (b) Agriculture.
  (c) Engineering.
  (d) Forestry.

BIBLIOGRAPHY

Avrahami, Y. and Y. Doytsher. 2010. "Algorithms for Automatic Matching of Polygons or Closed Curves Derived from Different Images. *Surveying and Land Information Science* 70 (No. 4): 175.

Campbell, J. et al. 2012. "Using a Geographic Information System to Resolve a Riparian Boundary." *Surveying and Land Information Science* 72 (No. 4): 195.

Greenfeld, J. 2011. "GIS Body of Knowledge." *Surveying and Land Information Science* 71 (No. 3–4): 103.

Miima J. and L. Potts. 2015. "Integrated Teaching and Assessment of GIS/LIS in Geomatics for Improving Student Performance." *Surveying and Land Information Science* 73 (No. 1): 15.

Nagarajan, S. 2014. "A Linear Feature-Based Recovery of Registration Parameters Using Area Minimization." *Surveying and Land Information Science* 73 (No. 2): 61.

Nettleman, C. A., et al. 2012. "GIS-Based Modeling of Sea Level Rise Effect on Coastal Property Management Policies." *Surveying and Land Information Science* 72 (No. 4): 173.

Roberge D. and B. Kjellson. 2009. "What Have Americans Paid (and Maybe the Rest of the World) for Not Having a Public Property Rights Infrastructure?" *Surveying and Land Information Science* 69 (No. 3): 135.

A Tape Correction Problems

A.1 CORRECTING SYSTEMATIC ERRORS IN TAPING

As discussed in Section 6.14, tape measurements are subjected to error sources caused by instrumental and environmental factors. All tape problems develop from the fact that a tape is either longer or shorter than its graduated "nominal" length because of manufacture, temperature changes, tension applied, or some other reasons. There are only two basic types of taping tasks: an unknown distance between two fixed points can be *measured*, or a required distance can be *laid off* from one fixed point. Since the tape may be too long or too short for either task, there are four possible versions of taping problems, which are: (1) measure with a tape that is too long, (2) measure with a tape that is too short, (3) lay off with a tape that is too long, and (4) lay off with a tape that is too short. The solution of a particular problem is always simplified and verified by drawing a sketch.

Assume that the fixed distance AB in Figure A.1 is measured with a tape that is found to be 100.03 ft as measured between the 0- and 100-foot marks. Then (the conditions in the figure are greatly exaggerated) the first tape length would extend to point 1; the next, to point 2; and the third, to point 3. Since the distance remaining from 3 to B is less than the correct distance from the true 300-ft mark to B, the *recorded* length AB is too small and must be increased by a correction. If the tape had been too short, the *recorded* distance would be too large, and the correction must be subtracted.

In laying out a required distance from one fixed point, the reverse is true. The correction must be subtracted from the desired length for tapes longer than their nominal value and added for tapes that are shorter. A simple sketch like Figure A.1 makes clear whether the correction should be added or subtracted for any of the four cases.

In taping linear distances, several types of systematic errors often occur simultaneously. The following examples illustrate procedures for computing and applying corrections for the two basic types of problems, *measurement* and *layoff*.

Figure A.1
Taping between fixed points, tape too long.

Example A.1

A 30-m steel tape standardized at 20°C and supported throughout under a tension of 5.45 kg was found to be 30.012 m long. The tape had a cross-sectional area of 0.050 cm² and a weight of 0.03967 kg/m. This tape was held horizontal, supported at the ends only, with a constant tension of 9.09 kg, to measure a line from A to B in three segments. The data listed in the following table were recorded. Apply corrections for tape length, temperature, pull, and sag to determine the correct length of the line.

(a) The tape length correction by Equation (6.3) is

$$C_L = \left(\frac{30.012 - 30.000}{30.000}\right) 81.151 = +0.0324 \text{ m}$$

(b) Temperature corrections by Equation (6.4) are (Note: Separate corrections are required for distances observed at different temperatures.):

Section	Measured (Recorded) Distance (m)	Temperature (°C)
A-1	30.000	14
1-2	30.000	15
2-B	21.151	16
	Σ81.151	

$$C_{T_1} = 0.0000116(14 - 20)30.000 = -0.0021 \text{ m}$$
$$C_{T_2} = 0.0000116(15 - 20)30.000 = -0.0017 \text{ m}$$
$$C_{T_3} = 0.0000116(16 - 20)21.151 = -0.0010 \text{ m}$$
$$\sum C_T = -0.0048 \text{ m}$$

(c) The pull correction by Equation (6.5) is

$$C_P \left(\frac{9.09 - 5.45}{0.050 \times 2{,}000{,}000}\right) 81.151 = 0.0030 \text{ m}$$

(d) The sag corrections by Equation (6.6) are (Note: Separate corrections are required for the two suspended lengths.):

$$C_{S_1} = -2\left[\frac{(0.03967)^2(30.000)^3}{24(9.09)^2}\right] = -0.0429 \text{ m}$$

$$C_{S_2} = -\frac{(0.03967)^2(21.151)^3}{24(9.09)^2} = -0.0075 \text{ m}$$

$$\sum C_S = -0.0504 \text{ m}$$

(e) Finally, corrected distance AB is obtained by adding all corrections to the measured distance, or

$$AB = 81.151 + 0.0324 - 0.0048 + 0.0030 - 0.0504 = 81.131 \text{ m}$$

Example A.2

A 100-ft steel tape standardized at 68°F and supported throughout under a tension of 20 lb. was found to be 100.012 ft long. The tape had a cross-sectional area of 0.0078 in.^{2} and a weight of 0.0266 lb/ft. This tape is used to lay off a horizontal distance CD of exactly 175.00 ft. The ground is on a smooth 3% grade; thus, the tape will be used fully supported. Determine the correct slope distance to layoff if a pull of 15 lb is used and the temperature is 87°F.

Solution

(a) The tape length correction, by Equation (6.3), is

$$C_L = \left(\frac{100.012 - 100.000}{100.000}\right)175.00 = +0.021 \text{ ft}$$

(b) The temperature correction, by Equation (6.4), is

$$C_T = 0.00000645(87 - 68)175.00 = +0.021 \text{ ft}$$

(c) The pull correction, by Equation (6.5), is

$$C_P = \frac{(15 - 20)}{0.0078(29,000,000)}175.00 = -0.0004 \text{ ft}$$

(d) Since this is a layoff problem, all corrections are subtracted. Thus, the required horizontal distance to layoff, rounded to the nearest hundredth of a foot, is

$$CD_h = 175.00 - 0.021 - 0.021 + 0.0004 = 174.96 \text{ ft}$$

(e) Finally, a rearranged form of Equation (6.2) is used to solve for the slope distance (the difference in elevation d for use in this equation, for 174.96 ft on a 3% grade, is $174.96(0.03) = 5.25$ ft):

$$CD_s = \sqrt{(174.96)^2 + (5.25)^2} = 175.04 \text{ ft}$$

B

Example Noteforms

MEASURING DISTANCES

Sta.	Fwd. Dist.	Back Dist.	Mean	Error	Ratio
A					
	321.18	321.22	321.20	.04 / 321.20	1 / 8,000
B					
	276.54	276.60	276.57	.06 / 276.57	1 / 4,600
C					
	100.30	100.29	100.30	.01 / 100.30	1 / 10,000
D					
	306.77	306.81	306.79	.04 / 306.79	1 / 7,700
E					
	255.47	255.50	255.48	.03 / 255.48	1 / 8,500
A					

Plate B.1

BORROW-PIT LEVELING

Point	+ Sight	HI	− Sight	Elev.	Cut
BM Road	4.22	364.70		360.48	
A,0			5.2	359.5	1.5
B,0			5.4	359.3	1.3
C,0			5.7	359.0	1.0
D,0			5.9	358.8	0.8
E,0			6.2	358.5	0.5
A,1			4.7	360.0	2.0
B,1			4.8	359.9	1.9
C,1			5.2	359.5	1.5
D,1			5.5	359.2	1.2
E,1			5.8	358.9	0.9
A,2			4.2	360.5	2.5
B,2			4.7	360.0	2.0
C,2			4.8	359.9	1.9
D,2			5.0	359.7	1.7
A,3			3.8	360.9	2.9
B,3			4.0	360.7	2.7
C,3			4.6	360.1	2.1
D,3			4.6	360.1	2.1
A,4			3.4	361.3	3.3
B,4			3.7	361.0	3.0
C,4			4.2	360.5	2.5
BM Road			4.22	360.48	

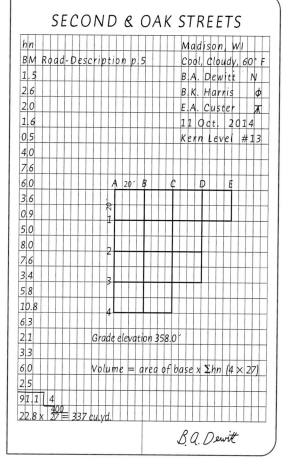

Plate B.2

TRAVERSING WITH A TOTAL STATION INSTRUMENT

Sta. Sighted	D/R	Horiz. Circle	Zenith Angle	Horiz. Dist.	Elev. Diff.
Instrument at sta 101					
$h_e = 5.8$	$h_r = 5.3$				
104	D	0°00′00″	86°30′01″	324.38	+19.84
102	D	82°18′19″	92°48′17″	216.02	−10.58
104	R	180°00′03″	273°30′00″		
102	R	262°18′18″	267°11′41″		
Instrument at sta 102					
$h_e = 5.5$	$h_r = 5.5$				
101	D	0°00′00″	87°11′19″	261.05	+10.61
103	D	95°32′10″	85°19′08″	371.65	+30.43
101	R	180°00′02″	272°48′43″		
103	R	275°32′08″	274°40′50″		
Instrument at sta 103					
$h_e = 5.4$	$h_r = 5.4$				
102	D	0°00′00″	94°40′48″	371.63	−30.42
104	D	49°33′46″	90°01′54″	145.03	−0.08
102	R	180°00′00″	265°19′14″		
104	R	229°33′47″	269°58′00″		

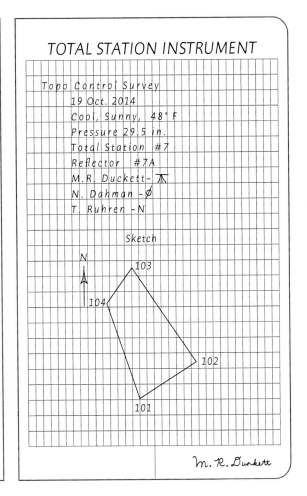

Topo Control Survey
19 Oct. 2014
Cool, Sunny, 48° F
Pressure 29.5 in.
Total Station #7
Reflector #7A
M.R. Duckett - ⚐
N. Dahman - ø
T. Ruhren - N

Sketch

M. R. Duckett

Plate B.3

DOUBLE DIRECT ANGLES

Hub	Dist.	Single ∠	Double ∠	Avg. ∠
A		38°58.0′	77°56.8′	38°58.4′
	321.31′			
B		148°53.6′	297°47.0′	148°53.5′
	276.57′			
C		84°28.1′	168°56.2′	84°28.1′
	100.30′			
D		114°40.3′	229°20.9′	114°40.4′
	306.83′			
E		152°59.4′	305°58.6′	152°59.3′
	255.48′			
A				
	1260.49′			539°59.7′
		Misclosure		0°00.3′
	interior	∠s =	(N-2) 180°	
		=	(5-2) 180°	
		=	540°00′	

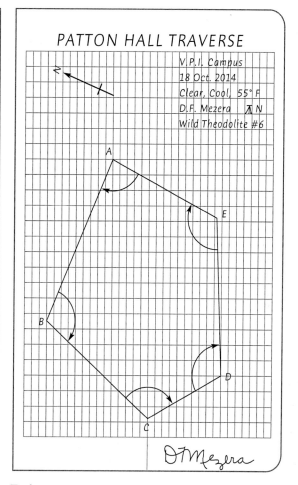

PATTON HALL TRAVERSE

V.P.I. Campus
18 Oct. 2014
Clear, Cool, 55° F
D.F. Mezera
Wild Theodolite #6

Plate B.4

CROSS-SECTION LEVELING

Sta.	+Sight	H.I.	−Sight	Elev.
5+00			9.5	
4+00			12.6	
TP 1	10.25	106.61	1.87	96.36
3+00			2.1	
2+50			5.8	
2+00			7.4	
1+35			9.7	
1+00			5.6	
0+50			7.6	
0+00			8.5	
BM Pod	8.51	98.23		89.72

HONOLULU-KAILUA HIGHWAY

Diamond Highway
25 Oct. 2014
Warm, Sunny 70°
A.C. Chun
R.E. Neilan N
W.E. Grube ΦC
M.L. Hagawa C
Lietz level #10

99.2	101.5	97.4	97.1	95.8	97.0	103.8
7.4	5.1	9.2	9.5	10.8	9.6	2.8
52	30	10		12	28	45

102.3	99.9	98.4	94.0	100.1	101.5	98.7
4.3	6.7	8.2	12.6	6.5	5.1	7.9
48	24	8		10	25	50

95.2	95.8	96.6	96.1	94.4	91.1	95.7
3.0	2.4	1.6	2.1	3.8	7.1	2.5
50	25	10		8	31	48

95.1	92.8	89.5	93.8	92.4	90.7	93.4	96.6
3.1	5.4	8.7	4.4	5.8	7.5	4.8	1.6
48	32	15	8		10	25	50

92.3	90.0	90.8	90.8	91.3	93.2	95.9
5.9	8.2	7.4	7.4	6.9	5.0	2.3
54	30	10		9	25	40

85.4	88.9	85.7	88.5	89.8	91.7	94.1
12.8	9.3	12.5	9.7	8.4	6.5	4.1
48	25	10		8	15	45

88.6	97.2	92.2	92.6	95.8	93.6	95.4
9.6	10	6.0	5.6	2.4	4.6	2.8
52	28	12		10	28	50

90.0	97.0	92.7	90.6	94.4	95.4	95.5
8.2	1.2	5.5	7.6	3.8	2.8	12.7
50	25	8		9	24	42

88.6	96.1	92.0	89.7	93.5	97.0	91.5
9.6	2.1	6.2	8.5	4.7	1.2	6.7
50	25	10		8	25	50

BM Pod—Kalini Valley, Oahu, Ewa-makai corner Hibiscus and Kiawe Drives. Spike in 30″ monkey pod tree, 2 ft. above ground.

R.E. Neilan

Plate B.5

8″ SEWER STAKEOUT

(1) Station	(2) +Sight	(3) HI	(4) −Sight	(5) Ground Elev.	(6) Pipe Flow Line
BM 16	2.11	102.76		100.56	
0+00			6.21	96.55	96.55
+00 ℄			3.20	99.56	96.55
+50 ℄			3.91	98.85	95.95
1+00 ℄			4.07	98.69	95.34
+31 ℄			8.22	94.54	94.97
+50 ℄			4.01	98.75	94.74
2+00 ℄			4.52	98.24	94.14
+33.7 ℄			5.03	97.73	93.73
+33.7			9.03	93.73	93.73
BM 16			2.11√	100.65	

Flowline Calculations

Line drops 50′ (1.206%) = 0.60′ per 50′

Example
Sta. 0+50 = 96.55 − 0.60 = 95.95
1+00 = 96.55 − 1.21 = 95.34
1+31 = 96.55 − 131 (1.206%) = 94.97

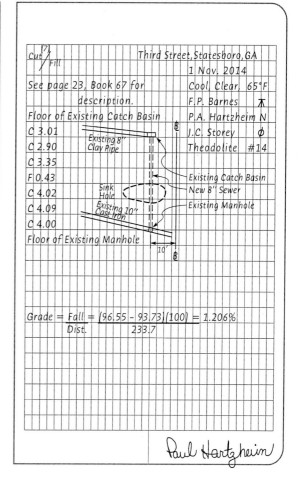

Cut (7) / Fill

Third Street, Statesboro, GA
1 Nov. 2014
See page 23, Book 67 for description.
Cool, Clear, 65°F
F.P. Barnes
P.A. Hartzheim N
Floor of Existing Catch Basin
J.C. Storey
C 3.01
Theodolite #14
C 2.90   Existing 8″ Clay Pipe
C 3.35
F 0.43   Existing Catch Basin
C 4.02   Sink Hole   New 8″ Sewer
C 4.09   Existing 10″ Cast Iron   Existing Manhole
C 4.00
Floor of Existing Manhole   10′

Grade = Fall / Dist. = (96.55 − 93.73)(100) / 233.7 = 1.206%

Paul Hartzheim

Plate B.6

C
Astronomic Observations

■ C.1 INTRODUCTION

Astronomical observations in surveying (geomatics) consist of observing positions of the Sun or certain stars. The principal purpose of astronomical observations in plane surveying was to determine the direction of the astronomic meridian (astronomic north). The resulting azimuth was needed to establish directions of new property lines so parcels could be adequately described; to retrace old property boundaries whose descriptions include bearings that were determined by astronomical methods; to specify directions of tangents on route surveys; to orient map sheets; and for many other purposes. These procedures have today been replaced with GNSS surveys where the coordinates of two points are established on the ground using either static and kinematic GNSS methods as discussed in Chapters 14 and 15. From these coordinates either geodetic or grid azimuths can be determined for the line.

The latitudes and longitudes of points can also be determined by making astronomical observations. However, this is seldom, if ever, done today for two reasons: (1) the field procedures and computations involved, especially for longitude, are quite difficult and time consuming especially if accurate results are expected; and (2) the use of the global navigation satellite systems has now made the determination of latitudes and longitudes a rather routine operation. Thus in this appendix, only astronomical methods for determining azimuth are discussed. For a more thorough presentation of this subject, readers are directed to the 11th or earlier edition of this book.

To expand on the definition of the astronomic meridian given in Section 7.4, at any point it is a line tangent to, and in the plane of, the great circle that passes through the point, and the Earth's north and south geographic poles. This is illustrated in Figure C.1, where P and P' are the poles located on the Earth's axis of rotation. Arc PAP' is the great circle through A, and line AN, the astronomic meridian (tangent to the great circle at A in plane $POP'A$). With an

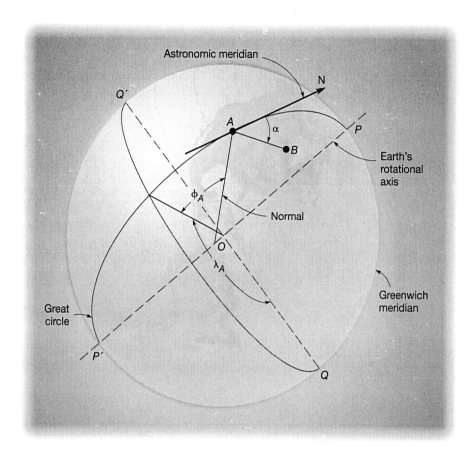

Figure C.1
Astronomic meridian, astronomic azimuth, latitude, and longitude.

astronomic meridian established, the astronomic azimuth α of any line, such as *AB* of Figure C.1, can readily be obtained by determining horizontal angle *NAB*.

Astronomical observations are not necessarily required on every project where azimuths or bearings are needed. If a pair of intervisible control monuments from a previous survey exists in the area, and the azimuth or bearing is known for that line, new directions can be referenced to it. Also, as noted earlier, GNSS survey methods, as described in Chapters 13 through 15, can be used to establish the positions of the two points of a project line from which the azimuth is determined.

C.2 OVERVIEW OF USUAL PROCEDURES FOR ASTRONOMICAL AZIMUTH DETERMINATION

In Figure C.2, imagine that *P* is on the extension of the Earth's polar axis, and that *S* is a star, which appears to rotate about *P* due to the Earth's rotation on its axis. Point *N* is on the horizon and vertically beneath *P*, and therefore line *AN* represents true north. For the situation shown in this figure, the general field procedures employed by surveyors to define the direction of astronomic north consist of the following steps: (1) a total station is set up and leveled at one end of the line whose azimuth is to be determined, like point *A* of Figure C.2; (2) the station at

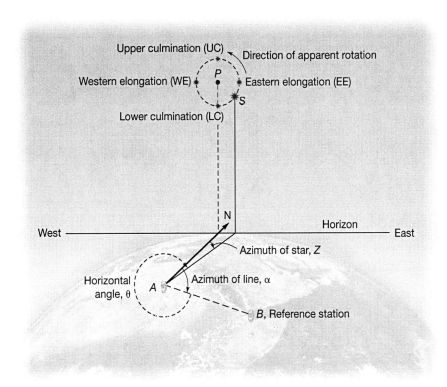

Figure C.2
Azimuth of a line on the ground from the azimuth of a star.

the line's other end, like B of Figure C.2, is carefully sighted and the instrument's horizontal circle indexed to 0°00′00″; (3) the telescope is turned clockwise and the star S carefully sighted; (4) the horizontal, and sometimes vertical, circles of the instrument are read at the instant of pointing on the star; (5) the precise time of pointing is recorded; and (6) the horizontal angle is recorded from the reference mark to the star, like angle θ of Figure C.2 from B to S. Office work involves (a) obtaining the precise location of the star at the instant sighted from an ephemeris (almanac of celestial body positions); (b) computing the star's azimuth (angle Z in Figure C.2) based on the observed and ephemeris data; and (c) calculating the line's azimuth by applying the measured horizontal angle to the computed azimuth of the star as

$$\alpha = 360° + Z - \theta \quad \text{(C.1)}$$

Any visible celestial body for which ephemeris data are available can be employed in the procedures outlined. However, the Sun and, in the northern hemisphere, *Polaris* (the north star) are almost always selected.[1] The Sun permits observations to be made in lighted conditions during normal daytime working hours; Polaris is preferred for higher-order accuracy. In the southern hemisphere a star in the *Southern Cross* is often used.

---

[1] In the southern hemisphere, the star *Sigmus Octantis* and the stars in the constellation *Southern Cross* are commonly used for astronomical observations.

Accuracies attainable in determining astronomical azimuths depend on many variables, including (1) the precision of the instrument used, (2) ability and experience of the observer, (3) weather conditions, (4) quality of the clock or chronometer used to measure the time of sighting, (5) celestial body sighted and its position when observed, and (6) accuracy of ephemeris and other data available. In the northern hemisphere Polaris observations provide the most accurate results and, with several repetitions of measurements utilizing first-order instruments, accuracies to within $\pm 1''$ are possible. Sun observations yield a lower order of accuracy but values accurate to within about $\pm 10''$ or better can be obtained if careful repeated measurements are made.

C.3 EPHEMERIDES

As noted previously, ephemerides are almanacs containing data on the positions of the Sun and various stars, versus time. Nowadays, ephemeris data are most conveniently obtained through the Internet. Jerry Wahl of the U.S. Bureau of Land Management, for example, maintains an ephemeris of the Sun and Polaris on his website.[2] Table C.1, which applies to December, 2000, was taken from this website. The data in this table is used in connection with some of the example problems presented later in this chapter.

A variety of ephemerides are also published annually and are available to surveyors for astronomical work. One of those most useful to surveyors is the *The Sokkia Celestial Observation Handbook and Ephemeris*, published annually by Sokkia Corporation.[3] It not only contains tabulated data for the Sun and Polaris, but also for several other of the brighter stars in the heavens. This booklet also includes a substantial amount of explanatory material, plus worked examples that demonstrate the use of the tabulated data and the computational procedures. Other published ephemerides are *The Apparent Place of Polaris and Apparent Sidereal Time*, published by the U.S. Department of Commerce; *The Nautical Almanac*, published by the U.S. Naval Observatory; and *Apparent Places of Fundamental Stars*, published by Astronomisches Rechen-Institut, Heidelberg, Germany.

In addition to published ephemerides, computer programs are also available, which solve for the positions of celestial bodies. Their major advantages are that they provide accurate results without tables, and can be used year after year. However, they must occasionally be updated.

Values tabulated in ephemerides are given for *universal coordinated time* (UTC), which is also Greenwich civil time, so before extracting data from them, standard or daylight times normally recorded for observations must be converted. This topic is discussed further in Section C.5.

---

[2] The Internet address for obtaining the ephemeris is http://www.cadastral.com/.

[3] The Sokkia ephemeris is authored by Drs. Richard Elgin, David Knowles, and Joseph Senne, and is available from the Sokkia Corporation, 9111 Barton, Box 2934, Overland Park, Kansas 66021; telephone: (800) 255-3913.

TABLE C.1 EPHEMERIS DATA FROM THE CADASTRAL SURVEY INTERNET SITE AT HTTP://WWW.CADASTRAL.COM/.

2000 Date	SUN Declination			..... For 0 hrs Universal Time							..... Polaris				0 hrs UT					
				... GHA ...			Eq of Tm			Semi-Di		Declination				... GHA ...			... TUC-*		
	d	M	s	d	m	s	m	s	m	s	d	m	s	d	m	s	h	m	s
Dec 1 FR	−21	48	47.9	182	45	19.8	+11	01.32	16	13.2	89	16	08.98	31	36	42.7	21	49	58.0
Dec 2 SA	−21	57	53.4	182	39	38.3	+10	38.55	16	13.4	89	16	09.35	32	36	02.1	21	46	01.3
Dec 3 SU	−22	06	33.5	182	33	47.8	+10	15.18	16	13.5	89	16	09.70	33	35	22.8	21	42	04.6
Dec 4 MO	−22	14	47.9	182	27	48.5	+09	51.23	16	13.7	89	16	10.04	34	34	44.4	21	38	07.8
Dec 5 TU	−22	22	36.4	182	21	40.9	+09	26.72	16	13.8	89	16	10.36	35	34	06.4	21	34	11.0
Dec 6 WE	−22	29	58.7	182	15	25.3	+09	01.69	16	14.0	89	16	10.67	36	33	28.2	21	30	14.2
Dec 7 TH	−22	36	54.7	182	09	02.1	+08	36.14	16	14.1	89	16	10.96	37	32	49.4	21	26	17.4
Dec 8 FR	−22	43	24.0	182	02	31.7	+08	10.12	16	14.3	89	16	11.24	38	32	09.4	21	22	20.7
Dec 9 SA	−22	49	26.5	181	55	54.5	+07	43.63	16	14.4	89	16	11.52	39	31	28.1	21	18	24.1
Dec 10 SU	−22	55	02.0	181	49	10.9	+07	16.72	16	14.5	89	16	11.81	40	30	45.7	21	14	27.6
Dec 11 MO	−23	00	10.4	181	42	21.1	+06	49.41	16	14.6	89	16	12.12	41	30	02.9	21	10	31.1
Dec 12 TU	−23	04	51.5	181	35	25.7	+06	21.71	16	14.7	89	16	12.45	42	29	20.6	21	06	34.6
Dec 13 WE	−23	09	05.2	181	28	24.9	+05	53.66	16	14.8	89	16	12.80	43	28	40.0	21	02	37.9
Dec 14 TH	−23	12	51.3	181	21	19.2	+05	25.28	16	14.9	89	16	13.16	44	28	01.7	20	58	41.1
Dec 15 FR	−23	16	09.7	181	14	08.9	+04	56.59	16	15.0	89	16	13.51	45	27	26.0	20	54	44.1

(Continued)

TABLE C.1 Ephemeris Data from the Cadastral Survey Internet Site at http://www.cadastral.com/.

2000 Date	SUN Declination			..... For 0 hrs Universal Time							..... Polaris ..... Declination			0 hrs UT					
				... GHA ...			Eq of Tm			Semi-Di					... GHA ...			.. TUC-*		
	d	M	s	d	m	s	m	s	m	s	d	m	s	d	m	s	h	m	s
Dec 16 SA	−23	19	00.2	181	06	54.6	+04	27.64	16	15.1	89	16	13.84	46	26	52.2	20	50	47.0
Dec 17 SU	−23	21	22.9	180	59	36.5	+03	58.44	16	15.2	89	16	14.13	47	26	19.5	20	46	49.9
Dec 18 MO	−23	23	17.6	180	52	15.4	+03	29.02	16	15.3	89	16	14.43	48	25	46.9	20	42	52.7
Dec 19 TU	−23	24	44.1	180	44	51.5	+02	59.43	16	15.3	89	16	14.68	49	25	13.7	20	38	55.6
Dec 20 WE	−23	25	42.5	180	37	25.5	+02	29.70	16	15.4	89	16	14.92	50	24	39.5	20	34	58.5
Dec 21 TH	−23	26	12.7	180	29	57.8	+01	59.85	16	15.5	89	16	15.16	51	24	04.3	20	31	01.5
Dec 22 FR	−23	26	14.7	180	22	29.1	+01	29.94	16	15.5	89	16	15.39	52	23	28.2	20	27	04.5
Dec 23 SA	−23	25	48.4	180	14	59.9	+00	59.99	16	15.6	89	16	15.64	53	22	51.6	20	23	07.6
Dec 24 SU	−23	24	53.8	180	07	30.7	+00	30.05	16	15.6	89	16	15.89	54	22	15.2	20	19	10.7
Dec 25 MO	−23	23	31.1	180	00	02.2	+00	00.15	16	15.7	89	16	16.15	55	21	39.4	20	15	13.7
Dec 26 TU	−23	21	40.1	179	52	34.8	−00	29.68	16	15.7	89	16	16.42	56	21	04.8	20	11	16.7
Dec 27 WE	−23	19	21.0	179	45	09.3	−00	59.38	16	15.8	89	16	16.69	57	20	31.5	20	07	19.6
Dec 28 TH	−23	16	33.8	179	37	46.0	−01	28.93	16	15.8	89	16	16.96	58	19	59.7	20	03	22.3
Dec 29 FR	−23	13	18.5	179	30	25.7	−01	58.29	16	15.8	89	16	17.23	59	19	29.5	19	59	25.0
Dec 30 SA	−23	09	35.4	179	23	08.8	−02	27.41	16	15.9	89	16	17.47	60	19	00.4	19	55	27.6
Dec 31 SU	−23	05	24.4	179	15	55.8	−02	56.28	16	15.9	89	16	17.70	61	18	32.3	19	51	30.1

(Courtesy Jerry Wahl—Cadastral Survey, Bureau of Land Management.)
*Universal time of upper culmination at Greenwich.

C.4 DEFINITIONS

In making and computing astronomical observations, the Sun and stars are assumed to lie on the surface of a *celestial sphere* of infinite radius having the Earth as its center. Because of the Earth's rotation on its axis, all stars appear to move around centers that are on the extended Earth's rotational axis, which is also the axis of the celestial sphere. Figure C.3 is a sketch of the celestial sphere and illustrates some terms used in field astronomy. Here, O represents the Earth and S a celestial body, as the Sun or a star whose apparent direction of movement is indicated by an arrow. Students may find it helpful to sketch the various features on a basketball, or globe. Definitions of terms pertinent to the study of field astronomy follow.

The *zenith* is located where a plumb line projected upward meets the celestial sphere. In Figure C.3, Z designates it. Stated differently, it is the point on the celestial sphere vertically above the observer.

The *nadir* is the point on the celestial sphere vertically beneath the observer and exactly opposite the zenith. In Figure C.3, it is at N.

The *north celestial pole* is point P where the Earth's rotational axis, extended from the north geographic pole, intersects the celestial sphere.

The *south celestial pole* is point P' where the Earth's rotational axis, extended from the south geographic pole, intersects the celestial sphere.

A *great circle* is any circle on the celestial sphere whose plane passes through the center of the sphere.

A *vertical circle* is any great circle of the celestial sphere passing through the zenith and nadir, and represents the line of intersection of a vertical plane with the celestial sphere. In Figure C.3, $ZSS'N$ is a vertical circle.

The *celestial equator* is the great circle on the celestial sphere whose plane is perpendicular to the axis of rotation of the Earth. It corresponds to the Earth's

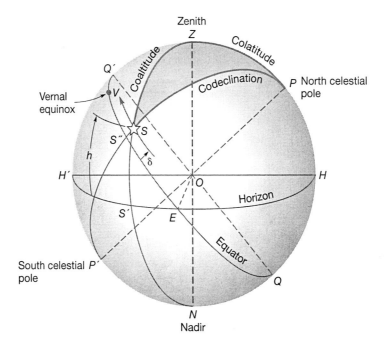

Figure C.3
Celestial sphere.

equator enlarged in diameter. Half of the celestial equator is represented by $Q'EQ$ in Figure C.3.

An *hour circle* is any great circle on the celestial sphere that passes through the north and south celestial poles. Therefore, hour circles are perpendicular to the plane of the celestial equator. They correspond to meridians (longitudinal lines) and are used to observe hour angles. In Figure C.3, $PSS''P'$ is an hour circle.

The *horizon* is a great circle on the celestial sphere whose plane is perpendicular to the direction of the plumb line. In surveying, the plane of the horizon is determined by a level vial. Half of the horizon is represented by $H'EH$ in Figure C.3.

A *celestial meridian*, interchangeably called *local meridian*, is that unique hour circle containing the observer's zenith. It is both an hour circle and a vertical circle. The intersection of the celestial meridian plane with the horizon plane is line $H'OH$ in Figure C.3, which defines the direction of true north. Thus, it is the astronomic meridian line used in plane surveying. Since east is 90° clockwise from true north, line OE in the horizon plane is a true east line. The celestial meridian is composed of two branches; the *upper branch* contains the zenith and is the semicircle $PZQ'H'P'$ in Figure C.3, and the *lower branch* includes the nadir and is arc $PHQNP'$.

A *diurnal circle* is the complete path of travel of the Sun or a star in its apparent daily orbit about the Earth. Four terms describe specific positions of heavenly bodies in their diurnal circles (see Figure C.2): (1) *lower culmination*—the body's position when it is exactly on the lower branch of the celestial meridian; (2) *eastern elongation*—where the body is farthest east of the celestial meridian with its hour circle and vertical circle perpendicular; (3) *upper culmination*—when it is on the upper branch of the celestial meridian; and (4) *western elongation*—when the body is farthest west of the celestial meridian with its hour circle and vertical circle perpendicular.

An *hour angle* exists between a meridian of reference and the hour circle passing through a celestial body. It is measured by the angle at the pole between the meridian and hour circle, or by the arc of the equator intercepted by those circles. Hour angles are measured westward (in the direction of apparent travel of the Sun or star) from the upper branch of the meridian of reference.

The *Greenwich hour angle* (GHA) of a heavenly body at any instant of time is the angle, measured westward, from the upper branch of the meridian of Greenwich to the meridian over which the body is located at that moment.[4] In the ephemeris of Table C.1, it is designated by GHA. Local hour angle (LHA) is similar to GHA, except it is observed from the upper branch of the observer's celestial meridian.

A *meridian angle* is like an LHA, except it is measured either eastward or westward from the observer's meridian, and thus its value is always between 0° and 180°.

The *declination* of a heavenly body is the angular distance (measured along the hour circle) between the body and the equator; it is plus when the body is north of the equator, and minus when south of it. Declination is usually denoted by δ in formulas, and represented by arc $S''S$ in Figure C.3.

---

[4] The meridian of Greenwich, England, is internationally accepted as the reference meridian for specifying longitudes of points on Earth and for giving positions of celestial bodies.

The *polar distance* or *codeclination* of a body is equal to 90° minus the declination. In Figure C.3, it is arc *PS*.

The *position* of a heavenly body with respect to the Earth at any moment may be given by its GHA and declination.

The *altitude* of a heavenly body is its angular distance measured along a vertical circle above the horizon, $S'S$ in Figure C.3. It is generally obtained by measuring a vertical angle with a total station, (or theodolite), and correcting for refraction, and parallax if the Sun is observed. Altitude is usually denoted in formulas by h.

The *coaltitude* or *zenith distance* is arc ZS in Figure C.3 and equals 90° minus the altitude.

The *astronomical* or *PZS triangle* (darkened in the figure) is the spherical triangle whose vertices are the pole P, zenith Z, and astronomical body S. Because of the body's movement through its diurnal circle, the three angles in this triangle are constantly changing.

The *azimuth* of a heavenly body is the angle observed in the horizon plane, clockwise from either the north or south point, to the vertical circle through the body. An azimuth from north is represented by arc HS' in Figure C.3, and it equals the Z angle of the *PZS* triangle.

The *latitude* of an observer is the angular distance, measured along the meridian, from the equator to the observer's position. In Figure C.3, it is arc $Q'Z$. It is also the angular distance between the polar axis and horizon, or arc HP. Depending on the observer's position, latitude is measured north or south of the equator. Formulas in this book denote it as φ. Colatitude is $(90° - \varphi)$. As a side note and as can be seen in Figure C.3, since Polaris is less than 1° from the north celestial pole, the angular distance PH is within 1° of the latitude for the observing station. Thus, the altitude of Polaris in the northern hemisphere is within 1° of the latitude of the observing station.

The *vernal equinox* is the intersection point of the celestial equator and the hour circle through the Sun at the instant it reaches zero declination and is proceeding into the northern hemisphere (about March 21 each year). For any calendar year, it is a fixed point on the celestial sphere (the astronomer's origin of coordinates in the sky), and moves with the celestial sphere just as the stars do. In Figure C.3, V designates it.

The *right ascension* of a heavenly body is the angular distance measured eastward from the hour circle through the vernal equinox to the hour circle of a celestial body. It is arc VS'' in Figure C.3. Right ascension frequently replaces GHA as a means of specifying the position of a star with respect to the Earth. In this system, however, the GHA of the vernal equinox must also be given.

C.5 TIME

Four kinds of time may be used in making and computing an astronomical observation.

1. *Sidereal time*. A sidereal day is the interval of time between two successive upper culminations of the vernal equinox over the same meridian. Sidereal

time is star time. At any location for any instant, it is equal to the local hour angle of the vernal equinox.

2. *Apparent solar time.* An apparent solar day is the interval of time between two successive lower culminations of the Sun. Apparent solar time is Sun time, and the length of a day varies somewhat because the rate of travel of the Sun is not constant. Since the Earth revolves about the Sun once a year, there is one less day of solar time in a year than sidereal time. Thus, the length of a sidereal day is shorter than a solar day by approximately 3 min 56 sec. The relationship between sidereal and solar time is illustrated in Figure C.4. (*Note:* The Earth's orbit is actually elliptical, but for simplicity it is shown circular in the figure.)

3. *Mean solar*, or *civil*, *time.* This time is related to a fictitious Sun, called the "mean" Sun, which is assumed to move at a uniform rate. It is the basis for watch time and the 24-h day.

   The *equation of time* is the difference between "apparent" solar and "mean" solar time. Its value changes continually as the true or apparent Sun gets ahead of, and then falls behind, the mean Sun. Values for each day of the year are given in ephemerides (see Table C.1). If the apparent Sun is ahead of the mean Sun, the equation is plus, if behind, it is minus. *Local apparent time is obtained by adding the equation of time to local civil time.*

4. *Standard time.* This is the mean time at meridians 15° or 1 h apart, measured eastward and westward from Greenwich. Eastern Standard Time (EST) at the 75th meridian differs from universal time (UT), or Greenwich civil time (GCT), by 5 h (earlier, since the Sun has not yet traveled from the meridian of Greenwich to the United States). Standard time was adopted in the United States in 1883, replacing some 100 local times used previously. *Daylight*

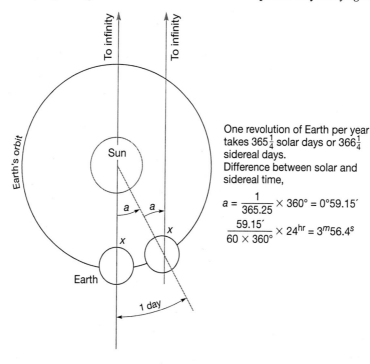

One revolution of Earth per year takes $365\frac{1}{4}$ solar days or $366\frac{1}{4}$ sidereal days.
Difference between solar and sidereal time,

$$a = \frac{1}{365.25} \times 360° = 0°59.15'$$

$$\frac{59.15'}{60 \times 360°} \times 24^{hr} = 3^m 56.4^s$$

Figure C.4
Comparison of sidereal and solar time.

TABLE C.2 LONGITUDES OF STANDARD MERIDIANS IN THE UNITED STATES AND TIME DIFFERENCES FROM GREENWICH

Standard Time Zone (and Abbreviation)	Longitude of Standard Meridian	Corrections in Hours, to Add to Obtain UT	
		Standard Time	Daylight Time
Atlantic (AST)	60°	4	3
Eastern (EST)	75°	5	4
Central (CST)	90°	6	5
Mountain (MST)	105°	7	6
Pacific (PST)	120°	8	7
Yukon (YST)	135°	9	8
Alaska/Hawaii (AHST)	150°	10	9
Bering Sea (BST)	165°	11	10

saving time (DST) in any zone is equal to the standard time in the adjacent zone to the *east*; thus, central daylight time is equivalent to Eastern Standard Time.[5]

As previously noted, Sun and star positions tabulated in ephemerides are given in UT. Observation times, on the other hand, may be recorded in the standard or daylight times of an observer's location and must therefore be converted to UT. Conversion is based on the longitude of the standard meridian for the time zone. Table C.2 lists the different time zones in the United States, the longitudes of their standard meridians, and the number of hours to be added for converting standard and daylight time to UT.

In making civil time conversions based on longitude differences, the following relationships are helpful:

$$360° \text{ of longitude} = 24 \text{ h}$$
$$15° \text{ of longitude} = 1 \text{ h}$$
$$1° \text{ of longitude} = 4 \text{ min (of time)}$$

C.6 TIMING OBSERVATIONS

In the United States, the *National Institute of Standards and Technology* (NIST), formerly the National Bureau of Standards, broadcasts time signals from station WWV in Fort Collins, Colo., on frequencies of 2.5, 5, 10, 15, and 20 MHz. These signals can be received with short-wave radios, including inexpensive *time kubes*

---

[5] Daylight saving time officially begins at 2:00 A.M. on the first Sunday of April and ends at 2:00 A.M. on the last Sunday of October each year.

especially designed for this purpose. They can also be received over the telephone by dialing (303) 499-7111. To broaden coverage, signals are also transmitted from station WWVH in Hawaii on the same frequencies. These signals are broadcast as audible ticks with a computerized voice announcement of UT at each minute. In Canada, EST is broadcast from station CHU on frequencies of 3.33, 7.335, and 14.67 MHz. This can be converted to UT by adding 5 h.

The time that is broadcast by WWV is actually *coordinated universal time* (UTC), whereas the time used for tabulating Sun and star positions in ephemerides is a corrected version known as UT1. UTC is a uniform time at Greenwich that, unlike UT1, does not vary with changing rates of rotation and other irregular motions of the Earth. Leap seconds are added to UTC as necessary to account for the gradual slowing of the Earth's rotation rate, and thus keep UTC within ∓ 0.7 sec of UT1 at all times. For precise astronomical work, a difference correction (DUT) can be added to UTC to obtain UT1. The required DUT correction is given by means of double ticks broadcast by WWV and CHU during the first 15 sec following each minute tone. Each double tick represents a 0.1 sec correction. A plus correction is applied for double ticks that occur during the first 7 sec after the minute tone, while a negative correction is made for double ticks that are heard during seconds 9 through 15. Thus, if double ticks occurred for the first 5 sec after the minute tone, +0.5 sec would be added to broadcast UTC to obtain UT1. If double ticks were heard during the ninth and tenth seconds, -0.2 sec would be added to UTC. Because the DUT correction is quite small, it can be ignored for most observations on Polaris or other stars with very high declinations. The correction should be considered, however, for more accurate observations on the Sun and stars of lower declination. The current DUT correction is also available in Bulletin A on the International Earth Rotation and Reference System Service at http://www.iers.org/nn_10968/IERS/EN/DataProducts/EarthOrientationData/eop.html on the Internet.

Digital watches, watches with sweep-second hands, and stopwatches are all suitable for timing most astronomical observations in surveying. Hand calculators and data collectors equipped with time modules are especially convenient since they can serve not only as timing devices, but also for recording data and computing. Regardless of the timepiece used, it should be checked against WWV before starting observations, and either set to agree exactly with UTC, or the number of seconds it is fast or slow recorded. The time a check is made should also be recorded. When all observations are completed, the clock check should be repeated and any change recorded. Then intermediate observation times can be corrected in proportion to the elapsed time since the original check. With a stopwatch, checks can be made before and after each individual observation.

■ C.7 COMPUTATIONS FOR AZIMUTH FROM POLARIS OBSERVATIONS BY THE HOUR ANGLE METHOD

In this method, only the horizontal circle reading and precise time need to be recorded when the star is sighted. A vertical circle reading for at least one pointing is recommended, however, to ensure that the correct star has been sighted. To make the observations, the instrument is set up and leveled on one end of a line whose azimuth is to be determined. In the usual field procedure, the line's other end is first sighted, and then the horizontal angle measured to the star. To eliminate the effects of instrumental errors, equal numbers of direct and reversed observations are taken and the results averaged.

Computations after fieldwork require the solution for angle Z in the astronomical (PZS) triangle (see Figure C.3). Two formulas for Z that apply in the hour angle method, derived from laws of spherical trigonometry, are:

$$Z = \tan^{-1}\left(\frac{\sin t}{\cos \varphi \tan \delta - \sin \varphi \cos t}\right) \quad \text{(C.2)}$$

and

$$Z = \tan^{-1}\left(\frac{-\sin(LHA)}{\cos \varphi \tan \delta - \sin \varphi \cos(LHA)}\right) \quad \text{(C.3)}$$

The geometry upon which these equations are based is shown more clearly in Figure C.5, using the PZS triangle. The latitude φ of the observer's position is arc HP; thus arc PZ is $(90° - \varphi)$, or *colatitude*. Declination δ of the star is arc $S''S$, so SP is $(90° - \delta)$, or *polar distance*. Angle ZPS in Figure C.5 is t, the *meridian angle*, which is related to the LHA of the star. Diagrams such as those of Figure C.6 are helpful in understanding and determining t angles and LHAs. These diagrams show the north celestial pole P at the center of the star's diurnal circle *as viewed from the observer's position within the sphere*. On the diagrams, west is to the left of the pole, east is to the right, and the apparent rotation of the stars is counterclockwise. Angle λ between the meridian of Greenwich (G) and the local meridian (L) through the observer's position is the longitude of the station occupied. The star's Greenwich hour angle (GHA) for the observation time is taken from an ephemeris. Sketching λ and GHA approximately to scale on diagrams such as those of Figure C.6 immediately makes clear the star's position. From Figure C.6, it can be seen that the LHA in the western hemisphere can be computed as

$$\text{LHA} = \text{GHA} - \lambda \quad \text{(C.4a)}$$

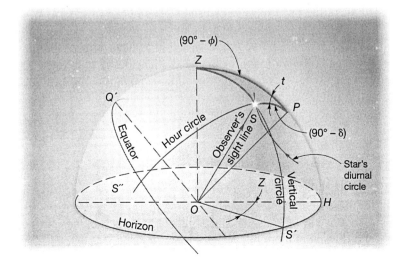

Figure C.5
The PZS triangle for Polaris at any hour angle.

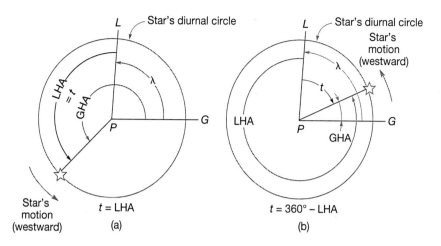

Figure C.6
Computation of the meridian angle t.

TABLE C.3 RELATIONSHIP BETWEEN LHA, SIGN OF Z, AND AZIMUTH OF STAR

LHA from	Z < 0	Z > 0
0° to 180°	Azimuth = 360° + Z	Azimuth = 180° + Z
180° to 360°	Azimuth = 180° + Z	Azimuth = Z

For the eastern hemisphere Equation (C.4a) becomes

$$\text{LHA} = \text{GHA} + \lambda \quad \text{(C.4b)}$$

As shown in Figure C.6(a), the LHA is between 0° and 180° when the star is west of north, and as seen in Figure C.6(b), it is between 180° and 360° if the star is east of north. Also, $t = \text{LHA}$ if the star is west of north, and $t = (360° - \text{LHA})$ if the star is east of north. The relationships between the LHA of a star, the sign of Z that is obtained using Equation (C.5), and the azimuth of the star are shown in Table C.3.

Note that the latitude of an observer's position is used directly in Equations (C.2) and (C.3), and that the station's longitude is also needed to compute either t or LHA. These values can both be scaled from a USGS quadrangle map and, with reasonable care, obtained to within ±2 sec. Declinations for use in these equations are extracted from an ephemeris for the instant of sighting.

C.8 AZIMUTH FROM SOLAR OBSERVATIONS

Reduction of solar observations uses the same equations as star observations. However, at least one and possibly two major differences exist in the computations. Since the Sun is relatively close to the Earth, the simple linear interpolation of the declination for stars is inadequate for the Sun. The interpolation formula for the declination of the Sun is

$$\delta_{\text{Sun}} = \delta_0 + (\delta_{24} - \delta_0)(\text{UT1}/24) + 0.0000395\, \delta_0 \sin(7.5 \times \text{UT1}) \quad \text{(C.5)}$$

where δ_0 is the tabulated declination of the Sun at 0-h UT1 on the day of the observation, δ_{24} is the tabulated declination of the Sun at 24-h UT1 on the day of the observation (0-h UT1 of the following day), and UT1 is the universal time of the observation.

Observations on the Sun can be made directly by placing a dark glass filter (designed for solar viewing) over the telescope objective lens. *A total station instrument should never be pointed directly at the Sun without an objective lens solar filter. Failure to heed this warning may result in serious damage to sensitive electronic components in the total station. Additionally, the observer should never look at the Sun without the solar filter in place, or permanent eye damage will occur.*

The second major difference depends on the method of pointing on the Sun. Because the objective lens filter only allows the observer to see the instrument's cross hairs in the Sun's illuminating circle, the most precise pointing will occur at the trailing edge of the Sun as shown in Figure C.6. Thus, the observer should place the trailing edge of the Sun near the vertical cross hair as in Figure C.6(a), then wait as the Sun moves and record the time when the trailing edge of the Sun just touches the vertical cross hair as shown in Figure C.6(b). When using this field procedure, or any field procedure that involves the edges, a correction must be made to the horizontal angle for the Sun's semidiameter. The Sun's semidiameter varies with the distance from the Earth to the Sun, and values are tabulated for each day in the ephemeris (see Table C.1). The correction that must be applied for semidiameter is computed as

$$C_{SD} = \frac{\text{Sun's semidiameter}}{\cos h} \tag{C.6}$$

The correction for the Sun's semidiameter can be avoided by using a *Roelof solar prism*. The Roelof solar prism is a device designed specifically for sighting the Sun. It is easily mounted on the objective end of the telescope, and by means of prisms, produces four overlapping images of the Sun in the pattern shown in Figure C.7. While viewing the Sun, an observer can accurately center the vertical cross wire in the small diamond-shaped area in the middle of the field of view. Because of symmetry, this is equivalent to sighting the Sun's center.

To compute the azimuth of a line from observations on the Sun, the hour angle equation, [either Equation (C.2) or Equation (C.3)], can be employed. These are the same ones used for Polaris observations. Again, latitude and

Figure C.7 View of the Sun (a) just prior to coincidence of the vertical cross hair and (b) at coincidence.

(a)

(b)

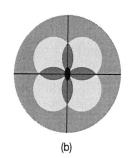

Figure C.8 View of the Sun using a Roelof solar prism (a) just prior to and (b) at the time of the observation.

longitude can be taken from a USGS quadrangle map, and declination extracted from an ephemeris for the time of observation.

C.9 IMPORTANCE OF PRECISE LEVELING

As discussed in Section 8.20.1, precise leveling is extremely important to horizontal directions whenever vertical angles are large. For example, at an altitude angle of 40° with a leveling error of 15" (1/2 of division of a 30" bubble), the estimated error in the horizontal direction observed by a total station is given by Equation (8.4) as

$$15'' \tan(40°) = \pm 12.6''$$

Typically the level on the vertical circle is more sensitive than the plate bubble. This fact can be used to precisely level a total station when performing an astronomical observation. The procedure involves aligning two leveling screws on the tripod in the direction of the astronomical body. After performing a typical leveling procedure, point the instrument in the general direction of the third leveling screw, clamp the vertical circle, read, and record the zenith angle. Now turn the instrument 180° opposite this position leaving the vertical circle clamped, read, and record the zenith angle again. Precise level is achieved by averaging the two zenith angles and slightly adjusting the third leveling screw to read this average on the vertical circle. This procedure only achieves precise level in the direction of the celestial body. However, precise leveling to the ground station is not as critical since the vertical angle to the target is small typically.

D Using the Worksheets from the Companion Website

D.1 INTRODUCTION

The Mathcad worksheets on the companion website at www.pearson.com/ghilani demonstrate many of the computational exercises presented in this book. The 42 worksheets allow you to modify the values of variables in the equations and see instantaneous changes in the results. Additionally, they further discuss topics presented in this book. These sheets require Mathcad version 14.0 or higher. For readers who do not own Mathcad 14.0 or higher, these worksheets have been converted to html files and can be viewed with a web browser. However, the html files are not computationally dynamic; that is, they can only display the equations at the time of creation, and do not allow changes to the variables or equations. To use either the worksheets or html files, you must copy them on your computer with the installation program provided on the companion website.

As shown in Figure D.1, if the Mathcad® worksheets are unzipped in the *handbook* directory under Mathcad, you will find the link to the electronic book (E-book) in the Mathcad help menu. Select the menu item entitled "Support files for Elementary Surveying: An Introduction to Geomatics" to open the electronic book. If the menu item is missing, use the "Open Book..." menu item to manually browse for the file "ElemSurv.hbk." Additionally, individual files in the *ElemSurv* subdirectory can be opened directly in Mathcad and modified as desired.

D.2 USING THE FILES

Sixteen of the 28 chapters in this book have associated Mathcad worksheets and html files. In some chapters, such as Chapter 20, there are several associated worksheets shown in Figure D.2. These additional worksheets provide further

Figure D.1 Opening support files in Mathcad.

Figure D.2 Worksheets demonstrating map projection computations.

information on the map projections that are briefly mentioned in Section 20.13 of this book. They allow you to explore map projections that are not commonly used in the United States.

Other chapters with more than one worksheet include 11, 16, 19, and 27. In Chapter 11, besides demonstrating coordinate geometry problems presented in the chapter, the reader can view a least-squares solution of a two-dimensional conformal coordinate transformation, discussed in Section 11.8. In Chapter 16, several Mathcad worksheets demonstrate the least-squares method. The first worksheet shows the least-squares method for fitting points, first to a line, and then to a circle using the equations discussed in Section 11.2. These adjustments, along with the two-dimensional conformal coordinate transformation, are not formally covered in Chapter 16, but instead the theory of the least-squares method is applied. Additionally, both the horizontal plane survey and differential-leveling network adjustments are demonstrated in separate worksheets. In Chapter 19, there are worksheets discussing the basics of geodesy, geodetic reductions of traditional observations, direct and inverse geodetic problems, three-dimensional geodetic computations, and transformation of ITRF 2000 coordinates with velocity vectors to NAD 83 coordinates for a particular epoch in time. Finally, in

Chapter 27, not only are the varying photogrammetric problems in the chapter demonstrated, but there are also two worksheets that cover the two-dimensional affine and projective coordinate transformations, which are commonly used in photogrammetric computations.

While some worksheets obtain their data from values assigned to variables directly in the worksheet, others, such as the least-squares worksheets, obtain their data from text files generated using a text editor such as Notepad. For example, Figure D.3 shows the first screen in the support file for Chapter 3: Theory of Errors in Observations. Note the line that states "data:= data3-1.txt" (there is an accompanying disk icon). This indicates that the contents of the file "data3-1.txt" are being read into the variable "data." The resultant variable data are partially listed on the right side of the window. (The values for this file come from Table 3.1 in this book.) The data file contains one observed value per line.

As previously stated, you can use a text editor such as Notepad or those in WOLFPACK, MATRIX, and STATS to create your own data files for other statistical problems. Once a file is created and saved to disk, you can change the worksheet's selected data file by right-clicking the "data" variable and selecting "Properties" in the resulting pull-down menu (see Figure D.4). This will display the "Component Properties" dialog box shown in Figure D.5. Click the "Browse"

CHAPTER 3

Theory of Errors in Observations

Statistics

This worksheet computes and plots data from the file data3-1. This is the data in Table 3.1 in **Elementary Surveying: An Introduction to Geomatics**. This worksheet can be used with different data files. Simply right-select the following command and select Choose File.

data :=

data3-1.txt

data := sort(data)

n := length(data$^{\langle 0 \rangle}$)

sum := $\sum$ data

μ := mean(data)

med := median(data)

mod := mode(data)

S := Stdev(data)

Partial listing of data ->

n = 100.000

sum = 2494.000

μ = 24.9400

med = 24.850

mod = 24.000

S = 2.18484

data =

	0
0	19.500
1	20.000
2	20.500
3	20.800
4	21.200
5	21.300
6	21.500
7	22.100
8	22.100
9	22.300
10	22.400
11	22.500
12	22.500
13	22.600
14	22.800

Figure D.3 Statistical computations for data from Table 3.1 of this book.

Figure D.4
Data entry pop-up menu.

Figure D.5
Component Properties box displaying the file name in the middle of the box.

button to locate the desired data file and then click "OK." The worksheet will then automatically update its computations and graphical plots to match the data in the newly specified file.

In Figure D.3, the left side of the worksheet shows the calls to the statistical functions contained in Mathcad under the heading "Computations." Near the right side of the window, the column labeled "Results" displays the values that are assigned to each variable. For example, the mean of the data is 24.90, the median is 24.85, and the mode is 24.0. You can refer to the Mathcad Help system to learn more about Mathcad variables, functions, and their use in worksheets.

Figure D.6 shows the top of a worksheet that obtains its data from variables entered directly in the worksheet. This listing demonstrates the use of tape-correction formulas from Example 6.1. The variables near the top of the worksheet contain the calibration data for a 30-m tape as given in the example. Immediately following the calibration data is the field data for the length of 21.151 m. Once the calibrated and field data are entered into the appropriate variables, corrections are computed using Equations (6.3) through (6.6) in the book. Finally, the sum of the corrections is determined in the variable Ctotal. Similar problems can be solved using this worksheet by modifying the field and calibration data as appropriate.

Chapter 6
Distance Measurement

Table of Contents

- Taping Corrections
- Propagation of Electromagnetic Energy
- Slope Reductions of Short Lines
- Error Propagation

LIBRARY FILE Reference:C:\Program Files (x86)\Mathcad\Mathcad 15\Handbook\ElemS

Taping Corrections

Calibration data

$l' := 30$ $\quad\quad A := 0.050$ $\quad\quad P := 5.45$

$l := 30.012$ $\quad\quad w := 0.03967$ $\quad\quad T := 20$

$k := 0.0000116$ $\quad\quad E := 2000000$

Field data (single tape measurement)

$L_s := 21.151$ $\quad\quad P_1 := 9.09$ $\quad\quad T_1 := 16$

Corrections

Length $\quad\quad C_L := \dfrac{l - l'}{l} \cdot L_s \quad\quad C_L = 0.0085$

Temperature $\quad\quad C_T := k \cdot (T_1 - T) \cdot L_s \quad\quad C_T = -0.0010$

Tension $\quad\quad C_P := (P_1 - P) \cdot \dfrac{L_s}{A \cdot E} \quad\quad C_P = 0.0008$

Sag $\quad\quad C_S := 0$

+ Total: $\quad\quad C_{total} := C_L + C_T + C_P + C_S \quad\quad C_{total} = 0.0082$

Figure D.6 Tape corrections for last taped distance in Example 6.1 of book.

■ D.3 WORKSHEETS AS AN AID IN LEARNING

You should not use these worksheets to solve assigned homework problems, since you will only truly learn by solving problems on your own. Instead, you should use these worksheets as a method for testing your understanding and checking your computations. As can be seen in Figure D.6, a major advantage of using the worksheets is that intermediate calculations can be viewed and compared with hand-computed results. These comparisons allow you to determine the location of computational errors.

Another advantage of the worksheets is that they demonstrate some of the common programming routines used in surveying. For example, the worksheets on least-squares adjustments demonstrate the parsing of values from data files, computation of coefficients, building of matrices, and matrix methods used to solve the problem and determine post-adjustment statistics. Many of these routines can be emulated in higher-level programming languages such as Basic, C, Fortran, or Pascal. Additionally, many of these programming sheets can be modified to solve other problems that may be encountered in future studies.

E Introduction to Matrices

E.1 INTRODUCTION

Matrix algebra enables users to express complicated systems of equations in a compact and easily manipulated form. It also provides a systematic mathematical method for solving systems of equations that can be easily programmed. Throughout this book, matrices have been used to solve systems of equations. Nowhere is this more evident than in Chapter 16 on least-squares adjustments. Matrices are frequently encountered in surveying, geodesy, and photogrammetry. This appendix provides readers with a basic understanding of matrices and their manipulations.

■ E.2 DEFINITION OF A MATRIX

A matrix is a set of number or symbols arranged in an array with m rows and n columns. This arrangement allows users to systematically express large systems of equations. For example, assume we have the three equations with three unknown parameters $x, y,$ and z. The system of equations may appear as

$$3x + 5y - 7z = -24$$
$$2x - y + 6z = 33 \qquad \text{(E.1)}$$
$$9x + 4y - 2z = 12$$

This system of equations can be represented in matrix form as

$$\begin{bmatrix} 3 & 5 & -7 \\ 2 & -1 & 6 \\ 9 & 4 & -2 \end{bmatrix} \begin{bmatrix} x \\ y \\ z \end{bmatrix} = \begin{bmatrix} -24 \\ 33 \\ 12 \end{bmatrix} \qquad \text{(E.2)}$$

Equation (E.2) can be represented in compact matrix notation as

$$AX = L \qquad \text{(E.3)}$$

As can be seen in Equation (E.2), each coefficient from Equation (E.1) has been placed in order in the first matrix called A; each unknown parameter has been placed in a single row of the second matrix called X; and similarly, each constant has been placed in a single row of the last matrix called L. Thus, the A matrix is often called the *coefficient matrix*, the X matrix the *unknown matrix*, and the L matrix the *constants matrix*. Once in this form, a system of equations can be manipulated and solved algebraically using matrix methods.

E.3 THE DIMENSIONS OF A MARIX

The number of rows and columns in each matrix expresses the dimensions or size of a matrix. For example, the A matrix in Equation (E.2) has three rows and three columns. It is said to have dimensions of 3 by 3 and is known as a *square matrix*. The X and L matrices have three rows, but only one column. Their dimensions are 3 by 1 and are also known as *vectors*. In general, a matrix can have m rows and n columns. When m is not equal to n, the matrix is known as a *rectangular matrix*. As previously stated by example, a *square matrix* is formed when the number of rows m equals the number of columns n.

Individual elements of a matrix can be designated by their row and column locations in the matrix. The row-column identifiers are known as *indices*. For example, the A matrix in Equation (E.2) has a value of 3 in row 1 and column 1. The index of 3 is thus 1,1 indicating that 3 is in the first row and first column of A. The elements of matrices are generally written in lowercase letters with subscripts representing their index. The indices of the elements are generally written without the intervening comma. For example, $a11$ has a value of 3. With this in mind, the entire matrix A can be written as

$$A = \begin{bmatrix} a_{11} & a_{12} & a_{13} \\ a_{21} & a_{22} & a_{23} \\ a_{31} & a_{32} & a_{33} \end{bmatrix} \qquad \text{(E.4)}$$

In Equation (E.4), each element of the A matrix from Equation (E.2) has been replaced by its elemental name. Thus, in reference to Equation (F.2), a11 is 3, a12 is 5, a13 is −7, and so on.

When a matrix is square such as Equation (E.4), the elements that have equal row and column indices are known as *diagonal elements*. Thus, a11, a22, and a33 are the diagonal elements of the A matrix in (E.4). In their entirety, they are known as the *diagonal* of matrix A. Their sum is known as the trace of matrix A. Only square matrices have diagonals. Matrices that are not square in dimensions, such as X and L in Equation (E.2), do not have diagonals.

E.4 THE TRANSPOSE OF A MATRIX

The *transpose* of a matrix is a process where each column of the transpose matrix is a row in the original matrix. That is, column 1 of the transpose matrix is row 1 of the original matrix, column 2 of the transpose matrix is row 2 of the original matrix, and so on. The transpose of the A matrix in Equation (E.2) is

$$A^T = \begin{bmatrix} 3 & 2 & 9 \\ 5 & -1 & 4 \\ -7 & 6 & -2 \end{bmatrix} \quad \text{(E.5)}$$

Notice Equation (E.5) that placing a "T" as a superscript indicates the transpose of A. Also note that the first column of A^T is the first row of the A matrix in Equation (E.2). Similarly, the second column is the second row and the third column in the third row. As seen in Chapter 16, the transpose of the coefficient matrix is used to create the *normal equations*.

E.5 MATRIX ADDITION

Two matrices can be added or subtracted when they have the same dimensions. As an example, assume that we have two matrices, A and B, which have dimensions of 3 by 2. The addition or subtraction of the two matrices is performed element by element. The following example illustrates this procedure:

$$A + B = \begin{bmatrix} -1 & 4 \\ 2 & 3 \\ 4 & 8 \end{bmatrix} + \begin{bmatrix} 5 & -3 \\ 6 & 7 \\ -2 & -5 \end{bmatrix} = \begin{bmatrix} -1+5 & 4-3 \\ 2+6 & 3+7 \\ 4-2 & 8-5 \end{bmatrix} = \begin{bmatrix} 4 & 1 \\ 8 & 10 \\ 2 & 3 \end{bmatrix} = C \quad \text{(E.6)}$$

Notice in Equation (E.6) that the resulting matrix has the same dimensions as the original two matrices, A and B. Also note that addition is performed element by element. The difference between A and B is written as

$$A - B = \begin{bmatrix} -1 & 4 \\ 2 & 3 \\ 4 & 8 \end{bmatrix} - \begin{bmatrix} 5 & -3 \\ 6 & 7 \\ -2 & -5 \end{bmatrix} = \begin{bmatrix} -1-5 & 4-(-3) \\ 2-6 & 3-7 \\ 4-(-2) & 8-(-5) \end{bmatrix} = \begin{bmatrix} -6 & 7 \\ -4 & -4 \\ 6 & 13 \end{bmatrix} = C \quad \text{(E.7)}$$

In Equation (E.7), each element of the C matrix is found by subtracting the individual elements of the B matrix from the A matrix.

E.6 MATRIX MULTIPLICATION

Matrix multiplication requires that the two matrices being multiplied have the same *inner dimensions*. That is, if A has dimensions of m rows by i columns and is to be multiplied by B, then B must have dimensions of i by n. Notice that A has

i columns and B has i rows. These are the inner dimensions of the product AB. Their resulting product, AB, will have dimensions of m rows and n columns. The *outer dimensions* of the product AB are m and n. This can be expressed as

$$_mA_i^iB^n = {}_mP^n \tag{E.8}$$

where P is the product of AB having dimensions of m by n.

When the number of rows of A does not equal the number of columns in B, then the product BA can not be performed. Thus, matrix multiplication is not commutative. That is, the product of AB does not necessarily equal BA. In fact, when m does not equal n in Equation (E.8), it can't even be performed. The following matrix multiplications are possible:

$$_3A_2^2B^4 = {}_3P^4$$
$$_1A_3^3B^2 = {}_1P^2$$
$$_6A_2^2B^6 = {}_6P^6$$

The following matrix multiplications are not possible:

$$_6A_6^3B^3$$
$$_2A_4^3B^2$$

The reason why these multiplications are not possible is best explained by understanding how matrix multiplication is performed. To obtain the first element of the product matrix P, we must multiply the first row of A by the first column B. This is best demonstrated with an example. Suppose we wanted to find the product AB using the following two matrices:

$$A = \begin{bmatrix} 1 & -2 \\ 3 & 4 \end{bmatrix} \quad B = \begin{bmatrix} 1 & 2 & 3 \\ 4 & 5 & 6 \end{bmatrix} \quad P = \begin{bmatrix} p_{11} & p_{12} & p_{13} \\ p_{21} & p_{22} & p_{23} \end{bmatrix}$$

where P is the product of AB. Then $p11$ is computed as

$$p_{11} = 1 \times 1 - 2 \times 4 = -7$$

Note how each element in the first row of matrix A is multiplied by each element in the first column of matrix B and their sums accumulated. The remaining elements of P are computed as

$$p_{12} = 1 \times 2 - 2 \times 5 = -8$$
$$p_{13} = 1 \times 3 - 2 \times 6 = -9$$
$$p_{21} = 3 \times 1 + 4 \times 4 = 19$$
$$p_{22} = 3 \times 2 + 4 \times 5 = 26$$
$$p_{23} = 3 \times 3 + 4 \times 6 = 33$$

Thus, the product of AB is

$$\begin{bmatrix} -7 & -8 & -9 \\ 19 & 26 & 33 \end{bmatrix}$$

It should be understandable now that the product of BA can't be performed since there are not enough elements in the first column of A to pair with the elements in the first row of B.

Using matrix multiplication, the representation of Equation (E.1) as Equation (E.2) can now be verified. That is, the product of the first row of the A matrix in Equation (E.2) with the elements of the X matrix results in the first equation of Equation (E.1). Similarly, the product of the second row of the A matrix in Equation (E.2) with the X matrix results in the second equation, and the use of the third row of the A matrix results in the third equation.

E.7 MATRIX INVERSE

A *matrix inverse* is similar to division when working with numbers. When A is multiplied by its inverse, the resulting matrix is known as the *identity matrix I*. The identity matrix has values of 1 for the diagonal elements and zeros for all other elements. Thus, an identity matrix with dimensions of 3 by 3 is

$$I = \begin{bmatrix} 1 & 0 & 0 \\ 0 & 1 & 0 \\ 0 & 0 & 1 \end{bmatrix}$$

When the identity matrix of appropriate dimension is multiplied by another matrix, say B, the resulting product is the same as matrix B. Thus, the identity matrix shares the properties of 1 in simple arithmetic multiplication. To solve Equation (E.2), we need to determine the inverse of A and multiply it by L to obtain X, or

$$X = A^{-1}L \tag{E.9}$$

There are several methods of inversing a matrix. While these methods are beyond the scope of this book, these methods often employ the same *elementary row transformations* that are used in mathematics to solve a system of equations. Software is readily available that can perform this operation. Once the inverse of the matrix is known, it can be used to solve a system of equations. For example, the inverse of the A matrix in Equation (E.2) expressed to five decimal places is

$$A^{-1} = \begin{bmatrix} -0.20952 & -0.17143 & 0.21905 \\ 0.55238 & 0.54286 & -0.30476 \\ 0.16190 & 0.31429 & -0.12381 \end{bmatrix}$$

When the inverse of A is multiplied times L, the resulting matrix X is

$$X = A^{-1}L = \begin{bmatrix} x \\ y \\ z \end{bmatrix} = \begin{bmatrix} 2 \\ 1 \\ 5 \end{bmatrix}$$

Thus, solution of Equation (E.2) is where x is 2, y is 1, and z is 5. The reader should check this by inserting these values into Equation (E.1) and confirming that the constants on the right side of the equation are determined.

The software program MATRIX is available for download on the companion website at http://www.pearsonhighered.com/ghilani. The reader can use software to solve other matrix problems presented in this book.

U.S. State Plane Coordinate System Defining Parameters

F.1 INTRODUCTION

As discussed in several chapters, the U.S. State Plane Coordinate System (SPCS) is used in many applications in surveying/geomatics. In Chapter 20, the equations necessary to compute SPCS map projection coordinates are presented. This appendix provides the defining SPCS parameters for use in the United States. Section F.2 provides the defining parameters for states using the Lambert conformal conic map projection. Section F.3 provides the defining parameters for states using the Transverse Mercator map projection.

F.2 DEFINING PARAMETERS FOR STATES USING THE LAMBERT CONFORMAL CONIC MAP PROJECTION

State Zone	AR North	AR South	CA 401	CA 402	CA 403	CA 404	CA 405
φ_S	34°56'	33°18'	40°00'	38°20'	37°04'	36°00'	34°02'
φ_N	36°14'	34°46'	41°40'	39°50'	38°26'	37°15'	35°28'
φ_0	34°20'	32°40'	39°20'	37°40'	36°30'	35°20'	33°30'
λ_0	92°00'W	92°00'W	122°00'W	122°00'W	120°30'W	119°00'W	118°00'W
N_b	0.000	400,000.000	500,000.0	500,000.0	500,000.0	500,000.0	500,000.0
E_0	400,000.000	400,000.000	2,000,000.0	2,000,000.0	2,000,000.0	2,000,000.0	2,000,000.0

F.2 Defining Parameters for States Using the Lambert Conformal Conic Map Projection

State Zone	CA 406	CO North	CO Central	CO South	CT 600	FL North	IA North
φ_S	32°47'	39°43'	38°27'	37°14'	41°12'	29°35'	42°04'
φ_N	33°53'	40°47'	39°45'	38°26'	41°52'	30°45'	43°16'
φ_0	32°10'	39°20'	37°50'	36°40'	40°50'	29°00'	41°30'
λ_0	116°15'W	105°30'W	105°30'W	105°30'W	72°45'W	84°30'W	93°30'W
N_b	500,000.0	304,800.6096	304,800.6096	304,800.6096	152,400.3048	0.0	1,000,000.0
E_0	2,000,000.0	914,401.8289	914,401.8289	914,401.8289	304,800.6096	600,000.0	1,500,000.0

State Zone	IA South	KS North	KS South	KY North	KY South	LA North	LA South
φ_S	40°37'	38°43'	37°16'	37°58'	36°44'	31°10'	29°18'
φ_N	41°47'	39°47'	38°34'	38°58'	37°56'	32°40'	30°42'
φ_0	40°00'	38°20'	36°40'	37°30'	36°20'	30°30'	28°30'
λ_0	93°30'W	98°00'W	98°30'W	84°15'W	85°45'W	92°30'W	91°20'W
N_b	0.0	0.0	400,000.0	0.0	500,000.0	0.0	0.0
E_0	500,000.0	400,000.0	400,000.0	500,000.0	500,000.0	1,000,000.0	1,000,000.0

State Zone	LA Offshore	MD 1900	MA Mainland	MA Island	MI North	MI Central	MI South
φ_S	26°10'	38°18'	41°43'	41°17'	45°29'	44°11'	42°06'
φ_N	27°50'	39°27'	42°41'	41°29'	47°05'	45°42'	43°40'
φ_0	25°30'	37°40'	41°00'	41°00'	44°47'	43°19'	41°30'
λ_0	91°20'W	77°00'W	71°30W	70°30'W	87°00'W	84°22'W	84°22'W
N_b	0.0	0.0	750,000.0	0.0	0.0	0.0	0.0
E_0	1,000,000.0	400,000.0	200,000.0	500,000.0	8,000,000.0	6,000,000.0	4,000,000.0

State Zone	MN North	MN Central	MN South	MT 2500	NE 2600	NY Long Island	NC 3200
φ_S	47°02'	45°37'	43°47'	45°00'	40°00'	40°40'	34°20'
φ_N	48°38'	47°03'	45°13'	49°00'	43°00'	41°02'	36°10'
φ_0	46°30'	45°00'	43°00'	44°15'	39°50'	40°10'	33°45'
λ_0	93°06'W	94°15'W	94°00'W	109°30'W	100°00'W	74°00'W	79°00'W
N_b	100,000.0	100,000.0	100,000.0	0.0	0.0	0.0	0.0
E_0	800,000.0	800,000.0	800,000.0	600,000.0	500,000.0	300,000.0	609,601.2199

U.S. STATE PLANE COORDINATE SYSTEM DEFINING PARAMETERS

State Zone	ND North	ND South	OH North	OH South	OK North	OK South	OR North
φ_S	47°26'	46°11'	40°26'	38°44'	35°34'	33°56'	44°20'
φ_N	48°44'	47°29'	41°42'	40°02'	36°46'	35°14'	46°00'
φ_0	47°00'	45°40'	39°40'	38°00'	35°00'	33°20'	43°40'
λ_0	100°30'W	100°30'W	82°30'W	82°30'W	98°00'W	98°00'W	120°30'W
N_b	0.0	0.0	0.0	0.0	0.0	0.0	0.0
E_0	600,000.0	600,000.0	600,000.0	600,000.0	600,000.0	600,000.0	2,500,000.0

State Zone	OR South	PA North	PA South	SC 3900	SD North	SD South	TN 4100
φ_S	42°20'	40°53'	39°56'	32°30'	44°25'	42°50'	35°15'
φ_N	44°00'	41°57'	40°58'	34°50'	45°41'	44°24'	36°25'
φ_0	41°40'	40°10'	39°20'	31°50'	43°50'	42°20'	34°20'
λ_0	120°30'W	77°45'W	77°45'W	81°00'W	100°00W	100°20W	86°00'W
N_b	0.0	0.0	0.0	0.0	0.0	0.0	0.0
E_0	1,500,000.0	600,000.0	600,000.0	609,600.0	600,000.0	600,000.0	600,000.0

State Zone	TX North	TX North Central	TX Central	TX South Central	TX South	UT North	UT Central
φ_S	34°39'	32°08'	30°07'	28°23'	26°10'	40°43'	39°01'
φ_N	36°11'	33°58'	31°53'	30°17'	27°50'	41°47'	40°39'
φ_0	34°00'	31°40'	29°40'	27°50'	25°40'	40°20'	38°20'
λ_0	101°30'W	98°30'W	100°20'W	99°00'W	98°30'W	111°30'W	111°30'W
N_b	1,000,000.0	2,000,000.0	3,000,000.0	4,000,000.0	5,000,000.0	1,000,000.0	2,000,000.0
E_0	200,000.0	600,000.0	700,000.0	600,000.0	300,000.0	500,000.0	500,000.0

State Zone	UT South	VA North	VA South	WA North	WA South	WV North	WV South
φ_S	37°13'	38°02'	36°46'	47°30'	45°50'	39°00'	37°29'
φ_N	38°21'	39°12'	37°58'	48°44'	47°20'	40°15'	38°53'
φ_0	36°40'	37°40'	36°20'	47°00'	45°20'	38°30'	37°00'
λ_0	111°30'W	78°30'W	78°30'W	120°50'W	120°30'W	79°30'W	81°00'W
N_b	3,000,000.0	2,000,000.0	1,000,000.0	0.0	0.0	0.0	0.0
E_0	500,000.0	3,500,000.0	3,500,000.0	500,000.0	500,000.0	600,000.0	600,000.0

State Zone	WI North	WI Central	WI South	PR VI 5200
φ_S	45°34'	44°15'	42°44'	18°02'
φ_N	46°46'	45°30'	44°04'	18°26'
φ_0	45°10'	43°50'	42°00'	17°50'
λ_0	90°00'W	90°00'W	90°00'W	66°26'W
N_b	0.0	0.0	0.0	200,000.0
E_0	600,000.0	600,000.0	600,000.0	200,000.0

F.3 DEFINING PARAMETERS FOR STATES USING THE TRANSVERSE MERCATOR MAP PROJECTION

State Zone	AL East	AL West	AK 5001/O.M.	AK 5002	AK 5003	AK 5004	AK 5005
$1:k_0$	25,000	15,000	10,000	10,000	10,000	10,000	10,000
φ_b	30°30'	30°00'	57°00'	54°00'	54°00'	54°00'	54°00'
λ_b	85°50'W	85°50'W	133°40'W	142°00'W	146°00'W	150°00'W	154°00'W
E_0	200,000.0	600,000.0	5,000,000.0	500,000.0	500,000.0	500,000.0	500,000.0
N_b	0.0	0.0	0.0	−5,000,000.0	0.0	0.0	0.0
S_0	3,375,406.7112	3,319,892.0570		5,985,317.4367	5,985,317.4367	5,985,317.4367	5,985,317.4367

State Zone	AK 5006	AK 5007	AK 5008	AK 5009	AZ East	AZ Central	AZ West
$1:k_0$	10,000	10,000	10,000	10,000	10,000	10,000	15,000
φ_b	54°00'	54°00'	54°00'	54°00'	31°00'	31°00'	31°00'
λ_b	158°00'W	162°00'W	166°00'W	170°00'W	110°10'W	111°55'W	113°45'W
E_0	500,000.0	500,000.0	500,000.0	500,000.0	213,360.0	213,360.0	213,360.0
N_b	0.0	0.0	0.0	0.0	0.0	0.0	0.0
S_0	5,985,317.4367	5,985,317.4367	5,985,317.4367	5,985,317.4367	3,430,631.2260	3,430,631.2260	3,430,745.5918

State Zone	DE 700	FL East	FL West	GA East	GA est	HI 5101	HI 5102
$1:k_0$	200,000	17,000	17,000	10,000	10,000	30,000	30,000
φ_b	38°00'	24°20'	24°20'	30°00'	30°00'	18°50'	20°20'
λ_b	75°25'W	81°00'W	82°00'W	82°10'W	84°10'W	155°30'W	156°40'W
E_0	200,000.0	200,000.0	200,000.0	200,000.0	700,000.0	500,000.0	500,000.0
N_b	0.0	0.0	0.0	0.0	0.0	0.0	0.0
S_0	4,207,476.9816	2,692,050.5001	2,692,050.5001	3,319,781.3865	3,319,781.3865	2,083150.1655	2,249,193.4045

U.S. STATE PLANE COORDINATE SYSTEM DEFINING PARAMETERS

State Zone	HI 5103	HI 5104	HI 5105	ID East	ID Central	ID West	IL East
$1:k_0$	100,000	100,000	0	19,000	19,000	15,000	40,000
φ_b	21°10'	21°50'	21°40'	41°40'	41°40'	41°40'	36°40'
λ_b	158°00'W	155°30'W	155°30'W	112°10'W	114°00'W	115°45'W	88°20'W
E_0	500,000.0	500,000.0	500,000.0	200,000.0	500,000.0	800,000.0	300,000.0
N_b	0.0	0.0	0.0	0.0	0.0	0.0	0.0
S_0	2,341,506.4725	2,415,321.4658	2,396,891.1333	4,614,370.6555	4,614,370.6555	4,614,305.8890	4,059,417.9793

State Zone	IL West	IN East	IN West	ME East	ME West	MS East	MS West
$1:k_0$	17,000	30,000	30,000	10,000	10,000	20,000	20,000
φ_b	36°40'	37°30'	37°30'	43°40'	42°50'	29°30'	29°30'
λ_b	90°10'W	85°40'W	87°05'W	68°30'W	70°10'W	88°50'W	90°20'W
E_0	700,000.0	100,000.0	900,000.0	300,000.0	900,000.0	700,000.0	700,000.0
N_b	0.0	250,000.0	250,000.0	0.0	0.0	0.0	0.0
S_0	4,056,280.6721	4,151,863.7425	4,151,863.7425	4,836,302.3615	4,744,046.5583	3,264,526.0416	3,264,526.0416

State Zone	MO East	MO Central	MO West	NV East	NV Central	NV West	NH 2800
$1:k_0$	15,000	15,000	17,000	10,000	10,000	10,000	30,000
φ_b	35°50'	35°50'	36°10'	34°45'	34°45'	34°45'	42°30'
λ_b	90°30'W	92°30'W	94°30'W	115°35'W	116°40'W	118°35'W	71°40'W
E_0	250,000.0	250,000.0	250,000.0	200,000.0	500,000.0	800,000.0	300,000.0
N_b	0.0	0.0	0.0	8,000,000.0	6,000,000.0	4,000,000.0	0.0
S_0	3,966,785.2908	3,966,785.2908	4,003,800.5632	3,846,473.6437	3,846,473.6437	3,846,473.6437	4,707,019.0442

State Zone	NJ/NY East 2900	NM East	NM Central	NM West	NY East	NY Central	NY West
$1:k_0$	10,000	11,000	10,000	12,000	10,000	16,000	16,000
φ_b	38°50'	31°00'	31°00'	31°00'	38°50'	40°00'	40°00'
λ_b	74°30'W	104°20'W	106°15'W	107°50'W	74°30'W	76°35'W	78°35'W
E_0	150,000.0	165,000.0	500,000.0	830,000.0	150,000.0	250,000.0	350,000.0
N_b	0.0	0.0	0.0	0.0	0.0	0.0	0.0
S_0	4,299,571.6693	3,430,662.4167	3,430,631.2260	3,430,688.4089	4,299,571.6693	4,429,252.1847	4,429,252.1847

State Zone	RI 3800	VT 4400	WY East	WY East Central	WY West Central	WY West
$1:k_0$	160,000	28,000	16,000	16,000	16,000	16,000
φ_b	41°05'	42°30'	40°30'	40°30'	40°30'	40°30'
λ_b	71°30'W	72°30'W	105°10'W	107°20'W	108°45'W	110°05'W
E_0	100,000.0	500,000.0	200,000.0	400,000.0	600,000.0	800,000.0
N_b	0.0	0.0	0.0	100,000.0	0.0	100,000.0
S_0	4,549,799.4141	4,707,007.8366	4,484,768.4357	4,484,768.4357	4,484,768.4357	4,484,768.4357

Answers to Selected Problems

CHAPTER 2

2.4(a)	13,548.44 sft	**2.5(a)**	163.84 m
2.6(a)	668.6 sft	**2.7(a)**	11.19 ac
2.10(a)	9.76 ac.	**2.12(a)**	9756.14 ift
2.13(a)	122°24′00″; 2.13628 rad	**2.14(a)**	161

2.16(a) 0.692867, 1.59705, and 0.851672 where sum = π rad.

CHAPTER 3

3.7(a) 65.401 **(b)** ±0.003 **(c)** ±0.001
3.11(a) 65.3957 ≤ m ≤ 65.4067, 100%
3.15(a) 23°30′04″ **(b)** ±7.5″ **(c)** ±3.8″
3.18 ±0.09 ft
3.22(a) 146.13 ± 0.023 ft
3.24(a) 32.049 ft
3.27(a) 50,888 ± 14 ft^2
3.28(a) $A = 49°24′28″; B = 39°02′28″; C = 91°33′04″$

CHAPTER 4

4.2 0.068 m; 1.688 m; 6.750 m
4.7 38,160 ft or 7.23 mi
4.12 −0.003 ft
4.15 273.075 m
4.21 0.048 ft
4.28 854.02 ft; 846.18 ft

CHAPTER 5

5.8	0.000 ft	**5.13**	First order, Class I
5.19	1.62 ft; 6.80 ft	**5.21**	16.7 m
5.23	2259.694 m	**5.31**	2.1%
5.35	±13.7 mm		

CHAPTER 6

6.2(a)	2.18 ft/pace, (b) 186 ft	**6.5**	See Section 6.14 and Table 6.1
6.8	253.41 ft	**6.12**	2.7 μsec
6.14(a)	9.9543 m	**6.18**	408.41 ft
6.21	1653.860 m	**6.25**	±8.7 mm

CHAPTER 7

7.10 190°57′36″
7.13 S60°20′57″E
7.16 Az_{CD}: 212°01′13″; Brg_{CD}: S32°01′13″W
7.26(a) 11.6° W   **7.30** 2°15′W
7.33 N22°03′E

CHAPTER 8

8.12(a)	21″	**8.14**	124″
8.20	275°46′04″	**8.24**	±27″
8.26(a)	56″	**8.28**	16″

CHAPTER 9

9.9(a)	720°	**9.15**	13″
9.16	−10″	**9.19**	15.3″
9.22	14″	**9.24**	−15″; Third Order, Class I

CHAPTER 10

10.2 −21″; + 3″
10.5 −2″ per angle; CD = 201°32′40″
10.7 0.063 ft; 1:26,800   **10.16** Distance for line DE
10.17 AC = 403.854 m; Az = 17°29′26″

CHAPTER 11

11.3	$m = -1.2993$; $b = 2445.548$ m		
11.5	51°18′26″	**11.7**	1.425 m
11.9	(6932.18, 4868.39)	**11.13**	(3560.56, 2791.19)
11.15	(1509.50, 1160.43) or (180.08, 248.39)		
11.19	(1284.92, 114.94)	**11.21(a)**	0.3048277

CHAPTER 12

12.1	280,800 ft² or 6.447 ac.	**12.5**	25,220 sq. ft.
12.9	96,894 m² or 9.6894 ha.	**12.16**	8060 ft² or 0.185 ac.

CHAPTER 13

13.4	See Section 13.3		
13.19	See Section 13.9.2	**13.21**	$X = -2,249,118.734$ m
13.24	$\varphi = 40°26'11896''$	**13.28**	65.640 m
13.30	95.888 m		

CHAPTER 14

14.2	80 min.	**14.9(b)**	5 seconds
14.11	2	**14.18**	4 sessions
14.24	4th order, class II	**14.34(a)**	2.40 ppm
14.36(a)	0.71 ppm		

CHAPTER 15

15.1	PPK, RTK, and RTN	**15.6**	±3.8 mm
15.10	See Section 15.4, paragraph 8		
15.18	450 – 470 MHz	**15.28**	2160

CHAPTER 16

16.4	532.686	**16.9**	$x = 135.469$
16.10	±0.033 ft	**16.12(a)**	942.16
16.30	$-0.2092 dx_{Steve} - 0.9779 dy_{Steve} + \cdots = 0.005$		
16.33	$t = 171°26'19.7''$		
16.35	0.045 ft		

CHAPTER 17

17.6	6 ft	**17.10**	1.25 in.
17.15	No overhead obstructions or multipath conditions		
17.27	(5111.76, 4943.11, 818.74)	**17.34**	0.47 in.

CHAPTER 18

18.2	160 in.	**18.10**	±20 ft
18.15	See Section 18.4, paragraph 2	**18.18**	1 in. = 30 ft or 1:360
18.24	12.5 mm	**18.25**	AB: (5431.445, 4472.812, 250)

CHAPTER 19

19.4	436.0 m	**19.10**	6,364,725.399 m; 6,387,949.711 m
19.12	272.624 m	**19.15**	42°36′52.8″
19.19	−33.880 m	**19.22**	D_2: 2456.100 m
19.26	268°19′43.1″	**19.31**	85°56′00.1″; 204°32′47.3″

CHAPTER 20

20.11 592.304 m
20.14 1,937,305.70 sft; 228,864.06 sft
20.17 (177,084.383, 211,996.039)
20.23 N 41°12′23.20370″, W 78°26′30.33402″
20.27(a) 567.82 sft
20.31 0.9995 8406; 1:2400
20.38 205°49′03.1″

CHAPTER 21

21.12 e, c, b, d, a
21.17(b) See Section 21.7, paragraph 1
21.20 11,800 ft^2
21.31 See Section 21.7, paragraph 1

CHAPTER 22

22.1	2726 ft	**22.6(a)**	777.0 ft
22.8(a)	30 mi	**22.14(a)**	240 rods
22.21	Single proportion; Single proportion		

CHAPTER 23

23.10	−0.60%	**23.17**	4000 ft^2
23.18	5.61 ft	**23.29**	1.00%

CHAPTER 24

24.6(a)	R = 1432.68 ft	**24.22**	1392.04 ft
24.23(a)	$I/2$	**24.28**	370.8 ft

CHAPTER 25

25.2(a)	Crest	**25.4**	44+50.0 = 3389.02 ft
25.10	L = 6 sta	**25.18**	612.26 ft
25.20	3389.02 ft @ Station 44+60.29		
25.25	593.64 ft	**25.28**	833.99 ft
25.30	3.92 ft		

CHAPTER 26

26.5	708 yds^3	**26.10**	6168.5 yd^3
26.19	761.8 yd^3; 759.1 yd^3	**26.23**	3.1136 ac-ft
26.26	419.3 ft^3/sec		

CHAPTER 27

27.8(a)	1/2000 in./ft	**27.11(a)**	1 in./1015 ft
27.13	69.19 ac	**27.16**	7988 ft
27.19(a)	116 ft	**27.22**	3409 ft
27.24	672.5 ft	**27.32**	12
27.34	30%		

CHAPTER 28

28.8(a) 15,360,000 pixels
28.16(a) See Section 28.7.3
28.21(b) 4-1/6 in.

Commonly Used Conversions and Abbreviations

■ CONVERSION FACTORS

Length

1 millimeter (mm) = 1000 micrometers (μm)
1 centimeter (cm) = 10 mm
1 meter (m) = 100 cm
1 m = 39.37 inches (in) [U.S. Survey Foot]
1 kilometer (km) = 1000 m
1 km = 0.62137 miles
1 in. = 25.4 mm *exactly* [International Foot]
1 ft = 304.8 mm *exactly* [International Foot]
1 mile = 5280 ft
1 nautical mile = 6076.10 ft = 1852 m
1 rod = 1 pole = 1 perch = 16.5 ft
1 Gunter's chain (ch) = 66 ft = 4 rods
1 mile = 80 ch
1 vara = about 33 inches in Mexico and California and 33-1/3 inches in Texas
1 fathom = 6 ft

Volume

1 m^3 = 35.31 ft^3
1 yd^3 = 27 ft^3 = 0.7646 m^3
1 litre = 0.264 gal [U.S.]
1 litre = 0.001^3
1 gal [U.S.] = 3.785 litres
1 ft^3 = 7.481 gal [U.S.]
1 gal [Imperial] = 4.546 litres = 1.201 gal [U.S.]

Area

1 mm^2 = 0.00155 in.^{2}
1 m^2 = 10.76 ft^2
1 km^2 = 247.1 acres
1 hectare (ha) = 2.471 acres
1 acre = 43,560 ft^2
1 acre = 10 ch^2, i.e., 10 (66 ft $\times$ 66 ft)
1 acre = 4046.9 m^2
1 ft^2 = 0.09290 m^2
1 ft^2 = 144 in.^{2}
1 in.^{2} = 6.452 cm^2
1 mile^2 = 640 acres (normal section)

Angles

1 revolution = 360 degrees = 2π radians
1° (degree) = 60' (minutes)
1' = 60" (seconds)
1° = 0.017453292 radians
1 radian = 57.29577951° = 57°17'44.806"
1 radian = 206,264.8062"
1 revolution = 400 grads (also called gons)
tan 1" = sin 1" = 0.000004848
π = 3.141592654

Other Conversions

1 gram (g) = 0.035 oz
1 kilogram (kg) = 1000 g = 2.20 lb
1 ton = 2000 lb = 2 kips = 907 kg
1 m/sec = 3.28 ft/sec
1 km/hr = 0.911 ft/sec = 0.621 mi/hr

GPS SIGNAL FREQUENCIES

Code	Frequency (MHz)
C/A	1.023
P	10.23
L1	1575.42
L2	1227.60
L5	1176.45

ELLIPSOID PARAMETERS

Ellipsoid	Semimajor Axis (a)	Semiminor Axis (b)	Flattening (1/f)
Clarke, 1866	6,378,206.4	6,356,583.8	294.97870
GRS80	6,378,137.000	6,356,752.314	298.257222101
WGS84	6,378,137.000	6,356,752.314	298.257223563

SOME OTHER IMPORTANT NUMBERS IN SURVEYING (GEOMATICS)

Errors and Error Analysis

68.3 = percent of observations that are expected within the limits of one standard deviation
0.6745 = coefficient of standard deviation for 50% error *(probable error)*
1.6449 = coefficient of standard deviation for 90% error
1.9599 = coefficient of standard deviation for 95% error *(two-sigma error)*

Electronic Distance Measurement

299,792,458 m/sec = speed of light or electromagnetic energy in a vacuum
1 Hertz (Hz) = 1 cycle per second
1 kilohertz (kHz) = 1000 Hz
1 megahertz (MHz) = 1000 kHz
1 gigahertz (GHz) = 1000 MHz
1.0003 = approximate index of atmospheric refraction (varies from 1.0001 to 1.0005)
760 mm of mercury = standard atmospheric pressure

Taping

0.00000645 = coefficient of expansion of steel tape, per 1°F
0.0000116 = coefficient of expansion of steel tape, per 1°C
29,000,000 lb/in.^{2} = 2,000,000 kg/cm^2 = Young's modulus of elasticity for steel
490 lb/ft^3 = density of steel for tape weight computations
15°F = change in temperature to produce a 0.01 ft length change in a 100 ft steel tape
68°F = 20°C = standard temperature for taping

Leveling

0.574 = coefficient of combined curvature and refraction (ft/miles^2)
0.0675 = coefficient of combined curvature and refraction (m/km^2)
20.6 m = 68 ft = approximate radius of a level vial having a 20″ sensitivity

Miscellaneous

6,371,000 m = 20,902,000 ft = approximate mean radius of the earth
1.15 miles = approximately 1 minute of latitude = approximately 1 nautical mile
69.1 miles = approximately 1 degree of latitude
101 ft = approximately 1 second of latitude
24 hours = 360° of longitude
15° longitude = width of one time zone, i.e., 360°/24 hr
23°26.5′ = approximate maximum declination of the sun at the solstaces
$23^h 56^m 04.091^s$ = length of sidereal day in mean solar time, which is 3m55.909^s of mean solar time short of one solar day
5,729.578 ft = radius of 1° curve, arc definition
5,729.651 ft = radius of 1° curve, chord definition
100 ft = 1 station, English system
1000 m = 1 station, metric system
6 miles = length and width of a normal township
36 = number of sections in a normal township
10,000 km = distance from equator to pole and original basis for the length of the meter

ABBREVIATIONS

Construction Surveys

BB	batter boards	esmt	easement
BL	building line	F	fill
CB	catch basin	FG	finish grade
CG	centerline of grade	FH	fire hydrant
CL	centerline	FL	fence line
C	cut	FS	finished surface
CS	curve to spiral	GC	grade change

GP	grade point		
GR	grade rod (ss notes)		
L, Lt	left (X-sect notes)		
MH	manhole		
PC	point of curvature		
PI	point of intersection		
PL	property line		
PP	power pole		
PT	point of tangency		
pvmt	pavement		
R, Rt	right (X-sect notes)		
R/W	right-of-way		
SC	spiral to curve		
SD	storm drain		
SG	subgrade		
spec	specifications		
Sq	square		
ss	slope stake; side slope		
Std	standard		
Str Gr	straight grade		
X sect	cross-section		

Control Surveys; GPS Surveys

BM	benchmark
BS	backsight
CORS	continuously operating reference station
DGPS	differential GPS
EDM	electronic distance measurement
FS	foresight
GPS	global positioning system
HARN	high accuracy reference network
HDOP	horizontal dilution of precision
NGRS	National Geodetic Reference System
OTF	on-the-fly initialization (kinematic GPS)
PDOP	positional dilution of precision
PPP	precise point positioning
RTDGPS	real-time differential GPS
RTK	real-time kinematic (GPS)
SNR	signal-to-noise ratio
VDOP	vertical dilution of precision

Property Surveys

A	area
CF	curb face
ch "X"	chiseled cross
CI	cast iron
diam	diameter
Dr	drive
ER	end of return
Ex	existing
H & T	hub and tack
HC	house connection sewer
IB	iron bolt; iron bar
IP	iron pipe; iron pin
L & T	lead and tack
MHW	mean high water
MLLW	mean low low water
MLW	mean low water
Mon	monument
P	pipe; pin
PS	professional surveyor
Rec	record
St	street
Std Surv Mon	standard survey monument
2″ × 2″	two-inch square stake
"X"	crosscut in stone
yd	yard

Public Lands Surveys

AMC	auxiliary meander corner
bdy, bdys	boundary; boundaries
BT	bearing tree
CC	closing corner
ch, chs	chain; chains
cor, cors	corner; corners
corr	correction
decl	declination
dist	distance
frac	fractional (sec, etc.)
GM	guide meridian
lk, lks	link; links (Gunter's chain)
mer	meridian
mkd	marked
Mi Cor	mile corner

MC	meander corner	hi	height of instrument above station
MS	mineral survey	HI	height of instrument above datum
Prin Mer, PM	principal meridian	hor	horizontal
		IS	intermediate sight
R, Rs	range; ranges	IFS	intermediate foresight
R 1 W	range 1 west	ISS	inertial surveying system
SC	standard corner	lat	latitude
Sec, Secs	section; sections	LHA	local hour angle
SMC	special meander corner	LIS	land information system
Stan Par, SP	standard parallel	long	longitude
		MDT	mountain daylight time
T, Tp, Tps	township; townships	MST	mountain standard time
T 2 N	township 2 north	N	nadir point
USMM	U.S. mineral monument	NAD27	North American Datum of 1927
WC	witness corner	NAD83	North American Datum of 1983
		NAVD88	North American Vertical Datum of 1988

Miscellaneous

		NGVD29	National Geodetic Vertical Datum of 1929
alt	altitude		
Chf	chief of party	NSRS	National Spatial Reference System
CI	cast iron		
Con Mon	concrete monument	obs	observer
CDT	central daylight time	obsn	observation
CS	corrugated steel	orig	original
CST	central standard time	PDT	Pacific daylight time
Delta (Δ)	central angle (of curve)	PST	Pacific standard time
dep	departure	red	reduction
Dir, D	direct	RP	reference point
EDT	eastern daylight time	rev, R	reversed
Elev	elevation	sta	station
EST	eastern standard time	stk	stake
FB	field book	TBM	temporary benchmark
FL	face left	TIN	triangulated irregular network
FR	face right	TP	turning point
GHA	Greenwich hour angle	tel	telescope
GIS	geographic information system	temp	temperature
GPS	global positioning system	UTC	universal coordinated time
HA	hour angle	Z	zenith

■ SOME COMMONLY USED SURVEYING SYMBOLS

Symbol	Description	Symbol	Description
✕	Chiseled monument	⊕	Power pole
⬢	Concrete monument	℔	Property line
═══▭═══	Curb inlet	▣	Stake (or hub) with tack
●―●―●―●	Fence, chain link	‒ ‒ 60'–15" RC ‒ ‒	Storm sewer (length, size, type)
○ ○ ○ ○	Fence, wood w/posts	―⊥―	Telephone line (buried)
―G―	Gas line	―T―	Telephone line (suspended)
⊥	Guy wire	•	Telephone pole
═══⊏	Headwall	✥	Traffic signal
―W―♦―	Hydrant on water line	―▽―⊠―⊠―	Valves on water line
⊙	Iron pipe	―W―	Water line
―⋀⋀⋀―	Power line		

Index

A

Abney hand level, 138
Accidental errors. *See* Systematic errors
Accuracy, 47–48
  ASPRS, 513
  circular map accuracy standard (CMAS), 463
  control surveys, 561–564
  definition of, 47
  FGDC, 513
  of field notes, 30
  mapping, 512–516
  public land surveys, 701–702
Acre, definition of, 24
Acre-foot, definition of, 797
Actual group refractive index, 145
Addition
  matrix, 924
  significant figures, 26–28
Add tape, 135, 136
Adjacency, 870, 882
Adjustments
  conditional adjustment of observations, 64
  least-squares, 66–67
  levels, 95–99
  for parallax, 96
  simple level circuits, 112–113
  total station instruments, 212–216
  traverse computations, 246–249
  weights of observations, 65–66
Adverse possession, 672
Aerial cameras, 820–822
  body, 820
  cone, 820
  fiducial marks, 820
  focal plane, 820
  lens, 820, 821
  magazine, 821
  principal point, 821
  shutter, 820, 821
  single-lens frame, 820
Aerial photo
  flying height from, 832–833
  ground coordinates from, 828–830
  relief displacement from, 830–832
  stereoscopic parallax, 833–836
  vertical, 824–825
Aerotriangulation, 825
Agonic line, 176
Airborne laser-mapping, 848–849
Air, standard, 145
Airy Transit Circle, 543
Algorithms, smoothing, 530
Alidade, 189
Alignments, horizontal. *See* Circular curves
Alignment surveys, 12
Aliquot part, 658
ALTA/NSPS Land Title Survey Standards, 671–672
Altitude angles, 79–80, 205
Ambiguities, integer, 341
Analysis
  of fixed baselines measurements, 385–387
  of loop closures, 388–389
  of repeat baselines, 387–388, 389
  of static surveys, 381–391
Analytical functions, GIS, 880–884
Analytical photogrammetry, 839
Analytical plotters, 839
Angles
  adjustment of, 240–242
  altitude, 79–80, 205
  angles to the left, 165
  angles to the right, 165–166, 228
  azimuths, 167–168, 229
  bearings, 168–169
  clockwise, 165, 166
  closing the horizon, 201–202
  computing, 746–748
  deflection, 166, 202–204, 229, 744–748
  of depression, 205
  direction of lines, 166–167
  distances, relationships, 196–197
  of elevation, 205
  exterior, 165, 166
  horizontal, by direction method, 200–201
  horizontal, total station instruments, 197–200
  of inclination, 331
  instrumental errors, 216–220
  interior, 165, 166, 228
  to the left, 165
  measurements, 196–200
  misclosure, 233–234
  natural errors, 220–221
  observation equations for, 452–454
  parallactic, 836
  phase, 146
  propagation of random errors in, 222–223
  reduction of, 585–586, 622–632
  reduction of distance observations, 579–585
  repetition method, 199, 200
  to the right, 165–166, 228
  three-point resection, 282–285
  units of measurement, 164
  vertical, 204–206
  zenith, 154, 187, 219
Angular units of measure
  gons, 24
  grads, 24

mils, 24
radians, 24, 164
sexagesimal system, 164
Annual change in declination, 176–178
Annual variation, 178
Antenna reference point (ARP), 396, 404, 506
Anti-spoofing (A-S), 327–328
Apparent solar time, 909
APPS. *See* Automatic Precise Positioning Service (APPS)
Arcs, 738
Arc-to-chord correction, 628–630
Area, 301–320
   of circular boundaries, 312–313
   by coordinates, 305–309
   by counting squares, 317
   by digitizing coordinates, 318
   by double-meridian distance (DMD), 309–312
   from map measurements, 317–319
   by offsets from lines, 303–305
   partitioning, 313–317
   by planimeters, 318–319
   by simple figures, 302–303
Argument of perigee, 331
Arrangement of field notes, 30
Arrow, meridian, 525
As-built circular alignments, 771–774
As-built surveys, 734
Assumed bearings, 660
Assumed meridian, 167
Astrolabe, 5
Astronomical observations, 900–901
   azimuths, 901–903, 913–915
   definitions, 906–908
   ephemerides, 903
   Polaris observations, computing, 911–913
   time, 908–910
   timing, 910–911
Astronomical zenith, 548
Astronomic positioning *vs.* geodetic positioning, 543
Atomic clocks, 325–326

Attribute data, 871
Automated contouring systems, 482–483
Automated mapping, 528–533
Automatic compensator, 83, 88, 190
Automatic levels, 83
Automatic Precise Positioning Service (APPS), 412
Autonomous coordinates, 368
Autonomous position, 411
Average-end-area formula, 800–801
Axis
   of the level vial, 86
   of the plate-level vial, 212
   of sight, 186
Azimuth, 167–168
   assumed, 167
   astronomical, 901–903
   from astronomical observations, 913–915
   back, 168
   bearings, comparing, 169–171
   computation of, 171–172, 242–243
   forward, 168
   geodetic, 167
   magnetic, 167
   observation equation, 451–452
   observation of, 204
   reduction of, 627–632

B

Backsight (BS), 77
Balancing angles, 240–242, 250–252
Balancing-in, 209
Barometric leveling, 78–79
Baselines
   fixed measurements, 385–387
   network adjustments, 389–390
   nontrivial, 374–375
   public land surveys, 688–690
   repeat measurements, 387–388
   trivial, 375–376, 398
   vectors, 442–446
Bases of total stations, 191
Base station, 350

Batter boards, 713–716, 718–720, 728
Bearings, 168–169
   assumed, 660
   azimuths, comparing, 169–171
   balanced angles, 250–252
   computation of, 173–174
   magnetic, 176, 178
   traverse computations, 242–243
BeiDou, 359
Benchmark (BM), 72, 73
Biases. *See* Systematic errors
BIH 1984.0, 543, 544
Bivariate distribution, 462
Blazing trees, 698
Block-and-lot system, 663–664
Borrow-pit method, 809–810
Boundary operations, 882
Boundary surveys, 657–658
   adverse possession, 672
   block-and-lot system, 663–664
   categories of land surveys, 658–659
   condominium surveys, 672–679
   coordinates, 665
   easement, 672
   error, sources of, 679
   geographic information systems (GIS), 679
   historical perspectives, 659–660
   metes and bounds, 660–663
   mistakes, 679–680
   partitioning land, 670–671
   registration of title, 671–672
   retracement surveys, 665–668
   subdivision surveys, 668–670
Bounds, 660–663
Bowditch rule, 247–249
Breaking tape, 136
Breaklines, 483
Broken-back curve, 738
Bubbles
   circular level, 216
   leveling errors, 122–123
Buffering, 881
Bureau International de l'Heure (BIH), 358, 543
Bureau of Land Management (BLM), 15, 658, 684, 703–704

C

Cadastral surveys, 11
Calculations. *See* Computations
Call for surveys, 666
Cameras
   aerial, 820–822
   of field notes, 32
Canadian Spatial Reference System-Precise Point Positioning (CSRS-PPP), 412
Canopy restrictions, 369
Caribbean Terrestrial Reference Frame of 2022 (CATRF2022), 556
Carrier, frequencies, 326
Carrier phase-shift measurements, 341
   ionospheric free model, 344
Carrier signal, 147
Carrying level, 102–104
Cartographic map elements, 525–527
Cartography, 511
Categories of land surveys, 658–659
Celestial equator, 906
Celestial meridian, 907
Cells (grid) as spatial data, 866
Center-point coding, 872, 873
Centimeter (cm), 24
Central meridian, 167, 608–609
Chaining pins, 133
Chains, 870
Charge-coupled device (CCD), 821, 822
Chords, computing, 746–748
Circles
   area of parcels, 312–313
   coordinate geometry, 271–273
   lines, intersections, 275–280
Circuits
   leveling, least-squares, 436–440
   levels, 110
   simple level, adjustments, 112–113
Circular arcs, 738
Circular curve elements
   deflection angles, 744–746, 748–753
   degree of curvature, 739–741
   external, 741
   incremental chord, 745, 746, 748–753
   length, 740, 741
   long chord, 741
   middle ordinate, 761
   point of compound curvature (PCC), 766
   point of curvature, 741
   point of intersection, 741
   point of tangency, 741
   point on curve, 741
   point on tangent, 741
   total chord, 753
   vertex, 741
Circular curves
   arc definition, 739
   as-built surveys of, 771–774
   chord definition, 739–741
   compound, 765–766
   computation of coordinates on, 753–755
   equation for, 742
   intersection with a line, 765
   intersection with another, 765
   layout, 748–749
   layout by coordinates, 755–760
   layout by incremental chords, 745
   layout by offsets, 761–764
   layout with total chords, 753
   metric, 751–753
   passing through a fixed point, 764–765
   reverse, 765–766
   setups on, 750–751
   sight distance on, 766
   sources of error in, 774
   special problems, 764–765
   stakeout using GNSS, 760–761
   stakeout using robotic total stations, 760–761
   stationing, 743–744
Circular error probable (CEP), 463
Circular level bubbles, adjustment of, 216
Circular map accuracy standard (CMAS), 463
Circular tangent length, 757–758
Civil time, 909
Clamp handles, 133
Clarity
   of field notes, 30
   in maps, 517–519
Class G airspace, 849
Classification, GISs, 865
Class interval, 52
Clocks, 325–326
   bias, 341–342
Closed traverses, 165, 226
Closing the horizon, 201–202
Closure conditions, 245
Cloud service, 391
Coarse/acquisition code (C/A), 327
Code-based GNSS receivers, 490, 875
Codeclination, 908
Code ranges, 339–341
   ionospheric free model, 344
   mathematical model, 341
Coding, 871
Coefficient matrix, 923
COGO
   direction intersection problem, 275–277
   inaccessible point problem, 290–292
   intersection of circle and line, 277–280
   intersection of two distances, 280–282
   perpendicular distance of point from line, 273–275
   three-dimensional resection problem, 292–294
   three-point resection, 282–285
Coincidence-type tube level, 87
Colatitude, 908
Collimation correction factor, 121
Collimation error, 576
   adjustment of, 97
Color in maps, 473
Combined factor, 626
Combined networks, 573–574
Common identifiers, 871
Communication ports, 192

Comparison of azimuths and bearings, 169–171
Compass, 174–176
Compass (Bowditch) rule, 247–249
Compensating errors, 46
Compound curves, 738, 765–766
Computations
   area, 307–308
   astronomical observations for azimuth, 913–915
   azimuths, 171–172
   bearings, 173–174
   curves by proportion, 787
   geodetic position computations, 591–594
   magnetic declination, 181
   slope intercepts, 804–806
   SPCS 83, 611–616
   three-dimensional coordinate computations, 596–598
   traverse in state plane coordinates, 632
   traversing (See Traverse computations)
   unequal tangent vertical curves, 788–790
   vertical curves, 782–786
   volumes, 808–809
Computer-aided drafting and design (CADD)
   mapping, 528–533
   mapping procedures, 516–517
   plotting using, 521–522
Computer-based systems, 4
   traverse computations, 263
Concrete nails, 227
Conditional adjustment of observations, 64
Condition equation, 428
Condominium Act, 673
Condominium surveys, 11, 672–679
Cones, cameras, 820
Conformal map projections, definition of, 604, 608–609, 928–930
Conformal projections, 605
Connections, data collectors, 39

Connectivity, 870, 882
Constants matrix, 923
Constants, zone, 611–612, 616–617, 628
Construction, machine control, 419–421, 732–734
Construction surveys, 707–735
   as-built surveys, 734
   batter boards, 713–716, 718–720, 728
   building layout, 717–721
   computing the bend angles in pipelines, 716–717
   control, 712–713
   equipment, 708–711
   grade rod, 715
   highway layout, 721–724
   horizontal and vertical control, 712–713
   machine control, 732–734
   mistakes in, 735
   other types, 726–727
   pipelines, staking grades, 714–716
   sources of error, 735
   using GNSS, 730–732
   using laser scanners, 734
   using total stations, 727–730
Continuing education units (CEUs), 17
Continuously Operating Reference Station (CORS), 18, 368, 379–380
Contour-area method, 811–812
Contouring
   automated methods, 482–483
   characteristics of, 479–480
   controlling-point method of locating, 480–481
   direct method of locating, 480
   index, 518
   indirect method of locating, 480
   methods of locating, 480–481
   plotting of, 523–524
   smoothing factor, 530
Contours
   definition of, 476–479
   interval, 477, 522–523
   mapping survey, 476–479

Contrast, 518
Control
   barometer, 78
   for mapping survey, 475–476
   monumentation, 734
   point descriptions, 565–569
   segment, 324, 325
Control surveys
   accuracy standards, 561–564
   control point descriptions, 565–569
   conventional terrestrial poles, 542–544
   deflection of the vertical, 547–549
   differential leveling, 540, 574–579
   ellipsoid and geoid, 540–542
   field procedures for, 569–574
   geodetic position, 544–546, 591–594
   geoid undulation, 547–549
   GNSS, 540
   horizontal, 565, 569–574
   local geodetic coordinate system, 594–596
   National CORS Network (NCN), 368–380, 411
   National Horizontal Control Network, 564
   National Spatial Reference System (NSRS), 564
   National Vertical Control Network, 565
   networks, 573–574
   NSRS horizontal control network, 564
   precise differential leveling, 540, 574
   reduction of field observations, 579–591
   three-dimensional coordinate computations, 596–598
   transforming coordinates between frames, 556–560
   triangulation, 569–570
   trilateration, 573
   U.S. reference frames, 549–556
   vertical, 565, 574–579

Conventional horizontal plane surveys, 447–456
Conventional terrestrial pole (CTP), 331, 542–544
Convergence angle, 612
Convergence of meridians, 234
  equation for, 572–573
Conversion
  data format, 871–874
  raster-to-vector, 873–874
  significant figures, 26
  vector-to-raster, 871–873
Coordinated universal time (UTC), 911
Coordinate geometry, 270–296
  circles, 271–273
  inaccessible point problem, 290–292
  intersections, 275–282
  lines, 271–273
  perpendicular distances, 273–275
  three-dimensional two-point resection, 292–294
  three-point resection, 282–285
  transformation, 285–290
Coordinates
  adjusting, 252–254
  area by, 305–309
  boundary surveys, 665
  end areas by, 803–804
  ground, 828–830
  horizontal curves, 755–760
  local geodetic coordinate system, 594–596
  manually plotting by, 521
  methods for determining area, 316–317
  property descriptions by, 665
  reference coordinate systems, 329–339
  state plane coordinate systems, 249
  three-dimensional coordinate computations, 596–598
  transforming between reference frames, 556–560
Coordinate transformation
  three-dimensional conformal, 496–498
  two-dimensional, 285–290

Corners
  lost and obliterated, 699–701
  marking, 698
  meander, 698–699
  witness, 698
Corrections
  arc-to-chord, 628
  ground-surveyed distances, 622
  lines, 690
  second-term, 628
Covariance matrix, 435, 442–443, 445
Crest curve, 783
Cross hairs, 85
  leveling errors, 122
  testing, 96–97
Cross-section leveling, 120–121
Cross-section method, 798–799
  types of, 799–800
Cross-sections
  five-level section, 799, 800
  irregular section, 799, 800
  side-hill section, 799, 800
  three-level, 799, 800
  transition section, 799, 800
Cubic feet, 24, 797
Cubic meter m^3, 24, 797
Cubic yards, 24, 797
Cultural features, 472
Cumulative errors, 46
Cumulative volumes, 808–809
Curvature
  differential leveling, 119
  ellipsoidal radii of, 544–546
  errors, 119
  leveling, 74–76, 122
Curvature and refraction, 74–76
Curves. *See also* Circular curves
  day-lighting, 809
  easement, 738
  horizontal (*See* Horizontal curves)
  normal density, 52
  vertical (*See* Vertical curves)
Cutoff lines, 265, 670
Cut tape, 135–136
Cycle slips, 355

D
Data collectors, 29
  advantages of, 41–42
  disadvantages of, 41–42
  electronic distance measurement (EDM), 143–144
  file management, 40–41
  manual entry in, 29
  mapping surveys, 498–501
  transferring files, 38–40
Data files, safe storage of, 40–41
Data format conversions, 871–874
Data latency, 403
Data processing, static surveys, 381–384
Data sources, GISs, 865
Datum, regional, 418
Datum, transformation of, 556–560
Decimals, 23
Decimeter (dm), 24
Declination, annual change in, 176–178
Deeds, 658
  Register of Deeds office, 669
Definite corners, 661
DEFLEC18, 586
Deflection angles, 202–204, 749–753
  traversing, 228–229
Deflection of the vertical, 547–549
Degrees, 24
  of circular curves, 739–741
  of freedom, 462–463
Density, normal curves, 52
Departures
  compass (Bowditch) rule, 247–249
  coordinates, adjusting, 252–254
  traverse computations, 245
Descriptions, 29
  public land surveys, 702
Descriptive data, 871
Designation of townships, 692
Design, mapping, 517–519
Detailing, 483
Developable surfaces, 604

cone, 604
cylinder, 604
plane, 604
Deviation of the compass. *See* Variations of the compass
Deviation of the vertical. *See* Deflection of the vertical
Diagrams, mass, 809
Diapositives, 840
Differences, elevation, 211–212
Differential leveling, 77–78, 105–110
   adjustment by least-squares, 436–440
   collimation error, 97–98
   equipment used in, 70–99
   loop, 110, 461
Differential positioning, 349–351
Differential rectification, 844
Digital cameras, 38
Digital data file management, 40–41
Digital elevation models (DEMs), 473, 482–483, 732, 798, 877, 878, 885, 886
Digital image correlation, 843
Digital image processing, 844
Digital levels, 83, 90–91
Digital line graphs (DLGs), 512, 878
Digital ortho quarter quadrangles (DOQQs), 886
Digital raster graphics (DRGs), 885
Digital terrain models (DTMs), 482, 530
Digitizing
   aerial photographs, 875–876
   coordinates, area by, 318
   graphic materials, 876–877
Digits, rounding off, 28–29. *See also* Numbers
Dilution of precision (DOP), 348–349
Dimensions of matrices, 923
Direct geodetic problem, 592
Direction-direction problem, 275
Direction-distance problem, 277
Direction method, 200–201
Direction of turning, 164

Directions, 870
   azimuths, 167–172
   bearings, 168–169, 173–174
   final adjustments, 255–257
   from latitudes and departures, 243–244
   reduction of, 585–587
Direct observations, 44–45
Discrepancies, 47
Dispersion, 50
Displacement from refraction, 75
Distance
   angles, relationships, 196–197
   angular, 164
   backsight (BS), 97
   double-meridian distance (DMD), 309–312
   geodetic, 579–580
   geodetic reduction of, 579
   mark-to-mark, 579, 581
   perpendicular, 273–275
   reduction of, 622–632
Distance-distance intersection, 280–282
Distance measurement, 129–161
   electronic, 76, 143–161
   foresight (FS), 97
   linear, 129–130
   odometer readings, 130
   optical rangefinders, 130–131
   pacing in, 130
   subtense bar, 131
   tacheometry, 131
   taping, 131–143
Distance observation equation, 449–451
Distance root mean square (DRMS), 463
Diurnal circle, 907
Doppler Ranging Integrated on Satellite (DORIS), 338
Doppler shifts, 323
DOP spikes, 370
DOQQs. *See* Digital ortho quarter quadrangles (DOQQs)
Double centering, 208
Double-meridian distance (DMD), 309–312

Double-rodded lines of levels, 110
Drafting procedures, 516–517
Drawing designators, 500
Drawing profiles, 120
DTREE, 500
Dual-frequency receivers, 343
Duties of a rod person, 104–105

E

Earth, 109
   grid reduction of distances, 623–627
   magnetic field and compasses, 174–176
   position of poles, 167
Earth-Centered Earth-Fixed (ECEF), 338
Earthwork sections, 800
Easement curves, 738
   spirals, 738
Eccentricity of centers, 219
Electromagnetic energy, 144–147
Electromagnetic interference (EMI), 356
Electromagnetic spectrum, 851, 852
Electronic distance measurement (EDM), 76, 143–161
   centering error, 155–156
   errors in, 154–161
   instrumental errors in, 157–160
   natural errors in, 160–161
   personal errors in, 156–157
   principles in, 147–148
   reflector constant offset, 158–159
   reflectorless mode, 152
   system measurement constant, 158
Electronic field book, 37
Electronic laser levels, 84
Elevation, 71, 547
   differences, 211–212
   digital elevation models (DEMs), 482–483
   factor, 624
   leveling, methods of determining, 76–83

Elevation of the instrument above datum, 80
Ellipsoid, 9, 540–542
  Clarke 1866, 541, 547
  eccentricities, 541
  flattening factor, 541
  Geodetic Reference System of 1980 (GRS 80), 541
  semiaxes, 541
  World Geodetic System of 1984 (WGS 84), 541
End areas
  by coordinates, 803–804
  determination of, 801–804
Endlap, 825
Engineer's tapes, 132
Ephemeris, 345, 902, 903, 907
Epoch rate, 366
Epochs of time, 350
Equal tangent property of parabola, 786
Equal tangent vertical parabolic curve equation, 780–781
Equation of time, 909
Equations
  angles observation, 452–454
  azimuth observation, 451–452
  distance observation, 449–451
  equal tangent vertical parabolic curve, 780–781
  linearizing nonlinear, 447–449
  matrix, 434–436
  tangent offset, 782–786
  vertical parabolic curve, 779–781
Equipment
  construction surveys, 708–711
  development of, 7
  for differential leveling, 70–99
  electronic distance measurement (EDM), 143–144
  hydrographic surveys, 501–506
  taping, 132–133
  testing, 95–99
Equipotential surfaces, 71–72
Error formulas, applications of, 59, 60, 62
Errors, 44
  area, 320
  boundary surveys, 679
  circular curves, 774
  construction surveys, 735
  curvature, 109
  ellipses, 456–461
  indexes, 188, 206
  leveling, 121–123
  mapping, 535
  mapping surveys, 506
  of the mean, 63–64
  photogrammetry, 858
  of product, 62–63
  public land surveys, 704
  refraction, 109
  residuals, 50
  of series, 60–62
  sources, 46
  static surveys (GNSS), 396–398
  of sum, 59–60
  taping, 139–143, 890–892
  three-sigma, 58
  total station work, 216–222
  traversing, 237
  two-sigma, 57
  types of, 46–47
  volumes, 814
Euler pole, 554
Euler Pole Parameters 2022 model (EPP2022), 554–555
Existing digital data sets, 878
Expansion of fill volumes, 809
Exterior angles, 165, 166
Exteriors
  survey, 660, 669
  township, 690–691
External measurement beam, 149
Eyepiece, 84

F
FAA. *See* Federal Aviation Administration (FAA)
Factors
  combined, 626
  elevation, 624
  grid, 626
  scale, 605
Father Point, 71, 73, 552
Fault lines. *See* Breaklines
Federal Aviation Administration (FAA), 849
Federal Communications Commission (FCC), 414
Federal Emergency Management Agency (FEMA), 886
Federal Geodetic Control Subcommittee (FGCS), 111, 234, 374, 384, 465, 561
Federal Geographic Data Committee (FGDC), 464, 513, 880
Fédération Internationale des Géométres. *See* International Federation of Surveyors (FIG)
Fiberglass tape, 133
Fiducial marks, 820
Field books, types of, 31–32
Field notes, 29–30
  arrangement of, 30
  erasures in, 35–38
  general requirements of, 30–31
  horizontal curves, 748, 749
  noteforms, 893–899
  procedures for control surveys, 569
  profile leveling, 117
  recording, 34–35
  suggestions for recording, 34–35
  survey controllers, 35–38
  three-wire leveling, 114
  traversing, 232–233
  types of, 31–32
Field procedures
  for control surveys, 569–574
  for profile leveling, 117–119
Field-to-finish systems, 39, 144
Files
  management, 40–41
  transfers from data collectors, 39
Fill volume, 808
Five-level section, 799, 800
Fixed baseline measurements, 385–387
Fixed points
  taping problem, 890–892

vertical curves, passing through, 790–792
Flight lines, 825
Flight planning, 846–848
Flood insurance rate maps (FIRMs), 886
Flying height, 832–833
Focal length, 85, 820, 822, 824–826
Focal plane, 821
Foot point latitude, 621
Foresight (FS), 78, 117
  distances, 97
Formulas, circular curves, 741–742
Forward tapeperson, 134, 135
Fractional sections, 695–696
Frequency, 50
  carrier, 326
  polygons, 52
  reference, 147
Full station, 116

G

Gal, 588
Galileo system, 358
Gauss, Karl, 66
Gaussian mean radius, 546
Gauss–Krüger map projection, 604
GEIOD03, 553, 554
General Land Office, 15, 684, 695
General laws of probability, 54
Geocentric coordinate system, 329–331
  computation from geodetic coordinates, 331–337
  conventional terrestrial pole, 542–544
Geodetic coordinates, 329, 331–337
  computation from geocentric coordinates, 331–337
  height, 335–337
  latitude, 331
  longitude, 331
Geodetic height, 336
Geodetic meridian, 167
Geodetic position computations, 591–594
Geodetic problem
  direct, 592
  inverse, 593–594
Geodetic radii
  at any azimuth, 545–546
  meridian, 544
  normal, 544
Geodetic reductions, 579–591
Geodetic Reference System of 1980 (GRS 80), 541, 547, 550
Geodetic surveys, 9
  accuracy standards, 561–564
  networks, 226
  precise traversing, 571–573
  triangulation, 569–570
Geographic Coordinate Data Base (GCDB), 703, 704
Geographic information systems (GISs), 14–15, 534, 862–886
  applications, 884–885
  boundary surveys, 679
  databases, creating, 874–880
  data format conversions, 871–874
  data sources and classifications, 865
  functions, 884
  Land Information System (LIS), 864–865
  mapping, 534–535
  metadata, 880
  nonspatial data, 871
Geoid, 71
  models for, 336, 414
  undulation, 336, 547–549
GEOID09, 553, 554
GEOID12B, 553, 554
GEOID18, 74, 390, 414, 553, 554
GEOID96, 554
GEOID99, 554
GEOID2022, 74
Geoidal heights, 574
Geomatics, 3–4
  definition of, 1
Geometric ranges, 340
Geometry
  coordinate (See Coordinate geometry)
  of observed satellites, 347–349
Geopotential number, 72
Geopotential unit, 589
Georgia West Zone, 649
Geospatial Positioning Accuracy Standards, 513
Gigahertz (GHz), 144
GIS analytical functions, 880–884
  boundary operations, 882
  line buffering, 881
  logical operations, 884
  overlaying, 882
  point buffering, 881
  polygon buffering, 881
  proximity analysis, 881–882
  spatial joins, 882–884
GIS applications, 884–885
GIS data
  from aerial photos, 875–876
  areas, 866
  from existing digital data, 876
  from field surveys, 874–875
  format conversions, 871–874
  from graphic materials, 876–877
  grid cells, 866
  from keyboard entry, 878
  lines, 866
  metadata, 880
  nonspatial, 871
  points, 866
  raster format, 867–869
  from scanners, 878–880
  spatial, 865–871
  topology, 869–871
  vector format, 867–869
GIS database, 14, 703
  creation, 874–880
GISs. See Geographic information systems (GISs)
Global navigation satellite system (GNSS), 129, 323–361
Global positioning system (GPS). See GPS
GLONASS, 357–358
GNSS Analysis and Positioning Software (GAPS), 412
GNSS antenna
  antenna reference point (ARP), 346
  phase center offsets, 347
  slant height, 346

GNSS baseline vectors,
    adjustment of, 442–446
GNSS initialization, 405–406
    antenna swapping, 405
    known baseline method, 405
    on-the-fly (OTF), 405
GNSS receiver, 10, 323, 364,
    406–407, 489–490
GNSS surveys, 323–361
    analysis of fixed baselines,
        385–387
    analysis of loop closures,
        388–389
    analysis of repeat baselines,
        387–388
    baseline network adjustment,
        389–390
    base station, 350
    canopy restrictions, 369
    determined coordinates of
        stations (2022), 554–555
    developing an observation
        scheme, 374–377
    DOP spikes, 370
    factors affecting accuracy, 403
    fast static method, 367, 376
    FGCS specifications, 374,
        384–385
    field procedures, 364–369
    field reconnaissance, 373–374
    kinematic method, 352, 367, 370,
        402–422
    observation windows, 369
    obstruction diagrams, 371
    planning, 369–380
    post-processed kinematic
        (PPK), 409
    precise point positioning (PPP),
        355–357
    pseudokinematic method, 367
    rapid static method, 367, 373
    RTK, 402, 408, 417–419
    selecting a field procedure,
        364–365
    sky plot, 370
    static positioning, 366–367
    static surveys, 363–399
Gons, 164, 187
    definition of, 24

GPS, 7, 323–361
    anti-spoofing (A-S), 327–328
    base station, 350
    C/A code, 327
    carrier phase-shift
        measurements, 339, 341
    centering errors, 155–156,
        397–398
    code ranging, 339–341
    control segment, 324, 325
    differential, 349–351
    DOP factors, 348
    ephemerides, 903
    fundamental frequency, 326
    fundamentals in positioning,
        339–341
    geometric ranges, 340
    hand-over word (HOW), 328
    initialization techniques,
        405–406
    introduction to, 323–324
    ionospheric refraction, 342
    kinematic surveys, 402–422
    L1 signal, 326
    L2 signal, 326, 343
    L5 signal, 347, 360
    master control station, 325
    M codes, 327
    monitoring stations, 325
    multipathing errors, 345, 397
    orthometric heights from,
        394–396
    P code, 327
    PDOP, 348, 409
    PPK, 402, 409
    precise positioning service
        (PPS), 326
    PRN codes, 327, 328
    pseudorange, 340, 375, 396
    radio spectrum, 410
    ranges, 324
    refraction errors, 342–344
    relative positioning, 352–355
    repeater stations, 410, 415
    RTK, 351, 402
    satellite clock bias, 341–342, 354
    satellite geometry, 344, 347
    satellite reference coordinate
        system, 329–330

satellite vehicle number (SVN),
    325
signals, 317–320, 323–328
solar activity, 397, 404
space segment, 324
standard positioning service
    (SPS), 326
static surveys, 363–399
stop-and-go method, 489
tapped feedback shift registers,
    327
time of week (TOW), 328
user equivalent range error
    (UERE), 348
W code, 327
Y code, 327
GPS week number, 360
Grad, definition of, 164
Grades
    construction surveys, 707
    machine control, vertical
        curves, 788
    pipelines, staking, 714–716
    rate of, 120
    rod, 715
    stakes, 722
    vertical curves, 780
Gradient, 120
Grantee, 660
Grantor, 660
Graphic data, 865
Gravity for the redefinition of the
    American Vertical Datum,
    553–556
Great circle, 906
Greenwich hour angle (GHA),
    331, 907
Greenwich Royal Observatory,
    543
Grid factor, 626
Grid leveling, 120
Grid reduction of distances,
    623–627
Grids
    cells as spatial data, 866
    meridian, 167
    method, 482
    reduction of azimuths and
        angles, 627–632

state plane coordinates, 622–632
three-dimensional perspective, 530
Ground control for photogrammetry, 845
Ground plane, 346
Ground-surveyed distance corrections, 622
Ground surveys, 12
Ground versus grid problem, 641–646
Group refractive index, 145
Group refractivity, 145
GRS 80, 541, 547, 550
Guide meridians (GM), 690
Gunter's chain, 686–687

H

Hachures in mapping, 473
Half-marks, 837, 838
Hand levels, use of, 84
Hand-over word (HOW), 328
Hard disk drive (HDD), 391
Harmony, 519
HARN. *See* High Accuracy Reference Networks (HARNs)
HDOP. *See* Horizontal dilution of precision (HDOP)
Hectares, definition of, 24
Height of instrument (HI), 78, 80, 235, 728
Height, orthometric, 335
Helmert transformation, 556–557
Hertz (Hz), 144
High Accuracy Reference Networks (HARNs), 377, 551
High-definition surveying, 492
High point on vertical curves, 782
Highway layout, full stations, 721–722
Highway surveys
  point of intersection, 722
  points on tangent, 722
Histogram, 50, 52
Horizontal angles, 22
  by direction method, 200–201
  three-point resection, 282–285

total station instruments, 197–200
Horizontal circles, 187
Horizontal control, hierarchy of, 564
Horizontal control surveys, 539, 569–574
Horizontal curves, 738–739. *See also* Circular curves
  as-built circular alignments, 771–774
  chords and angles, computing, 746–748
  circular curve formulas, 741–742
  circular curve stationing, 743–744
  compound and reverse, 765–766
  coordinates, 755–760
  degree of circular curves, 739–741
  field notes, 748, 749
  layout by deflection angles, 749–750
  metric, 751–753
  mistakes, 775
  offsets, 761–764
  setups on curve, 750–751
  sight distance, 766
  spirals, 767–771
  staking out, 760–761
Horizontal dilution of precision (HDOP), 348–349
Horizontal line, 71
Horizontal plane, 71
Horizontal stations, 462–465
Horizontal time-dependent positioning (HTDP), 368–369, 379, 392, 557
Hour angle method, 911–913
Hour circle, 907
HTDP. *See* Horizontal time-dependent positioning (HTDP)
Hubs, 227
Hybrid geoid model, 74
Hydrogen maser clocks, 325
Hydrographic mapping, 506
Hydrographic surveys, 501–502
  echo sounder, 502

lead lines, 502
sounding pole, 502
soundings, 502

I

Identifiers, 867
  common, 871
Identity matrix, 926
IFVM2022, 554
Inaccessible point problem, 290–292
Inclination angle, 331
Inclination line, 331
Index contours, 518
Indexes
  errors, 188, 206
  field books, 33
  of refraction, 144
Indexing errors, 188
Index of refraction, 144
Indices, 923
Indirect method, 480
Indirect observations, 44–45
Inertial measurement units (IMUs), 407, 540, 849
Inertial surveying systems (ISSs), 539
Initial control, establishing, 367–369
Initialization techniques in kinematic survey, 405–406
Initial point, 687–688
Instantaneous position of the pole, 544
Instrumental errors, 46, 121–122, 139, 157, 229, 396
  electronic distance measurement (EDM), 157–160
  leveling, 121–122
  static surveys, 396
  taping, 139
  total station work, 216–222
Integer ambiguity. *See* Ambiguities, integer
Integrity, of field notes, 30
Intent
  of parcels, 661
  of the parties, 666

Intercepts, slopes, 804–806
Interior angles, traversing, 228
Interior areas as spatial data, 866
Intermediate sights, 118
International Earth Rotation and Reference Systems Service (IERS), 544
International Federation of Surveyors (FIG), 1, 17
International System of Units (SI), 24–26
International Terrestrial Reference Frame (ITRF), 74, 544, 552. *See also* Individual entries
International Terrestrial Reference Frame of 2018 (ITRF2018), 554
Internet, surveying on the, 18
Interpolation, taping errors, 142–143
Interpretative photogrammetry, 818
Intersection
   hydrographic surveys, 504
   of a line with a circle, 277–280
   of two circles, 280–282
   of two lines with known directions, 275–277
   vertical point of intersection (VIP), 780
Intersection stations, 570
Inter-system bias (ISB), 357
Intervals
   classes, 52
   contours, 522–523
   stadia interval factor, 115
Intra-Frame Velocity Model (IFVM), 554
Invar tapes, 132
Inverse geodetic problem, 593–594
Inverse matrix, 926–927
Inverse problem, 614–616
Ionosphere, 416
Irregular section, 799, 800
Irregular variation, 178
ISO 17123-3, 631
Isogonic chart, 176

Isogonic line, 176
ITRF90, 338
ITRF91, 338
ITRF92, 338
ITRF93, 338
ITRF94, 338
ITRF95, 338
ITRF96, 338
ITRF97, 338
ITRF2000, 338
ITRF2005, 338
ITRF2008, 338
ITRF2014, 338, 554–555

J
Jacob's staff, 175
Jefferson, Thomas, 25, 73
Jog/shuttle mechanism, 190
Junior rights, 664

K
Kilohertz (KHz), 144
Kilometer (km), 24
Kinematic surveying methods
   semikinematic mode, 409
   stop-and-go method, 409
   true kinematic mode, 409
Kinematic surveys
   accuracy of, 403
   errors in, 421–422
   methods used, 408–411
   mistakes in, 422
   planning of, 403–404

L
LAMBDA method, 405
Lambert conformal conic projection, 608–609
   central meridian, 609
   common functions in, 611
   direct problem, 611
   inverse problem, 611
   scale changes in, 608
   scale factor, 605
   SPCS 83, computing, 611–616
   standard parallels, 690
   using tabular method, 613
   zone constants, 611–612
Land, 683

   partitioning, 670–671
   titles, 671–672
Land Information System (LIS), 14–15, 679, 703–704, 864–865
Landsat satellites, 852, 853, 855
Land surveys, 11
   categories of, 658–659
Land tenure system, 657
LandXML exchange format, 533–534
Laplace stations, 549
Large errors, 54
Large-scale maps, 473, 475, 819
Laser plummets, 192
Laser scanner, resolution of, 491–494, 710–711
Laser-scanning survey, 494–496
Latitude, 243–249
   compass (Bowditch) rule, 247–249
   coordinates, adjusting, 252–254
   expressed with 4 or 5 decimals, 545
   traverse computations, 243–244
Latitudinal (township) lines, 690–691
Law of cosines, 271
Law of sines, 271
Least squares, 46, 249
   adjustment of GNSS baseline vectors, 442–446
   adjustment of horizontal plane survey data, 447–456
   angles observation equation, 452–454
   azimuth observation equation, 451–452
   conventional horizontal plane surveys, 447–456
   covariance matrix, 435, 442
   differential leveling adjustment, 105–112
   errors ellipses, 456–461
   fundamental condition of, 427–428
   horizontal distance observation equation, 293
   leveling circuits, 436–440
   matrix methods, 432–434

measures of precision, 462–465
minimally constrained
    adjustment, 461
observation equation method,
    428–432
traverse adjustment, 249
weight matrix, 442, 445
Least-squares adjustments. *See*
    Least squares
Leenhouts, P., 464, 561
Left thumb rule, 103
Legends, 525
Legibility of field notes, 30
Length, 22
    final adjustments, 255–257
    tape, reading, 131
    traverse computations, 263
    traversing, 229–230
Lengths of lines, from latitudes
    and departures, 244
Lenker level rod, 94
Lens, 820
Lettering mapping, 524–525
Leveling
    adjustment of, 73
    allowable misclosure in, 112
    barometric, 78–79
    borrow-pit, 120–121
    circuits, least-squares, 436–440
    cross-section, 120–121
    curvature and refraction errors,
        81, 109
    definitions, 70–73
    differential, 77–78
    differential leveling loops, 462
    double-rodded lines, 110
    elevation, methods of
        determining, 76–83
    equipment for differential,
        83–84
    errors, 121–123
    field procedures, 117–119
    grids, 120–123
    loop misclosure, 110
    minus sight, 78
    mistakes in, 123–124
    networks, 576
    North American vertical
        datum, 73–74, 552–553

notes in, 110, 125
orthometric heights, 587–591
plate, 94–95
precision in, 111–112
profile, 116–120
reciprocal, 113–114
refraction, 74–76
section misclosure, 110
setting up levels, 102–104
sources of errors in, 121–123
three-screw arrangement, 88
three-wire, 114–116
trigonometric, 79–82
weights in, theory and methods,
    70
Level line, 71
Levels
    adjustment collimation error, 97
    adjustment of, 95–99
    automatic type, 88–90
    categories of, 83–84
    digital type, 90–91
    handling of, 192–193
    hand type, 133
    laser, 84, 709
    testing of, 95–99
    tilting type, 88
Level vial, errors caused by
    temperature variation, 140
Light Detection and Ranging
    (LiDAR), 503–504, 848
Linearization of a nonlinear
    equation, 447–449
Linear measurement, 129–130
Linear misclosure, 245–246
Line buffering, 881
Line of apsides, 329
Line of sight, 84
    adjusting of, 97–98
    leveling errors, 121–122
    minimum ground clearance in,
        112
    testing of, 97–98
    total station instruments, 186
Lines
    agonic, 176
    azimuths, 166
    bearings, 168–169
    buffering, 881

circles, intersections, 277–280
coordinate geometry, 271–273
cutoff, 265, 670
direction of (angles), 166–167
double-rodded, 110
equation for, 805, 806
flight, 825
intersections, 275–277
isogonic, 176
latitudinal (township), 690–691
meridional (range), 690–691
offsets from, 487–488
points to, perpendicular
    distances, 273–275
prolonging, 208–209
public land surveys, 690
reference, staking, 116–117
as spatial data, 865–871
Liquid crystal display (LCD), 186
LIS. *See* Land Information
    System (LIS)
Local attraction, 179–180
Local geodetic coordinate system,
    594–596
Localization, 418
Logical operations, 884
Longitude, 539
    definition of, 544–545
    expressed with 4 or 5 decimals,
        545
Loop misclosure, 110
Loops
    closures, 388–389
    differential leveling, 462
Lost corner, 699
Lovar tape, 132
Lower culmination, 907
Low distortion projections,
    642–645
Low point on vertical curves, 782

M

Machine control, 419–421
    using GNSS, 420
    using total stations, 420–421
Magnetic declination
    software to compute, 178–179
    typical problems in, 180–182
    variations in, 176–178

Magnetic field, local attraction, 175
Maintenance, taping equipment, 133
Management of files, 40–41
Manual of Surveying Instructions, 684, 685, 687, 689, 690
Map drafting
  automated methods, 528–533
  in CADD, 521–522, 528–533
  clarity, 517
  contrast, 518
  graphical scales, 525
  harmony, 519
  meridian arrow, 525
  mistakes in, 535–536
  order, 517
  title block, 527
  unity, 518–519
Mapping, 510–511
  accuracy, standards for, 512–516
  airborne laser-mapping systems, 851–852
  angle, 612
  automated, 528–533
  availability of maps, 511
  cartographic elements, 525–527
  circular map accuracy standard (CMAS), 463
  computer-aided drafting and design (CADD), 516–517, 528–533
  contour intervals, 522–523
  design, 517–519
  drafting procedures, 516–517
  errors, 535
  geographic information systems (GISs), 534
  layout, 519–520
  lettering, 524–525
  manual procedures, 516–517
  mistakes, 535–536
  plotting contours, 523–524
  plotting procedures, 520–522
  projections, 604–605, 637–641
  Thematic Mapper (TM), 852, 854
  types of, 472
  U.S. government agencies, 15–16
Mapping surveys, 472–507
  aerial, 473
  automated contouring systems, 482–483
  contours, 476–479
  data collectors, 498–501
  digital elevation models (DEMs), 482–483
  digital terrain models (DTMs), 482
  field, 473
  grid method, 482
  hydrographic, 506
  laser scanners, 491–494
  methods, 473
  mobile, 493
  offsets from reference lines, 487–488
  radiation by total station, 504
  scale, 474–475
  three-dimensional conformal coordinate transformation, 496–498
  topographic details, locating, 483–494
  using GNSS, 489–490
Map projections
  Gauss–Krüger projection, 604
  Lambert conformal conic, 608–609
  oblique Mercator, 641
  oblique stereographic, 637–641
  state plane coordinate system (SPCS), 610–611
  Transverse Mercator, 609–610
  Universal Transverse Mercator (UTM), 512, 636–637
Maps
  accuracy standards, 512–516
  cartographic elements, 525–527
  contours, 522–523
  cultural features, 472, 511, 512, 517
  design of, 517–519
  enlargements, 475
  large-scale, 474, 514
  layout of, 519–520
  legends, 525
  lettering, 524–525
  migrating between software packages, 533–534
  overlaying, 534
  planimetric, 472
  plotting of, 520–521
  relief, 472
  scale, 474–475
  spot elevations, 479
  topographic, 472
Map scales, 474–475
Mariana Terrestrial Reference Frame of 2022 (MATRF2022), 556
Marking, 206–208
  corners, 698
  taping errors, 143
Mark-to-mark distance, 579
Mass diagrams, 809
MATHCAD worksheets, 295, 916
  as a learning aid, 916–917
Matrix, 922–927
  addition, 924
  algebra, 922
  coefficient of, 923
  constants, 923
  definition of, 922–923
  diagonal elements, 923
  diagonal of, 923
  identity, 926
  indices, 923
  inverse, 926–927
  multiplication, 924–926
  square, 923
  transpose of, 924
  unknown, 923
Meanderable bodies of water, 688, 689
Meander corner (MC), 698–699
Mean higher high water (MHHW), 72
Mean high water (MHW), 72
Mean lower low water (MLLW), 72
Mean low water (MLW), 72
Mean sea level (MSL), 72
Measurements
  angles, 164, 196–200
  area, 317–319
  carrier phase-shift, 339, 341

electronic distance
    measurement (EDM), 76,
    129, 143–144, 147–148
  of parallax, 837–838
  types of, 22
  units of, 22–24
  vertical distances, 47
  volumes, 797
  water discharge, 812–813
Measures of precision, 55–57,
    462–465
Mechanical projection
    stereoplotters, 844
Megahertz (MHz), 144
Meridians
  arrows, 525
  assumed, 167
  astronomic, 167
  azimuths, 167–168
  bearings, 168–169
  central, 167
  deed, 167
  geodetic, 167
  grids, 167
  record, 167
Meridional (range) lines,
    690–691
Metadata, 391, 880
  importance of, 338, 535
Meter
  conversion of, 23
  definition of, 25, 550
Metes and bounds, 660–663
Methods
  condition equation method, 428
  controlling point, 480, 798
  coordinate squares, 486
  cross-sections, 798–799
  directional, 200–201
  least-squares, 249
  matrix methods in least-squares
    adjustment, 432–434
  observation equation, 428–432
  repetition, 199, 200
  trace-contour, 480
  trial and error, 313–315
  volume measurement, 797
Metonic cycle, 72
Meton of Athens, 72

Metrical photogrammetry, 818, 819
Metric circular curves, 751–753
Metric system, 22
  horizontal curves, 751–753
Meyer, Thomas H., 644
Microclimate, 161
Micrometers, 145, 838
Microprocessors, 144
  total station instruments, 192
Millimeter (mm), 24
Mils, definition of, 24
Minimally constrained
    adjustment, 461
Minus sight, 78
Minutes, 24
Misclosure
  loop, 110
  section, 110
Mississippi East Zone, 649
Mistakes, 45
Mobile mapping system, 493, 874
Models
  digital elevation models
    (DEMs), 473, 482–483
  digital terrain models (DTMs),
    482
  three-dimensional perspective,
    473
  triangulated irregular network
    (TIN), 528
  World Magnetic Model, 178
Modes
  post-processed kinematic
    (PPK), 402
  real-time kinematic (RTK), 402
  total station instruments, 187
  tracking, 151
Modulation, 149
Monuments, 666
Most probable value, 49–50
Multipathing errors, 344, 345
Multiple circles, intersections,
    280–282
Multiplication, matrices, 924–926

N

NAD 27 and NAD 83, in state
    plane coordinates, 610–611
NAD83 (2007), 553

NAD 83 (2011), 391, 392, 553
NAD 83 (HARN), 553
NAD 83, realizations of, 551–552,
    553
Nanometer (nm), 504
National CORS Network,
    368–380, 411
National Flood Insurance
    Program (NFIP), 886
National Geodetic Survey (NGS),
    15, 71, 73, 74, 167, 262, 368, 377,
    392, 394, 416, 564
National Geodetic Vertical Datum
    of 1929 (NGVD 29), 552
National Geophysical Data
    Center (NGDC), 176
National Geospatial-Intelligence
    Agency (NGA), 338–339
National Horizontal Control
    Network, 564
National Institute of Standards
    and Technology (NIST), 139,
    910
National Integrated Land System
    (NILS), 703
National Map Accuracy
    Standards (NMAS), 512, 522
National Mapping Program,
    511–512
National Oceanic and
    Atmospheric Administration
    (NOAA), 878
National Spatial Data
    Infrastructure (NSDI), 878
National Spatial Reference
    System (NSRS), 377, 556, 564
National Tidal Datum Epoch, 72
National Vertical Control
    Network (NVCN), 565
National Weather Service, 160
Natural errors, 46
  electronic distance
    measurement (EDM),
    160–161
  leveling, 122–123
  static surveys (GNSS), 398–399
  taping, 139
  total station instruments,
    220–221

Natural features, 472
Natural Resources Conservation Service (NRCS), 886
Nautical mile, 545
Navigation Satellite Timing and Ranging Global Positioning System. *See* GPS
NCN station, 368, 411
Near-vertical photographs, 824
Nestedness, 870
Network, 226
　horizontal control, 564
　leveling, 576
　triangulated irregular networks (TINs), 798
New Mexico Central Zone, 647–649
NGS antenna calibration data, 383
NGS Geodetic Tool Kit, 394
NILS. *See* National Integrated Land System (NILS)
NOAA Continuously Operating Reference Station (CORS) Network (NCN), 368, 379, 554
NOAA National Geophysical Data Center (NGDC), 179
Nodes, 870
Nominal tape length, 139
Nonlinear equations, linearizing, 447–449
Nonspatial data, GISs, 871
Normal distribution curve, 52
Normal equations, 428, 924
North American Datum of 1927 (NAD 27), 549
North American Datum of 1983 (NAD 83), 549–550
North American-Pacific Geopotential Datum of 2022 (NAPGD2022), 74
North American Terrestrial Reference Frame of 2022 (NATRF2022), 556
North American Vertical Datum of 1988 (NAVD 88), 71, 73–74, 552
North celestial pole, 906

Noteforms, 893–899
Notes. *See* Field notes
NSDI. *See* National Spatial Data Infrastructure (NSDI)
NSRS Horizontal Control Network, 564
Numbers
　rounding off, 28–29
　transposition of, 45
　turning points (TPs), 106
Nutation, 542

O
Objective lens, 84
Oblique Mercator projection, 641
Oblique stereographic projection, 637–641
　defining parameters, 638
Oblique triangle, 271
Obliterated corner, 699–701
Observations
　astronomical, 900–901 (*See also* Astronomical observations)
　conditional adjustment of, 64
　distance observation equation, 449–451
　least squares, 46
　measurements of precision, 55–57
　occurrence of random errors, 50–54
　redundant, 49, 435
　solar, azimuths from, 913–915
　sources of errors, 46
　types of, 46–47
　weights of, 65–66
　windows, 369
Obstruction diagrams, 371
Occupational Safety and Health Administration (OSHA), 13
Occurrence of random errors, 50–54
Odometer readings, 130
Off-line, taping errors, 142
Offsets
　area, 303–305
　horizontal curves, 761–764
　lines, 13
　from a reference line, 487–488

tangent offset equation, 782–786
Online Positioning User Service (OPUS), 392, 411–412
Online processing services, 368
On-the-fly (OTF) ambiguity resolution, 405
Open traversing, 226, 259–262
Operational Land Imager (OLI), 855
Optical plummet, 189, 191
　adjustment of, 215
Optical rangefinders, 130–131
Optical-reading theodolites, 897
OPUS. *See* Online Positioning User Service (OPUS)
Order, 517
Original surveys, 11, 658
Orthometric correction, 590
Orthometric height. *See* Elevation
Orthophotos, 844–845
OSHA. *See* Occupational Safety and Health Administration (OSHA)
OTF. *See* GNSS initialization
Outlines, public land surveys, 696–698
Overlay maps, 862, 882–884

P
Pacific Terrestrial Reference Frame of 2022 (PATRF2022), 556
Pacing, 130
Page check, 107, 110
Parabola
　equal tangent property of, 786
　equations, 779–780
Parallactic angles, 836
Parallax, 123
　adjustments, 96
　leveling errors, 123
　measurements of, 837–838
Parallelogram, 315
Parallel-plate micrometer, 89
Parol evidence, 658
Partitioning of lands, 313–317
　by coordinates, 316–317
　by simple figures, 315–316

trial and error, 313–315
Passive marks, 553
Passometer, 130
Pentagonal prism, 487, 488
Percent grade, 120
Perigee, 329
Permanent identifier (PID), 569
Perpendicular distance from point to line, 273–275
Personal errors, 123
Phase angles, 146
Phase-based technology, 492
Phase center offsets, 347
Phase shift, 148
Philadelphia rod, 92
Photodetectors, 190
Photogrammetry, 12, 818–858
   aerial cameras, 820–822
   airborne laser-mapping systems, 848–849
   analytical, 839
   flight planning, 846–848
   flying height, 832–833
   ground control for, 845
   ground coordinates, 828–830
   interpretative, 818
   mapping surveys, 473–474, 476
   metrical, 818, 819
   mistakes, 858
   orthophotos, 844–845
   relief displacement, 830–832
   remote sensing, 851–856
   scale of vertical photographs, 825–828
   small unmanned aerial systems (sUAS), 849–851
   software, 856–858
   sources of error in, 858
   stereoscopic parallax, 833–836
   stereoscopic plotting instruments, 839–844
   stereoscopic viewing, 836–837
   types of aerial photographs, 822–824
   uses of, 819
   vertical aerial photographs, 824–825
Photographic interpretation, 818
Photographs
   aerial, 317
   terrestrial, 12
   types of aerial, 822–824
   vertical aerial, 824–825
Pipelines
   bend angles in, 716–717
   grades, staking, 714–716
   staking out, 713–714
Pixels, 821
   as spatial data, 866
Planimeters, 318–319
Planimetric map, 513
Planning
   flight, 846–848
   kinematic surveys, 403–404
   laser-scanning survey, 494–496
Plate fixed frame, 554
Plate-level vial, 214
Plots of profile leveling, 119
Plotting
   contours, 523–524
   instruments, stereoscopic, 839–844
   procedures, mapping, 520–522
PLSS, 658
   accuracy of, 701–702
   baseline, 688–690
   bearing trees, 698
   closing corner, 690
   corner markings, 698
   correction lines, 690
   descriptions, 702–703
   fractional sections, 695–696
   Geographic Coordinate Data Base (GCDB), 703, 704
   guide meridians, 690
   initial point, 687–688
   lost corner, 699
   meander corner, 698–699
   mistakes, 704
   obliterated corner, 699–701
   principal meridian, 688
   quadrangles, 688, 690
   range lines, 690–691
   ranges, 692
   sources of errors, 704
   standard corner, 690
   standard parallels, 690
   subdivision of quadrangle, 692–693
   subdivision of sections, 695
   subdivision of townships, 692–695
   topographic calls, 701
   witness corner, 698
Plumb bob, use of, 133
Plumbing
   rod levels, 104
   taping errors, 142
Plus sight, 77
Point buffering, 881
Point cloud, 492
Point code, 500
Point identifiers (PIDs), 759
Point of beginning (POB), 660, 661
Point of commencement (POC), 660, 661
Point of intersection (PI), 722
Points
   corresponding principal, 833
   high/low on vertical curves, 782
   inaccessible point problem, 290–292
   to a line, perpendicular distances, 273–275
   referencing, 231
   as spatial data, 866
Polar distance, 908
Polaris observation, 234
   computing, 911–913
Polar stereographic projection, 638
Polygon buffering, 881
Polygons, 165
   buffering, 881
   frequency, 52
Positional dilution of precision (PDOP), 348, 349
Positioning, astronomic vs. geodetic, 543
Positive errors, 54
Post-process kinematic (PPK), 402
   surveys, 404, 409–411
Practical location, 659–660

Precedence coding, 872
Precession, 542
Precise code, 327
Precise ephemeris, 345, 383
Precise leveling, 915
Precise point positioning (PPP), 355–357, 368, 412
Precise traverse, 571–573
Precision, 47–48
 of adjusted quantities, 434–436
 definition of, 47
 leveling, 111–112
 matrix equations, 434–436
 measures of, 55–57, 462–465
 relative, 245–246
Predominant type coding, 871
Preferences, field notes, 32–33
Principal meridian, 688
Principal point, 821
Principle of reversion, 208, 209
Prismoidal formula, 806–807
Probability, 48–49
 general laws of, 54–55
Probable error, 57
Professional engineer (PE), 16
Professional land surveyor (PLS), 16
Professional organizations, 17
Professional qualifications, 16
Profile leveling, 116–120
 field procedures in, 117–119
 intermediate sights, 117–118
Profile, use of, 119–120, 798
Projections
 equatorial stereographic, 638
 Lambert conformal conic, 608–609
 map, 652–654
 oblique Mercator, 637, 641
 oblique stereographic, 637–641
 polar stereographic, 638
 state plane coordinates, 604–608
 Transverse Mercator, 609–610
Prolonging a line, 208–209
Propagation of electromagnetic energy, 144–147
Propagation of errors, 440–441

Propagation of random errors, 59, 222–223
Property descriptions
 block-and-lot system, 663–664
 by coordinates, 665
 by metes and bounds, 660–663
Proportion, vertical curve computation, 787
Proximity analysis, 881–882
Pseudokinematic surveys, 367
PseudoRandom noise (PRN), 325, 327
Pseudorange, 340, 372, 396
Pseudorange corrections (PRCs), 350–352
Public lands
 surveys of, 683–704
Public Land Survey System (PLSS). See PLSS
Pythagorean theorem, 138
PZS triangle, 908, 912

Q
Quadrangles, 686, 688, 690, 691
 subdivision of, 692–693
Quadratic equation, 271, 277, 295, 316

R
Radial traversing, 236–237
Radian, definition of, 24
Radiation
 hydrographic surveys, 504
 by total station, 484–486
Radiometers, 851
Radio Technical Commission for Maritime Services Special Committee 104 (RTCM SC-104), 351
Radius in the prime vertical, 332
Radius searching, 881
Random errors
 occurrence of, 50–54
 propagation of, 59, 222–223, 234
Random traverse, 209–211
Range rate corrections (RRCs), 350
Ranges diurnal, 78
 geometric, 340

 poles, 133
Rapid ephemeris, 345
Rapid static relative positioning, 366–367
Rapid static survey, 373
Raster formats, 867–869
Raster-to-vector conversion, 873–874
Rate of grade, 120
Ratio, 474
Reading
 leveling errors, 124
 odometer, 130
 rods, 106
 tapes, 135–136
Real-time differential GPS (RTDGPS), 350
Real-time kinematic surveys. See RTK surveys
Real-time network (RTN) surveys, 356, 402, 403
Real-time networks, 415–417
Rear tapeperson, 134–136
Reciprocal leveling, 113–114
Recording
 field notes, 34–35
 tape distance, 136
Rectangular coordinates, 249–250
Rectification, 844
Reduction
 azimuths and angles, 627–632
 distances and angles, 622–632
 field observations, 579–591
Redundant observations, 49, 435
Reference coordinate systems, 329–339
Reference frames, 549–556. See also WGS84 reference frame
 Helmert transformation, 556–557
 International Terrestrial Reference Frame, 544
 NAD 27, 549
 NAD 83, 549–550
 NAD 83 (2007), 552
 NAD 83 (CORS93), 551
 NAD 83 (CORS94), 551
 NAD 83 (CORS96), 551
 NAD 83 (HARN), 551

NAVD 88, 73, 552–553
NGVD 29, 73, 552
transforming coordinates between, 556–560
Reference frequency, 147
Reference gravity, 589
Reference line, 116–117, 487–488
Reference stations, 377–380
Referencing points, 270
Reflector constant offset, 158–159
Reflectorless EDMs, 710
Refraction
    differential leveling, 109
    errors, 109
    GNSS errors, 342–344
    group index of, 145
    index of, 145
    leveling, 74–76
Register of Deeds office, 669
Registration of title, 671–672
Relative Positional Accuracy, 671–672
Relative positioning, 352–355
    double differencing, 354–355
    single differencing, 354
    triple differencing, 355
Relative precision, 245–246
Relative relationships, 869
Relative weights, 65
Reliability of weight, 49
Relief displacement, 830–832
Remote-controlled survey, 503
Remote positioning unit (RPU), 759
Remote sensing, 12, 818, 851–856
Remote total station instruments, 195–196
Repeat baseline measurements, 387–388
Repeater radio, 411
Repeater stations, 408, 410
Repetition method, 199, 200
Reports, static surveys (GNSS), 390–391
Resection
    three-dimensional two-point, 292–294
    three-point, 282–285
Residuals, 50

Resolution, 868
Resolving power, 86
Resurveys, 686
Reticle, 84–85
Retracement surveys, 665–668
Reverse curves, 738, 765–766
Right ascension, 908
    of ascending node, 331
Right, traversing angle to, 228
RINEX file format, 379
Robotic total stations, 760–761
Rod person, duties of, 104–105
Rods
    Chicago type, 92, 94
    fixed height, 398, 406
    Lenker type, 94
    level, 92–94
    Philadelphia type, 92, 93
    precise level types, 575
Rollins, Craig M., 644
Root mean square error, 513–516
Rotating-beam lasers, 708–709
Rotation
    calculations, 287
    coordinate transformation, 286–287
    total station instruments, 191
Rotor, 190
Rounding off numbers, 28–29
Roving instruments, 78
RPU. See Remote positioning unit (RPU)
RTK surveys, 402, 403, 408, 410–411, 417
    communication in, 414–415
    performing, 417–419
Rubidium clocks, 325

S

Sanitary sewers, 713
Satellite availability plot, 370
Satellite geometry, DOP factors, 348
Satellite Laser Ranging (SLR), 338
Satellite orbits
    apogee, 329
    argument of perigee, 331
    inclination angle, 331

line of apsides, 329
perigee, 329
reference coordinate systems, 329–339
right ascension of ascending node, 331
Satellite reference coordinate systems, 329
Satellite signals, signal to noise ratio (SNR), 355
Satellite systems
    BeiDou, 370
    Galileo, 358–359
    GLONASS, 357–358
    GPS, 325–326
Satellite vehicle number (SVN), 325
Scaled lengths, area by, 318
Scale, equivalence, 474
Scanners, laser, 710–711
Secant method, 689
Secant planes, 637
Sections
    fractional, 695–696
    misclosure, 110
    subdivision of townships, 693–695
Secular change in declination, 176
Senior rights, 661, 666
Series, errors of, 60–62
Servo-driven total stations, 195–196
Setting up
    levels, 102–104
    total stations, 192–195
Settlement, leveling errors, 123
Setups on curve, 750–751
Sexagesimal system, 164
Shading in maps, 473
Shape of parcels, 661
Short lines, 80
SI (International System of Units), 24–26
Side-hill section, 799, 800
Sidelap, 825
Side-looking airborne radar (SLAR), 851
Sidereal time, 908
Sides, property, 661

962   Index

Sight distances, 792–794
   horizontal curves, 766
Signals
   carrier, 147
   global positioning system (GPS), 323–328
Significant figures, 26–28
   rounding of, 28–29
Simple curves, 738
Simple figures, end area by, 802–803
Simple level circuits adjustments, 112–113
Simple spatial objects, 866
Single-beam lasers, 708, 709
Single differencing, 354
Single-lens frame cameras, 820, 822
Site calibration, 418
Size of parcels, 661
Sky plot, 370–371
Slant measurements, 346
Slopes
   correction, 28
   distances, reduction of, 579
   intercepts, 722, 804–806
   measurements, 137–138
   stakes, 722, 723
   taping, 137–138
Small unmanned aerial systems (sUAS), 849–851
Small unmanned aerial vehicles (sUAVs), 473, 490, 849
Smoothing algorithms, 530
Snyder, J. P., 604, 642, 644
Softcopy photogrammetric system, 842
Softcopy plotter, 7, 9, 842
Software
   coordinate geometry, 295–296
   field-to-finish, 498–501
   least-squares adjustments, 571
   magnetic declination, 178–179
   between packages, 533–534
   projection, 652–654
Soil Conservation Service (SCS), 886
Solar activity, 391, 404
Solar method, 689

Solar observations, 913–915
Spads, 727
Spatial correction parameter (FKP), 415
Spatial joins, 882–884
Specifications
   for control surveys, 561–564
   static surveys (GNSS), 385–387
Spectrums, wavelength, 851
Spheres, 342
Spheroid. *See* Ellipsoid
Spirals, 738, 739, 767–771
   calculation and layout, 769–771
   geometry, 767–768
Spot elevations, 479
Square matrix, 923
Stadia, 106, 107, 131
   interval factor, 115
Stakeless construction, 732
Stakeout, 151
Staking
   buildings, 717–721
   curves, 782
   highways, 721–726
   horizontal curves, 758, 759
   pipelines, grades, 714–716
   reference lines, 116–117
   vertical curves, 782
Standard air, 145
Standard deviation, 55
   interpretation of, 57
Standard parallels, 608
Standards
   circular map accuracy standard (CMAS), 463
   control surveys, 561–564
   field notes, 31
   mapping, 510
   total station instruments, 186
Standard time, 909
State plane coordinate system (SPCS), 249, 928
   arc-to-chord correction, 628
   combined factor, 626
   convergence angle, 612
   conversions SPCS27 and SPCS83, 635
   defining parameters, 928–930
   direct problem, 617–619

   elevation factor, 624
   inverse problem, 614–616
   Lambert conformal conic map projection, 608–609, 928–930
   reduction of distances and angles, 622–632
   SPCS 83, computing, 611–616
   Transverse Mercator map projection, 616–617, 931–932
   traverse computations, 263
   Universal Transverse Mercator (UTM), 636–637
   zones, 635–636
State Plane Coordinate System of 1927 (SPCS 27), 610
   conversions, 635
State Plane Coordinate System of 1983 (SPCS 83), 610–611
   computing, 611–616
   conversions, 635
   foot, definitions of, 606–607
State Plane Coordinate System of 2022 (SPCS 2022), 647–652
   designed by NGS, 647
   GIS applications, 652
   Lambert conformal conic projections, 647, 649–651
   latitude and longitude parameters, 652
   preliminary SPCS2022 statewide zone, 647–648, 652
   projection axis, 647
   transverse Mercator, 647
Static GNSS surveys, 10
Static relative positioning, 365–366
Static surveys (GNSS), 363–399
   analysis, 381–391
   data processing, 381–391
   errors, 396–398
   field procedures, 364–369
   mistakes, 398–399
   obtaining orthometric heights from, 394–396
   performing, 380–381
   planning, 369–380
Stationing, 116–117
Stations
   full, 116

half, 117, 118
quarter, 117, 118
referencing, 231–232
Stator, 190
Stereoplotters, 839
aerial photos with, 875–876
basic concepts in, 840–842
softcopy type, 842–844
Stereoscopic
measurement of parallax, 837–838
parallax, 833–836
plotting instruments, 839–844
viewing, 836–837
Stop-and-go method. *See* Kinematic surveying methods
Storm sewers, 713
Straight lines, prolonging, 208–209
Strides, 130
Subdivision
quadrangle into townships, 692–693
of sections, 695
surveys, 658, 668–670
of townships, 692–695
Subtense bar, 131
Subtraction, significant figures, 27
Sum, errors of, 59–60
Sun's semidiameter, 914
Superelevation, 739
Surfaces
developable, 604
state plane coordinates, 604
Survey controllers. *See* Data collectors
Surveying
alignment, 12
as-built, 12, 734
boundary, 257–259
construction, 12, 707–735
control, 11, 539–598
definition of, 9–10
geodetic, 9–10
ground, 12
history of, 4–8
hydrographic, 12, 501–506
importance of, 10–11
industrial, 12
on Internet, 18

mapping (*See* Mapping surveys)
mining, 12
plane, 9–10
professional organizations, 17–18
profession of, 16–17
public land (*See* PLSS)
safety, 12–14
satellite, 12, 331, 347, 363–365
topographic, 11, 499
types of, 11–12
U.S. government agencies, 15–16
Surveyor's Certificate, 674
Surveyor's compass, 164, 174
Surveyor's tapes, 132
Survey report, 390–391
Surveys
as-built, 12, 734
construction, 12, 707–735
control, 11, 539–598
highway, 171
laser-scanning survey, 494–496
mapping, 472–507
public lands, 683–687
underground, 727
Symbols, cartographic, 525
Systematic errors, 46
eliminating, 48
Systeme Pour d'Observation de la Terre (SPOT), 855
System measurement constant, 158

T
Table of contents, field books, 33
Tabular form of volume computation, 808
Tabulations, 32
Tangent, 190
offset equation, 782–786
plane, 637
screw, 190
Tapes
add type, 135
correction for support, 141
correction problems, 890–892
cut, 135

length correction in, 139–140
sag correction in, 141–142
surveyor's, 132
temperature correction in, 133
tension correction in, 133
Taping, 131–143
applying tension, 134
combined corrections, 142
equipment, 133
error correction, systematic, 890–892
field procedures, 135–136
on level ground, 134–136
maintenance, 133
marking lengths, 134
plumbing, 134
reading, 135–136
slope measurements, 137–138
on sloping ground, 136–137
sources of errors, 139–143
Taping pins, 133
Tapped feedback shift registers, 327
Targets, leveling errors, 123
Taylor series, 447
Telescopes, 84–85
total station instruments, 189–192
Temperature
taping errors, 143
variations, 122, 145
Terrestrial frame of reference, 330
Terrestrial photographs, 12
Terrestrial poles, 542–544
Testing
collimation, 97
levels, 95–99
Thematic Mapper (TM), 852, 854
Theodolites, 908
Three-dimensional conformal coordinate transformation, 496–498
Three-dimensional coordinate computations, 596–598
Three-dimensional perspective grid, 530
Three-dimensional perspective models, 473

Three-dimensional resection problem, 292–294
Three-dimensional two-point resection, 292–294
Three-level section, 799, 800
Three-point resection, 282–285
Three-screw arrangement, leveling of, 103
Three-sigma errors, 58
Three-wire leveling, 114–116, 122
Tidal datum, 72
Tilted photographs, 824
Tilting levels, 83, 86
Time
  apparent solar, 909
  astronomical observations, 910–911
  civil, 909
  daylight saving time (DST), 909–910
  equation of, 909
  mean solar, 909
  sidereal, 908–909
  standard, 909
Time dilution of precision (TDOP), 348
Time of flight (TOF), 491–492
Time of week (TOW), 328
Timing, astronomical observations, 910–911
TIN, 483, 798, 877
  breaklines, 483
Title, registration of, 671–672
Topographic details, locating, 483–494
Topographic maps, 472
Topology, 869–871
  adjacency, 870
  connectivity, 870
  direction, 870
  nestedness, 870
Total station instruments, 151–152, 185
  adjustments, 212–216
  angle measurement system, 190
  azimuths, 204
  balancing-in, 209
  characteristics of, 185–188
  construction surveys, 727–730
  direct mode, 200, 203
  elevation differences using, 211–212
  handling and care of, 192–193
  horizontal angles, 197–200
  horizontal motion, 191
  indexing errors, 188
  laser plummets, 192
  parallax in, 189–190
  parts of, 189–192
  plate bubble adjustment, 193
  reflectorless, 152, 708
  reverse mode, 198, 199
  robotic, 188, 761
  servo-driven, 195–196
  setting up, 192–195
  sights and marks, 206–208
  sources of errors, 216–222
  straight lines, prolonging, 208–209
  telescopes, 189–192
  tracking mode, 187
  traversing with, 234–235
  tribrach, 191–192
  vertical angles, 204–206
  vertical circle, 190
Townships, 690–695
  designation of, 692
  subdivision of, 692–695
Trace-contour method, 480
Tracking mode, 151
TRANSIT system, 323, 337
Transpose of a matrix, 924
Transverse Mercator projection, 609–610
  central meridian, 609
  convergence angle, 612, 617
  defining parameters, 611
  direct problem, 611
  inverse problem, 611
Traverse
  angular misclosure, 233–234
  closed, 226, 227
  definition of, 226
  field notes, 232–233
  link, 226
  open, 226
  polygon, 226, 228
  random, 222–223
  selection of stations, 230–231
Traverse computations, 239
  adjustments, 246–249
  azimuth, 242–243
  balanced angles, 240–242
  bearings, 242–243
  boundary surveys, 257–259
  closure conditions, 245
  compass (Bowditch) rule, 247–249
  departures, 245
  errors, 263–266
  final adjustments, 255–257
  inversing, 254–255
  latitude, 243–244
  least-squares adjustment, 249
  linear misclosure, 245–246
  locating blunders in, 263–265
  misclosure, 245–246
  mistakes in, 266
  open traverses, 259–262
  rectangular coordinates from, 249–250
  relative precision, 245–246
  state plane coordinate system, 262
  steps in, 239
Traversing, 226–237
  angles, 228–229
  angular misclosure, 233–234
  by azimuths, 229
  by deflection angles, 229
  field notes, 232–233
  by interior angles, 228
  least-squares adjustments, 447–456
  lengths, 229–230
  mistakes in, 237
  radial, 236–237
  sources of errors, 237
  stations, 231–232, 632–635
  total station instruments, 234–236
  Universal Transverse Mercator (UTM), 262
Triangle, oblique, 271
Triangulated irregular network. See TIN
Triangulation, 130
  control surveys, 569–570
  mapping surveys, 476

Tribrachs, 191–192
  adjustment of, 215
Trigonometric leveling, 79–82
Trilateration
  control surveys, 573
  mapping surveys, 476
Triple difference, 355
Tripod, 91
  adjustment of, 214–215
  errors caused by settlement, 123
  fixed-height, 347, 406
  sidehill setup of, 103
Tropospheric refraction, 342
Tube vials, 86–87
Turning points (TPs), 94–95, 105, 106
  numbers, 106
Turtle, 95
Two-dimensional conformal coordinate transformation, 285–290
Two plus one approach, 557–560
Two-sigma errors, 57
Types
  of aerial photographs, 822–824
  of azimuths, 167–168
  of cross sections, 799–800
  of errors, 46–47
  of field books, 31–32
  of field notes, 30–31
  of horizontal angles, 165–166
  of levels, 83–84
  of maps, 472
  of measurements, 22
  of observations, 46–47
  of surveys, 11–12
  of traversing, 230

U

UAS. *See* Unmanned aerial systems (UAS)
Undulation, geoid, 547–549
Unequal tangent vertical curves, computing, 788–790
Unit-area method, 809–810
Units of measurements, 22–24
International System of Units (SI), 24–26
Universal Coordinated Time (UTC), 358, 903
Universal Transverse Mercator (UTM), 262, 636–637, 886
Unmanned aerial systems (UAS), 540
Unmanned systems, 490–491
Upper culmination, 902, 907, 908
U.S. Coast Survey, 73
User equivalent range error (UERE), 348
User-friendly CORS option, 380
User segment, 324, 326
U.S. Geological Survey
  quadrangle maps, 477
U.S. government agencies, 15–16
U.S. Public Land Survey System, 684
U.S. reference frames, 549–556
U.S. State Plane Coordinate System (SPCS). *See* State plane coordinate system (SPCS)
U.S. survey foot, definition of, 23, 26, 607–608

V

Variance, 55
Variations of the compass, 176
  annual, 178
  daily, 178
  magnetic declination, 176–178
  secular, 178
  temperature, 122
Vectors, baselines, 442–446
Vector-to-raster conversion, 871–873
Vernal equinox, 331, 908
Verniers, 93
Vertical aerial photographs, 824–825
  flying height, 832–833
  ground coordinates, 828–830
  relief displacement, 830–832
  scale of, 825–828
Vertical alignments. *See* Vertical curves
Vertical angles
  observation of, 204–206
  reduction of distance observations using, 582–585
Vertical axis, 186
Vertical circle, 190
Vertical control, 73, 476
  hierarchy of, 565
Vertical curves, 778–794
  computing, 782–786
  design templates, 798
  elevation at center point, 783–784
  equal-tangent property of, 786
  equal tangent vertical parabolic curve equation, 780–781
  general equation of, 779–781
  high or low point on, 782
  machine control, 788
  mistakes, 794
  passing through a fixed point, 790–792
  rate of change of grade, 778, 781
  sight distance, 792–794
  sources of error in, 794
  staking, 787–788
  tangent offset equation, 782–786
  unequal tangent, 788–790
  vertical parabolic curve equations, 787–788
Vertical datum, 71
  differential leveling, 110
  North American, 73–74, 549
Vertical dilution of precision (VDOP), 348
Vertical distances, 22
  by differential leveling, 78
  measurements, 76
Vertical exaggeration, 837
Vertical line, 70
Vertical photographs, 824–825
  flying height of, 832–833
  ground coordinates from, 828–830
  relief displacement on, 830–832
  scale of, 825–828
  tilt of, 844

Vertical point of intersection (VIP), 780
Very Long Baseline Interferometry (VLBI), 338, 544
Vials
   level, 86–87
   testing, 96
Viewing, stereoscopic, 836–837
Virtual reference station (VRS), 415, 416
Visible laser-beam instruments, 708–710
Volumes, 22
   average end-area formula, 800–801
   borrow-pit method, 809–810
   computations, 808–809
   contour-area method, 811–812
   cross-section method, 798–799
   end areas, determining, 801–804
   measurement methods, 797
   mistakes, 814
   prismoidal formula, 806–807
   software, 813–814
   sources of error, 814
   types of cross sections, 799–800
   unit-area method, 809–810
   water discharge measurements, 812–813

W

Warranty deed, 658
Water discharge measurements, 812–813
Wavelengths, 145
   spectrums, 851
Weight
   matrix, 442, 445
   of observations, 65–66
Western elongation, 902, 907
   distance, 908
WGS84 reference ellipsoid, 332
WGS84 reference frame
   Doppler Ranging Integrated on Satellite (DORIS), 338
   International Earth Rotation and Reference Systems Service (IERS), 337
   ITRF90, 338
   ITRF91, 338
   ITRF92, 338
   ITRF93, 338
   ITRF94, 338
   ITRF95, 338
   ITRF96, 338
   ITRF97, 338
   ITRF2000, 338
   ITRF2005, 338
   ITRF2008, 338
   ITRF2014, 338
   satellite laser ranging (SLR), 338
   very long baseline interferometry (VLBI) stations, 338
Wide Area Augmentation System (WAAS), 351
Wiggling-in, 209
Witness corner, 698
WOLFPACK computation, 593
World Geodetic System of 1984. *See* WGS84 reference ellipsoid
World Magnetic Model, 178

Z

Zenith angles, 22, 190
   trigonometric leveling, 79–82
Zones, constants, 611–612, 616–617